The Hour of Our Delight

The Hour of Our Delight

Cosmic Evolution, Order, and Complexity

Hubert Reeves

W. H. Freeman and Company
New York

Library of Congress Cataloging-in-Publication Data

Reeves, Hubert.
  [Heure de s'enivrer. English]
  The hour of our delight : cosmic evolution, order, and complexity
  / Hubert Reeves.
      p.   cm.
  Translation of: L'heure de s'enivrer.
  Includes bibliographical references (p.   ) and index.
  ISBN 0-7167-2220-8
  1. Cosmology. I. Title.
QB981.R3813   1991
523.1—dc20   90-24117
                                                             CIP

Printed in the United States of America

1 2 3 4 5 6 7 8 9 0   VB   9 9 8 7 6 5 4 3 2 1

*To all the feeling hearts
who "hate emptiness dark and cold,"
I dedicate this book*

Contents

Acknowledgments

Preliminary drafts of the French edition of this book were read, discussed, and corrected by many people whom I would like to thank again: Michel Cassé, Georges Charpak, Jean-Paul Meyer, Alice Milkin, Isabelle Montpetit, Lyette Plourde, Evelyne and Nicholas Reeves, Camille Scoffier, and Marianne Verga. As in the writing of my earlier books, I was happy to receive encouragement and valuable advice from Jean-Marc Lévy-Leblond, my editor at Éditions du Seuil.

PART
1

DEATH URGE

Prologue: Of Seals and Soldiers

I love the wildlife programs shown on television. The variety of animal behaviors is an endless source of fascination. I clearly recall the images of a battle. Two male seals are fighting over a female. Although they seem to be struggling in earnest, the narrator's voice is reassuring: "This is merely a mating ritual. *There is no killing here.* The combat is over at the first sight of bloodshed. The winner allows the loser to escape once he has given up.

"This process," adds the biologist, with growing enthusiasm, "ensures the selection of strong mates who will sire healthy and vigorous young seals." We see them on the screen, cavorting on the sunny cliffs whitened by the sea spray.

"Thanks to the advancement of scientific knowledge," the commentator continues, "we now know how to interpret such seemingly cruel behavior. It is merely a manifestation of the process of natural selection, the mechanism by which each species, from the amoeba to *Homo sapiens*, evolves."

A flick of the channel away, the situation changes. World War I is the subject of a historical program. We see a muddy landscape strewn with barbed wire and corpses. *There is killing here.* Long lines of stretchers are borne to and from the trenches. The narrator speaks of the desperation of these men, forced at gunpoint to expose themselves to the fire of enemy machine guns. Dominating everything, the sense of the absurdity of this useless, interminable war, which has no more reason to end than it had to begin.

An innocent question: All these handsome soldiers marching off to slaughter, aren't they, like the young seals, the result of the admirable

mechanism of natural selection? Is this the product of biological evolution? From the primeval ocean to the trenches of Verdun . . . The
reassurances of the other program are cruelly absent. Is it possible to
discern some purpose behind the fighting? Or did something go wrong
at some point along the way?

TWO VIEWS OF THE WORLD

This brief anecdote about flipping channels contains some of the
questions facing us today. Each of us is daily confronted with two
entirely different types of report about the nature of reality, about the
state of our world. These reports lead to two very different views.

Through scientific research we learn more each year about the
marvels of our universe. We discover the wonderful deeds of cosmic
evolution. We are given to see the gradual transformation of the chaos
of primordial matter into highly organized structures. For billions of
years the stars have been burning in the sky. Each photon that reaches
our eyes tells of new atoms generated by the nuclear processes in a
blazing stellar core. In February 1987 the news of a violent stellar
death in the Large Magellanic Cloud came to us in the form of photons
and neutrinos. Large crops of newly made atoms had just been liberated into space, remnants of a supernova. Similar events in past eras of
our own galaxy had been responsible, we know, for the existence of the
atoms of our solar system—and in turn for all matter, from solid
planets to living bodies.

At its birth, 4.5 billion years ago, our planet was hot and sterile.
Geologic exploration tells us that a billion years later the early oceans
were already thriving with living cells. It tells us also that in the
following billions of years a large variety of organisms evolved and
specialized. The human brain, with its fantastic capacity, is a recent
development, reaching its present state within the past few million
years.

Science and technology are our own creations. Through them we
acquire knowledge. Gradually our understanding improves regarding

the mechanisms and processes by which the primordial stew of electrons, quarks, photons, and other elementary particles gave birth to nuclei, atoms, molecules, living cells, and plants and animals.

This is the first type of account. It leads to the pleasant view of the world that we proudly teach or gladly learn in our university lectures. We receive it from popular science programs on TV, such as the one on the behavior of mating seals. This view might tempt us to embrace unwarranted optimism. The complexity of the universe, derived from such simple constituents, might suggest that the universe makes sense, that its existence has meaning, even if we do not know what that meaning is.

But there is another type of account that assails us. It reaches us through all the information and entertainment media. Every day we learn about the unending, cruel war in Lebanon, the outrageous oppression in South Africa, children dying of starvation in various parts of the world. We read and hear about the deteriorating natural condition on our planet, the spread of industrial pollution, the thinning of the ozone layer, the growing concentration of carbon dioxide, the dangers of radioactive contamination threatening the whole biosphere, threatening the very habitability of our earth. We know of the stock of nuclear warheads hidden underground. With all this in mind, can our optimism survive? Do we dare speak of the meaning of reality?

One popular way out of this troublesome dilemma is to invoke *chance*: We, and all organized systems, are simply the product of random phenomena. Therefore, reality is basically nonsensical, and any appearance to the contrary must be chalked up to anthropomorphic illusion. According to this view, reality is simply *absurd.*

However, on second thought, this conclusion does not appear entirely satisfactory. The success of Darwinian theory has indeed confirmed the importance of chance in biological evolution: It is the success of a randomly expressed trait that ensures that trait's survival in the succeeding generation, shaping the development of a species over time.

We also know that physical laws play a fundamental role in the evolution of our universe. The lawfulness of our universe is clear: Without laws, random processes lead nowhere. Pure chance is sterile.

Yet matter has come out of primordial chaos, reaching ever-higher levels of complexity.

There are more surprises in store on this subject. As we gain greater understanding of the physical processes associated with the evolution of matter, we find in them a number of so-called coincidences (discussed in detail below). Complexity grows through multiple random processes governed by physical laws. But just *any* physical law will not do. Each law is expressed by a set of numbers. For complexity to arise, these numbers have to be *just right*. Slightly modifying the numerical values associated with physical laws in the very early universe (that is, dreaming up alternative laws) or altering the initial conditions, howsoever slightly, leads to a sterile universe, one in which chance cannot generate complexity.

This astonishing discovery has caused much controversy. The expression "anthropic principle" has been introduced to name the purposive organization of matter. I personally prefer a less chauvinistic expression: *a principle of complexity.*

Metaphorically speaking, our universe is animated by a *life urge.* Initially it had exactly the right conditions to produce organization, complexity, and ultimately life. Throughout its history, the urge has done just that.

In the chapters to follow, I will attempt to present, in simple terms, the clues that science can provide for understanding this quest, as it were, for material organization. Random processes indeed play a fundamental role. Given the appropriate context of specific laws and physical conditions, chance becomes *creative.* The concept of evolution, first developed by biologists, has spread to all fields of scientific thought. For fifteen billion years, matter has been evolving toward ever higher states of organization, complexity, and performance. Out of the primordial chaos, nucleons, atoms, molecules, cells, and living organisms, including the human race and its marvelous brain, have been engendered.

However, that is where our problem begins. A switching of channels — from the healthy seals to the handsome soldiers — illustrates this situation vividly. For millenia the human race has used its intelligence and knowledge to manufacture increasingly efficient and deadly weapons. With the advent of the atomic bomb and the escalation of the nuclear arms race, we have reached the point at which extinction of our species has become a possibility.

Metaphorically speaking, again, our species appears to be animated by a *death urge* that leads us to perform, as rapidly and intelligently as possible, the very actions that could bring about our own elimination. As we examine the mechanisms leading to the growth of complexity, we will find that, far from being foreign to one another, the life urge and the death urge appear to be bound together inseparably.

We are faced with a second dimension of the absurd. We encountered the first in the existentialist view of the role of chance and random processes: If all that exists is the result of chance, then life is essentially nonsensical. Looking closer, however, we were led to question this conclusion, reappraising the role of chance in the growth of complexity. The situation was described in terms of a life urge at work in the universe. It is this urge, at work for fifteen billion years, that has engendered the contentious human brain that has caused so many of the problems we face today.

The discovery that the nuclear activity of stars, the electromagnetic hum of interstellar nebulae, the biochemical frenzy of the early oceans, and the ritual fighting of seals could have no other outcome than the trenches of Verdun or the nuclear holocaust might be called an awareness of *the second degree of absurdity:* Intelligence, consciousness, and knowledge emerge, after fifteen billion years of development, only to eliminate themselves in fifteen minutes. Complexity has a purpose: to fling us back into nothingness!

Is our doom inevitable? The fate of *meaning* is in our hands.

DESIGN OF THE BOOK

In order to achieve historical perspective, I have concentrated, first, on one of the most important events of the past few decades: the birth and development of nuclear arsenals on our planet. I have personally known many of the scientists involved in the building of the first atomic bomb. I have often tried, through conversations, to reconstruct the emotional context of this period. Behind the words, the question was always the same: *Was the atomic bomb inevitable?*

In the second section the metaphorical life urge of the universe is analyzed in scientific terms. An effort is made to understand the physical elements involved in the growth of complexity accompanying the expansion of the universe. The role of natural forces, stability, entropy, information, and states of disequilibrium in the elaboration of organized systems are discussed at length.

In the third section I consider from the same point of view the higher levels of complexity in the world of the living. The aim is to identify the sources of our present problems.

In the fourth section I present some personal views on possible solutions to the present crisis. They all come to the same point: The awakening of the sense of wonder and delight is the best antidote to absurdity at all levels. We should not try to escape reality, but *live it passionately*. This was the meaning of the original title of this book in French, *L'heure de s'enivrer*, from a poem by Charles Baudelaire. (Translations are always difficult. The "time of elation" is probably the most literal rendering; but it can also be translated as the "hour of delight.")

Some readers of the original French version of this book have expressed disappointment at my concluding remarks. To problems of cosmic scope I offer no grandiose solutions, but almost trifling suggestions concerning wonder and delight. I do not have much faith in grandiose solutions. Those presented in the past have generally failed miserably. What is needed, I believe, is a change in the state of mind of human beings. This can only take place gradually, in the context of everyday life.

The search for possible solutions to the present crisis is everybody's business. I have felt the need to present my own views, with their shortcomings. I can only invite the interested reader to bring his or her own mind to bear on the subject and to formulate personal solutions. At this point we certainly need all possible contributions.

1

The Bomb Is with Us

From my window, I can watch the sun setting over Paris. Graceful building facades reflect its golden light into the streets already cloaked in shadow. Between the displays of fruit and vegetables lined up for purchase, people choose, chat, pay, and go on their way lugging heavy market baskets.

The sight of such a calm and peaceful scene belies the existence of a looming threat. Stored in nuclear silos, thirty thousand atomic bombs are ready to explode in a matter of minutes. *One single bomb* would suffice to annihilate the entire city, leaving a glassy, lifeless crater, like those seen on the moon, stretching from the Place de l'Etoile to the Place de la Nation, and from the Porte d'Orléans to the Porte de Clignancourt.

One billion victims, one billion seriously injured: This is the evaluation of the immediate losses in the case of a generalized nuclear war. The after-effects could be even more terrifying. The survivors might wish they had perished in the initial blast. According to the best estimates, millions of tons of dust and soot suspended in the atmosphere would plunge a significant part of the earth's surface into a night lasting several months. Solar radiation would no longer heat the earth. The temperature might drop to several tens of degrees below the freezing point of water and remain there; hence the term *nuclear winter.*

Extremely violent storms over the two hemispheres would spread toxic substances, whose radioactive content would neutralize the immunogenic defense systems of people and animals. Agriculture, medical care, and transportation would be reduced to zero. According to some studies, famine, cold, and disease could lead to the extinction of human race. (These findings have been debated. The point is, however, that although the range of possibilities is wide, it does not exclude the possibility of the extermination of our species.)

What brought us to this point? Why did we allow this Trojan Horse to enter our gates? What fatal flaw in our makeup led us to manufacture the instruments of our own destruction? This chapter is a reflection on the sorrowful theme that *humanity is doing all it can to bring about, as quickly as possible, its own destruction.*

THE BOMB SPEEDS INTO EXISTENCE

It seems to me that the advent of the atomic bomb is better described in the epic style of myth than in the cold, impersonal style of contemporary historiography. Epic language appears to be better suited to and more enlightening about the actual dimension of the elements in play.

A myth is far more than a belief easily demonstrated as false. Traditionally, myths have been a means for transmitting wisdom or a way of life. A myth cannot be judged right or wrong, but must be examined in terms of its effectiveness as a teaching technique.

The theme of a spirit from the Beyond who takes physical shape and disrupts the human world appears frequently in mythic lore. Oracles, prophets, and high priests and priestesses are often designated to announce and prepare for its coming. Of all the deities, the atomic bomb is without a doubt the most tyrannical and the most cruelly demanding. Like the vestal virgins, the technicians had to dedicate their entire lives to that weapon's service. The sense of duty, competence, efficiency, and scientific honesty required to accomplish the complex tasks that preceded the birth of the atomic bomb had to be of the highest order.

The bomb is unable to tolerate the slightest hesitation, weakness, or lack of faith. Those whose love faltered for a moment lived to regret it. They were quickly replaced by scores of other, more ardent worshippers impatient to be given the opportunity to serve.

In less than ten years, the atomic bomb went from the state of pure speculation to that of terrifying reality, nurtured by one of the densest concentrations of gray matter in human history. In 1942 in Los Alamos, a tiny town isolated in the vastness of the New Mexico desert, the world's foremost scientists were assembled — physicists, mathematicians, and chemists.

The United States Army had installed a super-laboratory on the site, equipped to provide every available means to the researchers. The atmosphere was electrifying. Science was racing ahead faster than ever. People were working day and night, never leaving the site.

This long and difficult period of labor gave birth to the bomb. The first one was exploded in secret in July 1945 at Alamogordo, New Mexico. A short time later, in a demonstration seizing the attention of the whole world, it proved to be no ordinary weapon. Two Japanese cities, Hiroshima and Nagasaki, were annihilated. In a few seconds tens of thousands of people literally went up in smoke. In all there were more than one hundred fifty thousand victims.

The bomb grew more powerful. On the Bikini Island atoll and in the Siberian snows of New Zemble it reached a force equivalent to tens of millions of tons of dynamite. And it proliferated. According to the latest count, more than thirty thousand of its offspring lie scattered in the arsenals of the world.

Installed on ballistic missiles of awesome accuracy, several of them bear the sweet name of the city of Paris. Others are aimed at targets called New York, Moscow, Beijing. To the engineers who care for them and maintain them daily, these cities suggest nothing more than the name of one of their missiles.

"NONSENSE"

Let us return to the genesis of the nuclear weapon. The earliest rumors of the possibility of making an atomic bomb began to circulate in the scientific community several years before the Second World War.

It was the beginning of the nuclear age. The explosive properties of uranium were still a matter for conjecture. Its atom was known to be radioactive, splitting (fissioning) easily and releasing energy in the process. Laying one's hand upon a lump of uranium ore, one can feel the heat that is continuously released. Inside the lump millions of nuclei are fissioning every second.

This fissioning process can be accelerated artificially by subjecting the atoms to a neutron flow. When a neutron is absorbed, the uranium atom becomes much more vulnerable to fission. It splits rapidly, releasing several neutrons. Hence the possibility of a chain reaction: A uranium nucleus absorbs a neutron, fissions, and releases other neutrons, which are immediately absorbed by the surrounding atoms; these in turn fission and release more neutrons, and so on.

The energy released by each of these fissions adds up and can attain gigantic proportions. Hence the idea of a bomb. One kilogram of uranium gives off more heat than one thousand tons of dynamite. Enough to devastate a small town. One ton of uranium could easily wipe out any of the largest cities.

But early in the nuclear age no one knew whether this project was physically practicable. The technical difficulties seemed insurmountable. Most scientists were skeptical about the possibility. It was an idle speculation to be left on the shelf beside the perpetual motion machine and the time travel machine. "The merest moonshine," said Ernest Rutherford, one of our century's greatest physicists.

IN A HOTEL ROOM BATHTUB

In *Les heures étoilées de l'humanite* Stefan Zweig writes of specific historic events (the cries of the geese at Capitoline, the writing of Handel's *Messiah*) that, in spite of their brief duration — and in some cases, their unassuming appearance — had a profound effect on the destiny of mankind. I would like to add an event to Zweig's collection.

It occurred in London in 1935. A learned Hungarian Jew, freshly arrived from Budapest, had just rented a hotel room and was hurriedly transforming the bathroom into a laboratory.

He filled the bathtub with water and used it as a tank in which to experiment with small radioactive sources smuggled out of the university at which he had been teaching and brought to London, hidden in his luggage. The man's name was Leo Szilard, and he was haunted by a vivid anxiety. He firmly believed in the possibility of liberating the energy of the atom. He intended to accomplish the necessary steps as quickly as possible.

He cannot be written off as another "mad scientist," drunk with power at the thought of such a fantastic project. From the start he was well aware that the success of his project could have dire consequences for the human race; from the start he was lucid about the danger for posterity inherent in the development of an atomic bomb. "Mankind is on the brink of destruction," he kept repeating as the technical difficulties proved surmountable.

At the same time the rapid progress of the Nazi movement over the past several years terrified him. Having left his professorship at the University of Budapest, he had fled the continent because of the rise of anti-Semitism. Adolf Hitler's militaristic ambitions were obvious to him, and threatened by barbarism, the stability of Europe was also in the balance.

Imagine him, leaning over his hotel room bathtub, a political refugee without a laboratory, obsessed with the need to invent a bomb as quickly as possible, at any price, hoping to circumvent the German physicists serving the Nazi cause.

In 1935 Leo Szilard was alone in bearing the burden of anguish caused by an as yet embryonic bomb. Most of his colleagues had given little thought to the possibility. But the bomb would not have to wait in the wings much longer. Little by little, it came to fascinate and obsess an entire generation of physicists and engineers. Like Szilard, they, too, were subject to ambivalent and contradictory feelings: *excitement* at the idea of the forces to be released, *consciousness of the fatal risk* implied by the idea, and given the political situation, a sense of *imperious necessity*, forcing them to spare no energy in bringing it into the world as soon as possible.

SOME MODEL DISCIPLES

In 1935 a mere rumor; in 1990 a terrifying reality. The historians who, after some future nuclear cataclysm, wish to trace the roots of such a catastrophe, will be correct in citing Los Alamos as one of the original nerve centers. In the logistics of the sabotage upon humanity, this laboratory held a key position.

Several project members of the Los Alamos laboratory were later interviewed in a film entitled *The Day After Trinity*. They spoke not only of their professional roles, but also of their personal feelings during the Los Alamos years. This film, both moving and instructive, provides ample material for reflection. Another film, Stanley Kubrick's *Doctor Strangelove*, familiarized the public with an image of the atomic scientist as a paranoid genius, infatuated with ever more destructive machines. Although this image is not altogether devoid of a factual basis (it has been interpreted by some as a portrait of Edward Teller, another Hungarian refugee, responsible in large part for the development of the hydrogen bomb), it definitely does not apply to the majority of scientists who participated in the Manhattan Project (the code name for the secret Los Alamos atomic operations). As a student at Cornell University in the late 1950s, I was personally acquainted with several of the heads of this project. They were all ardent pacifists, actively opposed to the anticommunist fury stirred up at the time by Senator Joseph McCarthy.

Hans Bethe coordinated the theoretical research at the Manhattan Project from 1943 to 1946. Of German Jewish extraction, he had fled Europe only a few years earlier. Said to be "one of the most valuable gifts the United States ever received from Nazi Germany," he had already achieved reknown as a nuclear physicist before the war. At an early age, he had solved a century-old problem, that of the source of the sun's energy. In 1938, with a few coworkers, he demonstrated that nuclear reactions are taking place in the cores of stars. The energy released is largely sufficient to account for the light given off by stars. He received the Nobel Prize for this work in 1967.

I remember him well: tall, dignified, and serene, striding through the corridors of Rockefeller Hall, the university's physics building, surrounded by a coterie of young researchers.

During the physics department's weekly conferences, he usually sat in the back of the room and pursued his own research. He appeared to be paying little attention to the exchange of ideas going on around him. However, should the discussion grow vague and imprecise, he would clear his throat gravely. This was the signal for a long moment of silence. Then, in a few carefully chosen sentences, he dispelled the fog and everything became clear. I can still remember the admiration we felt for his intelligence.

His attention to each and every laboratory student was friendly and exacting. He made regular rounds in the lab and, though he appeared to be paying a casual visit, submitted us to a thorough interrogation on our progress.

His arrival used to create a certain amount of apprehension: "He's in So-and-so's lab now. I saw him at the window. He didn't look very pleased with the work." Yet his attention to our research and the advice he gave were invaluable to us. We were borne up by his vigor and discipline, and we gladly accepted his high standards of excellence.

Well aware of the scientist's responsibility, he devoted a great deal of his activity after Los Alamos to opposing the arms race, arguing that the only effective way to defuse the threat of war was through dialogue with the adversary and political understanding. He was a member of the U.S. President's Scientific Council, and he is to be thanked for the ban on atmospheric testing of nuclear weapons signed in 1963.

I do not know exactly which role Robert Wilson played at Los Alamos, although it is most likely that he worked in the experimental section. When I was a student at Cornell, he used to share the management of the Newman Nuclear Laboratory with Hans Bethe. His straightforward and jovial ways delighted us as much as Bethe's dignity and seriousness. As a lecturer, he never missed an opportunity to poke fun at the theoretical physicists drowning in their equations. To his mind, everything was simple, clear, and efficient.

A pioneer of modern particle accelerators, Wilson left Cornell after building the betatron there and later directed the Fermi Laboratory in Chicago, the site of one of the world's largest particle accelerators, comparable to that operated by the CERN in Geneva. Architecture is one of his great passions, and his affinity for the Gothic period prompts him to speak of it with warmth and authority. To his mind, the giant accelerators are the Gothic cathedrals of today. He brings to his work the fervor of the medieval craftsman.

I should now like to introduce Philip Morrison, who, I must admit, is my favorite. His courses and lectures were not to be missed at any cost. I can still hear his halting step as he limped to the podium. I can still see the slightly pathetic movements he made in preparing to address the gathering, and I can still see the smile on his face, intelligent and a bit mischievous.

Morrison's lecture would get off a running start. The flow of ideas, leaping ahead in a torrent of brilliance, was tinted with truculence and an overriding enthusiasm. A masterful blend of logical discipline, flights of fancy, and insolent jabs at "institutions" lifted us into an ecstatic state from which we returned with difficulty. We clamored for more! One of his best routines began with an introduction to the wonders of telecommunications and wound up with biting satire of the idiocy of the messages conveyed over the waves.

Blacklisted by U.S. government authorities in the 1950s because of his liberal sympathies, Morrison's security clearance was refused, and he was denied access to secret defense documents. At this time, his telephone was tapped and his mail systematically opened before delivery. As a joke, he sent a proposal to the Pentagon in Washington for a system to detect nuclear explosions by observing their reflection on the dark side of the moon. The panic it created in government circles brought him a written warning and a reminder that he was considered a security risk. His response was, "Well, if you don't want it, where would you suggest I send it?"

I must also mention Richard Feynman, a man engrossed in philosophical and religious problems. He often had lunch with students. We were captivated by the man's genius, which allowed him to "do" physics as instinctively as if he were playing the bongo drums.

THE ROAD TO THE BOMB IS PAVED
WITH GOOD INTENTIONS

I never met Robert Oppenheimer, but those who knew him say he was an exceptional person. In addition to his perfect command of physics, he possessed a vast knowledge of literature and philosophy, equally well acquainted with the intricacies of Hindu mythology and medieval French literature.

Oppenheimer paid little attention to international political events before the war. Nevertheless, no one could remain aloof from politics in the early 1940s, least of all a Jew. Hitler was winning every battle he undertook. His ambition was boundless. The German army was spreading over all of Europe, enslaving entire populations. More and more extermination camps were operating. The rising order was a serious menace to civilization itself. A return to barbarism was in the air, and the death of the Jewish people was imminent.

Moreover, the rumor was afoot that the Nazis were interested in the atomic bomb. Today we have proof that the Germans did indeed attempt to develop nuclear weapons. But they did not get very far. Luckily, Hitler was convinced that scientists were more useful on the battlefield than in the laboratory. A German atomic bomb could have changed the course of history.

At the same time an alliance as unlikely as it was significant was about to take place. The U.S. Army designated a staunchly conservative career officer, General Leslie Groves, to command the nuclear project. This man, whose extreme right-wing leanings and love of military discipline made him allergic to intellectuals and liberals, had to appoint a scientist to head the vital research section of the project. The man he chose was Robert Oppenheimer. This pair of minds — to all appearances the worst, most volatile match possible — was to function in close collaboration for several years.

News of German victories and revelations concerning the extermination camps stimulated and energized the Los Alamos team. The moral necessity for the bomb was clear: it *had* to be developed, as *quickly* as possible. The rescue of civilization depended on it. Had I been the right age, and had I been invited to do so, I would have hurled myself into the adventure wholeheartedly.

On May 8, 1945, the Germans surrendered: Victory day for the Allied forces in Europe. In the Pacific the Japanese were still fighting, but they were visibly losing the war. The atomic bomb was not yet ready.

. . . AND WITH NOT-SO-GOOD INTENTIONS

How did the researchers at Los Alamos react? In retrospect Bob Wilson has said, "That day, I should have turned in my badge, locked up my lab, and left the site for good. Why didn't I do it? That's the one thing I can't understand. It's the one thing in my life I regret the most."

Several scientists timidly discussed the possibility of abandoning the research at that point. However, they spoke up in order to assuage their own consciences, more than anything else. Of course, some arguments in favor of the bomb could still be produced, but in a singularly diluted form. The bomb was no longer necessary to rescue civilization, but it could still eliminate the need for a land invasion of Japanese territory, saving the lives of several million soldiers and civilians. It would be better to sacrifice the lives of one hundred thousand Japanese people. A simple matter of accounting. However, these figures failed to take into account the long-term losses. The hundreds of millions of deaths that may result from a future nuclear war should also have been put in the balance. "Once I'm here, I'm here to stay," said the bomb.

We have now arrived at the crux of our discussion: Was the bomb inevitable? Let us suppose that in May 1945 the researchers at Los Alamos had gone ahead with the decision to shut down the project, had burned all the findings and had torn down the installations. Even so, the Soviets were already interested in the bomb, and sooner or later the

United States would have been forced to start research anew, only hindered by the delay.

Like a large ship cruising at top speed, Oppenheimer later said, the project was irresistibly propelled forward by its own momentum. "When you see something that is technically sweet, you go ahead and do it and you argue about what to do about it after you have had your technical success. That is the way it was with the atomic bomb." Work on the atomic bomb wasn't delayed in the least by the neutralization of the Nazi threat; it found other arguments to justify itself. Work continued without interruption, and the first bomb was tested in New Mexico in the summer of 1945.

From then on the question of the bomb's future had to be addressed. Should it be used? Under what circumstances? There was serious talk of inviting the enemy generals to New Mexico, to witness a demonstration of nuclear capability. Others felt it would be sufficient just to drop a nuclear bomb on an uninhabited Japanese island.

"A fine time for conscience pangs," said the hawks of the time, "you should have mentioned it earlier. Let's annihilate a couple of cities that have never been bombed by conventional weapons, in order to accurately measure the damage." The other alternatives were rejected. After Hiroshima and Nagasaki, the Japanese begged for mercy.

How did Hans Bethe feel at the time? In the beginning he was quite worried whether the machine would function properly. When the operation was declared a success, he felt relieved at first. The sheer horror of what they had done didn't strike him until afterward. "What have we done? What have we done?" From that moment on, he made a decision, which has never wavered, to oppose any further nuclear testing.

Philip Morrison gave the bomb its final inspection, on Tinian Island, just before the final take-off for Hiroshima. He spoke willingly on the subject at Cornell, adopting as his own the official viewpoint, that millions of lives were saved by avoiding the invasion of Japan by conventional means. But to me he seemed to be far more ill at ease with himself than he would admit.

As for Bob Wilson, his wife remembers that on Hiroshima day, he came home from the laboratory and vomited. "I still vomit," he adds, "every time I think of it."

. . . AND WITH ILLUSIONS: THE TWO FACES
OF PROMETHEUS

Regarding the feelings shared by the scientists at Los Alamos, Oppenheimer later described them with his usual lucidity, irritating many: "In some sort of crude sense which no vulgarity, no humor, no over-statement, can quite extinguish, the physicists have known sin; and this is a knowledge which they cannot lose." Following the example of Prometheus ripping fire out of the sky, they had dared to control, subdue, and dominate natural forces to a degree never reached before. One can imagine their elation at the first explosion. Oppenheimer tells of how, at that very instant, the words of Krishna in the *Mahabharata*, one of the sacred books of Hinduism, leaped to his mind. A stanza of a prophetic nature:

> *. . . if the radiance of a thousand suns*
> *were to burst into the sky,*
> *that would be like the splendor of the mighty one . . .*
> *I am become death,*
> *the shatterer of worlds.*

In his book *Disturbing the Universe*, physicist Freeman Dyson appropriately terms the Manhattan Project a Faustian pact. In order to attain a higher level of knowledge and power, physicists surrendered their consciences to the military, as Faust clinched his deal with Mephistopheles. Faust alone, however, suffered the consequences of his action; the deeds of Los Alamos are a burden for all humanity.

There are two sides to the Prometheus myth. The first relates to Faust again: the infatuation with knowledge and power. The second is messianic: Prometheus as the benefactor of humanity.

The myth of the "benevolent force" is a timeless concept, one of the archetypes deeply engraved in the collective human psyche. Examples of it abound in the world's literature: Gilgamesh, for the Assyrians; Samson, for the Jews; Hercules, for the ancient Greeks; and in our time, Superman, Tarzan, or Zorro. A power that is inherently good, rushing to the rescue of the widow and the orphan.

In the same spirit, Oppenheimer mentions another passage from the *Mahabharata*. It is an event from the life of Shiva, who is both the creator of the universe and, when the cycle is over, its destroyer. Shiva has been trying to reason with a despotic warrior prince, who refuses to listen. In order to scare him into behaving, Shiva takes on his terrifying aspect as the Destroyer of Worlds. "Each of us at Los Alamos was, at some point, influenced by such an image," adds Oppenheimer.

It was the hope of many scientists that the nearly infinite power about to be born in the laboratories would serve to check the spread of harmful intentions and would force humanity to behave reasonably. They wished to present humanity with a tool that, in the hands of an organization like the United Nations, would guarantee world peace. These sentiments, according to Oppenheimer, sustained the researchers in their efforts and served to justify them at times of doubt or self-reproach.

In the same vein, several years later the U.S. delegate Bernard Baruch brought before the United Nations a proposal for a nuclear arsenal to be held by the United Nations, which would be empowered to punish any nation that, having accepted the terms of new agency, dared to violate its regulations. "Spare the rod and spoil the child," the proverb says, in fewer words, but for this a rod is needed. The proposal was unanimously rejected.

Who would still doubt the effectiveness and power of mythical archetypes, knowing that for many years this elite group of scientists relied on such images to justify their quest and appease their consciences? Theirs was an illusion. As history took shape, such high hopes were proven vain. This new "force" was never benevolent. The bomb is a weapon like any other in the history of war, only infinitely more powerful.

Later on politicians also made use of mythic language to justify the bomb, but the myth had become more narrowly focused. Truman hoped that the United States, entrusted with something close to divine right, would be the only country ever to possess this "sacred gift."

In 1946 he asked Oppenheimer, "When will the Russians be able to build the bomb?"

"I don't know."

"I know," said Truman.

"When?"

"Never," he replied, sure that he had heaven on his side. Three years later, a nuclear device was exploded in the Soviet Union.

Inspired by the same religious spirit, after the Japanese surrender, Senator Brian McMahon affirmed that the bombing of Hiroshima was "the greatest event in world history since the birth of Jesus Christ." He added that the United States had to stay ahead in the nuclear arms race. "If the Russians should, by some mishap, overtake the U.S., total power in the hands of total evil will equal total destruction."

So the holy war against communism replaced the holy war against fascism. The bomb held all the winning cards. It was developing, improving, and multiplying at an infernal pace. The weapons were piling up. When would the voice of reason be heard?

REASON'S INAUDIBLE MURMUR

The effort expended by Niels Bohr, the father of quantum physics, to put an end to the nuclear process must be one of the most moving tales in this dark saga. Even before the bomb was built he declared that it was of the most urgent necessity to share the secret with Stalin. He was convinced that if the Allies failed to make this gesture of confidence and goodwill, it would never be possible to establish international agreements about nuclear weapons. He correctly predicted an escalation of terror.

Bohr spent several months trying to arrange meetings with heads of state. At last, Winston Churchill agreed to see him briefly. Bohr presented his deep concerns to an uninterested prime minister, who soon turned his attention to another visitor. Crushed, Bohr asked if he might forward a memorandum to Churchill on the subject. Churchill replied positively, hoping that it would deal with "a subject other than politics." Later on, Churchill wrote: "I did not like the man . . . with his hair all over his head. It seems to me Bohr ought to be confined or at any rate made to see that he is very near the edge of mortal crimes."[1]

Roosevelt was more sociable, but the end result was the same. For the Allies, the bomb was a powerful weapon, not a "benevolent force." This was true even as early as 1943, long before it was built.

On two occasions, Leo Szilard asked Albert Einstein to speak to the U.S. government on his behalf. In 1939, when he was asking for

Roosevelt's help on the atomic project, his opinion was valuable. After the victory over the Germans, when Einstein and Szilard pleaded for the discontinuation of the Manhattan Project, the White House turned a deaf ear.

Another British physicist, Dr. Patrick Blackett, a Nobel Prize winner, approached Prime Minister Clement Attlee, Churchill's successor, with a report opposing the development of nuclear weapons in England. He was brutally reprimanded. "The author, a distinguished scientist, speaks on political and military problems of which he is a layman." To drive the point home, he was excluded from the committee set up to study national defense alternatives. Clearly, the myth of the benevolent force had a weaker influence on politicians than on scientists.

LOVE POTIONS AROUND THE WORLD

Bending over the bathtub in his London hotel room, Leo Szilard can be cast as the messenger angel of the bomb goddess. General Leslie Groves and Robert Oppenheimer, acting under the spell of two different but equally effective love potions, were the high priests of the cult.

When Groves was summoned to serve the bomb, he was wasting his life constructing artillery hangars. Like every true soldier, he dreamed of military glory. "The secretary of war has designated you to lead an extremely important mission. If you do the work assigned to you correctly, victory and glory will be ours."

Promoted to the rank of brigadier general, Groves went to work immediately. He organized the transportation of his "troops" — the most brilliant scientists of the day, including several Nobel Prize laureates — to an isolated spot in New Mexico. He planned to have them wear U.S. Army uniforms, use the military salute, and work in total secrecy. To his great disappointment these three demands were refused.

Groves was furious when he learned that several scientists were opposed to dropping the bomb on Japanese cities. In a report entitled "Handling of Undesirable Scientists," he wrote: "The Manhattan Project has been plagued since its inception by the presence of certain scientists of doubtful discretion and uncertain loyalty." As victory in Europe neared, he sent out a memorandum that mirrored his devotion to the cause: "The Manhattan Project will continue and increase after V E Day [Victory in Europe], with Japan as the objective. The avoidance of lost time on riotous celebration of V E Day itself should also be stressed if deemed necessary."

But the bomb owes allegiance to no nation. It is above such considerations. It serves itself and itself alone. Groves' mistake was in thinking and spreading the idea that the Russians would never be able to make an atomic bomb. Drunk with success, convinced of absolute American supremacy, he requested a technical report on the Manhattan Project, ordering many copies to be printed and distributed. The report was meant to be a hymn of praise to Yankee superiority. It has been of great use to Soviet engineers.

Groves was never forgiven for this indiscretion. His unbridled zeal had made him undesirable as a member of the nuclear club. He was soon replaced by other men, more modest, more discreet, more competent. His role was over. The bomb would go on without him.

The bomb wooed soldiers with promises of military glory. To entice scientists to serve, it gave promises of omnipotence, with an aura of good intentions.

Oppenheimer is a tragic figure. His fall from grace was more dramatic. He had dedicated most of his career to the development of the first nuclear weapon. At his first sign of hesitation, he was cast out. As Edith Piaf sings, "Life gives you everything, only to take it all back."

Robert Oppenheimer had been a prodigy. The Oppenheimer who had stepped up to the podium to make a presentation on his work in geology to the New York Academy of Sciences had been a child of twelve . . . "I never met anyone as quick to grasp a line of reasoning," said Hans Bethe, who knew what he was talking about. "In a few seconds he was able to understand something that it had taken hours for the rest of us to absorb."

This was a finely tuned mind, navigating with ease in the realms of abstraction but also able to solve concrete problems in directing a laboratory of several hundred staff members. Top this off with keen literary and artistic judgment and also remarkable culinary talent that was highly appreciated by his colleagues.

Oppenheimer was the perfect candidate for the Faustian challenge called for by the political situation: to bring the atom bomb into the world. He accomplished the task and was carried to the heights of glory. After the war he held a post in Washington where utmost attention was given to his opinions. As long as he kept his end of the bargain with the bomb, his life was a success story of glory, honor, and official recognition. The wheel of fortune turned against him when he showed signs of doubt and expressed objections on ethical grounds. Military men and scientists never forgave him for revealing their hidden motives.

He was "suspected" of trying to peddle atomic secrets when he demanded radioactive isotopes for medical research. His real transgression, however, was to have opposed the priority given to the strategy of massive nuclear bombardment. He was framed during a trial designed to disgrace him. His personal secrets were dragged out in public. The fall from grace was a brutal one.

He was excluded from defense committees; he was denied access to the scientific equipment he needed. He would never recover. The last images of the film *The Day After Trinity* show him after the fall: prematurely aged, a shadow of his former self.

Although Hans Bethe also opposed the escalation of the atomic madness, destiny was not as cruel to him as it was to Oppenheimer. Nevertheless, he has been the target for biting criticism from the young movers in the arms race: "You were enthusiastic about the atomic bomb when you were building it, in spite of opposition from your elders, who felt it was an impossible task. So stop harping on the same old objections, and let us have our chance." These words are enough to suggest the level of ethical reflection and moral responsibility of the new generation of nuclear weapons builders. Who could speak more eloquently for the seductive qualities of the bomb's potions?

THE BOMB PROLIFERATES

The U.S. bomb was born in a euphoric rush of determination and enthusiasm. The Soviet bomb appeared in a climate of terror.

Truman's belief that the Russians would be unable to develop a thermonuclear device was not based on faith alone. The Soviet Union had been ravaged by the Germans. An immense field of ruins was all that remained. The development of an atomic bomb in the United States had demanded every possible effort from an enormous technical and industrial infrastructure. If we compare the economic situations of the two nations at the time, it is indeed difficult to imagine how Stalin could have achieved his goals.

Now we know how it came about. In spite of the poverty of the land, the bomb demanded a higher priority than social reconstruction. Free labor was available in the prison camps. Armed guards forced hundreds of thousands of people to go to work, day and night, often under horrifying working conditions.

The installations were constructed as quickly as possible, without wasting time for safety measures. Later, a German engineer qualified the process as "criminal." The dangers of fire and flood were matched by the certainty of absorbing high levels of radioactivity.

The explosion of a dynamite dump in 1947 left seventy dead and one hundred seventy wounded. Nothing caused the work to slow down. Even the physicists worked in terror. "What would have happened to us, had we failed?" one of them wrote to his family. "They simply would have shot us."

Certain parallels can be drawn between the destiny of Soviet physicist Andrei Sakharov and that of Oppenheimer. As a pioneer of the Soviet hydrogen bomb, he was for many years the darling of the military authorities. The persecution he suffered later were not entirely unrelated to his opposition to atmospheric nuclear testing. Khrushchev never forgave him for it.

How did the bomb fare in France? Taking advantage of the weakness of the Fourth Republic, with its frequently changing leaders, the French atomic bomb was the brainchild of a small group of technocrats. It never obtained an official Parliamentary order and was never submitted to a process of democratic debate. When it was tested in 1960, it was too late to do anything except glorify the bomb. The shamelessness of its "blooming" weighs upon the nation, leaving stinging memories. The latest reaction is called Greenpeace.

In England Churchill, returned to power in 1951, was stunned to find that under the preceding, socialist government, English engineers had received, without any Parliamentary discussion, one million pounds for the bomb.

THE WINNING HAND

The curse is that the bomb holds all the winning cards of the game. Bob Wilson spoke of "the irresistible momentum of technological genius coupled with the bureaucratic structure" when he sought to understand why German surrender did not even slow down work on the bomb.

Add to this the paranoid fear and hysteria in response to news of a Soviet bomb, which, experts have judged, "never should have caused such a panic." The hawks aren't selective about the arguments they use. Any excuse will do, as long as it serves their purposes; political responsibility is not one of their concerns.

"Throughout the history of the atomic age, man has been presented with decisions that have been surrounded with an aura of inevitability, obscuring the freedom of choice. . . . [T]he course of history was . . . allowed to be determined by transitory fears and enthusiasms," write Peter Pringle and James Spigelman in *The Nuclear Barons*, a book I urge everyone to read.[2]

The following passage by Peter Sloterdijk beautifully expresses the situation:

> The Atom Bomb is the real Buddha of the West, a
> perfect, detached, sovereign apparatus. Unmoving it rests in
> its silo, purest actuality and purest potentiality. It is the
> embodiment of cosmic energies and the human share in
> these, the highest accomplishment of the human race and
> its destroyer, the triumph of technical rationality and its
> dissolution into paranoia. . . .
> Its repose and its irony are endless. It is the same to the
> Bomb how it fulfills its mission, whether in silent waiting
> or as a cloud of fire; for it the change of conditional status
> does not count. As with a Buddha, all there is to say is said

by its mere existence. It is not a bit more evil than reality
and not a hair more destructive than we are. It is not only
our unfolding, a material expression of our ways. It is
already completely incarnate while we in comparison are
still divided. In the face of such an instrument great listening
is called for, rather than strategic considerations. The Bomb
requires from us neither struggle nor resignation, but
experience of ourselves. We are it.[3]

PROLIFERATION IN 1990

After the United States, the Soviet Union, France, and Great Britain,
China and India tested their own thermonuclear devices. Seven other
countries are also running in the race: Argentina, Brazil, Iraq, Israel,
Libya, Pakistan, and South Africa. These countries may not yet possess
a complete nuclear arsenal, but they are believed to be well on the way
to attaining this goal.

Several years ago a nuclear nonproliferation treaty was put before
the United Nations. But for various reasons there were objections to it.
France and China, two countries that already belonged to the nuclear
pool, refused to sign, followed by most of the states eager to develop a
bomb of their own, as would be expected.

In the last few years, however, the tension on this front has diminished considerably. Nuclear disarmament appears to be becoming a
reality. There is hope for the future, but the threat is still present. The
conflict between India and Pakistan and the political stand of Iraq are
matters of great concern in this respect.

THE ANCESTORS OF THE BOMB

In mythicizing the bomb and seeing it as the incarnation of a diabolical
power, it is easy to forget that it has notorious ancestors. It is the
newest member of an old family, the murderous weapons created by
the fertile human imagination throughout the history of mankind.

As far back as we can trace history, every invention and every new form of energy has systematically been used to wage war. Darts, arrows, fire, and horses lent themselves to the arsenals of conquering armies. Lucretius, our "correspondent in Rome" at around 100 B.C., provides vivid testimony to this effect:

> It was found that wild horses could be tamed and harnessed to carry a man. Then the chariot, drawn by two horses, was tested on the battlefields. Then a team of four horses was tried, and chariots armed with scythes were invented. Then the Carthaginians tamed elephants and trained them for warfare.
> Thus, cruel discord invented more and more murderous weapons for the human race, and the horrors of war increased as time went on.

We have much to learn from these words written more than two thousand years ago. The last sentence could have been written yesterday. In spite of technological progress, human nature has remained faithful to ancient tradition.

Hannibal and his elephants threatened only the Roman legions. Gunpowder and dynamite were able to inflict far more damage. Today, armed with nuclear energy, "cruel discord" can seriously envisage wiping out the entire human race.

ONE OF NATURE'S ERRORS?

The human species, so well provided for by nature, upsets the natural balance of the planet. We seem to be ill-adapted precisely because of these wonderful gifts. Are we one of nature's errors?

Currently, it is estimated that more than a million different plant and animal species are actually living on the planet. Over the course of biological evolution, a much larger number of species have existed. Most have died out.

No species is sacred. Each evolves out of nature's process, the product of random biological mutation. In order to last, a species has to adapt. To elaborate a form of exchange with the environment. To give and receive. To find a niche in an ecosystem. If it fails, elimination is inexorable.

Sixty-five million years ago, dinosaurs, giant ferns, and ammonites were wiped out of the biosphere. No one knows for sure how this catastrophe occurred. The earth may have encountered a huge shower of extraterrestrial matter (giant meteorites or an interstellar cloud). In any case, it is highly unlikely that these animals were responsible for their own extinction. Nature did not give them any choice in the matter. Conversely, when we contemplate the likelihood of the extinction of our own species, we have no one to blame but ourselves. We are threatening our very existence.

Should humanity perish in a nuclear blast, many plant and animal species would die in the same conflagration. If today's weapons are not yet capable of such massive slaughter, tomorrow's will be. Once again, we must take our hats off to the efficiency of human intelligence.

At this point we must recognize the fact that Western civilization has played a major role in these events. Were we to measure the degree of civilization of a human group in terms of the harmony of its relationship with the natural environment, ours would be, I believe, at the bottom of the scale. As evidence to this effect, I would like to offer the testimony of an old Indian from Canada.

The White people never cared for land or deer or bear.
When we Indians kill meat, we eat it all up. When we dig
roots, we make little holes. When we build houses, we make
little holes. When we burn grass for grasshoppers, we don't
ruin things. We shake down acorns and pinenuts. We don't
chop down trees. We only use dead wood. But the White
people plow up the ground, pull down the trees, kill
everything. The tree says, "Don't. I am sore. Don't hurt
me." But they chop it down and cut it up. The spirit of the
land hates them. They blast out trees and stir it up to its
depths. They saw up the trees. That hurts them. The
Indians never hurt anything, but the White people destroy

all. They blast rocks and scatter them on the ground. The
rocks says, "Don't. You are hurting me." But the White
people pay no attention. How can the spirit of the earth like
the White man? . . . Everywhere the White man has
touched it it is sore.[4]

Our planet is home to many different cultures. Each has developed
its own subsistence strategies into a life-style that is adapted to its
natural habitat. Eskimos don't fish the same way that people in Benin
fish. The massive cultivation of the Canadian plains differs greatly
from the gardening in India. Humanity's relationship to nature varies
according to habitat and to life-style. Many traditional societies have
developed a deep respect for nature, tinged with animism.

Science and technology developed in the Western world precisely
when the mystical relationship to nature was first questioned. This is
certainly not accidental. Once again, we encounter the image of
Prometheus ripping fire out of the heavens, the "sin," according to
Oppenheimer, that the physicists came to know at Los Alamos.

If a relationship between the rejection of ancestral reverence for
nature and the development of science can be discerned, how did it
operate? Did irreverence foster science, or did science foster irreverence? It probably happened both ways, either alternately or simultaneously.

Our concern at this point is the historic fact of the appearance of
Western technological culture. Once it got a start, its strengths allowed
it to rule the planet and dominate all other cultures. Commercial needs
and modern transportation prevent cultural isolation. In the nineteenth century the Japanese were forced to open their gates to the
West. At the end of the twentieth century the last Amazonian tribes
are fast disappearing.

Is a technological society centered on harnessing energy the inevitable outcome of intelligence and curiosity? This question is often raised,
but I feel that it is poorly expressed — and misses the point. Imagine a
planet Lambda, like our earth, where a multitude of different cultures
have developed many distinct ways of relating to nature. Even if the
great majority of these human groups show only slight interest in
science and technology, if only a single one possesses a passion for such

knowledge, that is sufficient for its eventual domination of all the
others. Technology is invasive; it brings about its own territorial
expansion.

THE NATURE OF THE SCORPION

By the sandy banks of an African river, a lion is dozing. The sun is high
in the sky, and there isn't a breath of wind.

A scorpion creeps up to the lion. "Wake up. I need your help," he
says, nudging the lion with his elbow. "I have to get on the other side
of the river. Let me climb on your back and swim across with you."

The lion is flabbergasted. "I, carry a scorpion on my back? You
would sting me, and I would die."

The scorpion's pleas are eloquent. "Don't be silly. If I sting you, we
both drown. Nothing will happen to you."

Stubbornly, the lion resists, but the scorpion's agile wit, coupled
with the implacable logic of his argument, finally persuades the lion.
"Climb aboard."

Slowly and suspiciously, the lion steps into the water. He begins to
swim. In the middle of the river, a sharp pain paralyzes him instantly.
They are both swept away with the current.

"Look what you've done," gasps the lion. "We will drown!" "I
know," says the scorpion. "I'm terribly sorry. But you can't overcome
nature."

The events of the past few decades make this fable particularly
pertinent. Is it human nature to prepare, as quickly and efficiently as
possible, the weapons of our own elimination? If this is the case, is it
possible for us to overcome our nature?

THE COSMIC STAKES

In this first chapter we have taken stock of a particularly alarming
situation — the dubious future of the human species. The frantic race
to accumulate thermonuclear weapons and the proliferation of atomic
devices give us reason to expect the worst.

History shows us that weapons always end up being used. Self-defense is systematically turned into an excuse to violate the rights of others. If the past is proof of the future, who would bet on the chances for world peace? And once the first shot is fired, who would bet on the chances for the survival of the human race?

Viewed from interstellar space, what sort of effect would a pyrotechnical display of atomic explosions — our grand finale — produce? Practically none at all. The inhabitants of even the closest planetary systems would be unable to detect it! On the galactic or cosmic scale, no one would bat an eyelid. So why make so much fuss?

All the same, if life exists on planets revolving around other stars and if technological civilizations have appeared, aren't they, too, threatened by "cruel discord"? How many planetary populations have already met the same crisis? How many have blasted themselves into nothingness? And how many have passed the test of peaceful coexistence with their own power?

A FRIGHTENING SILENCE

In the seventeenth century the French philosopher Blaise Pascal wrote: "The silence of these infinite spaces frightens me." Today we have some keys to the mystery of the heavens. Out there, in the center of stars and nebulae, nuclei, atoms, and molecules are being generated. These particles may later be assembled in a brain, may lead to consciousness.

Does life exist on other planets, in other planetary systems, somewhere in the billions of galaxies of our universe? We have reasons to surmise that matter does organize itself and reach the higher states of complexity when the physical conditions are appropriate.[5]

Then why don't we ever receive messages from the sky, via radio waves or any other means? There are several possible answers to this question. We'll examine four:

1. Contrary to the opinion expressed above, ours is the only form of life to have developed in the universe. This is possible. But in

view of our current scientific knowledge I find this rather unlikely.

2. Extraterrestrial civilizations communicate via means that we cannot detect. This hypothesis cannot be refuted.
3. Our closest neighbors are too far away for current reception equipment. The next generation of radiotelescopes may have surprises in store for us.
4. Unable to cope with their own power and aggressivity, technological civilizations exterminate themselves as soon as they reach the capacity to do so.

If the fourth response is correct, the silence of infinite spaces should frighten us far more than it did Pascal.

PART
2

LIFE URGE

The pyramid of complexity rises over the course of
the ages

2

The Pyramid of Complexity

In the first part of this book I summarized the difficult situation in which humanity now finds itself. Since there is always hope that by understanding how we got into trouble on the first place, we may find paths leading out of trouble, the rest of this book will be devoted to just that understanding. In the following chapters we shall dip into the immense reservoir of knowledge accumulated by scientific investigation, in order to try to understand the origin of the contemporary crisis.

This is not a new story. War, suffering, and death have always been with us. The primate's evolving intelligence and the child's awakening consciousness soon apprehend the fact that reality is frightening. In order to defend and reassure themselves, human beings try to understand the world they live in. I think we can find in this vast endeavor of seeking reassurance the root of all explanatory systems of reality, whether mythological or scientific.

Scientific investigation, in the current sense of that term, appeared for the first time, to our knowledge, in ancient Greece. That approach to reality sustained intense activity for several centuries before petering out during the fall of the Roman Empire. In the Middle Ages, science was confined mostly to the Arab world. It returned to Europe during the Renaissance. Since then, its progress has continued unimpeded in several areas on our planet.

Is scientific knowledge cumulative? I believe it is. We know far more about nature than the Greek and Latin philosophers or the Renaissance scholars. But indeed, how is our knowledge superior to theirs? Can the gain in knowledge be expressed in a few concise sentences?

We know that life has not always existed. What is more, we know that organization has not always existed. Matter arose from a primordial chaos and produced, step by step, increasingly complex beings. Living things (in the usual sense of the term) are the latest arrivals in this evolutionary sequence.[1]

The aim of the following chapters is to take apart the set of mechanisms responsible for the generation of organization and life. First, we explain the pattern that forms what we call the "pyramid of complexity." Fifteen billion years ago, this pyramid did not exist, although the potential for it did. It was formed over the course of time.

The dawn of the organization of primordial chaotic matter raises fascinating questions. Modern cosmology can sketch the outlines of some answers. Organization implies the existence of *information*. What is the source of this information? Where and how did it hide in the primordial chaos? How did the events that engendered material structures cope with the celebrated exigencies of the increasing entropy? What is the entropy of the universe? Does it obey the laws of thermodynamics?

These are the themes that we will now address.

NATURE IS STRUCTURED LIKE A LANGUAGE

Is it possible to sum up in a few brief sentences the scientific knowledge acquired since the Renaissance? To paraphrase an observation by the French psychologist Jacques Lacan, we could begin by saying that nature is structured like a language. But what does that mean?

Written language uses an alphabet as its basic element. An alphabet is a group of *letters* everyone has agreed to adopt. We use the Latin alphabet, made up of twenty-six letters, but many other alphabets exist: Greek, Hebrew, Armenian, and so forth. Dozens of different alphabets are being used on our planet.

When the letters are arranged in a predetermined order, they form *words*. The key concept here is that of combination. Let's say that a typesetter has four empty spaces and can fill them with any one of the twenty-six letters of our alphabet. How many different "words" can be written? The answer is four hundred thousand! With seven spaces, more than ten billion different combinations could be printed! Of course, only a very small fraction of these would correspond to dictionary spellings of words from our language.

For example, take the letters E, L, B, and U. By rearranging them, we can spell the word BLUE. This sequence bears a meaning: It immediately conjures up an image of the color of the sky or sea. Yet none of the letters, taken singly, contains even a fragmentary reference to this color. The image emerges from the combination of these four letters in the proper order. This is analogous to the chemist's idea of emerging properties. Variations on the combination, such as BULE or EBLU, are entire foreign to the concept of BLUE.

Sentences are obtained by combining *words*, which in turn are combinations of *letters*. Again, there appears a meaning that did not exist before. The sentence "The sky is blue" bears a message that is absent from any one of the four words it uses.

A set of sentences forms a *paragraph*. As we climb the pyramid of the written language, we encounter *chapters, volumes*, and *collections*. The same structural pattern is repeated at every level: The units are composed of elements from the lower level, and they themselves will combine to compose the units of the next higher level. Letters are the alphabets of words; words are the alphabets of sentences.

Our planet's written languages are thought to be five to six thousand years old. Today, we recognize the fact that nature has been using a structure similar in every way to the pyramid of language (see the illustration on the following page). The history of science could conceivably be written as the unfolding of this fundamental discovery.

The chemists of the late 1700s, faced with the huge variety of natural substances catalogued by the alchemists, took a major step

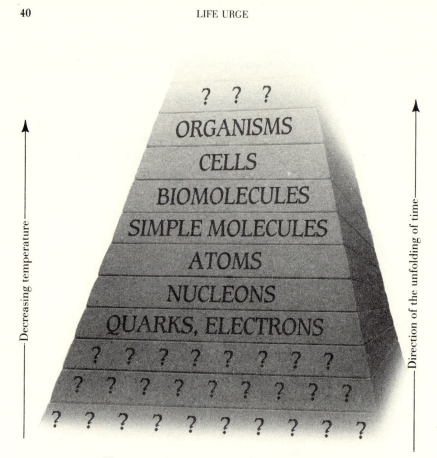

The pyramid rises over the course of time

The Pyramid of Complexity

The organization of matter uses the strategy of "superimposed alphabets." In the same way that letters combine to form words, elements from one level of the pyramid combine to form the elements of the next-higher level. Atoms are thus the "letters" that compose molecules.

The question marks on the lower levels represent the limits of contemporary research. The problem of the upper levels is discussed in the text.

Astronomy tells us that the pyramid did not exist fifteen billion years ago. It rises gradually over the course of eons, in parallel with the cooling of the universe.

toward simplification. They discovered that by combining the *atoms* of various chemical elements, a great number of *molecules* could be formed. These molecules were the substances displayed in the colorful bottles in the apothecary's shop.

In chemistry, water is a *word* composed of two *letters*, called *hydrogen* and *oxygen*. Daily experience shows us that water possesses remarkable properties. For example, it is an excellent solvent. Yet hydrogen (H) and oxygen (O) do not possess this property. It is the juxtaposition of two H atoms and one O atom, in an appropriate geometrical arrangement, that produces the property. The properties of hydrogen peroxide (two H atoms and two of O) are quite different. It would not be a good idea to use the two substances interchangeably. Likewise, by combining hydrogen and carbon (C), we obtain several different hydrocarbons (methane, propane, butane, octane), which are useful in various ways.

The immense variety of gems, minerals, and rock formations derives from the combination of only a small number of atoms: oxygen, silicon, iron, calcium, aluminum, magnesium, and a few others. With these letters we can *write* the whole of geology, be it terrestrial, lunar, or martian.

The work of chemists has made it possible to list more atoms. We now know of about one hundred. Various combinations of them account for all the chemical substances on earth. As we explore our solar system, the stars in our Milky Way, and those in more exotic galaxies, we find that the same atoms occur everywhere and combine according to the same laws. This *atomic language* is resolutely universal. This is, briefly, the message of chemistry.

The observations of Lucretius on the same subject still ring surprisingly true:

> *Why, yes, in the very verse I write, you see*
> *dozens of letters shared by dozens of words,*
> *and yet you must admit that words and verses*
> *differ one from another in sound and meaning,*
> *Such power have letters when order alone is changed,*
> *but atoms, our basic stuff, can claim more patterns*
> *of change whence countless things may be created.*[2]

PHYSICS

In the early 1900s physicists picked up where the chemists had left off. Atoms, previously thought to be indivisible (that is the meaning of the Greek word *a-tom*), were found to be composed of more elementary particles: electrons, protons, and neutrons. These three types of particle are the *letters* that spell atoms. The nucleus of an atom is made up of protons and neutrons (called in combination nucleons). The number of protons determines the chemical nature of the element. Six protons together make a carbon atom, twenty-six make iron, and ninety-two make uranium.

The number of neutrons, however, does not affect the atom's chemical identity. The carbon most commonly found in nature is made up of six protons and six neutrons. Occasionally, in about 1 percent of all cases, the carbon contains six protons and seven neutrons. That is how we differentiate between two carbon isotopes: carbon-12 and carbon-13.

To date, physicists have discovered more than a thousand different nuclei. The largest ones contain one hundred protons and about two hundred neutrons. Most are unstable. Their life expectancies range from a billionth of a second to billions of years.

The nucleon family is not limited to the two members mentioned above, protons and neutrons. The large particle accelerators built after World War II made it possible to detect dozens more. They last a very short time, less than a billionth of a second, before becoming protons or neutrons.

When I was a graduate student at Cornell University in the late 1950s, physicists considered protons and neutrons to be the "true" elementary particles. They assumed that nucleons were not composed of even smaller units. George Gamow, the physicist who announced, twenty years before its detection, the existence of microwave background radiation, used to say that he would stake half his fortune on the "elementariness" of the nucleons. Luckily for him, no one took him up on it. In the past twenty years experimenters in high-energy physics have detected, within the nucleons, smaller units called quarks.

A proton is made up of the combination of two u (for "up")-type quarks and one d (for "down")-type quark. A neutron is made of two d quarks and one u quark. Four other types of quarks are known: they are called s (for "strange"); c (for "charmed"), t (for "top" or "truth"), and b (for "bottom" or "beauty"). Naturally, whimsical names like "truth" and "beauty" reflect nothing more than the sense of humor of the physicists.

THE BASE OF THE PYRAMID

All nucleons are "words," made up of quark "letters." This is perhaps the most significant finding of high-energy physics in the past twenty years. The next question to arise is whether quarks and electrons are finally the true elementary particles, or whether they contain even smaller units, like the Russian doll that opens to reveal a smaller replica inside? Have we reached the lowest step of the pyramid of alphabets, or can we descend even further? And if so, how many steps? Is there such a thing as the last step? Or must we consider the hypothesis of an endless descent?

To the question "Are electrons and quarks composite bodies?" today's physicists reply, "We have no reason to think so." Scientists are above all pragmatic people averse to unnecessary complications. They will consider a new hypothesis only if it promises some hope: that of explaining something that was not understood before. However, many theoretical physicists are seeking to construct composite models according to which quarks and electrons would be made up of even simpler units (called preons or rishons, depending on the author). It is true that these models encounter great obstacles. They have not yet proved themselves to be viable descriptions of reality. But this situation could change.

THE LEVELS OF BINDING ENERGY

We note that the pyramidal steps explored successively by nineteenth-century chemists and twentieth-century physicists correspond to different levels of energy. The strength of the links that bind physical systems like molecules, atoms, nuclei, and nucleons can be measured in terms of binding energy units. This quantity specifies the amount of energy that must be injected into a system in order to dissolve the bonds. The unit currently in use is the electron volt (ev).[3]

To separate a molecule into its constituent atoms requires only a few electron volts. To disintegrate a nucleus into nucleons, millions of electron volts must be injected. To affect a nucleon's inner structure, the energy required is measured in billions of electron volts (see Appendix 4).

Increasingly powerful accelerators were needed to explore the inner structure of the nucleus. The earliest machines were barely able to attain the several million electron volts needed to split atomic nuclei. Energies of more than one billion electron volts (10^9 ev) were obtained in the late 1950s. Today, accelerators in Geneva, Chicago, and Stanford are probing the state of matter at energies around 10^{12} ev. These powerful machines accelerate particles almost to the speed of light (three hundred thousand kilometers, or one hundred eighty-six thousand miles, per second). They are the scalpels required to "dissect" a nucleon and reveal its inner structure, the quarks. The accumulated evidence gives ample confirmation of the validity of the present theories of matter.

The contest to build a bigger particle accelerator continues. Projects underway, for instance the SSC to be built in Texas, will attain several tens of trillions of electron volts. With such tools, physicists will obtain important new results in the decades to come. We may learn whether the pyramid of alphabets stops at quarks or goes down even lower.

The Upper Steps: Biology

Having discussed the contributions of chemistry and physics to the elucidation of the superimposed alphabets of nature, we turn now to the contribution of the biological sciences. Over the past two centuries, the importance of chemical phenomena in plants and animals has been demonstrated. The archaic distinction between inert and living matter has gradually faded. Frogs, like rocks, are made up of atoms that can be identified by chemists.

A major breakthrough occurred when it became possible to synthesize in the laboratory substances normally secreted by living glands. This marked the appearance of a new science, biochemistry, the study of the chemical processes taking place in the structure and functions of biological systems. Research in this field led to the discovery of the existence of extremely important new entities: the giant molecules (or biomolecules) of the living world. These are really impressive in size; some proteins include several million atoms (mainly carbon, nitrogen, oxygen, and hydrogen) arranged according to extremely complex architectural patterns. The mobile forms thus produced possess the properties needed to accomplish vital functions.

When, in the 1950s, the genetic code was cracked, the alphabet metaphor was validated more literally than ever before. The cells of living things contain the information essential to the maintenance and reproduction of life. These instructions are written using an alphabet consisting of four letters: *A*, *C*, *G*, and *T*. Each letter stands for a nucleotide, a molecule made up of a few dozen atoms. The "words" obtained when the four letters are arranged in the proper order are the genes. A single gene may require several thousand nucleotide letters. The gene sequences then assemble to form chromosomes. Several dozen chromosomes are present in the nucleus of every living cell.

In the human body each cell contains forty-six chromosomes, each of which is made up of millions of genes, that is, billions of nucleotides incorporating, in all, more than a trillion quarks and electrons. The molecular sequences, curled tightly within the cell nucleus, would measure several yards in length if unwound. Every living species and every individual within the species possesses its own particular sequence, the key to its identity.

Let us move up another step. Living cells become the letters of a new combination. Approximately two hundred different types of cells suffice to describe the physical makeup of any member of the mammalian class. Assembled in various ways, they spell words like *dog*, *cat*, *elephant*, and *kangaroo*. By including a few additional kinds of cells, we could account for all the living things on earth, both plants and animals.

On the pyramid of complexity, humans are on the same level as seagulls and sardines. The "emerging properties" of each are what

The Rising of the Pyramid in Time

An illustration of the various processes by which the pyramid of complexity rises during the course of cosmic history, as discussed in the text.

Note in particular how the strength of binding progressively decreases as we climb up the steps. The nuclear force manifests itself in a strong version (binding quarks in nucleons) and later in a weaker version (binding nucleons in nuclei). The (much weaker) electromagnetic force manifests itself in a strong version (binding electrons to nuclei, forming atoms) and in several progressively weaker versions (binding atoms in molecules and molecules in life processes).

I adopt here, for convenience, the "model" age calendar based on the hypothesis of a "zero" age of the universe.

Step A: The Birth of the Nucleons (protons and neutrons)

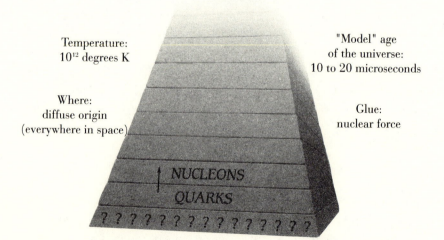

Temperature: 10^{12} degrees K

Where: diffuse origin (everywhere in space)

"Model" age of the universe: 10 to 20 microseconds

Glue: nuclear force

NUCLEONS

QUARKS

? ? ? ? ? ? ? ? ? ? ? ? ? ? ? ?

Nucleons are composed of quarks bound together by the nuclear force. The glue is obtained by the continual exchange of particles called "gluons," giving rise to the strongest bind in nature. This statement is essentially a summary of high energy physics in the second half of our century.

Step B: The Birth of the Atomic Nuclei

Temperature:
10^9 to 10^6 degrees K

"Model" age
of the universe:
from 10^2 seconds
until now

Where:
first, diffuse origin;
later, stellar interiors

Glue:
nuclear force

ATOMS

NUCLEONS

QUARKS

? ? ? ? ? ? ? ? ? ? ? ? ? ? ? ?

In the first part of our century, physicists showed that all atoms are composed of nucleons. The binding is again provided by the nuclear force, although in an appreciably weaker version than for the case of the quarks.

Step C: The Birth of the Atoms and the Molecules

Temperature:
10^4 to 10^2 degrees K

"Model" age
of the universe:
from 10^6 years
until now

Where:
supernovae remnants,
interstellar clouds

Glue:
electromagnetic
force

MOLECULES

ATOMS

NUCLEONS

QUARKS

? ? ? ? ? ? ? ? ? ? ? ? ? ? ? ?

The history of chemistry in the eighteenth and nineteenth centuries is essentially the story of how scientists discovered that molecules are made of atoms. Atoms are made of electrons bound to atomic nuclei by a strong version of the electromagnetic force. This force is obtained by the exchange of photons between charges. A weaker version of the same force provides the binding of atoms in molecules.

Step D: The Birth of the Living Cells (on earth)

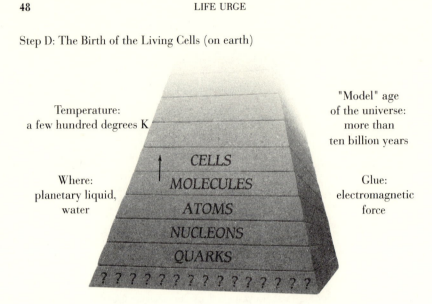

Temperature:
a few hundred degrees K

"Model" age
of the universe:
more than
ten billion years

CELLS
MOLECULES
ATOMS
NUCLEONS
QUARKS
? ? ? ? ? ? ? ? ? ? ? ? ? ? ?

Where:
planetary liquid,
water

Glue:
electromagnetic
force

Biochemistry can be related as the story of the discovery of the chemical nature of living processes. In a still weaker version than at lower levels, the electromagnetic force governs structures and evolution at this level.

Step E: The Birth of the Organisms (on earth)

Temperature:
a few hundred degrees K

"Model" age
of the universe:
more than
ten billion years

ORGANISMS
CELLS
MOLECULES
ATOMS
NUCLEONS
QUARKS
? ? ? ? ? ? ? ? ? ? ? ? ? ? ?

Where:
planetary surfaces

Glue:
electromagnetic
force

Here the situation is about the same as for the birth of the living cells.

distinguish it from other living things. Fundamentally, the human and the canine brains are both systems of neurons.

What level of organization is at the next step? There, we could place social systems: an ant colony, a hive of bees. These insect communities have often been likened to a single organism with coordinated behavioral patterns; taken from its colony, an ant is as helpless as a cell separated from a living body. In its natural habitat, however, it fulfills one of the multiple roles needed for the good functioning of the anthill.

The preceding has been an attempt to sum up the recent history of the science (see the illustrations on pages 46–48). Biologists "spell" living organisms in terms of cells and cells in terms of biomolecules; chemists decipher the biomolecules as combinations of smaller molecules and the latter as association of atoms; physicists smash the atoms into nuclei and electrons and the atomic nucleus into nucleons, which themselves are found to contain smaller entities called quarks. All physical systems are cemented together by natural forces. At the lower levels, *nuclear* force keeps the nucleons from flying apart; on the higher rungs, *electromagnetic* force plays a similar role.

The relative strengths of these forces have a great bearing on our analysis. Our pyramid is characterized by *a decrease in the strength of its bonds* as we ascend its steps. Nucleons are more strongly bound than nuclei, nuclei more strongly bound than atoms, and atoms more strongly bound than molecules. Biological systems follow the same trend. The higher the pyramidal step, the more fragile and vulnerable the structure. The base of the pyramid is like reinforced concrete, sturdy and practically immutable (powerful accelerators are needed to break nuclei). On the upper levels, we find weak bonds and delicate structures, easily disrupted, in need of protection. *This fragility of complex systems will play a fundamental role in our story.*

Our written alphabets are five to six thousand years old; those of nature are fifteen billion years old. Once again, human inventors have been outdone. It's as though Sir Edmund Hillary, when reaching the peak of Mount Everest, had found someone else's footprints in the snow. Instead of saying, "Nature is structured like a language," we should say, "Language is structured like nature."

We know for certain the "ground level" of our pyramid of human language: Letters, conventionally adopted for historical reasons, cannot be decomposed into smaller units.[4] In nature, however, the concept of elementary particles is provisional, continually challenged by researchers' insatiable curiosity. An object is called an elementary particle merely because physicists do not yet have reason to believe that it contains an inner structure and constituent parts. As inconclusive as this definition may be, we have to be satisfied with it.

COMPLEXITY AND DIVERSITY

Why the image of a pyramid to describe the scale of complexity in the universe? By definition, the width of a pyramid decreases progressively as we move up. While specimens from the lower levels abound throughout the universe, only a small fraction of matter ever reaches the upper rungs.

Almost all quarks in the universe are associated in trios in the form of nucleons. (A lone quark has never been observed.) No more than two out of every ten nucleons, however, are integrated in a complex atomic nucleus. The eight others wander about in the form of free protons or as hydrogen atoms. Likewise, the number of molecules that consists of several atoms is infinitesimal in comparison with the number of single atoms. And only an extremely tiny fraction of the molecular matter is incorporated in cells. Finally, the mass of all specialized, evolved multicellular organisms is nothing compared to the mass of the microscopic cell life pullulating on the planet. The pyramidal form therefore provides an excellent analogy for the gradual rarefaction of systems along the ascent.

If we turn the pyramid upside-down and stand it on its point, we can illustrate another property of the structure of alphabet: the increase in the number of *species* inhabiting each level. At the tip of the upside-down pyramid, we find six species of quarks (of which only two are used in natural conditions, *u* and *d*). The nucleons catalogued to date number in the dozens (but again, only two types can be found in nature, the proton and the neutron). More than two thousand types of atomic nuclei (isotopes) can be counted; they give rise to nearly one hundred different chemical species, called atoms. On the next level, to get some idea of the potential number of molecules, we could leaf through a catalogue of the medicines sold by a large pharmaceutical firm: thousands of pages covered with fine print. Even so, this is far from being an exhaustive list of all the possible combinations of atomic building blocks.

As for cells, let us recall that each of our chromosomes is made up of a chain of a billion letters, or nucleotides, and that each letter may be *A*, *C*, *G*, or *T*. How many different combinations are possible, if each nucleotide can take on any of the four possible values? A number of six hundred million digits is required to write the answer (ten to the six-hundred-millionth power) (see Appendix 1).

In terms of these quantities the proliferation of animal species (one hundred fifty thousand mushroom species, for example, or one hundred ten thousand sorts of butterflies) does not seem exaggerated. Nor does the existence of five billion human beings, each one entirely unique. Nature's capacity for variation is still alive and strong.

To act upon their environment and to interact with the rest of the universe are properties all natural systems share. Atoms absorb and emit light. Molecules change shape, form compounds, and dissociate. Bacteria move toward their food sources. The diversity, specificity, and scope of the activities increase as the level of complexity increases. Termites control the temperature of their colony to within one degree of optimum. Human beings build stereo systems, perform microsurgery, and analyze the soil of planets in their solar system. The scale of performance corresponds to the scale of complexity.

But we need a precise definition of the word *complexity*. Although we each have an intuitive notion of what the word means, the meaning is not easy to pinpoint. For a start, let us say that a complex system is one that contains a large number of component particles. However, the mere inclusion of the large number of particles does not automatically mean that a system is complex. The arrangement of the components is of key importance. We will soon discuss the idea of *information* in a given system, which is related to the nonrandom arrangement of components. The more complex the system, the more information it requires to form itself and to pursue its specific activities. This will be clear when appropriate illustrations are given later.

We may also associate the concepts of specificity, personality, and individuality to the idea of complexity. Elementary particles are identical, indistinguishable from one another. The higher we climb the pyramid of complexity, the more ways we find to differentiate between forms sharing the same level. This individuality is marked by unique, specific types of behavior.

THE GROWTH OF COMPLEXITY

At the beginning of this chapter we attempted to summarize the whole of scientific knowledge in a few keys sentences. To our first statement, "Nature is structured like a language," we will now add a second: "The pyramid of complexity rises over the course of time." In Chapter 4 I will treat in greater depth the observations and arguments justifying this statement. For the moment, however, I will simply touch on this history of the growth of complexity.

Fifteen billion years ago the universe was at the base of the pyramid. All matter was maintained by extremely high temperatures in a state of complete and permanent dissociation. Or more precisely, any bond was immediately destroyed. This was the primordial chaos; there were neither structures nor organizations.

But the universe is expanding, and the expansion causes cooling. Temperature and density decrease with time, and at certain moments, the cooling process is marked by associative events that allow matter to organize. Successively, matter will engender all the natural systems in the pyramid of complexity.

In the beginning these events took place simultaneously everywhere in the universe. Later, they were confined to specific regions at given times. When the temperature dropped below one trillion degrees Celsius, trios of quarks began to assemble, giving rise to nucleons. As the temperature cooled to one billion degrees, a smaller number of the nucleons merged to engender the first helium nuclei. At that point, about one minute has passed on the cosmic clock.[5]

A million years later, when the temperature had fallen enough to allow electrons to stabilize in orbit around protons, the first atoms and molecules of hydrogen formed. This was approximately the time when microwave background radiation was emitted.[6]

Several hundred million years passed before the first galaxies appeared. Mystery still shrouds the circumstances that brought about their birth. Later, after successive fragmentation, galactic matter gave birth to the stars. After this epoch, associative events no longer took place everywhere at the same time; they became localized in specific regions of the cosmos and stretched over billions of years.

Stars, at their incandescent core, induce nucleons to associate, forming nuclei (helium, carbon, silicon, and iron). Upon the death of a star, these materials are hurled into interstellar space, where they capture electrons and form atoms. The atoms bind to one another, making molecules, as well as minuscule crystalline structures, the dust grains of sidereal space.

Still later, some dust grains clung to one another, generating extended solid bodies, the asteroids and planets. On the surface of the largest ones atmospheres and oceans may be deposited. More molecular interactions taking place in these fertile fluid layers may produce chains of molecules, forming biomolecules, living cells, and organisms.

Every year, new stars appear in our Milky Way. Stars born soon after the birth of a galaxy do not possess solid planets: An infant galaxy is made only of hydrogen and helium, but iron, oxygen, and silicon nuclei are the links in nature's rigid structures. Several generations of stars must live and die — over billions of years — in order to generate planets. When our sun was born, the galaxy had been functioning for ten billion years. The eldest stars had already set free into space their crop of heavy nuclei, so solid planets could form.

Our telescopes are not yet powerful enough to ascertain whether other stars have planets. However, we have reasons to believe that among the hundred billion stars in our Milky Way there are many — perhaps younger, perhaps older, than our sun — that possess planetary systems similar to our own. Cell life appeared on our planet about four billion years ago; intelligent life, some million years ago. Other planets may present different evolutionary history and state of development.

What matters is the fact that the scale of complexity corresponds to a historical sequence linked to the cooling of the universe. Step by step, matter ascends the pyramid when corresponding systems are no longer threatened with dissociation because of the heat. In future eras, will there be new steps above those we know? Who can say? Seeing the blue-green algal cell floating in the primeval terrestrial ocean, could we suspect that someday, given four billion years of additional evolution, a new level of the pyramid of complexity would appear over its head?

Is cosmic evolution still going on? We have no reason to believe it is not. Over time, it seems to have accelerated, rather than slowed down. In only four million years the average weight of the humanoid brain grew from five hundred grams (about a pound) to its present fourteen hundred grams. And four million years is only a tiny fraction of the total duration of biological evolution on earth.

In order to grasp this notion, imagine that the life of our planet could be squeezed into a single twenty-four-hour day. On this scale each minute represents three million years. The earth is formed at midnight. At six A.M. algae and bacteria are proliferating in the tepid waters. The earliest mollusks and crustaceans do not appear until the evening, around six or seven P.M. Dinosaurs make their entrance at 10:30 P.M., only to leave at 11:40. In the last twenty minutes of this day, mammals swarm over the planet and rapidly differentiate. Our primate forebears arise at ten minutes to midnight, and in the final two minutes their brains triple in size. In the light of an accelerating sequence, therefore, it seems wiser to top our pyramid with a question mark.

3

At the Sources of
Cosmic Organization

Certain texts give pause for reflection; they force us to take a stand. We realize that we disagree, but we don't know exactly why. Confronted with a well-stated argument, we have nothing but vague discomfort. Such writings are valuable. They force us to react. They open the mind to a new train of thought.

This is how I felt when I first read a certain passage from Claude Lévi-Strauss's *Tristes Tropiques*. I feel that it summarizes the vision of the world held true by many scientists, before modern cosmology came into existence. In the chapters to come we will refer to this passage often, almost line by line; the phrases to be commented upon in greater detail are printed in italics:

The world began without man and will end without him.
The institutions, morals and customs that I have spent my life noting down and trying to understand are the transient efflorescence of a creation in relation to which they have no meaning, except perhaps that of allowing mankind to play its part in creation. But apart from the independent position given to man and to his endeavours — even if they are

doomed to failure — in opposition to universal decline, he
himself appears as perhaps the most effective agent working
towards the disintegration of the original order of things and
accelerating powerfully organized matter *towards an ever
greater inertia, an inertia which, one day, will be final.*
From the time he first began to breathe and eat, until the
invention of atomic and thermonuclear devices, by way of
the discovery of fire — and except when he has been engaged
in self-reproduction — what else has man done except
blithely break down billions of structures and reduce them
to a state in which they are no longer capable of integration?
Of course he has built towns and cultivated the land; yet, on
reflection, urbanization and agriculture are themselves
instruments destined to create inertia, at a rate in an
infinitely higher proportion than the amount of organization
they involve. As for the creations of the human mind, their
significance only exists in relation to the mind itself, and
they will merge into the general chaos, as soon as the
human mind has disappeared. Thus civilization, taken as a
whole, can be described as an extraordinarily complex
mechanism, which we might be tempted to see as offering
an opportunity of survival for the human world, if its
function were not to produce what physicists call entropy,
that is inertia. Every verbal exchange, every line printed,
establishes communication between people, thus creating an
evenness of level, where before there was an information gap
and consequently a greater degree of organization.
"Anthropology" could with advantage be changed into
"entropology," a discipline concerned with the study of the
highest manifestations of this process of disintegration.[1]

In brief, the text states more or less the following: It is natural for
everything to fall apart, and the presence of humanity serves only to
accelerate this destructive process. A nuclear holocaust would be the
next logical step. After all, why not bring this absurd race to the void
to its swiftest possible conclusion? The author does not mince words
when he comes to the question of the "meaning" of reality and human

life: Both to him are pointless, and any impression to the contrary should be chalked up to a vast illusion, bound to be refuted on deeper examination.

This reasoning is based on the cosmological vision formulated by physicists in the late nineteenth century. The findings of contemporary astronomy have radically altered that viewpoint. But before we can realize the full scope of this transformation, we must understand the elements forming the basis of Lévi-Strauss's thesis. Such will be our goal in the next few pages.

THERMAL DEATH

Warm bodies tend to communicate their heat to cooler ones. If enough time is allowed to elapse, temperatures will equalize. When nineteenth-century physicists extended this observation to the universe as a whole, they came up with the image of the thermal death of the world. Today, they said, the objects in the universe, such as stars and planets, present great differences in temperature. Such a state of affairs can only be temporary. Temperature differences will gradually shrink and disappear, inevitably bringing about the death of the universe. But why the *death*?

Imagine a ship becalmed on a smooth sea. Its fuel tanks are empty. The weather is sunny, and the water is warm. The sea is packed with "thermal energy," enough to fuel the boat for millions of years. But the captain cannot use this energy; for him it is worthless. He is unable to transform the sea's thermal energy (stored in the disorderly movements of the water molecules) into mechanical energy (which would move the ship).

Why? Because the sea temperature is (essentially) uniform. Temperature differences (so-called thermal disequilibrium) must be created in order to set the boat in motion.

How? Bring the captain some fuel. The explosion of the fuel droplets in the cylinders creates heat. This heat pushes the pistons, which in turn activates the engine shaft and propeller. The ship can continue its journey.

The earth does not possess an inexhaustible supply of fuel, as the oil crisis reminded us in 1973. Nor does it have endless sources of nuclear fuels. The sun and the other stars are not immortal. When the heat sources are spent, motion will stop, even if the universe is chock-full of energy. If there should be a universal thermal death, it would be caused by the progressive decrease in the thermal differences of natural bodies and therefore by the deteriorating quality of existing energies.

ORGANIZATION, CHAOS, INFORMATION, ENTROPY

Mathematics can help us flesh out the intuitive image we have of the above concepts. For those who wish to pursue the subject, I advise such gymnastics. Appendix 1 contains a few exercises that should shed light on the premises put forth in the main body of this book, should those premises seem a bit vague. We can become better acquainted with the mathematical content of the concepts of information and entropy by rolling dice, flipping a coin, or drawing lots, which will demonstrate that, contrary to popular belief, they can be strictly defined and often (but not always) formulated in quantitative terms.

The problems to which we shall apply these findings are of interest to all. Even a reader with little taste for mathematics should be able to participate. And simplified summaries have been added here and there to help readers of different scientific backgrounds follow the thread of my discussion. However, if so inclined, the hurried reader may immediately jump ahead to Chapter 11, which provides a synthesis of the preceding chapters, enabling one to tackle the book's conclusions.

ENTROPY AND QUALITY OF ENERGY

The concept of entropy is rich and complex. It applies to reality on many different levels. It is an idea that may assume varying forms and appearances. I intend to treat only the qualitative aspects of entropy by

using examples from daily life. Since these examples are mainly illustrative, they must be used with care.

The becalmed captain showed us that the mere presence of energy is
not enough to create movement. Energy varies in quality. *Entropy is a
measurement of the value of an energy, its "usability."* But (and this is
important!) this measurement is expressed in an inverse proportion:
the *higher* the entropy, the *less usable* the energy.

In the ocean entropy is at a peak; the thermal energy stored in the
water is useless for the captain. Conversely, gasoline is a low-entropy
energy source. A large portion of the chemical energy contained in
petroleum hydrocarbons can be transformed into mechanical motion
(or into electricity, through the use of a generator). A large portion of
the energy is usable, but *not all* the energy: The hot gases emitted from
the exhaust pipes take away part of the gasoline's energy, and nothing
is gained. Is this wastefulness? No, it's an inevitable loss. The phenomena of real life always involve a deterioration of the quality of energy. It
is impossible for us to use one hundred percent of the quantity of
energy available at the outset.

ENTROPY AND WASTE

Living, eating, working, washing: All these processes create waste. We
toss onion peels, cellophane wrappers, and disposable diapers into
trash cans. What happens to them? Who cares! We trust the garbage
collectors to get rid of them, that's all.

When the sanitation workers go on strike, the trash piles up and
paralyzes our existence. In 1963 some New Yorkers had to evacuate
their homes because of the accumulation of trash in their streets.

Like the hot gases of the exhaust pipes, waste is part of our effort to
live and to organize our homes, meals, and work. The excrement we
pass after absorbing the nutrients in the food are part of the same
phenomenon. All organizational processes generate entropy, to be
evacuated as soon as possible.

A STRANGE CASINO

Steam engines were important during the industrialization of our society in the nineteenth century. Inventors sought ways to maximize their efficiency. Their studies produced concepts like the laws of thermodynamics and the ideas of energy and entropy.

The first law of thermodynamics rules out the possibility of perpetual motion, an ever-popular dream in previous centuries. No one has ever found a bottomless well. A ton of coal will produce a certain quantity of energy, no more. You might as well accept the harsh truth: You get what you pay for. There's no such thing as a free lunch.

When the Superphoenix breeder reactor went on-line in France, television newscasters triumphantly spread the word that this reactor produced more electrical energy than it consumed. This information, trickling into the living rooms, must have left more than one person perplexed. Energy is not *produced* by the reactor; it is already present in the nuclei of uranium-238 atoms. These nuclei are reluctant to release their energy, however. The reactor merely makes the energy usable. We should take note of this example, because it will be quite helpful later.

Let us return to the laws of thermodynamics. The second law is as frustrating as the first, if not more so. The energy available to us cannot be used entirely for our benefit. Some loss is always involved. Any event — a ship's navigation, an animal's life, a hurricane in the Pacific — brings about a deterioration of energy. The total entropy of the world can only increase. Considered in these terms, nature is a casino where the odds are always against us. Not only can we never win (first law), but every time we play, we lose a few chips (second law).

The second law of thermodynamics leads us quite naturally to the idea of the world's thermal death. If every natural phenomenon inevitably adds to universal entropy, through the necessary deterioration of some preexisting forms of energy, we are tempted to perceive each event as a step closer to the final state, in which all motion would cease due to an absence of usable energy. Hence the image to which Lévi-Strauss refers: inexorable progress toward ever greater inertia, an inertia that will one day be final. As the cause of so many events, movers and doers as they are, the human race makes an ideal executioner for its own universe. In this context nuclear annihilation is almost too normal, too foreseeable.

How can modern astronomy and physics challenge what appears to be such irrefutable logic? The subject will be discussed at length in the following chapters. Here I give only a brief overview. Cosmological observation has introduced the image of an ever-expanding and cooling universe. This process of expanding and cooling increases temperature differences between star cores and the interstellar void. It is this vast movement of all the galaxies that rescues us from an inevitable rendezvous with final inertia; indeed, every passing day takes us farther from it.[2] Energy forms that are unusable in certain physical conditions may be released by a modification of these conditions. Breeder reactors provide a good example. By transforming the practically unusable uranium-238 into plutonium-239, they give access to nuclear energy stored billions of years ago in the uranium nuclei. These unfamiliar concepts will be easier to grasp when we introduce the concept of information. Information is what provides energy with its quality and value.

ENTROPY AND INFORMATION

For millennia, human beings have marveled at nature's order. But does such an order really exist, or is it just a figment of human imagination? In 1749 Denis Diderot wrote *la Lettre sur les aveugles* (letter on the blind), an essay so controversial that it landed him in prison. In response to a letter from Voltaire, Diderot took up again his pen and wrote a reply that includes the following excerpt:

What am I to think of this marvelous order visible all
around me, the infinite interrelatedness of all the things I see?
    That they are metaphysical entities that exist in the mind
alone. Imagine a field covered with litter, some of which
happens to make comfortable homes for worms and ants.
What would you say if these insects, mistaking the random

scattering of litter for some great design, admired the beauty
of the architecture and worshiped the higher intelligence of
the gardener who had placed the objects at their disposal?

Like the ants that sees order in a trash heap where, according to
Diderot, there is nothing but chaos, we, who admire the beauty of
natural order, could be merely the victims of an anthropomorphic
illusion. Such is the conclusion of Diderot.

Two centuries after Diderot, we know that order and organization
are not purely "metaphysical entities" existing merely in the eye of the
human beholder. By counting the number of specific instructions
needed to build a system, we can give an objective estimation of its
"level of organization." This leads us to another aspect of the concept
of entropy. Very loosely stated, entropy is a measure of the amount of
"disorganization" or "absence of organization" within a given system.
The more organized a structure is, the lower its entropy. Entropy
reaches its maximum value when the system reaches a state of total
disorganization, or chaos.

Entropy can be opposed to information. For example, the information needed to assemble a television set is printed in the manufac-
turer's manual. The more complicated the set, the greater the amount
of information. But no one needs a manual to assemble a pile of
junkyard scrap, using the same elements as those in the set. The
entropy of the scrap would be much higher than that of the television
set.

The entropy could be even higher. But no accident, no matter how
violent, can reduce a TV set to a state of total chaos. Some of its
components would still be recognizable. The same is true of Diderot's
trash dump. The broken dishes, smashed-up furniture, and twisted
metal are useless to humans, but they contain a sufficiently high level
of organization to suit the needs of ants and worms; on their scale the
litter is useful as shelter. It would be more judicious for Diderot's
animals to feel grateful to the potters and cabinet makers who fashioned the objects littering their field than to the gardener, who merely
scattered them. Even when broken, these objects retain enough organization (contain enough information) to make decent homes for worms

and ants. In other words, when human users broke the objects and scattered them as trash, they did not eliminate every bit of information introduced by the craftsworkers into their handiwork.

Can the amount of information in a given object be calculated? Yes, we might define it as the number of items required to provide a complete description of an object to someone who has absolutely no knowledge of or contact with the object.

If you have you ever played Twenty Questions, in which one player tries to guess what another player has in mind by asking a series of "yes" or "no" questions, you are aware of the fact that the number of questions needed is greater if the object in mind is highly specific. For instance, fewer questions are necessary if the secret is "a bird" than if it is "the cardinal nesting behind the barn."

Let's imagine we're trying to transmit the information contained in a painting or on a television screen. If the surface is all white, all we need to say is "White everywhere." If it is not all white, we have to divide the surface into little squares, and provide specific information for each square. For example, the brightness of an image could be expressed in certain units on a scale, from faintest to strongest, each step subject to a "yes or no"–type binary response. The same would be true for colors. If the image is detailed, the size of the squares has to be accordingly reduced, and the number of questions, which pertain to individual squares, is necessarily greater than for a simple image.

The photocopy of a drawing is always less accurate than the original. If the system is at its best, the copied image may be almost as good, but it will never be the same as the original, let alone better. A photocopy of the photocopy brings about another deterioration. Each further photocopy diminishes information. The drawing gradually blurs. Soon retrieving information from it becomes impossible. The concept of entropy reappears here as a measurement of "unusability" caused by information loss.

A Ferrari smashing into a wall becomes a heap of scrap metal (loss of information). But, of course, no collision can transform scrap metal into a race car. This example and that of the nth-generation photocopy illustrate a well-known fact about reality: Organization tends to diminish, and entropy tends to increase. With due credit to folk wisdom, let's say that "things" left to their own devices go from bad to worse.

"Things are bad, things are bad," says the pessimist; "they couldn't be worse."

"Of course they could!" the optimist replies.

Scientific observation shows us nature as an incredibly organized system. The regularity of planetary orbits and the precision of biochemical mechanisms reveal that matter can be structured to a very high level, corresponding to a very low level of entropy.

Nineteenth-century physicists, conscious that systems tend to decay over the course of time, were forced to assume that organization had been even greater in the past. From this assumption they conceived the ideas of an "original order" and "strongly organized matter" that would have existed at the dawn of time. In addition, they concluded that the present order of the world was inevitably bound to deteriorate and disappear. Thus, Lévi-Strauss's text refers us to a distant past, a sort of Garden of Eden, from which we are continually being expelled. What exactly is the original order in this context? In more modern terms, what was the nature of the primordial "information" that allows us to use the energies of the universe today?

The development in the past decades of a genuine "scientific cosmology" is still largely unknown outside the small community of physicists and astronomers. Many people still confuse cosmology and mythology. This is why I've decided to devote much of the next chapter to the *history of the history of the universe*. In the following pages I will summarize the events that secured this vision of the world in scientific thought, anchoring it as solidly as had, for instance, Darwin's theory of evolution. This history will also be helpful when we introduce another resource vital to our quest for cosmic information. In 1965 two American radio astronomers, Arno Penzias and Robert Wilson, discovered the existence of a "background radiation" evenly spread throughout the universe. Emitted fifteen billion years ago, this radiation gives us an image of the most ancient past of the universe. The information contained in this image will be of major importance for our story.

SUMMARY

Nature's organized structures tend to deteriorate over time. Energies become less and less usable. It would thus appear that the universe is doomed to "die"; that is, it will eventually reach a state in which all movement and life are impossible. This is how physicists saw the cosmos in the late 1800s. Modern astronomy, however, provides an entirely different image of the history of the universe.

4

The Entropy of the Universe

THE HISTORY OF THE HISTORY OF THE UNIVERSE

The scientific method for exploring the past is based on celestial observation. Sometimes very simple questions take us quite far (or quite far back).

What keeps the heavenly bodies from falling out of the sky? No doubt this question has been asked for many thousands of years. The ancient Greeks replied that heat tends to rise. Empedocles of Agrigento, who had observed lights leaping around ships' masts in stormy weather (Saint Elmo's fire), believed that the flames escaping from the masts ascended into the heavenly vault and gave birth to the stars in the sky.

Two thousand years later, Newton discovered why the moon doesn't fall onto the earth[1]: It's because the moon *revolves* around our planet. The centrifugal force created by the rotation opposes the gravitational force that attracts bodies to one another. The balance between these two forces is what determines our satellite's orbit. A similar principle governs the rotation of the earth and the other planets around the sun.

What about the stars? Newton knew nothing of the structure of galaxies. He assumed that stars were evenly sprinkled throughout space. What keeps them away from each other despite their mutual attraction? Why don't they all converge at some point in space? And what would determine the location of such a point? These were troubling questions that Newton was unable to answer. He never found a satisfactory explanation for the problem of the stars' movement.

It was later discovered that stars revolve around the center of our galaxy. Again, centrifugal force is what counters the gravity of the billions of celestial bodies in the Milky Way and maintains the sun and other stars in their galactic orbit. But the next question is, What keeps the galaxies themselves away from each other?

When Einstein announced the general theory of relativity in 1915, he explained — luminously — the relations among time, space, matter, and gravitational force. Nature still held a surprise for Einstein, however: *A universe of matter cannot be static.* The theory clearly expresses the idea that "cosmic matter" has to be in some state of "motion."

If I showed you a photograph of a stone suspended in midair, you would say that this is a "snapshot," something that "froze" the action. You know that this state can only be temporary. There are only two possibilities: Either the stone is falling, or it is rising because someone has just thrown it. In both cases, however, it is moving. Your common sense tells you that.

Einstein's equations say essentially the same thing about galaxies in space. They cannot remain immobile: they draw either nearer to or farther from one another. Einstein found this conclusion unacceptable. Why? I admit that I've never clearly understood his reason. Einstein seems to me a thinker in the Greek Apollonian tradition. He has faith in the *intelligibility* of the universe. Reality must fundamentally be simple, clear, and harmonious.

His God, inherited from the age of Enlightenment and from the deterministic vision of Newton and Laplace, is a strict mathematician who "does not play dice." But are we to suppose that, for Einstein, stability and permanence are the essential attributes of any self-respecting creation? For Einstein, perfection and transformation are apparently irreconcilable.

In 1915 Einstein thought of a possible way to avoid this unpleasant situation. It was mathematically acceptable to add a new term to his equations. This term, dubbed the cosmological constant, could in principle be given any numerical value. Einstein decided to add such a term and to assign to it the numerical value needed to stop the movement of the galaxies. This assumption was entirely arbitrary. Its only justification, its only excuse, lay in the aesthetic of immutability. And it soon became obvious that this was altogether a mistake. The artificially induced arresting of galactic movement could not hold. Sooner or later, galaxies would start to move again.

When, ten years later, Edwin Hubble discovered that the galaxies move away from one another at speeds proportional to their distances, Einstein bitterly regretted his blunder. In 1915, his immobilist aesthetic had deprived him of an astonishing prediction — that *the cosmos cannot be stationary*. Einstein would have been unable at any rate to determine whether the overall trend of galactic motion is outward or inward, expansion or contraction because his equations allowed for either possibility. By measuring the reddish glow of the galaxies, however, Hubble determined that our universe is expanding.[2]

In the following decades other authors — George Gamow, Georges Lemaître, and Alexander Alexandrovitch Friedman — drew even more earthshaking conclusions from these observations. Physics teaches us that as matter expands, it cools; therefore, in the past, the universe was hotter. Heat and light are intimately bound together (another physical fact); thus, in the past the universe glowed. Gamow pursued the argument further: This ancient radiation could not have disappeared entirely; traces of it must remain. In what form? His research led him to predict the existence of a glow — invisible to our eyes but detectable with the proper radio equipment — that is evenly distributed throughout the cosmos.

A conservative hides in the soul of every scientist. He is always skeptical of new ideas and always insists on a solid demonstration of their validity. The more outlandish a theory, the more it departs from accepted views, the more reticent he will be, the more numerous and serious the justifications will have to be. The fact that Gamow's prediction was almost universally ignored for thirty years, until fossil radiation was detected *by accident* in the course of unrelated research, shows to what extent the concept of a *dynamic universe* is foreign to conventional scientific thought.

In the same context, we need only to recall how difficult it was for Darwin's theory of evolution to shake the firm belief that animal forms were fixed and unchanging.

THE OLDEST IMAGE OF THE WORLD

When, in 1965, the astronomers Arno Penzias and Robert Wilson discovered the existence of the universal glow predicted by Gamow (sometimes referred to as "fossil radiation," "microwave background radiation," or "3 K radiation"),[3] they could hardly imagine how profoundly their findings would influence human thought. Their discovery gave clear evidence to support the hypothesis of an expanding universe. Scientists may be conservative, but they do know when, under the weight of accumulating evidence, it's time to accept new ideas and change their ways of thinking. This discovery meant that physicists and astrophysicists had to give serious thought to the idea of a *history of the universe.*

Since that time, there have been several other findings supporting the theory of universal expansion (the relative abundance of the chemical elements hydrogen, helium, and lithium and the number of types of elementary particles)[4] giving the theory even more credibility.

What do we know about this fossil radiation? Its photons, detected by radiotelescopes, were emitted fifteen billion years ago, when the temperature of the universe was three thousand degrees. Ever since, this radiation has been traveling through space, progressively cooling to its present three degrees, as the universe expands. It is the bearer of the oldest image of our world: a portrait of the cosmos at the most remote past.

Examination of this radiation reveals a surprising property: its intensity is the same in every direction. Whether it is measured from the north, south, east, or west, whether we compare two very distant or very close regions of the sky, it shines with exactly the same glow. This implies that the matter that emitted this radiation was everywhere at the same temperature (to within one ten-thousandth of a degree).[5]

What does this observation tell us about the ancient cosmos? At the time when this radiation was emitted, some fifteen billion years ago, the universe was an extremely hot, uniform and isothermic fluid (a fluid with the same temperature throughout). In other words, a vast, incandescent, high-entropy chaos. It was entirely devoid of contrast, structure, and organization. There was no trace of an original order, of "strongly organized matter," as Lévi-Strauss puts it.

Here we are faced with a paradox: If differences in temperature tend to *decrease* over time, how can we explain the transformation from the ancient isothermic universe to today's world, which is so full of temperature contrasts? The truth is, the thermal death, the uniformity of temperature discussed by nineteenth-century physicists, should no longer be located in the future, but in the remote past. In fact, it would be more appropriate to speak of a *thermal limbo* in which matter slumbered, like Sleeping Beauty until she was awakened by the prince's kiss. But who played the role of Prince Charming?

A CAVEAT ABOUT THE COSMIC CLOCK

Before we go any further, I want to stress the need for caution about the chronology of these events. The discovery of universal expansion and microwave background radiation[6] laid the basis for the image of a "beginning of the universe" or a "creation of the world" in scientific literature. But are these expressions justified? Can they be supported by rigorous scientific arguments? Nothing is less certain. In Chapter 7 I will go over the reasons for this uncertainty and present more sober views of what we do and do not know.

For the purpose of discussion, we will adopt the conventional chronology of the Big Bang theory, which assumes the existence of a "time zero" when the cosmic clock began ticking. However, it's better to maintain a certain intellectual distance from this way of counting the time. No one knows what there was before the conventional time zero.

INCREASING THE ENTROPY OF THE UNIVERSE

Lévi-Strauss seems to have everything backwards — thermal lethargy is not lurking in the future, but in the remote past; this is the message borne by fossil radiation, a fossil of the early universe. The history of the universe is not characterized by the deterioration of the initial, strongly organized order; on the contrary, it is the story of the construction over the eons, of the pyramid of complexity, which astronomy, physics, chemistry, and biology reveal in all its splendor.

How can this increase in material organization be reconciled with the laws of thermodynamics? There is a striking contrast between the spontaneous tendency to disorder that we expect, and the slow ascent toward the higher levels of complexity. Some are tempted to see this contrast as proof of divine intervention, of "a little boost from the Great Beyond." Today, however, there is no need to believe in miracles. In the following pages, I will try to show that, from the earliest times, matter possessed all the information it required to begin and then to continue climbing the pyramid of complexity.

All the same, science is not (yet) able to supply the answer to the question, Why did matter contain this information?

Entropy and Light: The Birth of a Star

In order to bring the situation into focus, let's pay close attention to one of the events most representative of the cosmic economy: the birth of a star.

Somewhere in the Milky Way, a mass of gaseous matter slowly begins to collapse inward upon itself under the effect of its own gravity. The matter contracts, becomes opaque, and heats up; it is a new stellar embryo.

For the sake of our discussion, we will arbitrarily divide the universe into two regions: (a) the contracting mass and (b) the rest of space (see the illustration on the following page). A star contains a certain degree of organization. It is not shapeless like most nebulae; instead, it forms an almost perfect sphere, slightly flattened by its own rotation. The

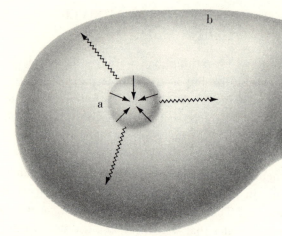

Evaluation of Entropy at the Birth of a Star

The universe is divided into two parts: (a) the volume, where the star is located, and (b) the rest of the universe. During the contraction of matter, which gives birth to the star, the entropy in region (a) *decreases*. The light emitted by the star spreads through the rest of the universe — region (b) — and increases its entropy. Calculating the two contributions, we find that the increase in (b) is greater than the decrease in (a). Net result: The entropy of the whole universe, (a) plus (b), *increases* at the birth of a star.

birth of the star corresponds to a decrease of entropy in region (a), the stellar volume. According to the second law of thermodynamics, however, total entropy cannot decrease. Therefore, the entropy of region (b), the rest of space, must increase by an equal or greater amount.

How does this happen? The star emits light. This radiation disperses and travels through the vastness of space. Physics teaches us that the light emitted by a hot body is one of the most disorganized or entropic entities that can exist (see Appendix 1). Thus stellar radiation contributes its share to universal entropy, to the disorganization of the outer world. This is the price the star must pay to organize and decrease its own entropy in obedience to the second law of thermodynamics.

Organize while emitting light. This is a universal recipe, serving at every level of the pyramid of complexity. Nuclei, atoms, molecules, cells, plants, and animals all make constant use of it. At the upper rungs of the ladder, organization is not created at once and forever as in the case of nuclei or atoms. It must be maintained throughout an existence. The human body is a source of infrared light. To keep our wonderful machine alive, we must give off as much radiation as a one-hundred-watt lightbulb. At death, the light goes out.

Information in the Solar System

The mathematical exercises laid out in Appendix 1 illustrate a very useful principle: The entropy assigned to a system is proportional to the number of elements the system contains. Though true in the cases we shall examine here, this principle is not invariably exact. It will, however, be our golden rule.

When a star shines, it creates photons. These new photons increase the population of particles in the universe and consequently increase its entropy.

Let's look closer at the activity of the famous triad, information, organization, and entropy. Information comes first. Without it, there would be no organization. Entropy is the price that must be paid for an increase in complexity. As an exercise to illustrate the situation, let us analyze the mechanism that allows the triad to operate. There is no need to look far; the sun–earth system will provide an excellent example.

The surface of the sun is hotter than the surface of the earth. Thanks to this temperature difference, or thermal disequilibrium, the sun's energy is usable on earth. This thermal gradient is the source of the *information* by which the biosphere creates and supports life. This temperature difference meets the entropic price that must be paid to the universe. How?

The earth constantly receives a large amount of light energy from the sun. This energy arrives in the form of a shower of yellow photons, rapidly absorbed by the surface of our planet. It is then converted into heat, and reemitted skyward in the form of infrared light (see the illustration on the following page).

The color of the photons is related to the temperature of their source. At 6,000 K, the solar surface is twenty times hotter than the terrestrial one (about 300 K, or 25° C, or 77° F, see note 3). The energy

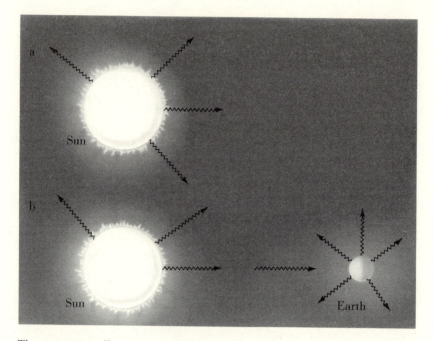

The sun emits yellow photons that increase the entropy of the universe (a). A part of these solar photons is absorbed by the earth and reemitted in the form of infrared photons (b). Each yellow photon is worth *twenty* infrared photons. The entropy released is proportional to the number of photons (our golden rule). Thus the presence of the earth near the sun is a powerful source of entropy.

of the photons is proportional to their temperature. A yellow photon carries twenty times more energy than an infrared photon. The earth reradiates as much energy, on average, as it absorbs from the sun. Hence the number of infrared photons sent out into space by our planet is twenty times greater than the number of yellow photons it receives from the sun.

In its partnership with the hot sun, the earth is a photon-multiplying machine. What does our golden rule say? That entropy is proportional to the number of particles. By increasing the number of photons, the earth–sun duo creates entropy. If the earth did not exist, solar photons would continue to travel through space without stopping. We might say that the earth "shatters" solar photons into particles that are twenty times lower in energy, but twenty times greater in number. That is, they carry twenty times as much entropy.

The infrared photon flow might be pictured as a current of entropy emanating from earth into space, increasing universal entropy. The phenomena responsible for organization and life use this current to their advantage.

Sailing Downwind

On the day of a race, sailors rejoice to see the wind ripple the calm surface of the sea. As the breeze rises, the erratic chaotic movements of the air molecules become gradually coordinated toward a given direction. The coordination of movement corresponds to an entropic decrease of the atmospheric mass (see Appendix 1). It is possible only if accompanied by an increase in the total entropy of the universe. Hence the need for a source of entropy, the current of infrared photons emitted by the earth. (The entropy flow is present, whether or not it is tapped out. A similar flow is emitted by the moon, but, deprived of the appropriate physical conditions, the lunar surface hosts neither wind nor life).

The birth of the wind is a complex phenomenon that meteorologists would like to understand more fully in order to make more accurate predictions. First, an atmosphere is needed. Mars and the Earth have one, but the Moon and Mercury do not. Next, temperature differences are required: hot and cold masses. As the solar energy heats up the equatorial regions, the air expands upward and propagates horizontally to higher latitudes. There, it receives less radiation, cools, and sinks to the ground. A current is thus established that, because of the earth's rotation, circulates clockwise in the southern hemisphere and counterclockwise in the northern hemisphere.

If the sun's surface were not hotter than that of the earth, the North and South poles would be at the same temperature as the equator. No wind would arise, no matter how much or how little energy the sun was giving off. The energy would not be usable.

Yellow solar photons are not all immediately converted into infrared terrestrial photons. For example, when a plant carries out photosynthesis, it intercepts and stores solar energy, using it in order to grow and organize. The "food chain" of the biologist is the sequence of events whereby plants absorb solar energy, herbivorous animals digest the plants, and carnivores devour the animals that have eaten the plants. In physical terms this sequence can be seen as the process by which a tiny amount of solar energy is recycled by increasingly efficient systems. Human abilities like consciousness, intelligence, and intellectual activity require the growth and maintenance of a highly organized structure, the brain. The quantity of information stored and manipulated by the human mind is enormous. The creation of language, arts, and sciences is the product of a system of extremely low entropy.

But deterioration is inevitable. Every organism, be it plant or animal, dies and decomposes. Sometimes, when fossilized, the plant retains its high-quality energy for a long time before releasing it; this is the case with coal and oil. Inside an automobile engine, the combustion of fossilized plants creates the temperature difference required to move the pistons. If the atmosphere was the same temperature as the engine's gases, nothing at all would happen.)

"From the time he first began to breathe and eat, until the invention of atomic and thermonuclear devices, by way of the discovery of fire — and except when he has been engaged in self-reproduction — what else has man done except blithely break down billions of structures and reduce them to a state in which they are no longer capable of integration?" Although this question by Lévi-Strauss, quoted in Chapter 3, touches on a truth, it must be tempered by two observations. First, Beethoven's quartets and the Gothic cathedrals were acquired through this dissociation of structures, and we're glad to have them. Second, on the cosmic scale, information is not diminishing, but continually increasing.

To sum up, human beings, like animals and plants, prey on solar information. Not only do we use the daily crop of light energy, but we also avail ourselves of fuel stored millions of years ago in coal and oil. (We even use information from dead stars, in the form of nuclear fuel.) Taking advantage of the transformation of solar photons into terrestrial infrared photons, we generate increasingly organized structures that are farther and farther in nature from the chaos of the early universe. Scientific research and artistic creations are among the finest examples of this human endeavor.

The Limits of Solar Information

The farther a planet is from the sun, the lower its temperature. On Jupiter, for example, permanent hurricane winds blow in the red spot. These gales take advantage of the enormous flow of entropy emitted when the planet's glacial surface faces the torrid sun. (Is there life on Jupiter? Some authors have imagined a population of strange "birds" gliding through the upper layers of the atmosphere. Perhaps.)

Is there a limit to the multiplication of entropy? The temperature of fossil radiation uniformly spread throughout space is 3 K. It is therefore two thousand times cooler than the surface of the sun. Theoretically, each of the sun's photons could sire two thousand new photons.

The universe has been cooling ever since the Big Bang occurred; cosmic temperature has been decreasing. Year by year, the temperature difference between the surface of our sun and the background radiation increases. And, of course, the same is true of the other stars in the sky.

Yet the ability to use this information is lacking. Only an infinitesimal fraction of the photons emitted by the sun is absorbed by the surrounding planets and nebulae. Nearly all stellar radiation is spread through an increasingly empty universe, only to blend with background radiation.[7]

This is a state of affairs we will encounter again and again. Although the information is present, it remains untapped for lack of an appropriate mechanism. If artificial planets were placed in orbit around the earth by a technologically advanced civilization, they would maximize the harvest of solar information by reflecting back to earth the photons they had captured.

Let us close this section with a visit to the center of the sun. The temperature is sixteen million degrees Kelvin. We call the high-energy photons emitted by thermonuclear reactions gamma rays. Solar matter is opaque to these photons. Absorbed and reemitted several times, these particles multiply as they travel toward the surface of the star, which is much cooler than its center: only 6,000 K. The amount of entropy generated by this trip between the center and the surface is approximately twenty-six hundred (sixteen million divided by six thousand). The information associated with the temperature difference is immediately put to good use: It is instrumental in opposing the strong gravitational pull of the upper layers pressing on the central core. Near the surface, this creates strong convective movements, like the bubbles in a boiling stew.

The earth is also boiling. Earthquakes, volcanic eruptions, and continental drift remind us of this fact, even when we'd prefer to forget it. In this case, information is contained in the temperature difference between the center and the surface of our planet. In the absence of sufficiently deep exploration wells, it is difficult to evaluate the temperature of the earth's core, although we think it is around 10,000 K. Why so high? Because, unlike the moon and the asteroids, the earth has not yet released into space all the heat it accumulated when it formed.

We have noted that temperature differences are the source of various types of movement. How, then, does the existence of these thermal variations fit in with the isothermic portrait of ancient times revealed by background radiation?

Before we explore the source of thermal gradients (or temperature disequilibria) in the cosmos, however, we must answer a preliminary question: Is it possible to give a quantitative meaning to the expression "entropy of the universe"?

The Entropy of the Universe: A Barely Perceptible Increase

If we approach the subject properly, we can give an accurate and useful definition for the expression "entropy of the universe," despite the evidence suggesting that our universe is infinite. In fact, two definitions can be given.

Version 1

The first definition, by introducing the notion of the *covolume*, takes into consideration the fact that the universe is expanding. Imagine a gigantic cube, somewhere in extragalactic space, containing a great number of galaxies, including one at each corner (see the illustration on the facing page). Under the influence of the general cosmic movement (sometimes called the Hubble flow) the galaxies are gradually

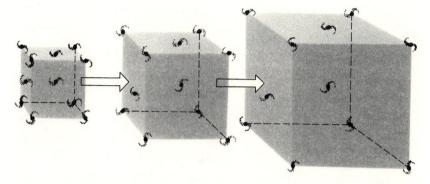

"Covolume" is the name for an imaginary volume growing along with cosmic expansion. Imagine a cube with a galaxy at each of its eight corners. Over the course of time the galaxies gradually move away from one another, and the volume of the cube increases. The number of stars within the covolume does not change as time passes.

moving away from each other. Without changing shape, the volume of the cube increases, while its inner population of celestial objects, that is, galaxies, stars, and atoms, remains the same. This expanding cube, called a covolume, is the unit most appropriate to the study of the behavior of matter in an expanding universe. (All covolumes are identical, so we need to examine only one of them. We can thus avoid the sticky problem of working with infinite quantities.)

In addition to the galaxies, stars, and atoms mentioned above, our covolume hosts myriads of photons. Most of these photons belong to the universal fossil radiation. They number about four hundred per cubic centimeter and, according to our golden rule, can be used to measure the entropy contained in the covolume.

Over the eons space has been dilating. The galaxies have been receding from each other, and the background radiation has been cooling as it has become increasingly diluted. When it was emitted fifteen billion years ago, its temperature was around 3,000 K. Today its temperature is 3 K. In fifteen billion years its temperature will be 2 K. However, at the same time, the covolume has been gradually increasing. The net result of these two effects is that the number of background radiation photons in the covolume remains the same.[8] In other words, the cosmic entropy per covolume (borne mainly by the background radiation photons) remains practically unchanged over the course of time. The expansion of the universe does not generate entropy.

But we also have to consider the effects of the billions of stars shining in the sky. Torrents of new photons are pouring out of their surfaces. As discussed before, any terrestrial, planetary, or stellar event that creates organization must eject its ransom of entropy-bearing photons into space. The contribution of this celestial cosmic entropy can be calculated, and the result is surprising: The totality of photons emitted by all the stars and planets in the multitudes of galaxies over the past fifteen billion years accounts for *less than one part in one thousand* of the total number of photons in space. In other words, the entropy produced by the joint effort of all those stars does not amount to more than one part in one thousand of the cosmic entropy that existed fifteen billion years ago when background radiation was emitted. Hence, the meaning of the expression given above: The universe evolves under conditions of almost constant entropy.

Version 2

In comparison with the other particles found in nature, such as protons and electrons, photons display a behavior quite unique, a particular way of existence. When I switch on my desk lamp, I create photons that formerly did not exist. Let's follow one of these newborns. What happens to it? It travels toward the wall, where the atoms immediately absorb it. It no longer exists. Conversely, protons and electrons, except under very unusual circumstances, can be neither created nor destroyed. Whether they combine and regroup as atoms and molecules or wander about unattached, their number remains stable. In nearly every known case, it is the phenomenon of photon creation that nature uses to increase the entropy of the universe, thereby to ascend the levels of the pyramid of complexity.

In version 1 we estimated cosmic entropy by counting the photons in a covolume. Now we will use the permanence of the nucleon population (protons and neutrons), compared with the ephemeral quality of the photons, to provide a new definition of cosmic entropy.

There are approximately one billion photons for every proton or neutron in the universe. As time goes on, this number is increased by every photon emitted by the sun, the moon, our own warm bodies, and everything shining in cold, dark space. This number measures the cosmic entropy per nucleon, giving us another way of calculating the entropy of the universe. Since the emission of background radiation fifteen billion years ago, this number has increased by about one million additional photons per nucleon. The ratio, one million to one billion, gives us, once again, one in a thousand, the same answer as in version 1. Cosmic entropy has practically not changed since the farthest past we have been able to perceive.

But why was the number of photons per nucleon already so high at that early time? What unknown phenomena, creators of entropy, were able to do a thousand times better than all the stars, past and present? These questions, for the moment, escape our understanding.

Entropy and Gravity

Is your soup too cold, and you want to reheat it? Nothing is easier: Turn on the stove and heat it up. Too hot now? Be patient. If you allow it to sit for awhile, it will lose its excess heat and become cool enough to eat. This scene taken from everyday life shows familiar behavior: When heated, a body's temperature increases; when a body loses heat, its temperature decreases. However, this is not the case with stars. We have to address a crucial point, that stars display an inverse behavior: Loss of heat increases their temperature, and a gain in heat cools them off.

A star's internal temperature increases by steps throughout its life, as it throws off thermal energy into space. It begins its life as a cold, gaseous nebula, gradually warming up to temperatures of millions of degrees; prior to its death it may reach billions of degrees. *Because of this particular thermal behavior, stars are the main factors responsible for the appearance of thermal gradients (uneven temperatures) in the universe.*

Why do the soup and the sun react to heat in such different ways? Chalk it up to gravitational force. The sun is big. It contains a lot of matter. For our sun, gravity plays a dominant role. It maintains our star's spherical shape and keeps it from flying apart into space. Gravity is its glue. Conversely, the mass of the soup is too small for gravity to play a significant role. Electromagnetic force is the governing force, binding the molecules of fluid together.

A star, like the soup, behaves in such a way as to increase the entropy of the universe. The soup does so by cooling, and the star by heating (see Appendix 2). To shine, for a star, means to create entropy-bearing photons and to release them into space.

The augmenting of the number of particles is an application of the golden rule of entropy. This increase can be obtained in many ways. One method was outlined in the preceding pages: Because the earth is cooler than the sun, it gives off many times more infrared photons than the number of yellow photons it receives in solar radiation. Here, we present a second method: *the transformation of matter into light.*

To create light, the sun sacrifices a fraction of its mass: nearly four million tons (equivalent to the mass of ten supertankers) every second. The number of photons created by this disappearance of matter is in the forty-five-digit range (2×10^{45}). Einstein's famous formula, $E = mc^2$, describes the details of this metamorphosis. It tells us, with great accuracy, the amount of light energy released by the transmutation of a given quantity of matter.

Entropy and Natural Forces

Physicists know of four different forces at work in nature. They are called gravity, electromagnetism, the strong nuclear force, and the weak nuclear force. The first three are in a position to create stable structures. Gravity welds planets, stars, and galaxies; electromagnetic force binds atoms and molecules; and the strong nuclear force glues together nucleons within the nucleus of the atom. Weak nuclear force does not contribute to any of nature's stable structures.

The binding principle is always the same. In joining elements, the force transforms a part of their mass into energy. This energy, called binding energy, is ejected into space, usually in the form of photons. The resulting system is thus lower in mass than the sum of its *compressing*

component elements.[9] As a rule, the bound system is also more complex; that is, it occupies a higher level of the pyramid. The photons carry off the entropy to pay for this higher level of organization.

Thus, the first three forces play a triple role in the growth of complexity:

1. They enable particles to combine, and they weld together organized structures (nuclei, atoms, molecules, cells, and organisms, as well as planets, stars, and galaxies).
2. The bonds thereby created are accompanied by the transformation of a fraction of the matter into luminous photons. These photons carry energy that may be used in later exchanges for the construction of even more complex systems.
3. The photons also act to release the entropy that must accompany any growth of complexity.

The Forces in the Primordial Stew

> Gather every vestige of the luminous past!
> — *Charles Baudelaire*

Equipped with what we have just learned, let us take a deep breath before returning to the remote past of the universe. Our mission: to identify the source of the organization and diversity in the cosmos.

Did the natural forces that are the prime movers in the elaboration of complexity exist in this far-off time? If so, did they have the same intensity as today? These questions lend themselves to tests on various events that took place a very long time ago.

Here are three examples:

1. More than a billion years ago in what is now Gabon, spontaneous nuclear reactions occurred in uranium ore (a sort of natural reactor).
2. We receive from distant quasars, photons that were generated by atomic phenomena more than five billion years ago.
3. The helium atoms we now find in the cosmos were produced during nuclear reactions that took place fifteen billion years ago.

Consideration of these three phenomena and comparison with data from observation lead us to the conclusion that *the physical laws have not changed over the past fifteen billion years* — more precisely between now and the time when the universe was at ten billion degrees Kelvin. In other words, this analysis reveals to us the remarkable constancy of the natural laws since at least the first minutes of the universe. What happened before that will be discussed in Chapter 7.

Such a sweeping conclusion raises the question of the *enacting* of physical laws. What is the origin of the laws? And how, contrarily to all that exists in the universe, were they unaffected by change? Unable to answer these fascinating questions, we'll have to be satisfied by noting that nature, by this token, is sending us an as yet undecipherable message.

Because of the high temperatures that dominated the early days of the universe, the forces, although present, were inoperative. Any attempt at combination was immediately defeated by the great heat. But the disorganized, chaotic primordial stew hid a secret that would later be revealed. Forces present amid the particles of the stew had the potential to create bonds and transform matter into light. These potentialities would be fulfilled when, due to cosmic expansion, the temperature had dropped sufficiently to allow complex systems to survive.

Should we see the existence of these forces as the ultimate source of cosmic organization? No, another ingredient is necessary. Although natural forces were essential, they were not sufficient for the elaboration of complexity. The sad fate of the horses at Lake Ladoga will provide us with an illustration of the nature of this extra ingredient.

SUMMARY

The laws of physics dictate that any organization of matter must be accompanied by a simultaneous increase in the global disorganization (also called entropy) of the universe.

How can matter be organized and life created in a universe of continually increasing disorder? In practice, entropy is best measured in terms of quantity of light. To conform to the law of increasing disorder, the quantity of light in the universe must increase continuously. That is, new photons must be created. This can happen in one of two ways:

By placing a hot body near a cold body (for example, the sun is the hot body near the earth).

By transforming matter into light (this is what natural forces accomplish when they bring about the generation of stars, atoms, or nuclei).

The entropy of the universe can be calculated. It does increase over time, as the laws of physics require. However this increase is very small, practicably immeasurable.

5

The Fertility of Disequilibrium

THE HORSES OF LAKE LADOGA

In his postwar novel *Kaputt*, the novelist Curzio Malaparte tells of the dramatic death of a thousand Russian horses, frozen in the ice of Lake Ladoga in the winter of 1942. To escape a forest fire, following intense German bombing, the horses plunged into the lake. The water had not frozen, despite the cold wave that had recently swept the area. As the horses swam toward the bank, with their heads lifted high, a loud noise was heard. The water froze in a flash, imprisoning the animals in ice. The next day the sun glittered on the stiff manes covered with jewel-like ice crystals. Made immobile, each head was a sculpture whose beauty anyone, under other circumstances, would have admired. What had happened?

Supercooling, the phenomenon that caused the event, is familiar to physicists. If a body of water cools slowly, it should solidify, that is, *freeze*, when the temperature reaches 0° C (32° F). But if the water cools too quickly, as it had that December night, the ice will be late in forming. The water may remain liquid well below the theoretical freezing point. The liquid is said to be supercooled. This state is unstable. A few grains of sand thrown into the water could trigger a

flash freeze. In our story the thin hairs of the Russian horses stampeding into the lake were enough to bring on the sudden formation of an
icy grasp from which there was no escape.

Around the grains of sand or fine hairs, ice forms and spreads
rapidly, eventually immobilizing the whole mass of water. The tiny
bodies, playing the role of "nucleation centers," speed the solidifica-
tion process considerably. In their absence the formation of ice cannot
keep up with the rate at which the water cools. Extremely pure water
can remain in a supercooled state for quite a long time.

The same is true for water that is heated rapidly. It will not necessarily boil the minute the temperature reaches 100° C (212° F). As
everyone knows, the water in the kitchen pan does not change suddenly into steam. The process is gradual, beginning with tiny bubbles
clinging to the sides of the pan. Microscopic flaws in the smooth
surface of the metal act as centers of nucleation. If the sides of the pan
are extremely smooth, formation of the steam can be considerably
delayed.

A TANGLE OF CRYSTALS

Have you ever admired the beauty of crystals left over by cooling salty
water? The best way to observe it is with a microscope and a source of
polarized light. Long, thin needles appear as a result of the crystalliza-
tion (see the illustration on the following page). Moored to the sides of
the container, the needles intersect and overlap in a jumble of intricate
patterns, visible in the polarized light as a variety of shimmering hues.
The faster the cooling process, the more detailed and delicate the
crystalline structures. No matter how many times you perform the
experiment, the results is always new and unique.

If, on the other hand, the solution is cooled *very slowly*, the result is
very disappointing. Instead of an endless variety of multicolored gems,
you obtain a single block of salt that fills the whole container. One
block of salt looks like any other block of salt. The result is always the
same, perfectly predictable.

a   b

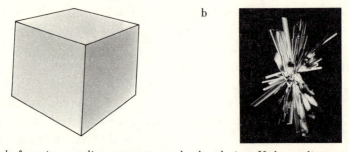

Crystals form in a cooling supersaturated salt solution. If the cooling proceeds very slowly, the result is always the same: a big lump of salt (a). If the cooling is more rapid, we obtain a jumble of gemlike needle shapes. Every experiment produces a different result that is both unique and unpredictable (b).

(Photograph by Claude Nurisdany.)

Ice formation involves myriad reactions, in the course of which the water molecules join with their neighbors. These reactions occur at their own rate. They require time. When the water is cooled slowly, the molecules have enough time to bind together in an orderly way. We say in this case that the ice is developing in a state of equilibrium. The result is the predictable block of salt.

Supercooling takes place when the liquid cools very quickly in comparison with the time needed for the ice-generating reactions. The resulting disequilibrium introduces unpredictability and randomness into the ice. Each of the frozen manes of the horses of Lake Ladoga was unique.

Snowflakes acquire their lacy patterns as they fall through thick layers of atmospheric clouds. As they travel, they capture water molecules that hook onto their intricate geometry. The chemical reactions involved in these processes are said to take place in state of disequilibrium. A single universal trait characterizes all snow crystals: They always have six points. But beyond this single uniformity, each snow crystal of the billions that fall during a blizzard is unique. Diversity arises from the action of a force in a context of disequilibrium. In the case of the snowflakes, the force requires that all crystals have six points. With that satisfied, any other combination is allowed. The lacy effect is generated by the random path of the snowflake through the supercooled moisture of the clouds. Conversely, in the context of equilibrium of the slowly cooled container, randomness is banished. All salt blocks look alike.

VARIETY AND MONOTONY

I love botanical gardens and aquariums. When my work takes me abroad, I enjoy visiting the local parks in my free time. What enchants me is the extraordinary variety of flora and fauna to be found on our planet. Complexity and diversity are the essential characteristics of nature's handiwork. A great number of elementary particles must be combined in order to form a complex system. For example, a water molecule contains seventy-two; a chimpanzee, approximately 10^{29} (a twenty-nine-digit number).

These combinations are effected by natural forces. However, the mere combination of particles does not suffice to engender complexity and variety. An appropriate context is required for this to be realized. When the cosmic temperature dropped low enough, many billion years ago, the forces went to work and generated successively stable structures: galaxies, stars, atoms, nuclei, and molecules. In most cases, these systems are capable of further bonding, becoming part of structures that are more stable, more powerfully welded by the appropriate force. The recombining process can continue until the most stable state has been reached, that is, the state where the binding possibilities of the force are exhausted.

If red giant stars obtain light energy by fusing helium nuclei into carbon, it is because the carbon nucleus is a more stable nuclear structure, more strongly bound by the nuclear force, than the helium nucleus. Later in life the same stars combine carbon nuclei to obtain an even more strongly bound structure: magnesium. And this fusion process continues, until it reaches iron, which is the most stable structure that can be generated by nuclear force. Still more massive nuclei (uranium, plutonium, and so forth) are relatively less strongly bound. It is their fission into lighter nuclei that releases energy in nuclear reactors.

If a bit of anthropomorphism is permitted, we might say that nuclear force cherishes an ambition: to change all matter into iron nuclei. If nuclear force had carte blanche, that is, given the appropriate conditions, that's what it would do.[1]

An analogous situation exists with electromagnetic force. Atomic systems (atoms or molecules) are of different stability. The most stable atomic structures are the atoms of the "rare," or "noble," gases (helium, neon, argon, krypton, xenon). The most stable molecular structures are small molecules, such as water and carbon dioxide. I shall categorize as "noble structures" all these systems that are powerfully bound by electromagnetic force. Electromagnetic force "aspires" to create these noble structures, just as nuclear force "aspires" to create iron.

Similarly, gravity can condense stars into increasingly massive bodies. It "aspires" to turn them all into "black holes," superconcentrated celestial objects, so strongly bound that even light cannot escape from their surfaces. It even strives to bind together all existing black holes into one supermassive black hole that would include the entire universe.

Such reflections upon the "ambitions" of natural forces give us an important clue to our history. Even if the organized structures of our universe owe their existence to natural forces capable of combining elementary particles, these combinations do not *necessarily* generate the variety that is so characteristic of our world. All variety and novelty would be absent from a world in which the forces had been allowed to reach their ultimate goal and exercise their binding power to its utmost. The atomic landscape would consist only of monotonous iron; the stellar one, entirely of monotonous black holes.

AN UNFINISHED SYMPHONY

Luckily, our universe is not composed only of iron and black holes. Of any one hundred thousand atoms plucked at random from cosmic matter, only one is an iron atom. In varying numbers, all the atomic species listed on Mendeleev's table of chemical elements are present in nature. Their existence, their multiplicity, and their combinational possibilities are the basis for the universe's marvelous diversity.

Likewise, we believe that there are many black holes in the sky. But they make up no more than 1 percent of the mass of all galaxies. Space is swarming with blue, yellow, and red stars. Nature is never guilty of uniformity. Every star family is brewing batches of nuclei that, after its death, will further enrich the atomic landscape of the universe.

Hence, the following questions: Why didn't the forces accomplish their ends, once they went to work? Why don't they exhaust their binding possibilities? What makes them give a free rein to the emergence of variety without reducing it to monotony?

As we observed in Chapter 2, the pyramid of complexity, placed upside-down, represents a pyramid of multiplicity. Through multiple combinations, the number of different species increases with every higher step: Atomic species outnumber nuclear ones, molecular species outnumber atomic, and so on up. What keeps the forces from stifling, at each level, this fragile and delightful multiplicity?

A REPRIEVE FOR COSMIC DIVERSITY

Before we try to answer the questions we've posed, let us stay for a moment more on the subject of increasing diversity. The illustration on the following page represents schematically the situation encountered so far. The vertical axis is the time axis. In the distant past, intense heat completely blocked the activity of the nuclear force. Consequently nucleons (protons and neutrons) behaved merely as free particles. There were as yet no nuclei in existence.

As time went on, the temperature dropped and thanks particularly to the activity of the stars, nuclear force was able to begin working. Throughout the eons the phases of thermonuclear fusion within stars succeeded in generating a great variety of nuclei. These nuclei can be described as systems in intermediate states of stability. In our drawing, the arrows show various routes leading from a state of minimum stability (that of free nucleons), on the bottom, to one of maximum stability (iron), at the top.

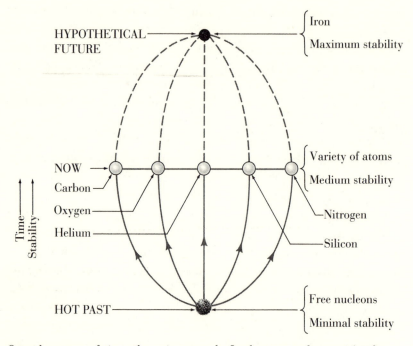

Over the course of time, the universe cools. In the remote, hot past (at the
bottom of the figure), it was composed of free nucleons (protons and nucleons).
Gradually, due to stellar activity, a variety of atomic nuclei appeared in the
cosmos (arrows leading upward). If material conditions allow the nuclear force
to give full rein to its binding powers, all atoms would become iron. It does
not seem that this will <u>ever happen</u>; hence the dashed lines at the top of the
figure.

Free nucleons represent nuclear matter in its unbound state, while iron
nuclei represent the same matter in its most stable bound state. <u>Other nuclei
(carbon, oxygen, helium, nitrogen, silicon, etc.) are at intermediate states of
stability, which, it would seem, will last for a very long time.</u> Nuclear force
"seeks" to engender increasingly stable nuclei, but its <u>quest for stability is
hindered by the expansion rate of the universe.</u> In the figure on page 88, the
single lump of salt shown in part (a) is analogous to iron; the jumble of
needles there in part (b) is analogous to the variety of nuclei dispersed
throughout the universe.

We observe that, although nuclear force has already acted, produc-
ing the large diversity of nuclei found in nature, it is still far from
having exhausted the possibilities open to it. There is still plenty of
hydrogen and little iron. This state of intermediate stability is what
gave birth to the variety of the atomic landscape. At the hypothetical

final phase, when maximum stability would be reached, monotony would dominate, even though the production of an atom of iron requires the combination of one hundred ninety-four nuclear particles. This is proof that the mere combination of particles does not suffice to create diversity.

We are in suspension between two types of monotony: the bottom one, in the distant past, and the top one, into a future that seems ceaselessly to recede. In the illustration the improbable event of reaching maximum stability is represented by the converging dotted lines.

ELECTROMAGNETIC VARIETY

The two forces that team up to build the pyramid of complexity are nuclear and electromagnetic forces. The first crafts the lower steps (nucleons and nuclei), while the second acts as the upper levels (atoms, molecules, organisms). (Remember that in the economy of cosmic organization gravitational force acts to give birth to the hot stars that generate atomic nuclei.)

Although the nuclear force is about a hundred times more powerful than the electromagnetic force, its sphere of influence is much smaller in spatial terms. Likewise, there is a significant difference between the types of structures that these forces are capable of building (see Chapter 2).

Nature's biggest nuclei combine fewer than three hundred nucleons. Beyond this number, they fission. Nuclei are usually spherical in shape, although some of the heavier ones may be shaped like cigars, disks, or pears. Nevertheless, the variety of geometric forms available to nuclei is quite limited. Conversely, the structural and combinative potential of the electromagnetic force appear infinite. Food proteins sometimes combine millions of atoms. DNA molecules are chains of billions of atoms. The process by which atomic species combine to form simple molecules that then recombine to make giant molecules could go on almost endlessly. Likewise, the variety of shapes assumed by the molecules is almost limitless; whether the molecular geometry resembles a coiled string, a tightly wound skein, or a twisted cable depends upon the way it interacts with its surroundings.

Toward the end of its life a star incorporates vast quantities of different nuclei (distributed in successive strata, like the layers of an onion), such as carbon, nitrogen, oxygen, and iron. Hurled into space upon the death of the star, these nuclei find themselves in a very different condition. The nuclear force, paralyzed by the cold, can no longer act upon them. But the electromagnetic force is now in a good position to do so. In a "buzz" of interstellar chemistry, electrons are set in orbit around the nuclei, thereby forming atoms that then combine to make molecules. (As shown in the illustration below.)

Here, again, we meet a situation that is now familiar to us. Electromagnetic force deploys itself in the context of disequilibrium. It gives birth to a myriad of molecular structures but is kept from expressing

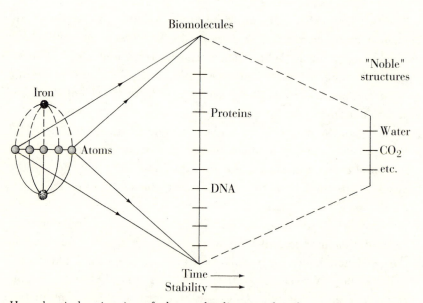

Here chemical variety is grafted onto the diagram of nuclear variety illustrated in the previous figure. At the left, the free atoms ejected by stars are in a state of minimal stability relative to the electromagnetic force. They may combine to form the immense variety of biochemical molecules, which are in a state of intermediate stability between free atoms and the "noble" structures (neon, krypton, water, carbon dioxide), shown at the right of the diagram. Matter in our biosphere is in an intermediate state between these two extremes. This is how the variety of chemical species grafts itself onto nuclear variety to engender the wealth and diversity of our universe.

its full potential and exhausting the possibilities for making combinations. The fact that its work is unfinished is the reason for the exis- ✳
tence of a wide variety of molecules in our universe.

The noble structures (helium, neon, argon, krypton, xenon, carbon
dioxide, water, and so forth) are systems in a state of maximum
stability, the structures most strongly bound by electromagnetic force.
Carbon dioxide is the main component of the atmosphere on Venus
and Mars; water (in a frozen state) makes up a large part of comets and
of the moons of Jupiter and Saturn. Like carbon atoms relative to iron
ones, giant molecules are systems of intermediate stability; they
abound in our colorful biosphere, where the monotony of noble structures has not imposed its dull tyranny.

The drawing shows how molecular variety can be "grafted" onto the
diagram of atomic variety. Time is read from left to right. The nuclei
released into space by dying stars spawn a nearly infinite variety of
complex molecules. The forked paths symbolize the multiplicity of
intermediate molecular states that exist between single atoms (on the
left) and the noble structures (on the right).

When, in about five billion years, the sun dies, it will immolate the
earth in the process. The proteins of living organisms will break down
into more stable structures. Later, perhaps, as part of new planets, they
will combine with other structures to form, once again, highly complex
systems. Nature does not require that they remain for eternity in a
state of maximum stability.

SUMMARY

Natural forces bind elementary particles together, thereby opening the
possibility for organized systems — nuclei, atoms, molecules, cells,
stars, galaxies. "Pursuing their ambition," these forces continuously
bind more and more particles. The decrease in the cosmic temperature
allowed the forces to go to work, leading to the combination of elementary particles in complex systems. These combinations, however, do
not by themselves lead necessarily to the diversity of structures and
interactions; monotony could just as easily result.

Gravity endeavors to transform all matter into black holes. Nuclear force tends to turn matter into iron. Electromagnetic force wants to make matter into stable structures like helium, neon, or water. Hence, our question: Where does diversity come from? Disequilibrium is the answer.

The horses of Lake Ladoga have not perished in vain. Their sad story is an excellent illustration of the concepts of equilibrium and disequilibrium. Newly formed crystals and the precipitation during spring hailstorms also guide our understanding of these concepts. If events occur in a context of perfect equilibrium, monotony follows monotony in an entirely predictable succession. Newness, variety, and the unexpected cannot arise in contexts of equilibrium (see Appendix 2). To conserve a wealth of different forms, the activity of the natural forces must take place under conditions of disequilibrium akin to those in the waters at Lake Ladoga in December 1942.

6

A Supercooled Universe

NUCLEAR SUPERCOOLING

Because of the extreme heat of the early universe, nuclear interactions in the cosmic stew were extremely rapid. Nuclei were constantly forming only to dissolve immediately in the sterility of "equilibrium." Temperature had a profound effect on the course of these reactions of capture and dissociation. As the stew cooled, the reaction time lengthened. At the first minute, when the temperature fell below one billion degrees Kelvin, some nuclei were able to resist the great heat. For the first time in the universe, helium nuclei were able to survive.

This event, called the primordial nucleosynthesis, resulted in the transmutation of approximately one-quarter of the cosmic matter into helium. The other three-quarters, which had remained in a proton state, was later to become universal hydrogen. The production of iron was negligible.

Chemistry textbooks teach that water freezes at zero degrees Celsius. Similarly, nuclear matter "freezes" at one billion degrees Kelvin.[1] At higher temperatures nucleons behave as free particles, like water molecules in a gaseous state. At cooler temperatures the nuclear force binds the nucleons together. In our comparison, the iron nucleus is the equivalent of the block of ice.

Which brings us to the following question: Once the temperature was low enough to permit nuclear matter to "freeze" into atomic nuclei, why did nuclear force spare three-quarters of the available particles, transforming only one-quarter? The answer lies the speed of the temperature change. The universe, which is still largely composed of hydrogen, despite the fact that the nuclear freezing point was reached long ago, can be compared to Lake Ladoga, still liquid on that cold winter night in 1942: Had the cold weather arrived more gradually, the lake would have frozen, and the horses would have escaped both the forest fire and the flash freeze; similarly, had the universe expanded more slowly, cosmic matter would have been entirely transmuted, or frozen, into iron nuclei.

In the stellar ovens the transformation of hydrogen into iron requires millions of years.[2] In the Big Bang, however, matter remained at the appropriate temperatures for only a few minutes. No wonder the work remained unfinished and nuclear force did not exhaust its potential!

Hydrogen, saved by the onset of nuclear supercooling, was to become the main star fuel. In stellar interiors it may burn for billions of years. It may keep a star shining long enough to ensure the emergence of life on associated planets.

If the universe were made of iron, star life-spans, instead of lasting billions of years, would only amount to millions of years.[3] The elaboration of giant molecules would be jeopardized in two ways. First, there would be an absence of hydrogen, carbon, nitrogen, and oxygen atoms. Second, even if a small quantity of these elements were available, the life expectancy of the mother-stars would be much shorter than the planetary gestation period required. The universe would have been entirely different, and mankind would probably never have appeared. Supercooling, which took the lives of the Russian horses, gave us ours!

NUCLEAR INFORMATION

The rate of expansion of the universe is responsible for the onset of disequilibrium, which in turn eliminates the possibility of reaching maximum nuclear stability. The failure of nuclear force to remain in a state of equilibrium is what gave rise to the nuclear information existing in the cosmos today. Let's examine that more closely.

First, recall that, as the nuclei were forming by the process of nucleon association, a few photons were emitted, and they carried away the energy and entropy produced by these reactions. In further discussion it will be convenient to use the term *nuclear entropy* to describe this contribution to cosmic entropy.

Let's imagine briefly that something entirely different had happened. What if a very slow expansion had enabled all the hydrogen to be transformed into iron? All the energy liberated by these reactions would have been emitted immediately in the form of photons. *They would have made the maximum possible nuclear contribution to the universal entropy.* The term "nuclear contribution" expresses the idea that these new photons came from nuclear reactions; the word "maximum" means, in this case, that the nuclear resources of the universe would then be exhausted, that the nuclear "well" would have run dry. To express it in yet another way, cosmic entropy would have *immediately* gained the maximum value, corresponding to the entropy released by the difference in mass between the iron nucleus and its fifty-six nucleons.

Now let's go back to the way things actually happened. Before the period of nucleosynthesis (about one billion degrees Kelvin), it was too hot for the nuclear force to express itself by binding nucleons and releasing photons. Nuclear entropy was therefore nil. After this period, the cosmic temperature could no longer neutralize the effects of the nuclear force; nuclei were able to survive.

The nuclear reactions that created helium nuclei during the first minutes of the universe emitted a certain number of photons into space. Upon being absorbed and reemitted many times by the surrounding matter, these photons did indeed increase cosmic entropy, although they did so to a much lesser degree than they would have if all the nucleons had been changed into iron (as in our first hypothetical example).

The difference between the two values of emitted entropy — the real one and the maximum possible one — represents the quantity of *nuclear information* generated by these events.

This example can help us grasp an essential characteristic of the relationship between entropy and information. *Information appears in nature when a source of entropy becomes available but its entire entropy content is not emitted immediately.* Underlying this is the concept of the reprieve. Information is "delayed entropy"; it is proportional to the amount of entropy available (see Appendix 1).

Like nuclear entropy, nuclear information was nil before the first minute. It grew thanks to the failure of nuclear force to match the rapid pace of cosmic expansion and to fulfill its entire binding potential. Today, nuclear information is carried by all existing nuclei that are not iron nuclei — mostly hydrogen, with a small amount of helium.

Billions of years later, in the stellar cores, high temperatures were again attained, and the nuclear information engendered in the early minutes of the universe could progressively be released. On page 75 we saw how information and entropy function in the genesis of winds and living organisms.

We will learn later how the birth of stars, essential to the release of nuclear information, depends on *gravitational information*, the other major source of organization in the cosmos. We will discover how this information also resulted from a phase of supercooling, yet another variation on the theme of instability and disequilibrium.

NUCLEAR ACCOUNTING

We will close this section with an estimate of the amount of nuclear information currently stored in the universe. So far, all the stars in all the galaxies have not consumed more than 5 percent of the protons in the cosmos. We have barely scratched the surface of the supply of nuclear information!

These processes of stellar fusion have generated photons that are now roaming the vastness of space, thereby increasing universal entropy. This increase, however, is quite small. It amounts to only one million photons per nucleon, compared with the billion photons per nucleon already present in fossil radiation; it corresponds to an added contribution of only one part in one thousand over fifteen billion years.

To complete the picture, we should add that the continuing cooling down of the universe permits the ever-greater multiplication of photons from stellar sources by further absorptions and reemissions. The example of the sun – earth system gave us an appropriate illustration of this process. When all the accounts are tallied, nuclear information, rather than decreasing, turns out to be continuously increasing. So much for Lévi-Strauss's "inertia, one day to be final."

We've traveled a long way. Let's pause for a moment, catch our breath, and give ourselves the benefit of a brief recapitulation.

Before the first minute, the intensity of the heat immediately disintegrated any forming nucleus. Nuclear energy was inaccessible, and nuclear information was nil. In the first minute the temperature fell enough to allow nuclei to form and survive, transforming their binding energy into photons. Nuclear reactions combined some of the protons into helium, but most of the primordial nucleons escaped this fate.

This incomplete transformation, due to the quick cooling-down of the cosmic stew, was the source of the nuclear information in the universe. Today the stars take advantage of this nuclear information to remain hot in an ever-cooling universe.

ELECTROMAGNETIC INFORMATION

Analogous to nuclear entropy, the entropy carried away by the photons emitted at the birth of atoms and molecules is called *electromagnetic entropy*.

Atomic and molecular bonds are much weaker than those of nuclei; therefore, they are far more vulnerable to the high temperatures of the early cosmos. The first atoms of hydrogen did not emerge until the universe was nearly a million years old; before that, they had been destroyed as soon as they'd combined.

Unlike the production of the helium nuclei in the first minutes, the formation of the hydrogen atoms occurred in the context of equilibrium. All the protons and electrons combined at once, and the energy released by these reactions was dispersed into space. As a result, the electromagnetic entropy related to the birth of these atoms immediately reached its maximum value, and no electromagnetic information was generated by the process.

Electromagnetic information appeared in the universe when, thanks
to stellar activity, heavy atoms and molecules came into being. When
stars die, they project the heavy atoms they have produced into space,
and these combine to form molecules. With our radiotelescopes we are
able to explore the molecular populations of interstellar clouds. Molecules incorporating several atoms abound within them. These fragile
systems, which can be opposed to the stable sturdy structures of the
noble substances were probably the first storehouses of electromagnetic information in the universe.

Here on earth, the complex molecules generated by living plants,
such as cellulose and starch, are combustible. When ignited, they are
transformed into noble molecules (carbon gas, water, and so forth); at
that moment they release the electromagnetic information that had
earlier been accumulated, for example, by trees or giant ferns. The key
factor allowing them to store this information was the difference in
temperature between the surfaces of the sun and the earth. Thanks to
the yellow solar photons, the biosphere gradually moved away from a
state of chemical monotony. Were the sun to go out, our planet would
inevitably revert to that state.

You may have noticed that we have described organic material not
as a source of energy, as it is usually done, but as a source of information. Unless it is usable, energy is of little interest to us. The higher the
information, the greater the possibility of using the energy. For any
fuel, the measure of usability is the information it contains, that is, the
difference between its present entropy and the maximum entropy it
reaches when it is burned.

GRAVITATIONAL INFORMATION

The nuclear force exerts its effect over extraordinarily small distances,
no bigger than the size of an atomic nucleus (10^{-13} centimeters, or less
than one-trillionth of an inch). The electromagnetic force extends its
influence a considerably greater distance: over the structure of giant
molecules. The influence of gravitational force, however, seems limit-

less: It keeps the moon in orbit around the earth, the earth around the sun, the sun around the center of the Milky Way, and the galaxies within their clusters. What is more, as described by cosmological models based on Einstein's theory of relativity, it regulates the whole universe's rate of expansion. This is why, when we speak of gravitational force, we must consider activity on two different scales:

1. Universal scale: Gravitational force acts upon cosmic matter as a whole. It is the force governing the universe's vast expansion movement.
2. Local scale: In a more localized manner it acts upon clouds of dispersed matter and assembles them into stars and galaxies.

We know of no force capable of neutralizing gravity's effects on the universal scale. Conversely, the intense heat of early times discouraged any efforts to condense matter on a local scale. Whenever a part of the primordial stew manifested a tendency to condense, a process that could have made it an embryonic star or galaxy, it was immediately repressed and flattened out by what we might call the steamroller of the initial heat. Any lump forming somewhere in cosmic matter would have soon dissolved into homogeneity. The thermal limbo of the universe's infancy paralyzed gravity on the local scale in the same way that it paralyzed the nuclear and electromagnetic forces, which existed but were unable to gather and associate particles to form nuclei, molecules, or stars.

It is at the moment of the emission of fossil radiation that gravity first became able to initiate the formation of galaxies and stars. Extending our analogy of freezing points, zero degrees Celsius for water, one billion degrees Kelvin for nuclear matter. The temperature at which gravity "froze" was 3,000 K. On the conventional cosmic clock, approximately a million years had elapsed since time zero.

This is the era when electrons and protons in the cosmic stew combined to create the first hydrogen atoms. Several cosmological effects accompanied this event. One was the emission of fossil radiation, which sends us the oldest image of the cosmos. Another was a fundamental change in the state of the stew.

In the terminology of physicists, a *plasma* is an ensemble of electrically charged particles (electrons, ions), as found, for instance, in fluorescent tubes. In the earth's atmosphere, the nitrogen and oxygen molecules are neutral; they form a *gas*. The cosmic stew had been in the state of plasma before electrons had been captured by protons to form neutral hydrogen atoms. Now it entered the gaseous state.

Laboratory experiments demonstrate that light interacts much more with a plasma than with the neutral atoms and molecules of a gas. Consequently, at this particular moment, the universe became *transparent* to radiation. After this moment, there is practically no chance that a photon traveling through space would ever be absorbed. That is why the photons emitted at this time can travel fifteen billion years to reach us and be detected by our measuring apparatus. These are the photons that constitutes the fossil, or background, radiation.

Like Samson, who lost his immense strength when Delilah cut his hair, the steamroller of the initial heat lost nearly all its power when electrons began combining with protons. Until then, it had exerted its leveling influence through the effects of the intense interaction between light and free electrons. Now that those free electrons were no longer to be found, the galaxies were thus able to condense.

Much time is required for the formation of heavenly bodies. In fact, this brings us to a riddle that contemporary astrophysicts have been unable to solve: How did galaxies form? Scientists theorize that the birth of galaxies was initiated by fluctuations in the density of the cosmic stew, once it was freed from the leveling influence of the thermal steamroller. Beginning at a very slow pace, the "seeds" of galaxies condensed due to their own gravity, which itself increased because of the contraction process. As it continued, this mechanism snowballed, picking up speed and magnitude.

Fossil radiation gives us the image of a perfectly homogeneous early universe. Thus the beginning of the condensation process could only have been *very slow* — too slow, it seems, to account for the formation of the galaxies we see today. Current theories have a very hard time explaining satisfactorily how galaxies were able to "sprout" in the relatively brief interval between the moment that fossil radiation was emitted and the present.

According to some physicists, our universe contains, in addition to the particles with which we are already acquainted, populations of exotic particles that answer to sweet names like "photinos" and "gravitinos," whose special properties would have enabled them to accelerate galactic germination.

This book is not the forum for further conjecture on a question that still teases contemporary cosmology. I merely observe that galaxies do exist. We'll have to admit that nature has more than one trick up her sleeve.

After the emission of fossil radiation, the universe embarked upon an era of gravitational supercooling. Matter could then condense, release energy, and progress toward increasingly stable states. For this, though, galaxies and stars — in other words, nucleation centers of gravitational force, would need to form. And that takes billions of years . . .

Like that of a nucleus or an atom, the mass of a star is less than the mass of the sum of its components.[4] The excess mass is emitted in the form of photons, which carry away energy and entropy. Following our convention, we will call _gravitational entropy_ the entropy that is carried by photons from contracting celestial bodies. Hence, to be sure, the simultaneous appearance in this period of gravitational information. As was the case with nuclear and electromagnetic information, it was caused by the stars' delay in forming and releasing light. It is proportional to the entropy made available by the formation of stars.

THE BOTTOM OF THE STELLAR WELL

As a star shines and contracts, it gradually increases the thermal gradient within its own body. The central core becomes hotter and hotter as the stellar luminosity becomes brighter. Is a star an inexhaustible source of entropy? A bottomless well? No. Like nuclear and electromagnetic entropies, gravitational entropy is not limitless. When the star dies, the flow stops and the current is cut off.

Many types of dead stars have been observed. Those approximately equal in mass to the sun die a slow death as white dwarfs before burning out and becoming black dwarfs, while those with a mass a few times greater than the sun become neutron stars. In the case of the more massive stars, gravity is so powerful that nothing can stop the collapse that occurs as they die. Beyond a certain point, light no longer escapes from the star, but falls back upon it, like water in a fountain. For the outside world, the star has ceased to shine. Hidden behind its own horizon, it collapses with increasing speed, hurtling toward an unknown destiny. We do know, however, that it no longer sends its energy- and entropy-bearing radiation into space. Like the iron nucleus and the noble molecules, it has emptied its well.[5]

If one day celestial matter were to be entirely transformed into black holes, the universe would then have reached its state of maximum gravitational entropy. This may never happen; at any rate, we are still a long way from such a grim state.

> It is the stars,
> The stars above us, govern our conditions. . . .
> —*King Lear,* act IV, scene 3

When, on a clear night, we look up at the twinkling skies and see the multitude of stars sending out their colored photons, we are witnessing the delayed nuclear and gravitational freezing of matter. Since the emission of fossil radiation, the cosmic stew, in a state of gravitational supercooling, has slowly been progressing toward the stable state embodied by dead stars.

Stars use gravitational information to contract and to create the local temperature gradients so essential to the growth of cosmic complexity. When the core temperature reaches several million degrees Kelvin, nuclear reactions get under way. They take over the role of providers of the stellar luminosity. By the same token, they slow down stellar heating and contraction.[6]

When, in five billion years, our sun has transformed all its central hydrogen into helium, solar contraction will begin again due to ever-present gravitational force. The temperature will rise to about one hundred million degrees Kelvin, at which point, through a new phase of nuclear reactions, the helium nuclei will be transformed into carbon and oxygen nuclei. This will bring a new halt to the solar contraction and heating. All through the star's life, periods of thermonuclear fusion alternate with periods of gravitational contraction.

Water, cooled quickly, reaches the state of ice through the presence of centers of nucleations (grains of sand, manes of the horses of Lake Ladoga). With respect to the gravitational force, cosmic matter "froze" (that is, it reached its most stable state) by means of the formation and heating of stellar embryos. This heating, in turn, brings about the "freezing" of matter with respect to the nuclear force (nucleons are transformed into the more stable state of heavy nuclei). As they prime the pump of the gravitational well, the stars also prime the pump of the nuclear well.

Stars could have existed without the nuclear information generated by primordial nucleosynthesis when the universe was a few minutes old, but their life spans would have been much shorter. Without the gravitational information created when fossil radiation was emitted, nuclear energy would be inaccessible for lack of appropriate stellar "crucibles."

THE MOMENT IS UNIQUE

In sum, the physical structures inhabiting the various levels of the pyramid of complexity owe their cohesion to the existence of natural forces. In order to create variety and diversity, these forces had to exert their building activity under conditions of disequilibrium. Snow crystals are a good example.

Disequilibria exist because today's universe is in a state of supercooling relative to the natural forces responsible for its architecture. We might say that the source of cosmic information resides in the fact that, although the whole of matter is exposed to natural forces, so far only a fraction of it has yielded to their binding capacities. Over the eons, more and more matter is bonded, but only extremely slowly and at an ever-slowing pace. The state of maximum stability that Lévi-Strauss characterized as final inertia becomes more and more remote.

The results of natural events taking place out of equilibrium are partially unpredictable, thus giving rise to the uniqueness of each moment.

NEWNESS IS IRREVERSIBLE

The laws of thermodynamics tell us that no natural event can result in a net decrease of the entropy of the whole universe. But in certain cases this entropy does not increase, either. It remains at the same level.

Imagine that we have filmed the earth's yearly orbit around the sun. In the center of the screen, we see the sun; around it, our planet makes its majestic way. But what if we project the film backward? Would the audience be able to see a difference? In the true version the earth moves in one direction; in the reverse version, it moves in the opposite direction. Either direction is possible; that is why it is impossible to tell which version is the correct one. In other words, the earth's movement is reversible. By looking at the film, we cannot say if time is running backward or forward. (See the illustration on the facing page.) In the standard terminology, we say that the "arrow of time" is not inscribed into the movement. Events in which the arrow of time is not inscribed do not increase the entropy of the universe.

Is the expansion of the universe a reversible phenomenon? Yes, a contraction phase is just as possible. In this case Edwin Hubble would have observed a "blueing" in far-off galaxies, instead of a reddening. But a priori neither he nor Einstein, ten years earlier, would have been able to predict the direction of the cosmic movement.

All reversible events have something in common: They create nothing new; past and future are interchangeable. This is just another way of saying that, as far as past and future are concerned, time is reversible.

The expansion of the cosmos is reversible, but the conditions it imposes on matter are the sources of irreversibility and of the unprecedented in nature. By causing phases of disequilibrium, expansion introduces irreversibility and, as a result, enables the growth of complexity and variety on a cosmic scale.

(a) Let us film the earth's movement around the sun (1). Then let us project the film backward from end to beginning (2). The only difference is that the path of the orbit is reversed. There is no way for us to guess, while watching the film, whether it is a normal or a backward unfolding of the process.

(b) Let us record the same event, this time taking into account the infrared radiation the earth emits into space. In the normal sequence (1), while the earth follows its orbit, red photons escape from it into outer space. In the backward sequence (2), these photons *converge on earth from outer space* — an extremely improbable situation. It is now easy to identify the normal sequence. The thermal disequilibrium between the earth (at 300 K) and outer space (at 3 K) is what makes this identification possible. Expansion, by causing disequilibrium, inscribes the "arrow of time" into the universe.

The arrow of time is inscribed in the cosmos by the expansion and through the disequilibrium states thereby generated. To illustrate the situation, let's return to the previous example. To the nineteenth-century physicist, the earth's movement around the sun was a perfect example of the reversibility of all natural movements. Closer examination of the film, however, reveals a different situation. The earth emits infrared radiation into space; as it orbits around the sun, these photons go *from* the earth *toward* space. If the sequence is reversed, the infrared photons come *from* space *toward* the Earth, as shown in the illustration on page 109. From this viewpoint, the sequences are not at all equivalent. It is easy to see which is true and which is reversed. Why? Because extragalactic space is too cold to emit infrared photons toward our planet. The thermal imbalance between space and the surface of our planet is a result of cosmic expansion. This demonstrates the key role of this expansion in relation to the direction of time and the emergence of newness in the universe.

WHAT CAUSES SUPERCOOLING?

Thus we go from one question to another. We have discovered that the occurrence of supercooling phases is a basic requirement for the emergence of complexity. Insatiable, now we ask: What causes the supercooling? I must warn you straightaway that we have only incomplete answers to this question. The reader will undoubtedly remain unsatisfied.

From the frozen horses of Lake Ladoga we have learned that supercooling is due to rapid freezing. Cosmology, on the other hand, relates the rate of universal cooling to the rate of cosmic expansion. We are led to this question: Why was the expansion fast enough to ensure the supercooling of the universe? Or more precisely, why did the rate of nuclear and atomic reactions, which were initially more rapid than the rate of cooling (thus maintaining an equilibrium), later slow down and lag behind the cooling, causing the onset of supercooling?

The two phenomena, cooling time and particle reaction time, are governed by entirely different factors.

What determines the rates of the expansion and the cooling of the universe? They are governed by the action of the gravitational force (on the universal scale) on cosmic matter. So far, Einstein's general theory of relativity has provided the best explanation of this whole process. According to this theory, matter influences the geometry of space. Both the rate of expansion and the rate of cooling resulting from this influence are proportional to the density of matter and energy in space. On the other hand, the rate of particle reaction is governed by the intensity of the nuclear and electromagnetic forces. We should thus seek the answers to our questions in the subtleties of the confrontation between gravitational force, which governs cosmic expansion, and the nuclear and electromagnetic forces, which govern reaction times (see Appendix 2).

For several years, physicists have been endeavoring to elaborate a unified theory of the forces of nature. That is, they are trying to show how the four forces are actually different manifestations of a single force that underlies all natural phenomena. Such a theory would, in principle, specify relationships of intensity between the forces (as they manifest themselves to us) and thus explain the onset of the phases of supercooling. Although much progress has been made in determining the relationship between the nuclear, weak, and electromagnetic forces, the unity with gravitational force eludes our grasp.

The answer is most likely bound to the solution of a very difficult problem. Today, we apply quantum theory to obtain an accurate description of the behavior of matter and light. However, this theory is no longer valid if the atoms and photons are immersed in an intense gravity field, as in the vicinity of a black hole or in the cosmic stew of the earliest times. Until now, no one has been able to formulate a complete quantum theory of gravity, for a correct description of the behavior of atoms and radiation in very strong gravitational fields. This gap in our knowledge is a major handicap to our exploration of reality. It blocks, in particular, any possibility of explaining the occurrence of the phases of supercooling.

That's where we stand at present. At some point in cosmic history the natural forces quit their initial period of equilibrium to enter overlapping phases of nuclear, gravitational, and electromagnetic supercooling. Although contemporary physics cannot give complete answers to the origin of these phases, we can at least be certain that, without them, we would not be here to ask such questions.

SUMMARY

On each step of the pyramid of complexity we found a group of structures. The forces of nature are the glue by which these structures are bound. In order to engender variety and diversity, however, these forces must exert their activity under rather special physical conditions, which we call a state of disequilibrium. The formation of snowflakes is a good example. When natural events occur in a state of disequilibrium, their results are partially unpredictable. That is what gives rise to the uniqueness of the moment.

The appearance of states of disequilibrium in the universe is linked to two key dates in its history:

1. Before the first minute of time on the cosmic clock, the intensity of the heat caused nuclei to disintegrate immediately. Nuclear energy was inaccessible. During this first minute, the temperature dropped enough to allow nuclei to form and survive, transforming part of their mass into photons. A certain number of nuclear reactions took place at this time, by which a fraction of the protons were transformed into helium. However, the majority of the original nucleons escaped this fate and survived this period of nuclear activity. The rapid cooling of the cosmic stew is responsible for the remaining nucleons, which provide us with the nuclear energy available in our universe.

2. Likewise, until the first million years had elapsed, galaxies and stars were unable to start their formation processes. Hundreds of millions of years later, the cosmic matter had cooled enough to begin contracting and giving birth to heavenly bodies. Since then, stars have shone and, by this very activity, have progressively increased the entropy (a measure of the disorder) of the universe.

We should be grateful for these events. They ensured the creation and longevity of the stars. Had they not occurred, terrestrial life would not have been able to appear and prosper, availing itself of the billions of years of beneficial warmth the sun has bestowed on its planetary system.

The universe expands in a certain way that provokes, at given moments in cosmic history, the onset of phases of disequilibrium. Thanks to these phases, the natural forces are able to generate the complexity and variety of natural forms.

At this writing, in 1990, physics cannot provide an entirely satisfactory explanation for the occurrence of these phases of disequilibrium.

7

An Anthropic Principle?

In a burst of enthusiasm the French author Bernardin de Saint-Pierre once wrote: "[S]ince cantaloupes are divided into equal parts . . . they were meant to be shared by families." To tease him, his friends added, "And the great rivers were laid out to supply water to the cities. . . ."

Is man the center and the destiny of the universe? This old question comes up again and again. The concept of anthropocentrism, repeatedly advanced only to be rejected, ridiculed, and trampled, continually resurfacing, surrounded by a halo of religious and mystical connotations.[1]

To the casual observer, it would seem that modern science had definitively crushed any anthropocentric pretentions by discovering the immensity of the cosmos. Here we are on a minuscule planet orbiting around an average star on the outskirts of an ordinary galaxy that is merely one of billions of other galaxies. Nevertheless, in today's scientific literature, the latest findings of modern astronomers are rekindling the controversy. Scientists are finding some justification for a so-called anthropic principle.

It all started in the 1920s, with Hubble's observation of the movement of the galaxies. Instead of wandering at random in any direction, all participate in a vast movement carrying them further and further away from one another. In relation to us, the most distant ones are moving away at almost the speed of light. Hence the idea of a cosmological horizon, that is, a distance beyond which the galaxies "disappear" from our field of view.

The cosmological horizon is located about fifteen billion light years away. A light-year (ten trillion kilometers, or six trillion miles) is, like the meter or the foot, a conventional unit of measure. We can replace it with a more natural unit of length, for instance, the radius of a proton (equivalent to 10^{-13} centimeters). In 1930 the British astronomer Sir Arthur Eddington noticed that if one measures the distance to the cosmological horizon (the radius of the "observable" universe) in proton radius units, one obtains a number already familiar in another context. The number is approximately 10^{40} (one followed by forty zeros). Physicists are well acquainted with this number.

Within a hydrogen atom, two different forces are exerted between the electron and proton: electromagnetic force and gravitational force. The former is far more powerful than the latter; the ratio of intensity between these two forces is approximately 10^{40}, that is, a number in the vicinity of the distance ratio given above. I use the phrase "in the vicinity of" to express the fact that, even though these numbers are not equal, they are quite similar. Moreover, at the scale of magnitude we are working with, their similarities are much more striking than their differences. We are speaking here in terms of orders of magnitude: These numbers are of the same order of magnitude.

Furthermore, using our knowledge of astronomy, we can estimate the number of nucleons existing within the cosmological horizon. The answer is 10^{80}, that is, the square of 10^{40}. Fascinating coincidences . . .

We might reject these coincidences as silly numerology. Perhaps. But what if they were not? We might be turning our backs on a valuable source of insight. If we pay attention to these findings, we might discover a connection between the "infinitely" large and the "infinitely" small, between macro and micro, between the universe and a single atom. This is worth a second glance.

Many of our century's great minds have pondered this question. We may learn even more from examining it from other angles. For instance, instead of measuring the age of the universe in years (again, a conventional unit of time), let's use a more natural unit of time: the time it takes for light to cross a proton. In this unit, which corresponds to approximately 3×10^{-24} seconds; the age of the universe is close to 10^{40}. (This, however, is not a *new* coincidence, different from the previous one. We have simply expressed the relationship between cosmological horizon distance and the size of an atom in a new way. The new coincidence simply comes from the fact that the cosmological horizon is, by definition, moving away from us at the speed of light.)

The age of the universe is always changing, and therefore this coincidence can only be momentary. Today, the age of the universe, expressed in this natural units, is given by the same number (10^{40}) as the ratio of intensity of two natural forces. In the past, of course, this would not hold true — unless the forces of nature change with the passage of time. This idea was put forth by the British physicist Paul Dirac in 1938. What if, in the past, the force of gravity had been more intense, such that the ratio between it and the age of the universe (expressed in natural units) had remained constant?

Unfortunately, this hypothesis, for all its elegance, is untenable. If, in the past, gravity had been more intense, the sun would have been hotter at its birth than it is today. The presence of life on earth, which covers the past four billion years, is incompatible with the hypothetical increase in heat of the early sun. Moreover, most of the evidence (see page 85) points to the remarkable constancy of the natural forces, at least since the first minutes of the universe.

In the 1950s, the U.S. astronomer Robert Dicke found a much more interesting trail. "The human being," he pointed out, "is a newcomer to the universe." The manufacture of the atoms and molecules of the human body is a long-term process. Several billion years of cosmic evolution were required to generate a brain capable of inquiring about the age of the universe. In other words, the age of the universe has not always been equal to the ratio of intensity between the electromagnetic and the gravitational forces, but in the past no one was there to compare them . . .

Dicke thus pointed out, quite rightly, that the time (in natural units) needed to evolve this questioning brain is similar to the ratio of the intensities. Thus the human race enters the cosmological scheme merely because *the universe becomes observable when there is someone to observe it*. This assertion will be of great use to us in our future investigations.

EXPLORERS OF THE PAST

The physicist in the quest for facts about the remote past of the universe sets off like an explorer charting unknown realms. Physical theories like quantum physics and Einstein's of general theory relativity are his maps. His navigational landmarks are the cosmological observations: the recession of the galaxies, fossil radiation, and the relative populations of atoms and particles.

"Tell us what you saw and heard," people beseeched Marco Polo when he returned from China, "and don't make anything up just because you've been so very far away." Strictly speaking, the only thing the astrophysicist-explorer can affirm is that, in the past, the universe was denser, hotter, and more disorganized than it is today.

What if we ask the astrophysicist the highest temperature ever reached by cosmic matter in the ancient times? Basing his answer on the existence of fossil radiation, he will conclude that matter reached a temperature of at least several thousand degrees. Seeking an explanation for the relative populations of hydrogen and helium atoms, he must raise his bid to several billion degrees. Searching for a reason for the absence of antimatter in the sky,[2] within the context of the unified theory of natural forces (page 111), he can go as high as 10^{28} degrees. Yet he cannot say anything about temperatures over 10^{32} degrees. Why not?

Because, as we mentioned on page 111, we do not know how to describe the behavior of atoms and radiation in a field of high-intensity gravity. This problem was first pointed out by the German physicist Max Planck. The point at which our physical theories run into most serious difficulties is that where matter reaches a temperature of approximately 10^{32} degrees, also known as Planck's temperature. The extreme density of radiation emitted at this temperature creates a disproportionately intense field of gravity.

To go even further back in the past, a quantum theory of gravity would be necessary, but such a theory has yet to be written. For the explorer, these conditions mark the limits of his journey; he has reached the wall of ignorance. Planck's wall hides the rest of the remote past and keeps us from seeing what there was "before."

This is why, strictly speaking, we cannot speak in a rigorous sense of a "birth of the universe" or a "creation of matter." These terms are impossible to define. Although they are colorful expressions, often used by today's authors, we should be aware that there is no solid scientific argument to back them up.

The universe before this time is hiding behind Planck's wall. In scientific literature this wall is often used as the "zero hour" of the cosmic clock, and the age of the universe is then measured from that point. But we should always keep in mind that we have no reason to believe that this wall is really a "beginning."

WILLIAM'S ARROW

Faced with this obstacle, must we give up our exploration, go home, and bide our time until the problem of quantum gravity is solved? It would be better to adopt a second course, well known to physicists: the method of "initial data." It's an excellent means of navigating in the fog of ignorance.

An archer is preparing to shoot an arrow. The physicist wishes to calculate in advance the arrow's trajectory. To do so, he needs two types of information: on one hand, knowledge of the physical laws governing motion (in this case, Newton's law of gravity will do, because the earth attracts the arrow), and on the other hand, the initial data concerning the arrow's flight, the angle at which the archer is taking aim and the tension he is applying to the bowstring.

These two types of information are entirely different. Physical laws are fundamental — and theoretically unchangeable — properties of universal matter. The initial data of the arrow's flight (the angle at which it is directed, the tension on the bowstring), on the other hand, are arbitrary and anecdotal. They are subject to the archer's whims. Nevertheless, knowledge of both the laws and the initial data is required in order to calculate the path of the arrow.

We might imagine that the archer's decision is arbitrary only in appearance, that a deeper understanding of his psychological makeup and background might enable us to predict accurately the direction and speed with which he will shoot his arrow. In this case, the initial data themselves would depend on other factors and laws that we would otherwise have neglected, out of pure ignorance.

Let us now state the event in slightly different terms: An arrow springs from a bow and traces a path directly into an apple placed on a small boy's head. The archer is William Tell.

Suppose that our physicist can observe only the final result. He measures the angle at which the arrow pierced the apple. Physics enables him to reverse the course of time and calculate the initial data. He will then know the angle at which the arrow was aimed and the speed of its flight, without having observed them. If he is unfamiliar with Swiss folktales, his findings may lead him to believe that the archer chose his target entirely at random.

THE INITIAL DATA OF THE UNIVERSE

William's arrow is a metaphor for many aspects of our problem. In view of our ignorance of the laws governing the cosmic fluid on the other side of Planck's wall, we will follow the ensuing strategy. We will ask what the properties of the early cosmic matter could have been just after Planck's wall. We will call these properties the initial data of the universe. Just as the initial arrow angle and bowstring tension are responsible for the hole in the apple, these initial data are responsible for the present appearance of the cosmos after fifteen billion years of evolution.

Of course, these properties are not arbitrary. There should be a way to find an explanation of them, perhaps in terms of even earlier events. But those, for the moment, are outside the realm of our understanding. Hence our decision to neglect them.

THE SPHERE OF CAUSALITY

Thanks to Einstein and the theory of relativity, we know that the speed of a particle cannot exceed the speed of light (300,000 kilometers, or 186,000 miles, per second). This limit also applies to the propagation of "messages" between various parts of the universe, because such messages are transmitted by moving particles.

The cosmological horizon of an observer or of any particle is the maximum distance from which messages can be received. The observer is, out of necessity, unaware of emitters (or causes) located beyond the limits of a sphere whose surface is formed by the cosmological horizon.[3] This is called the *sphere of causality*. Today, the radius of the sphere is fifteen billion light-years. Naturally, its size increases with time. Every day new sources enter our horizons, from which no information could have been received before.[4]

When the electrons and protons of the primordial stew combined and emitted fossil radiation (see Chapter 4), the cosmic clock had ticked off one million years. The sphere of causality of these particles was approximately one million light-years. Compared to its current dimensions, that sphere was tiny. The photons that reach us from various parts of the sky were emitted by atoms *that have never been in contact with one another.*

Whence the prize question: Why is their temperature exactly the same? How did they arrive at such a perfect "unanimity" without ever knowing of each other?

Recently cosmologists have offered various explanations to solve this awkward problem. Some have called upon "inflationary phases," in which supercooling phenomena would play a major role.[5] In my opinion, these solutions are still incomplete and not entirely convincing. Many questions are left pending. We do not properly understand the origin of the extraordinary uniformity of the temperature of fossil radiation. In my opinion we can consider it yet another "coincidence" for our collection.

THE GEOMETRY OF THE UNIVERSE

The world . . . is an infinite sphere of which
the center is everywhere and the surface nowhere.
— *Blaise Pascal*

In this section, we will examine the receding motion of the galaxies, responsible for the expansion of the universe. Space engineers know that an initial speed of more than eleven kilometers (6.8 miles) per second is required to launch a rocket outside the earth's field of gravity. If the object is traveling more slowly, it will fall back to earth; if it is traveling more quickly, it will recede indefinitely from our planet. Eleven kilometers per second is the earth's "escape velocity." The moon has a weaker field of gravity; on its surface astronauts weigh less and are able to jump higher. The moon's escape velocity is two kilometers (1.2 miles) per second. Each planet, each star, has its own escape velocity in relation to its surface gravity.

Like the rocket attracted by the gravity of the planet from which it is launched, the attraction diminishing gradually as it climbs, so each galaxy is attracted by all the other galaxies, and this collective attraction slows down its motion. If the galaxies are moving at a high enough speed in relation to each other, they will continue to recede from each other indefinitely. This is called an *open universe*. Otherwise, following a period of recession, their motion will reverse itself, and they will close in on one another, like the rocket falling back to earth. This is the case in a *closed universe*.

As the galaxies recede, the universe gradually cools. If, one day, the movement reversed itself (closed universe), the current cooling phase would be replaced by a heating phase and would eventually bring us back to the initial inferno.

What are the facts about our universe? Which course will it follow? Here is a first observational result, solidly anchored in observational data: The speed at which the galaxies are expanding is close to their escape velocity (as defined above). A second result is the following: Current data indicate that the galactic speed is slightly higher; in other words, it seems that the galaxies will recede indefinitely.

I have made a distinction between these two results to point out that the second is not as well established as the first. This is because the observed expansion speed differs very little from the escape speed. In fact, this is another coincidence that deserves our attention.

The same coincidence can be expressed differently, in terms of the mean density of the universe. In a (fictitious) cosmos where the density was extremely high, the mutual attraction of galaxies would be quite strong (in other words, the escape speed would be very high, as it is for a large planet), and the reversal of expansion to contraction would occur rapidly (closed universe). If, on the other hand, the density was low, gravity would be too weak to first stop and then reverse the expansion (open universe).

The intermediate case would consist of a universe where the expansion speed was exactly the same as the escape velocity. This cosmos would be at the so-called *critical density*. Under such conditions, the galaxies would gradually slow down; their speed would approach zero, but their movement would never be reversed.

It is possible to calculate the value of the critical density for our present universe (the figure changes with time). The density obtained is about ten nucleons (protons or neutrons) per cubic meter, that is, the equivalent of a mass of ten protons in a one-cubic-meter volume. If our universe had this density of matter, the galaxies would slow down for eternity but never move back toward each other.

Now the key question: What is the real density of our universe? The mean density of visible matter (galaxies and stars) is equivalent to about one-tenth of a nucleon per cubic meter. More matter exists as "dark matter," observed only through its gravitational attraction upon visible matter, which amounts to two nucleons per cubic meter. However, in addition to these objects, it may be necessary to consider the contribution of unknown matter that might exist in the form of exotic particles we are not yet able to detect. Is there any way of knowing?

There is. Because of gravitational attraction, all matter, visible and invisible, known and unknown, slows down the galaxies. If we were to measure this deceleration, we could evaluate the total density of the cosmos. However, the galactic slowdown is so slight that, as yet, it has escaped all attempts to detect it. This fact, in itself, is significant: It enables us to set an upper limit on the total density of the universe. The upper limit is, by this argument, estimated at twenty nucleons per cubic meter. A higher amount would have made the galactic deceleration already detectable.

The real density, probably close to two nucleons per cubic meter (and definitely lower than twenty), is seen to be in the vicinity of the critical density (ten protons per cubic meter). Here we have a new coincidence.

Let's pose another question: How many protons would there be in a universe measuring fifteen billion light-years (that is, 10^{40} atomic units of length), with a density equal to critical density? The answer: Approximately 10^{80}, that is, the square of the ratio of intensity of electromagnetic and gravitational forces. This finding may explain the mysterious coincidence noted on page 115. It means that the density of our universe is more or less equal to the critical density.

If today, the real density is in the vicinity of critical density, the cosmological model tells us that in the past, the two densities were even closer together! At the time that fossil radiation was emitted, real density differed from critical density by less than one part per thousand! And at the moment of primordial nucleosynthesis, by less than one part per quadrillion! They were really very close! Here, the expression "in the vicinity" takes on its full meaning . . .

For the reader familiar with the subtleties of geometry, we might add that the situation can also be expressed in terms of space-time curvature. If the real density is greater than the critical density, the curvature is positive, analogous to a sphere in three-dimensional space (see the illustration on the following page); in this case parallel lines eventually cross each other. Density less than critical density would imply a negative, or saddle-shaped, type of curvature; parallel lines always diverge from each other. The intermediate case (expansion speed equal to escape speed) corresponds to a flat space, described by the plane (Euclidean) geometry of our school days; in this case parallel lines neither meet nor diverge.

Our discussion indicates that the current geometry of the universe is almost plane (flat space), with a slight negative curvature. In the remote past, it was extremely close to being absolutely plane.

This long and difficult section has given us some information on the initial data of our world. The ancient universe was extremely isothermic; its expansion rate was extremely close to its escape speed. In other words, its density was extremely close to critical density; that is, its geometry was extremely plane. Why were there so many "extremes" in the choice of initial data?

TOY UNIVERSES

It might be fun to use our equations and computers to calculate how the universe could be evolved, had the initial data been different. Let's play the game of toy universes.

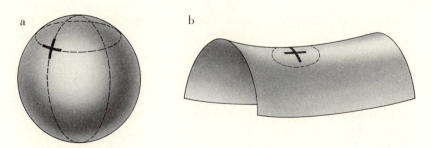

Let us draw two lines intersecting one another at right angles on the surface
of a sphere (a). These two lines curve inward in the same direction: toward the
center of the sphere. This is called a positive curvature. The sphere's surface
closes itself in; it is of finite dimension. On a horse's saddle or mountain pass
(b), the perpendicular lines curve in the opposite direction. This is called a
negative curvature. The surface does not close itself in. It is (probably but not
certainly) of infinite dimension.

If the density of the universe is greater than critical density, the space–time
curvature is positive, like the curvature of the sphere. The volume of the
universe is finite. If the density is less, the curvature is negative and the
volume is probably, but not certainly, infinite. If the density is equal to critical
density, the curve is nil and the universe is infinite.

Beware of analogies. The surface of the sphere is two-dimensional, while the
volume of space–time is four-dimensional. To pursue the comparison, since
the surfaces of the sphere or saddle (represented in two dimensions) curve into
a third dimension. the space–time volume should curve into a fifth dimension.

Again, we can use William Tell's arrow to illustrate the situation:
When it was shot, we might ask, how was the arrow aimed, and how
fast did it have to travel, in order to hit the apple? Similarly we might
ask: When the universe was forming, what initial conditions were
required in order to engender diversity and complexity? Remember
that these conditions could be considered as resulting from even earlier
events that are hidden from us by the wall of ignorance (page 118).

How will our toy universes develop? What is the state of matter
within them? Will they be expanding, contracting, or stationary? How
long will they last? Will galaxies be able to form? Will stars? Will
stellar nucleosynthesis generate the atoms encountered in the real
world? Will there be planets with oceans and chemical reactions? Will

life, intelligence, and consciousness ever develop? Will there be any people around to ask questions?

Behind these questions lies a hypothesis that could be contested: *We associate intelligence with biochemical activity carried out by giant molecules containing hydrogen, carbon, oxygen, and so forth.* Some scientists have considered the possibility of life forming in entirely different structures, for example, crystals or molecular clouds. In my opinion, however, such hypotheses run into serious difficulties. Living things must not only be capable of storing huge quantities of information, they must also have rapid access to it. Life and survival require quick, efficient reactions. At present, no one has been able to demonstrate how crystals or nebulae could possess a comparable reaction speed. In any case, the living crystal hypothesis implies that atoms able to fit into the crystalline matrix had already been formed. They, in turn, are dependent on the prior formation of galaxies and stars. (Incidentally, we might also wonder why such living crystals have never been observed if indeed they are possible). Because of the above reasons, our discussion will concern only the biochemical path to consciousness, the only one that we know to have succeeded in generating life on our planet.

Our toy universes have surprises in store for us. On the whole, they show that even slight alterations in the choice of the initial data of the universe would have resulted in a sterile cosmos unable to engender even the atoms required to form organic molecules.

Let us first consider the life-span of the universe, which depends upon its density. We have already examined the concept of critical density (page 122), which, if exceeded, would limit the expansion time of the universe, causing it eventually to contract and collapse inward. To carry out our calculations, we must first choose a set of initial data, analogous to direction and speed in the story of William Tell's arrow. Imagine that we select them at random. For the majority of choices, the life-span of the universe is extremely short: On the order of a second or even less. No star would have the time to form. As a result, neither would the atoms of carbon, nitrogen, or oxygen, so essential to our biochemistry. (Similarly, an arrow shot at random would have very little chance of hitting the apple.)

A universe at a density lower than critical is not confronted with the problem of time: It would theoretically be eternal. However, in practically every case, the thermal steamroller would be so powerful that it would prevent any condensation from starting; thus no star would be able to form.[6] No stars, no atoms, and above all no long-term source of warmth to provide a nest for the genesis of life.

Only a universe with an initial density exactly equal to critical density would be capable both of engendering motherly stars and of lasting long enough to provide a home for the nuclear, chemical, and biological reaction required for life to subsist. The gestation period would last billions of years.

This "fertile" density corresponds, in the remote past, to a universe in which space possesses no curvature and where geometry is extremely plane. This is the first condition our toy universes must meet before they can hope to produce their own observer.

PROPERTIES OF THE UNIVERSE

In ancient times no one had an idea of the actual size of the world. The sky seemed a hollow sphere; circling within it, the planets and stars. The Renaissance astronomers burst the sphere open. Little by little, the human race become aware of the dimensions of the universe it inhabited. Realizing each of us is a mere grain of dust in a gigantic volume, we are awed by its silent immensity.

Why is the universe so big? The initial data essential to the emergence of complexity implies that the cosmos existed for several billion years. Thus, the volume of the observable universe (the distance to the cosmological horizon) must be measured in terms of billions of light years! A smaller world would never have been able to generate an observer to be amazed by its dimensions.

The surface area of the San Diego Zoo is so huge that a gazelle can roam the park freely and rarely ever reach its boundaries. The animal lives and flourishes as if it were in its native savannah. Of course, the park provides these natural conditions to a large number of other animals, each of which takes advantage of the immense territory in its own way.

Thus, it seems to me, a universe that is big enough to engender one civilization may have engendered several. The conditions that are good for one are not necessarily bad for the others.

Fossil radiation shows us that, when the cosmic clock struck one million years, the universe was extremely isothermic. This isothermy (which raised the problem of causality for us) indicates that the cosmic stew was extremely homogeneous on a large scale. Here we have another instance of our narrow escape from sterility. Large condensations in the form of, say, giant black holes scattered here and there in space would have caused major perturbations, most likely detrimental to the growth of complexity (see Appendix 3).

Homogeneous, yes, but not too much so. Galaxies formed, like lumps, in the cosmic batter. For this to happen, the "seeds" of the galaxies, had to be "sown" in space to act as centers of nucleation (see Chapter 5). We know very little about the birth of galaxies, but the infant universe must have possessed both homogeneity on a large scale and material condensations on a small scale; in time, the latter were able to ensure the gestation of galaxies and stars.[7]

In "The Horses of Lake Ladoga" in Chapter 5 we pointed out another basic ingredient in the elaboration of cosmic complexity: the cooling and expanding mode of cosmic matter. To it, we owe the sequence of disequilibria that allowed nuclear and gravitational information to emerge.

PHYSICAL "CONSTANTS"

Scientists describe the behavior of matter in terms of laws of physics. Newton's law of gravity, for instance, states that the force between massive bodies is proportional to the product of their masses and inversely proportional to their distances. This proportionality is specified by a number physicists call G, the gravitational constant. As a matter of fact, every physical law contains in its expression some definite numbers that are called physical constants. In addition to G, the gravitational constant, there are c, the speed of light, h, the action constant of Planck (the German physicist who pointed out the wall of ignorance), e and m, the charge and the mass of the electron, and many other numbers. These parameters have been measured in laboratory experiments, with great precision. But the reasons why light travels at one hundred eighty-six thousand miles per second, rather than, for example, one hundred eighty-six miles per minute, are still entirely unknown.

The Swiss archer could decide to change the angle and speed of his arrow. But he had no control over the law of gravity, which determines the earth's influence on his projectile. The numerical value of G has been measured experimentally, but no one knows why it has that numerical value, rather than another. Perhaps one day this will be explained within the framework of a more general theory. At the moment, however, its wherefore is much like that of the initial data. The value has to be taken as it is, when used to calculate the path of the arrow. We cannot give any justification for this choice.

This is the case, in general with all physical laws providing a "local" description of particle behavior. These laws can be thought of as microscopic initial data, parallel to the macroscopic initial data (density, temperature, and homogeneity) we used earlier in calculating the fate of the toy universes. In either case, the ultimate explanation of why these data are what they are is hidden from us by Planck's wall.

How many numbers, or mathematical parameters, are needed to describe the cosmos? Current physical and cosmological theories already require more than twenty — not a satisfactory situation at all. Theoretical physicists are attempting to reduce this number. This is the secret hope behind the theory of the unification of forces. Unifying two forces would consist of showing that they represent two aspects of a single force; they could thus be described using a single parameter, instead of two.

This search has yielded encouraging results in the past several years. In 1984 partial reunification of the electromagnetic and weak nuclear forces was strikingly confirmed by the detection of three new particles, called intermediate bosons, at CERN in Geneva. The data obtained in 1990 has limited to three the number of families of elementary particles, thereby giving some reassurance against the possible discovery of even more unknown parameters.

Naturally, it would be useful to know how far we could go in reducing the number of mathematical parameters necessary to describe our universe. Is the minimum ten? Three? One? Perhaps even zero. None would be required if there were one fundamental principle *like topology* from which all values could be calculated.[8]

But we are still a long way from formulating such a principle, if indeed it exists. In fact, we do not even know whether we have identified all the physical parameters of our world. For example, if it were shown that electrons and quarks were not really elementary particles, but composed of even smaller particles, we would then have to determine the properties of these new particles. This would produce a set of new mathematical parameters to be explained.

Of course, like the Russian dolls that contain increasingly smaller dolls inside them, the new particles themselves might be made of smaller elements. Could the regression go on forever? Hats off to anyone who can answer such questions.

STILL MORE COINCIDENCES

To conclude this chapter, I would like to mention two additional elements that deserve our attention in this context. We have already discussed the two different types of nuclear force: the strong nuclear force, which binds the nucleons in the nucleus, and the weak nuclear force, which converts protons into neutrons, and vice versa. In this section we will deal only with the former, which, to simplify our discussion, will be referred to simply as nuclear force, without further qualification.

Nuclear force is far more powerful than electromagnetic force. The intensity ratio, around one hundred to one, is of prime importance in the shaping of our world. It is the basis of chemistry and biochemistry and gives atoms and molecules their characteristic shapes.

The core of an atom is formed by a massive tiny nucleus welded by powerful nuclear bonds. Huge (on this scale) electron clouds in varied shapes revolve around the nucleus. Their flexibility and versatility is partly due to the relative weakness of the electromagnetic bond tying them to the nucleus. These structures, highly mobile and adaptable, are well suited for the innumerable combinations required by the activity of organic molecules.

We have already agreed to consider the numbers associated with the physical constants as the initial data of our universe. Let us now explore the kind of toy universes that are created when we substitute other numbers. What would happen if the factor of one hundred characterizing the ratio of intensity between nuclear and electromagnetic forces were increased or decreased? Before we look at the findings, I should introduce a note of caution. We still know little about nuclear force, and calculations in which it is involved may not give very accurate results. In qualitative terms, however, I feel they can be trusted.

Had the strength of the nuclear bond been increased, hydrogen would have vanished from the universe at the time of primordial nucleosynthesis, taking with it nearly all nuclear information (see page 98). Stars would be able to form, but their life-spans would be short due to a lack of long-burning fuel. In a few million years they would become stellar corpses, leaving no time for any hypothetical planets to reach biological maturity. Moreover, the very possibility of biological maturity would no doubt be jeopardized by the absence of hydrogen atoms, since they are a major component of the giant molecules of biochemistry.

Were the strength of the bond decreased, we would no longer recognize most of the atoms that make up our world. A large number of atomic species would become radioactive. Through nuclear fission most would disappear quickly, before the beginning of the era of biological evolution. Of the nuclei able to survive, a majority would be poisonous for living metabolisms because of their radioactivity.

Undeniably, the factor of one hundred was optimal for the emergence of complexity.

What is more, theories unifying the forces seem to be able to explain this factor of one hundred. As mentioned earlier, the forces of nature appear to have remained highly constant from the creation of fossil radiation until today (see page 83). This was not true at earlier times when the universe was extremely hot, despite the fact that the laws themselves were apparently the same as they are now.

At temperatures over one trillion degrees Kelvin the weak and strong nuclear forces both gradually weaken. This decrease becomes significant before the first millisecond. Electromagnetic force behaves in a similar manner, but contrary to both nuclear forces, its strength increases at very high temperatures, gradually becoming comparable in intensity to the nuclear forces. At a cosmic temperature of approximately 10^{28} K, electromagnetic and nuclear forces were of equal intensity. The time on the cosmic clock was then 10^{-35} seconds.

Later on, as the temperatures cooled and the forces subsequently differentiated, nuclear force increased in intensity in relation to electromagnetic force, until it became one hundred times stronger in our "cold" world. We can thus have two different ways to explain the factor of one hundred that now links the electromagnetic and strong nuclear forces. The first explanation originates in our current knowledge of physics, relative to a plausible theory that may soon be confirmed: the theory of unified forces, which describes the behavior of the intensities as a function of the cooling of the universe. The second explanation, which is quite popular today, is a so-called *a posteriori* (after the fact) explanation. In a hypothetical universe in which the ratio of the two forces would be different, no biomolecules that would allow the evolution of complex organisms could not have formed. This explanation tends to support the statement of the anthropic principle, to which we will soon turn our consideration.

A PRINCIPLE OF COMPLEXITY

How should we react when confronted with all these amazing coincidences? Some authors have supposed that there exists, outside the range of our observation, an infinite number of universes, each characterized by different initial conditions. As we saw before, these universes are most likely sterile. In such inhospitable worlds, of course, no one is around to ask questions. We ourselves are able to wonder about our existence precisely because our universe happened to provide us with the right, fertile parameters.

Perhaps. However, this solution not only seems unsatisfactory to me, it also seems wasteful. To explain a problem that concerns only ourselves, we invent an infinite number of unobservable universes. That is truly the stuff of science fiction.

Instead, let's reconsider a hypothesis we touched on earlier. Let us suppose that, far from being arbitrary, all the numerical parameters of the world's initial data, macroscopic (temperature, density, and homogeneity) and microscopic (the constants of physics), were *necessarily* predetermined to take on the values they have in our universe. Let us further assume that this determination is related to the existence of a universal principle (the physicist's dream) from which all these numbers can be derived. It would then be logical to conclude that this ultimate hypothetical principle has contained, from the beginning, the recipe for intelligence and consciousness.

We are then confronted with two possibilities: Either (1) we accept the idea that the values, chosen from a set of arbitrary parameters, just happened to be those that would lead to the birth of consciousness or (2) we acknowledge the existence of an initial principle that contained the seeds of consciousness. Personally, I do not know which of these two possibilities is the most amazing.

If the first possibility is correct (that is, the values were chosen from a set of arbitrary parameters), we may ask whether this choice was made "at random" or "intentionally" (whatever these words may mean in this context). At the present time, there is no scientific answer to this question.

In the light of modern cosmological findings, our cosmos thus seems to be quite special. Are we once again the victims of our own naiveté? Who knows? At any rate, it seems difficult to deny that such fine-tuning of physical and cosmological parameters is awe-inspiring, even if we are far from knowing what it really means.

Lévi-Strauss writes, "The universe was born without man." Understood in chronological terms, this affirmation states the obvious: The heat of the initial stew was incompatible with the existence of biochemical structures. The conclusion of our discussion, however, invites us to adopt a different reading of the history of the universe. The temporal sequence of the chapters of nuclear, atomic, molecular, and biological evolution links human existence to the physical conditions reigning in the very remote past of the universe.

This statement, though, does not, to my mind, justify the term "anthropic principle" currently used in cosmological literature. If sparrows could talk, they would accuse us of chauvinism. The same fine-tuning was required to bring about their own existence. Thus a chorus of amebas, jellyfish, cantaloupes, and other living things could rightly claim the existence of a protozoan, coelenterate, or violalate principle governing the universe. The only way to deal with their protestations is to use another, less chauvinistic terminology. I should like to suggest the term *"complexity* principle." This principle could be stated as follows: *Since the earliest times accessible to our exploration, the universe has possessed the properties required to enable matter to ascend the pyramid of complexity.* Such a statement gives equal consideration to cantaloupes and human beings.

A SUCCESSFUL PRINCIPLE?

Is this principle of complexity sterile and useless? Or can it be fruitful? I believe the latter is the case, if one applies the principle correctly. Here is an interesting example.

In the 1950s the English astrophysicist Fred Hoyle speculated that the chemical species of our world formed in the heat of stellar ovens, that nucleons and nuclei effected successive captures, thereby giving birth to atoms. But there is a catch: The combination of two helium nuclei form a highly unstable structure that breaks apart almost immediately. How could helium nuclei undergo successive combinations to form carbon (composed of three heliums) or oxygen (four heliums).

On the one hand, Hoyle knew that carbon is a very abundant cosmic element; on the other hand, he was aware that, because of its instability, the production of this element in stellar ovens would be very inefficient unless there were a fine-tuned adjustment between some parameters of nuclear physics. Hoyle foresaw that only an extraordinary numerical coincidence between the parameters describing the structures of helium and carbon nuclei could make this reaction occur and generate carbon in appropriate abundance.[9]

He assumed that this adjustment occurred in nature and greatly facilitated production of carbon atoms. When Hoyle suggested his hypothesis, these parameters were still unknown. However, based on the observation of the abundance of carbon in the universe, Hoyle had a hunch that this coincidence existed. Only later did experiments in nuclear physics prove the validity of his intuition. This is an example of a successful *a posteriori* explanation.

In a manner of speaking, Hoyle "discovered" the existence of those special fine-tuned properties of carbon required to explain its abundance. Yet mere assumption, or an educated guess, was not enough. Laboratory experiments were needed to provide proof. Hoyle's hindsight was profitable in that it guided the planning of the laboratory experiment needed to measure and verify the intuition. Similarly, the principle of complexity is successful if one respects the duality of the situation: (1) an educated guess based *a posteriori* on what is already observable, and (2) experimental verifications.

A physicist enters the laboratory. She wants to perform an experiment in order to increase our knowledge of reality. The complexity principle tells her that her very presence in the laboratory is an extremely important clue to the nature of reality. By the fact that she exists, she knows that all the conditions required for the emergence of intelligence are assembled in our universe. But an educated guess is not enough. As in the case of carbon, the physicist must go further and confirm these intuitions by a proper study of these conditions and proper laboratory experimentation.

Strong nuclear force is one hundred times more intense than electromagnetic force. This factor of one hundred can be "explained" in retrospect by showing that it was required for the emergence of complexity. Moreover, within the framework of contemporary physics, we hope it will be possible to demonstrate that this factor is a result of the properties of very hot matter as it cools.

These two explanations (*a priori* and *a posteriori*) of the factor of one hundred existing between the intensity of electromagnetic and strong nuclear forces are a good example of a successful application of the complexity principle.

SUMMARY

To invent theoretical universes, a physicist hypothesizes that the initial properties of the universe (temperature, density, homogeneity, mode of expansion), as well as the numbers that specify physical laws, can be selected at random, in other words, assumes that these numbers can be given values different from those in our world. With the help of powerful computers, the physicist calculates the evolution of these fictitious universes, arriving at the surprising conclusion that, in just about every possible case, they would be sterile; matter in them would fail to ascend the pyramid of complexity and generate life. The interpretation of this conclusion is presently a matter of animated discussions. It suggests *a posteriori* a principle of complexity underlying the fecundity of the universe we inhabit.

The Future of Life in the Universe

The physicist is often tempted to act as a prophet. His knowledge of natural laws gives him an understanding of the past and the present and can also reveal some aspects of the future. What will happen to energy, entropy, and information? Will life continue, and if so, for how long?

A word of caution before we begin. Our discussion is, out of necessity, based upon *current* scientific knowledge. We make no claim to having catalogued all the forces of nature. Remember, at the end of the nineteenth century only two forces were known to physicists: gravity and electromagnetism. Within less than a century, physicists added two others: the strong and weak nuclear forces. Have they exhausted the list? The discovery of a new force could modify our predictions for the future. (Although their interpretation is highly controversial, several experimental findings indicate that such a possibility exists.)

The question of the density of the universe and its influence on universal expansion was discussed in Chapter 7. The best current observations imply that the cosmic density is lower than critical density. The gravity field of cosmic matter appears to be too weak to bring cosmic expansion to a halt. Thus the galaxies will not be drawn inward on top of one another in a contractive movement accompanied by the heating of the cosmic matter (the so-called Big Crunch following the Big Bang). Although this conclusion is not yet definitive, the theory of a "closed" universe (with recontraction) seems less and less plausible. We will therefore confine our discussion to the framework of an "open" universe (continually expanding and cooling).

Stars live for a long time, but they are not immortal. When they exhaust their nuclear fuel, they die. Our sun will perish in around five billion years. Smaller stars last longer; some (the red dwarfs) may last for a trillion years, but then they, too, go out.

New stars are continually being formed from the nebular matter of the galaxies. However, as this matter grows scarce, stars form at a slower pace. For example, in our own Milky Way, the birth rate is now about one or two new stars per year, while in the past, it was several times higher. Much of the galaxy's nebular matter has already been changed into various stars.

When they die, stars eject a major part of their component atoms into space. However, the core of the star collapses inward on itself, forming a stellar corpse—either a white dwarf, a neutron star, or a black hole, depending upon the mass of the star. The number of stellar corpses increases regularly over the years. A trillion years from now, the sky will be dimly lit by old stars (red dwarfs and white dwarfs) slowly burning out amid the neutron stars and black holes. In those remote days, extragalactic space will be much emptier than it is today. Expansion will have increased the average distance between galaxies from 1 to 20 million light years.

COSMIC FARMING

Is life doomed to disappear when the stellar sources of energy, information, and entropy are exhausted?

Our distant ancestors were hunters and gatherers who survived by availing themselves of what nature had to offer, season after season. Being limited and unreliable, their food sources were not satisfactory for the needs of increasingly large groups of people. The early people, we believe, turned to agriculture and animal husbandry to avoid shortages and famine. After some time they probably wondered about the optimal size of a herd. Maintaining two animals does not involve much more work than maintaining only one, but the yield in meat, milk, and skins is doubled. The number of animals can be adapted to the size of the tribe. As the population grows, new animals can be captured and domesticated.

At this time, about eight to ten thousand years ago, mankind stopped being a *passive predator* and started controlling the availability of food sources. This transformation, from the Paleolithic to the Neolithic era, enabled the human population on our planet to grow considerably.

Similarly, with respect to the sky, we are passive predators. We use whatever the heavens offer us in the way of energy and information. But our progress toward the celestial Neolithic Age has already begun. Astronauts have set foot on the moon and brought back hundreds of pounds of lunar rocks. Unmanned laboratories have landed on the surface of the planet Mars in order to analyze its soil and tell us about its chemical composition.

There are plenty of projects in store. Some of the asteroids orbiting near the earth contain large quantities of technologically precious metals whose terrestrial supplies are growing scarce. In the decades to come we might find a way to draw the asteroids away from their natural orbit and bring them gently to our planet. The goal of other projects is to increase our harvest of solar photons by launching orbiting captors. Solar wind is also being eyed with interest; the energy and atoms it contains would be extremely useful. How could they be extracted? I leave it up to our engineers, who are never at a loss for imagination.

Imagination will be essential in one trillion years when the stars are burning out. We may already have one solution.

PUMPING ENERGY FROM BLACK HOLES

We have good reasons to believe that our universe contains a certain number of black holes, highly condensed celestial objects with an extremely large gravity field. As Stephen Hawking likes to say, black holes ain't so black (see Appendix 3). As they evaporate, they emit energy and information that might be tapped by surviving civilizations when ordinary stars are burned out. For a black hole of a mass equal to our sun this process will last for quite a long time, much longer than the star's life-span: 10^{64} years. But its luminosity, for most of this duration, will be extremely weak.

In contrast, a black hole of lesser mass will emit much more intense radiation but for a shorter period of time. We may speculate that, like the herder of the Neolithic Age, each planetary civilization will have to make a judicious decision about the mass of black holes to be captured and domesticated, that is, maintained in orbit at an appropriate distance.

There's one other problem: The evaporation that reduces the mass of the black hole increases its temperature and the intensity of its luminous flow. This flow would have to be controlled by compensating for losses in mass, that is, by feeding the beast continuously. What would it eat? Stars, planets, and asteroids would be a decent diet. Moreover, a tame black hole would take care of all our trash disposal problems. Radioactive or not, this undesirable matter would disappear from our environment to reappear in the form of radiation.

And where does information come from? From the endless supply acquired when gravitational overcooling began, at the moment of the emission of fossil radiation (see page 103).

The capture of stars, placing black holes in orbit — those certainly sound like grandiose projects, straight out of a science fiction thriller. Nevertheless, serious scientific journals publish articles on these subjects. Like Leonardo da Vinci with his airplane models, we know the basic theories involved but are not yet in possession of the material means required to initiate these projects.

And after that? Other threats loom in the distant future. Some theories of the unification of physical forces indicate that nucleons may disintegrate into lighter particles (electrons, neutrinos, and photons) after a mean period of about 10^{32} years. Up to now, efforts to confirm this hypothesis have failed, but it has not been abandoned by every physicist. Would intelligence still be present if cosmic matter no longer contained nucleons?

In the nineteenth century, because they knew nothing of nuclear energy, physicists set the sun's age at only a few million years, provoking vehement protestations from contemporary geologists and paleontologists. By studying the behavior of uranium salts, Henri Becquerel

discovered nuclear phenomena in 1898. The work of Ernest Rutherford, Jean-Baptiste Perrin, Enrico Fermi, Irène and Frédéric Joliot-
Curie and many others contributed to our present knowledge of
nuclear energy sources. In this way they lengthened considerably our
estimate of the life-span of the sun (to ten billion years) and the stars
(to one trillion years). The phenomenon of black hole evaporation,
discovered about twenty years ago, may have added yet another period
to the possibility of life in the universe.

No one can say how long the universe will be habitable, since our
posterity may make amends for any shortages. In doing so, they will
merely be imitating their Neolithic ancestors, although on a far
grander scale.

"Anthropology could with advantage be changed into 'entropology,'
a discipline concerned with the study of the highest manifestations of
this process of disintegration," says Lévi-Strauss.[1] On the strength of
the preceding discussion, we would be tempted to contradict this
assertion and state that anthropology is instead the study of the ways
that human beings harness new sources of information and entropy in
order to delay the disintegration process. "The world will end without
mankind," the text asserts. But we counter that the long-term goal
would be *to fight off such an end to life in the universe*.

In the first chapter of this book we encountered the death drive,
manifest in the inexorable development of war-making techniques, to
the point that the extinction of our species is a possibility. If, in the
spirit of the nineteenth-century cosmological vision, the result of
human activity is, inevitably, to bring about the destruction of the
universe, then the build-up of the nuclear arsenal becomes the proto-
type of every human endeavor. Annihilation and doom are prime
objectives, not mere unwanted side effects.

On the contrary, contemporary cosmology gives us a very different
vision of the world. We know now that the universe provides us with
ever-greater quantities of information. The mission of humanity is to
figure out how to use and manage this information. Nuclear suicide
would bring a swift end to this quest, brutally terminating all efforts to
prolong life in the universe. But since the French edition of this book
was written, in 1984, the situation has considerably improved. The
superpowers have shown definite interest in nuclear disarmaments.
There is now reason for prudent optimism about our future.

3

ORIGINAL SIN

The death urge is inscribed in the life urge

9

Out of Eden

The other day I was browsing in a bookshop when I saw on display an *Auschwitz Album*. My heart shrank. I rushed away, unable to bring myself to open the cover. Later, I read about its subject matter in a literary review; it was just what I had expected: "total horror."

Horror is not confined to bookshop displays. It is an integral part of reality. As I write these words, as you read them, horror is raging in Lebanon, Ethiopa, Kuwait, on the streets of our own cities, in many corners of our human world; some people face it every day.

One reader criticized my *Atoms of Silence* on the ground that I talked only about the "wonders" of the universe. He was right. Any vision of the world worthy of the name must integrate every facet of reality.

When and how did the death urge permeate the hum of gestating cosmic matter? Is it irrevocably entangled with the life urge? Is preservation of the life urge any guarantee that we will avoid the fatal effects of the death urge?

To answer these questions, we must start at the beginning. In the pages that follow, we will take apart the mechanisms of complexity. This will bring us to the origin of the elements by which the death urge introduced itself into the structure of the pyramid.

Complexity is costly. It does not flourish without damage. However, without completely avoiding this damage, we can hope to limit it. Thus we can minimize the threat that complexity poses to itself. *Understanding* will help us to *prevent* or at least to *act*.

THE ARTICULATIONS OF COMPLEXITY

We must first reconsider certain elements from the preceding chapters.

Matter is perceptible to us in a variety of forms. We can classify these forms according to a scale of complexity. At the bottom are elementary particles (quarks, electrons, and photons), next come nucleons (protons and neutrons), next atomic nuclei, and then, in ascending order, atoms, molecules, cells, and multicellular organisms. Among the levels there are analogies and resemblances. For example, the "alphabet" structure is repeated on every level. But differences also exist. Characteristics that are absent from lower levels gradually develop on higher ones. Increasing complexity articulates itself through them and uses them to move up from one level to the higher one. In this section, we will try to identify these key characteristics. They will enable us to understand how the pyramid grew over the course of time.

Such an approach will lead us to reexamine and reformulate an ever-popular question: Do extraterrestrial civilizations exist? Our first step will be to determine whether the emergence of intelligence is a "normal," "natural" phase of the growth of complexity. If it is, we can be fairly sure that there are plenty of inhabited planets and interstellar cultures. If, on the contrary, the emergence of intelligence is a freak event in the normal flow of organizational currents, we may speculate on our solitude in the universe.

Our analysis tends to support the first alternative. The reasoning can be summarized as follows: In the same way that the concept of *stability* governs the course of complexity as it ascends through the lower levels, the concept of *competitiveness* takes over at higher ones, pointing the way to the summit of the pyramid. As we shall discuss later, living systems are constantly threatened. They have to compete in order to survive. Under pressure they develop strategies at an ever-higher level of efficiency. Because it is a particularly effective survival strategy, intelligence naturally finds its way into the pyramid.

THE FUNDAMENTAL STATE

Lower-level systems like nucleons, atoms, and molecules, because of the stability conferred on them by natural forces, are relatively invulnerable. Large amounts of energy are required to destroy them.

Smaller supplies of energy, usually in the form of radiation, enable them to interact with other matter without placing them at greater risk of disintegration. Upon absorption of luminous energy, they leave their *fundamental state* and go to one of their several *excited states*. The number of possible excited states increases as we climb the scale of complexity. (Appendix 4 contains illustrations of the spectra of the various types of states; nucleonic, nuclear, atomic, and molecular.)

Absorption or emission of radiation causes the systems to migrate between states. Usually, access to a given excited state is accompanied by the acquisition of new properties. For example, if a carbon nucleus is excited in a certain way, it enters a phase of high-amplitude collective vibration called giant resonance, in which the movements of all its nucleons are coordinated.

When interaction with the radiation stops, the nucleus or atom returns to its fundamental state. It doubles back upon itself until another source of excitation appears. Like a bear in winter, it "hibernates." For this reason, the elements at the lower levels of the pyramid can be said to be immortal. Their identity is ensured by the mere juxtaposition of their component particles.

A carbon nucleus is created when six protons and six neutrons combine in a small space. The removal of a proton or a neutron is enough to change the nature of the nucleus. On the other hand, once it has been created, no activity is vital to its survival. Its continued existence does not rely on interaction with the outside world. This situation is reversed on the pyramid's upper levels. Living things are vitally dependent on their exchanges. How did such a state of affairs come about?

Throughout cosmic evolution, transformations and emergences took place gradually. Primitive or rudimentary forms and behaviors developed, diversified, and enriched themselves, although they always retained some vestige of their origin. Close examination of any behavior will reveal ancestral roots reaching to the humblest levels of reality.

In fact, dependence upon exchanges is not confined to living organisms. The physical world is full of systems that must rely on an energy supply for their existence. Deprived of this supply, they have no fundamental state to which they may withdraw in hibernation. They cease to exist when their energy source is exhausted.

A hurricane is a striking example of a complex system that is wholly dependent on energy. It can be thought of as an air mass in a collective state. The motion of the molecules is preferentially orientated in a single direction; this is called a wind (see Appendix 2). Every fall, many hurricanes appear in the Caribbean. You may remember Hugo, which devastated South Carolina. When the wind subsides, the hurricane ceases to exist. For lack of activity, Hugo died.

VULNERABILITY AND DEPENDENCE

A living organism is the site of many exchanges. Molecules and photons enter through various orifices in the form of food, breath, and heat. The organism must accommodate these deliveries, extract what is useful, integrate that into its structure, and dispose of the waste.

Unlike atoms and molecules, living organisms are better identified by this set of exchange-carrying functions than with the sum of their component particles. If you lose a hair, or even an arm or leg, you are still yourself. If your exchanges cease, you regress on the pyramid of complexity: You are only a set of molecules.

From this viewpoint, we might define life as a highly excited state of matter in which inert and independent particles enter into a collective state that enables them to coordinate their activity and pursue exchanges with the environment. We've seen that collective states exist even for inert matter, be they giant resonances of nuclei (as on page 145) or Caribbean hurricanes, but at a far more elementary level. In living things these collective states display themselves as the many amazing feats we call behavior: moving around, reproducing, writing symphonies, and so on.

In this respect, it is interesting to note the ambiguous status of the tobacco mosaic virus. It can be crystallized, like any other salt. Maintained in this state, it is entirely inoffensive, independent of the outside world. But if it is dissolved, it recovers its full powers. It becomes simultaneously dependent, virulent, and mortal.

DEATH AND FILIATION

An excited atom can transmit its energy to another atom. Similarly, by the process of filiation a new animal appears and perpetuates the mortal animal from which it issued. This process provides some permanence to the vulnerable structures inhabiting the upper levels of complexity.

At the beginning of terrestrial life, this filiation was carried out through cell division: One cell split into two individuals ready to repeat the process. Later, filiation took the far more efficient form of an encounter between two individuals, in which their genes are combined.

THE LIFE STRUGGLE

The energy required to maintain life is not always available. Sometimes it must be sought. The necessity for vital exchanges introduced the concept of predation.

It is probable that the first complex systems that were forced to become predators lived in the primeval ocean. At first, there was no lack of food in that environment. For hundreds of millions of years, the interplay of molecular combination—energized by the yellow solar photons and the purple lightning of atmospheric storms—generated and accumulated vast amounts of energy-rich molecules in the seas: sugars, fats, and amino acids. Laboratory simulations of the primeval sea set up by the U.S. chemists Harold Urey and Stanley Miller give us reasons to believe that the primeval oceans were indeed well stocked with nutritive substances. Food was there for the taking.

Thomas Malthus was the first to note the problem of providing adequate nourishment, given the fact that food reserves are not infinite. Under the proper conditions, twenty minutes suffices for a cell to divide and produce two identical daughter cells. At such a pace, how long would it take before the mass of living cells equaled the mass of the earth? The answer is easy: a little more than one day!

COMPETITION STRATEGIES

To reproduce, each cell begins by splitting into two identical parts. The two halves then seek to mature by using the necessary elements from the nutritive substances accumulated in the sea.

The sea abounded with food for the early cells. Their population thus grew rapidly. But soon there was not enough to go around. Famine and shortage limited the wild population increase. At this point life had reached the gates of the Garden of Eden. Food became scarce. All wanted to eat, but only a few could manage to get a piece of the pie. Hence the appearance of a concept that was to have a major influence on the upper rungs of the pyramid: competitiveness.

As living organisms competed for limited food supplies, they invented and developed thousands of strategies. Some invested in movement; speed can be a valuable asset both for capturing prey and for avoiding predators. Others tried protective armor or chemical poisons. Each family, each genus, each species thus defined the terms under which it would engage in the harsh business of life.

ADVANTAGES AND MUTATIONS

In a gas chemical reactions tend to produce increasingly stable molecules (Chapter 5). At the upper reaches of complexity, however, the influence of stability fades. Differences in stability among the giant molecules are negligible. A new factor takes up where stability left off:

Efficient performance in survival strategies is the reward for adaptational advantages.

Tough skin, sharp teeth, a long beak: like stability on the lower levels, these assets contribute to population increase by improving an individual's chances to survive and bear young. Where do these advantages come from? How do organisms acquire them? Modern genetics holds that they are the result of mutation. Genes, which contain the information necessary to reproduction and life, are subject to modification. Coiled in the cell nucleus, they are continually affected by external influences (radiation) and internal ones (transcription errors) that change their composition. These modifications can, in turn, alter the properties or behavior of the organism (coloration, resistance to cold, and so on). Under certain conditions, the changes are an improvement; for example, new coloring may provide camouflage. The breed with improved protection from predators will proliferate more quickly.

At a still uncertain historical moment nature invented sexuality. Two living things meet and blend their genes; this blending provides a fantastic source of newness and diversity. Creatures no longer merely produced exact replicas of themselves as they had through cell division. Now they could engender new beings endowed with original genetic combinations, potentially leading to immense varieties of adaptational advantages.

DEATH AND COMPLEXITY

Lichens feed on rock; they are capable of subsisting on the meager nutrients the rain leaches out of the stone surfaces. After a volcanic eruption, when all life vanishes within a radius of several kilometers, these minuscule organisms, able to eke out an existence from the atoms of rain and rock, are the first to reappear. As the years pass, the strata of lichen accumulate and decompose, forming a rich soil for plant seeds sown by the winds. Gradually, vegetation takes hold: grasses, shrubs, and then tall trees. The animals can then return. First come the herbivores, who need only eat plants. Later, the carnivorous animals will find the food types essential to their metabolism.

I've narrated this simple tale to illustrate a very important point. As time passed, predation became more specialized. Unlike lichens and plants, animals cannot live on what they derive from inert matter. *They must kill.* Herbivores kill plants; carnivores kill animals. A herd of cows grazing a pasture seems less cruel to us than three lions ripping apart a gazelle. But all things considered, is the difference that great?

In a way, it is. The carnivorous diet introduces and emphasizes the idea of aggression. A wolf cannot afford to be a pacifist. To possess long incisors is not enough for the tiger; it must also use them. Attack and killing emerge naturally from the gradual specialization of dietary sources all along the food chain.

We may dream of another world. Life would be sweeter if each being, emulating the lichen, could feed on stone and water. But we have to be realistic. Highly evolved organisms cannot survive on such a restricted diet. Death is the price of complexity.

10

The Poisoned Gift of Knowledge

PERCEPTION

By means of their interlocking structure, crystals are able to recognize and reject atoms that do not fit. Antibodies in the blood identify proteins foreign to their home organism. A rudimentary form of perception is present even at the pyramid's lower levels. For living things perception ranks as one of the most effective competition strategies.

Channels of perception are many and varied. In plants, as well as in animals, the identification of chemical substances is involved (olfactory perception). The ear perceives air vibrations, while the eye detects photons emitted by luminous objects. The information (food in sight, danger imminent, possible mate nearby) gathered by the nose, ear, and eye lead to decisions: Hide in a hole until the hawk flies away; fold leaves to conserve moisture; send a signal to the female.

Where is the deciding done? In plants and lower species of animals this operation does not appear to be localized in a specific part of the organism. A cutting from a plant or an earthworm cut in two continues to perceive and act in each of its parts. The same is true of beehives and termite colonies: Following a disaster, the workers busily rebuild the edifice, without (to our knowledge) orders sent from a central authority. At this level, behavior consists of reflexes. A bacterium, for example, responds automatically and immediately to light stimulation. Although the phenomena involved in bacterial migration are infinitely more complex than those governing the deflection of an electron in an electrical field, to our eyes they appear to occur in just as mechanical a fashion, without any possibility of alternative or choice.

In more complex animals, the situation is quite different. When an organism possesses several means of perception, or senses (hearing, sight, touch, smell), we note the appearance of a data integration center, more often referred to as a brain. Information conveyed by the senses is collected, weighed, and integrated in the brain. These operations yield an image, an inner representation of the outside world. Instinctive behavior, less automatic than reflexive behavior, can be identified with the set of reactions triggered by these images.

In the absence of any apparent stimuli, some vertebrates explore the world around them almost as if in amusement. An animal thus enriches its knowledge of the environment and collects images it stores in its memory. These data allow it to increase its efficiency and make its life more comfortable. A bird knows where it is most likely to find food. Bees explore their territory in search of nectar sources; using a set of codified movements, they communicate to their fellows the course and distance to fly, in order to find the meadow or field in flower. It seems that images and symbols appeared first as survival strategies in a world where life is vulnerable.

Thanks to biological evolution, members of the mammalian class have learned to associate images. Realizing that a banana is hanging just out of his reach, the monkey sees itself in the process of knocking it down with a stick, and goes off in search of the right stick. Image association as a problem-solving tool is the most elementary form of intelligence.[1]

SELF-AWARENESS

The rabbit "knows" that it is the eagle's prey and must hide. We can assume that perception of a self arose from instinctive behavior related to survival. But that is about all we can say; doubtlessly, we lack an appropriate vocabulary.

I recently watched a television program where researchers painted a chimpanzee's face red while it was sleeping. Upon waking, the chimp looked in a mirror and immediately put its hand to its face. Not only did the creature know that the monkey in the mirror was itself, it also had an accurate enough idea of its normal appearance to react to the red paint. This, to my mind, can be taken as an indication that self-awareness is not confined to humans. It also suggests that human behavior arises from analogous manifestations in our animal ancestors, even if the latter are more rudimentary. Plants also supply many examples.

A WELL-BLAZED TRAIL

Three major conclusions can be drawn from the above discussion:

1. Complex systems are vulnerable. Their existence requires exchanges with the outer world. Resources are limited. A winning strategy will beat the competition.
2. A particularly effective strategy is the perception of the world by mental images.
3. Intelligence emerges from the association of images.

The first point applies to all complex systems, from viruses to orchids, to human beings. The range of the second point is somewhat more limited: As far as we know, plants have not invested in the mental image strategy, although they reach peaks of organization and sophistication in structure as well as behavior. The application of the third point is even more limited; nevertheless the path of intelligence, traced by competitiveness and the valuable benefits to be obtained from the association of mental images, has been followed by several animal lines. Along this trail the human race hobnobs with not only its primate cousins (chimpanzees and gorillas), but also with cetaceans (dolphins, whales, and walruses), and many other mammals (rats, elephants, and dogs). Nor are mammals alone: Certain birds (crows) and insects (bees) are skillful manipulators of mental images.

The multiplicity of intelligence lines is, I believe, highly significant. Far from being rare and improbable, this strategy appears to be a major thoroughfare in the ascension of complexity. Species from highly varied evolutionary backgrounds have developed some form of intelligence.

These ideas bring us to the principle of complexity. From the above analysis two points are apparent. First, intelligence is not the only path to complexity; the plant kingdom does very well without it. Second, the human being is not the only animal using intelligence as a survival strategy; at best, we might say that the human is the one who has progressed the farthest. But the human race has only a small headstart on the other animals — a few million years, that is, only a thousandth of the total duration of biological evolution. As far as we know, our ancestors of three million years ago were not terribly more intelligent than present-day chimpanzees. (These comments are, I feel, two more reasons to prefer the term "principle of complexity" to the chauvinistic expression "anthropic principle.")

THE RISKS

In previous chapters we noted that the ascension of the pyramid's lower levels is governed by the quest for stability. But this quest is fraught with threats for complexity. Stability, pushed too far, is synonymous with monotony.

The danger of physical monotony was circumvented by the universe's particular mode of expansion. Dominated by gravitational force on the universal scale, quick cooling brought cosmic matter into a succession of disequilibrium phases called phases of supercooling. Later, on the local scale, gravity generated hot stars in an increasingly cold universe. These phases of disequilibrium curbed stability's tyrannical aspirations, enabling a marvelous variety of structures to appear and survive.

Vulnerability and the obligation to kill appear at the upper echelons of complexity. They bring competitiveness and adaptational advantages to the forefront. Taken to an extreme, competitiveness could be harmful as the overriding quest for stability: One supremely gifted species, able to eliminate every rival, would bring monotony on the biological world. Luckily, this is not the case in nature. The advent of new species does not necessarily involve the extinction of previous lines. Tortoises are still here; they have not changed much in the past three hundred million years and are quite able to coexist with reptiles and mammals. The same is true of sea urchins, mosses, and ferns.

The wide diversity of climatic conditions on our planet restrains the despotic ambitions of competitiveness. Habitats in deserts, prairies, mountains, oceans, tropical regions, and polar zones require very different survival strategies. Something that constitutes a physical advantage in one climate may be a serious handicap elsewhere. Thick fur is indispensable in the tundra, but in the savannah you're better off without it. The variety of ecological habitats results in the diversity of living beings.

It is worth recalling here that the variety of climates ultimately originated with the effect of thermal disequilibria on cosmic matter (Chapter 5). Rain, snow, and hot and cold air are the results of solar radiation on our planet. They are caused by the temperature difference between the surfaces of the sun and earth. Supercooling in the universe's remote past resulted in these thermal disequilibria; they were also responsible for the diversity of nuclei, atoms, and molecules, imperiled by the various forces' tendency to instill monotony. Thus we come to the following remarkable conclusion: Cosmic supercooling, which, on the lower levels of the pyramid of complexity, limits the dangers inherent in stability (Chapter 6), also limits the risks inherent in competitivity.

THE HUMAN DIMENSION

Nature apparently has little use for compassion. One's life requires another's death. Animals feed on the flesh of other animals. Eat and avoid being eaten; catch and don't be caught: Those are the two poles of animal life. Even plants kill each other.

Animal evolution could be schematized as refinement in the art of killing others and protecting oneself, as offense and defense. Each advance in the offensive panoply brings about an advance in defensive protection, and vice versa. Sharks have long teeth, but sea urchins are covered with spines. Wrens evolved long beaks they can poke into rotten wood, but certain insects eject toxic substances to defend themselves.[2] Camouflage and bait are techniques widely used in the animal kingdom. Our analysis of the articulation of complexity (Chapter 9) revealed the reasons for this state of conflict. The fragility and vulnerability inherent to the complexity of higher systems, associated with the requirement for a specific food, does not allow them to relax their vigilance. Existence becomes a full-time job.

Judicious doses of aggressiveness and brains appears to be the perfect christening gift for a new animal line. But in order to survive, it also needs a sense of family, a drive to protect and care for the young. The ritual battle between male seals (Chapter 1) does not end with the death of the loser: Natural inhibitions preserving the species are engraved in the animals' instinctive behavior. Today, biologists explain these inhibitions in terms of natural selection. A seal line that did not inherit such behavior with respect to the lives of its own members would quickly become extinct.

With the human race, the situation takes on a new dimension. In the case of mankind, advantage over predators lies not in physical superiority, but in intelligence and knowledge. Thanks to these advantages, the human race has become an organized structure capable of the highest performance. The human being is a formidable competitor, able to adapt to any climatic conditions. We have neither claws nor sharp teeth, neither the tiger's speed nor the tortoise's protective armor, but we can make weapons: Knives are our claws, and shields our hard shell. Long ago, we conquered our natural enemies. No animal is a match for our wits. Even the strategies of bacteria and viruses, hidden by their microscopic size, are, we hope, on the way to being overcome by our intelligence. No other organism controls the matter surrounding it to the degree that we do so. We build bridges, dig canals, drain lakes, flood entire provinces. We change the face of our own planet and will soon be mining the riches of outer space (see Chapter 8).[3] Problems arise precisely from the *scope* of our power. The human being's adaptational skills have gone well beyond their goal. Now it is the human race itself that threatens nature's balance and, by the same token, its own existence.

As a result of the activity of the human race, many animal species have disappeared without leaving any descendants. The cases of the dodo and the trumpet swan are well known. But little is said of the two hundred other species of mammals and birds and the forty varieties of butterflies that have vanished in the past two hundred years. To this total, we should add another thousand animal species and thirty thousand plant species that are on their way to extinction (as shown in the map below).

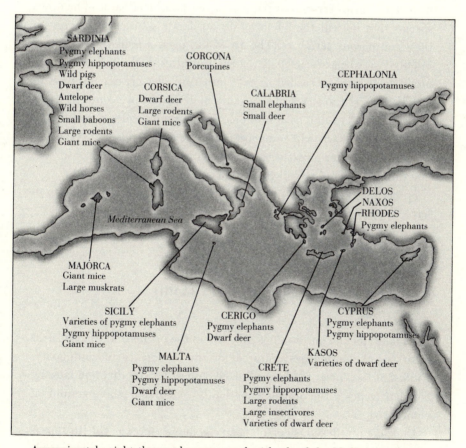

Approximately eight thousand years ago, the islands of the Mediterranean were populated with a variety of strange animals. They were probably hunted to extinction by our ancestors. There were giant mice, pygmy elephants and hippopotamuses, small baboons, antelopes, and many others, listed on the map. (From *New Scientist*, January 1985, p. 29.)

Our power is turning against us. The damage we inflict on our environment endangers our own survival. Massive clearing of wooded land transforms the forests into barren wastes. Even more tragic is our behavior toward our fellow human beings. Competition is racing along in high gear, but the train is running off the track.

By overriding the automatism of instinctive behavior, intelligence neutralizes the natural system of checks and balances that protected living things for millions of years. Unlike seals and wolves, humans kill one another. Killing is, in fact, one of our major activities. Human history is a grim series of wars, massacres, and bloodbaths. The other species that do kill their fellows devote far less time and energy to these murderous activities. The art of killing is a field where human intelligence surpasses itself. Over the centuries, techniques have improved and wars succeeded in exterminating ever-greater numbers of people. We used to speak of thousands of dead; now we count in terms of millions and even tens of millions. The human species is prolific: "A single night in Paris will compensate for our human losses," Napoleon callously said after a costly battle. But after a full-size nuclear attack, any number of nights will be insufficient.

Sometimes I wonder if it wouldn't have been better for evolution to have stopped with the butterflies . . .

A SISYPHIAN PYRAMID

Let's take stock of our findings. Already in the primordial chaos matter possessed the properties required for the emergence of complexity. Guided at the outset by the (unfinished) quest for stability and later by competitiveness, the universe patiently elaborated the pyramid of complexity.

On some privileged islets (planetary systems), life was able to appear and develop. Certain animal species evolved intelligence and knowledge; certain cultures invented science and technology. These very achievements — the crowning glory of complexity — may put an end to all the others.

To illustrate the fundamental absurdity (as he saw it) of human existence, the philosopher Albert Camus evoked the image of unhappy Sisyphus, condemned indefinitely to push his boulder to the top of a mountain, only to see it roll back down ineluctably again. If the tip of the pyramid of complexity must crumble every time it reaches the level of intelligence and knowledge, wouldn't an apter name for it be the pyramid of Sisyphus?

Let us repeat here the key question of the current day, as formulated in the prologue of this book. Should we conclude that all the amazing coincidences, all the fine-adjustment of initial conditions (physical constants, isotropy, density, homogeneity), the succession of super-coolings, the infinitely fertile combinations, the nuclear activity of the stars, the electromagnetic hum of the interstellar nebulae, the exuberant biochemical fever of the primordial ocean, the trials, errors, and successes of natural evolution — the purpose of *all this* is to culminate in nuclear annihilation? After fifteen billion years in the making, could intelligence emerge only to commit suicide in a few minutes? Is knowledge such a poisoned gift?

If this turns out to be necessarily and fatally the case, then it would be fair to say that "meaning" is a disastrous illusion and God a pitiful cheat.

PAWNS ON A CHESSBOARD

The idea of a ludicrous deity is not a new one. Literature contains many references to such a god. Contrary to the principles of atheism, the existence of a god is accepted, but the deity is perceived as a foolish, irresponsible character, well beneath human expectations, a being unworthy of our "metaphysical anguish."

Heraclitus of Ephesus, known as Heraclitus the Gloomy, lived in Greece in the fifth century B.C. He deserved his nickname. Certain passages of his writings have received widely varying interpretations over the course of the centuries, and scholars are unable to agree on their true meaning. We are further hampered by the problem of a questionable translation from the archaic Greek. At any rate, Heraclitus writes:

> *Time is a child*
> *playing backgammon*
> *Royalty of a child*

Backgammon is a game using dice. The players move their pieces around the board according to the throw of the dice. Hence a popular interpretation of Heraclitus's words: "We are nothing more than pawns. The events and dramas in our lives, as well as nature's catastrophes, are merely the reflection of the aimless play of a child king who is rolling dice."

Omar Khayyám, the twelfth-century Persian poet and astronomer, like the English poet Robert Herrick ("Gather ye rosebuds while ye may"), sang of the fleetingness of life, but in a much more bitter tone. Inexorably, death awaits us, bearing grim proof of the foolishness of our lives. Ours is to enjoy life, without wondering why, for fear of discovering that

> *We are no other than a moving row*
> *Of Magic Shadow-shapes that come and go*
>   *Round with the Sun-illumined Lantern held*
> *In Midnight by the Master of the Show;*
>
> *But helpless Pieces of the Game He plays*
> *Upon this Chequer-board of Nights and Days;*
>   *Hither and thither moves, and checks, and slays,*
> *And one by one back in the Closet lays.*

A child rolling dice, a chess player who annihilates his pieces when he tires of his play — these images would do quite well to illustrate the absurd cynicism of a universe fated to destroy its crowning glory.

SENSE AND NONSENSE

The question of "sense" penetrated the life and work of the Swiss psychologist Carl Gustav Jung. He valiantly explored the most elusive aspects of reality. At the age of eighty-three, in his autobiography (1961),[4] he summed up his search for meaning:

The world into which we are born is brutal and cruel, and at the same time of divine beauty. Which element we think outweighs the other, whether meaninglessness or meaning, is a matter of temperament. If meaninglessness were absolutely preponderant, the meaningfulness of life would vanish to an increasing degree with each step in our development. But that is—or seems to me—not the case. Probably, as in all metaphysical questions, both are true: Life is—or has—meaning and meaninglessness. I cherish the *anxious hope* that meaning will preponderate and win the battle.

I have italicized the words *anxious* and *hope*. What would Jung add today, twenty-eight years later? The cosmological findings that revealed the principle of complexity to us were unknown to him. No doubt he would have interpreted them as a confirmation of the "sensible aspect of life." Yet the intensification of the nuclear peril and planetary pollution since Jung's death can only serve to underscore a "nonsensical aspect."

One point for hope, one point for anxiety . . .

PART
4

A NOTE OF HOPE

But of man, when will it be a question?
— *St. John Perse*

11

Man's View of the Universe

"Where am I?" "Why am I here?" People have been asking themselves these questions for millennia, everywhere on the planet (and perhaps elsewhere, too). "Does science teach us to die better?" a novelist asked. "Does the discovery of a new particle or the dissection of a frog teach us the meaning of life?"

Science, of course, cannot give answers to these fundamental questions. But science does provide elements of knowledge about reality that any search for meaning must recognize and integrate. These elements of knowledge have been accumulated over several centuries, and our way of seeing the world has been profoundly altered by them. In the pages that follow I shall try to retrace the main steps of this transformation.

For the purposes of my exposition, I have chosen to divide the history of man's view of the universe into three major periods. We will call them:

The Ancient Dialogue,
The Universe No Longer Answers, and
Outline of a New Vision.

THE ANCIENT DIALOGUE

In the cosmogony of almost all primitive cultures, as in that of most great traditional wisdom, the earth is the center of the world. Above it lies heaven, the starry vault where the gods cavort. The universe is not very big. The planets and stars are just out of reach of human hands, but within reach of human voices.

The heavens are inhabited by paternal figures who take a direct and personal interest in human affairs. In fact, they often dwell on the peaks of high mountains.

Mankind is not afflicted with loneliness. Quite the contrary: Humans are under constant surveillance and must behave accordingly. Heavenly manifestations like thunder, lightning, and comets are the language of an awe-inspiring "beyond." The ancestors live there. They are the intermediaries between heaven and men.

In several mythologies human beings are the offspring of celestial beings, in the genetic sense of the term as well as the symbolic one. This relationship, to which the tale of mankind's divine birth attests, gives existence purpose and meaning. Man is responsible for his fate. If he respects the commandments, he will be admitted to the land of the ancestors. If he transgresses, he will be punished.

My goal in presenting this oversimplified version of our first period was to emphasize the fact that certain beliefs are common to several religious traditions:

1. The world is small.
2. Humans are in direct contact with a "beyond."
3. Humanity is the offspring of divinity.
4. Life has meaning, and the responsibilities are clear.

In the past centuries this traditional vision of the world has been severely undermined by a number of scientific discoveries. I have borrowed from Sigmund Freud his discussion of these attacks, in which he outlines three major blows.

THE UNIVERSE NO LONGER ANSWERS

The Astronomical Shock

The Greeks were the first to perceive the true dimensions of the sky. To Heraclitus, jokingly asserting that the sun is no bigger than his foot, Anaxagoras answered that it was larger than the Peloponnesus.

With the development of telescopes during the Renaissance, the sky suddenly assumes gigantic proportions. The earth is no longer the immobile pedestal around which the stars revolve. Like the moon and the planets, the earth itself is a heavenly body. All these celestial objects revolve around the sun, which, during this period, emerges as the new center of the world.

But not for long. Soon it becomes clear that our triumphant sun is a star, like the multitude of others, sparkling in the sky on a moonless night. In fact, it is a very ordinary star, located in the suburbs of the Milky Way. It seems brighter than the others only because they are much, much more distant. (Anaxagoras already had a hunch this was the case.) The light emitted by the sun reaches us in eight minutes, but the light from the stars in the night sky takes several years to travel to us.

In the eighteenth and nineteenth centuries they identified the universe with what we call our galaxy today: the Milky Way, a volume extending over one hundred thousand light-years. Then, at the dawn of the twentieth century, there is a new revelation: Our galaxy is not unique! New galaxies, similar to the Milky Way, are continually being discovered. Hundreds of millions of galaxies have already been detected, spread over billions of light-years. In the context of contemporary cosmology we seriously envisage the possibility that the number of galaxies is infinite, that the universe is without limits.

These dizzying dimensions and the new awareness of our insignificance in the measureless void of space gradually influenced philosophical thought. As early as the mid-seventeenth century, Blaise Pascal voiced a reaction. This man was both a scientist — thus well informed on astronomical findings — and a philosopher — therefore capable of appreciating their impact. His famous saying, "The eternal silence of these infinite spaces frightens me," succinctly summarizes the torment of this era. In the chilling immensity of this universe, man is alone, an alienated stranger. The void is silent. Comforted by his Christian faith and deep attachment to the "God of Abraham, Isaac, and Jacob, rather than the one of philosophers and scholars," Pascal was not as deeply traumatized as later thinkers by the solitude of the human race in the sidereal immensity. He was nevertheless profoundly shaken by the impact of the development of astronomy on the traditional vision of man's relationship to the cosmos.

The Biological Shock

A second shock came from the field of biological sciences, with the Darwinian theory of evolution. Doubt was now cast upon the divine origin of mankind. It was no longer possible to believe that Adam and Eve were created by God in the Garden of Eden on the sixth day of Creation. And the brilliant Athenians had not sprung in full battle dress from Jupiter's thigh. Our ancestry is much more humble: Our forefathers emerged from between the legs of primates.

The further we search into our past for the ancestors of our ancestors, the closer we nuzzle the primitive, lower animal species. Before the primates appeared, we find reptiles, amphibians, fish, and invertebrates; ultimately, our family tree has its roots in the microscopic world of primitive cellular organisms much like those we can observe swimming in a murky pool of stagnant water. It is easy to see why Darwin's theory was unpopular for quite a long time.

To add to the discomfiture, biological research revealed the important role of *chance* in the mechanism of the evolution of living things. The genetic code of every being, stored in the cell nucleus, is continually submitted to accidental mutations that will affect its posterity. Random processes are essentially involved in the emergence of new animal forms, which may or may not remain in the realm of living species.

These two biological discoveries—the animal ancestry of humans and the intervention of accidental phenomena in the evolution of terrestrial life — dealt another blow to the ancient dialogue. Man is not the offspring of gods but the child of chance. Can you talk to chance?

The Psychological Shock

During the nineteenth century the traditional harmony between man and the universe was to receive two more major blows.

By exploring the depths of the human psyche and discovering the existence of the unconscious mind, Freud and his followers profoundly altered our perception of human responsibility. Handicapped by infantile conflicts, we are far less free than we like to think. Moreover, psychoanalysis has placed into evidence mental processes by which the father-image is projected in the form of divine figures. Having lost the status of both genetic and spiritual parent, God becomes a product of infantile fantasy. "God does not exist since we now know how He was invented by human beings," wrote Nietzsche.

A few decades before the birth of psychoanalysis, Karl Marx had cast doubt upon the image of God as the spiritual leader of the great nations. Social power is not a divine right, as the royal rulers would have had us believe. The power to govern is, first and foremost, the power of the strongest. Human history is not the stately accomplishment of a divine project, but the tale of the ups and downs of the class struggle.

These multiple shocks engendered what has been called the anguish of modern civilization, faced with the silence of the heavens and our own solitude. "The absurd," writes Camus, "arises from the confrontation of the human call with the silence of the world." Without contact with any "beyond," alienated in our own universe, Western man has arrived at an uncharted position in the history of humanity: the essential loner.

OUTLINE OF A NEW VISION

We come now to the third part of our history, presenting some of the features through which a new vision of the world appears to be emerging. In this vision, man is no more a stranger in the universe. Quite the contrary: he discovers that he belongs here, that he is inextricably involved in a long story that involves as well the stars, the galaxies, and the universe as a whole. Ironically, through a rereading and a reappreciation of this story, a new vision emerges from the very same scientific facts that had given such lethal blows to the ancient dialogue.

For a long time the various sciences progressed independently, each on its own track, without any attempt at exchange or contact. Instead of expanding their horizons and touching edges, these disciplines became more and more specialized, narrowing their scope, and concentrating on smaller and smaller areas of reality. The expert on beetles had as little to say to his colleague the spider specialist as to the volcanologist or to the thermodynamicst. George Bernard Shaw summarized the situation by stating that in contrast with the so-called humanist who knows "precious nothings about everything," the scientific expert is one who knows "everything about nothing."

Only in the twentieth century, as a result of the accumulation of knowledge, did the scientists begin to remove their blinders. Scientists rediscovered something they had overlooked in their zeal: Their common object is the universe, inhabited by man, the author of science.

The Waltz of History

Properly placed side by side, the teachings of the various scientific disciplines form a vast tapestry of knowledge. Just as the dabs of paint in a pointillist painting take on shape and meaning when viewed from a certain distance, the tapestry of knowledge adds depth, meaning, and perspective to the results of scientific research.

The history of the universe is the central theme around which the various branches of knowledge are best harmonized. For many centuries scientists studied a reality they perceived as eternal and fundamentally unchanging: the laws of stars, the laws of the atoms. Then, gradually, the historic dimension entered the field of science and began to exert an influence. It started with geology and the exploration of deep soil strata. The skeletons of animals buried two hundred million years ago were found to bear little resemblance to the species we know today. In the process of trying to account for these differences, biology encountered history in the form of the theory of evolution: Living animal and plant species have been elaborated from primitive cells that existed four billion years ago.

Astronomy, chemistry, and physics joined in the waltz of history with the discovery that atoms and stars have not always existed. Stars are born in space from interstellar clouds. Atoms are born in stellar interiors, thanks to a propitious combination of gravitational, electromagnetic, and nuclear forces.

Cosmology extended the historical vision to the whole universe. Not only are the universe's inhabitants subject to change, but the very structure of the cosmos is in constant evolution.

Fifteen billion years ago the whole universe was chaos: extreme temperature, density, and disorganization. It did not contain a single structure or organized system. There were neither stars nor galaxies; there were not even any molecules, atoms, or atomic nuclei. At that time the universe was composed of a homogeneous and isothermic stew of elementary particles that today's physicist can identify as electrons, photons, quarks, neutrinos, and several other exotic ones that haunt the giant modern accelerators.

The Tapestry of Knowledge

Modern sciences teach us how matter organized itself, step by step and bit by bit, as the cosmic temperature, continuously decreasing due to universal expansion, lowered enough to allow the various natural forces to come into play.

Astronomy tells us of the formation of galaxies and stars due to the influence of gravitational force on the primordial stew.

Physics endeavors to understand how, in the stars' incandescent cores, elementary particles were able to fuse and generate atomic nuclei.

Chemistry reconstructs the steps by which nuclei and electrons, ejected by dying stars, associated with one another in interplanetary space, forming atoms, molecules, and grains of dust.

Gravity went to work again to collect these small dust grains and pack them together as solid planets. The study of these events, as yet not fully explained, is called planetology.

Some of the planets possessed atmospheres and oceans, in which biochemistry, in concert with geology, attempts to understand the formation of the first living cells.

At this point biology joins with paleontology to lead us from the roots of the animal family tree to the higher organisms and hominoids.

Human behavior, including the development of scientific activity, is the field of study we call psychology.

Thus, instead of observing a fundamentally unchanging world, the scientist gradually becomes a historian of nature, only to realize that he or she is writing an autobiography of the human race.

The key concept in each of these phases of history is the organization of matter. Science teaches us that everything that exists — atom, star, frog, or human being — is made of the same matter, the same elementary particles. The only difference is the state of organization of these particles in relation to one another. It lies in the number of steps climbed on the pyramid of complexity.

At the lowest levels the huge diffuse clouds of matter floating in sidereal space are in a state close to primordial chaos. Stars, which are born in these nebulae, are relatively more structured. Gravity has made them spherical, but their inner organization is still quite primitive. Encased in rigid crystalline shells, the atoms of stones, asteroids, and solid planets are much more organized than the atoms of a gaseous star, yet they are incomparably less organized than the atoms in a frog's foot. The human brain, with its billions of interconnected neurons, is the most formidably structured sample of matter known to us.

The activities, accomplishments, and achievements of organized systems are proportional to their degree of complexity and inner organization. Stars are content merely to shine, bacteria, to move and reproduce, but we human beings perform the most outstanding feat ever realized in the universe: taking consciousness of our own existence and that of the outer world. Science emerges from this activity of the thinking brain.

Geographically, almost all cosmic matter possesses a rudimentary, if not nonexistent, organizational level. Nevertheless, in this ocean of chaos, in certain special regions, on certain islets blessed with the proper physical conditions, matter was able to yield to its organizational drive and conceive the wonders that are within its power.

Man's Place in the Universe

The synthesis of scientific teachings, woven together into the coherent fabric of the tapestry of knowledge, leads us to redefine humankind's place in the universe and our relationship to nature.

In terms of mass and volume, a person is nothing, an infinitesimal grain of dust in a limitless void. According to the much more meaningful criterion of organization, however, man is quite important. As far as we know, mankind occupies the highest level of the pyramid of complexity, a height from which we can gaze upon the universe and wonder about its origins and future. Before us, no one (at least on our planet) was able to formulate an answer to these questions.

With the nebulae, the stars, the stones, and the frogs, with all that exists, we are engaged in this immense experiment, the organization of matter. Far from being strangers in the universe, we are an integral part of an adventure that extends over billions of light-years. We are the children of a cosmos that gave birth to us after a fifteen-billion-year gestation period. As the Hindu tradition has it, the stones and stars are our sisters.

THE THREE SHOCKS REVISITED

How could science, after its early repudiation of the ancient alliance between man and the gods, now suggest this new vision of man's place in the universe? By considering one by one the shocks described above and integrating scientific findings into the larger context of contemporary cosmology, we can come to assess them differently, and give them new meaning. In short, these shocks were healthy for humankind's vision of the universe. By ridding us of a primitive infantilism, they give us access to the real attributes of the cosmos.

What about the astronomical shock at the new awareness of the immensity of space? We now know that this immensity is no idle wastefulness (page 167). Given a smaller world, on the scale of that of Dante or the Greek cosmologies, matter would never have been able to give birth to life and consciousness. Our brain is made of atoms, which are generated within massive stars. Many generations of stars burning over billions of years in galaxies like our Milky Way are needed. Countless stellar explosions have released fertile atoms and sown them in the immensity of outer space.

Furthermore, it is now apparent that the very expansion of the universe is vital to the emergence of life on earth, for several reasons.

First, because expansion results in cosmic cooling, which allows particles to associate and bind. Moreover, by continually increasing the space between galaxies, expansion acts to dilute the entropy produced by natural processes, enabling complex systems to elaborate.

Life takes several billion years to emerge. Meanwhile, expansion spreads the galaxies over billions of light-years. The astronomical shock replaced the tiny heaven populated with anthropomorphic characters with a limitless space humming with the fever of the cosmic gestation to which we owe our existence. Space not only surrounds our planets, as our forefathers believed; it also begets life. And that task required enormous volumes of space . . .

The discovery of biological evolution has created the second shock. Placing it side by side with chemical evolution in the early oceans and with nuclear evolution in stars, we discover progressively the steps of our cosmic gestation.

Far from being the descendants of gods, we emerge at the summit of an arduous ascent. Our genealogy goes back, step by step, to include primates, reptiles, fish, and protozoans; before that, giant molecules, simple molecules, atoms, nuclei, nucleons, and the elementary particles of the Big Bang. The combination of letter-elements and superimposed alphabets are nature's favorite ways of constructing complexity. An acceptance of the real roots of our family tree reveals our close kinship with everything that exists, our place in the vast organizational tendency of universal matter.

At the same time, it becomes a little clearer how organization and complexity were able to emerge from the primordial chaos. Forces acting on the particles of cosmic stew fashioned matter as soon as the decreasing temperatures gave them the opportunity to do so. Chance was a part of this interplay of forces, but its role must be reinterpreted. Evidence that random phenomena influenced animal evolution is often given as a proof that nature has no "master plan." (I do not know if it has one or not.) In the context of the theory of evolution, however, chance can be seen as playing an essential role in the actualization of matter's immense potentialities.

The wealth and diversity of life forms in the plant and animal kingdoms are related to the fact that physical laws allow the existence of a nearly infinite variety of different entities. Nuclei colliding in the stellar crucibles, molecular captures in the primeval ocean, and the impact of cosmic rays on the genes of living cells are examples of random phenomena that perpetually engender the new and unique. These accidental events are the key that unlocks nature's fantastic possibilities. A universe without chance, a world where every event was entirely predetermined, would produce nothing but dullness and monotony.

Here reside several of the characteristics of our cosmos that are still a mystery to us. Our analysis pointed out the "fine adjustment" of many mathematical parameters governing the behavior of matter. We were especially aware of the advent of phases of disequilibrium and their important contribution to the emergence of complexity.

Organization is constantly threatened by "thermal death," that is, the establishment of states of equilibrium. Fortunately, thanks to universal expansion, matter escapes these deadly states. This is our reprieve, and it is continually being extended, always providing more time for innovation and improvisation.

The foregoing is a new interpretation of the elements that caused the biological shock. We are the manifestations of the potentialities of extremely rich, inventive matter, whose actualization requires the intervention of chance. According to this developmental scheme, our place is not in the descent of the gods, but in the ascent of the apes and so-called lower animals. All along this evolution matter behaved in a way we do not yet grasp; nevertheless, we have good reasons to believe that, had it been different, we would not be here to discuss it.

The psychological shock was important in purifying our conception of the cosmos. It enlightened us on the unconscious processes that lead us to project into the heavens the agents of our own inner conflicts. But this explanation in itself does not help us very much in dealing with our problems of survival. When children argue too heatedly, an adult intervenes to prevent them from hurting each other. We now know that in case of war, no deity — no product of human psychological projection — will intercept the nuclear missiles. Nothing will halt the holocaust once we set it in motion.

12

The Delivery of Meaning

Pride of man marching on beneath his burden
of humanity. Pride of man marching on beneath
his burden of eternity.
— *St. John Perse*

MEANING IS NOT AUTOMATICALLY GRANTED

The universe engenders complexity. Complexity engenders efficiency. But efficiency does not *necessarily* engender meaning; it can just as well engender nonsense. This, in brief, is what comes out of the preceding chapters.

To avoid nonsense, an entirely new element is required: a conscious, active *decision* in favor of meaning. We could very well say that the decision itself *is* meaning.

Neither blue-green algae nor meadowlarks have the ability to make such a decision. But there was no need for such a decision before we entered the scene. Indeed, if the need for such a decision did not arise until human beings appeared, it is because major threats to complexity hardly existed until we arrived.

From the earliest times, conflict has been present in living processes and played an important role in establishing complexity. With mankind, conflict acquires a tragic dimension; at the same time, however, a solution becomes possible: It is the duty of man to give meaning to reality.

We (the current generation of human beings) are both witnesses of and actors in the period of history when the problem enters its crucial phase. If there is indeed a purpose for our presence in the universe, it must be to help nature give birth to herself. The most threatening of nature's creations is also the only one who can make a successful delivery.

NATURE SHOWS THE WAY

Let us recall that, despite the power of natural forces, the universe does not contain only iron atoms, noble gases, and black holes. The variety of nuclear, chemical, and stellar species convinced us that the natural forces have not exhausted their possibilities, have not completed their job. Their despotic ambitions were curbed by cosmic supercooling, whereby matter escaped the pitfalls of sterility and monotony.

Atoms endure. They are invulnerable. Conversely, living organisms are in perpetual danger of losing their lives. They must fight for survival. In the long run, however, every individual will die. The only hope for survival is through posterity. If the species is to be preserved and offspring to be permitted to reach puberty, the young must be protected, nurtured, and taught the difficult art of living. Death is a cruel reality of the living world. But it can have meaning: *Parental care and filial love emerge from the will to survive in the face of individual death.*

Sexual reproduction allows nature to proceed ever further in her favorite activity: *the creation of new forms.* The blending of genes is a magnificent source of new combinations and untried biological experiments. In the sexual encounter and in the more or less elaborate emotional exchange that ensues, feelings play a role. The family, whether stable or temporary, is the natural context for emotional, social, and sexual relations. It is a social structure nature has chosen for many animal species, especially birds and mammals.

When butterflies gather nectar from flowers, they transport the fecundating pollen that provides plants with a means of reproduction.

Symbiotic relationships, or ecosystems based on mutually beneficial exchanges, are common in nature. Thus systems of mutual support have evolved alongside weapons for survival. Nature's strategy of conflict, drawing on strength and aggressiveness, is accompanied by a strategy of cooperation, involving in some cases affection.

Solidarity and feelings, curbing ruthless competition, come to the forefront. Nature has already blazed the trail; it's up to us to follow. Wise management of our legacy of survival strategies is the only way to ensure that nature will have a chance to continue surpassing herself.

HOMINIZATION AND HUMANIZATION

The sequence of events through which our primate ancestors gradually converted themselves into *Homo sapiens* is called hominization.[1] The acquisition of intelligence as a survival strategy (in the Darwinian sense of the term) was an essential part of this transition. To save mankind from itself, intelligence must transcend its earlier role and participate in the humanization of our species.[2] But how? We may very well be dumbstruck before the magnitude of the problem and the urgency of the situation. What recommendations, other than pious clichés and unworkable resolutions, could we offer? I am not cut out to be a preacher. Nevertheless, in the words of René Dubos, it is important to "think globally and act locally."

Syntheses are not sufficient, and political stunts aimed at "changing society" have proven vain. Each of us acts within a sphere of influence that is limited but not nil. In my opinion, only personal initiatives have some chance of making an effective contribution to the project.

I would like to outline a few ideas for contemplation. Anyone can try to invent solutions or share suggestions. This is a subject that is best handled gently. There will inevitably be a repetition of ideas familiar to some. In any case, I make no claim to either novelty or originality. We'll recognize concepts that have already been propagated thousands of times. Perhaps the hugeness of this book's themes will give them new dimensions.

I am aware that, in several cases, I'm standing on shaky ground. In no case are my opinions definitively committed. If the reader disagrees, if his or her outlook differs from mine, so much the better. A productive discussion may ensue, and perhaps show us a new perspective from which to view the urgent problems facing us all.

DARWIN AT THE STAKE

The increase in wantonly violent behavior is, without a doubt, one of the most disturbing phenomena of the past few decades. Brutal aggression is not confined to deprived regions, where the need to fight for survival could explain, if not justify, the use of force. The wealthiest places in the world are also the scene of much murderous violence and crime. It would seem that, despite the best efforts of individual, family, and social programs, a free society is powerless to contain the wave of violence.

Educators and social workers are seeking the underlying reasons for this problem and the ways to instill more ethical conduct. Dr. Daniel Robinson, a professor of psychology at Georgetown University, expressed this need in an interview in *U.S. News and World Report*: "I don't think there's any question but that people, generically, sooner or later must find what can only be called a moral reason for their lives — provided in the past by governments, monarchies and powerful institutions."[3]

The past few years have seen a resurgence of the theory of "scientific creationism" in the United States. Defenders of this theory contest the Darwinian outline. They demand that more than one explanation of the origin of the human species be taught in the schools. The biblical version of the birth of Adam and Eve in the Garden of Eden (which, according to them, is no less scientific than the theory of biological evolution) should also be presented. Ronald Reagan, two-term president of the United States, has given some support to creationism. Asked his view on the teaching of evolution, he responded: "It [the theory of evolution] is a scientific theory only, and it is not believed in the scientific community to be as infallible as it once was believed. But if it is going to be taught in the schools, then I think the biblical story of creation should also be taught."[4]

If the fundamentalists have not always received the success they counted on in legal battles engaged in fourteen U.S. states, they have achieved a considerable moral victory. Many instructors present the two "theories" on an even footing. The more cautious ones avoid the subject entirely. Science teaching has taken a serious blow.

As *scientific explanations*, how can one compare the biologists' evolutionary reasoning and the symbolic account given in Genesis (or any other traditional mythology)? Evidently, beyond this ridiculous debate, something else is at stake. In a torrent of alliterative eloquence, a Georgia judge put his finger on the real problem:[5] "This monkey mythology of Darwin is the cause of permissiveness, promiscuity, pills, prophylactics, perversions, pregnancies, abortions, pornotherapy, pollution, poisoning, and proliferation of crimes of all types." Or, to quote a different source, "The theory of evolution is essentially an immoral idea that gives rise to immoral conduct."[6]

Let's be more explicit. According to that interpretation, the degradation of human behavior is the result of the contemporary scientific vision, which reduces the universe to the level of physiochemical machinery obeying blind laws. Mankind is the accidental product of a series of incoherent processes. Total selfishness is the one rational behavior if human existence has no meaning. Looking out for number-one is the only thing that counts. People and other objects are meant to be used or consumed. Behind the fuss and foolishness stirred up by the creationists, a real problem lurks: What should be the basis for ethical behavior?

In ancient tribal societies the rules of conduct appear to have been intimately linked to the vision of the world. The Native Americans are a good example. The attitudes were impregnated with reverence for nature and what nature represents. The acts of life — eating, lovemaking, hunting, warmaking — were performed with a sense of communion with the ancestors and the immortals.

Similarly, the major religions, each in its own way, have codified teachings and laws. What created the world? What is the relationship between heaven and earth? How does it affect human behavior? Christian philosophy, for example, is founded on the redemption of the human sinner by the Son of God. The Christian believer, by integrating this faith, perceives the attitudes it is appropriate for him to adopt.

But the influence of religion has diminished noticeably in the past few centuries. Religious truths have given way to scientific findings,

Is Anyone There?

(Drawing by J. F. Batellier, from the collection *Y'a quelqu'un?* (Is anyone there?), published by the author, January 1983.)

which carry no moral motivation or obligation. Hence the vacuum deplored by the supporters of creationism (see the illustration on the following page). Hence their efforts to return to the past, by intimidation if persuasion should fail, and to introduce Bible stories in the natural science curriculum.

In my opinion, the problem is not limited to the field of ethics. The experience of a relationship with the "beyond" — in the vaguest and most general sense of the term — seems to be one of the most basic human needs. The definition for human being used to be "the laughing animal," but "the religious animal" would be more appropriate. Anthropologists make it clear: No human group, no matter how isolated, no matter how primitive, has failed to establish and codify its relations with an intangible divine reality.

The disarray of modern civilization, deprived of such a relationship and in need of relief for the ache of emptiness, is evidenced by the proliferation of new religions, cults, and teachings (often originating in countries that also harbor the leading centers of scientific research).

After the discoveries of Galileo and Newton, astrology, like its sister alchemy, should have been eclipsed, only kept for its historical interest. Yet today there is a massive renewal of enthusiasm for astrology (this upsets certain scientists, who are trying to fight it off, without, apparently, understanding what is at stake).

But there is no way to return to the past. The gates of the earthly paradise are locked forever. One can no longer bury one's head, like the ostrich, in the sands of traditional certainties. Our attempt to find a new vision of the world must integrate all the facts of modern science, or it will fail. The survival of humanity, the victory of the life urge over the death urge — this is the purpose determined for us by the analysis of reality presented in this book. This is the purpose that could form a basis for our code of ethics.

The most significant message modern science has for us is perhaps the evidence of the connection between humanity and the universe. This new assertion appears to me to be easier and more satisfying for the critical mind to accept. Mathematics is nature's language. Atoms and molecules are our very substance; the discovery of how they combine is the key to our present as well as our past. It is a way to improve our understanding of the vast enterprise of organization to which we owe our very existence.

"I feel that nothing human is foreign to me," wrote the Latin poet Terence two thousand years ago. We can now add: "Nor is anything physical, chemical, or biological foreign to me." In a nutshell, that is the sum of our progress since the Roman Empire.

Becoming adult means that, without too much pain, one can accept the idea that there is no Santa Claus. An adult learns to cope with doubt and uncertainty. Science cannot answer questions like Does God exist? Does life have a meaning? Is there life after death? But scientific knowledge does allow us to define our place in the cosmos in relation to stars, plants, and animals. Science retraces our past, uncovers our cosmic roots, and describes the adventure of matter's organization, of which our existence is a part. In order to live and interact with our fellow beings, each of us develops a personal philosophy of life, a personal world vision. It is in the elaboration of this world vision that scientific knowledge plays a primordial role.

HUMAN DIGNITY

Often, when a young man misbehaves, his entire family feels ashamed.
When I see reports of violence against South African blacks or Cambodians, I also feel a nauseating sense of shame. I am ashamed that I
belong to a species capable of such crimes.

A photo extracted from World War II archives was recently published, showing a particularly cruel scene: a charred human head sticks
out from a half-burnt wooden fence. Just before the Allies arrived to
liberate the prisoners of an extermination camp, the SS set the bar-
racks on fire and fled. This extra instance of gratuitous horror brings
us to the tragic question: How could human beings stoop so low?

In their struggle against sexism, feminists attacked it at its roots,
exposing prejudice in hitherto unsuspected ground: the words, everyday language, and, to all appearances, harmless sentences innocently
uttered at home and at school. In solving our problem, we must be as
thorough as the feminists.

Over the years, certain expressions come to provide a solid foundation for abominable attitudes. When the circumstances allow it, these
attitudes cause us humans to behave disgracefully. When my children
were in school, I often attended their graduation ceremonies. I remember the warm June mornings fragrant with lilac and cherry blos-
soms. After the traditional awards ceremony, students, instructors, and
parents would lift their voices in song, intoning "La Marseillaise," the
French national anthem, wishing with one heart that their fields might
be "watered with impure blood." What disturbed me even more than
the racism of the expression "impure blood" was the carelessness with
which the words were sung. With the smiling approval of their parents
and teachers, the children were ingesting a kind of poison.

Let's not exaggerate. Millions of French children have sung the
"Marseillaise" without turning into bloodthirsty monsters. Nevertheless, our history reminds us — painfully — how easy it is for the "common folk," in certain political situations, to applaud the prophets of
racism. The waves of anti-Semitism that swept through France during
the Nazi occupation, as well as the current vogue for fascist-influenced,
nationalistic ideas, are sufficient proof.

A new vision of humanity emerges from contemporary scientific knowledge. Though mankind can no longer pretend to be the "center of the world," our new position gives us our real dignity. With animals and plants, we occupy the top level of the pyramid of nature's organized entities. We reached this level after a gestation period of fifteen billion years, in which all the cosmic phenomena participated.

All human beings, regardless of their origin, have an equal claim to this dignity. The respect for human rights implies also an awareness of the importance of every individual in the history of the universe.

ANIMAL RIGHTS

In Roman times slaves had no rights or property. Subject to the whims and vagaries of their owners, they were not even the masters of their own bodies. At times slaves were mutilated as babies to be displayed as circus freaks. Seneca, a humanist of the era, tolerated these practices. "Ideally," he would have said, "slavery should be abolished. But let's be realistic. Roman citizens are attached to their privileges, and free labor is essential to the general welfare of the Empire." The man on the street would have had an even simpler viewpoint: "Some men are born free; some are born slaves. Can you change a cat into a dog? Can you change a slave into a citizen?"

Who speaks for the slaves? What hope have they of a better life? With the citizen acting as judge, jury, and prosecution, the slave cannot expect a fair trial. Spartacus, crucified on the Roman Way, learned his lesson the hard way.

Nevertheless, *slavery was abolished.*

That does not necessarily mean that all people are free. Our century has witnessed the proliferation of concentration camps and gulags. I like to believe, however, that there has been some improvement. The United Nations did sign and ratify the Charter of Human Rights. Violations, although frequent, are decried and condemned. A collective awareness has been awakened, and it is dismayed by the fate of the Cambodians tortured by Pol Pot and that of the Romanians under

Ceausescu's boot. This awareness did not exist, I believe, in the Assyrian, Egyptian, or Roman Empires.

But how do we stand in our relation with animals? At the moment, we are the strongest and (so it pleases us to believe, at least) the most intelligent of the animals. In our relations with other species, we have generally used our intelligence to exploit them as much as possible. Anyone who doubts it should visit a farm where calves are raised for veal. Observe the refinement of the technique used to obtain this delicacy. And help yourself to another slice, if you still have any appetite.

"Why did Nature create the hideous race of ferocious beasts, bitter enemies of human beings?" wondered Lucretius in the first century B.C. Today, we wonder how to protect the few specimens that remain. Like the awareness of the injustice of slavery, the awareness of animals' feelings and rights has evolved. Ecology preoccupies our fellow citizens. It is generally accepted that care for the natural environment is vital to human comfort and happiness. Nevertheless, this observation both justifies and limits environmental awareness. In a communiqué, the Soviet Union clearly expressed its official position on the question: "Environmental concerns are valid only inasmuch as they contribute to the well-being of human society." But who will speak for the animals?

The discovery, through contemporary scientific research, of cosmic evolution, the vast movement in which stars, frogs, and humans are manifestations of matter's fantastic potentialities, is the basis for *every* being's right to existence and dignity. To question established personal privileges in order to reach a higher level of objectivity is the sign of true intelligence.

What cases and what circumstances give a person the right to deprive an animal of life? At the outset, we must realize that we are dealing with an act of *murder*. The act therefore requires justification. Animal murder becomes a necessity under certain circumstances. It can be seen as an act of self-defense on the part of the human race: Eating and protecting oneself from the cold are essential to survival. These vital needs can justify animal slaughter.

Killing for the purpose of eating is admissible; torturing for the purpose of gourmet delights is not. Brutal treatment, such as force-feeding geese, confining calves in dark stalls, and so on, is unworthy of human beings. The death of a polar bear is justified if its fur is used to protect an Eskimo from the Arctic freeze. The urban socialite, however, has no right to the bear's skin; in our industrialized cities protection from cold no longer requires the use of animal hides.

The Romans enjoyed circus entertainment. Watching gladiators in battle or slaves wrestling lions was a pleasant way to spend a Sunday afternoon. Certainly, we humans have become less bloodthirsty. Yet a bullfight, with or without the death of the bull, is a gory sight beneath the sunny Spanish skies.

"Trifling, ridiculous concerns, when there are so many more pressing problems": Those who defend animal rights often hear this argument when they object to fox hunting or to the massacre of baby seals. "Why worry about baby seals when children are dying in Palestinian refugee camps?" But a single code of ethics applies to animals and humans alike. Sensitivity to the act of killing has to be developed. I am not certain that a trigger-happy hunter will react humanly and refuse to shoot prisoners if, during a war, he is ordered to do so by his superiors.

AT WHAT PRICE KNOWLEDGE?

Is there anyone who has forgotten Joseph Mengele's medical experiments on prisoners in Hitler's concentration camps? After the war, no one protested that science would be crippled if his research was stopped. Today, many people feel that vivisection and the use of live animals in medical laboratories are reprehensible.

Curiosity is a basic urge, driving humans as well as kittens to explore how things work. Justification for it lies chiefly in the pure pleasure of discovery. How far can one go in the name of knowledge? Can knowledge become a goal in itself, justifying any means? If pure knowledge is not the ultimate end, where should the line be drawn? What risks are admissible? The same questions apply to technical exploits. The idea held by certain scientists, that anything that *can* be done *must* be done, is rightly called technological folly.

We are currently forced to give thought to these questions by the controversy surrounding genetic experimentation. Those who wonder about the long-term risks involved in playing sorcerer's apprentice with human genes are sometimes satisfied with the usual alibi, "If I don't do it, someone else will." The problem is not a new one. Lucretius grappled with it, addressing the idea of using ferocious animals in combat. The Carthaginian generals experimented with a tactic whereby maddened bulls and crazed elephants armored with sharp blades were released onto the battlefield. The strategists were unpleasantly surprised to note that, in the heat of the battle, the stampeding animals attacked friend and foe alike. This technological folly was quickly abandoned. Likewise, and for similar reasons, the use of poisonous gases was dropped during World War I. Nevertheless we have information that gas production is still going on actively in many countries. Toxic weapons were used in the recent war between Iraq and Iran.

A moratorium exists among the planet's major powers on the development of weapons for germ warfare. Representatives of these nations drew up and signed a treaty whereby each promised to halt all research aimed at designing and perfecting those lethal weapons. After all, this is another case where the common interest of all the participants is at stake. Like the Carthaginian elephants, the viruses would be unable to tell the "good guys" from the "bad guys."

These examples give us a gauge for deciding where technological folly begins: It is when an experiment threatens the future of the experimenter's own species. This is a new dimension to Lucretius's problem. The possibility for humanity to exterminate itself, which did not exist until recently, is one of the few positive effects of the power of destructive means. It gives pause to even the boldest technician. The idea of ethics has *forced* its way into the sanitized field of scientific research.

For the kitten, exploration is pure profit. Knowing his territory will be a valuable advantage in leading his life. Mankind, however, is in a much more ambiguous position in relation to scientific research. "What really matters is not so much the "truth," but rather, what helps in living. That humane sentence by Nietzsche contains much wisdom, I believe. Isn't the ultimate goal of all knowledge and experimentation to make each life a little more bearable? And the words "each life" embrace all living beings. Legislation on scientific experimentation must respect the rights of all concerned, people and animals.

ON DELIGHT

> In the woods there are
> Trees wild with birds.
> — *Paul Eluard*

As a student I sometimes hitched rides on boats during the summer vacation. "Where are you headed? Do you have room for a passenger?" These two questions carried me the entire length of the Saint Lawrence River, from Montreal to the sea.

I spent whole nights stretched out on deck, gazing at the lights on the top of the masts as they swung back and forth amid the stars of the Milky Way. I watched the dawning of a new day from beginning to end, from the first gray glimmers until the sky was ablaze with scarlet rays.

Dolphins often accompanied us, swimming past the boat and then turning back to rejoin us. Their upturned bellies reflected the sun's golden light on the blue sea dappled with white foam. One day, overjoyed by the beauty of this sight, I called one of the sailors with whom I had made friends the day before. He disappeared down a hatch and emerged with three men carrying rifles. In spite of my objections, the sea turned red. Farewell to the dolphins . . .

I saw this sailor again later the same day, and reminded him of our long conversation the night before. He had told me of his miserable childhood, his drunkard father, and the sadness of his current situation. He had added, "All I ever get from this world is trouble and pain." Was he aware that he had destroyed a source of pleasure? Apparently, these words meant nothing to him, and perhaps that was his real misfortune. How can one restrain the drive to kill when delight is absent?

"Music often transports me like the sea." When, with Beethoven or Wagner, I climb aboard for a journey, those lines by Baudelaire sometimes come to mind. With a genius at the helm, as I drift along, rising and falling with the swelling waves, I am overcome by an irresistible surge of exaltation and gratitude for life and the universe that engendered it. Sounds, colors, and words are the alphabets of artists. From their combinations arise new emotions that reveal within us unknown oceans and hidden treasures.

It is difficult to imagine that barely three centuries ago, the works of Bach, Haydn, and Schubert and the paintings of Turner, Monet, and Van Gogh did not exist. We should rejoice in their creation as yet another revelation of the marvelous potentialities of the primordial matter. Thanks to the efforts of artists, reality acquires new dimensions; the universe takes on splendor and majesty. New directions open to transform the moments of our lives into instants of exultation.

"The more the universe seems comprehensible, the more it also seems pointless," writes the physicist Stephen Weinberg, in *The First Three Minutes*.[7] I challenge him to repeat those words while listening to Mozart's *Marriage of Figaro*, as I am now.

In *Epousailles*, Annie Leclerc writes, "Very early, as a child, I had the idea — but it was also the embrace of a desire — that knowing and asserting had to come from the strongest and most forceful place, that our faculty for attaining what is truly desirable was neither understanding, nor reasoning, nor intelligence, but solely *delight*." Goethe says the same thing in fewer words: "Theories are gray, but trees are always green." These sentences are the best summaries I could find for this last section, indeed for the whole book . . .

From a poster on the wall of my room, Baudelaire is watching me as I write these lines. In his eyes, I read a message, the one from "Spleen of Paris":

And if sometimes on the steps of a palace, or on the green grass of a ditch, or in the dreary solitude of your own room, you should awaken and find the inebriety half or entirely gone, ask of the wind, of the wave, of the star, of the bird, of the clock, of all that flies, of all that sighs, of all

that moves, of all that sings, of all that speaks, ask what
hour it is; and wind, wave, star, bird or clock will answer
you: *It is the hour to be drunken!*

With wine, with poetry or with virtue, as you please.

APPENDICES

1

Games of Chance

The concepts of entropy and information have played a major role throughout our story. Vivid definitions have been applied to them, but these are still imprecise. It is much easier to grasp them by examining the mathematical structure on which they are based. By studying simple examples in detail, we will be able to understand what entropy and information mean and what they do not mean; that is, we will demystify them.

The number of elementary particles involved in the tiniest of reality's events is always extraordinarily high. This is a crucial factor in our discussion. The blue spot on a butterfly wing or the thin column of smoke rising from a cigarette are the result of arrangements composed of billions and billions of atoms. Of course, it is impossible to describe the individual behavior of each particle in such a huge population, or set. Like opinion pollsters at election time, we must therefore compute probabilities.

The way to become familiar with these calculations and concepts is to apply them to sets that are much smaller and therefore easier to work with. A die, for example, is made up of six sides, and a roulette wheel (if I remember correctly) is composed of thirty-seven positions. By studying these games of chance, one can see the mathematical link between the concepts of randomness and probability and the entropy–information pair.

We will see that high-information (or low-entropy) systems are very improbable systems. Conversely, the higher the entropy, the greater the probability that a system is caused by uncontrolled, random phenomena, in other words, that it is the product of chance.

HEADS OR TAILS

A coin that is flipped in the air can land either heads (H) or tails up (T). We will say that there are two possible configurations: H and T. If the coin has not been weighted, the probabilities of these two configurations are equal. The probability of each configuration is thus one out of two, that is, one half (0.5).

It is assumed that it is impossible to know beforehand which of the two configurations, H or T, will occur. Using this example to define the chance, we will say that an event takes place "at random" if it is not predetermined to give one result and not another. (I shall not discuss the question of whether chance really exists or is merely an excuse for our ignorance. I simply mention that modern physics admits and incorporates in its structure the idea of true randomness. Classical physics' concept of absolute determinism has been replaced by the concept of statistical causality: Reality is only partially determined by the laws of nature.)

Using **two** coins, we could obtain the following **four** (2×2) results:

HH HT TH TT

The probability of obtaining one of these four events is ¼ or 0.25.

Using **three** coins, we would obtain the following **eight** $(2 \times 2 \times 2)$ results:

HHH HHT HTH HTT THH THT TTH TTT

The probability of each event is now ⅛ or 0.125.

With **four** coins, we would obtain **sixteen** $(2 \times 2 \times 2 \times 2)$ events. I leave it to the reader to write them out. The probability of obtaining one of them is $\frac{1}{16}$, or 0.0625.

The above examples suffice to prove a few general rules. We have learned to calculate the probability of each event. A wager consists in opting for a single event among n number of possible events. The probability of the optimal event is 1 over n. The problem consists in calculating the value of n.

For those who may have forgotten it, a brief review of basic mathematics follows:

The number 4 can also be written 2^2. That can be read as (2×2) or "two to the second power" or "two squared."

The number 8 is also written as 2^3 or $(2 \times 2 \times 2)$ or "two to the third power" or "two cubed."

The number 16 is written as 2^4 or $(2 \times 2 \times 2 \times 2)$ or "two to the fourth power."

In the above examples the number 2 is called the *base* number while the smaller figures written a bit higher and to the right are the *powers* or *exponents*. The figure 2 (the base) expresses the number of configurations possible with a single coin (heads or tails). The figure 4 (the exponent) in the last example stands for the number of coins required to produce sixteen possible events. If we decided to use ten coins, the number of events would be 2^{10}, or 1,024.

Now let's substitute a die for the coin. Each throw can produce six configurations: **1** or **2** or **3** or **4** or **5** or **6**. How many events could be obtained with two dice? The answer is six times six: 6^2, or 36. With three dice? Answer: 6^3, or 216.

This gives us a simple rule: To calculate the probability of a certain event in a game of chance, one must first know the number of configurations assigned to each of the game's elements (two for a coin, six for a die, and so forth). This number will serve as the base. The number of different elements will become the exponent. The total number of possible events, n, will be obtained by bringing the base to the power indicated by the exponent. The inverse value, $1/n$, will then give us the probability of any one event.

Without expending any particular effort, while amusing ourselves with these innocent games, we have just established a basic result that will lead us very far in our understanding of the physics of the cosmos.

Here comes a little more terminology: the *logarithm*. Let's use as our example the number sixteen, written 2^4. **Four** is the exponent that must be given to the base, **two**, to obtain the result, **sixteen**. That can also be expressed: "**four** is the logarithm with the base **two** of **sixteen**." By definition, the logarithm is the exponent that must be assigned to a number (the base) in order to obtain a given number. Here's another example: **Three** is the logarithm with the base **ten** of **one thousand**, since $1,000 = 10^3$.

What is the logarithm with the base two of sixty-four? The answer is six.

Let us return to our games of chance. To increase the number of possible events, we only need to increase the number of elements (dice or coins) used in the game. The number of events increases very rapidly (exponentially) with the number of elements.

As we shall soon see, the entropy assigned to a physical substance is proportional to the total number of configurations in which this substance can be found.

Like the dice and the coins, each physical particle can be found at a given moment in one of a certain number of possible physical states (see Appendix 4). Hence the opportunity to apply the concepts expressed in studying the behavior of sets of particles. It is not hard to understand that the total number of configurations that a substance can take on increases exponentially with the number of component particles. Whence our golden rule: *To increase entropy, merely increase the number of particles.* This is exactly what the earth does in relation to solar radiation: It increases the number of photons (see Chapter 5). This increase is the source of all terrestrial life.

GAMES WITH DICE

The (complex) events of our life can be analyzed in terms of a large quantity of (simple) events at the scale of atoms and molecules. It is thus appropriate to make a distinction between the macroscopic scale (the one to which we have access) and the microscopic scale (the one of which we perceive only the effects). Our guide here will be an analysis of the results of some games of dice.

Two dice are thrown at the same time and their sum is calculated. The result can be anywhere from 2 (if 1 comes up on both dice) to 12 (if we roll a double 6). The question is, which result is the most probable? To analyze the situation, we can refer to the square in the illustration. The first row and first column give the results possible with each die — all the numbers from 1 to 6. In the center square, we find the sum of the corresponding results. Note that the number 7 appears most frequently (a total of six times), followed by 6 and 8 (five times), and so forth. The essential concept here is that the probability of any of the individual squares occurring is the same: All results — (1 + 1 = 2), (1 + 6 = 7), (6 + 6 = 12), and so forth — have exactly the same chance of appearing. We count the thirty-six squares (6 × 6) and thus say that the probability of any one square is $\frac{1}{36}$.

Let us now differentiate between two types of events, called single and composite events. A *single event* corresponds to a single little square. It might be, for example, the result (3 + 4). We count the thirty-six small squares that illustrate thirty-six single results. The probability of any given single result is $\frac{1}{36}$. The *composite event* is defined by the fact that we are interested in the sum of two numbers, rather than in which two numbers give the sum. In our diagram a

Second die

	1	2	3	4	5	6
1	2	3	4	5	6	7
2	3	4	5	6	7	8
3	4	5	6	7	8	9
4	5	6	7	8	9	10
5	6	7	8	9	10	11
6	7	8	9	10	11	12

First die

composite event can correspond to several of the little squares. For example, (7) corresponds to a whole diagonal (six little squares). The probability of the composite event (7) is thus $6/36 = 1/6$. For the events (5) and (8) the probability is $5/36$, while the rarest events, (2) and (12), have a probability of only $1/36$.

Now, for the purposes of our study, we are going to change the language slightly. The results of the events will become configurations. Instead of "single," we shall say "microscopic"; instead of "composite," we shall say "macroscopic." The probability of a macroscopic event will be the sum of the probabilities of all the individual microscopic events corresponding to that macroscopic event. Finally, the large square containing all possible configurations will be called the sample field.

We will now attribute a certain entropy to each of the macroscopic configurations. *By definition* this entropy will be proportional to the number of microscopic configurations entering into the macroscopic configuration. (More precisely, it will be proportional to the logarithm of this number.) The macroscopic configuration (7) thus possesses the greatest entropy, while (2) and (12) possess the least. (Note that entropy is associated only with macroscopic events.)

We can thus begin to see the relationship between probability and entropy. The higher the entropy of a certain configuration, the greater the probability of it occurring in the course of a game.

To illustrate the situation, let us imagine that, in a space-age casino, a specially designed machine rolls two dice once per second and displays the composite result on a screen. The spectators watch as the figures change. The one that appears most often is (7). But the others also flash by. Approximately once every thirty-six seconds, the players note the appearance of a (2) or a (12).

Let's begin a new game, this time using three dice. The sample field is now a cube. It contains 216 individual small cubes representing each of the microscopic configurations. The probability of $1/216$ can be assigned to each. Events go from (3) to (18). The most probable are (10) and (11), each of which has a probability of $27/216$, that is, approximately 12 percent (you can easily work this out), while the (3) and (18) have the minimum probability, $1/216$.

If greater numbers of dice are used, increasingly improbable events are encountered. For example, with ten dice, the probability of throwing (10) or (60) is only 1/60,000,000. The casino screen will display the number (60) only about once every two years, if the dice are rolled once per second, as in the preceding examples. Let us imagine that upon entering the casino one day the gamblers see the number (60) on the screen. They would then wonder if (60) had been the last number rolled the day before or some mischievous casino employee had tampered with the game in order to arouse the gamblers' curiosity. The first possibility cannot be dismissed, although the second seems far more likely. The machine is started up again. The number (60) could, of course, come back, but the gamblers expect to see a figure around thirty. If yesterday's last figure was (33), the chances of remaining in the neighborhood of this number are much greater than those of seeing (60) appear.

We can summarize this study by saying that the states of greatest entropy (the most common) are those that are repeated most often over a certain period of time. In other words, if an improbable state is observed, it is probably the result of some tampering with chance. We are much more likely to pass from an improbable state to a probable one than from a probable to an improbable one. Nevertheless, the opposite is not at all impossible: If enough time elapses, it will occur, sooner or later.

THE NORTH WIND

These statements describe not only the behavior of dice and roulette wheels, but also that of natural phenomena.

Air, for example, is composed of billions of molecules in motion. In the absence of wind, this movement occurs in all possible directions, that is, "at random" or "by chance." The expression "The wind starts blowing" means that the molecular movements are beginning to be oriented in a certain direction. When the wind dies, the movements again become random. What does that have to do with dice and their configurations? Each molecule corresponds to a die and each direction to a face of a die. This comparison is flawed, however, because a die has only six faces, while the number of different directions is infinite. The analogy would be more exact if we played with new dice composed of a huge number of different faces.

Let us now associate the phenomenon north wind (movements in a single direction) with a configuration in which the same figure (for example, 1) would come up on all the dice. In this case nearly all the other configurations would correspond to a state of calm.

Wind can be thought of as a sort of organization of air, while calm air is characterized by the absence of organization. On its own, with no outside influences and thus no source of information, an air mass has a much greater probability of remaining in a state of calm than of changing to a state of wind. A rise in the wind with no external cause is not an impossible event (it violates no physical law), but it is an extremely improbable one.

In case a meteorologist is reading this, I hasten to admit that I distorted the situation somewhat, for the sake of simplicity. A wind in which all molecules were oriented in the same direction would blow at nearly a thousand kilometers an hour—if any instruments were left standing to measure it! Luckily, the winds we encounter in daily life are of a much less violent sort. Even the most destructive Caribbean hurricanes are the result of a slight orientation of the speed of the particles.

Wind is nevertheless an excellent illustration of the ways the concepts of probability, entropy, and organization can work together to produce a natural event on a human scale.

A DROP OF INK

A drop of ink is thrown into a glass of water. Little by little, the ink mixes with the liquid; in the end, we have a glass of colored water.

What if we filmed this event and then ran the film backward? We would see a glass of colored water purify itself bit by bit while a drop of ink formed somewhere on its surface. The backward sequence seems impossible, but that is not altogether true. Nothing prevents the molecular movements from having precisely the speeds and directions that would cause them to converge as a bubble. Given all the movements possible, however, it is an excessively improbable one.

We could say that the entropy of the initial state "drop of ink in water" is much lower than that of its final state. The first state is characterized by a restriction: All the ink molecules are located together in a small space. In the second state, this limitation no longer holds. Here, the microscopic configurations are the precise positions of the molecules; the macroscopic ones have to do with the average color of the liquid. There are many more ways for the ink molecules to arrange themselves in the whole glass of water than in the small volume represented by the drop of ink.

THE ENTROPY OF LIGHT: OUR GOLDEN RULE

We maintain the walls of an enclosed space at constant temperature. This heat creates light in the enclosure. The photons inside the space, like the molecules of the calm air, are moving in all directions. This is a case of thermal radiation. As we did with the other phenomena, we could associate a certain entropy with this radiation. As it was for the dice and coins, the entropy is proportional to the number of photons in the enclosure.

The rule that entropy is proportional to the number of particles is valid not only for photons, but also, under certain conditions, for many other types of particles, such as neutrinos. The rule is approximately valid for all kinds of particles in nearly all the conditions that interest us in this book. We have adopted it as our golden rule.

Why are photons so vital to the events of natural organization? There are several reasons.

First, it is extremely easy to increase their number. All you have to do is turn on a light. My aquarium light is a source of entropy essential to the well-being of my pretty tropical fishes.

In the second place, photons have no mass. Let's look at a specific example. We would like to generate the most entropy possible using a limited quantity of energy. We know the recipe: The number of particles must be increased. What type of particles would be the best ones for us to use?

Would electrons do? They have a mass. To generate them, we would be forced to dip into our energy budget to manufacture the mass of these particles (as always, $E = mc^2$).

Photons, because they have no mass, are much more economical to produce. For the same price, we could have much more! Thermal radiation is the substance that can store the greatest quantity of entropy; this is why it is widely used in phenomena in which matter acquires structure.

THE EPIPHANY CAKE

We shall now explore the mathematical structure that connects the ideas of entropy and information. The increase of entropy always corresponds to a decrease of information.

Let us turn again to our game with two dice. We add the sum of the numbers shown on each die. If the result obtained is 2, I know exactly what occurred: The dice showed, 1 and 1. In the diagram back on page 199 we are in the upper left-hand corner. If told that the result was 7, I know only that we are somewhere along the diagonal; I do not know exactly where we are on this diagonal. The 7, you may recall, is a high-entropy state (see page 200). The 2 is the opposite, a low-entropy state corresponding to greater information about the nature of the event that produced it. That is, we have more information about the exact position of this event in the grid of possible results of the experiment with the sum of two dice.

Similarly, at the beginning of the ink dilution experiment I can say that the ink molecules are located within the volume of the drop of ink. At the end of the experiment I have lost this information. The ink molecules could be anywhere in the volume of the container. The state of higher entropy is the state about which I know the least.

In France families traditionally celebrate Epiphany by sharing a cake that has been baked with a bean somewhere inside it. The person who gets the slice with the bean is king for a day. This year, an imaginative pastry cook has baked a rather special Epiphany cake. First, he baked eight little cubical cakes. In one of them he hid the bean. He then put them together to make one big cube-shaped cake, as shown in the illustration below. The game consists of guessing where the bean is by asking questions that can be answered only by either *yes* or *no*. The person who finds it using the fewest questions wins.

To facilitate the discussion, we have given each cake a number between 1 and 8.

The first question could be "Is the bean in the bottom layer (cakes 1, 2, 3, or 4)?" Depending upon the answer, we will know whether it is in the top or bottom layer. Let's assume that the answer was *no*. We are now limited to cakes 5, 6, 7, or 8.

The second question will be "Is the bean in the cakes nearest to us?" If the answer is *yes*, we will know that it is in 5 or 6.

The third question will be. "Is the bean in the cakes on the right?" If the answer is *no*, we will find the bean in cake 5.

It is clear that these questions enable us to reduce our ignorance and, as a result, increase the information we have about the system. Let us

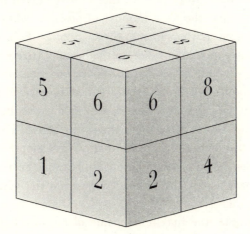

The Epiphany Cake

assume for a moment arbitrarily that each question and answer gave us one unit of information. At the start information was nil (0); it then went from (1) to (2). At (3) it was complete. (We know all that we want to and can.)

The reduction of ignorance can be presented in another way. In the beginning it is measured by the fact that the whole of our knowledge is that the bean is hidden in one of the eight cakes. After the first question, we know that it is hidden in one of four cakes. After the second question, in one of two cakes. After the third question, we can identify which cake.

Let us recall a point we made earlier. The number eight is obtained by multiplying $2 \times 2 \times 2$. This can also be written as $8 = 2^3$ (read aloud as "two to the third power." Three is the number of times that two must be multiplied by itself to make eight; it is the exponent that must be given to two to make eight. In technical terms, three is the logarithm with the base two of eight.

Eight cakes, three questions — that is the key. Each question reduces by half the number of possible cakes. Therefore the ignorance decreases by half each time the information increases by one unit.

We shall now find our concept of entropy in a new light: Entropy can be seen as a measure of our ignorance about a given situation. At the beginning there is maximum entropy. The bean is in one of the eight cakes, and we are three questions away from knowing where it is. We define entropy as the exponent that must be given to the number two in order to find the total number of possible cakes $(2 \times 2 \times 2)$. The answer, three, is the same as the answer to the question "How many questions will it take to locate the bean?" After we ask the first question, four possible cakes and two more questions remain, or $2 \times 2 = 4$. In the table on page 207 the situation is diagrammed in detail.

As we read the table, we note a remarkable result: The sum of entropy plus information is always the same (three). This convenient rule justifies the definition of information we adopted earlier. The pastry cook has shown us the relationship between entropy and information: One decreases as the other increases, and vice versa. Their sum remains constant.

In this example, the maximum value of entropy is three. It goes from three to zero during the series of questions. The seventh (and last)

	Number of cakes where the bean might be	Entropy	Number of questions to come	Information	Entropy plus information	Maximum entropy minus actual entropy
Before the first question	8 (2×2×2)	3	3	0	3	0
Before the second question	4 (2×2)	2	2	1	3	1
Before the third question	2	1	1	2	3	2
At the end	1	0	0	3	3	3

column 1   2   3   4   5   6   7

column of the chart shows maximum entropy minus entropy, that is, three minus the number shown in the third column. The values are the same as in the "information" column. This is yet another illustration of our definition of information as the difference between maximum and real entropy, or, in other words, as the amount of the entropy that can still be released.

The example of the Epiphany cake shows that the concept of information can be established on a quantitative basis. A series of questions and answers can enable us to calculate the information contained in a system. In the second part of this book I described a way of analyzing a painting according to this method. The painting would be subdivided into as many little squares as it would take for each square to represent a single color, like a paint-by-number sketch. Then the color spectrum would be subdivided into a scale of hues as accurate as necessary to match the painter's palette. To inform someone who had never seen the painting of exactly how it looked, we would have to answer a great number of *yes* and *no* questions, the same way we did in the preceding example.

I would now like to touch on an objection that no doubt occurred to the reader during the Epiphany cake discussion. There, entropy was a function of the observer's degree of ignorance about the bean's position. The pastry chef, of course, knew where it was from the outset. Is entropy entirely relative to the observer?

Although the simple examples I have used have definite advantages as teaching tools, they also present some drawbacks: To illustrate certain aspects of reality, they distort others. The entropy of physical systems certainly can be defined without consideration to the observer; in order to illustrate the mathematical relationship between entropy and information, I have made use of an ambiguous example.

THE RIGHT DIGITS IN THE WRONG ORDER: MOLECULES AND CRYSTALS

Let's imagine a lottery where one thousand tickets numbered from one to one thousand have been sold, and the winning number is **134**. The probability of this number being drawn is one in one thousand; the holder of the ticket will receive a handsome prize. As a rule, smaller prizes are awarded to people holding tickets with the same digits in the wrong order. How many are there? Five in all:

143 413 431 314 341

Recalling the dice game discussed earlier, we recognize the concepts of single configuration (in the proper order) and composite configuration (in the wrong order). The probability (and therefore the entropy) of the "wrong order" result is greater than the probability of the "proper order" result. The information associated with them is therefore less. In other words, someone who is apprised of not only of which digits are on the winning ticket, but also of the order in which they occur, "knows" more than someone who is apprised of only the digits and not their order.

We can conclude that sequences where order is important contain far more information than sequences where order does not matter.

Let us suppose that we want to make linear molecules (where the atoms are connected end-to-end to form a train) of four atoms using carbon and hydrogen. Sixteen combinations are possible (as shown below). If we then specify that within the molecule, the C and H atoms must alternate, only two possibilities remain: HCHC and CHCH. It would be enough to describe the string and name the lead atom to know the composition of the molecule.

By this example we can grasp the difference in information between a crystal and a complex molecule. A crystal is composed of an infinity of patterns that are always repeated. Table salt, for example, contains atoms of chlorine (Cl) and sodium (Na) that always alternate in a three-dimensional structure, in such a way as to surround each Na

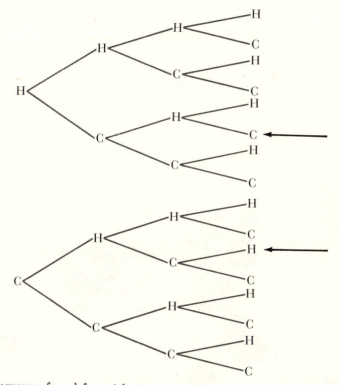

Each sequence, from left to right, is a possible species. If the order does not matter, we have sixteen species. If the C and H must alternate, then we have only two species, as indicated by the arrows.

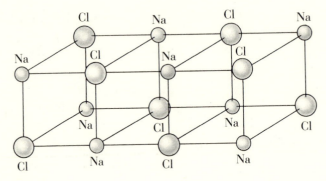

A Crystal of Salt

atom with eight Cl atoms, and vice versa (as shown above). A grain of salt may contain a trillion atoms; a sugar molecule contains only thirty-six (as shown below). Nevertheless, much more information is needed to describe a single sugar molecule than to describe a grain of salt. Moreover, for sugar it is imperative that I specify not only the atoms, but also their positions relative to one another and the angles they form together.

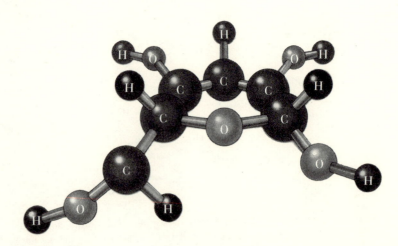

A Molecule of Sugar

THE ENTROPY OF LIFE

Living cells use an alphabet made up four letters, **A, C, G,** and **T,** which stand for the molecules adenine, cytosine, guanine, and thymine. Genes are composed of long strings (called DNA) on which these molecules follow each other in a certain specific order. In bacteria the number of links is somewhere near one thousand; in mammals it reaches one billion.

How many different combinatory sequences can be obtained for a billion-link string if each link can take any one of the values A, C, G, or T? The answer is $10^{600,000,000}$ — that is, a six-hundred-million-digit number. The approximately five billion people in our planetary population represent only a tiny fraction of all the different combinations possible. These *astronomical* figures help us understand the amazing diversity of living things.

Are we equal to the task of calculating the entropy and information assigned to living things? Certainly not. Even though we now know the alphabet of living cells, we are still extremely far from being able to provide accurate descriptions of all of life's atomic processes. This ignorance has two major impacts on our discussion.

First, the growth of complexity in living things, from bacterium to jellyfish to baboon to human being, appears to correspond qualitatively to a scale of increasing information and decreasing entropy. The fact that bacterial genes contain only thousands of links, while the genes of larger organisms represent billions, tends to support this assumption. But no one, today, can show quantitatively that the baboon possess less information than a human being. These calculations are still well beyond our abilities.

The second impact is relative to the problem of the origin of life. Several scientists claim to have calculated the probability of the appearance of life in the earth's primitive sea. Can we believe them?

Let's return to our dice games. To calculate the probability of a selected event (for example, the 7 in a game with two dice), it is necessary to know both the number of possible events (in this case, sixty-four), and the number of events corresponding to the one selected (here, the six events along the diagonal running from the upper right-to lower left-hand corners). Likewise, to calculate the probability of the appearance of life in the primitive sea, it would be necessary to know both the number of chemical reaction sequences that result in life (the events chosen) and the total number of possible chemical reaction sequences, regardless of their result (sample field). These operations are well beyond our current possibilities.

The preceding paragraph demonstrates an obvious aspect of probability calculations: It is impossible to calculate the probability of an event without knowing exactly how it happens. Similarly, it is impossible to calculate the probability of obtaining a winning number in a casino without prior knowledge of the number of spaces where the roulette wheel may stop.

2

Disequilibria

In physics the word *equilibrium* can be applied in several different circumstances. We will be discussing thermal equilibria first, and then associative equilibria. In our cosmic saga the two play parallel roles and are of equal importance.

THERMAL DISEQUILIBRIA: THE PARADOXES OF GRAVITY

The behavior of stars is altogether paradoxical. As a rule, when an object emits heat, it cools, and its temperature tends to equalize with that of the surrounding environment. Conversely, however, when a star shines, it heats up. Throughout its lifetime, it increases the thermal disequilibrium between its core and the interstellar environment.

As we know, a natural system, if left alone, has a tendency to seek its state of greatest entropy.

For objects on our human scale (cabbages, pebbles, teddy-bears, and so forth), maximum entropy corresponds to isothermy, that is, the disappearance of temperature differences. This is the reasoning that led earlier physicists to write of thermal death (see Chapter 3).

In a system whose mass and density are high enough for gravity to play a major role, however, the state of maximum entropy is not a state of uniform temperature. Spontaneously matter tends to collapse inward; the atoms located in the core react to this collapse by heating up, thus generating *thermal pressure* that counterbalances the collapse.

The equilibrium, however, cannot hold very long. All hot bodies emit light; for a star, the light emitted signifies a loss of heat, and therefore a loss of pressure. Hence further collapse. Hence additional heating to counteract the weight of the outer layers. Hence an increase in luminosity. And so forth. These interrelated phenomena occur continuously throughout the star's life-span until it becomes a dead star.

What causes this behavior in stars? It is due to the strange properties the force of gravity possesses in relation to other forces. The properties of the forces are reflected in the behavior of the energies they generate.

The chemical energy (of electromagnetic origin) that can be stocked by an object is proportional to the mass of the object. Two gallons of gasoline contain twice as much energy as one gallon of gasoline. The same is true of nuclear energy; two pounds of uranium are equivalent to two times one pound of uranium. In technical terms these energies are said to be *extensive*.

Gravitational energy, however, is proportional to the square of the mass. A star two times the mass of the sun incorporates four times as much energy as the sun (if the radii of these stars are equal). Gravitational energy is not extensive. As a result massive objects, such as stars, are the best choice for storing very large amounts of energy. For instance, they are the best candidates, as super-massive black holes, to be the power-houses of the fantastically luminous quasars.

Why is there such a difference in behavior among the forces? For two reasons: First, gravity acts over very large distances; second, there is no "screen" against gravity. Electromagnetic force can also act on large distances, but its effective range is attenuated by the fact that there are two types of electrical charge, positive and negative. Within a plasma, an ion and electron gas, these charges neutralize one another, except over very short distances. Both types of nuclear force, weak and strong, are very short-range forces. Gravity, in contrast, is a long-range force and is always attractive. Thus its effect can be felt even over a distance of millions of light-years. All the stars in a cluster and all the particles of a star interact together on the entire volume of the system, and not only with the nearest elements of matter. This long-distance interaction is what is ultimately responsible for the fact that gravitational energy is proportional to the square of the mass.

Entropy behaves in a way similar to energy, and for quite analogous reasons. The entropy of radioactive or perfect gases is proportional to the number of particles included in these substances (hence, for example, our golden rule, page 74). In these cases, entropy is extensive. In a massive, concentrated object, however, entropy is proportional to the square of the mass. That is why black holes are the universe's biggest storehouses of entropy.

ASSOCIATIVE DISEQUILIBRIUM: THE ROLE OF TIME

In the atomic world, for every possible process there exists an inverse process. If a nuclear reaction occurs, causing the association of two free particles and the formation of a new nucleus, the corresponding inverse reaction would be the dissociation of the nucleus, liberating the two initial particles.

Let's imagine we could film a sequence in which a proton captures an electron. A hydrogen atom is formed, and an ultraviolet photon is released. Now let's run the film backward. A photon is absorbed by a hydrogen atom that proceeds to separate into a proton and an electron. An uniformed observer would be unable to see which of the two sequences was the backward-running one. The two phenomena correspond to two possible events. The sequence is said to be *reversible*.

When a proton–electron gas is maintained at constant temperature, a great number of combinations (forming hydrogen atoms) and dissociations (back to the electron–proton pairs) take place. If the number of combinations per second is equal to the number of dissociations per second, we speak of a state of associative equilibrium. The relative populations of the various species of particles remains the same.

In the infancy of the universe the cosmic stew was in just such a state of equilibrium; all sorts of particles were busily engaged in a hum of perfect associative equilibrium. But the new systems generated by all these reactions did not last; they immediately dispersed into their constituent particles. The net result in terms of organization was absolutely *nothing*. The whole state could be called primordial thermal limbo.

The temperature, of course, was not constant; it was dropping quickly. In such a hot environment, however, the capture and dissociation reactions were extremely rapid, far more rapid than the cooling rate. As far as these phenomena are concerned, it was as though the temperature were constant. We apply the term *quasi-equilibrium* to a situation where the number of associations and dissociations is practically equal and uniform.

Now we have reached the crux of our discussion: *In a substance where the reactions are in equilibrium, the states of maximum stability are always attained.* Consequently, in a universe cooling slowly enough to allow for the equilibrium of nuclear reactions, all matter would ineluctably become iron. As the temperature drops, states of equilibrium are never perfectly fulfilled, of course, but the quasi-equilibria are enough to yield the same result.

To neutralize the tyrannical ambitions of the natural forces, to engender the diversity of structures and shapes, and to banish monotony, it is essential to introduce into the universe situations of violent associative (as well as thermal) disequilibrium. How could this be brought about? Here is where the concept of a time-scale is important.

First, there is the time characterizing the cooling of the universe, that is, the time it took for the temperature to drop by half. Although this cooling time was quite short in the beginning, it gradually lengthened as the universe expanded. Today, it is calculated at fifteen billion years, about the same as the age of the universe according to our cosmic clock.

We also have to take into account the capture and dissociation time. Nuclear and atomic phenomena are not instantaneous. It takes time for them to occur. Reaction times vary widely, depending upon the ambient temperature. The hotter it is, the shorter they are; they lengthen as the temperature falls. Moreover, captures are usually quicker than dissociations.

The expansion rate decreases as a function of the square of the temperature (T^2), while reaction rates are proportional to higher powers $(T^5$, generally). Although initially, at extremely high temperatures, they are more rapid than the expansion rate, they gradually become slower. The "freezing point" would be reached when the mathematical curves tracing the two rates cross: at one billion degrees Kelvin for the nuclear force and three thousand degrees Kelvin for the electromagnetic force.

It now becomes clear how the concepts of associative equilibrium and disequilibrium are linked to our history. In the beginning there is equilibrium: Reaction times are short relative to cooling time. A little later, the situation is reversed, and states of disequilibrium are instilled. This means that the number of captures is no longer equal to the number of dissociations. There is less and less chance that the nuclear and atomic systems, once they have formed, will be destroyed. They might survive forever (more precisely, as long as expansion continues).

Natural structures bound together by physical forces (nuclei, atoms, and stars, for example) have reaction times (captures and dissociations) that differ greatly. After a phase of equilibrium during which systems cannot survive the effects of the extreme heat, the decreasing temperature triggers the transition to disequilibrium and the durability of structures, at lengths of time that vary according to the natural force involved.

3

Black Holes and Radiation

What would happen to the sun if it was thrust into a field of high-temperature radiation like the one that existed in the beginning of the universe? It would absorb light continually instead of emitting it. It would heat up, swell, and dilute its mass; in the end, it would dissolve itself entirely into the cosmic fluid.

This situation is analogous to that of the iron nucleus, which, if subjected to intense heat, breaks down into nucleons despite the fact that, in the absence of radiation, iron is the most stable of nuclear configurations. The same is true of ice, which melts in hot weather even though ice is more stable than water.

But what would happen if we brought a black hole, instead of a normal star, into this primordial radiation?

According to Einstein's physics, a black hole is a region of space where gravity is so tremendously intense that nothing is able to escape from it, not even light. Such an object should suck in and absorb radiation, gradually increasing its own mass ($E = mc^2$, still) and increasing the entropy of the universe until it brought about the end of gravitational supercooling.

After Einstein, however Niels Bohr, Werner Heisenberg, and quantum physics came along and changed everything. The Einsteinian version of a black hole boils down to the affirmation that matter located within the black hole is definitively "sentenced to life imprisonment" in that particular volume of space. The absolutist character of this statement goes against the spirit of quantum physics. Nothing is ever definitively stuck somewhere; there is always the probability of an outlet. If the prison wall is too high, the prisoners will dig a tunnel; the day of escape will come, if they are patient. It's just a question of time.

Thanks to the work of Stephen Hawking, we now believe that black holes "evaporate" and "shine": Matter is constantly escaping from them in the form of radiation. A black hole's surface behaves like that of any body heated to a certain temperature. The greater the mass of the black hole, the lower its temperature and the slower its evaporation rate. This evaporation process, however, reduces the mass of the black hole; the temperature thus increases, and the loss of matter speeds up.

When the mass has been lowered to about fifty micrograms (the equivalent of a very fine grain of sand), the moment of climax is reached. The evaporation becomes an incredibly powerful explosion. The flash of light it causes is equivalent to that of ten quadrillion galaxies.

This fifty-microgram mass is called Planck's mass. It is the smallest black hole that can exist in the universe. Its surface temperature would be 10^{32} K (Planck's temperature). It evaporates in 10^{-43} seconds (Planck's time).

The reader may recognize some of the figures we discussed in the main body of the book. Planck's temperature is the temperature of the universe at the point in the past where we lose track of it, where we explorers are blocked by the "wall of ignorance."

A black hole with a mass equivalent to that of the sun would have a very low surface temperature, approximately one one-millionth of a degree Kelvin. Its evaporation process would take more than 10^{64} years, provided that the temperature of the fossil radiation in which it was immersed were lower than its own. But this is not the case today; fossil radiation is at 3 K. At current rates of cosmic expansion and cooling, such a black hole will have to wait more than 10^{20} years before it can begin evaporating.

4

States of Natural Systems

In Chapter 2 we studied the ways in which matter organizes itself. The strategy of superimposed alphabets was present at every level of complexity. Quarks are combined to obtain nucleons, nucleons are combined to obtain nuclei, and so on. The words spelled out by the component letters are nature's organized systems. They can exist in a wide variety of different states.

In the absence of interaction with the outer world, the system remains in its *fundamental state*, that is, the one in which its inner energy is at the lowest. The transition to other states, known as *excited states*, requires the use of energy obtained from other systems. The exchanges that result from the transition are the manifestation of the activity specific to the system.

NUCLEONIC STATES

By combining u and d quarks, protons (two u and one d) and neutrons (one u and two d) are formed. But that isn't all. The association of quarks may occur in any of a large number of different states. Proton and neutron states are the fundamental states, while the other configurations are excited states.

As shown in the illustration below, a sample of these states has been mapped on a scale of energy (the vertical scale). The energy is measured in billions of electron-volts. For example, to bring a proton to its first excited state, it is necessary to subject it to nearly two hundred million ev. Left alone, it will reemit this energy into space and regain its fundamental state. If interaction with the outer world continues,

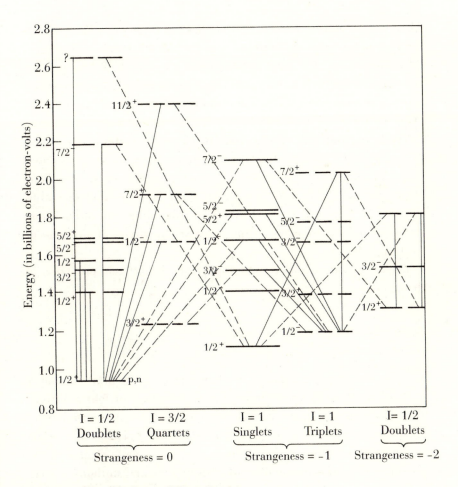

A Variety of Possible States of Three Quarks

(From V. Weisskopf, *Knowledge and Wonder*, MIT Press.)

however, the proton will reach all the states described in the diagram, one after another.

The horizontal scale shows other properties of these particles, one of which is its spin, or the angular momentum. Spin is measured in units of Planck's constant, h. The only possible values of this unit are one-half, three-halves, and so on.

NUCLEAR STATES

The diagram on the facing page describes the variety of states possible for eleven nucleons (protons or neutrons). Each block represents a different nucleus. The lithium-11 nucleus is made up of three protons and eight neutrons. The beryllium-11 is composed of four protons and seven neutrons; boron-11 of five protons and six neutrons; carbon-11 of six protons and five neutrons.

The horizontal lines crossing each block are the states of the nuclei. At the bottom we see the fundamental state; above, we see the excited states. The figures on the left show the quantities of energy, in millions of electron-volts, that must be injected to get from the fundamental to an excited state. (Bear in mind that in the preceding scale of nucleonic states we measured energy differences in hundreds of millions of electron-volts.)

The figures on each horizontal line specify the different properties of these states.

ATOMIC STATES

A lithium atom is composed of three electrons in orbit around a nucleus, itself made up of three protons and three or four neutrons. The illustration on page 224 shows a small part of the variety of different states in which this atom may be found. The vertical scale specifies the energies of these states, in electron-volts. Scales in the preceding diagrams, remember, were in billions and millions of electron-volts, respectively . The change in scale is an excellent illustration of the difference in strength between electromagnetic force (which

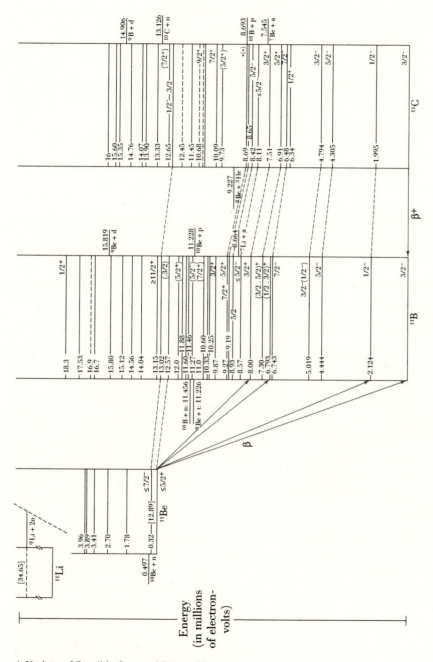

A Variety of Possible States of Eleven Nucleons

(From *Energy Levels*, USAEC, 1954.)

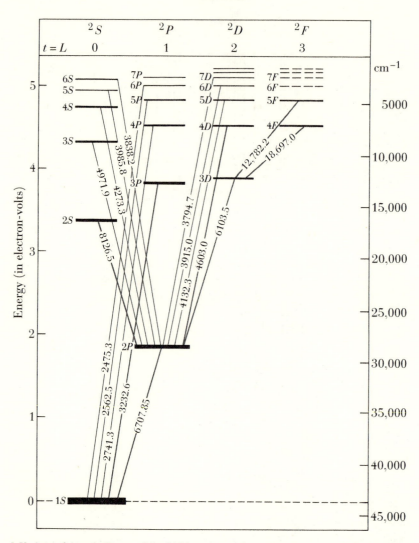

A Variety of Atomic States of the Lithium Atom (three electrons around the nucleus)
(From G. Hertzberg, *Atomic Spectra.*)

binds atoms and molecules) and nuclear force (which binds nuclei and nucleons).

The horizontal scale shows the spin properties of these atoms. The slanted lines describe the events by which the lithium atom changes energy state and spin as it emits or absorbs photons of certain frequencies. The frequencies are written along the lines.

MOLECULAR STATES

In the illustration below, the molecule is one called OH, made up of
one hydrogen atom and one oxygen atom. Each horizontal line represents an energy state. These states specify the quantities of vibration
and spin of the molecule. The differences in energy between one state
and the next (indicated by the forked lines) are lower than one one-
millionth of an electron-volt.

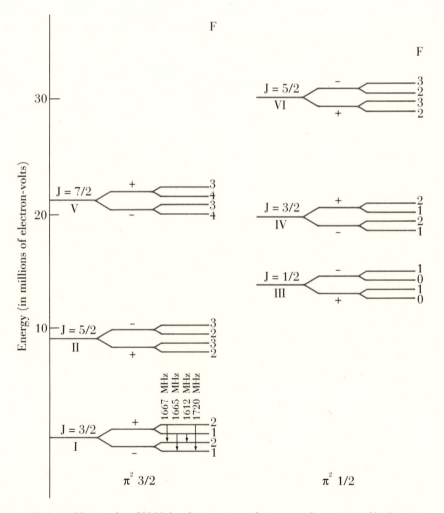

A Variety of States of an OH Molecule (one atom of oxygen and one atom of hydrogen)

Notes

Chapter 1

1. M. J. Sherwin, *A World Destroyed* (New York: Knopf, 1975), p. 110.
2. Peter Pringle and James Spigelman, *The Nuclear Barons* (New York: Avon Books, 1983), p. 23.
3. Peter Sloterdijk, *Critique of Cynical Reason* (Ann Arbor, MI.: University of Michigan Press, 1988), p. 120.
4. T. C. McLuhan, *Touch the Earth: A Self-Portrait of Indian Existence* (New York: Touchstone Books, 1976), p. 15.
5. Eric Chaisson, *Cosmic Dawn: The Origin of Matter and Life* (New York: Berkley, (1984).

Chapter 2

1. Unlike the majority of ancient philosophers, Lucretius has opinions on this question that are well ahead of his time:

> . . . As I think, our universe is quite new,
> our world is young; it started not long ago.
> Some arts are just now taking final polish
> or still developing: daily, our ships gain new
> tackle; today, musicians produce sweet sound.
> Nature and nature's law have been revealed just now. . . .

The Nature of Things, translated by Frank O. Copley (New York: Norton, 1977), p. 20.
2. Ibid.

3. The electron volt is traditionally defined as the kinetic energy of an electron accelerated by a one-volt difference in potential. It is also the energy of an infrared photon. The mass of a proton is equivalent to an energy of 10^9 ev (one billion electron volts).

4. The human imagination is never idle. The letters of the alphabet themselves inspire compositions and associations, as Arthur Rimbaud so winningly demonstrates:

> *A* black, *E* white, *I* red, *U* green, *O* blue — I'll tell,
> One day, you vowels, how you come to be and whence.
> *A* black, the glittering of flies that form a dense
> Velvety corset round some foul and cruel smell,
>
> Gulf of dark shadow; *E*, the glaciers' insolence,
> Steams, tents, white kings, the quiver of a flowery bell;
> *I*, crimsons, blood expectorated, laughs that well
> From lovely lips in wrath or drunken penitence;
>
> *U*, cycles, the divine vibrations of the seas,
> Peace of herd-dotted pastures or the wrinkled ease
> That alchemy imprints upon the scholar's brow;
>
> *O*, the last trumpet, loud with strangely strident brass,
> The silences through which the Worlds and Angels pass;
> —*O* stands for Omega, His Eyes' deep violet glow!

5. The definition and meaning of this cosmic clock are discussed Chapters 3 and 8.

6. See my *Atoms of Silence* (Cambridge, Mass.: MIT Press, 1983; original French title: *Patience dans l'azur* [Paris, Le Seuil, 1981]), p. 41.

Chapter 3

1. Claude Lévi-Strauss, *Tristes Tropiques*, translated by John and Doreen Weightman (London: Jonathan Cape, 1973), p. 103, reprinted by permission.

2. As we shall see in Chapter 7, the total entropy generated by all the stars and all the natural events that have occurred so far is negligibly small in comparison to the entropy made available by expansion.

Chapter 4

1. In fact, we could say that the moon is indeed falling on the earth but it always misses its target because of its initial movement.

2. A light source moving away from us appears to be redder than the same source at a standstill. Hubble observed that galaxies redden as they grow more distant. The reddening has been interpreted as a sign of their movement (Doppler – Fizeau effect) away from us. Today, that interpretation is difficult to contest; it is accepted by nearly all contemporary astrophysicists.

3. A word about the temperature scales. The Kelvin scale (K) of physicists corresponds to the Celsius (C) or (centigrade) scale shifted by 273 degrees. In other words, the freezing point of water (zero degrees in C, 32 degrees in Fahrenheit [F]) is at 273 K, while its boiling point ($100°$ Celsius, $212°$ F) is at 373 K. Thus the temperature of the microwave background (3 K) corresponds to $-270°$ C, or $-454°$ F. Throughout this book I use the Kelvin scale. Needless to say, when very high temperatures (millions of degrees) are involved, the difference between the scales becomes irrelevant.

4. See Joseph Silk, *The Big Bang*, rev. ed. (New York: W. H. Freeman and Co., 1988).

5. In fact, there is a slight variation in the temperature of fossil radiation. It's cyclical: Over a period of exactly one year it varies by approximately one one-thousandth of a degree. This variation is not due to any real variation in the fossil glow, but rather to our movement through space (the earth moves around the sun; the sun turns around the galactic center; the galaxy moves in its cluster). We move in relation to fossil radiation. It is as though we had finally discovered the fabled aether of nineteenth-century physics. This aether of ours, however, does not alter the speed of light.

6. See my *Atoms of Silence* (Cambridge, Mass.: MIT Press, 1983).

7. At least, according to the open universe hypothesis, which current astronomical observations seem to confirm (see Chapter 7).

8. Let us call the distance between two galaxies (or between any two points in space) R. This distance increases with time. The volume of a covolume increases with the cube of R (R^3). Consequently the particle density (number of particles per unit of volume) decreases with the inverse of the cube (R^{-3}). In addition, the number of photons in fossil radiation is proportional to the cube of the cosmic temperature (T^3), and the cosmological model shows us that this temperature decreases in inverse proportion to the expansion (T is proportional to R^{-1}). Therefore, the number of fossil radiation photons in a covolume (the product of the number of photons per unit of volume multiplied by the volume of the covolume) remains constant over the course of time: $R^3 \times R^{-3} = 1$.

9. See *Atoms of silence*, p. 172.

Chapter 5

1. A comment is in order, to avoid confusion. Using the analogy of an alphabet, the operations we are discussing here combine letters to make new *letters*, not words in Chapter 2. We remain on the same level of the pyramid. It is clear that the letters of these physical alphabets are not all equal. One property distinguishes one from another: stability.

Chapter 6

1. The exact value of these temperatures depends upon density. I have selected the densities appropriate to our discussion.

2. There are two phases of this transmutation that are always slow, whatever the temperature. The first one is the capture of two protons to form deuterium. This reaction implies the simultaneous transformation of a

proton into a neutron, and thus it requires the intervention of the weak nuclear force, an extremely improbable event.

The second phase, the fusion of helium into heavier nuclei, is hindered by the instability of the beryllium-8 nucleus (consisting of two heliums). To generate a carbon nucleus, *three* heliums must meet at a single point, which is a very rare event at any temperature.

3. Had the sun, for example, been made up of iron atoms, it would have passed directly from the state of interstellar nebula to that of white dwarf due to the effect of its own gravity. This transformation would have taken about twenty million years. Because it is the most stable state for nuclear material, iron cannot emit energy and support a star.

4. See my *Atoms of Silence* (Cambridge, Mass.: MIT Press, 1981), p. 183.

5. This last affirmation is not entirely accurate. After the work of Stephen Hawking, quantum physics allows black holes to "evaporate." They permanently emit radiation, but for a star whose mass is comparable to our sun's this luminosity is extraordinarily weak. See Appendix 3).

6. Contrary to widespread opinion, nuclear reactions, rather than *causing* high stellar temperatures, act as *thermostats* for them, maintaining a constant inside temperature until the fuel is exhausted.

Chapter 7

1. "To pretend that the gods wanted to prepare the world and its wonders for humans; that, consequently, their admirable handiwork deserves all our praise, and that it can be believed to be eternal and destined for immortality . . . all these ideas . . . are nothing but pure nonsense." Lucretius, *The Nature of Things*, vol. 1.

2. "Antimatter" is the name of a type of matter altogether similar to the matter we know, except that the signs of electrical charge are reversed: electrons are positive, and protons are negative.

Particle accelerators manufacture antimatter (at a high price), but it does not appear to exist naturally, except in very small amounts in cosmic rays. See Joseph Silk, *The Big Bang*, rev. ed. (New York: W. H. Freeman and Co., 1988), for more details.

3. See my *Atoms of Silence* (Cambridge, Mass.: MIT Press, 1983), p. 202.

4. The quantity of matter contained in the sphere of causality, when the universe reached the age of one year, was about equal to that at present of our galaxy. In a stationary universe (one without expansion), the radius of the sphere of causality would measure exactly one light-year. Expansion, by enlarging the space, added volume to the sphere. The radius became two light-years. In that era, there was no way for an exchange to occur between the matter that was to become our galaxy and, for example, the future matter of the Andromeda galaxy.

Every year, hitherto unobservable galaxies appear. The sphere of causality (the cosmological horizon) progresses more quickly than expansion. The present "radius of the observable universe" extends fifteen billion light-years.

5. See Silk, *The Big Bang*.

6. The lower the density, the later the transition from the state of plasma (ionized) to the state of gas (neutral). A field of gravity that had been greatly weakened by dilution of cosmic matter would no longer be able to precipitate the condensation of matter into galaxies and stars.

7. A word about another observational fact: the absence of "turbulence" on the cosmic scale. In our cosmological models we have compared the universe to a vast fluid (of infinite dimensions, if it is an open model; of limited ones, if it is a closed one). The fluids we know of—the ocean, the atmosphere—are generally agitated by more or less disorderly movement: currents, turbulence, ripples, whirlpools, cyclones, and so forth. Conversely, the observation of galaxies reveals a universe-fluid so extremely calm that it appears to be entirely free of turbulence. Has it always been so?

   Let us return to our toy universes and arbitrarily assign to them turbulent movements with a variety of amplitudes. Then let us compute the course of their evolution. In nearly every case, the turbulence progressively *increases*, making the formation of galaxies impossible. Only in a universe in which the density is very close to critical density does the turbulence subside so that the fluid reaches the calm state we observe today.

8. An assumption that would be to physics as topology is to geometry.

9. The transmutation of three heliums into a carbon is greatly abetted by the fact that carbon possesses a level of energy—at 7.656 million electron volts (MeV)—with just the properties required for the fusion, only very slightly (0.372 MeV) above the mass of three heliums. I have written more extensively on this subject in *Stellar Evolution and Nucleosynthesis* (New York: Gordon and Breach, 1969).

   Were it not for this extraordinary coincidence, carbon would be almost entirely absent from the universe.

Chapter 8

1. Claude Lévi-Strauss, *Tristes Tropiques*, translated by John and Doreen Weightman (London: Jonathan Cape, 1973), p. 103.

Chapter 10

1. "An open system conserves itself only by means of constant exchanges with the environment, relative to the requirements of nourishment and protection from predators. It is unceasingly threatened by its limitations, and, even if its current environment suffices momentarily in terms of its immediate requirements, the development of the most elementary precautionary and anticipatory conducts results in an enlargement of the environment as surroundings that are known and that may or may not be used for current physiological exchanges.

   "This process of exploration, which can be observed even in the lower orders of animal life, is naturally reinforced when, in addition to simple precautions, the curiosity of the higher vertebrates joins in, and then when the actions are internalized as symbols, a process culminating in the infinite expansion of the need to know, proper to human thought." Jean Piaget, *Problèmes de psychologie génétique* (Paris: Denoël/Gonthier, 1972).

2. See, for example, G. A. Rosenthal, "The Chemical Defenses of Higher Plants," *Scientific American*, January 1986, p. 76.

3. Numberless are the world's wonders, but none
More wonderful than man; the stormgrey sea
Yields to his prows, the huge crest bear him high;
Earth, holy and inexhaustible, is graven
With shining furrows where his plows have gone
Year after year, the timeless labour of stallions.

The lightboned birds and beasts that cling to cover,
The lithe fish lighting their riches of dim water,
All are taken, tamed in the net of his mind;
The lion on the hill, the wild horse windy-maned,
Resign to him; and his blunt yoke has broken
The sultry shoulders of the mountain bull.

Words also, and thought as rapid as air,
He fashions to his good use; statecraft is his;
And his the skill that deflects the arrow of snow,
The spears of winter rain: from every wind
He has made himself secure — from all but one:
In the late wind of death he cannot stand.

From the *Antigone* of Sophoclés, in *Four Greek Plays* translated by Dudley Fitts and Robert Fitzgerald (New York: Harcourt Brace, 1960), p. 25.

4. Carl Gustav Jung, *Memories, Dreams, Reflections*, translated by Richard and Clara Winson (London: Collins and Routledge & Kegan Paul, 1963), p. 330.

Chapter 12

1. Albert Jacquard, *Endangered by Science?*, translated by Margaret Moriarty (New York: Columbia University Press, 1985; original French title: *Au peril de la science* [Paris: Editions du Seuil, 1981]); idem, *In Praise of Difference: Genetics and Human Affairs*, translated by Margaret Moriarty (New York: Columbia University Press, 1984; original French title: *Eloge de la différence* [Paris: Editions du Seuil, 1979]).

2. Peter Pringle and James Spigelman, *The Nuclear Barons* (New York: Avon, 1983).

3. Daniel Robinson, "Behind Violence," *U.S. News and World Report*, May 25, 1981, p. 25.

4. "Putting Darwin Back in the Dock," *Time Magazine*, March 16, 1981, p. 82.

5. Ibid.

6. Ibid.

7. Stephen Weinberg, *The First Three Minutes* (New York: Basic, 1977), p. 134.

Bibliography

Bateson, Gregory. *Mind and Nature*. New York: Bantam, 1979.

Careri, Giorgio. *Order and Disorder in Matter*. Translated by Kristin Jarrat. Menlo Park, Calif.: Benjamin/Cummings, 1984.

Chaisson, Eric. *Cosmic Dawn: The Origin of Matter and Life*. New York: Berkley, 1984.

Colinveaux, Paul. *Why Big Fierce Animals Are Rare*. Princeton: Princeton University Press, 1978.

Davies, Paul. *The Cosmic Blueprint*. New York: Simon and Schuster, 1988.

Dyson, Freeman. *Disturbing the Universe*. New York: Harper & Row, 1979.

————. *Weapons and Hope*. New York: Harper & Row, 1984.

Fang Li Zhi and Li Shu Xian. *Creation of the Universe*. Translated by T. Kiang. Teaneck, N.J.: World Scientific, 1989.

Ferris, Timothy. *Coming of Age in the Milky Way*. New York: Morrow, 1988.

Gould, Stephen Jay. *The Mismeasure of Man*. New York: Norton, 1981.

————. *The Flamingo's Smile*. New York: Norton, 1985.

Hawking, Stephen. *A Brief History of Time*. New York: Bantam, 1988.

Jacquart, Albert. *In Praise of Difference: Genetics and Human Affairs.* Translated by Margaret Moriarty. New York: Columbia University Press, 1984. (Original French title: *Eloge de la différence.* Paris: Editions du Seuil, 1979.)

———. *Endangered by Science?* Translated by Margaret Moriarty. New York: Columbia University Press, 1985. (Original French title: *Au peril de la science.* Paris: Editions du Seuil, 1981.)

Jung, C. G. *Memories, Dreams, Reflections.* Translated by Richard and Clara Winston. London: Collins and Routledge & Kegan Paul, 1963.)

Lévi-Strauss, Calude. *Tristes Tropiques.* Translated by John and Doreen Weightman. New York: Atheneum, 1974.

Lucretius. *The Nature of Things.* Translated by Frank O. Copley. New York: Norton, 1977.

Pringle, Peter, and James Spigelman. *The Nuclear Barons.* New York: Avon, 1983.

Silk, Joseph. *The Big Bang*, rev. ed. New York: W. H. Freeman, 1988.

Thomas, Lewis. *Late Night Thoughts on Listening to Mahler's Ninth Symphony.* New York: Bantam, 1983.

Symon, Arthur, translator. *Baudelaire, Rimbaud, Verlaine. Selected Verse and Poems.* New York: Citadel Press, 1947.

Weinberg, Stephen. *The First Three Minutes.* New York: Basic, 1977.

Zeldovich, Ya. B., A. A. Ruzmaikin, and D. D. Sokoloff. *The Almighty Chance.* Teaneck, N.J.: World Scientific, 1990.

Index

Mathematics for Physicists

Mathematics for Physicists

Susan M. Lea
San Francisco State University

THOMSON
™
BROOKS/COLE

Australia • Canada • Mexico • Singapore • Spain
United Kingdom • United States

THOMSON

BROOKS/COLE

Publisher: David Harris
Acquisitions Editor: Chris Hall
Assistant Editor: Rebecca Heider
Editorial Assistants: Lauren Raike, Seth Dobrin
Technology Project Manager: Sam Subity
Marketing Manager: Kelley McAllister
Marketing Assistant: Sandra Perin
Advertising Project Manager: Stacey Purviance
Project Manager, Editorial Production: Karen Haga

Print/Media Buyer: Karen Hunt
Permissions Editor: Sarah Harkrader
Production Service: Lifland et al., Bookmakers
Copy Editors: Sally Lifland, Gail Magin
Illustrator: Scientific Illustrators
Cover Designer: Bill Stanton
Text and Cover Printer: Phoenix Color Corp
Compositor: ATLIS Graphics

For more information about our products,
contact us at:
**Thomson Learning
Academic Resource Center
1-800-423-0563**

For permission to use material from this text,
contact us by:
Phone: 1-800-730-2214
Fax: 1-800-730-2215
Web: http//www.thomsonrights.com

Library of Congress Control Number: 2002117220

ISBN 0-534-37997-4

Brooks/Cole—Thomson Learning
10 Davis Drive
Belmont, CA 94002
USA

Asia
Thomson Learning
5 Shenton Way #01-01
UIC Building
Singapore 068808

Australia/New Zealand
Thomson Learning
102 Dodds Street
Southbank, Victoria 3006
Australia

Canada
Nelson
1120 Birchmount Road
Toronto, Ontario M1K 5G4
Canada

Europe/Middle East/Africa
Thomson Learning
High Holborn House
50/51 Bedford Row
London WC1R 4LR
United Kingdom

Latin America
Thomson Learning
Seneca, 53
Colonia Polanco
11560 Mexico D.F.
Mexico

Spain/Portugal
Paraninfo
Calle/Magallanes, 25
28015 Madrid, Spain

For my father,
who started me down this road

Preface

*Philosophy is written in this grand book, the universe, which stands
continually open to our gaze. But the book cannot be understood unless one
first learns to comprehend the language and read the letters in which it
is composed. It is written in the language of mathematics, and its
characters are triangles, circles and other geometric figures without which
it is humanly impossible to understand a word of it; without these one
wanders about in a dark labyrinth.*

— GALILEO GALILEI

The words of Galileo are just as true today as when he wrote them in the 17th century.
A serious student of physics must first learn the language of mathematics. This is a more
complex task for today's student than it was for Galileo. The characters include integrals,
matrices, groups, tensors, and other concepts unheard of in Galileo's time. A physics major
is also faced with a list of physics courses that may include not only the mechanics that
Galileo studied, but also electromagnetic theory, quantum mechanics, and relativity. With
a push to include more physics in the undergraduate major, there is even less time to study
mathematics. We usually attempt to remedy this by putting a lot of mathematics instruction
into the physics classes, introducing the mathematical tools as they are needed.

A beginning graduate student usually needs to gather together tools that have been accrued
piecemeal and develop a deeper understanding of their use. She or he also needs to review
and reinforce material learned in undergraduate mathematics classes and become more
secure in using mathematical tools. That is the purpose of this book. A student who has
studied the material in this book should be prepared for graduate classes in mechanics,
quantum mechanics, or electricity and magnetism.

The text is an outgrowth of a course at San Francisco State University that is taken by
seniors in their final year of a physics BS or, even more commonly, by incoming graduate
students. These students have taken the usual undergraduate courses in mathematics: calculus, vector calculus, differential equations, and linear algebra. They have also already taken
upper-level courses in mechanics and electricity and magnetism. While these students have
seen some of the material in the first few chapters of this book, they are rarely comfortable,
confident, and *accurate* using these tools. At San Francisco State, incoming graduate students who already have a degree in physics take a placement test, which shows that most
of them cannot perform vector calculations correctly, especially when asked to abandon
Cartesian components! Thus, the early chapters include some basic material as a review,

and for completeness. At the same time, this material can now be approached in a more sophisticated manner.

There are many books that discuss the mathematics that graduate students in physics need to know. Some of them are extensive reference works in two volumes. Every student needs one of these books, but not as a textbook. Most textbooks cover too much material for a one-semester course, and most of them are too mathematical in their approach for today's students, who usually do not have an extensive undergraduate preparation in mathematics. The goal of this book is to offer a one-semester course aimed at *physics* students—a book that provides students with the tools that they need and shows them how to use those tools in physics problems. Because these students have some experience in physics, I draw on that knowledge in examples and exercises.

The book is organized as follows. The eight chapters cover the basic material that I have taught every year. Then five optional topics provide additional material, from which instructors may select to round out the semester or which students may read on their own. Some of the chapters include material normally covered somewhere in the undergraduate curriculum, but pushed to more depth and/or breadth. Instructors may wish to skip some of this material or assign it as required reading. Chapter 1 (on vectors and matrices) and Chapter 3 (Differential Equations) fall into this class. Some chapters cover material not normally explored in detail until the graduate program. Chapter 6 (on generalized functions) and Chapter 7 (Fourier Transforms) are in this category. The optional topics are at a more advanced level than the material in the chapters. Within each chapter, the material becomes progressively more difficult. Instructors who want a lower-level presentation can simply omit the final sections of each chapter and the optional topics. Instructors with more well-prepared students may want to use the early sections of each chapter as assigned reading and concentrate on the more difficult material in the later sections.

Material introduced in one chapter is reinforced in later chapters. Thus, Chapter 2 (Complex Variables) uses material on vectors (Chapter 1); the discussion of inverse transforms (Chapters 5 and 7) makes extensive use of contour integration (Chapter 2); and Chapter 7 (Fourier Transforms) uses material on delta functions (Chapter 6). Material on special functions (Chapter 8) draws on series solutions of differential equations (Chapter 3) and prior experience with Fourier series (Chapter 4) and Fourier transforms (Chapter 7). After a tool such as the Fourier transform has been introduced, I immediately give examples of how that tool is used in the solution of physics problems. Even though there is no single chapter labeled "Partial differential equations," many techniques and examples involving the solution of PDEs are presented throughout the text.

Mathematical proofs are kept to a minimum in the text, but many have been placed in the appendices. Students who plan a career in theoretical work will need to study the proofs in more detail at some time. My goal here is to show students how to use the mathematics to get results (see, for example, the sections on using contour integration in Chapter 2). Similarly, many of the examples are applications in physics.

I debated whether to include numerical methods in this text. I finally decided to omit computer applications almost entirely. (There is a brief discussion of numerical solution of differential equations in Chapter 3 and some discussion of computer methods in solving matrix equations in Chapter 1.) Of course, in the 21st century every physics student needs competence in using a computer, and many of the problems suggest using a computer for graphics or numerical answers. But the student also needs to understand what the computer is doing, and why. And that same student still needs to have the basic mathematical tools and to understand why and how they work. The computer material belongs in a different course, and in a different book.

A *Student Solutions Manual*, containing solutions to about 25 percent of the problems in the book, is available for sale to students. A box around the problem number in the text indicates a problem solved in the *Student Solutions Manual*. An *Instructor's Solutions Manual*, containing solutions to all of the problems, is available to instructors. For further information about either of these ancillaries, instructors should contact their Thomson Learning sales representative.

ACKNOWLEDGMENTS

The material has been extensively class tested—with a favorable response—at San Francisco State University, and I thank the many graduate students who have taken this class and have shaped my vision of what this book should be. I am also grateful to my colleagues at SFSU who read the numerous drafts of the text. Special thanks go to Dr. John Burke (Physics and Astronomy Department) and Dr. David Ellis (Department of Mathematics). My department chair, Dr. James Lockhart, also provided useful criticism of the manuscript and supported my efforts. I would also like to thank the following reviewers for their helpful comments on various drafts of the manuscript:

Albert Altman, *University of Massachusetts, Lowell*
Giles Auchmuty, *University of Houston*
Thomas Beatty, *Florida Gulf Coast University*
Paul L. DeVries, *Miami University*
Nevin Daniel Gibson, *Denison University*
Porter W. Johnson, *Illinois Institute of Technology*
David Kastor, *University of Massachusetts, Amherst*
Igor Kogoutiouk, *Minnesota State University–Mankato*
Daniel P. Lathrop, *University of Maryland*
Romulo Ochoa, *The College of New Jersey*
Daniel Phillips, *Ohio University*
George R. Plitnik, *Frostburg State University*
Asok K. Ray, *University of Texas at Arlington*
John F. Reading, *Texas A&M University*
Peter S. Riseborough, *Temple University*
Sergei Shandarin, *University of Kansas*
Charles Stanton, *CSU San Bernardino*
Krzysztof Szalewicz, *University of Delaware*
Robert L. Zimmerman, *University of Oregon*

The responsibility for any remaining errors, however, is entirely my own.

I thank Rebecca Heider, Assistant Editor for Physics and Chemistry at Brooks/Cole, for her tireless efforts on my behalf in getting this project completed. It has been a pleasure to work with George Morris at Scientific Illustrators, who completed the art with his usual skill and good humor, and Sally Lifland, who did a masterful job on the production. Finally, I thank my husband, Dr. Michael Lampton, for his love and support.

Contents

Mathematics for Physicists

CHAPTER 1

Describing the Universe

1.1. A UNIVERSAL LANGUAGE

The wonder of physics is that its laws are universal. So far as we can tell, the same laws describe the behavior of things everywhere in the universe. We need a language for these laws that is equally universal; that language is mathematics. The laws of physics are written using mathematical terms that are independent of the reference frame that we use; that is, the laws should be independent of the coordinate system that we choose, and also of any uniform motion of the reference frame. Ensuring that all physical laws obey the velocity rule was Einstein's inspiration for the theory of special relativity. Here, we'll begin by considering the first constraint—that the physical laws be stated in a form that is independent of the coordinate system used.[1]

1.1.1. Common Coordinate Systems

Coordinate systems are used to describe the position and orientation of objects in space. The three coordinate systems in most common use are the Cartesian, cylindrical, and spherical coordinate systems.

Cartesian Coordinates

The Cartesian coordinate system is the familiar x, y, z system (Figure 1.1a). The coordinate axes are mutually perpendicular and are generally chosen to be right-handed. That is, if you pick a point to be the origin and then choose the orientation of two of the axes, the third axis is determined by the right-handed convention.

---

[1] See Optional Topic A for more on the mathematics of special relativity.

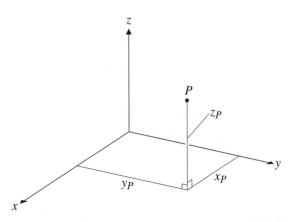

FIGURE 1.1a. Cartesian coordinates: the position of a point is described by its perpendicular distances from three mutually orthogonal planes, as shown here.

The distance ds between two neighboring points with coordinates x, y, z and $x + dx$, $y + dy$, $z + dz$ is given by Pythagoras' theorem (Figure 1.1b):

$$ds^2 = dx^2 + dy^2 + dz^2 \tag{1.1}$$

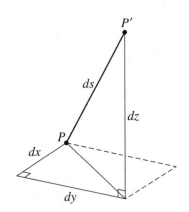

FIGURE 1.1b. The distance between two neighboring points (the line element) is found using the Pythagorean theorem.

The absence of cross terms like $dx\, dy$ is characteristic of an *orthogonal* coordinate system. The Cartesian system is orthogonal because the axes are mutually perpendicular. We'll give a more careful definition of orthogonality in Section 1.3.1.

Cylindrical Coordinates

The cylindrical coordinates are ρ, the distance from the z-axis; ϕ, an angle measured counterclockwise from a reference line (usually the positive x-axis); and z (Figure 1.2a).

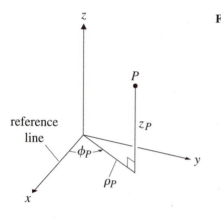

FIGURE 1.2a. Cylindrical coordinates: the position of a point is described by its perpendicular distance ρ from one coordinate axis, its perpendicular distance z from a plane perpendicular to that axis, and an angle. To obtain the angle ϕ, we drop a perpendicular from P to the plane $z = 0$, draw a line from the origin to the foot of the perpendicular, and measure the angle ϕ between that line and a reference line in the plane. Traditionally angles are measured counterclockwise from the reference line.

In terms of these coordinates,

$$x = \rho \cos \phi, \quad y = \rho \sin \phi \tag{1.2}$$

and, conversely,

$$\rho = \sqrt{x^2 + y^2}, \quad \phi = \tan^{-1} \frac{y}{x} \tag{1.3}$$

This is an orthogonal coordinate system because the unit vectors $\hat{\rho}$ (pointing outward from the z-axis, parallel to the x-y plane), $\hat{\phi}$ (perpendicular to $\hat{\rho}$, pointing in the direction in which ϕ increases), and $\hat{z}$ are mutually perpendicular. Most of the coordinate systems commonly used in physics are orthogonal coordinate systems.

The distance ds between two neighboring points with coordinates ρ, ϕ, z and $\rho + d\rho$, $\phi + d\phi$, $z + dz$ may be found by recognizing that the distance along an arc of a circle with radius ρ and angle $d\phi$ is $\rho \, d\phi$ (Figure 1.2b). Then, from Pythagoras' theorem,

$$ds^2 = d\rho^2 + \rho^2 \, d\phi^2 + dz^2 \tag{1.4}$$

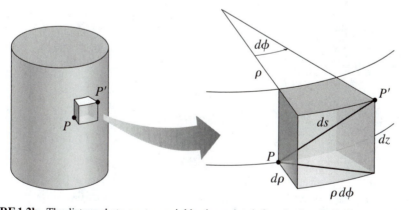

FIGURE 1.2b. The distance between two neighboring points is found using the Pythagorean theorem. The distance corresponding to the coordinate difference $d\phi$ is $\rho \, d\phi$.

Spherical Coordinates

The spherical coordinates of a point P are r, the distance from the origin; θ, the angle measured from the polar axis (usually the positive z-axis); and ϕ, an angle measured counterclockwise from the reference line (positive x-axis) (Figure 1.3a).

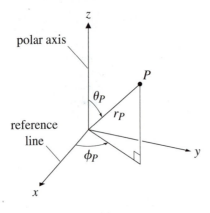

FIGURE 1.3a. Spherical coordinates: the position of a point is described by its distance from the origin and two angles. The angle ϕ is defined in the same way as in cylindrical coordinates. The *polar angle θ* is measured from the polar (z) axis to the position vector of the point.

In terms of Cartesian coordinates,

$$r = \sqrt{x^2 + y^2 + z^2}, \quad \theta = \cos^{-1}\frac{z}{r}, \quad \phi = \tan^{-1}\frac{y}{x} \tag{1.5}$$

and
$$x = r\sin\theta\cos\phi, \quad y = r\sin\theta\sin\phi, \quad z = r\cos\theta \tag{1.6}$$

When we increase the angle ϕ by $d\phi$, the point P moves around a circle of radius $r\sin\theta$ through a distance $r\sin\theta\, d\phi$ (Figure 1.3b), and so

$$ds^2 = dr^2 + r^2\, d\theta^2 + r^2\sin^2\theta\, d\phi^2 \tag{1.7}$$

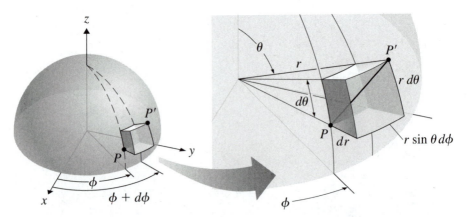

FIGURE 1.3b. The distance between two neighboring points is found using the Pythagorean theorem. The distances corresponding to the coordinate differences $d\theta$ and $d\phi$ are $r\, d\theta$ and $r\sin\theta\, d\phi$.

1.1.2. Representing Physical Laws[2]

Physical laws are represented by equations containing mathematical quantities that are independent of the reference frame or coordinate choice we use to describe them. We can write each physical law in a coordinate-independent way. For example, we may write Newton's second law without reference to specific coordinates:

$$\vec{F} = m\vec{a} = \frac{d\vec{p}}{dt} \tag{1.8}$$

The quantities appearing in this equation are scalars, vectors, and differential operators. No coordinates appear explicitly in the equation. (In special relativity theory, we also demand independence of inertial reference frame, and the time t becomes a fourth coordinate. Then we have to write Newton's second law somewhat differently. For now we are considering only coordinate independence and three spatial coordinates.)

Scalars

Scalars are mathematical quantities represented by a single number that is independent of the coordinate system used. Examples are mass of a particle, electric charge of a particle, and distance between two points. The quantity m in equation (1.8) is a *scalar.*

Vectors

Vectors are geometrical objects (arrows) with magnitude and direction. Thus, equation (1.8) is a relation between two arrows $\vec{F}$ and $\vec{a}$. These two vectors must have the same direction, and their magnitudes are related by the scalar m. Vectors are added geometrically by placing them head to tail. The sum has its tail at the tail of the first vector in the sum, and its head at the head of the last vector in the sum.

Some problems can be solved by using the vector representation (1.8). But often we find it easier to set up a coordinate system and work with the individual components of the equation. In any particular coordinate system, a vector is represented by *three* numbers— the three components of the vector. The vector may be written

$$\vec{v} = (v_x, v_y, v_z)$$

or, using index notation, v_i, where $i = 1, 2, 3$ represents the x, y, or z component of the vector. In this representation, we regard the vector $\vec{v}$ as the sum of *three* vectors, each parallel to one of the coordinate axes. The magnitude of each of these vectors is given by the magnitude of the corresponding component, while the sign of the component indicates the direction (in the direction of increasing or decreasing the coordinate).

When we change the coordinate system, these three numbers (v_x, v_y, and v_z) change. But they must change in a specific way in order for equation (1.8) to remain true. In order to see how the components change, let's restrict attention to Cartesian coordinate

---

[2]In this section it is assumed that the reader has had some previous experience with matrices. Readers who have not should read Sections 1.6.1–1.6.3 now.

systems.[3] The two ways to change the coordinates and still maintain a rectangular system are to (a) move the origin and (b) rotate the axes. Change of origin affects the position vector $\vec{r}$ but does not affect vectors like $\vec{F}$ and $\vec{a}$. Thus, the important changes are rotations of the coordinate axes.

Suppose that the system of coordinates x', y', z' is obtained from the original system x, y, z by rotating counterclockwise about the z-axis through an angle θ. Further, suppose that a vector $\vec{v}$ has components (v_x, v_y, v_z) in the original coordinate system. Let's find the components in the new system.

The x- and y-components may be constructed geometrically by projecting the vector onto the x-y plane and dropping perpendiculars from the end of the projected vector to the coordinate axes, as shown in Figure 1.4. Thus,

$$v_{x'} = OA + AB$$

$$= \frac{v_x}{\cos \theta} + (v_y - v_x \tan \theta) \sin \theta$$

$$= \frac{v_x}{\cos \theta}(1 - \sin^2 \theta) + v_y \sin \theta$$

$$= v_x \cos \theta + v_y \sin \theta \tag{1.9}$$

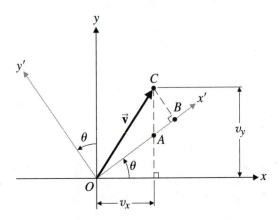

FIGURE 1.4. Rotation of coordinate axes. Here the prime axes are obtained from the unprime axes by rotating counterclockwise through an angle θ about the z-axis. OC is the projection of the vector $\vec{v}$ onto the x-y plane.

Similarly,

$$v_{y'} = BC$$

$$= (v_y - v_x \tan \theta) \cos \theta$$

$$= -v_x \sin \theta + v_y \cos \theta \tag{1.10}$$

[3]In any coordinate system we need three numbers to completely describe a vector, but in a non-Cartesian system the description is more complicated. See, for example, Problem 10 and Optional Topic A.

and

$$v_{z'} = v_z$$

Each new component is a linear combination of the old components. The two representations are related by a set of nine numbers A_{ij}:

$$v_i' = \sum_{j=1}^{3} A_{ij} v_j \tag{1.11}$$

where

$$A_{11} = \cos\theta, \quad A_{12} = \sin\theta, \quad A_{21} = -\sin\theta, \quad A_{22} = \cos\theta, \quad A_{33} = 1$$

and all other $A_{ij} = 0$. We may write these relations using matrix multiplication[4]:

$$\begin{pmatrix} v_x' \\ v_y' \\ v_z' \end{pmatrix} = \begin{pmatrix} v_x \cos\theta + v_y \sin\theta \\ -v_x \sin\theta + v_y \cos\theta \\ v_z \end{pmatrix} = \begin{pmatrix} \cos\theta & \sin\theta & 0 \\ -\sin\theta & \cos\theta & 0 \\ 0 & 0 & 1 \end{pmatrix} \begin{pmatrix} v_x \\ v_y \\ v_z \end{pmatrix} \tag{1.12}$$

or

$$\vec{v}' = \mathbb{A}\vec{v} \tag{1.13}$$

where

$$\mathbb{A} = \begin{pmatrix} \cos\theta & \sin\theta & 0 \\ -\sin\theta & \cos\theta & 0 \\ 0 & 0 & 1 \end{pmatrix} \tag{1.14}$$

is the rotation matrix that describes the coordinate transformation. This matrix allows us to find the components of *any* vector in the new coordinate system.

We write the vector components as v_i, $i = 1, 2, 3$. Similarly, the components of the matrix $\mathbb{A}$ are the numbers A_{ij}, where i labels the rows and j labels the columns[5]:

$$\mathbb{A} = \begin{pmatrix} A_{11} & A_{12} & A_{13} \\ A_{21} & A_{22} & A_{23} \\ A_{31} & A_{32} & A_{33} \end{pmatrix} \tag{1.15}$$

---

[4]See Section 1.6.1. To perform the matrix multiplication, take the dot product of the nth row of the matrix with the vector expressed in the original components to obtain the nth new component.
[5]This is the most commonly used convention for labeling the matrix components.

This matrix has several nice properties:

1. It is orthogonal; that is, each row, considered as a vector, is orthogonal to every other row:

$$\sum_j A_{ij} A_{kj} = \delta_{ik} \tag{1.16}$$

The same is true of the columns:

$$\sum_i A_{ij} A_{ik} = \delta_{jk} \tag{1.17}$$

In these expressions, δ_{ik} is the Kronecker delta, which equals 1 if $i = k$ and zero otherwise.

2. The determinant[6] of the matrix is $+1$.

3. Because of properties (1) and (2), the inverse[7] of matrix $\mathbb{A}$ equals its transpose:

$$\mathbb{A}^{-1} = \mathbb{A}^T = \begin{pmatrix} \cos\theta & -\sin\theta & 0 \\ \sin\theta & \cos\theta & 0 \\ 0 & 0 & 1 \end{pmatrix} \tag{1.18}$$

Matrices that share the properties (1) and (3) but with determinant -1 represent a transformation from a right-handed coordinate system to a left-handed system; that is, they represent a combination of a rotation and a reflection.

The matrix multiplication may also be written using index notation, as in equation (1.11). In a commonly used shorthand called the *summation convention,* we drop the $\sum$. Whenever an index is repeated once, we understand that we are to sum over that index. That is,

$$v_i' = A_{ij} v_j \quad \text{means} \quad v_i' = \sum_{j=1}^{3} A_{ij} v_j$$

The index i in this expression is not summed; it is called a free index, and it must appear once on each side of the equation. When using the summation convention, it is important to remember that, in any one term, an index may not be written *more than* twice. For example, $A_{jj} v_j$, with j repeated three times, is meaningless.

The identity matrix has components

$$\mathbb{I} = \begin{pmatrix} 1 & 0 & 0 \\ 0 & 1 & 0 \\ 0 & 0 & 1 \end{pmatrix} \tag{1.19}$$

---

[6]See Section 1.6.2.
[7]See Section 1.6.3.

and is written in index form as δ_{ij}. Multiplication by the identity matrix leaves a vector unchanged. In index notation, multiplication by the Kronecker delta δ_{ij} simply replaces the index label j with i:

$$\delta_{ij} v_j = v_i \tag{1.20}$$

We may express the matrix $\mathbb{A}$ in terms of the angles between the old and new axes. Let θ_{ij} be the angle between the ith new axis and the jth original axis. Then

$$A_{ij} = \cos\theta_{ij} \tag{1.21}$$

For the example above (rotation through θ about the z-axis), $\theta_{11} = \theta, \theta_{12} = \pi/2 - \theta$, $\theta_{21} = \pi/2 + \theta, \theta_{22} = \theta$, and $\theta_{33} = 0$. The angles $\theta_{3j} = \pi/2$ for $j = 1, 2$.

Invariance of Vector Equations

Now let's look at Newton's law again. In the first coordinate system, we may write each component of the equation as

$$F_i = ma_i \tag{1.22}$$

Each of the vectors may now be transformed to the new system. The new components are

$$F_i' = A_{ij} F_j \tag{1.23}$$

and

$$a_i' = A_{ij} a_j \tag{1.24}$$

We may invert each of these relations to obtain F_i in terms of F_i'. First multiply equation (1.23) on the left by the matrix $\mathbb{A}^{-1}$:

$$(\mathbb{A}^{-1})_{ki} F_i' = (\mathbb{A}^{-1})_{ki} A_{ij} F_j$$

Then use the results that $\mathbb{A}^{-1} = \mathbb{A}^T$ and that $\mathbb{A}^{-1}\mathbb{A}$ is the unit matrix:

$$(\mathbb{A}^T)_{ki} F_i' = (\mathbb{A}^{-1})_{ki} A_{ij} F_j = \delta_{kj} F_j$$

and so

$$A_{ik} F_i' = F_k \tag{1.25}$$

Equation (1.25) is the inverse of relation (1.23) and transforms the components in the primed system to the original system. We can use this relation to transform both vectors in equation (1.22):

$$A_{ji} F_j' = m A_{ji} a_j'$$

or

$$A_{ji}(F_j' - ma_j') = 0$$

Multiply on the left by $\mathbb{A}$:

$$A_{ki} A_{ji}(F_j' - ma_j') = 0$$

Use equation (1.16):

$$\delta_{kj}(F_j' - ma_j') = 0$$

Then use property (1.20):

$$F_k' - ma_k' = 0$$

So the equation is true in the primed system if it is true in the original system.

Thus, we see that the transformation law (1.23) is a necessary consequence of the geometrical nature of vectors and the fact that vector relations are independent of coordinate rotations. The transformation law (1.23) becomes part of the definition of a vector.

> A vector in three-dimensional space is represented in Cartesian coordinates by a set of three numbers (its components) that transform according to equation (1.23) when the coordinate axes are rotated.

Repeated rotations are represented by matrix multiplication. If the first rotation is represented by matrix $\mathbb{A}$, $v_i' = A_{ij}v_j$, and the second by $\mathbb{B}$, $v_k'' = B_{ki}v_i'$, then the two rotations together are represented by the matrix product[8] $\mathbb{B}\mathbb{A}$, $v_k'' = B_{ki}A_{ij}v_j$. Notice that the first rotation is on the right in this product. The order is important, because in general the result of doing rotation 2 followed by rotation 1 is not the same as that of doing 1 followed by 2. Thus, the matrix multiplication is not commutative; that is,

$$\mathbb{B}\mathbb{A} \neq \mathbb{A}\mathbb{B}$$

Example 1.1.   The coordinate axes are rotated about the z-axis by $90°$ and then about the new x-axis by $90°$. Find the matrix that describes this rotation. Compare it with the matrix for rotation about the x-axis followed by rotation about the new z-axis (that is, the same rotations but performed in the reverse order).

Using rule (1.21), we have for the first rotation $\theta_{12} = 0 = \theta_{33}$, $\theta_{21} = \pi$, all other $\theta_{ij} = \pi/2$. Thus,

$$\mathbb{A} = \begin{pmatrix} 0 & 1 & 0 \\ -1 & 0 & 0 \\ 0 & 0 & 1 \end{pmatrix}$$

---

[8]Note here how the index notation describes the order in which the matrices are multiplied. In the matrix product, the repeated index that indicates summation is in the center: $(\mathbb{B}\mathbb{A})_{kj} = B_{ki}A_{ij}$. We could also write the product as $A_{ij}B_{ki}$, where the meaning is indicated by the placement of the indices, independent of the order in which the terms are written. Then to perform the sum using matrix multiplication, we must first write the matrices with repeated indices next to each other: $A_{ij}B_{ki} = B_{ki}A_{ij} = (\mathbb{B}\mathbb{A})_{kj}$.

For the second rotation about the new x-axis, $\theta_{23} = 0 = \theta_{11}, \theta_{32} = \pi$, all other $\theta_{ij} = \pi/2$. So

$$\mathbb{B} = \begin{pmatrix} 1 & 0 & 0 \\ 0 & 0 & 1 \\ 0 & -1 & 0 \end{pmatrix}$$

The two successive rotations are represented by the matrix product:

$$\mathbb{B}\mathbb{A} = \begin{pmatrix} 1 & 0 & 0 \\ 0 & 0 & 1 \\ 0 & -1 & 0 \end{pmatrix} \begin{pmatrix} 0 & 1 & 0 \\ -1 & 0 & 0 \\ 0 & 0 & 1 \end{pmatrix} = \begin{pmatrix} 0 & 1 & 0 \\ 0 & 0 & 1 \\ 1 & 0 & 0 \end{pmatrix}$$

A unit vector along the x-axis with components $(1, 0, 0)$ in the original system has the following components in the rotated system:

$$\begin{pmatrix} 0 & 1 & 0 \\ 0 & 0 & 1 \\ 1 & 0 & 0 \end{pmatrix} \begin{pmatrix} 1 \\ 0 \\ 0 \end{pmatrix} = \begin{pmatrix} 0 \\ 0 \\ 1 \end{pmatrix}$$

That is, it lies along the new z''-axis. Similarly, the y-axis has become the x''-axis, and the z-axis has become the y''-axis (Figure 1.5).

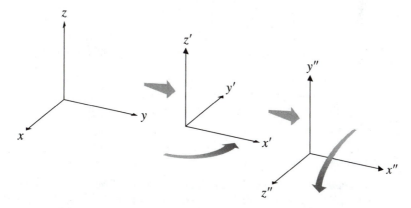

FIGURE 1.5.   The set of rotations in the first part of Example 1.1.

Doing the rotations in the opposite order, we get

$$\mathbb{A}\mathbb{B} = \begin{pmatrix} 0 & 1 & 0 \\ -1 & 0 & 0 \\ 0 & 0 & 1 \end{pmatrix} \begin{pmatrix} 1 & 0 & 0 \\ 0 & 0 & 1 \\ 0 & -1 & 0 \end{pmatrix} = \begin{pmatrix} 0 & 0 & 1 \\ -1 & 0 & 0 \\ 0 & -1 & 0 \end{pmatrix}$$

This set of rotations sends x to $-y''$, y to $-z''$, and z to x'' (Figure 1.6).

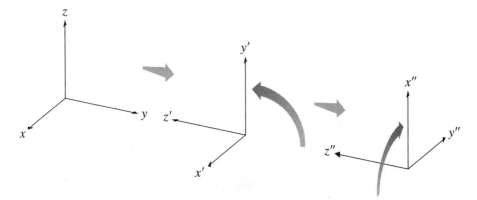

FIGURE 1.6. In the second part of Example 1.1, the same rotations performed in the opposite order give a different result.

However, two rotations about the *same axis* do commute. Thus, if we rotate about the z-axis through an angle θ_1 and then again about the z-axis through θ_2, the result is a rotation through $\theta_1 + \theta_2$.

$$
\mathbb{BA} = \begin{pmatrix} \cos\theta_2 & \sin\theta_2 & 0 \\ -\sin\theta_2 & \cos\theta_2 & 0 \\ 0 & 0 & 1 \end{pmatrix} \begin{pmatrix} \cos\theta_1 & \sin\theta_1 & 0 \\ -\sin\theta_1 & \cos\theta_1 & 0 \\ 0 & 0 & 1 \end{pmatrix}
$$

$$
= \begin{pmatrix} \cos\theta_2 \cos\theta_1 - \sin\theta_2 \sin\theta_1 & \cos\theta_2 \sin\theta_1 + \sin\theta_2 \cos\theta_1 & 0 \\ -\sin\theta_2 \cos\theta_1 - \cos\theta_2 \sin\theta_1 & \cos\theta_2 \cos\theta_1 - \sin\theta_2 \sin\theta_1 & 0 \\ 0 & 0 & 1 \end{pmatrix}
$$

$$
= \begin{pmatrix} \cos(\theta_2 + \theta_1) & \sin(\theta_2 + \theta_1) & 0 \\ -\sin(\theta_2 + \theta_1) & \cos(\theta_2 + \theta_1) & 0 \\ 0 & 0 & 1 \end{pmatrix} = \mathbb{AB}
$$

Multiplying Vectors

Product with a scalar   A vector may be multiplied by a scalar. The result is another vector. The product $m\vec{a}$ in Newton's second law is an example of this kind of multiplication. The new vector has the same direction as the original vector, but its magnitude is changed. The ith component of the new vector equals the ith component of the original vector multiplied by the scalar.

Scalar product of two vectors   Two vectors may be multiplied together to give either a scalar or a vector.[9]

_____

[9]Strictly speaking, this product is a pseudo-vector, as discussed below.

The *scalar* or *dot product* of two vectors is

$$\vec{\mathbf{v}} \cdot \vec{\mathbf{u}} = vu \cos\theta \qquad (1.26)$$

where v and u are the magnitudes of the two vectors $\vec{\mathbf{v}}$ and $\vec{\mathbf{u}}$, respectively, and θ is the angle between them.

Equivalently, we may write the product in terms of the Cartesian vector components:

$$\vec{\mathbf{v}} \cdot \vec{\mathbf{u}} = v_i u_i \qquad (1.27)$$

where, as usual, the repeated index means that we sum over the index i. That is,

$$\vec{\mathbf{v}} \cdot \vec{\mathbf{u}} = \sum_{i=1}^{3} v_i u_i = v_1 u_1 + v_2 u_2 + v_3 u_3$$

To show that this product is a scalar, let's start with the product in the prime frame and transform to the unprime frame:

$$\vec{\mathbf{v}} \cdot \vec{\mathbf{u}} = v_i' u_i' = A_{ij} v_j A_{ik} u_k$$

Notice that we needed to be careful not to write the index j more than twice, so in transforming $\vec{\mathbf{u}}$ we called the summed index k. Now

$$A_{ij} A_{ik} = A_{ji}^T A_{ik} = A_{ji}^{-1} A_{ik} = \delta_{jk}$$

and so

$$\vec{\mathbf{v}} \cdot \vec{\mathbf{u}} = v_i' u_i' = A_{ij} A_{ik} v_j u_k = \delta_{jk} v_j u_k = v_k u_k = \vec{\mathbf{v}} \cdot \vec{\mathbf{u}}$$

Since the product is the same in the two coordinate systems, it is a scalar. This demonstration actually shows that the product is the same in *any* two coordinate systems, since the transformation represented by the matrix $\mathbb{A}$ is an arbitrary rotation.

The dot product is also commutative:

$$\vec{\mathbf{v}} \cdot \vec{\mathbf{u}} = \vec{\mathbf{u}} \cdot \vec{\mathbf{v}}$$

The magnitude of a vector $\vec{\mathbf{v}}$ may be written in terms of the dot product of $\vec{\mathbf{v}}$ with itself:

$$v = |\vec{\mathbf{v}}| = \sqrt{\vec{\mathbf{v}} \cdot \vec{\mathbf{v}}}$$

Examples of dot products in physics include the definition of work ($dW = \vec{\mathbf{F}} \cdot d\vec{\mathbf{s}}$), the definition of electric flux ($d\Phi = \vec{\mathbf{E}} \cdot d\vec{\mathbf{A}}$), and the expressions $P = \vec{\mathbf{F}} \cdot \vec{\mathbf{v}} = \vec{\tau} \cdot \vec{\omega}$ for power.

Cross product of two vectors   The *vector* or *cross product* of two vectors is another vector whose magnitude is given by

$$|\vec{v} \times \vec{u}| = vu \sin \theta \tag{1.28}$$

The direction of the product is given by the right-hand rule, as shown in Figure 1.7a.

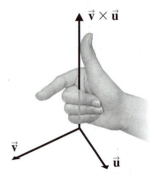

FIGURE 1.7a. The right-hand rule for cross products. Curl the fingers of the right hand from $\vec{v}$ toward $\vec{u}$ and the thumb gives the direction $\vec{v} \times \vec{u}$.

Its components are given by

$$(\vec{v} \times \vec{u})_1 = v_2 u_3 - v_3 u_2$$

$$(\vec{v} \times \vec{u})_2 = v_3 u_1 - v_1 u_3$$

and

$$(\vec{v} \times \vec{u})_3 = v_1 u_2 - v_2 u_1$$

We may express the result in index form by using the Levi-Civita symbol ε_{ijk}:

$$\varepsilon_{ijk} = \begin{cases} +1 & \text{if } i, j, k = \text{an even permutation of 1, 2, 3} \\ -1 & \text{if } i, j, k = \text{an odd permutation of 1, 2, 3} \\ 0 & \text{if any two of } i, j, k \text{ are equal} \end{cases} \tag{1.29}$$

(To get an even permutation, you may rotate the sequence of numbers $ijk \rightarrow kij \rightarrow jki$, but you may not interchange two of them. The permutation $ijk \rightarrow ikj$, obtained by interchanging j and k, is an odd permutation.) Then

$$(\vec{v} \times \vec{u})_i = \varepsilon_{ijk} v_j u_k \tag{1.30}$$

Strictly speaking, the result is not a true vector. It transforms as a vector under coordinate rotations but does not transform properly under reflections. (To see what this means, notice that your right hand looks like a right hand however you turn it, but in a mirror it looks like a left hand, as shown in Figure 1.7b.) The cross product is called a pseudo-vector. (See Appendix I and Optional Topic A for the transformation law.)

FIGURE 1.7b. When viewed in a mirror, the right hand becomes a left hand.

The cross product $\vec{v} \times \vec{u}$ represents the parallelogram formed by the vectors $\vec{v}$ and $\vec{u}$. The magnitude of the cross product equals the area of the parallelogram, and the direction is normal to the plane of the parallelogram. Examples of the cross product in physics include torque ($\vec{\tau} = \vec{r} \times \vec{F}$), angular momentum of a particle ($\vec{L} = \vec{r} \times \vec{p}$), and the Biot-Savart law for magnetic field ($d\vec{B} = \mu_0 \vec{j} \times \vec{r} dV / 4\pi r^3$).

Products of three vectors We may form a *triple scalar product* of three vectors by taking the dot product of a third vector $\vec{w}$ with the cross product $\vec{u} \times \vec{v}$. It is sometimes written with square brackets as

$$[\vec{u}, \vec{v}, \vec{w}] = (\vec{u} \times \vec{v}) \cdot \vec{w} = \vec{u} \cdot (\vec{v} \times \vec{w}) = (\vec{w} \times \vec{u}) \cdot \vec{v} \tag{1.31}$$

To see why these three products are equal, note that each is the volume of the parallelepiped with edges given by the three vectors $\vec{u}$, $\vec{v}$, and $\vec{w}$ (Figure 1.8). We may also derive the result algebraically:

$$[\vec{u}, \vec{v}, \vec{w}] = \vec{u} \cdot (\vec{v} \times \vec{w}) = \varepsilon_{ijk} u_i v_j w_k = \varepsilon_{kij} u_i v_j w_k = (\vec{u} \times \vec{v}) \cdot \vec{w} \tag{1.32}$$

and similarly for the third relation.

For example, a particle in circular motion is acted on by a force $\vec{F}$. The particle's angular velocity is $\vec{\omega}$, and its linear velocity is $\vec{v} = \vec{\omega} \times \vec{r}$, where the vector $\vec{r}$ is the position vector of the particle with respect to the center of the circle. The power delivered by the force $\vec{F}$ is

$$P = \vec{F} \cdot \vec{v} = \vec{F} \cdot (\vec{\omega} \times \vec{r}) = \vec{\omega} \cdot (\vec{r} \times \vec{F}) = \vec{\omega} \cdot \vec{\tau}$$

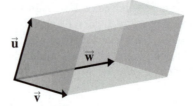

FIGURE 1.8. The triple scalar product equals the volume of the parallelepiped with the three vectors along its edges.

The *triple vector product* may be expressed in terms of dot products:

$$(\vec{u} \times \vec{v}) \times \vec{w} = \varepsilon_{ijk}(\vec{u} \times \vec{v})_j w_k = \varepsilon_{ijk}\varepsilon_{jlm}u_l v_m w_k \qquad (1.33)$$

To evaluate expression (1.33), we may use a nifty relation:

$$\varepsilon_{jki}\varepsilon_{jlm} = \delta_{kl}\delta_{im} - \delta_{km}\delta_{il} \qquad (1.34)$$

To write relation (1.34) correctly, we first put the repeated index j in the same position in both Levi-Civita symbols by a sequence of even permutations. Then we pair up matching indices in order (k with l and i with m). This gives the positive term $\delta_{kl}\delta_{im}$. Then we pair up the nonrepeated indices the other way to get the negative term.

Let's verify equation (1.34). We don't have to check all 3^4 possible combinations of $k, i, l,$ and m separately because we can use the properties of the ε and δ symbols to show that the relation is true for certain groups of values. First note that the ε symbol is zero if any two of its indices equal each other, so if $k = i$ or $l = m$, all three terms in the sum on the left-hand side are zero. But if $k = i$ or $l = m$, both terms on the right-hand side are identical and their difference is zero. The relation is true in this case.

If the indices $k, i, l,$ and m include all the values 1, 2, and 3, both sides also vanish. Suppose, for example, that $k = 1, i = 2, l = 2,$ and $m = 3$. Then on the left-hand side we have

$$\varepsilon_{j12}\varepsilon_{j23} = 0$$

since $\varepsilon_{j12} = 0$ unless $j = 3$, but then $\varepsilon_{j23} = \varepsilon_{323} = 0$. A quick check shows that in this case one of the Kronecker deltas in each term on the right-hand side is also zero. So we have verified the relation—both sides are zero—in all cases except $k = l, i = m$ or $k = m, i = l$. The first of these possibilities ($k = l, i = m$) gives

$$\varepsilon_{jki}\varepsilon_{jlm} = \sum_j \varepsilon_{jlm}\varepsilon_{jlm} = +1$$

Note. In the middle term, we use the summation sign and do not use the summation convention; that is, we do not sum over l and m.

On the right-hand side of equation (1.34), the first combination of δs is $1 \times 1 = 1$, while the second is zero. The other possibility is $k = m$, $i = l$, which leads to

$$\varepsilon_{jki}\varepsilon_{jlm} = \sum_j \varepsilon_{jml}\varepsilon_{jlm} = -1$$

This time it is the second set of deltas on the right-hand side that equals 1, so the right-hand side is also -1, as required, and relation (1.34) is proved.

Returning to the triple cross product (1.33), we have

$$(\vec{u} \times \vec{v}) \times \vec{w} = \varepsilon_{ijk}\varepsilon_{jlm}u_l v_m w_k$$

$$= \varepsilon_{jki}\varepsilon_{jlm}u_l v_m w_k$$

$$= (\delta_{kl}\delta_{im} - \delta_{km}\delta_{il})u_l v_m w_k$$

$$= u_k v_i w_k - u_i v_k w_k$$

$$\boxed{(\vec{u} \times \vec{v}) \times \vec{w} = \vec{v}(\vec{u} \cdot \vec{w}) - \vec{u}(\vec{v} \cdot \vec{w})} \qquad (1.35)$$

This result is sometimes known as the "bac-cab rule":

$$\vec{a} \times (\vec{b} \times \vec{c}) = \vec{b}(\vec{a} \cdot \vec{c}) - \vec{c}(\vec{a} \cdot \vec{b})$$

To be sure you get the signs right, always start with the *middle* vector ($\vec{v}$ or $\vec{b}$ in the examples above) times the dot product of the other two vectors. The second term has the other vector inside the original parentheses ($\vec{u}$ or $\vec{c}$) times a dot product.

For example, the torque exerted by the magnetic force on a charged particle is

$$\vec{\tau} = \vec{r} \times \vec{F} = \vec{r} \times q(\vec{v} \times \vec{B}) = q[\vec{v}(\vec{r} \cdot \vec{B}) - \vec{B}(\vec{r} \cdot \vec{v})]$$

Tensors

Some physical laws cannot be represented using only vectors and scalars—they require an object that is represented by more than three numbers. These objects are called tensors (see Optional Topic A).

1.2. SCALAR AND VECTOR FIELDS

Some physical quantities are associated with points in space. For example, a weather map shows the air temperature in different cities. In principle, we could measure the temperature at each point in space. Since temperature is a scalar quantity, the values of temperature constitute a *scalar field* $T(\vec{\mathbf{r}})$. We can represent the field visually by drawing contour lines of constant T. (On a weather map, these are called isotherms.) Other examples of scalar fields are density, electrostatic potential, and height above sea level (which is related to gravitational potential). We expect these physical quantities to be described by functions that are continuous at the least, and more often differentiable. In most cases, the derivatives are also continuous, and we shall make that assumption in what follows, except when explicitly noted otherwise.

When a vector quantity is associated with each point in space, we have a *vector field*. Examples are wind velocity, electric field, and magnetic field.

It is possible to derive one kind of field from another. For example, the wind speed is a scalar quantity derived from the wind velocity vector.

1.2.1. The Gradient

One vector field we can derive from a scalar field $\Phi(\vec{\mathbf{r}})$ is the gradient $\vec{\nabla}\Phi$. In Cartesian coordinates, the gradient operator is written

$$\vec{\nabla} \equiv \left(\frac{\partial}{\partial x}, \frac{\partial}{\partial y}, \frac{\partial}{\partial z} \right)$$

so

$$\vec{\nabla}\Phi = \left(\frac{\partial \Phi}{\partial x}, \frac{\partial \Phi}{\partial y}, \frac{\partial \Phi}{\partial z} \right) \tag{1.36}$$

The change in Φ when we move through an arbitrary displacement $d\vec{\mathbf{s}} = (dx, dy, dz)$ is

$$d\Phi = \frac{\partial \Phi}{\partial x}dx + \frac{\partial \Phi}{\partial y}dy + \frac{\partial \Phi}{\partial z}dz = \vec{\nabla}\Phi \cdot d\vec{\mathbf{s}} \tag{1.37}$$

If $d\vec{\mathbf{s}}$ is tangent to the constant-Φ contour, $d\Phi = 0$. Thus, the gradient vector is perpendicular to the contour of constant Φ at each point. In order to obtain the maximum change $d\Phi$ over a given *distance* $|d\vec{\mathbf{s}}|$, $d\vec{\mathbf{s}}$ must be parallel to $\vec{\nabla}\Phi$. Thus, the direction of $\vec{\nabla}\Phi$ at a point is the direction in which Φ changes most rapidly at that point (Figure 1.9). The maximum change in Φ is obtained by moving perpendicular to the lines of constant Φ.

The *directional derivative* of Φ in a direction described by a unit vector $\hat{\mathbf{l}}$ is $\hat{\mathbf{l}} \cdot \vec{\nabla}\Phi$.

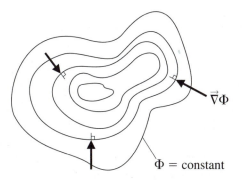

FIGURE 1.9. Contours of constant Φ. In this diagram, Φ is increasing inward. The gradient of a scalar field at a point is the steepest slope of the scalar field at that point. $\vec{\nabla}\Phi$, indicated by the arrows, is perpendicular to the constant-Φ contours.

1.2.2. Properties of Vector Fields

The Divergence

In Cartesian coordinates, the divergence of a vector field is

$$\vec{\nabla} \cdot \vec{u} \equiv \frac{\partial u_x}{\partial x} + \frac{\partial u_y}{\partial y} + \frac{\partial u_z}{\partial z} \tag{1.38}$$

In index notation, it is

$$\vec{\nabla} \cdot \vec{u} = \frac{\partial u_i}{\partial x_i} \tag{1.39}$$

The divergence is a scalar that indicates how much the vectors spread apart, or diverge, from each other around a point. The electric field due to a positive point charge is a good visual example of a vector field with positive divergence at the position of the charge. A vector field whose divergence is zero is called *solenoidal*.

The Curl

In Cartesian coordinates, the curl of a vector field $\vec{u}$ is given by

$$\vec{\nabla} \times \vec{u} \equiv \left(\frac{\partial u_z}{\partial y} - \frac{\partial u_y}{\partial z}, \frac{\partial u_x}{\partial z} - \frac{\partial u_z}{\partial x}, \frac{\partial u_y}{\partial x} - \frac{\partial u_x}{\partial y} \right) \tag{1.40}$$

In index notation, it is given by

$$(\vec{\nabla} \times \vec{u})_i = \varepsilon_{ijk} \frac{\partial u_k}{\partial x_j} \tag{1.41}$$

The curl, like all cross products, is a pseudo-vector. The curl is nonzero at a point when the vectors circulate around, or curl around, the point. The magnetic field due to a long straight wire is a good visual example of a vector field with nonzero curl at the location of the wire. A vector field whose curl is zero is called *irrotational*.

1.2.3. Integral Properties of Vector Fields

Line Integrals and Circulation

A line integral of the form

$$\int_A^B \vec{u} \cdot d\vec{l} \tag{1.42}$$

occurs frequently in physics.[10] The integral is the sum of the contributions $\vec{u} \cdot d\vec{l}$ for each differential displacement[11] $d\vec{l}$ along the path from A to B. For example, the work done by a force $\vec{F}$ between A and B is

$$W(A \rightarrow B) = \int_A^B \vec{F} \cdot d\vec{l}$$

In general, the result depends on the path taken between A and B (Figure 1.10).

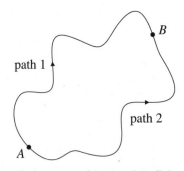

FIGURE 1.10. Two different paths between points A and B. Going from A to B along path 1 and returning to A along path 2, we form a closed curve C.

[10]See, for example, Lea and Burke, Section 7.1.4.
[11]In Cartesian coordinates, $d\vec{l}$ has components (dx, dy, dz).

However, when $\vec{\mathbf{u}} = \vec{\nabla}\Phi$ for some scalar field Φ, the integral (1.42) is independent of the path between A and B:

$$\int_A^B \vec{\mathbf{u}} \cdot d\vec{\mathbf{l}} = \int_A^B (u_x \, dx + u_y \, dy + u_z \, dz)$$

$$= \int_A^B \left(\frac{\partial \Phi}{\partial x} dx + \frac{\partial \Phi}{\partial y} dy + \frac{\partial \Phi}{\partial z} dz \right)$$

$$= \int_A^B d\Phi = \Phi(B) - \Phi(A) \tag{1.43}$$

When the integral is independent of path, the integral around the closed loop $A \to B \to A$ is zero:

$$\oint \vec{\mathbf{u}} \cdot d\vec{\mathbf{l}} = \int_{A,\text{ path }1}^B \vec{\mathbf{u}} \cdot d\vec{\mathbf{l}} + \int_{B,\text{ path }2}^A \vec{\mathbf{u}} \cdot d\vec{\mathbf{l}}$$

$$= \int_{A,\text{ path }1}^B \vec{\mathbf{u}} \cdot d\vec{\mathbf{l}} - \int_{A,\text{ path }2}^B \vec{\mathbf{u}} \cdot d\vec{\mathbf{l}} = 0$$

The integral around the closed loop $C = A \to B \to A$ is called the *circulation* of the vector field $\vec{\mathbf{u}}$ around the curve C. Thus, we have this result:

When a vector field $\vec{\mathbf{u}}$ is the gradient of a scalar field Φ in a region R, the circulation of $\vec{\mathbf{u}}$ around any curve C in R is zero.

Conversely, if the circulation of $\vec{\mathbf{u}}$ around *every* curve C is zero, then $\vec{\mathbf{u}}$ is the gradient of some scalar field Φ, and Φ may be determined (up to a constant) by equation (1.43).

The language that we use to describe vector fields comes from fluid theory, because fluid velocity was one of the first vector fields studied. Water flowing out the drain in a sink usually swirls around the drain. This is an example of a vector field with nonzero circulation around any curve that surrounds the drain.

Flux

The flux of a vector field $\vec{\mathbf{u}}$ through a surface S is

$$\Phi_{\mathbf{u}} = \int_S \vec{\mathbf{u}} \cdot \hat{\mathbf{n}} \, dA$$

where $\hat{\mathbf{n}}$ at any point on the surface is a unit vector normal to the surface at that point. For fluids, with $\vec{\mathbf{u}}$ equal to the fluid velocity, the flux describes the rate at which fluid flows through the surface.

Although flux is a scalar, it has a sign that is associated with the direction of the vector field $\vec{\mathbf{u}}$. Flux is positive if the field $\vec{\mathbf{u}}$ points generally in the direction of $\hat{\mathbf{n}}$ and negative if $\vec{\mathbf{u}}$ points generally opposite $\hat{\mathbf{n}}$.

The Divergence Theorem

The divergence theorem[12] relates the flux through a *closed* surface S to the divergence of the vector field in the enclosed volume. When integrating over a closed surface, it is conventional to choose the normal vector $\hat{\mathbf{n}}$ to point outward. Then

$$\oint_S \vec{\mathbf{u}} \cdot \hat{\mathbf{n}} \, dA = \int_V \vec{\nabla} \cdot \vec{\mathbf{u}} \, dV \qquad (1.44)$$

To prove this theorem, we cut the volume V up into a very large number of tiny cubes (Figure 1.11).

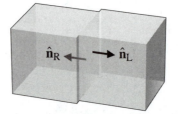

FIGURE 1.11. Two neighboring cubes inside volume V. On touching sides, the normals outward from the two cubes are opposite each other.

The volume integral is the sum of the integrals over all the cubes. Notice that the normal vectors on touching sides of two neighboring cubes are exactly opposite each other, and so the contributions to the surface integrals $\oint_S \vec{\mathbf{u}} \cdot \hat{\mathbf{n}} \, dA$ for the two cubes from these two sides exactly cancel. Thus, the sum of the surface integrals for all the cubes reduces to the integral over those surfaces that have no touching neighbors—that is, over the original closed surface. So if we can prove the result for one cube (Figure 1.12), we will have proved it for an arbitrary volume.

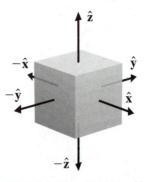

FIGURE 1.12. Integrating over a differential cube with sides of length dx, dy, and dz, respectively. The normals to each side are in the directions $\pm\hat{\mathbf{x}}$, $\pm\hat{\mathbf{y}}$, and $\pm\hat{\mathbf{z}}$.

[12]This theorem is also called Gauss' theorem.

The volume integral on the right-hand side of equation (1.44) is

$$\int_{\text{differential box}} \vec{\nabla} \cdot \vec{u}\, dV = \left(\frac{\partial u_x}{\partial x} + \frac{\partial u_y}{\partial y} + \frac{\partial u_z}{\partial z} \right) dx\, dy\, dz \tag{1.45}$$

To evaluate the surface integral on the left-hand side of (1.44), we first consider the pair of sides at x and at $x + dx$. For the side at x, $\hat{n} = -\hat{x}$ and

$$\int_S \vec{u} \cdot \hat{n}\, dA = -u_x(x)\, dy\, dz$$

where $u_x(x)$ is the x-component of $\vec{u}$ evaluated at x, while for the side at $x + dx$, $\hat{n} = +\hat{x}$ and

$$\int_S \vec{u} \cdot \hat{n}\, dA = u_x(x + dx)\, dy\, dz$$

Using the definition of the partial derivative $\partial u_x / \partial x$, we find that the sum of the two terms is

$$[u_x(x + dx) - u_x(x)]\, dy\, dz = \frac{\partial u_x}{\partial x}\, dx\, dy\, dz$$

We get similar contributions from the pairs of sides at y, $y + dy$ and z, $z + dz$. Thus, the left-hand side of equation (1.44) also equals the right-hand side of (1.45), and the theorem is proved.

To understand the meaning of the theorem, think of the electric field due to a positive point charge. If we surround the charge with a volume V, then the electric field is outward at each point of the surface S, and so the flux through the surface is positive. Similarly, the divergence of the electric field is positive at the position of the charge, as required by the theorem.

Example 1.2. Compute the divergence of the vector field $\vec{v} = k\vec{r}$, where $\vec{r}$ is the position vector and k is a constant. Find the flux of $\vec{v}$ through a cube of side a centered at the origin and show that it equals $\int \vec{\nabla} \cdot \vec{v}\, dV$ over the volume of the cube.

The divergence is

$$\vec{\nabla} \cdot \vec{v} = k\vec{\nabla} \cdot (x\hat{x} + y\hat{y} + z\hat{z}) = k\left(\frac{\partial}{\partial x}x + \frac{\partial}{\partial y}y + \frac{\partial}{\partial z}z \right) = 3k$$

Since the divergence is a constant, we can immediately compute the volume integral as $3k$ times the volume, or $3ka^3$.

The flux is

$$\oint_{\text{surface of cube}} k(x\hat{x} + y\hat{y} + z\hat{z}) \cdot \hat{n}\, dA$$

On each face of the cube, the normal is $\pm\hat{x}$, $\pm\hat{y}$, or $\pm\hat{z}$, so the flux through the face at $x = a/2$ is

$$\int_{\text{face}} k\frac{a}{2}\hat{x} \cdot \hat{x}\, dA = k\frac{a}{2}a^2 = \frac{ka^3}{2}$$

while at $x = -a/2$ we have

$$\int_{\text{face}} k\left(-\frac{a}{2}\right) \hat{\mathbf{x}} \cdot (-\hat{\mathbf{x}}) \, dA = k\frac{a}{2}a^2 = \frac{ka^3}{2}$$

The result is the same for each of the six faces, so the total flux is $3ka^3$ and equals the volume integral of the divergence.

Stokes' Theorem

Stokes' theorem relates the line integral of a vector field $\vec{\mathbf{u}}$ around a closed curve $\mathcal{C}$ to the curl of the vector field. First we define a surface that spans the curve $\mathcal{C}$ to be any surface whose edge is the curve $\mathcal{C}$. Think of blowing bubbles, using the curve $\mathcal{C}$ as the wire. Dip the wire in the soap solution. If curve $\mathcal{C}$ lies in a plane, you will get a planar film of soap solution attached to the wire (the curve $\mathcal{C}$) at its edges. This is the simplest surface that spans $\mathcal{C}$. Now as you blow the bubble, the surface expands but remains attached to the curve $\mathcal{C}$ at the edges. You obtain a sequence of surfaces, each of which spans the curve $\mathcal{C}$. Stokes' theorem applies to all such surfaces.

We define the normal to the surface using a right-hand rule. Curl your fingers in the direction of $\vec{dl}$, and the thumb gives the direction of $\hat{\mathbf{n}}$ (Figure 1.13). Then

$$\oint_{\mathcal{C}} \vec{\mathbf{u}} \cdot \vec{dl} = \int_{S} (\vec{\nabla} \times \vec{\mathbf{u}}) \cdot \hat{\mathbf{n}} \, dA \tag{1.46}$$

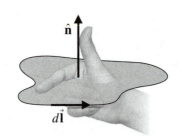

FIGURE 1.13. Definition of the normal vector used in Stokes' theorem. Curl the fingers of your right hand in the direction of $\vec{dl}$, and your thumb points in the direction of $\hat{\mathbf{n}}$.

To prove Stokes' theorem, we use a method similar to the one that we used for the divergence theorem. Divide the surface up into a mesh of little rectangles. The surface integral is just the sum of the integrals over all the rectangles. Summing the line integrals over all the rectangles, again we note that the contributions $\vec{\mathbf{u}} \cdot \vec{dl}$ from touching sides cancel, since the vectors $\vec{dl}$ have opposite directions. Thus, the sum of the line integrals over all the rectangles reduces to the integral over the curve $\mathcal{C}$, as shown in Figure 1.14.

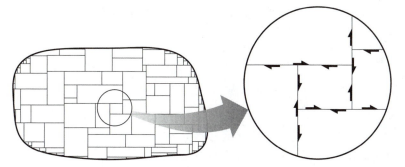

FIGURE 1.14. The surface divided into differential rectangles.

Let's look at one of the rectangles that is in a single plane, and choose that plane to be the x-y plane (Figure 1.15).

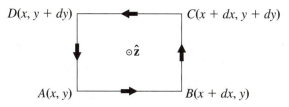

FIGURE 1.15. A single differential rectangle. We choose x- and y-axes parallel to the sides of this rectangle.

The integral on the right-hand side of equation (1.46) is

$$\int_{\text{differential rectangle}} (\vec{\nabla} \times \vec{u}) \cdot \hat{n} \, dA = (\vec{\nabla} \times \vec{u}) \cdot \hat{z} \, dx \, dy = \left(\frac{\partial u_y}{\partial x} - \frac{\partial u_x}{\partial y} \right) dx \, dy$$

while on the left we have

$$\int_{AB+BC+CD+DA} \vec{u} \cdot d\vec{l} = u_x(x, y)dx + u_y(x + dx, y)dy$$

$$- u_x(x, y + dy)dx - u_y(x, y)dy$$

$$= [u_y(x + dx, y) - u_y(x, y)]dy - [u_x(x, y + dy) - u_x(x, y)]dx$$

$$= \frac{\partial u_y}{\partial x}dx \, dy - \frac{\partial u_x}{\partial y}dy \, dx$$

which equals the right-hand side. Thus, the theorem is proved.

For the special case of a curve lying entirely in the x-y plane, Stokes' theorem reduces to

$$\oint_C u_x \, dx + u_y \, dy = \int_S \left(\frac{\partial u_y}{\partial x} - \frac{\partial u_x}{\partial y} \right) dx \, dy$$

Since any functions $f(x, y)$ and $g(x, y)$ may replace the x- and y-components of $\vec{u}$ in this proof

$$\oint_C (f \, dx + g \, dy) = \int_S \left(\frac{\partial g}{\partial x} - \frac{\partial f}{\partial y} \right) dx \, dy \tag{1.47}$$

In this form, the theorem is known as Green's theorem.

Example 1.3.   Find the circulation of the vector field $\vec{u} = x^2 y^3 (\hat{y} + \hat{x})$ around a square of side a in the x-y plane, centered at the origin. Also compute $\int (\vec{\nabla} \times \vec{u}) \cdot \hat{n} \, dA$ over the surface of the square, and show that the result is equal to the circulation.

We compute the circulation by calculating the line integral around each of the four sides of the square, starting at $x = a/2$. We have

$$I_1 = \int_{-a/2}^{a/2} \left(\frac{a}{2} \right)^2 y^3 \, dy = \left(\frac{a}{2} \right)^2 \frac{y^4}{4} \bigg|_{-a/2}^{a/2} = 0$$

We obtain the same result for the side at $x = -a/2$. At $y = a/2$,

$$I_2 = \int_{a/2}^{-a/2} \left(\frac{a}{2} \right)^3 x^2 \, dx = \left(\frac{a}{2} \right)^3 \frac{x^3}{3} \bigg|_{a/2}^{-a/2} = -\left(\frac{a}{2} \right)^3 \frac{2}{3} \left(\frac{a}{2} \right)^3 = -\frac{2}{3} \left(\frac{a}{2} \right)^6$$

while at $y = -a/2$,

$$I_4 = \int_{-a/2}^{a/2} \left(-\frac{a}{2} \right)^3 x^2 \, dx = -\left(\frac{a}{2} \right)^3 \frac{x^3}{3} \bigg|_{-a/2}^{a/2} = -\left(\frac{a}{2} \right)^3 \frac{2}{3} \left(\frac{a}{2} \right)^3 = -\frac{2}{3} \left(\frac{a}{2} \right)^6$$

Thus, the circulation is

$$\oint \vec{u} \cdot d\vec{l} = 0 - \frac{2}{3} \left(\frac{a}{2} \right)^6 + 0 - \frac{2}{3} \left(\frac{a}{2} \right)^6 = -\frac{4}{3} \left(\frac{a}{2} \right)^6$$

Now we calculate the curl. The normal to the square is in the $\hat{z}$ direction, so we only need the z-component:

$$(\vec{\nabla} \times \vec{u})_z = \frac{\partial u_y}{\partial x} - \frac{\partial u_x}{\partial y} = 2xy^3 - 3x^2 y^2$$

Then the surface integral is

$$\int \left(\vec{\nabla} \times \vec{u} \right) \cdot \hat{n} \, dA = \int_{-a/2}^{a/2} \int_{-a/2}^{a/2} \left(2xy^3 - 3x^2 y^2 \right) dy \, dx$$

$$= \int_{-a/2}^{a/2} \left(\frac{xy^4}{2} - x^2 y^3 \right) \Bigg|_{-a/2}^{a/2} dx$$

$$= 2 \int_{-a/2}^{a/2} \left(-x^2 \left(\frac{a}{2} \right)^3 \right) dx$$

$$= -2 \left(\frac{a}{2} \right)^3 \frac{x^3}{3} \Bigg|_{-a/2}^{a/2} = -\frac{4}{3} \left(\frac{a}{2} \right)^6$$

as we obtained from the circulation.

Now if curl $\vec{u} = 0$ everywhere on S, then $\oint_C \vec{u} \cdot d\vec{l} = 0$ for any curve C lying in S, and thus[13] we can express $\vec{u} = \vec{\nabla}\phi$ on S. Thus,

$$\vec{\nabla} \times \vec{u} = 0 \Rightarrow \vec{u} = \vec{\nabla}\phi$$

Similarly, it is easy to show[14] from the expression for curl that

$$\vec{u} = \vec{\nabla}\phi \Rightarrow \vec{\nabla} \times \vec{u} = 0$$

Notice that the divergence theorem and Stokes' theorem relate a locally defined property of the vector field (divergence or curl) to a globally defined property (flux or circulation). "Local" means that the property is defined at each point of the space, while "global" means that the property is defined over a region of the space (on a surface for the flux or around a complete curve for the circulation).

Numerous variants of these theorems exist, and they are all proved in a similar fashion. For example,[15]

$$\int_V \vec{\nabla}\Phi \, dV = \oint_S \Phi \hat{n} \, dA$$

1.2.4. Repeated Operations with $\vec{\nabla}$

We can define additional properties of vector fields using second derivatives. For example, we have already observed that

$$\text{curl (grad } \Phi) = \vec{\nabla} \times (\vec{\nabla}\Phi) \equiv 0 \tag{1.48}$$

---

[13] See Section 1.2.3.
[14] See Problem 19.
[15] See Problem 30. Problem 31 gives an additional variant.

Similarly, we can show that

$$\text{div (curl } \vec{\mathbf{u}}) = \vec{\nabla} \cdot (\vec{\nabla} \times \vec{\mathbf{u}}) \equiv 0 \tag{1.49}$$

for any vector field $\vec{\mathbf{u}}$. Let's evaluate the expression in Cartesian coordinates:

$$\frac{\partial}{\partial x}\left(\frac{\partial u_z}{\partial y} - \frac{\partial u_y}{\partial z}\right) + \frac{\partial}{\partial y}\left(\frac{\partial u_x}{\partial z} - \frac{\partial u_z}{\partial x}\right) + \frac{\partial}{\partial z}\left(\frac{\partial u_y}{\partial x} - \frac{\partial u_x}{\partial y}\right)$$

$$= \frac{\partial^2 u_z}{\partial x \, \partial y} - \frac{\partial^2 u_z}{\partial y \, \partial x} - \frac{\partial^2 u_y}{\partial x \, \partial z} + \frac{\partial^2 u_y}{\partial z \, \partial x} + \frac{\partial^2 u_x}{\partial y \, \partial z} - \frac{\partial^2 u_x}{\partial z \, \partial y} = 0$$

Since the order in which we take the partial derivatives is irrelevant, the terms cancel in pairs.

The second derivatives (1.48) and (1.49) are identically zero, but many others are not. The combination

$$\text{div (grad } \Phi) = \vec{\nabla} \cdot (\vec{\nabla} \Phi) = \nabla^2 \Phi = \frac{\partial^2 \Phi}{\partial x^2} + \frac{\partial^2 \Phi}{\partial y^2} + \frac{\partial^2 \Phi}{\partial z^2} \tag{1.50}$$

appears frequently in physics, as does the combination

$$\text{curl (curl } \vec{\mathbf{u}}) = \vec{\nabla} \times (\vec{\nabla} \times \vec{\mathbf{u}})$$

We can evaluate this product using the bac-cab rule, being careful that each $\vec{\nabla}$ operates on everything to its right:

$$\vec{\nabla} \times (\vec{\nabla} \times \vec{\mathbf{u}}) = \vec{\nabla}(\vec{\nabla} \cdot \vec{\mathbf{u}}) - \nabla^2 \vec{\mathbf{u}} \tag{1.51}$$

This expression effectively *defines* the quantity $\nabla^2 \vec{\mathbf{u}}$.

In Cartesian coordinates, the scalar operator ∇^2, the *Laplacian* operator, is given by

$$\nabla^2 = \frac{\partial^2}{\partial x^2} + \frac{\partial^2}{\partial y^2} + \frac{\partial^2}{\partial z^2}$$

and the same expression holds whether ∇^2 operates on a scalar or a vector. The same is *not* true in other coordinate systems, as we shall see below.

1.3. CURVILINEAR COORDINATES

When solving a physics problem, we always want to choose a coordinate system that is well suited to the geometry of the problem. Thus, when studying the orbit of a satellite about the Earth, we would probably choose spherical coordinates with the origin at the center of the Earth. When studying the magnetic effects of a long current-carrying wire, we would choose cylindrical coordinates with the z-axis along the wire. And for studying the electric field due to a conducting disk, the best coordinates are spheroidal coordinates.[16]

1.3.1. Unit Vectors

In general, we may call the three curvilinear coordinates u, v, and w. The unit vector $\hat{\mathbf{u}}$ at any point indicates the direction in which u increases while v and w are held fixed. The vectors $\hat{\mathbf{v}}$ and $\hat{\mathbf{w}}$ are defined similarly. We almost always use an *orthogonal coordinate system*, in which $\hat{\mathbf{u}}$, $\hat{\mathbf{v}}$, and $\hat{\mathbf{w}}$ form a right-handed orthogonal set—that is, the triple scalar product

$$[\hat{\mathbf{u}}, \hat{\mathbf{v}}, \hat{\mathbf{w}}] = +1$$

We can relate these vectors to the Cartesian unit vectors $\hat{\mathbf{x}}$, $\hat{\mathbf{y}}$, and $\hat{\mathbf{z}}$ geometrically, using diagrams such as Figure 1.16, or we can find their components analytically. To see how, first write the new coordinates u, v, and w in terms of x, y and z and, conversely, express x, y, and z as functions of u, v, and w: $x = x(u, v, w)$, $y = y(u, v, w)$, and $z = z(u, v, w)$. Let point P be described by the position vector $\vec{\mathbf{r}}$ and have coordinates u_0, v_0, and w_0. A neighboring point Q has coordinates $u_0 + du$, v_0, and w_0. Then the vector from P to Q is in the direction of $\hat{\mathbf{u}}$:

$$d\vec{\mathbf{r}} = dx\,\hat{\mathbf{x}} + dy\,\hat{\mathbf{y}} + dz\,\hat{\mathbf{z}} \propto du\,\hat{\mathbf{u}} \tag{1.52}$$

and thus

$$\hat{\mathbf{u}} \propto \frac{\partial x}{\partial u}\hat{\mathbf{x}} + \frac{\partial y}{\partial u}\hat{\mathbf{y}} + \frac{\partial z}{\partial u}\hat{\mathbf{z}}$$

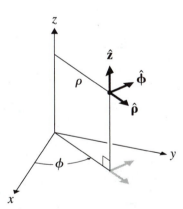

FIGURE 1.16. Unit vectors in cylindrical coordinates. As the coordinate ϕ changes, so do the directions of $\hat{\boldsymbol{\rho}}$ and $\hat{\boldsymbol{\phi}}$. We may express these vectors in Cartesian components: $\hat{\boldsymbol{\rho}} = \hat{\mathbf{x}}\cos\phi + \hat{\mathbf{y}}\sin\phi$, $\hat{\boldsymbol{\phi}} = -\hat{\mathbf{x}}\sin\phi + \hat{\mathbf{y}}\cos\phi$. The third unit vector in this system is $\hat{\mathbf{z}}$, the same as in Cartesian coordinates. Note that the three vectors are mutually perpendicular.

---

[16]See Chapter 2, Problem 13, for a definition of these coordinates.

Since $\hat{\mathbf{u}}$ is a unit vector, the x-, y-, and z-components of $\hat{\mathbf{u}}$ are

$$u_x = \frac{\partial x/\partial u}{\sqrt{(\partial x/\partial u)^2 + (\partial y/\partial u)^2 + (\partial z/\partial u)^2}} = \frac{1}{h_1}\frac{\partial x}{\partial u} \qquad (1.53)$$

$$u_y = \frac{\partial y/\partial u}{\sqrt{(\partial x/\partial u)^2 + (\partial y/\partial u)^2 + (\partial z/\partial u)^2}} = \frac{1}{h_1}\frac{\partial y}{\partial u} \qquad (1.54)$$

$$u_z = \frac{\partial z/\partial u}{\sqrt{(\partial x/\partial u)^2 + (\partial y/\partial u)^2 + (\partial z/\partial u)^2}} = \frac{1}{h_1}\frac{\partial z}{\partial u} \qquad (1.55)$$

where the constant of proportionality in equation (1.52) is

$$h_1 = \sqrt{\left(\frac{\partial x}{\partial u}\right)^2 + \left(\frac{\partial y}{\partial u}\right)^2 + \left(\frac{\partial z}{\partial u}\right)^2} \qquad (1.56)$$

Similarly, we can find $\hat{\mathbf{v}}$ and $\hat{\mathbf{w}}$:

$$\hat{\mathbf{v}} = \frac{1}{h_2}\left(\frac{\partial x}{\partial v}, \frac{\partial y}{\partial v}, \frac{\partial z}{\partial v}\right) \qquad (1.57)$$

where

$$h_2 = \sqrt{\left(\frac{\partial x}{\partial v}\right)^2 + \left(\frac{\partial y}{\partial v}\right)^2 + \left(\frac{\partial z}{\partial v}\right)^2} \qquad (1.58)$$

and

$$\hat{\mathbf{w}} = \frac{1}{h_3}\left(\frac{\partial x}{\partial w}, \frac{\partial y}{\partial w}, \frac{\partial z}{\partial w}\right) \qquad (1.59)$$

where

$$h_3 = \sqrt{\left(\frac{\partial x}{\partial w}\right)^2 + \left(\frac{\partial y}{\partial w}\right)^2 + \left(\frac{\partial z}{\partial w}\right)^2} \qquad (1.60)$$

In a general non-Cartesian coordinate system, the directions of the unit vectors change with position. This means that we have to be very careful when taking derivatives or integrals of vectors expressed in curvilinear components.

1.3.2. Metric Coefficients

The distance between two neighboring points described by the position vectors $\vec{r}$ and $\vec{r}+d\vec{r}$ may be expressed in terms of differential changes in the coordinates u, v, and w. In general, we may write

$$d\vec{r} = h_1 \, du \, \hat{u} + h_2 \, dv \, \hat{v} + h_3 \, dw \, \hat{w}$$

and so for an orthogonal system

$$ds^2 = d\vec{r} \cdot d\vec{r} = h_1^2 \, du^2 + h_2^2 \, dv^2 + h_3^2 \, dw^2 \qquad (1.61)$$

(compare with equations 1.1, 1.4, and 1.7). The coefficients h_i are called metric coefficients. They relate the length of a geometrical line segment to the differential coordinate element and can usually be obtained from geometry (for example, Figure 1.2b), or they can be calculated using equations (1.56), (1.58), and (1.60). We can see that such coefficients are necessary on dimensional grounds. For example, in cylindrical coordinates, the coordinate ϕ is an angle, and so the differential $d\phi$ is just a number and does not have the dimension of length. The corresponding $h_2 = \rho$ (compare equations 1.4 and 1.61) does have the required dimension of length.

Let's check the expression for h_2 using equation (1.58). First, from equations (1.2) we have

$$\frac{\partial x}{\partial \phi} = -\rho \sin \phi; \qquad \frac{\partial y}{\partial \phi} = \rho \cos \phi$$

and z is independent of ϕ, so

$$h_2 = \sqrt{\left(\frac{\partial x}{\partial \phi}\right)^2 + \left(\frac{\partial y}{\partial \phi}\right)^2 + \left(\frac{\partial z}{\partial \phi}\right)^2}$$

$$= \sqrt{(-\rho \sin \phi)^2 + (\rho \cos \phi)^2 + 0} = \rho$$

as obtained from the geometry in Section 1.1.1.

1.3.3. Div, Grad, and Curl in Curvilinear Coordinate Systems

Gradient

The differential change in a scalar field Φ over a displacement $d\vec{r}$ is (Section 1.2.1, equation 1.37)

$$d\Phi = \vec{\nabla}\Phi \cdot d\vec{r} = \vec{\nabla}\Phi \cdot (h_1 \, du \, \hat{u} + h_2 \, dv \, \hat{v} + h_3 \, dw \, \hat{w})$$

But we can also derive the differential from the fact that Φ is a function of u, v, and w, using the rules of calculus:

$$d\Phi = \frac{\partial \Phi}{\partial u} du + \frac{\partial \Phi}{\partial v} dv + \frac{\partial \Phi}{\partial w} dw$$

Comparing these two expressions, we find that the gradient is

$$\vec{\nabla} \Phi = \frac{1}{h_1} \frac{\partial \Phi}{\partial u} \hat{u} + \frac{1}{h_2} \frac{\partial \Phi}{\partial v} \hat{v} + \frac{1}{h_3} \frac{\partial \Phi}{\partial w} \hat{w} \qquad (1.62)$$

Again we see that the hs are needed on dimensional grounds.

Divergence

In curvilinear coordinates, the divergence is *defined* by the divergence theorem (1.44). Thus, the divergence of a vector field $\vec{f}$ at a point P is

$$\vec{\nabla} \cdot \vec{f} = \lim_{V \to 0} \frac{1}{V} \oint_S \vec{f} \cdot \hat{n} \, dA$$

where the point P is inside the volume V. We begin by evaluating the integral on the right-hand side. We can choose our volume V to be a curved "cube" with sides of length du, dv, and dw and each face lying along a coordinate surface (Figure 1.17a). Then the normals to the faces are $\pm\hat{u}$, $\pm\hat{v}$, and $\pm\hat{w}$. From the pair of faces at u, $u + du$ we get a contribution

$$-f_u(u, v, w) \, h_2 h_3 \, dv \, dw + f_u(u + du, v, w) \, h_2 h_3 \, dv \, dw$$

where f_u is the u-component of the vector $\vec{f}$.

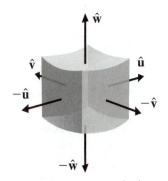

FIGURE 1.17a. A differential volume used to calculate $\vec{\nabla} \cdot \vec{f}$ in coordinates u, v, and w. The faces of the volume lie along surfaces of constant u, v, and w, respectively, and the edges have lengths $h_1 \, du$, $h_2 \, dv$, and $h_3 \, dw$.

We get similar contributions from the other two pairs of faces. Dividing by the volume $h_1 h_2 h_3 \, du \, dv \, dw$ and using the definition of each partial derivative, we have

$$\vec{\nabla} \cdot \vec{f} = \lim_{du \to 0} \frac{[-f_u(u, v, w) \, h_2 h_3 + f_u(u + du, v, w) \, h_2 h_3]}{h_1 h_2 h_3 \, du} + \text{ two similar terms}$$

$$\vec{\nabla} \cdot \vec{f} = \frac{1}{h_1 h_2 h_3} \left[\frac{\partial(h_2 h_3 f_u)}{\partial u} + \frac{\partial(h_1 h_3 f_v)}{\partial v} + \frac{\partial(h_1 h_2 f_w)}{\partial w} \right] \qquad (1.63)$$

Notice that the hs appear inside the derivatives, since they may be functions of the coordinates u, v, and w and so differ on opposite faces of the "cube."

In cylindrical coordinates (Figure 1.17b), with $h_1 = 1$, $h_2 = \rho$, and $h_3 = 1$, we obtain

$$\vec{\nabla} \cdot \vec{f} = \frac{1}{\rho} \left[\frac{\partial(\rho f_\rho)}{\partial \rho} + \frac{\partial f_\phi}{\partial \phi} + \frac{\partial(\rho f_z)}{\partial z} \right] = \frac{1}{\rho} \frac{\partial(\rho f_\rho)}{\partial \rho} + \frac{1}{\rho} \frac{\partial f_\phi}{\partial \phi} + \frac{\partial f_z}{\partial z} \qquad (1.64)$$

Notice that the result is dimensionally correct.

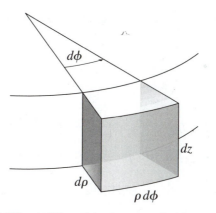

FIGURE 1.17b. A differential volume in cylindrical coordinates.

Example 1.4. Compute the divergence of the position vector, $\vec{\nabla} \cdot \vec{r}$, in cylindrical coordinates, and compare with the result in Cartesian coordinates (Example 1.2).

In cylindrical coordinates, the position vector[17] is $\vec{r} = \rho \hat{\rho} + z \hat{z}$. Thus, from equation (1.64), the divergence is

$$\vec{\nabla} \cdot \vec{r} = \frac{1}{\rho} \frac{\partial(\rho r_\rho)}{\partial \rho} + \frac{1}{\rho} \frac{\partial r_\phi}{\partial \phi} + \frac{\partial r_z}{\partial z} = \frac{1}{\rho} \frac{\partial(\rho^2)}{\partial \rho} + \frac{\partial z}{\partial z} = \frac{1}{\rho}(2\rho) + 1 = 3$$

This is the same result that we obtained in Example 1.2.

[17]There is no ϕ-component. Do you see why? See Problem 10.

Curl

In curvilinear coordinates, the curl of a vector $\vec{f}$ at a point P is defined via Stokes' theorem (1.46):

$$(\vec{\nabla} \times \vec{f}) \cdot \hat{n} = \lim_{A \to 0} \frac{\oint_C \vec{f} \cdot d\vec{l}}{A}$$

where A is the area of a surface S spanning the curve C, the point P lies on S, and $\hat{n}$ is the normal to that surface.

Again we begin by evaluating the integral. We may choose a curve that lies in a constant u surface (Figure 1.18). Then the normal is $\hat{n} = \hat{u}$, and we will obtain the u-component of the curl.

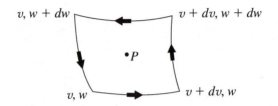

$v, w + dw$ $v + dv, w + dw$

$\bullet P$

v, w $v + dv, w$

FIGURE 1.18. Calculating the curl using a "rectangle" surrounding point P.

The line integral is

$$f_v h_2 \, dv|_w + f_w h_3 \, dw|_{v+dv} - f_v h_2 \, dv|_{w+dw} - f_w h_3 \, dw|_v$$

$$= -\frac{\partial}{\partial w}(f_v h_2) \, dv \, dw + \frac{\partial}{\partial v}(f_w h_3) \, dv \, dw$$

while the area is $dA = h_2 h_3 \, dv \, dw$, and so the u-component of the curl is

$$(\vec{\nabla} \times \vec{f}) \cdot \hat{u} = \frac{1}{h_2 h_3} \left[\frac{\partial}{\partial v}(f_w h_3) - \frac{\partial}{\partial w}(f_v h_2) \right]$$

The other two components may be derived similarly:

$$\vec{\nabla} \times \vec{f} = \frac{1}{h_2 h_3} \left[\frac{\partial}{\partial v}(f_w h_3) - \frac{\partial}{\partial w}(f_v h_2) \right] \hat{u} + \frac{1}{h_1 h_3} \left[\frac{\partial}{\partial w}(f_u h_1) - \frac{\partial}{\partial u}(f_w h_3) \right] \hat{v}$$

$$+ \frac{1}{h_1 h_2} \left[\frac{\partial}{\partial u}(f_v h_2) - \frac{\partial}{\partial v}(f_u h_1) \right] \hat{w} \tag{1.65}$$

We may also express this result using a determinant:

$$\vec{\nabla} \times \vec{f} = \frac{1}{h_1 h_2 h_3} \begin{vmatrix} h_1 \hat{u} & h_2 \hat{v} & h_3 \hat{w} \\ \dfrac{\partial}{\partial u} & \dfrac{\partial}{\partial v} & \dfrac{\partial}{\partial w} \\ h_1 f_u & h_2 f_v & h_3 f_w \end{vmatrix} \tag{1.66}$$

When using this expression, take care to keep the terms in the correct order. The operators in the second row operate on the functions in the bottom row, including the metric coefficients h_i.

In cylindrical coordinates, we obtain

$$\vec{\nabla} \times \vec{f} = \frac{1}{\rho} \left[\frac{\partial f_z}{\partial \phi} - \frac{\partial}{\partial z}(\rho f_\phi) \right] \hat{\rho} + \left[\frac{\partial f_\rho}{\partial z} - \frac{\partial f_z}{\partial \rho} \right] \hat{\phi} + \frac{1}{\rho} \left[\frac{\partial}{\partial \rho}(\rho f_\phi) - \frac{\partial f_\rho}{\partial \phi} \right] \hat{z}$$

$$= \left(\frac{1}{\rho} \frac{\partial f_z}{\partial \phi} - \frac{\partial f_\phi}{\partial z} \right) \hat{\rho} + \left(\frac{\partial f_\rho}{\partial z} - \frac{\partial f_z}{\partial \rho} \right) \hat{\phi} + \frac{1}{\rho} \left[\frac{\partial}{\partial \rho}(\rho f_\phi) - \frac{\partial f_\rho}{\partial \phi} \right] \hat{z} \tag{1.67}$$

Example 1.5. The magnetic field in a region of space is given by the function $\vec{B} = B_0(\rho/a^3)(a - \rho)^2 \hat{\phi}, 0 \le \rho \le a$, and $\vec{B} = 0$ for $\rho > a$. Find the current density in the region.

From Maxwell's equations,

$$\vec{j} = \frac{1}{\mu_0} \vec{\nabla} \times \vec{B}$$

Since $\vec{B}$ has only a ϕ-component, we have

$$\vec{\nabla} \times \vec{B} = \left(-\frac{\partial B_\phi}{\partial z} \right) \hat{\rho} + (0)\hat{\phi} + \frac{1}{\rho} \left[\frac{\partial}{\partial \rho}(\rho B_\phi) \right] \hat{z}$$

$$= B_0 \frac{1}{\rho a^3} \left[\frac{\partial}{\partial \rho} \left(\rho^2 (a - \rho)^2 \right) \right] \hat{z}$$

$$\vec{j} = \frac{B_0}{\mu_0} \frac{1}{\rho a^3} \left[2\rho(a - \rho)^2 - 2\rho^2 (a - \rho) \right] \hat{z}$$

$$= \frac{2B_0}{\mu_0 a^3} (a - \rho)(a - 2\rho)\hat{z}$$

Notice that the curl is in the positive z-direction for $\rho < a/2$ but is in the negative z-direction for $a/2 < \rho < a$. A plot of the vector field (Figure 1.19) helps us to understand this result. While all the vectors point counterclockwise around the origin,

$\oint_{\mathcal{C}} \vec{\mathbf{B}} \cdot d\vec{\mathbf{l}}$ is positive when $d\vec{\mathbf{l}}$ runs *clockwise* around a small curve $\mathcal{C}$ at $a/2 < \rho < a$ because the vectors get shorter as ρ increases in this region.

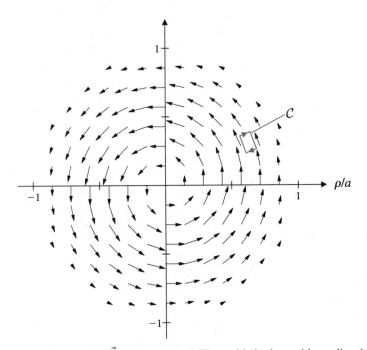

FIGURE 1.19. The vector field $\vec{\mathbf{B}}$ in Example 1.5. The curl is in the positive z-direction near the origin but is in the negative z-direction for $\rho > a/2$. Note that the integral $\oint_{\mathcal{C}} \vec{\mathbf{B}} \cdot d\vec{\mathbf{l}}$ is positive when $d\vec{\mathbf{l}}$ is taken *clockwise* around a small closed curve $\mathcal{C}$ placed at $\rho > a/2$, since the length of the vectors $\vec{\mathbf{B}}$ is decreasing as we go away from the origin in this region.

The Laplacian

With Φ a scalar function, the expression[18] for $\nabla^2 \Phi$ may be derived from the above expressions for gradient and divergence:

$$\nabla^2 \Phi = \vec{\nabla} \cdot (\vec{\nabla}\Phi) = \vec{\nabla} \cdot \left(\frac{1}{h_1} \frac{\partial \Phi}{\partial u} \hat{\mathbf{u}} + \frac{1}{h_2} \frac{\partial \Phi}{\partial v} \hat{\mathbf{v}} + \frac{1}{h_3} \frac{\partial \Phi}{\partial w} \hat{\mathbf{w}} \right)$$

$$\nabla^2 \Phi = \frac{1}{h_1 h_2 h_3} \left[\frac{\partial}{\partial u} \left(\frac{h_2 h_3}{h_1} \frac{\partial \Phi}{\partial u} \right) + \frac{\partial}{\partial v} \left(\frac{h_1 h_3}{h_2} \frac{\partial \Phi}{\partial v} \right) + \frac{\partial}{\partial w} \left(\frac{h_1 h_2}{h_3} \frac{\partial \Phi}{\partial w} \right) \right] \tag{1.68}$$

[18]To evaluate $\nabla^2 \vec{\mathbf{u}}$, with $\vec{\mathbf{u}}$ a vector function, you must use equation (1.51) at the end of Section 1.2.4.

In cylindrical coordinates, this is

$$\nabla^2\Phi = \frac{1}{\rho}\left[\frac{\partial}{\partial\rho}\left(\rho\frac{\partial\Phi}{\partial\rho}\right) + \frac{\partial}{\partial\phi}\left(\frac{1}{\rho}\frac{\partial\Phi}{\partial\phi}\right) + \frac{\partial}{\partial z}\left(\rho\frac{\partial\Phi}{\partial z}\right)\right]$$

$$= \frac{1}{\rho}\frac{\partial}{\partial\rho}\left(\rho\frac{\partial\Phi}{\partial\rho}\right) + \frac{1}{\rho^2}\frac{\partial^2\Phi}{\partial\phi^2} + \frac{\partial^2\Phi}{\partial z^2} \tag{1.69}$$

This operator occurs frequently in physics, and we shall need to refer to result (1.68).

1.4. THE HELMHOLTZ THEOREM

Any vector field $\vec{\mathbf{F}}$ may be expressed as the sum of two terms:

$$\vec{\mathbf{F}} = \vec{\mathbf{u}} + \vec{\mathbf{v}}$$

where

$$\vec{\mathbf{u}} = \vec{\nabla}\Phi$$

for some scalar field Φ and

$$\vec{\mathbf{v}} = \vec{\nabla}\times\vec{\mathbf{A}}$$

for some vector field $\vec{\mathbf{A}}$. The proof of this theorem[19] may be found in Appendix II. Then

$$\vec{\nabla}\cdot\vec{\mathbf{F}} = \vec{\nabla}\cdot\vec{\mathbf{u}} = \nabla^2\Phi$$

and

$$\vec{\nabla}\times\vec{\mathbf{F}} = \vec{\nabla}\times(\vec{\nabla}\times\vec{\mathbf{A}})$$

In particular, if $\vec{\nabla}\cdot\vec{\mathbf{F}} \equiv 0$, then we may choose $\Phi = $ constant and express $\vec{\mathbf{F}} = \vec{\nabla}\times\vec{\mathbf{A}}$, and if $\vec{\nabla}\times\vec{\mathbf{F}} \equiv 0$, then we may choose $\vec{\mathbf{A}} = 0$ and express $\vec{\mathbf{F}} = \vec{\nabla}\Phi$, as we showed in Section 1.2.3.

1.5. VECTOR SPACES

1.5.1. Properties of Vector Spaces

In physics we regard vectors both as geometrical objects (*arrows*) and also as algebraic objects (an ordered set of numbers: the components of the vector). It is important to have both descriptions, since each description proves most useful in certain classes of problems. Here we shall discuss aspects of the algebraic theory of vector spaces.

---

[19]See also Morse and Feshbach, *Methods of Theoretical Physics*, Section 1.5.

The number of components of a vector equals the dimension of the vector space. In most applications in this book, we shall be concerned with three-dimensional vector spaces, corresponding to the three dimensions of physical space. Infinite-dimensional vector spaces are also of importance—in quantum mechanics, for example—but we shall not discuss these spaces here. We shall also restrict attention, for the most part, to spaces in which the vectors are purely real.[20]

A vector space consists of a set of objects (the vectors) with the following properties[21]:

1. The sum of two vectors is another vector in the space:

$$\vec{a} + \vec{b} = \vec{c}$$

2. Addition is commutative:

$$\vec{a} + \vec{b} = \vec{b} + \vec{a}$$

3. Addition is associative:

$$(\vec{a} + \vec{b}) + \vec{c} = \vec{a} + (\vec{b} + \vec{c})$$

4. There is a zero vector, $\vec{0}$, which has the property that

$$\vec{a} + \vec{0} = \vec{a} = \vec{0} + \vec{a}$$

5. Each vector $\vec{a}$ has a corresponding negative vector $-\vec{a}$ with the property that

$$\vec{a} + (-\vec{a}) = \vec{0}$$

6. Vectors can be multiplied by scalars. The result is another vector:

$$c\vec{a} = \vec{b}$$

   This product is also associative:

$$k(c\vec{a}) = (kc)\vec{a}$$

7. Multiplication by scalars satisfies two distributive laws:

$$(k + c)\vec{a} = k\vec{a} + c\vec{a}$$

   and

$$c(\vec{a} + \vec{b}) = c\vec{a} + c\vec{b}$$

8. Multiplication by the scalar unity leaves a vector unchanged:

$$1\vec{a} = \vec{a}$$

---

[20] See Optional Topic B for additional material.
[21] The first five properties establish the vectors as elements of a group; see Optional Topic C.

1.5.2. Basis Vectors

Properties (1) and (6) together imply that the sum

$$\sum_{i=1}^{n} c_i \vec{v}_i$$

is a vector. In particular, we can make this sum equal to the zero vector by appropriate choice of the numbers c_i. If the only possible choice of values that makes the sum equal to zero is $c_i = 0$ for every number c_i, then the set of vectors $\vec{v}_i$, $i = 1$ to n, is said to be *linearly independent*. But if there is a set of numbers, at least two of which are not zero, such that

$$\sum_{i=1}^{n} c_i \vec{v}_i = \vec{0}$$

then the set of vectors $\vec{v}_i$ is linearly dependent.

In each vector space, there is some number N with the following property: there will be some vectors $\vec{v}_i$ such that $\sum_{i=1}^{N} c_i \vec{v}_i = 0$ only if each $c_i = 0$, but if we add one more vector, any vector, then we can find a set of constants such that $\sum_{i=1}^{N+1} c_i \vec{v}_i = 0$ with more than one of the c_i nonzero. The number N is the dimension of the vector space. The original N vectors are linearly independent and are said to *span* the space. They may be chosen as basis vectors $\vec{e}_i$ for the space. Then any vector in the space may be written as a linear combination of the basis vectors:

$$\vec{v} = \sum_{i=1}^{N} v_i \vec{e}_i$$

The numbers v_i are the components of the vector $\vec{v}$. Thus, we may write the vector algebraically as an ordered set of numbers, the components v_i.

In physics applications, we define an additional operation called the *inner product*[22] of two vectors with respect to the basis[23]:

$$\vec{a} * \vec{b} = \sum_{i=1}^{N} a_i b_i = \vec{b} * \vec{a} \qquad (1.70)$$

If the resulting value is zero, the two vectors are said to be orthogonal. The inner product of two basis vectors is

$$\vec{e}_k * \vec{e}_j = \sum_{i=1}^{N} \delta_{ik} \delta_{ij} = \delta_{jk}$$

and so the basis vectors are orthogonal by definition.

---

[22]An inner product may be defined in many different ways. This version is most useful for our purposes here.

[23]The inner product becomes a more important and useful concept when it is independent of the basis chosen. This proves to be the case when the operation that changes the basis is orthogonal. See Section 1.6.5 and Problem 50.

It is most convenient to have the inner product defined by equation (1.70) correspond to the dot product defined in Section 1.1.2 (equation 1.26). We can do this if we choose a set of basis vectors that are mutually orthogonal in the geometrical sense of being at right angles. From any initial choice of basis vectors, we can construct a set such that the inner product with respect to this basis is the geometrical dot product. Suppose we have one set of basis vectors $\vec{e}_i$ that are not mutually orthogonal; that is,

$$\vec{e}_i \cdot \vec{e}_j \neq 0$$

for some $i \neq j$. Then we can use a procedure called Gram-Schmidt orthogonalization to create an orthogonal set $\hat{f}_i$. First we normalize each vector by dividing by the square root of its dot product with itself:

$$\hat{e}_i = \frac{\vec{e}_i}{\sqrt{\vec{e}_i \cdot \vec{e}_i}}$$

Then

$$\hat{e}_i \cdot \hat{e}_i = 1$$

These vectors are called unit vectors. Now we start with the vector $\hat{f}_1 = \hat{e}_1$ and define

$$\vec{f}_2 = \hat{e}_2 - (\hat{e}_2 \cdot \hat{f}_1)\hat{e}_1$$

so that

$$\vec{f}_2 \cdot \hat{f}_1 = (\hat{e}_2 \cdot \hat{e}_1) - (\hat{e}_2 \cdot \hat{e}_1)(\hat{e}_1 \cdot \hat{e}_1) = 0$$

and

$$\vec{f}_2 \cdot \vec{f}_2 = [\hat{e}_2 - (\hat{e}_2 \cdot \hat{e}_1)\hat{e}_1] \cdot [\hat{e}_2 - (\hat{e}_2 \cdot \hat{e}_1)\hat{e}_1]$$
$$= \hat{e}_2 \cdot \hat{e}_2 - (\hat{e}_2 \cdot \hat{e}_1)^2 = 1 - (\hat{e}_2 \cdot \hat{e}_1)^2$$

Then we normalize to get

$$\hat{f}_2 = \frac{\vec{f}_2}{\sqrt{\vec{f}_2 \cdot \vec{f}_2}}$$

We may continue in this manner to construct each succeeding $\hat{f}_i$:

$$\vec{f}_i = \hat{e}_i - \sum_{k=1}^{i-1}(\hat{f}_k \cdot \hat{e}_i)\hat{f}_k; \quad \hat{f}_i = \frac{\vec{f}_i}{\sqrt{\vec{f}_i \cdot \vec{f}_i}}$$

The vectors $\hat{f}_i$ are both unit vectors and mutually orthogonal; they are called an *orthonormal* set.

With orthonormal basis vectors, the inner product defined by process (1.70) is identical to the geometrical dot product defined in Section 1.1.2. In what follows, we shall assume that the basis vectors are orthonormal, and we shall not distinguish between the dot product and the inner product.

In representing a vector algebraically by its components, we write the set of numbers as an array, either in a column such as

$$\vec{a} = \begin{pmatrix} a_1 \\ a_2 \\ a_3 \end{pmatrix}$$

called a column vector, or in a row such as

$$\vec{a} = (a_1, a_2, a_3)$$

called a row vector.[24] Clearly both represent the same vector.

We may now write properties (1)–(8) of vectors in terms of the vector components in an orthogonal basis. For example,

1. Two vectors are added by adding the corresponding components:

$$c_i = a_i + b_i$$

4. The components of the zero vector are all zeros.

5. The components of $\vec{b} = -\vec{a}$ are the negatives of the components of $\vec{a}$:

$$b_i = -a_i$$

6. The components of $\vec{b} = c\vec{a}$ are $b_i = ca_i$.

To work efficiently with vectors written in terms of components, we shall need to develop some additional mathematical tools—the mathematics of matrices.

1.6. MATRICES

1.6.1. Basic Properties

The list of properties that defines a vector space includes several procedures for creating one vector from another—for example, multiplication by a scalar. If the procedure is linear, each component of the new vector $\vec{v}$ is a linear combination of the components of the original vector[25] $\vec{u}$:

$$v_i = \sum_{k=1}^{N} a_{ik} u_k$$

---

[24] Strictly speaking, if the column vector is $\mathbf{a}$, the row vector is its transpose, $\mathbf{a}^T$. See Section 1.6.1.
[25] In this section we shall not use the summation convention introduced in Section 1.1.2. When summing over an index, we shall show the sum explicitly.

Thus, any linear operation that maps one vector in the space to another may be expressed as an array $\mathbb{A}$ of $N \times N$ numbers called a square matrix, whose elements are the set of numbers a_{ik}. The simple operation of multiplying the components of a vector by a number c is represented by the array

$$\mathbb{A} = \begin{pmatrix} c & 0 & 0 \\ 0 & c & 0 \\ 0 & 0 & c \end{pmatrix} = c\mathbb{I}$$

where $\mathbb{I}$ is the unit matrix

$$\mathbb{I} = \begin{pmatrix} 1 & 0 & 0 \\ 0 & 1 & 0 \\ 0 & 0 & 1 \end{pmatrix}$$

This example illustrates the rule:

> To multiply a matrix by a number c, multiply each element of the matrix by c.

The elements of the unit matrix may be represented by the Kronecker delta, δ_{ij}, whose components are zero for $i \neq j$ and 1 when $i = j$.

The usual convention[26] for writing matrices with indices, as in a_{ik}, is that the first index (here i) labels the row while the second (here k) represents the column. Thus, a_{21} is the element in the second row and the first column:

$$\begin{pmatrix} a_{11} & a_{12} & a_{13} \\ \boxed{a_{21}} & a_{22} & a_{23} \\ a_{31} & a_{32} & a_{33} \end{pmatrix}$$

An important example of a linear transformation is a rotation of the basis vectors (compare with Section 1.1.2). Other examples are the angular momentum of a rigid body, $\vec{L} = \mathbb{R}\vec{\omega}$, where the operator $\mathbb{R}$ is represented by a matrix containing the moments of inertia of the body and $\vec{\omega}$ is the angular velocity vector.

A symmetric matrix is invariant under reflection about the diagonal that runs from upper left to lower right; that is, $a_{ji} = a_{ij}$. An antisymmetric matrix has the property that $a_{ji} = -a_{ij}$. Thus, an antisymmetric matrix has zeros along the diagonal.

Two matrices are equal only if all of the corresponding elements are equal: $\mathbb{A} = \mathbb{B}$ if and only if $a_{ij} = b_{ij}$ for each i and j. The elements of the sum of two matrices are the sums of the corresponding elements: if $\mathbb{C} = \mathbb{A} + \mathbb{B}$, then $c_{ij} = a_{ij} + b_{ij}$.

When two transformations are applied one after the other, we can describe the result with matrix multiplication. If $\vec{v} = \mathbb{A}\vec{w}$ and $\vec{u} = \mathbb{B}\vec{v}$, then

$$\vec{u} = \mathbb{B}(\mathbb{A}\vec{w}) = (\mathbb{B}\mathbb{A})\vec{w}$$

The operations in a matrix product are performed in order, right to left. $\mathbb{B}\mathbb{A}$ means first do $\mathbb{A}$, then do $\mathbb{B}$.

[26]Any convention may be used, provided that you make it *really clear*.

Let's investigate how to perform this matrix multiplication. The components of the vector $\vec{v} = \mathbb{A}\vec{w}$ are

$$v_i = \sum_{j=1}^{N} a_{ij} w_j$$

Similarly,

$$u_k = \sum_{i=1}^{N} b_{ki} v_i = \sum_{i=1}^{N} \sum_{j=1}^{N} b_{ki} a_{ij} w_j = \sum_{j=1}^{N} \left(\sum_{i=1}^{N} b_{ki} a_{ij} \right) w_j = \sum_{j=1}^{N} c_{kj} w_j$$

Thus, the elements of the product $\mathbb{C} = \mathbb{B}\mathbb{A}$ are computed as

$$c_{kj} = \sum_{i=1}^{N} b_{ki} a_{ij}$$

The element c_{kj} is the inner product[27] of the row vector $b_{ki} = (b_{k1}, b_{k2}, b_{k3})$ and the column vector $a_{ij} = \begin{pmatrix} a_{1j} \\ a_{2j} \\ a_{3j} \end{pmatrix}$. This row and column intersect at the position labeled by the indices kj (Figure 1.20).

$$\begin{pmatrix} b_{11} & b_{12} & b_{13} \\ b_{21} & b_{22} & b_{23} \\ b_{31} & b_{32} & b_{33} \end{pmatrix} \begin{pmatrix} a_{11} & a_{12} & a_{13} \\ a_{21} & a_{22} & a_{23} \\ a_{31} & a_{32} & a_{33} \end{pmatrix} = \begin{pmatrix} c_{11} & c_{12} & c_{13} \\ c_{21} & c_{22} & \boxed{c_{23}} \\ c_{31} & c_{32} & c_{33} \end{pmatrix}$$

FIGURE 1.20. Matrix multiplication: $\mathbb{C} = \mathbb{B}\mathbb{A}$. Element c_{ij} is the dot product of the ith row of $\mathbb{B}$ with the jth column of $\mathbb{A}$. This row and column intersect at the position of element c_{ij}. In this diagram, $i = 2$ and $j = 3$.

Matrix multiplication is not, in general, commutative; that is, $\mathbb{B}\mathbb{A} \neq \mathbb{A}\mathbb{B}$.

Example 1.6. Let $\mathbb{A} = \begin{pmatrix} 1 & 2 & 3 \\ 1 & 0 & 1 \\ 0 & 1 & 2 \end{pmatrix}$ and $\mathbb{B} = \begin{pmatrix} 0 & 1 & 1 \\ 2 & 1 & 0 \\ 1 & 3 & 1 \end{pmatrix}$. Compute the products $\mathbb{A}\mathbb{B}$ and $\mathbb{B}\mathbb{A}$.

$$
\begin{aligned}
\mathbb{A}\mathbb{B} &= \begin{pmatrix} 1 & 2 & 3 \\ 1 & 0 & 1 \\ 0 & 1 & 2 \end{pmatrix} \begin{pmatrix} 0 & 1 & 1 \\ 2 & 1 & 0 \\ 1 & 3 & 1 \end{pmatrix} \\
&= \begin{pmatrix} 1 \times 0 + 2 \times 2 + 3 \times 1 & 1 \times 1 + 2 \times 1 + 3 \times 3 & 1 \times 1 + 2 \times 0 + 3 \times 1 \\ 1 \times 0 + 0 \times 2 + 1 \times 1 & 1 \times 1 + 0 \times 1 + 1 \times 3 & 1 \times 1 + 0 \times 0 + 1 \times 1 \\ 0 \times 0 + 1 \times 2 + 2 \times 1 & 0 \times 1 + 1 \times 1 + 2 \times 3 & 0 \times 1 + 1 \times 0 + 2 \times 1 \end{pmatrix} \\
&= \begin{pmatrix} 7 & 12 & 4 \\ 1 & 4 & 2 \\ 4 & 7 & 2 \end{pmatrix}
\end{aligned}
$$

---

[27] Clearly this operation makes sense only if the row and column have the same length. Matrices that do not meet this restriction cannot be multiplied.

On the other hand,

$$\mathbb{BA} = \begin{pmatrix} 0 & 1 & 1 \\ 2 & 1 & 0 \\ 1 & 3 & 1 \end{pmatrix} \begin{pmatrix} 1 & 2 & 3 \\ 1 & 0 & 1 \\ 0 & 1 & 2 \end{pmatrix} = \begin{pmatrix} 1 & 1 & 3 \\ 3 & 4 & 7 \\ 4 & 3 & 8 \end{pmatrix}$$

Clearly these products are not the same.

The *transpose* $\mathbb{A}^T$ of a matrix $\mathbb{A}$ with elements a_{ij} is formed by interchanging rows for columns. The components of the transpose are $\left(\mathbb{A}^T\right)_{ij} = a_{ji}$. In the event that the elements of a matrix are complex numbers, we can form the complex conjugate $\mathbb{A}^*$ of the matrix $\mathbb{A}$ by taking the complex conjugate[28] of each element. The matrix formed by performing both of these operations (complex conjugation and transposition) is called the hermitian conjugate or *adjoint matrix,* $\mathbb{A}^\dagger$. A matrix is described as *hermitian* if it equals its adjoint: $\mathbb{A} = \mathbb{A}^\dagger$.

The *trace* of a matrix is the sum of the elements on the diagonal:

$$\mathrm{Tr}\,(\mathbb{A}) = a_{11} + a_{22} + \cdots + a_{NN}$$

The trace is a linear operation: $\mathrm{Tr}\,(\mathbb{A} + \mathbb{B}) = \mathrm{Tr}\,(\mathbb{A}) + \mathrm{Tr}\,(\mathbb{B})$.

For a product, we find

$$\mathrm{Tr}\,(\mathbb{AB}) = \sum_i \sum_j a_{ij} b_{ji} = \sum_j \sum_i b_{ji} a_{ij} = \mathrm{Tr}\,(\mathbb{BA})$$

This result is true even when the matrix product is *not* commutative: $\mathbb{AB} \neq \mathbb{BA}$.

1.6.2. Determinants

A very important and useful property of a square matrix is its determinant. The determinant of a 2×2 matrix is a number formed by multiplying along the diagonals and subtracting:

$$\det \begin{pmatrix} a_{11} & a_{12} \\ a_{21} & a_{22} \end{pmatrix} = a_{11}a_{22} - a_{12}a_{21}$$

The products $a_{11}a_{22}$ and $a_{12}a_{21}$ are called elementary products. In an $N \times N$ determinant, each elementary product is a product of N elements, with no two being in the same row or the same column. We attach a sign to each product of the form $a_{1j}a_{2k}\cdots a_{Nm}$ according to whether the set $\{j, k, \ldots, m\}$ is an even $(+)$ or odd $(-)$ permutation[29] of $\{1, 2, \ldots, N\}$. The determinant of an $N \times N$ matrix is the sum of all the signed elementary products. For

---

[28] See Chapter 2, Section 2.1.

[29] To determine whether a permutation is even or odd, count the number of interchanges of two numbers that must be performed to execute the permutation. An even permutation needs an even number of interchanges, and an odd one needs an odd number.

a 3×3 determinant, this result may be written using the Levi-Civita symbol (Section 1.1.2, equation 1.29):

$$|\mathbb{A}| = \sum_{i,j,k=1}^{3} \varepsilon_{ijk} a_{1i} a_{2j} a_{3k} \qquad (1.71)$$

The determinant of a 3×3 matrix $\mathbb{A}$ may also be formed as follows. Take any row of the matrix—say, the top row. For each element a_{ij} in that row, take the determinant of the 2×2 matrix α_{ij} obtained by removing the row and column that contain the chosen element (Figure 1.21). Then the determinant is

$$\det(\mathbb{A}) = \begin{vmatrix} a_{11} & a_{12} & a_{13} \\ a_{21} & a_{22} & a_{23} \\ a_{31} & a_{32} & a_{33} \end{vmatrix} = \sum_{j} (-1)^{i+j} a_{ij} \det(\alpha_{ij})$$

$$= a_{11}(a_{22}a_{33} - a_{23}a_{32}) - a_{12}(a_{21}a_{33} - a_{23}a_{31}) + a_{13}(a_{21}a_{32} - a_{22}a_{31})$$

The quantity $(-1)^{i+j} \det(\alpha_{ij}) \equiv A_{ij}$ is called the *cofactor* of the element a_{ij}.

$$\begin{pmatrix} a_{11} & a_{12} & a_{13} \\ a_{21} & a_{22} & a_{23} \\ a_{31} & a_{32} & a_{33} \end{pmatrix} \Rightarrow \alpha_{12} = \begin{pmatrix} a_{21} & a_{23} \\ a_{31} & a_{33} \end{pmatrix}$$

FIGURE 1.21. Determining the elements of the submatrix α_{ij} whose signed determinant $A_{ij} = (-1)^{i+j} \det(\alpha_{ij})$ is the cofactor of the element a_{ij}. This figure shows how to find α_{12}. Eliminate the row and column that contain the chosen element. The remaining submatrix is α_{ij}.

We can now write a rule for finding the determinant of any matrix:

The determinant of a matrix $\mathbb{A}$ is the sum of the elements of any row or column of the matrix, each multiplied by its cofactor:

$$\det(\mathbb{A}) = |\mathbb{A}| = \sum_{j} a_{ij} A_{ij} = \sum_{i} a_{ij} A_{ij} \qquad (1.72)$$

Starting with the rule for a 2×2 determinant, we may use equation (1.72) to compute the determinant of a 3×3 matrix, then a 4×4 matrix, and so on. This is called the Laplace development.

These properties of determinants now follow[30]:

- If an $N \times N$ matrix is multiplied by a constant c, its determinant is multiplied by c^N.

---

[30] See Problems 37 through 39 and 41.

- A matrix and its transpose have the same determinant.
- If two rows or two columns of a matrix are identical, its determinant is zero.

The determinant proves very useful in understanding the properties of matrix products. We will need the following theorem[31]:

Product theorem for determinants.   *The determinant of the matrix* $\mathbb{C} = \mathbb{A}\mathbb{B}$ *is the product of the determinants of* $\mathbb{A}$ *and* $\mathbb{B}$.

Proof (for 3×3 matrices).   Matrix $\mathbb{C}$ has elements $c_{ij} = \sum_k a_{ik} b_{kj}$, and the determinant is thus (equation 1.71)

$$\det (\mathbb{C}) = \sum_{i,j,k=1}^{3} \varepsilon_{ijk} c_{1i} c_{2j} c_{3k}$$

$$= \sum_{i,j,k=1}^{3} \varepsilon_{ijk} \sum_{m} a_{1m} b_{mi} \sum_{n} a_{2n} b_{nj} \sum_{p} a_{3p} b_{pk}$$

$$= \sum_{m} \sum_{n} \sum_{p} a_{1m} a_{2n} a_{3p} \left(\sum_{i,j,k=1}^{3} \varepsilon_{ijk} b_{mi} b_{nj} b_{pk} \right)$$

Now if m, n, p are an even permutation of $1, 2, 3$, then the term in parentheses equals $\det (\mathbb{B})$, whereas if m, n, p are an odd permutation of $1, 2, 3$, then we get its negative. If any two of m, n, p are equal, the term is zero, since it equals the determinant of a matrix with two equal rows. Thus, we may write $\det (\mathbb{C})$ as

$$\det (\mathbb{C}) = \det (\mathbb{A}\mathbb{B}) = \sum_{m,n,p} a_{1m} a_{2n} a_{3p} \varepsilon_{mnp} \det (\mathbb{B}) = \det (\mathbb{A}) \det (\mathbb{B}) \qquad (1.73)$$

as asserted.

A matrix with determinant zero is said to be *singular*. Such matrices have some interesting properties. For example, if such a matrix is multiplied by any other matrix, the determinant of the resulting matrix is also zero. This fact allows us to understand why it is possible for two matrices, each nonzero, to have a product that equals the zero matrix[32]:

$$\mathbb{A}\mathbb{B} = \mathbf{0} \quad \text{does \textbf{not} imply} \quad \mathbb{A} = \mathbf{0} \quad \text{or} \quad \mathbb{B} = \mathbf{0}$$

As a consequence, in the mathematics of matrices, the square root of zero exists and is not necessarily zero.

---

[31]For additional proofs of this theorem, see, for example, Halmos, Section 53 or Tucker, p. 92.
[32]See Problems 42 and 43.

1.6.3. The Inverse of a Matrix

If the product of two matrices is the unit matrix, $\mathbb{A}\mathbb{B} = \mathbb{I}$, then $\mathbb{B}$ is the inverse of $\mathbb{A}$, $\mathbb{B} = \mathbb{A}^{-1}$. From the rule for computing the determinant of a matrix product (1.73), we conclude that the determinant of $\mathbb{A}^{-1}$ is $1/\det(\mathbb{A})$. Thus, the inverse exists only if the determinant of the matrix is not zero (the matrix is *nonsingular*).

Using the basic rule for matrix multiplication, we have

$$\sum_{j=1}^{N} a_{ij} b_{jk} = \delta_{ik}$$

This relation is actually $N \times N$ equations for the elements b_{jk}. The solution may be found by a process of successive elimination,[33] with the result that the inverse matrix equals the transpose of the matrix of cofactors of the elements of $\mathbb{A}$, divided by the determinant of $\mathbb{A}$:

$$\boxed{b_{ij} = \frac{A_{ji}}{|\mathbb{A}|}} \tag{1.74}$$

Let's check this result. The product of a matrix and its inverse must be the identity matrix. The product has elements

$$(\mathbb{A}\mathbb{B})_{ik} = \sum_{j} a_{ij} b_{jk} = \sum_{j} a_{ij} \frac{A_{kj}}{|\mathbb{A}|}$$

The first diagonal element is

$$\frac{\sum_{j} a_{1j} A_{1j}}{|\mathbb{A}|} = \frac{|\mathbb{A}|}{|\mathbb{A}|} = 1$$

where we used equation (1.72) for the determinant with $i = 1$. The result is the same for $i = 2$ and 3. The off-diagonal elements are of the form

$$\sum_{j} a_{1j} \frac{A_{2j}}{|\mathbb{A}|} \tag{1.75}$$

The numerator is the determinant of a matrix with two identical rows (here the first and second rows) and so is zero.

Example 1.7.   Find the inverse of the matrix $\begin{pmatrix} 1 & 0 & 0 \\ 2 & 1 & 0 \\ 0 & 0 & 2 \end{pmatrix}$.

---

[33] See below for an example of how to do this.

The determinant of the matrix is $D = 2$, so the inverse exists. The cofactors are $A_{11} = 2$, $A_{12} = -4$, $A_{13} = 0$, $A_{21} = 0$, $A_{22} = 2$, $A_{23} = 0$, $A_{31} = 0$, $A_{32} = 0$, and $A_{33} = 1$. Thus, the inverse is

$$\frac{1}{2}\begin{pmatrix} 2 & 0 & 0 \\ -4 & 2 & 0 \\ 0 & 0 & 1 \end{pmatrix} = \begin{pmatrix} 1 & 0 & 0 \\ -2 & 1 & 0 \\ 0 & 0 & \frac{1}{2} \end{pmatrix}$$

We can check the result by multiplying the original matrix and the inverse we have found:

$$\begin{pmatrix} 1 & 0 & 0 \\ 2 & 1 & 0 \\ 0 & 0 & 2 \end{pmatrix}\begin{pmatrix} 1 & 0 & 0 \\ -2 & 1 & 0 \\ 0 & 0 & \frac{1}{2} \end{pmatrix} = \begin{pmatrix} 1 & 0 & 0 \\ 0 & 1 & 0 \\ 0 & 0 & 1 \end{pmatrix}$$

as required.

A matrix whose transpose equals its inverse is called *orthogonal*.[34] A matrix whose adjoint equals its inverse is called *unitary*.

Computer mathematics packages such as *Mathematica* and *Maple* have matrix inversion routines. Even spreadsheets can invert matrices. Formula (1.75) for the elements of the inverse matrix is not usually the best computational tool.[35] Procedures such as Gauss-Jordan inversion are more efficient. To see how these methods work, first we note that we may devise a set of matrices, called elementary matrices, whose effect is to perform simplifying operations on a matrix. For example, a matrix with a 1 in each row and in each column and zeros elsewhere permutes the entries in the matrix:

$$\begin{pmatrix} 0 & 1 & 0 \\ 0 & 0 & 1 \\ 1 & 0 & 0 \end{pmatrix}\begin{pmatrix} a_{11} & a_{12} & a_{13} \\ a_{21} & a_{22} & a_{23} \\ a_{31} & a_{32} & a_{33} \end{pmatrix} = \begin{pmatrix} a_{21} & a_{22} & a_{23} \\ a_{31} & a_{32} & a_{33} \\ a_{11} & a_{12} & a_{13} \end{pmatrix}$$

Multiplication by this matrix has moved row 2 to row 1, row 1 to row 3, and row 3 to row 2. We can also add a multiple of one row to another:

$$\begin{pmatrix} 1 & 0 & 0 \\ -2 & 1 & 0 \\ 0 & 0 & 1 \end{pmatrix}\begin{pmatrix} a_{11} & a_{12} & a_{13} \\ a_{21} & a_{22} & a_{23} \\ a_{31} & a_{32} & a_{33} \end{pmatrix} = \begin{pmatrix} a_{11} & a_{12} & a_{13} \\ a_{21} - 2a_{11} & a_{22} - 2a_{12} & a_{23} - 2a_{13} \\ a_{31} & a_{32} & a_{33} \end{pmatrix}$$

Multiplication by this matrix subtracts two times row 1 from row 2. By putting the number x on the diagonal of the elementary matrix, we can multiply a row by any number x:

$$\begin{pmatrix} x & 0 & 0 \\ 0 & 1 & 0 \\ 0 & 0 & 1 \end{pmatrix}\begin{pmatrix} a_{11} & a_{12} & a_{13} \\ a_{21} & a_{22} & a_{23} \\ a_{31} & a_{32} & a_{33} \end{pmatrix} = \begin{pmatrix} xa_{11} & xa_{12} & xa_{13} \\ a_{21} & a_{22} & a_{23} \\ a_{31} & a_{32} & a_{33} \end{pmatrix}$$

[34] You should convince yourself that this definition is equivalent to the one given in Section 1.1.2.
[35] See Press et al., Chapter 2.

We can thus perform any of these operations on a matrix $\mathbb{A}$ until we get the unit matrix. Then the product of the elementary matrices that we used must be the inverse of $\mathbb{A}$:

$$\mathbb{E}_n \mathbb{E}_{n-1} \cdots \mathbb{E}_3 \mathbb{E}_2 \mathbb{E}_1 \mathbb{A} = \mathbb{I} \Rightarrow \mathbb{E}_n \mathbb{E}_{n-1} \cdots \mathbb{E}_3 \mathbb{E}_2 \mathbb{E}_1 = \mathbb{A}^{-1}$$

We may evaluate this product of elementary matrices as a single matrix by applying the same operations to the unit matrix:

$$\mathbb{E}_n \mathbb{E}_{n-1} \cdots \mathbb{E}_3 \mathbb{E}_2 \mathbb{E}_1 \mathbb{I} = \mathbb{A}^{-1} \mathbb{I} = \mathbb{A}^{-1}$$

Once we have established this result, we do not actually need to compute the elementary matrices; we can simply perform the operations on the matrix rows. The allowed operations are

1. Interchange two rows.
2. Multiply a row by any number.
3. Subtract a (positive or negative) number times one row from another row.

Example 1.8.   Invert the matrix $\begin{pmatrix} 2 & 3 & 4 \\ 1 & 3 & 2 \\ 0 & 2 & 3 \end{pmatrix}$.

First we check that the determinant is not zero, and thus that an inverse exists:

$$\det \begin{pmatrix} 2 & 3 & 4 \\ 1 & 3 & 2 \\ 0 & 2 & 3 \end{pmatrix} = 9$$

Next we perform simplifying operations.

1. Subtract row 2 from row 1:

$$\begin{pmatrix} 1 & 0 & 2 \\ 1 & 3 & 2 \\ 0 & 2 & 3 \end{pmatrix}$$

2. Subtract row 1 from row 2:

$$\begin{pmatrix} 1 & 0 & 2 \\ 0 & 3 & 0 \\ 0 & 2 & 3 \end{pmatrix}$$

3. Subtract 2/3 times row 2 from row 3:

$$\begin{pmatrix} 1 & 0 & 2 \\ 0 & 3 & 0 \\ 0 & 0 & 3 \end{pmatrix}$$

4. Subtract 2/3 times row 3 from row 1:

$$\begin{pmatrix} 1 & 0 & 0 \\ 0 & 3 & 0 \\ 0 & 0 & 3 \end{pmatrix}$$

5. Divide rows 2 and 3 by 3:

$$\begin{pmatrix} 1 & 0 & 0 \\ 0 & 1 & 0 \\ 0 & 0 & 1 \end{pmatrix}$$

Now we perform the exact same operations on the unit matrix.

1. Subtract row 2 from row 1:

$$\begin{pmatrix} 1 & -1 & 0 \\ 0 & 1 & 0 \\ 0 & 0 & 1 \end{pmatrix}$$

2. Subtract row 1 from row 2:

$$\begin{pmatrix} 1 & -1 & 0 \\ -1 & 2 & 0 \\ 0 & 0 & 1 \end{pmatrix}$$

3. Subtract 2/3 times row 2 from row 3:

$$\begin{pmatrix} 1 & -1 & 0 \\ -1 & 2 & 0 \\ \frac{2}{3} & -\frac{4}{3} & 1 \end{pmatrix}$$

4. Subtract 2/3 times row 3 from row 1:

$$\begin{pmatrix} \frac{5}{9} & -\frac{1}{9} & -\frac{2}{3} \\ -1 & 2 & 0 \\ \frac{2}{3} & -\frac{4}{3} & 1 \end{pmatrix}$$

5. Divide rows 2 and 3 by 3:

$$\begin{pmatrix} \frac{5}{9} & -\frac{1}{9} & -\frac{2}{3} \\ -\frac{1}{3} & \frac{2}{3} & 0 \\ \frac{2}{9} & -\frac{4}{9} & \frac{1}{3} \end{pmatrix}$$

Those 9s in the denominator are reassuring, since $9 = \det(\mathbb{A})$. Finally, we check our result by multiplying:

$$\begin{pmatrix} \frac{5}{9} & -\frac{1}{9} & -\frac{2}{3} \\ -\frac{1}{3} & \frac{2}{3} & 0 \\ \frac{2}{9} & -\frac{4}{9} & \frac{1}{3} \end{pmatrix} \begin{pmatrix} 2 & 3 & 4 \\ 1 & 3 & 2 \\ 0 & 2 & 3 \end{pmatrix} = \begin{pmatrix} 1 & 0 & 0 \\ 0 & 1 & 0 \\ 0 & 0 & 1 \end{pmatrix}$$

as required.

1.6.4. Matrices and Linear Equations

A set of N equations in N unknowns may be written in the language of matrices. The equations

$$a_{11}x_1 + a_{12}x_2 + \cdots + a_{1N}x_N = b_1$$
$$a_{21}x_1 + a_{22}x_2 + \cdots + a_{2N}x_N = b_2$$
$$\vdots$$
$$a_{N1}x_1 + a_{N2}x_2 + \cdots + a_{NN}x_N = b_N$$

may be written as

$$\mathbb{A}\vec{\mathbf{x}} = \vec{\mathbf{b}}$$

where the vector $\vec{\mathbf{x}}$ contains the unknowns x_i as its components and the vector $\vec{\mathbf{b}}$ contains the constants b_i. The solution may be found by multiplying both sides by the inverse matrix $\mathbb{A}^{-1}$:

$$\vec{\mathbf{x}} = \mathbb{A}^{-1}\vec{\mathbf{b}}$$
$$x_i = \frac{\sum_j A_{ji} b_j}{\det |\mathbb{A}|} \tag{1.76}$$

where A_{ji} is the cofactor of element a_{ji}. The numerator is the determinant of the matrix formed by removing the ith column of $\mathbb{A}$ and replacing it with the vector $\vec{\mathbf{b}}$. Result (1.76) is called Cramer's rule. It provides a formal solution to the equations, but it is rarely the best computational procedure.

In the Gaussian elimination method,[36] we use the first equation to eliminate x_1 from all of the remaining equations. Then we use the second equation to eliminate x_2 from all the equations below it, and so on. The final equation is then solved for x_N, and we then substitute back up the sequence of equations to find all the unknowns. This algorithm is easily programmed and is well suited to the characteristics of modern computers.

The homogeneous set of equations

$$\mathbb{A}\vec{\mathbf{x}} = 0$$

has a nontrivial solution for the vector $\vec{\mathbf{x}}$ if and only if the determinant of $\mathbb{A}$ is zero. One interpretation of the set of equations is that the row vectors with components a_{1j}, a_{2j}, a_{3j} are each perpendicular to the vector $\vec{\mathbf{x}}$. Thus, all three vectors must lie in the plane perpendicular to $\vec{\mathbf{x}}$. This means that the triple scalar product[37] of the three vectors must be zero, or, equivalently, that the determinant $|\mathbb{A}| = 0$.

1.6.5. Change of Basis

Similarity Transformations

The elements of a matrix that represents a linear operator in a vector space depend on the basis vectors chosen for the space. If we change the basis vectors, then the elements of the

---

[36] See, for example, Golub and van Loan, Chapter 4.
[37] See Section 1.1.2. Compare equation (1.32) with equation (1.71).

matrix change, even though the operator is unchanged. The new basis vectors are linear combinations of the old ones, so the operation of changing the basis is also represented by a matrix. Let's call this matrix $\mathbb{C}$. Thus, if a vector has components x_i in the original basis, its components in the new basis will be x_i', where

$$x_i' = \sum_{j=1}^{N} c_{ij} x_j \tag{1.77}$$

Now suppose an operator represented by $\mathbb{A}$ maps the vector $\vec{\mathbf{x}}$ to a new vector $\vec{\mathbf{y}}$:

$$y_i = \sum_{j=1}^{N} a_{ij} x_j$$

In the new basis, the operator is represented by $\mathbb{A}'$ with components a_{ij}' and

$$y_i' = \sum_{j=1}^{N} a_{ij}' x_j'$$

Thus,

$$y_i' = \sum_{k=1}^{N} c_{ik} y_k$$

$$= \sum_{k=1}^{N} c_{ik} \sum_{j=1}^{N} a_{kj} x_j$$

$$= \sum_{j=1}^{N} \left(\sum_{k=1}^{N} c_{ik} a_{kj} \right) x_j$$

or, in matrix notation,[38]

$$\vec{\mathbf{y}}' = \mathbb{C}\mathbb{A}\vec{\mathbf{x}}$$

Now the components x_j are related to the components x_m' by the inverse of $\mathbb{C}$, $x_j = \sum_j (\mathbb{C}^{-1})_{jm} x_m'$, so

$$\vec{\mathbf{y}}' = \mathbb{C}\mathbb{A}\mathbb{C}^{-1}\vec{\mathbf{x}}' = \mathbb{A}'\vec{\mathbf{x}}'$$

Thus,

$$\boxed{\mathbb{A}' = \mathbb{C}\mathbb{A}\mathbb{C}^{-1}} \tag{1.78}$$

[38] In this notation, $\vec{\mathbf{x}}$ and $\vec{\mathbf{x}}'$ represent the same vector, but the corresponding sets of components are expressed with respect to different bases.

Equation (1.78) describes a *similarity transformation*. It is sometimes written in terms of the matrix $\mathbb{T} = \mathbb{C}^{-1}$:

$$\mathbb{A}' = \mathbb{T}^{-1}\mathbb{A}\mathbb{T} \qquad (1.79)$$

The elements of the matrices $\mathbb{C}$ and $\mathbb{T}$ are the components of the new basis vectors in the original system. If a vector $\vec{\mathbf{x}}$ with components x_i is to be the first basis vector in the new system, then its prime components are $x_1' = 1$, $x_i' = 0$ if $i \neq 1$. Thus,

$$x_i = \sum_j t_{ij}x_j' = t_{i1}$$

Thus, the column vector x_i is the first column of the matrix $\mathbb{T}$. Similarly, the components x_i of the jth basis vector form the jth column of $\mathbb{T} = \mathbb{C}^{-1}$.

We often want to find the basis that makes the matrix $\mathbb{A}$ corresponding to a given operator as simple as possible. The simplest matrix is one with nonzero elements along the diagonal and all other elements zero. Then if $\vec{\mathbf{y}} = \mathbb{A}\vec{\mathbf{x}}$, the components of $\vec{\mathbf{y}}$ are simple multiples of the components of $\vec{\mathbf{x}}$: $y_i = \lambda_i x_i$. In particular, for a basis vector $\hat{\mathbf{e}}_i$ that has only one nonzero component, $\mathbb{A}'\hat{\mathbf{e}}_i = \lambda_i\hat{\mathbf{e}}_i$. This relationship among vectors must remain true in any basis, so there exist N vectors $\vec{\mathbf{x}}_i$ such that $\mathbb{A}\vec{\mathbf{x}}_i = \lambda_i\vec{\mathbf{x}}_i$. These N vectors are the eigenvectors of the matrix $\mathbb{A}$, and the constants λ_i are the corresponding eigenvalues. The matrix $\mathbb{A}'$ has the values λ_i along the diagonal when the vectors $\vec{\mathbf{x}}_i$ (suitably normalized) are chosen as the basis vectors.

To diagonalize the matrix, we must solve the equation

$$(\mathbb{A} - \lambda\mathbb{I})\,\vec{\mathbf{x}} = 0$$

where $\mathbb{I}$ is the unit matrix. This equation has a nontrivial solution for the vector $\vec{\mathbf{x}}$ only if[39]

$$\det(\mathbb{A} - \lambda\mathbb{I}) = 0 \qquad (1.80)$$

This equation is called the characteristic equation. If the matrix is a 2×2 matrix, the characteristic equation is a quadratic equation; if the matrix is a 3×3 matrix, the characteristic equation is a cubic; and so on. Thus, there are at most N real distinct eigenvalues. There may be fewer than N if the eigenvalues are not distinct or if the roots are complex.

The eigenvalues of a real symmetric matrix are real. Let's see why. If λ is an eigenvalue, then there exists a vector $\vec{\mathbf{x}}$ such that

$$\mathbb{A}\vec{\mathbf{x}} = \lambda\vec{\mathbf{x}}$$

Transpose both sides and take the complex conjugate[40]:

$$\vec{\mathbf{x}}\mathbb{A}^* = \vec{\mathbf{x}}\mathbb{A} = \lambda^*\vec{\mathbf{x}}$$

---

[39] See Section 1.6.4.
[40] Remember that we are working with vector spaces in which all the vectors are real.

(The process of transposing does not change the right-hand side, except that the column vector becomes a row vector.) Now take the inner product with $\vec{\mathbf{x}}$ in both equations:

$$\vec{\mathbf{x}} \mathbf{A} \vec{\mathbf{x}} = \lambda \vec{\mathbf{x}} \cdot \vec{\mathbf{x}} = \lambda^* \vec{\mathbf{x}} \cdot \vec{\mathbf{x}}$$

Thus,

$$(\lambda - \lambda^*) \vec{\mathbf{x}} \cdot \vec{\mathbf{x}} = 0$$

Since $\vec{\mathbf{x}} \cdot \vec{\mathbf{x}}$ cannot be zero unless $\vec{\mathbf{x}} = 0$, $\lambda = \lambda^*$ and the eigenvalues are real.

If two eigenvalues are distinct, then the corresponding eigenvectors are orthogonal. For if

$$\mathbf{A} \vec{\mathbf{x}}_1 = \lambda_1 \vec{\mathbf{x}}_1$$

and

$$\mathbf{A} \vec{\mathbf{x}}_2 = \lambda_2 \vec{\mathbf{x}}_2$$

then

$$\vec{\mathbf{x}}_1 \mathbf{A} \vec{\mathbf{x}}_2 = \lambda_2 \vec{\mathbf{x}}_1 \cdot \vec{\mathbf{x}}_2$$

and

$$\vec{\mathbf{x}}_2 \mathbf{A} \vec{\mathbf{x}}_1 = \lambda_1 \vec{\mathbf{x}}_2 \cdot \vec{\mathbf{x}}_1$$

Transposing the latter relation and subtracting from the former, we have

$$0 = (\lambda_2 - \lambda_1) \vec{\mathbf{x}}_1 \cdot \vec{\mathbf{x}}_2$$

Since the two eigenvalues are not equal, $\vec{\mathbf{x}}_1 \cdot \vec{\mathbf{x}}_2 = 0$ and the eigenvectors are orthogonal.

Example 1.9.   Diagonalize the matrix $\begin{pmatrix} 3 & 2 \\ 2 & 0 \end{pmatrix}$.

The characteristic equation is

$$\begin{vmatrix} 3 - \lambda & 2 \\ 2 & -\lambda \end{vmatrix} = 0$$

$$-\lambda (3 - \lambda) - 4 = 0$$

$$\lambda^2 - 3\lambda - 4 = 0$$

So the eigenvalues are

$$\lambda = \frac{3 \pm \sqrt{9 + 16}}{2} = \frac{3 \pm 5}{2} = -1, 4$$

The diagonalized matrix is thus

$$A' = \begin{pmatrix} -1 & 0 \\ 0 & 4 \end{pmatrix}$$

We have solved the problem, but we can learn more from this example.

The two eigenvalues are real and distinct, so the eigenvectors are orthogonal. Their components are found from the equation

$$2x_1 - \lambda x_2 = 0$$

Inserting the values of λ, we have

$$2x_1 = -x_2 \quad \text{or} \quad x_1 = 2x_2$$

So, with $x_1 = -1$,

$$\vec{x}_1 = (-1, 2)$$

and, with $x_2 = +1$,

$$\vec{x}_2 = (2, 1)$$

These vectors may be normalized to provide unit vectors as a basis. Let's verify the relation $A\vec{x} = \lambda\vec{x}$ for these two vectors:

$$\begin{pmatrix} 3 & 2 \\ 2 & 0 \end{pmatrix}\begin{pmatrix} -1 \\ 2 \end{pmatrix} = \begin{pmatrix} -3+4 \\ -2 \end{pmatrix} = \begin{pmatrix} 1 \\ -2 \end{pmatrix} = -1\begin{pmatrix} -1 \\ 2 \end{pmatrix}$$

and

$$\begin{pmatrix} 3 & 2 \\ 2 & 0 \end{pmatrix}\begin{pmatrix} 2 \\ 1 \end{pmatrix} = \begin{pmatrix} 6+2 \\ 4 \end{pmatrix} = \begin{pmatrix} 8 \\ 4 \end{pmatrix} = 4\begin{pmatrix} 2 \\ 1 \end{pmatrix}$$

as required.

We diagonalize the matrix by changing to a new basis with an operator (matrix) $\mathbb{C}$. Then $A' = \mathbb{C}A\mathbb{C}^{-1}$ (equation 1.78), and the elements on the diagonal of A' are the eigenvalues of A. The eigenvectors of A are the new basis vectors. Then from equation (1.77), the components of the kth eigenvector are related by

$$u_i^{(k)} = \sum_j (\mathbb{C}^{-1})_{ij} u_j^{(k)'} = \sum_j (\mathbb{C}^{-1})_{ij} \delta_{jk} = (\mathbb{C}^{-1})_{ik}$$

Thus, the elements in the kth column of $\mathbb{C}^{-1}$ equal the components of the kth eigenvector. If the matrix is orthogonal, then these are the elements of the kth row of $\mathbb{C}$.

Here, the matrix that effects the transformation is orthogonal and has elements

$$\mathbb{C} = \frac{1}{\sqrt{5}}\begin{pmatrix} -1 & 2 \\ 2 & 1 \end{pmatrix}$$

Let's verify that this does the job:

$$\mathbb{C}\mathbb{A}\mathbb{C}^{-1} = \frac{1}{\sqrt{5}} \begin{pmatrix} -1 & 2 \\ 2 & 1 \end{pmatrix} \begin{pmatrix} 3 & 2 \\ 2 & 0 \end{pmatrix} \frac{1}{\sqrt{5}} \begin{pmatrix} -1 & 2 \\ 2 & 1 \end{pmatrix}$$

$$= \frac{1}{5} \begin{pmatrix} -1 & 2 \\ 2 & 1 \end{pmatrix} \begin{pmatrix} 1 & 8 \\ -2 & 4 \end{pmatrix}$$

$$= \begin{pmatrix} -1 & 0 \\ 0 & 4 \end{pmatrix}$$

which is the matrix $\mathbb{A}'$.

In this example, $\mathbb{C}$ turns out to be orthogonal. In fact, since the diagonal matrix $\mathbb{A}'$ is symmetric and an orthogonal transformation maps a symmetric matrix to another symmetric matrix,[41] any symmetric matrix $\mathbb{A}$ is diagonalized by an orthogonal transformation.

If a matrix has two or more equal eigenvalues, then we cannot prove that the eigenvectors are orthogonal, but they may be.

Example 1.10.   Find the eigenvalues and eigenvectors of the matrix $\begin{pmatrix} 2 & 0 & -1 \\ 0 & 3 & 0 \\ -1 & 0 & 2 \end{pmatrix}$.

First we find the eigenvalues:

$$0 = \begin{vmatrix} 2-\lambda & 0 & -1 \\ 0 & 3-\lambda & 0 \\ -1 & 0 & 2-\lambda \end{vmatrix} = 9 - 15\lambda + 7\lambda^2 - \lambda^3 = -(\lambda-1)(\lambda-3)^2$$

Thus, the eigenvalues are $\lambda = 1$ and $\lambda = 3$, the latter being a repeated root.

Next we seek the eigenvectors.

$$\begin{pmatrix} 2 & 0 & -1 \\ 0 & 3 & 0 \\ -1 & 0 & 2 \end{pmatrix} \begin{pmatrix} x \\ y \\ z \end{pmatrix} = \lambda \begin{pmatrix} x \\ y \\ z \end{pmatrix}$$

$$\begin{pmatrix} 2x - z \\ 3y \\ -x + 2z \end{pmatrix} = \lambda \begin{pmatrix} x \\ y \\ z \end{pmatrix}$$

First we take the case of $\lambda = 1$:

$$2x - z = x \Rightarrow x = z$$

$$3y = y \Rightarrow y = 0$$

$$-x + 2z = z \Rightarrow z = x$$

Thus, this vector $\vec{v}_1$ has components $(x, 0, x)$ for any x.

---

[41] See Problem 48.

Then, with $\lambda = 3$, we have

$$2x - z = 3x \Rightarrow x = -z$$

$$3y = 3y \Rightarrow y \text{ is arbitrary}$$

$$-x + 2z = 3z \Rightarrow z = -x$$

Thus, these eigenvectors have components $(x, y, -x)$ for any values of y and x. Notice first that each of these vectors is orthogonal to $\vec{v}_1$. Second, we can find two such vectors that are orthogonal to each other:

$$(x, y, -x) \cdot (u, v, -u) = xu + yv + xu = 0 = 2xu + yv$$

Thus, with x, u, and y chosen, we simply pick $v = -2xu/y$. We can always do this, provided that y is not zero. For example, we might pick $\vec{v}_2 = (1, 1, -1)$ and $\vec{v}_3 = (1, -2, -1)$. Let's check that $\vec{v}_2$ is an eigenvector:

$$\begin{pmatrix} 2 & 0 & -1 \\ 0 & 3 & 0 \\ -1 & 0 & 2 \end{pmatrix} \begin{pmatrix} 1 \\ 1 \\ -1 \end{pmatrix} = \begin{pmatrix} 3 \\ 3 \\ -3 \end{pmatrix} = 3 \begin{pmatrix} 1 \\ 1 \\ -1 \end{pmatrix}$$

as required. You should check that $\vec{v}_3$ is also an eigenvector.

If two matrices commute, then if their eigenvalues are all distinct, they have the same eigenvectors. For example, suppose that $\mathbb{A}\mathbb{B} = \mathbb{B}\mathbb{A}$ and $\mathbb{A}\vec{x} = \lambda\vec{x}$. Then

$$\mathbb{B}\mathbb{A}\vec{x} = \mathbb{B}\lambda\vec{x} = \lambda\mathbb{B}\vec{x}$$

But

$$\mathbb{B}\mathbb{A}\vec{x} = \mathbb{A}\mathbb{B}\vec{x}$$

Thus,

$$\mathbb{A}\mathbb{B}\vec{x} = \lambda\mathbb{B}\vec{x}$$

Thus, $\mathbb{B}\vec{x}$ is also an eigenvector of $\mathbb{A}$ with eigenvalue λ. If the eigenvectors of $\mathbb{A}$ are distinct, then $\mathbb{B}\vec{x}$ must be a constant times $\vec{x}$, and thus $\vec{x}$ is also an eigenvector of $\mathbb{B}$.

Congruent Transformations

A set of linear equations results when we consider small oscillations of a physical system about its equilibrium. We can often analyze such a system most efficiently using Lagrangian mechanics. The Lagrangian of a system contains a kinetic energy term[42] K which is quadratic in the velocities and a potential energy term V which, in the case of

---

[42]In Lagrangian mechanics, the symbol T is often used for kinetic energy. But since we have already used $\mathbb{T}$ for the transformation matrix, we'll use K for the kinetic energy.

small displacements from equilibrium, may be expanded in a Taylor series with no linear term. (This is the equilibrium condition.) Thus, the potential energy term is a quadratic function of the coordinates. Each of these quadratic forms may be expressed in terms of matrices:

$$K = \frac{d\mathbf{x}^T}{dt} \mathbb{K} \frac{d\mathbf{x}}{dt}; \quad V = \mathbf{x}^T \mathbb{V} \mathbf{x}$$

where the vector $\mathbf{x}^T$ on the left is a row vector and the vector $\mathbf{x}$ on the right is a column vector. Furthermore, both matrices are symmetric. The motion of the system looks simplest when we choose our coordinates to correspond to the normal modes of the system. But we don't usually know what those modes are until we have solved the problem.

The normal modes of the system can be described by vectors $\mathbf{x}^{(N)}$. Thus, to simplify the system, we need to change our basis to make these vectors our basis vectors. Let's see what happens to a quadratic form when we change the basis. The energy is a scalar under coordinate transformations, so

$$V' = \mathbf{x}'^T \mathbb{V}' \mathbf{x}' = \mathbf{x}^T \mathbb{V} \mathbf{x} = V$$

But $\mathbf{x} = \mathbb{T}\mathbf{x}'$, so

$$\mathbf{x}'^T \mathbb{V}' \mathbf{x}' = (\mathbb{T}\mathbf{x}')^T \mathbb{V}\mathbb{T}\mathbf{x}' = \mathbf{x}'^T \mathbb{T}^T \mathbb{V}\mathbb{T}\mathbf{x}'$$

and thus

$$\boxed{V' = \mathbb{T}^T \mathbb{V}\mathbb{T}} \tag{1.81}$$

This transformation is almost a similarity transformation (equation 1.79), but here the transpose of $\mathbb{T}$ appears instead of the inverse. It is called a congruent transformation. If the transformation is orthogonal, $\mathbb{T}^T = \mathbb{T}^{-1}$ and the two transformations are identical.

Example 1.11. A system has two identical pendulums connected by a spring. Each pendulum has a point object of mass m on the end of a stiff but massless rod of length ℓ. Each pendulum is attached to a pivot on the roof. The two pivots are a distance s_0 apart. The two point objects are connected by a massless spring with spring constant k and relaxed length s_0. Find the possible motions of this system.

This system is most easily analyzed using Lagrangian methods.[43] In equilibrium, both rods hang straight down, and the spring exerts no horizontal force. Figure 1.22 shows the system when displaced from equilibrium. The kinetic energy is

$$K = \frac{1}{2}m\ell^2 \left[\left(\frac{d\theta_1}{dt}\right)^2 + \left(\frac{d\theta_2}{dt}\right)^2 \right]$$

---

[43] See Optional Topic E.

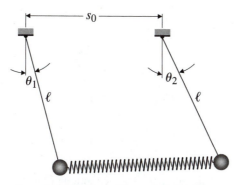

FIGURE 1.22. A system of coupled pendulums (Example 1.11). The two objects are connected by a spring.

The system has both gravitational potential energy and elastic potential energy. To compute the gravitational potential energy, we put the reference level at the level of the two pivots. Then

$$V_g = -mg\ell(\cos\theta_1 + \cos\theta_2)$$

The length of the spring in Figure 1.22 is a complicated function of the angles, but we can obtain a much simpler expression if the two angles θ_1 and θ_2 remain small. (This is the *small oscillation* approximation.) Then with $\theta_1 \ll 1$ and $\theta_2 \ll 1$, the two objects are separated by a distance $s_0 + \ell(\theta_2 - \theta_1)$ and the elastic potential energy is

$$V_e = \frac{1}{2}k\ell^2(\theta_2 - \theta_1)^2$$

Now applying the same constraint ($\theta \ll 1$) to the expression for the gravitational potential energy, we obtain the total potential energy:

$$V = -mg\ell\left(2 - \frac{\theta_1^2}{2} - \frac{\theta_2^2}{2}\right) + \frac{1}{2}k\ell^2(\theta_2 - \theta_1)^2$$

$$= -2mg\ell + \frac{1}{2}\ell\theta_1^2(k\ell + mg) + \frac{1}{2}\ell\theta_2^2(k\ell + mg) - k\ell^2\theta_2\theta_1$$

Then the Lagrangian is

$$\mathcal{L} = K - V = \frac{1}{2}m\ell^2\left[\left(\frac{d\theta_1}{dt}\right)^2 + \left(\frac{d\theta_2}{dt}\right)^2\right] + 2mg\ell$$

$$- \frac{1}{2}\ell\theta_1^2(k\ell + mg) - \frac{1}{2}\ell\theta_2^2(k\ell + mg) + k\ell^2\theta_2\theta_1$$

and Lagrange's equations are

$$\frac{d}{dt}\frac{\partial \mathcal{L}}{\partial \dot{\theta}_i} - \frac{\partial \mathcal{L}}{\partial \theta_i} = 0$$

$$m\ell^2 \frac{d^2\theta_1}{dt^2} - \ell(k\ell + mg)\theta_1 + k\ell^2\theta_2 = 0 \tag{1.82}$$

$$m\ell^2 \frac{d^2\theta_2}{dt^2} - \ell(k\ell + mg)\theta_2 + k\ell^2\theta_1 = 0 \tag{1.83}$$

These coupled linear equations show that the motion of one pendulum influences the motion of the other.

We shall now formulate this problem in the language of matrices. Note that both the kinetic and potential energies may be expressed as products of a matrix and a vector containing the velocities or the coordinates of the two pendulums.

$$K = \frac{1}{2}m\ell^2 \left(\frac{d\theta_1}{dt}, \frac{d\theta_2}{dt} \right) \begin{pmatrix} 1 & 0 \\ 0 & 1 \end{pmatrix} \begin{pmatrix} \dfrac{d\theta_1}{dt} \\ \dfrac{d\theta_2}{dt} \end{pmatrix} \tag{1.84}$$

and

$$V = -2mg\ell + \frac{1}{2}\ell\theta_1^2(k\ell + mg) + \frac{1}{2}\ell\theta_2^2(k\ell + mg) - k\ell^2\theta_2\theta_1$$

$$= -2mg\ell + \frac{1}{2}m\ell^2(\theta_1, \theta_2) \begin{pmatrix} \dfrac{k}{m} + \dfrac{g}{\ell} & -\dfrac{k}{m} \\ -\dfrac{k}{m} & \dfrac{k}{m} + \dfrac{g}{\ell} \end{pmatrix} \begin{pmatrix} \theta_1 \\ \theta_2 \end{pmatrix} \tag{1.85}$$

Here the coupling is indicated by the off-diagonal terms in the matrix. Notice that we have divided the symmetric term in $\theta_1\theta_2$ into two equal parts, making the off-diagonal terms equal and the matrix symmetrical.

To find the normal modes of the system, we should change the basis to obtain a diagonal matrix, indicating that the oscillations in the new coordinates are uncoupled. Since the potential energy matrix is symmetric, we know it can be diagonalized by an orthogonal transformation. A transformation of the type

$$\mathbb{V}' = \mathbb{T}^T \mathbb{V} \mathbb{T}$$

must also be applied to the matrix in the expression for kinetic energy, but since that is the unit matrix, it remains unchanged if the transformation is orthogonal:

$$\mathbb{K}' = \mathbb{T}^T \mathbb{I} \mathbb{T} = \mathbb{T}^T \mathbb{T} = \mathbb{T}^{-1} \mathbb{T} = \mathbb{I}$$

The characteristic equation is

$$\begin{vmatrix} \dfrac{k}{m} + \dfrac{g}{\ell} - \lambda & -\dfrac{k}{m} \\[2ex] -\dfrac{k}{m} & \dfrac{k}{m} + \dfrac{g}{\ell} - \lambda \end{vmatrix} = 0$$

$$\left(\dfrac{k}{m} + \dfrac{g}{\ell} - \lambda\right)^2 - \left(\dfrac{k}{m}\right)^2 = 0$$

$$\dfrac{k}{m} + \dfrac{g}{\ell} - \lambda = \pm\dfrac{k}{m}$$

and so the eigenvalues are

$$\lambda = \dfrac{g}{\ell}, \quad \dfrac{g}{\ell} + 2\dfrac{k}{m}$$

The eigenvectors are given by

$$\begin{pmatrix} \dfrac{k}{m} + \dfrac{g}{\ell} & -\dfrac{k}{m} \\[2ex] -\dfrac{k}{m} & \dfrac{k}{m} + \dfrac{g}{\ell} \end{pmatrix} \begin{pmatrix} \theta_1 \\ \theta_2 \end{pmatrix} = \lambda \begin{pmatrix} \theta_1 \\ \theta_2 \end{pmatrix}$$

In the first mode,

$$\begin{pmatrix} \left(\dfrac{k}{m} + \dfrac{g}{\ell}\right)\theta_1 - \dfrac{k}{m}\theta_2 \\[2ex] -\dfrac{k}{m}\theta_1 + \left(\dfrac{k}{m} + \dfrac{g}{\ell}\right)\theta_2 \end{pmatrix} = \dfrac{g}{\ell}\begin{pmatrix} \theta_1 \\ \theta_2 \end{pmatrix}$$

$$\begin{pmatrix} \dfrac{k}{m}(\theta_1 - \theta_2) + \theta_1\dfrac{g}{\ell} \\[2ex] -\dfrac{k}{m}(\theta_1 - \theta_2) + \theta_2\dfrac{g}{\ell} \end{pmatrix} = \dfrac{g}{\ell}\begin{pmatrix} \theta_1 \\ \theta_2 \end{pmatrix}$$

Thus,

$$\dfrac{k}{m}(\theta_1 - \theta_2) = 0 \quad \text{or} \quad \theta_1 = \theta_2$$

In this mode, the two pendulums oscillate in phase, and the distance between them stays fixed. The spring does not affect the system.

To find the other mode, we use the second eigenvalue:

$$\begin{pmatrix} \left(\dfrac{k}{m} + \dfrac{g}{\ell}\right)\theta_1 - \dfrac{k}{m}\theta_2 \\[2ex] -\dfrac{k}{m}\theta_1 + \left(\dfrac{k}{m} + \dfrac{g}{\ell}\right)\theta_2 \end{pmatrix} = \left(\dfrac{g}{\ell} + 2\dfrac{k}{m}\right)\begin{pmatrix} \theta_1 \\ \theta_2 \end{pmatrix}$$

$$-\dfrac{k}{m}(\theta_1 + \theta_2) = 0$$

Here $\theta_1 = -\theta_2$, the two pendulums are $180°$ out of phase, and the spring is alternately stretched and compressed.

We want to find a solution in which the system oscillates, so each angular displacement may be written in the form

$$x_i = A_i \cos(\omega_i t + \phi_i)$$

Then $d^2 x_i/dt^2 = -\omega_i^2 x_i$, and equations 1.82 and 1.83 take the following form:
Even mode: For $\theta_2 = \theta_1$,

$$-\omega_1^2 \theta_1 - \left(\frac{k}{m} + \frac{g}{\ell}\right)\theta_1 + \frac{k}{m}\theta_1 = 0$$

$$-\omega_2^2 \theta_2 - \left(\frac{k}{m} + \frac{g}{\ell}\right)\theta_2 + \frac{k}{m}\theta_2 = 0$$

The equations are identical, as expected, and the solution for the angular frequency is

$$\omega_e = \sqrt{\frac{g}{\ell}}$$

This is the oscillation frequency for a single pendulum of length ℓ, as expected.
Odd mode: For the second mode, $\theta_2 = -\theta_1$,

$$-\omega^2 \theta_1 - \left(\frac{k}{m} + \frac{g}{\ell}\right)\theta_1 - \frac{k}{m}\theta_1 = 0$$

$$\omega_o = \sqrt{2\frac{k}{m} + \frac{g}{\ell}}$$

This mode has a higher frequency because the spring provides an additional restoring force.

If we initially displace the two pendulums by unequal amounts θ_A and θ_B, then we must decompose the displacements into two parts, one corresponding to each normal mode:

$$\theta_1(0) = \theta_A = \frac{\theta_A + \theta_B}{2} + \frac{\theta_A - \theta_B}{2}$$

$$\theta_2(0) = \theta_B = \frac{\theta_A + \theta_B}{2} - \frac{\theta_A - \theta_B}{2}$$

The first term is the even mode, and the second is the odd mode. Thus, the system evolves as

$$\theta_1(t) = \left(\frac{\theta_A + \theta_B}{2}\right)\cos\sqrt{\frac{g}{\ell}}t + \left(\frac{\theta_A - \theta_B}{2}\right)\cos\sqrt{2\frac{k}{m} + \frac{g}{\ell}}t$$

and

$$\theta_2(t) = \left(\frac{\theta_A + \theta_B}{2}\right) \cos\sqrt{\frac{g}{\ell}}\, t - \left(\frac{\theta_A - \theta_B}{2}\right) \cos\sqrt{2\frac{k}{m} + \frac{g}{\ell}}\, t$$

We close by showing how the Lagrangian may be written in the new coordinates. Let $\phi = \frac{1}{2}(\theta_1 - \theta_2)$ and $\psi = \frac{1}{2}(\theta_1 + \theta_2)$. Then

$$\mathcal{L} = \frac{1}{2}m\ell^2 \left[\left(\frac{d(\phi + \psi)}{dt}\right)^2 + \left(\frac{d(\phi - \psi)}{dt}\right)^2\right] + 2mg\ell$$

$$- \frac{1}{2}\ell(\phi + \psi)^2(k\ell + mg) - \frac{1}{2}\ell(\phi - \psi)^2(k\ell + mg) + k\ell^2(\phi + \psi)(\phi - \psi)$$

$$= m\ell^2 \left[\left(\frac{d\phi}{dt}\right)^2 + \left(\frac{d\psi}{dt}\right)^2 + 2\frac{g}{\ell} - \phi^2\frac{g}{\ell} - \psi^2\left(2\frac{k}{m} + \frac{g}{\ell}\right)\right]$$

The absence of cross terms $\phi\psi$ shows that the motion is decoupled in these coordinates. In matrix language,

$$\mathcal{L} = m\ell^2 \left(\frac{d\phi}{dt}, \frac{d\psi}{dt}\right)\begin{pmatrix} 1 & 0 \\ 0 & 1 \end{pmatrix}\begin{pmatrix} \dfrac{d\phi}{dt} \\ \dfrac{d\psi}{dt} \end{pmatrix}$$

$$+ m\ell^2(\phi, \psi)\begin{pmatrix} \dfrac{g}{\ell} & 0 \\ 0 & 2\dfrac{k}{m} + \dfrac{g}{\ell} \end{pmatrix}\begin{pmatrix} \phi \\ \psi \end{pmatrix} + 2mg\ell$$

$$= m\ell^2(\phi, \psi)\begin{pmatrix} \dfrac{g}{\ell} - \omega^2 & 0 \\ 0 & 2\dfrac{k}{m} + \dfrac{g}{\ell} - \omega^2 \end{pmatrix}\begin{pmatrix} \phi \\ \psi \end{pmatrix} + 2mg\ell$$

The eigenvalues λ of the potential energy matrix are in fact the squares of the frequencies of the normal modes.

The transformation matrix has the components of the normalized eigenvectors as its columns,

$$\mathbb{T} = \frac{1}{\sqrt{2}}\begin{pmatrix} 1 & 1 \\ 1 & -1 \end{pmatrix}$$

and it is orthogonal. Thus, the transformed potential energy matrix is

$$
\mathbb{V}' = \frac{1}{2}
\begin{pmatrix} 1 & 1 \\ 1 & -1 \end{pmatrix}
\begin{pmatrix} \dfrac{k}{m} + \dfrac{g}{\ell} & -\dfrac{k}{m} \\[2ex] -\dfrac{k}{m} & \dfrac{k}{m} + \dfrac{g}{\ell} \end{pmatrix}
\begin{pmatrix} 1 & 1 \\ 1 & -1 \end{pmatrix}
$$

$$
= \frac{1}{2}
\begin{pmatrix} 1 & 1 \\ 1 & -1 \end{pmatrix}
\begin{pmatrix} \dfrac{g}{\ell} & \dfrac{2k\ell + gm}{m\ell} \\[2ex] \dfrac{g}{\ell} & -\dfrac{2k\ell + gm}{m\ell} \end{pmatrix}
$$

$$
= \begin{pmatrix} \dfrac{g}{\ell} & 0 \\[2ex] 0 & \dfrac{2k\ell + gm}{m\ell} \end{pmatrix}
$$

and it is diagonal, as expected. The diagonal elements are the eigenvalues of the matrix.

The Lagrangian contains two matrices, and we have succeeded in making them both diagonal simultaneously. The question then arises: Can we always do this? We can if both matrices are real symmetric matrices and at least one of them is positive definite. A matrix is positive definite if

$$
\mathbf{x}^T \mathbb{A}\mathbf{x} \geq 0 \tag{1.86}
$$

for an arbitrary vector $\mathbf{x}$. In the example above, one of our matrices was the unit matrix, and the eigenvalues were found from the equation

$$
\det(\mathbb{V} - \lambda\mathbb{I}) = 0
$$

More generally, the equation for the eigenvalues in a Lagrangian oscillation problem is

$$
\boxed{\det(\mathbb{V} - \lambda\mathbb{K}) = 0} \tag{1.87}
$$

where $\mathbb{K}$ is the kinetic energy matrix. This matrix is always positive definite, and both matrices are symmetric, so simultaneous diagonalization is possible in this case. (See Problem 53.) The resulting eigenvalues are called *generalized eigenvalues*. The eigenvectors are solutions of the matrix equation

$$
\mathbb{V}\mathbf{x} = \lambda\mathbb{K}\mathbf{x}
$$

The transformation in this case need not be, and usually is not, orthogonal.

1.6.6. Matrices and Quantum Mechanics

One very important application of matrices in physics is in quantum mechanics. The state of a system is represented by a vector in a finite or infinite-dimensional vector space. Any physically measurable property of the system may be obtained by allowing a matrix to operate on the state. The matrix represents the physical operator. The eigenvalues of the matrix represent the possible values of the physical measurement. The fact that matrices do not commute, in general, leads to some interesting physical consequences, such as the uncertainty principle.

PROBLEMS

1. **Circular motion.** Answer this question without using Cartesian components. A particle is moving around a circle with angular velocity $\vec{\omega}$. Write its velocity vector $\vec{v}$ as a vector product of $\vec{\omega}$ and the position vector $\vec{r}$ with respect to the center of the circle. Justify your expression using geometrical arguments. Differentiate your relation for $\vec{v}$, and hence derive the angular form of Newton's second law ($\vec{\tau} = I\vec{\alpha}$) from the standard form (equation 1.8).

2. Find two vectors, each perpendicular to the vector $\vec{u} = (1, 2, 2)$ and perpendicular to each other. *Hint:* Use dot and cross products. Determine the transformation matrix $\mathbb{A}$ that allows you to transform to a new coordinate system with x'-axis along $\vec{u}$ and y'- and z'-axes along your other two vectors.

3. Show that the vectors $\vec{u} = (15, 12, 16)$, $\vec{v} = (-20, 9, 12)$, and $\vec{w} = (0, -4, 3)$ are mutually orthogonal and right-handed. Determine the transformation matrix that transforms from the original (x, y, z) coordinate system to a system with x'-axis along $\vec{u}$, y'-axis along $\vec{v}$, and z'-axis along $\vec{w}$. Apply the transformation to find components of the vectors $\vec{a} = (1, 1, 1)$, $\vec{b} = (3, 2, 1)$, and $\vec{c} = (-2, 1, -2)$ in the prime system. Discuss the result for vector $\vec{c}$.

4. A particle moves under the influence of electric and magnetic fields $\vec{E}$ and $\vec{B}$. Show that a particle moving with initial velocity

$$\vec{v}_0 = \left(\frac{1}{B^2}\right) \vec{E} \times \vec{B}$$

is not accelerated if $\vec{E}$ is perpendicular to $\vec{B}$.

 A particle reaches the origin with a velocity $\vec{v} = \vec{v}_0 + \varepsilon\hat{e}$, where $\hat{e}$ is a unit vector in the direction of $\vec{E}$ and $\varepsilon \ll v_0$. If $\vec{E} = E_0(1, 1, 1)$ and $\vec{B} = B_0(1, -2, 1)$, set up a new coordinate system with x'-axis along $\vec{E} \times \vec{B}$ and y'-axis along $\vec{E}$. Determine the particle's position after a short time t. Determine the components of $\vec{v}(t)$ and $\vec{x}(t)$ in both the original and the new system. Give a criterion for "short time."

5. A solid body rotates with angular velocity $\vec{\omega}$. Using cylindrical coordinates with z-axis along the rotation axis, find the components of the velocity vector $\vec{v}$ at an arbitrary point within the body. Use the expression for curl in cylindrical coordinates to evaluate $\vec{\nabla} \times \vec{v}$. Comment on your answer.

6. Starting from conservation of mass in a fixed volume V, use the divergence theorem to derive the continuity equation for fluid flow:

$$\frac{\partial \rho}{\partial t} + \vec{\nabla} \cdot (\rho \vec{v}) = 0$$

where ρ is the fluid density and $\vec{v}$ its velocity.

7. Find the matrix that represents the transformation obtained by (a) rotating about the x-axis by $45°$ counterclockwise and then (b) rotating about the y'-axis by $30°$ clockwise. What are the components of a unit vector along the original z-axis in the new (double-prime) system?

8. Does the matrix

$$\begin{pmatrix} \cos\theta & \sin\theta & 0 \\ \sin\theta & -\cos\theta & 0 \\ 0 & 0 & 1 \end{pmatrix}$$

represent a rotation of the coordinate axes? If not, what transformation does it represent? Draw a diagram showing the old and new coordinate axes, and comment.

9. Represent the following transformation using matrices: (a) a rotation about the z-axis through an angle $\pi/3$, followed by (b) a reflection in the line through the origin and in the x-y plane, at an angle $2\pi/3$ to the *original* x-axis, where both angles are measured counterclockwise from the positive x-axis. Express your answer as a single matrix. You should be able to recognize the matrix either as a rotation about the z-axis through an angle α or as a reflection in a line through the origin at an angle α to the x-axis. Decide whether this transformation is a reflection or a rotation, and give the value of α. (*Note:* For the purposes of this problem, reflection in a line in the x-y plane leaves the z-axis unchanged.)

10. Solve this problem *without* using Cartesian components. Using polar coordinates, write the *components* of the position vectors of two points in a plane: P_1, with *coordinates* r_1 and θ_1, and P_2, with coordinates r_2 and θ_2. (That is, write each vector in the form $\vec{v} = v_r \hat{r} + v_\theta \hat{\theta}$.) What are the coordinates r_3 and θ_3 of the point P_3 whose position vector is

$$\vec{r}_3 = \vec{r}_1 + \vec{r}_2?$$

Hint: Start by drawing the position vectors.

11. A skew (nonorthogonal) coordinate system in a plane has x'-axis along the x-axis and y'-axis at an angle θ to the x-axis, where $\theta < \pi/2$.
    (a) Write the transformation matrix that transforms vector components from the Cartesian x-y system to the skew system.
    (b) Write an expression for the distance between two neighboring points in the skew system. Comment on the differences between your expression and the standard Cartesian expression.
    (c) Write the equation for a circle of radius a, with center at the origin, in the skew system.

12. Prove the Jacobi identity:

$$\vec{a} \times (\vec{b} \times \vec{c}) + \vec{b} \times (\vec{c} \times \vec{a}) + \vec{c} \times (\vec{a} \times \vec{b}) = 0$$

13. Evaluate the vector product

$$(\vec{a} \times \vec{b}) \times (\vec{c} \times \vec{d})$$

in terms of triple scalar products. What is the result if all four vectors lie in a single plane? What is the result if $\vec{a}$, $\vec{b}$, and $\vec{c}$ are mutually perpendicular? What is the result if $\vec{b} = \vec{d}$?

14. Evaluate the product $(\vec{a} \times \vec{b}) \cdot (\vec{c} \times \vec{d})$ in terms of dot products of $\vec{a}$, $\vec{b}$, $\vec{c}$, and $\vec{d}$.

15. Use the vector cross product to express the area of a triangle in three different ways. Hence prove the sine rule (see figure):

$$\frac{\sin \alpha}{A} = \frac{\sin \beta}{B} = \frac{\sin \gamma}{C}$$

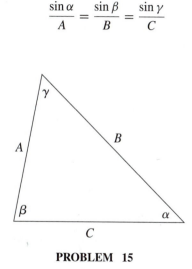

PROBLEM 15

16. Use the dot product $(\vec{a} - \vec{b}) \cdot (\vec{a} - \vec{b})$ to prove the cosine rule for a triangle:

$$c^2 = a^2 + b^2 - 2ab \cos \gamma$$

17. A tetrahedron has its apex at the origin and its edges defined by the vectors $\vec{a}$, $\vec{b}$, and $\vec{c}$, each of which has its tail at the origin (see the figure on the next page). Defining the normal to each face to be outward from the interior of the tetrahedron, determine the total vector area[44] of the four faces of the tetrahedron. Find the volume of the tetrahedron.

---

[44]A vector area is represented by a vector cross product as in Section 1.1.2.

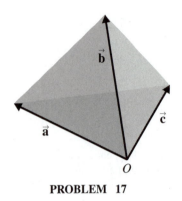

PROBLEM 17

18. A sphere of unit radius is centered at the origin. Points U, V, and W on the surface of the sphere have position vectors $\vec{\mathbf{u}}$, $\vec{\mathbf{v}}$, and $\vec{\mathbf{w}}$. Show that points P and Q on the sphere, located on a diameter perpendicular to the plane containing the points U, V, and W, have position vectors given by

$$\vec{\mathbf{r}} = \pm \frac{\vec{\mathbf{u}} \times \vec{\mathbf{v}} + \vec{\mathbf{v}} \times \vec{\mathbf{w}} + \vec{\mathbf{w}} \times \vec{\mathbf{u}}}{[\vec{\mathbf{u}}, \vec{\mathbf{v}}, \vec{\mathbf{w}}]} \cos \theta$$

where θ is the angle between the vectors $\vec{\mathbf{r}}$ and $\vec{\mathbf{u}}$.

19. Show that

$$\vec{\nabla} \times (\vec{\nabla}\Phi) = 0$$

for any scalar field Φ.

20. Find an expression for $\vec{\nabla} \times (\vec{\mathbf{a}} \times \vec{\mathbf{b}})$ in terms of derivatives of $\vec{\mathbf{a}}$ and $\vec{\mathbf{b}}$.

21. Prove the identity

$$\vec{\nabla}(\vec{\mathbf{a}} \cdot \vec{\mathbf{b}}) = (\vec{\mathbf{a}} \cdot \vec{\nabla})\vec{\mathbf{b}} + (\vec{\mathbf{b}} \cdot \vec{\nabla})\vec{\mathbf{a}} + \vec{\mathbf{a}} \times (\vec{\nabla} \times \vec{\mathbf{b}}) + \vec{\mathbf{b}} \times (\vec{\nabla} \times \vec{\mathbf{a}})$$

Hint: Start with the last two terms on the right-hand side.

22. Compute $\vec{\nabla} \cdot (\vec{\mathbf{a}} \times \vec{\mathbf{b}})$ in terms of curl $\vec{\mathbf{a}}$ and curl $\vec{\mathbf{b}}$.

23. Obtain an expression for $\vec{\nabla} \times (\phi\vec{\mathbf{u}})$, and hence show that $\vec{\nabla} \times (\phi\vec{\nabla}\phi) = 0$.

24. The equation of motion for a fluid may be written

$$\rho \left(\frac{\partial \vec{\mathbf{v}}}{\partial t} + (\vec{\mathbf{v}} \cdot \vec{\nabla})\vec{\mathbf{v}} \right) = -\vec{\nabla}P + \rho\vec{\mathbf{g}}$$

where $\vec{\mathbf{v}}$ is the fluid velocity at a point, ρ its density, and P the pressure. The acceleration due to gravity is $\vec{\mathbf{g}}$. Use the result of Problem 21 to show that for fluid flow that is incompressible ($\rho = $ constant) and steady ($\partial/\partial t \equiv 0$), Bernoulli's law holds:

$$P + \tfrac{1}{2}\rho v^2 + \rho g h = \text{ constant along a streamline}$$

Hint: Express the statement "constant along a streamline" as a directional derivative being equal to zero.

Under what conditions is $P + \frac{1}{2}\rho v^2 + \rho g h$ equal to an absolute constant, the same throughout the fluid?

25. Evaluate the integral

$$\oint_C \vec{u} \cdot d\vec{l}$$

where

(a) C is the unit circle in the x-y plane and centered at the origin and

$$\vec{u} = x^2 y \hat{x} - xy^2 \hat{y}$$

(b) C is a semicircle of radius a in the x-y plane with the flat side along the x-axis, the center of the circle at the origin, and

$$\vec{u} = xy^2 \hat{x} + yx^2 \hat{y}$$

(c) C is a 3-4-5 right-angled triangle with the sides of length 3 and 4 along the x- and y-axes, respectively, and

$$\vec{u} = x^2 \hat{x} + xy \hat{y}$$

(d) C is a semicircle of radius a in the x-y plane with the flat side along the x-axis, the center of the circle at the origin, and

$$\vec{u} = (2x - y^3)\hat{x} - (3y^2 + x^3)\hat{y}$$

26. Evaluate the integral

$$\int_S \vec{v} \cdot d\vec{A}$$

where

(a) S is a sphere of radius 2 centered on the origin and

$$\vec{v} = x^3 \hat{x} + 3yz^2 \hat{y} + 3y^2 z \hat{z}$$

(b) S is a hemisphere of radius 1, with the center of the sphere at the origin, the flat side in the x-y plane, and

$$\vec{v} = x^2 yz(\hat{y} + \hat{z})$$

27. Show that the vector

$$\vec{u} = x\hat{x} + y\hat{y} - 2z\hat{z}$$

has zero divergence (it is solenoidal) and zero curl (it is irrotational). Find a scalar function ϕ such that

$$\vec{\mathbf{u}} = \vec{\nabla}\phi$$

and a vector $\vec{\mathbf{A}}$ such that

$$\vec{\mathbf{u}} = \vec{\nabla} \times \vec{\mathbf{A}}$$

28. Show that the vector

$$\vec{\mathbf{v}} = \frac{\hat{\mathbf{r}}}{r^2}$$

has zero divergence (it is solenoidal) and zero curl (it is irrotational) for $r \neq 0$. Find a scalar function ϕ such that

$$\vec{\mathbf{v}} = \vec{\nabla}\phi$$

and a vector $\vec{\mathbf{A}}$ such that

$$\vec{\mathbf{v}} = \vec{\nabla} \times \vec{\mathbf{A}}$$

29. A surface S is bounded by a curve C. The solid angle subtended by the surface S at a point P, where P is in the vicinity of but not **on** the curve, is given by

$$\Omega = \int_S \frac{da_\perp}{R^2}$$

Here $da_\perp$ is an element of area of the loop projected perpendicular to the vector $\vec{\mathbf{R}} = \vec{\mathbf{x}} - \vec{\mathbf{x}}'$, $\vec{\mathbf{x}}$ is the position vector of the point P with respect to some chosen origin O, and $\vec{\mathbf{x}}'$ is a vector that labels an arbitrary point on the surface or the curve. Now let the curve be rigidly displaced by a small amount $d\vec{\mathbf{s}}$. Express the resulting change in solid angle $d\Omega$ as an integral around the curve. Hence, show that

$$\vec{\nabla}\Omega = -\vec{\nabla} \times \oint_C \frac{d\vec{\mathbf{l}}}{R}$$

30. Prove the theorems

(a)

$$\int_V \vec{\nabla}\Phi \, dV = \oint_S \Phi\hat{\mathbf{n}} \, dA$$

(b)

$$\int_V \vec{\nabla} \times \vec{\mathbf{u}} \, dV = \oint_S (\hat{\mathbf{n}} \times \vec{\mathbf{u}}) \, dA$$

31. Prove

(a)

$$\oint_C \Phi \, d\vec{l} = \int_S \hat{n} \times \vec{\nabla} \Phi \, dA$$

(b)

$$\int_S (\hat{n} \times \vec{\nabla}) \times \vec{u} \, dA = \oint_C d\vec{l} \times \vec{u}$$

32. Derive the expressions for gradient, divergence, curl, and the Laplacian in spherical coordinates.

33. In polar coordinates in a plane, the unit vectors $\hat{r}$ and $\hat{\theta}$ are functions of position. Draw a diagram showing the vectors $\hat{r}$ at two neighboring points with angular coordinates θ and $\theta + d\theta$. Use your diagram to find the difference $\Delta\hat{r}$ and hence find the derivative $\partial\hat{r}/\partial\theta$.

34. The vector operator

$$\vec{L} = \frac{1}{i}\vec{r} \times \vec{\nabla}$$

appears in physics as the angular momentum operator. (Here $i = \sqrt{-1}$ and $\vec{r}$ is the position vector.) Prove the identity

$$\vec{\nabla}(\vec{r} \cdot \vec{u}) = \vec{u} + \vec{r}(\vec{\nabla} \cdot \vec{u}) + i(\vec{L} \times \vec{u})$$

for an arbitrary vector $\vec{u}$.

35. Can you express the vector $\vec{a} = (1, 2, 3)$ as a linear combination of the vectors $\vec{u}_1 = (1, 1, 1)$, $\vec{u}_2 = (1, 0, -1)$, and $\vec{u}_3 = (2, 1, 0)$? Can you express the vector $\vec{b} = (1, 3, 2)$ as a linear combination of the vectors $\vec{u}_1$, $\vec{u}_2$, and $\vec{u}_3$? Explain your answers geometrically.

36. Show that an antisymmetric 3×3 matrix has only three independent elements. How many independent elements does a symmetric 3×3 matrix have? Extend these results to an $N \times N$ matrix.

37. Show that if any two rows of a matrix are equal, its determinant is zero.

38. Prove that a matrix with one row of zeros has a determinant equal to zero. Also show that if an $N \times N$ matrix is multiplied by a constant c, its determinant is multiplied by c^N.

39. Prove that a matrix and its transpose have the same determinant.

40. Prove that the trace of a matrix is invariant under change of basis—that is,

$$\text{Tr}\,(\mathbb{A}') = \text{Tr}\,(\mathbb{C}\mathbb{A}\mathbb{C}^{-1}) = \text{Tr}\,(\mathbb{A})$$

41. Show that the determinant of a matrix is invariant under change of basis—that is, $\det\,(\mathbb{A}') = \det\,(\mathbb{A})$. Hence show that the determinant of a real symmetric matrix equals the product of its eigenvalues.

42. If the product of two matrices is zero, it is not necessary that either one be zero. In particular, show that a 2×2 matrix whose square is zero may be written in terms of two parameters a and b, and find the general form of the matrix. Verify that its determinant is zero.

43. If the product of the matrix $\mathbb{A} = \begin{pmatrix} a & b \\ c & d \end{pmatrix}$ and another nonzero matrix $\mathbb{B}$ is zero, find the elements of $\mathbb{B}$. You may find it necessary to impose some conditions on matrix $\mathbb{A}$. If so, state what they are.

44. Diagonalize the matrix

$$\begin{pmatrix} 1 & 1 & 1 \\ 1 & 0 & 0 \\ 1 & 0 & 0 \end{pmatrix}$$

45. Show that a real symmetric matrix with one or more eigenvalues equal to zero has no inverse (it is *singular*).

46. Diagonalize the matrix $\begin{pmatrix} 1 & 2 \\ 3 & 4 \end{pmatrix}$ and find the eigenvectors. Are the eigenvectors orthogonal? If not, why not?

47. What condition must be imposed on the matrix $\mathbb{A}$ in order that $\mathbb{A}\mathbb{B} = \mathbb{A}\mathbb{C}$ with $\mathbb{B} \neq \mathbb{C}$? If

$$\mathbb{B} = \begin{pmatrix} 1 & 0 \\ 0 & 2 \end{pmatrix} \quad \text{and} \quad \mathbb{C} = \begin{pmatrix} 1 & 1 \\ 0 & 1 \end{pmatrix}$$

find a matrix $\mathbb{A}$ such that $\mathbb{A}\mathbb{B} = \mathbb{A}\mathbb{C}$.

48. Show that if $\mathbb{A}$ is a real symmetric matrix and $\mathbb{C}$ is orthogonal, then $\mathbb{A}' = \mathbb{C}\mathbb{A}\mathbb{C}^{-1}$ is also symmetric.

49. Show that $\mathbb{A}\mathbb{B} = \mathbb{B}\mathbb{A}$ if both $\mathbb{A}$ and $\mathbb{B}$ are diagonal matrices.

50. Prove that the inner product is invariant with respect to change of basis under orthogonal transformations.

51. A quadratic expression of the form $\alpha x^2 + 2\beta xy + \gamma y^2 = 1$ represents a curve in the x-y plane.
    (a) Write this expression in matrix form.
    (b) Diagonalize the matrix, and hence identify the form of the curve and find its symmetry axes. Determine how the shape of the curve depends on the values of α, β, and γ. Draw the curve in the case $\alpha = \beta = 2$, $\gamma = 3$.

52. Two small objects, each of mass m, are joined by a spring of relaxed length ℓ. Identical springs hold each mass to a wall. The walls are separated by a distance 3ℓ. Write the Lagrangian for the system, and find the normal modes and the oscillation frequency for each mode.

53. Find the normal modes of a jointed pendulum system. Two point objects, each of mass m, are linked by stiff but massless rods, each of length ℓ. The upper rod is attached to a pivot. The system is in equilibrium when both rods hang vertically below the pivot. The figure shows the system when displaced from equilibrium.

Write the Lagrangian for the system in matrix form. Simultaneously diagonalize the kinetic energy and potential energy matrices, and find the generalized eigenvalues. Determine the eigenvectors. Find the matrix that effects the transformation, and verify that both matrices are diagonalized.

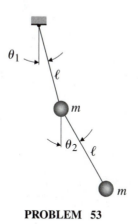

PROBLEM 53

CHAPTER 2

Complex Variables

2.1. ALL ABOUT NUMBERS

Numbers are classified emotionally! The Greeks liked integers and rationals (a rational number is the ratio of two integers).

Integers: $1, 2, 3, 4, \ldots$

Rationals: $\dfrac{1}{2}, \dfrac{3}{4}, \dfrac{107}{436}, \ldots$

But they did not like *irrationals* (irrationals are numbers that cannot be expressed as the ratio of two integers). Yet irrationals show up in lots of common circumstances, like the ratio of the circumference of a circle to its diameter.

Irrationals: $e, \pi, \sqrt{2}, \ln 2, \ldots$

In spite of the name, there is nothing wrong with these numbers, as we all know, and we have no trouble computing with them.

What happens when we try to solve the following quadratic equation?

$$f(z) = z^2 + z + \frac{5}{2} = 0$$

It doesn't factor, so we use the formula:

$$z = \frac{-1 \pm \sqrt{1 - 10}}{2} = -\frac{1}{2} \pm \frac{\sqrt{-9}}{2}$$

This result does not exist in the list of numbers written above. We can see this geometrically by plotting the function $f(z) = z^2 + z + \frac{5}{2}$ (Figure 2.1) and noticing that the function does not cross the z-axis. The solution to the dilemma is to add a dimension to our number system. We define the quantity[1] $\sqrt{-1} \equiv i$ so that the solution to the equation is

$$z = -\frac{1}{2} \pm \frac{3}{2}i = x + iy$$

---

[1] Engineers often use the notation $j = \sqrt{-1}$.

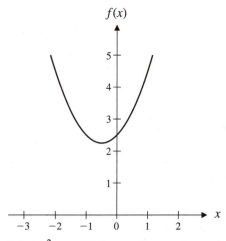

FIGURE 2.1. The curve $f(x) = x^2 + x + 5/2$ does not intersect the x-axis, so the equation $f(x) = 0$ has no real solutions.

We now need two real numbers, x and y, to describe the *complex number z*. The number y multiplying the i is called the imaginary part,[2] while x is the real part. Now the original expression becomes a function of the two real numbers x and y:

$$z^2 + z + \frac{5}{2} = (x + iy)^2 + x + iy + \frac{5}{2} = x^2 + 2ixy + i^2 y^2 + x + iy + \frac{5}{2}$$

Since $i^2 = -1$, this expression becomes

$$f(z) = f_1(x, y) + if_2(x, y) = x^2 - y^2 + x + \frac{5}{2} + iy(2x + 1)$$

To plot this expression we would have to plot the two functions $f_1 = x^2 - y^2 + x + \frac{5}{2}$ and $f_2 = y(2x + 1)$. To obtain solutions of $f(z) = 0$, we require that f_1 and f_2 be zero simultaneously. By allowing the extra dimension, we find that both figures intersect the plane $f = 0$ at two common points $x = -\frac{1}{2}, y = \pm\frac{3}{2}$.

By allowing ourselves to use complex numbers with both real and imaginary parts, we are able to solve an enormous number of equations that do not have real solutions, not just quadratics of the type we considered here.

Complex numbers satisfy all the usual algebraic rules, so long as we remember that $i^2 = -1$. Then, for example,

Addition:

$$(a + ib) + (c + id) = (a + c) + i(b + d)$$

$$(2 + 5i) + (3 - 4i) = (2 + 3) + (5 - 4)i$$

$$= 5 + i$$

---

[2]The name is unfortunate; these numbers are not "imaginary" in the normal sense of the word. In fact, they prove extraordinarily useful in solving problems in the real world of physics.

Subtraction:

$$(a + ib) - (c + id) = (a - c) + i(b - d)$$

$$(2 + 5i) - (3 - 4i) = (2 - 3) + (5 + 4)i$$

$$= -1 + 9i$$

Multiplication:

$$(a + ib) \times (c + id) = ac + i(bc + ad) + i^2 bd$$

$$= ac - bd + i(bc + ad) \tag{2.1}$$

$$(2 + 5i) \times (3 - 4i) = 26 + 7i$$

Division is trickier if we want to write the result in the standard $x + iy$ form. As a preamble, notice from equation (2.1) that, if we take $a = c$, the imaginary part of the product is zero for $d = -b$. Thus, the product of a number $z = a + ib$ with its *complex conjugate* $z^* = a - ib$ is purely real.

The complex conjugate z^* of any complex number z is formed by changing the sign of its imaginary part.

The product of a complex number and its complex conjugate is purely real and non-negative:

$$zz^* = (a + ib)(a - ib) = a^2 + b^2 \tag{2.2}$$

We can make use of this result when dividing two complex numbers. We multiply the top and bottom of the ratio by the complex conjugate of the denominator:

$$\frac{z_1}{z_2} = \frac{x_1 + iy_1}{x_2 + iy_2} = \left(\frac{x_1 + iy_1}{x_2 + iy_2}\right)\left(\frac{x_2 - iy_2}{x_2 - iy_2}\right)$$

$$= \frac{x_1 x_2 + y_1 y_2 + i(y_1 x_2 - x_1 y_2)}{x_2^2 + y_2^2} = \frac{x_1 x_2 + y_1 y_2}{x_2^2 + y_2^2} + i\frac{(y_1 x_2 - x_1 y_2)}{x_2^2 + y_2^2}$$

So we may write the result of the division in the standard form:

$$\frac{(2 + 5i)}{(3 - 4i)} = \frac{(2 + 5i)}{(3 - 4i)}\frac{(3 + 4i)}{(3 + 4i)} = \frac{-14 + 23i}{(3^2 + 4^2)} = -\frac{14}{25} + \frac{23}{25}i$$

2.1.1. The Argand Diagram

The Argand diagram (or complex plane) is a diagram in which we plot complex numbers as points with x and y coordinates equal to the real and imaginary parts of the number, respectively (Figure 2.2). Since we may equally well describe a point in the plane using polar coordinates (r, θ), we can express the number as

$$z = x + iy = r(\cos \theta + i \sin \theta) \tag{2.3}$$

where $r = \sqrt{x^2 + y^2}$ is the length of the line from the origin to P and is called the amplitude (or absolute value) of z: $r = |z|$. The angle measured counterclockwise from the x-axis, $\theta = \tan^{-1}(y/x)$, is called the argument (or phase) of z: $\theta = \arg(z)$. As $x \to 0$, $y/x \to +\infty$ for positive y, and $\theta \to +\pi/2$. Similarly, for negative y, $\theta \to -\pi/2$ as $x \to 0$. By convention, we choose a range of 2π for the angle θ. Usual choices are $0 \le \theta < 2\pi$ or $-\pi < \theta \le \pi$.

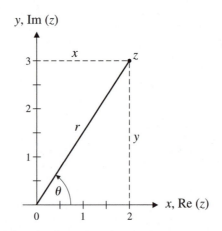

FIGURE 2.2. The Argand diagram allows us to plot complex numbers as points in a plane. The point is described by Cartesian $(z = x + iy)$ or polar $(z = re^{i\theta})$ coordinates.

There is a nifty way to express z in terms of the polar coordinates r and θ. Notice that

$$\frac{\partial z}{\partial \theta} = r(-\sin \theta + i \cos \theta) = iz$$

We can integrate this differential equation to get

$$z = g(r)e^{i\theta}$$

If we evaluate this result at $\theta = 0$ and compare with the original expression (2.3) in terms of $\cos \theta$ and $\sin \theta$, we see that we should take $g(r) \equiv r$. Thus,

$$\boxed{z = re^{i\theta}} \tag{2.4}$$

which is the *polar form* for z. In this form, multiplication and division are really easy:

$$z_1 z_2 = r_1 r_2 e^{i(\theta_1 + \theta_2)}$$

and

$$\frac{z_1}{z_2} = \frac{r_1}{r_2} e^{i(\theta_1 - \theta_2)}$$

Comparing relations (2.4) and (2.3), we see that we may write

$$e^{i\theta} = \cos\theta + i\sin\theta \qquad (2.5)$$

a relation called Euler's formula. Changing the sign of the angle, we find

$$e^{-i\theta} = \cos\theta - i\sin\theta$$

Then, adding the two relations, we have

$$\cos\theta = \frac{e^{i\theta} + e^{-i\theta}}{2} \qquad (2.6)$$

and subtracting, we have

$$\sin\theta = \frac{e^{i\theta} - e^{-i\theta}}{2i} \qquad (2.7)$$

These expressions are similar to the definitions of the hyperbolic functions:

$$\cosh x = \frac{e^x + e^{-x}}{2} \quad \text{and} \quad \sinh x = \frac{e^x - e^{-x}}{2}$$

Thus, we have the relation

$$\cosh(i\theta) = \cos\theta \qquad (2.8)$$

and, conversely,

$$\cos(i\theta) = \cosh\theta \qquad (2.9)$$

Similarly,

$$\sinh(i\theta) = i\sin\theta \qquad (2.10)$$

and

$$\sin(i\theta) = i\sinh\theta \qquad (2.11)$$

Using these relations, we note that the cosine of a complex number is not necessarily a real number between -1 and 1. When using complex numbers, we must recognize that functions may behave differently than they do when we work with numbers that are purely real.

Example 2.1. Evaluate $\sin(2i + \pi/4)$.

$$\sin(2i + \pi/4) = \frac{e^{i(2i+\pi/4)} - e^{-i(2i+\pi/4)}}{2i}$$

$$= \frac{e^{-2}(\cos \pi/4 + i \sin \pi/4) - e^{2}(\cos \pi/4 - i \sin \pi/4)}{2i}$$

$$= \frac{\sqrt{2}}{2}(i \sinh 2 + \cosh 2) = 2.6603 + 2.5646i$$

Notice that both the real part and the absolute value of this number are greater than 1.

Algebraic operations on complex numbers can be regarded as geometric operations in the complex plane (Figure 2.3, a, b, c). For example,

Addition of complex numbers $\longleftrightarrow$ Vector addition (Figure 2.3a)

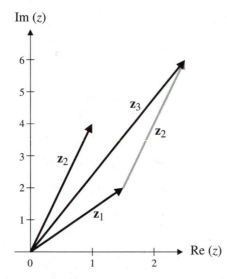

FIGURE 2.3a. Addition of complex numbers is equivalent to addition of position vectors in the plane. Here $z_3 = z_1 + z_2$.

Using this diagram, it is straightforward to prove the inequalities

$$|z_1 + z_2| \leq |z_1| + |z_2| \tag{2.12}$$

and
$$|z_1 - z_2| \geq \big||z_1| - |z_2|\big| \tag{2.13}$$

Complex conjugate $\longleftrightarrow$ Reflection in the real axis (Figure 2.3b)

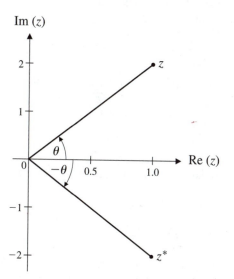

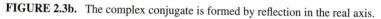

FIGURE 2.3b. The complex conjugate is formed by reflection in the real axis.

Multiplication of two numbers $\longleftrightarrow$ Magnification plus rotation

For example, in Figure 2.3c we show the product $z_1 \times z_2 = 2e^{i\pi/4} \times 3e^{i\pi/6} = 6e^{i5\pi/12}$.

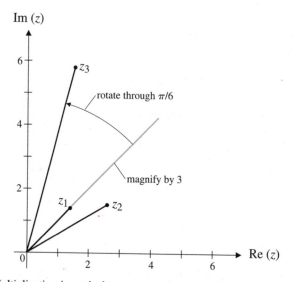

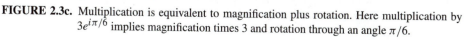

FIGURE 2.3c. Multiplication is equivalent to magnification plus rotation. Here multiplication by $3e^{i\pi/6}$ implies magnification times 3 and rotation through an angle $\pi/6$.

2.1.2. Roots

Since $e^{2\pi i} = \cos(2\pi) + i\sin(2\pi) = 1$, we can add 2π to the argument of any complex number without changing its value. That is,

$$z = re^{i\theta} = re^{i(\theta+2\pi)}$$

Now let's see what happens when we take the nth root of this number:

$$z^{1/n} = (re^{i\theta})^{1/n}$$
$$= r^{1/n}e^{i\theta/n}$$

This is the primary root. Now add 2π to the argument of z and take the root again:

$$z^{1/n} = r^{1/n}\exp(i[\theta + 2\pi]/n)$$
$$= r^{1/n}e^{i\theta/n}e^{2\pi i/n}$$

This root is distinct from the primary root. We can continue to add 2π $(n-1)$ times before we repeat a root. Thus, *there are n distinct nth roots of any nonzero number.* They are given by

$$z^{1/n} = r^{1/n}e^{i(\theta+2\pi m)/n} \tag{2.14}$$

$$= r^{1/n}\left[\cos\left(\frac{\theta+2\pi m}{n}\right) + i\sin\left(\frac{\theta+2\pi m}{n}\right)\right], \quad 0 \le m < n \tag{2.15}$$

Example 2.2. Find the cube roots of 1.
First write

$$1 = 1\exp(2\pi im), \quad m = 0, 1, 2$$

Then the roots are $\exp(2\pi im/3)$, with $m = 0$, 1, and 2.

$m = 0$: $1^{1/3}e^0 = 1$

$m = 1$: $1^{1/3}e^{2\pi i/3} = \cos(2\pi/3) + i\sin(2\pi/3) = -\dfrac{1}{2} + i\dfrac{\sqrt{3}}{2}$

and

$m = 2$: $1^{1/3}e^{4\pi i/3} = \cos(4\pi/3) + i\sin(4\pi/3) = -\dfrac{1}{2} - i\dfrac{\sqrt{3}}{2}$

Taking $m > 2$ does not give any new roots—we simply repeat the values we have already found. The roots fall at the vertices of an equilateral triangle in the complex plane (Figure 2.4).

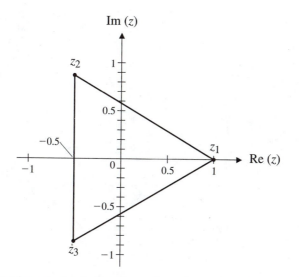

FIGURE 2.4. The cube roots of unity lie at the vertices of an equilateral triangle. One vertex lies on the real axis at the point $z = 1$.

The nth roots of a number always fall at the vertices of a symmetrical n-sided figure in the complex plane. For example, the fifth roots fall at the vertices of a pentagon.

2.1.3. Complex Functions

Mappings

A complex function $w(z)$ takes the number $z = x + iy$ and generates a new complex number $w = u + iv$. Since a plane is needed to plot a single number, complex functions cannot be represented in a single two- or three-dimensional diagram. We would need $2 \times 2 = 4$ dimensions. Thus, instead, we use two diagrams, and we think of a complex function as a mapping of the complex plane onto another copy of itself.

Example 2.3. Describe the mapping

$$w = \frac{1}{z}$$

In terms of the coordinates x, y or r, θ, we have

$$w = u + iv = \frac{1}{x + iy} = \frac{x - iy}{x^2 + y^2} = \frac{1}{r} e^{-i\theta} = \rho e^{i\phi}$$

We can map out in the w-plane the image of a curve in the z-plane. For example, the unit circle in the z-plane ($r = 1$) maps to a circle ($\rho = 1$) in the w-plane, but traversed in the opposite sense ($\phi = -\theta$). The inside of the circle ($r < 1$) in the z-plane maps to the outside of the circle ($\rho > 1$) in the w-plane, and vice versa (Figure 2.5). This function gives us the complete w-plane from the complete z-plane, one to one.

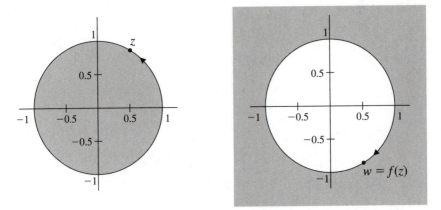

FIGURE 2.5. The function (mapping) $w = 1/z$ maps the interior of the circle $|z| = 1$ to the exterior of the circle $|w| = 1$. Traveling around the unit circle in the z-plane counterclockwise corresponds to traveling around the unit circle in the w-plane clockwise.

Multivalued Functions and Branch Cuts

Some functions don't work that way. The log function exhibits some of the possible curious behaviors:

$$w = \ln z = \ln (re^{i\theta}) = \ln r + i\theta = u + iv$$

If we take all possible points in the z-plane with $0 \le \theta < 2\pi$, we get only one strip of the w-plane, $0 \le v < 2\pi$. This is the first branch, or *principal branch*, of the function. To get the whole w-plane, we have to go around the z-plane more than once; that is, we have to add $\pm 2\pi n$ to the argument of z. (We can imagine the plane as a pad of paper containing many sheets, each labeled by the integer n.) Every time we cross the positive x-axis, θ increases by 2π, and thus v increases by 2π also. (We move to the next sheet in the pad and to a new branch of the function.) The positive x-axis forms the *branch cut* of the complex plane for this function. As we cross the positive x-axis, we move from one sheet to the next, or to a new copy of the plane. The branch cut—here, the positive x-axis—acts as a ramp that allows us to move from one sheet to the next. With the log function, we have to go around infinitely many times to generate the whole w-plane. Thus, this function has infinitely many branches. (Our pad of paper has infinitely many sheets.) With $n = 0$, we get the *principal branch* of the function.

If we choose to define θ such that $-\pi < \theta \le \pi$, then the branch cut lies along the negative x-axis at $\theta = \pi$. Thus, we can choose the position of the branch cut by choosing the range of values for θ. The choice that we make affects the value of $f(z)$. For example, consider the point z_0 that lies on the negative imaginary axis at a distance $|z_0| = 2$ from the origin (Figure 2.6, a, b). With the branch cut along the positive real axis, and hence $0 \le \theta < 2\pi$, z_0 has argument $\theta = 3\pi/2$, and so for the principal branch

$$\ln z_0 \,|_{\text{branch cut on positive real axis}} = \ln 2 + \frac{3\pi}{2}i$$

However, with the branch cut along the negative real axis, and hence $-\pi < \theta \le \pi$, z_0 has

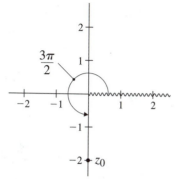

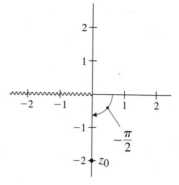

FIGURE 2.6a. Branch cut on the positive x-axis. The point z_0 $(x = 0,$ $y = -2)$ has argument $3\pi/2$.

FIGURE 2.6b. Branch cut on the negative x-axis. The point z_0 has argument $-\pi/2$.

argument $\theta = -\pi/2$, and so for the principal branch

$$\ln z_0 \text{ |branch cut on negative real axis} = \ln 2 - \frac{\pi}{2}i$$

Since the two values are different, so are the two functions.

The cube root function has three branches, since adding more multiples of 2π to the argument simply reproduces the same roots that we already found with $n = 0, 1, 2$.

For both these functions, the branch cut ends at the origin, which is the *branch point* for each function. Think of the branch cut as a tadpole with a fixed head (the branch point) and a movable tail (the branch cut).

2.1.4. An Example from Physics

To demonstrate the use of complex variables in physics, let's look at an electromagnetic wave. The general expression for a plane wave may be written in the form

$$y = y_0 \cos (\vec{k} \cdot \vec{x} - \omega t + \phi)$$

where y_0 is the amplitude of the wave; $\vec{k}$, the wave vector, gives the direction of propagation and also the wavelength, $\lambda = 2\pi/k$; ω is the angular frequency; and the wave speed is $v = \omega/k$. The phase constant, ϕ, is the wave phase at $\vec{x} = 0$, $t = 0$. For an electromagnetic wave, the electric wave amplitude is a vector:

$$\vec{E} = \vec{E}_0 \cos (\vec{k} \cdot \vec{x} - \omega t + \phi)$$

We can simplify the expression by choosing our x-axis along $\vec{k}$. Then, using equation (2.5) with $\theta = kx - \omega t + \phi$, we can write

$$\vec{E} = \vec{E}_0 \, \text{Re} \, (\exp (i[kx - \omega t + \phi]))$$

$$= \text{Re} \, \vec{E}_0 \, e^{i\phi} \exp (i[kx - \omega t]) = \text{Re} \, \vec{E}_a \exp (i[kx - \omega t]) \qquad (2.16)$$

where $\vec{E}_a = \vec{E}_0 e^{i\phi}$ is the complex amplitude of the wave.

This mathematical description makes it easier to investigate the propagation of a wave. Let's start with Maxwell's equations:

$$\vec{\nabla} \cdot \vec{D} = \rho_f \qquad (2.17)$$

$$\vec{\nabla} \cdot \vec{B} = 0 \qquad (2.18)$$

$$\vec{\nabla} \times \vec{E} = -\frac{\partial \vec{B}}{\partial t} \qquad (2.19)$$

$$\vec{\nabla} \times \vec{H} = \vec{j}_f + \frac{\partial \vec{D}}{\partial t} \qquad (2.20)$$

where $\vec{D} = \varepsilon \vec{E}$ is the electric displacement, the magnetic induction $\vec{B} = \mu \vec{H}$, and $\vec{H}$ is the magnetic field. Now assume a solution of the form (2.16) for both $\vec{E}$ and $\vec{B}$, and substitute into the equations. For waves in a region where the free charge and current densities are zero, $\rho_f = 0$ and $\vec{j}_f = 0$, Maxwell's equations take the following form:

Gauss' law (equation 2.17):

$$ik\varepsilon E_x = 0$$

Gauss' law for $\vec{B}$ (equation 2.18):

$$ikB_x = 0$$

Thus, $E_x = 0$ and $B_x = 0$; both $\vec{E}$ and $\vec{B}$ are perpendicular to the direction of propagation. Now let's simplify further by choosing our y-axis along the direction of $\vec{E}$. Then from Faraday's law (equation 2.19) we get

$$\frac{\partial E_y}{\partial x} = -\frac{\partial B_z}{\partial t}$$

$$ikE_y = i\omega B_z \qquad (2.21)$$

while from Ampere's law (equation 2.20) we get

$$-ik\frac{B_z}{\mu} = -i\omega\varepsilon E_y \qquad (2.22)$$

Combining equations (2.21) and (2.22), we get

$$kE_y = \omega\left(\frac{\omega\mu\varepsilon}{k}E_y\right)$$

Since we want a solution with $E_y \neq 0$, we must have

$$\frac{\omega}{k} = \frac{1}{\sqrt{\mu\varepsilon}} \qquad (2.23)$$

which is the wave phase speed. In a vacuum, relation (2.23) becomes $\omega/k = 1/\sqrt{\mu_0\varepsilon_0} \equiv c$.

Now let's see what happens if we allow the wave to propagate through a conducting medium with conductivity σ, but ε and μ real. There will be a current given by

$$\vec{\jmath} = \sigma \vec{E}$$

which must be included in Ampere's law. Equation (2.22) becomes

$$-ik\frac{B_z}{\mu} = \sigma E_y - i\omega\varepsilon E_y \Rightarrow B_z = E_y\frac{(\sigma - i\omega\varepsilon)\mu}{-ik}$$

Now when we combine this equation with (2.21), we get

$$kE_y = i\frac{\omega}{k}(\sigma - i\omega\varepsilon)\mu E_y$$

Again we may cancel the factor E_y to get a relation between ω and k:

$$k^2 = i\omega\mu(\sigma - i\omega\varepsilon) = i\omega\mu\sigma + \omega^2\mu\varepsilon \qquad (2.24)$$

We expect the frequency to be a real number,[3] but then the square of the wave number is complex! This means that k itself must be complex, so let's write it as

$$k = \kappa + i\gamma \qquad (2.25)$$

Before proceeding, we should ask whether this idea makes sense. If we put our expression for k back into equation (2.16), we find

$$\vec{E} = \text{Re } \vec{E}_a \exp\left(i[(\kappa + i\gamma)x - \omega t]\right)$$
$$= \text{Re } \vec{E}_a \exp\left(i[\kappa x - \omega t]\right) \exp\left(-\gamma x\right)$$
$$= \vec{E}_0 \cos\left(\kappa x - \omega t + \phi\right)e^{-\gamma x} \qquad (2.26)$$

Thus, the wave amplitude decreases exponentially as the wave propagates. The wave energy is converted to kinetic energy of electrons as the electric field drives an oscillating current in the conducting medium.

OK; let's solve for this complex k. With

$$k^2 = \kappa^2 - \gamma^2 + 2i\kappa\gamma$$

equation (2.24) becomes

$$\kappa^2 - \gamma^2 + 2i\kappa\gamma = i\omega\mu\sigma + \omega^2\mu\varepsilon$$

Equating real and imaginary parts, we get two equations for κ and σ:

$$\kappa^2 - \gamma^2 = \omega^2\mu\varepsilon \qquad (2.27)$$

---

[3]This is the usual convention, but it is sometimes useful to make the opposite assumption, choosing k real and ω complex.

and

$$2\kappa\gamma = \omega\mu\sigma \tag{2.28}$$

Then, from equation (2.28),

$$\gamma = \frac{\omega\mu\sigma}{2\kappa} \tag{2.29}$$

Substituting this into equation (2.27) gives

$$\kappa^4 - \omega^2\mu\varepsilon\kappa^2 - \left(\frac{\omega\mu\sigma}{2}\right)^2 = 0$$

We can solve this quadratic in κ^2 to get

$$\kappa^2 = \frac{\omega^2\mu\varepsilon}{2}\left(1 \pm \sqrt{1 + \left(\frac{\sigma}{\omega\varepsilon}\right)^2}\right)$$

Since κ is real by definition, κ^2 must be positive, so we take the $+$ sign:

$$\kappa^2 = \frac{\omega^2\mu\varepsilon}{2}\left(1 + \sqrt{1 + \left(\frac{\sigma}{\omega\varepsilon}\right)^2}\right) \tag{2.30}$$

which reduces to our previous result (2.23) when $\sigma \to 0$, as required.

Now let's solve for γ. We substitute our solution (2.30) for κ into equation (2.29):

$$\gamma = \frac{\omega\mu\sigma}{2\kappa}$$

$$= \sigma\sqrt{\frac{\mu}{2\varepsilon}\frac{1}{1 + \sqrt{1 + (\sigma/\omega\varepsilon)^2}}} \tag{2.31}$$

Now in the limit of low conductivity ($\sigma \ll \omega\varepsilon$), these results become

$$\kappa \simeq \omega\sqrt{\mu\varepsilon}\sqrt{1 + \frac{1}{4}\left(\frac{\sigma}{\omega\varepsilon}\right)^2} \simeq \omega\sqrt{\mu\varepsilon}\left[1 + \frac{1}{8}\left(\frac{\sigma}{\omega\varepsilon}\right)^2\right]$$

which is only slightly different from the result for a nonconducting medium, and

$$\gamma \simeq \frac{\sigma}{2}\sqrt{\frac{\mu}{\varepsilon}}$$

which is small and directly proportional to σ, as we might expect.

In the limit of high conductivity ($\sigma \gg \omega\varepsilon$), we get

$$\kappa = \frac{\omega\sqrt{\mu\varepsilon}}{\sqrt{2}}\sqrt{\frac{\sigma}{\omega\varepsilon}} = \sqrt{\frac{\omega\mu\sigma}{2}}$$

and

$$\gamma = \frac{\omega\mu\sigma}{2\kappa} = \frac{\omega\mu\sigma}{2}\sqrt{\frac{2}{\omega\mu\sigma}} = \sqrt{\frac{\omega\mu\sigma}{2}} = \kappa$$

Now what does this all mean? According to equation (2.26), when the conductivity is low, the amplitude decreases slowly ($\gamma \ll \kappa$), but when the conductivity is high, $\gamma = \kappa$ and the wave travels only a short distance into the medium, as shown in Figure 2.7.

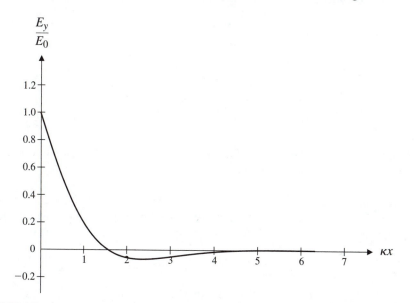

FIGURE 2.7. The magnitude of the electric field vector decreases rapidly with distance inside a conducting medium. Here we show E_y as a function of x at a fixed time t, with $\kappa = \gamma$ (the high conductivity limit).

2.2. FUNCTIONS OF COMPLEX VARIABLES

We have already discussed the notion of a complex function as a *mapping* of the complex plane onto itself. It is now time to explore some properties of these mappings.

2.2.1. Continuity

A real function $f(x)$ is *continuous* at $x = a$ if, for any positive ε, we can choose a positive δ such that whenever $|h| < \delta$, then also

$$|f(a + h) - f(a)| < \varepsilon$$

or, loosely speaking, $f(x)$ is close to $f(a)$ whenever x is close to a (for example, see Stewart, Section 2.5). We can extend this definition to complex functions simply by replacing x with z and interpreting the absolute value as the absolute value of a complex number. Then a function is continuous if $f(z + h)$ is close to $f(z)$ in the mapped plane.

Example 2.4. Show that the function $f(z) = 1/z = (1/r)e^{-i\theta}$ is continuous except at the origin.

Let $h = \eta + i\delta$. Then, for $z \neq 0$,

$$f(z+h) = f(x+iy+\eta+i\delta) = \frac{1}{x+\eta+i(y+\delta)} = \frac{x+\eta-i(y+\delta)}{(x+\eta)^2+(y+\delta)^2}$$

Thus, neglecting squares of the small quantities η and δ, we have

$$\begin{aligned}
f(z+h) - f(z) &= \frac{x+\eta-i(y+\delta)}{(x+\eta)^2+(y+\delta)^2} - \frac{x-iy}{x^2+y^2} \\
&= -\frac{(\eta+i\delta)(x-iy)^2}{(x^2+y^2+2x\eta+2y\delta)(x^2+y^2)} \\
&= -\frac{h(x-iy)^2}{(x^2+2x\eta+y^2+2y\delta)(x^2+y^2)} \to 0 \quad \text{as } h \to 0
\end{aligned}$$

Thus, the function is continuous. But the function is not defined at the origin, where $x = y = 0$, and so f cannot be continuous there.

But look at the cube root function (Section 2.1.2) with branch cut along the positive real axis ($0 \leq \theta < 2\pi$). If we consider only the first, or principal, branch, the function is not continuous. Let $z_0 = re^{i\delta}$ and the neighboring point $z_1 = re^{i(2\pi-\delta)}$ (Figure 2.8a). Then

$$w_0 = f_1(z_0) = r^{1/3}\exp(i\delta/3)$$

but
$$w_1 = f_1(z_1) = r^{1/3}\exp(i[2\pi-\delta]/3)$$

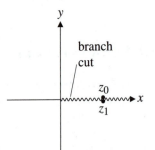

FIGURE 2.8a. The principal branch of the cube root function is not continuous across the branch cut, here chosen to be along the positive x-axis. The two points z_0 and z_1 are very close together, but on opposite sides of the branch cut.

The point w_1 is not close to w_0 (Figure 2.8b).

However, there are two more branches of this function:

$$f_2(z) = r^{1/3}\exp(i[2\pi+\theta]/3)$$

and

$$f_3(z) = r^{1/3} \exp\left(i\left[4\pi + \theta\right]/3\right)$$

Let's plot the points $w_2 = f_2(z_1) = r^{1/3} \exp\left[i(4\pi - \delta)/3\right]$ and $w_3 = f_3(z_1) = r^{1/3} \exp\left[i(6\pi - \delta)/3\right] = r^{1/3} \exp\left(-i\delta/3\right)$ in the w-plane also (Figure 2.8c). Then we see that $f_3(z_1)$ lies close to $f_1(z_0)$. That is, the function is continuous if we are willing to switch from one branch to another. Points that are on opposite sides of the branch cut are not *close* for a single branch of the function. We have to change from one branch to another as we cross the branch cut.

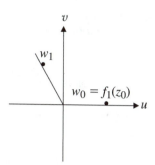

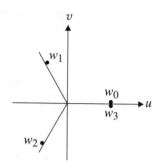

FIGURE 2.8b. The points $w_0 = f_1(z_0)$ and $w_1 = f_1(z_1)$ are not close together.

FIGURE 2.8c. The point $w_3 = f_3(z_1)$ is close to w_0.

Functions with More Than One Branch Point

The function $f(z) = \sqrt{1 + z^2}$ has two branch points, at $z = \pm i$. Thus, there are also two branch cuts, one starting at each of the branch points. We can understand the behavior of this function if we factor it as $f(z) = \sqrt{(z - i)(z + i)}$. Let's first choose each branch cut to run from its branch point upward along the imaginary axis (Figure 2.9a). Then, in polar coordinates with origin at the point $z = i$, we write

$$z - i = \rho_1 e^{i\phi_1}, \quad \text{with } \frac{\pi}{2} > \phi_1 > -\frac{3\pi}{2}$$

With origin at the point $z = -i$,

$$z + i = \rho_2 e^{i\phi_2}, \quad \text{with } \frac{\pi}{2} > \phi_2 > -\frac{3\pi}{2}$$

Then, for points on the right side of the imaginary axis above $z = i$, we have $\phi_1 = \pi/2$, $\phi_2 = \pi/2$, $\rho_1 = y - 1$, and $\rho_2 = y + 1$, so

$$f(z) = \sqrt{\rho_1 \rho_2} e^{i(\phi_1 + \phi_2)} = \sqrt{(y - 1)(y + 1)} e^{i\pi/2} = i\sqrt{y^2 - 1}$$

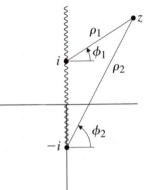

FIGURE 2.9a. Here we have chosen both branch cuts to run upward. We describe a point z by its distances ρ_1 and ρ_2 from the points i and $-i$ and the corresponding angles ϕ_1 and ϕ_2, as shown here.

For points on the left side of the imaginary axis above $z = i$, $\phi_1 = -3\pi/2$ and $\phi_2 = -3\pi/2$, so

$$f(z) = \sqrt{(y-1)(y+1)}e^{-i3\pi/2} = i\sqrt{y^2 - 1}$$

and the function is continuous across the imaginary axis. Below $z = -i$, there are no branch cuts and the function is continuous. However, for points between $-i$ and $+i$, we have

On the right:

$$f(z) = \sqrt{\rho_1 e^{-i\pi/2}\rho_2 e^{i\pi/2}} = \sqrt{\rho_1 \rho_2} = \sqrt{(1-y)(1+y)} = \sqrt{1-y^2}$$

On the left:

$$f(z) = \sqrt{\rho_1 e^{-i\pi/2}\rho_2 e^{-i3\pi/2}} = \sqrt{(1-y)(1+y)}e^{-i\pi} = -\sqrt{1-y^2}$$

The function is not continuous across the imaginary axis. This choice for the two branch cuts is equivalent to having a single branch cut that runs from one point to the other (Figure 2.9b).

On the other hand, if we choose the branch cut from $z = i$ to run upward while that from $z = -i$ runs downward (Figure 2.9c), then $\pi/2 > \phi_1 > -3\pi/2$ but $-\pi/2 < \phi_2 < 3\pi/2$. Above $z = i$, we have

On the right:

$$f(z) = \sqrt{\rho_1 e^{i\pi/2}\rho_2 e^{i\pi/2}} = e^{i\pi/2}\sqrt{\rho_1 \rho_2} = i\sqrt{y^2 - 1}$$

On the left:

$$f(z) = \sqrt{\rho_1 e^{-i3\pi/2}\rho_2 e^{i\pi/2}} = e^{-i\pi/2}\sqrt{\rho_1 \rho_2} = -i\sqrt{y^2 - 1}$$

and the function is discontinuous. A similar situation holds below $-i$. Between i and $-i$, there are no branch cuts and the function is continuous.

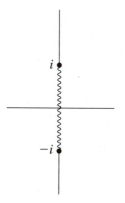

FIGURE 2.9b. The situation in Figure 2.9a is equivalent to having a single branch cut that runs from one branch point to the other.

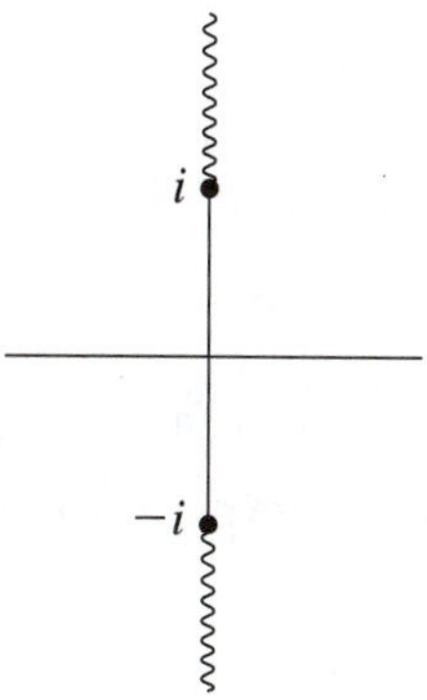

FIGURE 2.9c. Here we have chosen one branch cut to run upward and the other to run downward.

2.2.2. Differentiability

A real function $f(x)$ is differentiable at $x = a$ if, for $h > 0$,

$$\lim_{h \to 0} \frac{f(a+h) - f(a)}{h} \quad \text{exists and equals} \quad \lim_{h \to 0} \frac{f(a) - f(a-h)}{h}$$

(for example, see Stewart, Section 2.8). Thus, the function $f(x) = x^2$ (Figure 2.10a) is differentiable everywhere, while the function

$$f(x) = \begin{cases} x & \text{if } x < 0 \\ -x & \text{if } x > 0 \end{cases}$$

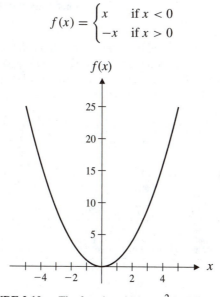

FIGURE 2.10a. The function $f(x) = x^2$ is differentiable.

(Figure 2.10b) is not differentiable at $x = 0$ because

$$\lim_{h \to 0} \frac{f(0+h) - f(0)}{h} = \lim_{h \to 0} \frac{-h}{h} = \lim_{h \to 0} -1 = -1$$

while

$$\lim_{h \to 0} \frac{f(0) - f(0 - h)}{h} = \lim_{h \to 0} \frac{0 - (-h)}{h} = \lim_{h \to 0} 1 = 1$$

The two limits are not equal, and thus the derivative does not exist.

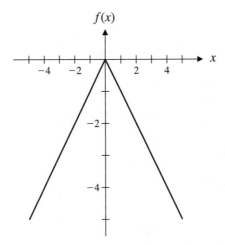

FIGURE 2.10b. This function has an abrupt change of slope at the origin: $\lim_{x \to 0+} df/dx = -1$ but $\lim_{x \to 0-} df/dx = +1$. The function is not differentiable at the origin.

Now with complex functions, we need to let one point approach the other not just along the real axis, in either direction, but along *any line in the complex plane* (Figure 2.11). In particular, we must get the same limit for the derivative if we approach z from a neighboring point along either the real or the imaginary axis[4]:

$$\lim_{dz \to 0} \frac{f(z + dz) - f(z)}{dz} \bigg|_{dz=dx} = \lim_{dz \to 0} \frac{f(z + dz) - f(z)}{dz} \bigg|_{dz=idy}$$

That is, with $f(z) = u(x, y) + iv(x, y)$,

$$\lim_{dx \to 0} \frac{u(x + dx, y) + iv(x + dx, y) - u(x, y) - iv(x, y)}{dx}$$

$$= \lim_{dy \to 0} \frac{u(x, y + dy) + iv(x, y + dy) - u(x, y) - iv(x, y)}{i \, dy}$$

[4]This condition turns out to be not only necessary but also sufficient.

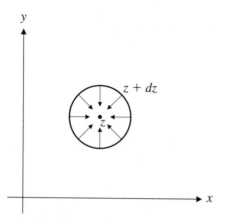

FIGURE 2.11. For a function to be differentiable at z, we must get the same limit when we approach the point z along any line in the complex plane.

Taking the limits, we obtain

$$\frac{\partial u}{\partial x} + i\frac{\partial v}{\partial x} = \frac{1}{i}\left(\frac{\partial u}{\partial y} + i\frac{\partial v}{\partial y}\right) = -i\frac{\partial u}{\partial y} + \frac{\partial v}{\partial y}$$

Equating real and imaginary parts, we get the *Cauchy-Riemann relations:*

$$\frac{\partial u}{\partial x} = \frac{\partial v}{\partial y} \qquad (2.32)$$

and

$$\frac{\partial v}{\partial x} = -\frac{\partial u}{\partial y} \qquad (2.33)$$

These relations must be satisfied by any function that is differentiable.[5]
   We may now write the derivative as

$$\frac{df}{dz} = \frac{\partial u}{\partial x} + i\frac{\partial v}{\partial x} = \frac{\partial v}{\partial y} - i\frac{\partial u}{\partial y} \qquad (2.34)$$

   Conversely, if the Cauchy-Riemann conditions are satisfied and the partial derivatives are continuous, then the function is differentiable. Consider a small displacement $\delta z = \delta x + i\,\delta y$. Then

$$\delta f = \left(\frac{\partial u}{\partial x} + i\frac{\partial v}{\partial x}\right)\delta x + \left(\frac{\partial u}{\partial y} + i\frac{\partial v}{\partial y}\right)\delta y$$

---

[5] See Problem 17 for an alternative statement of the differentiability condition.

Applying equations (2.32) and (2.33), we find

$$\delta f = \left(\frac{\partial u}{\partial x} + i \frac{\partial v}{\partial x} \right) (\delta x + i \, \delta y) = \left(\frac{\partial u}{\partial x} + i \frac{\partial v}{\partial x} \right) \delta z$$

and thus

$$\frac{df}{dz} = \lim_{\delta z \to 0} \frac{\delta f}{\delta z} = \left(\frac{\partial u}{\partial x} + i \frac{\partial v}{\partial x} \right)$$

for any δz.

Example 2.5.   Show that the function $f(z) = z^2$ is differentiable.
  First note that $z^2 = (x + iy)^2 = x^2 + 2ixy - y^2$, so the real and imaginary parts
of the function are

$$u(x, y) = x^2 - y^2$$

and

$$v(x, y) = 2xy$$

Then

$$\frac{\partial u}{\partial x} = 2x = \frac{\partial v}{\partial y}$$

and

$$\frac{\partial u}{\partial y} = -2y = -\frac{\partial v}{\partial x}$$

so the Cauchy-Riemann relations are satisfied and the partial derivatives are continuous. The derivative is

$$\frac{df}{dz} = \frac{\partial u}{\partial x} + i \frac{\partial v}{\partial x} = \frac{1}{i} \left(\frac{\partial u}{\partial y} + i \frac{\partial v}{\partial y} \right)$$
$$= 2(x + iy) = 2z$$

which is exactly the expression we might expect. In fact, complex derivatives obey
all the usual rules of real derivatives, such as the product rule.

  As is the case with real functions, complex functions that "blow up" are not differentiable,
and functions that are not continuous are not differentiable. Thus, functions that have branch
cuts are not differentiable at the cut. However, it is possible for a complex function to be
continuous everywhere but differentiable nowhere. The function $f(z) = z^* = x - iy$ is
one such function. With $\partial u / \partial x = -\partial v / \partial y = 1$, the Cauchy-Riemann conditions are never
satisfied.

2.2.3. Analyticity

A function f that is differentiable at $z = a$ and *within a neighborhood*[6]*of* $z = a$ is said to be *analytic* at $z = a$. Thus, a function analytic at a satisfies the Cauchy-Riemann relations within a small but finite region around $z = a$. The requirement that the function be differentiable within a neighborhood of a guarantees the existence of higher-order derivatives as well. A function that is analytic in the whole complex plane (except perhaps at infinity) is called an *entire* function. Analyticity is a powerful constraint on a function, as we shall see.

2.2.4. Integrals

Since a complex function $f(z)$ represents a mapping of the z-plane onto the w-plane, any integral

$$\int_{z_1}^{z_2} f(z)\, dz$$

is an integral[7] along a path that must be specified between the points z_1 and z_2. Expanding the integrand, we find

$$\int_{z_1}^{z_2} f(z)\, dz = \int_{z_1}^{z_2} (u + iv)(dx + i\, dy) = \int_{z_1}^{z_2} u\, dx - v\, dy + i(u\, dy + v\, dx)$$

Defining the vectors $\vec{\mathbf{u}}$, with components $(u, -v)$, and $\vec{\mathbf{v}}$, with components (v, u), we may write the integral as

$$\int_{z_1}^{z_2} f(z)\, dz = \int_{z_1}^{z_2} \vec{\mathbf{u}} \cdot d\vec{\mathbf{l}} + i \int_{z_1}^{z_2} \vec{\mathbf{v}} \cdot d\vec{\mathbf{l}} \tag{2.35}$$

The Cauchy Theorem

Next we shall prove[8] an extremely powerful theorem about integrals of analytic functions. Consider the integral from z_1 to z_2 along path C_1 and returning from z_2 to z_1 along path C_2 (Figure 2.12). The combined path forms a closed curve C in the complex plane, and result (2.35) applied to this path gives

$$\oint_C f(z)\, dz = \oint_C \vec{\mathbf{u}} \cdot d\vec{\mathbf{l}} + i \oint_C \vec{\mathbf{v}} \cdot d\vec{\mathbf{l}}$$

---

[6]A neighborhood of $z = a$ is a small region in the complex plane that contains the point $z = a$ strictly within it—for example, the circle defined by $|z - a| \le \varepsilon$ for any $\varepsilon > 0$.

[7]See Section 1.2.3 for information on path integrals.

[8]The proof presented here has the advantage of being relatively straightforward, but it requires continuity of the partial derivatives. An alternative proof due to Goursat that does not require the use of Green's theorem (see, for example, Jeffreys and Jeffreys, Section 11.052) requires only that $f(z)$ be differentiable and C have finite length.

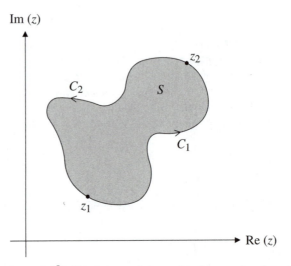

FIGURE 2.12. The integral $\int_{z_1}^{z_2} f(z)\,dz$ is a path integral in the complex plane. Here the two paths C_1 and C_2 can be combined to form a closed curve C from z_1 to z_2 and returning to z_1.

Then we may use Green's theorem (equation 1.47) to write each integral over C as a double integral over the surface S spanning C:

$$\oint_C f(z)\,dz = \int_S \left(-\frac{\partial v}{\partial x} - \frac{\partial u}{\partial y}\right) dx\,dy + i \int_S \left(\frac{\partial u}{\partial x} - \frac{\partial v}{\partial y}\right) dx\,dy$$

Now if the function is analytic everywhere on S, then both integrands are zero, by the Cauchy-Riemann relations, and so we find

---

If $f(z)$ is analytic in and on C, then

$$\oint_C f(z)\,dz = 0 \qquad\qquad (2.36)$$

a result known as the *Cauchy theorem.*

Cauchy's theorem applies to curves defined in a simply connected region. A curve in a simply connected region can be shrunk continuously to a point while remaining inside the region. If the region has a hole in it (Figure 2.13) and the curve surrounds the hole, then the region is not simply connected and the theorem does not hold. We will see regions like this in the following sections, and we will also find some ways to get around the restrictions.

An important corollary to Cauchy's theorem is that the integral of a function $f(z)$ between two points z_1 and z_2 is path independent if the function f is analytic everywhere in the region containing the paths of interest. This is important because it means you can often use a path that hugely simplifies the integration.

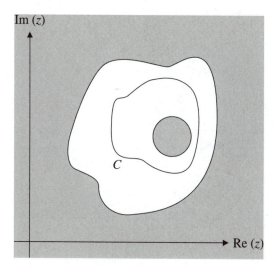

FIGURE 2.13. This white region has a hole (shaded circle), and so it is not simply connected. As the curve C shrinks, it surrounds the hole and can never shrink to a point.

2.2.5. The Cauchy Formula

Careful application of the Cauchy theorem leads to many interesting results. For example, consider the integral[9]

$$I = \oint_C \frac{1}{z - a} dz$$

where C is a closed curve in the complex plane. The function $f(z) = 1/(z - a)$ is analytic except at the single point $z = a$. Thus, by the Cauchy theorem, the integral is zero if the curve C does not enclose the point $z = a$. If the curve includes $z = a$, we consider the integral

$$I' = \oint_{C'} \frac{1}{z - a} dz$$

where the curve $C' = C + B_1 + \Gamma + B_2$ (Figure 2.14). The segments B_1 and B_2 are very close together. Note that this new curve C' does not enclose the point $z = a$, and thus the function f is analytic everywhere in and on C'. Thus, I' is zero by the Cauchy theorem. Since the function f is analytic everywhere except at $z = a$, it is continuous along B_1 and B_2 and so the contributions to the integral I' along B_1 and B_2 cancel. (The integrands are equal, but the path is traversed in opposite directions. Equivalently, the upper limit of the

[9]The symbol $\oint_C$ as used here means an integral *counterclockwise* around the closed path C. When you traverse such a curve, the interior of the curve is always to your left.

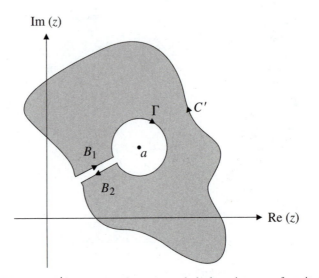

FIGURE 2.14. The curve C' is constructed so as to exclude the point $z = a$ from its interior. (When you travel counterclockwise around the curve, the interior, shown shaded, is on your left.)

first integral is the lower limit of the second, and vice versa.)

$$0 = \oint_{C'} \frac{1}{z-a} dz = \oint_{C+B_1+\Gamma+B_2} \frac{1}{z-a} dz = \oint_C \frac{1}{z-a} dz + \oint_{\Gamma,\text{clockwise}} \frac{1}{z-a} dz$$

and so

$$\oint_C \frac{1}{z-a} dz = -\oint_{\Gamma,\text{clockwise}} \frac{1}{z-a} dz$$

The integral around Γ is taken clockwise, the reverse of the usual direction. We may change the direction if we also change the sign in front of the integral:

$$\oint_C \frac{1}{z-a} dz = \oint_\Gamma \frac{1}{z-a} dz$$

Now[10] on Γ, $z = a + \rho e^{i\theta}$, so $dz = i\rho e^{i\theta} d\theta$ and then

$$\oint_C \frac{1}{z-a} dz = \int_0^{2\pi} \frac{1}{\rho e^{i\theta}} \rho e^{i\theta} i \, d\theta$$

$$= i \int_0^{2\pi} d\theta = 2\pi i$$

---

[10]Remember: Addition of complex numbers is equivalent to addition of vectors in the complex plane (Figure 2.3).

Thus, we obtain the first Cauchy formula:

$$\oint_C \frac{1}{z-a}dz = \begin{cases} 2\pi i & \text{if } z = a \text{ lies within } C \\ 0 & \text{otherwise} \end{cases} \tag{2.37}$$

We can easily modify the proof to show that for any function f that is analytic in and on C,

$$\oint_C \frac{f(z)}{z-a}dz = \begin{cases} \oint_\Gamma \frac{f(z)}{z-a}dz = \int_0^{2\pi} \frac{f(a+\rho e^{i\theta})}{\rho e^{i\theta}}\rho e^{i\theta}i\,d\theta & \text{if } z = a \text{ lies within } C \\ 0 & \text{otherwise} \end{cases}$$

Then we let $\rho \to 0$, to obtain

$$\oint_C \frac{f(z)}{z-a}dz = \begin{cases} if(a)\int_0^{2\pi} d\theta = 2\pi i f(a) & \text{if } z = a \text{ lies within } C \\ 0 & \text{otherwise} \end{cases} \tag{2.38}$$

We may extend this result to obtain the general Cauchy formula:

$$\oint_C \frac{f(z)}{(z-a)^n}dz = \frac{2\pi i}{(n-1)!}\left. f^{(n-1)}(z)\right|_{z=a} \qquad \text{if } z = a \text{ lies within } C \tag{2.39}$$

The proof of this more general result is in Appendix III. This result verifies the claim made in Section 2.2.3 that analytic functions possess derivatives of all orders.

Example 2.6. Evaluate the integral $\displaystyle\oint_{\text{circle } |z|=1} \frac{\cos z}{z^3}dz$ around the circle of unit radius centered at the origin.

The function $f(z) = \cos z$ is analytic within and on the circle $|z| = 1$. Using the Cauchy formula with $n = 3$, we have

$$\oint_{\text{circle } |z|=1} \frac{\cos z}{z^3}dz = \frac{2\pi i}{2!}\left. \frac{d^2}{dz^2}(\cos z)\right|_{z=0} = \pi i\left(-\cos z|_{z=0}\right) = -\pi i$$

2.3. COMPLEX SERIES

2.3.1. Real Sequences and Series—A Review

Sequences

A sequence of numbers a_n is an ordered set with a rule for computing the nth element in the set. The number a_n is the nth number in the sequence and can often be written as a function of n. Some examples are

$$S_1 = \{1, 2, 3, 4, \ldots\}$$

for which $a_n = n$, and

$$S_2 = \{1, -1, 1, -1, \ldots\}$$

for which

$$a_n = (-1)^{n-1}$$

A sequence S is unbounded if, for any real number $M > 0$, there is at least one member of the sequence a_n such that $|a_n| > M$. The sequence S_1 above is unbounded, since we need only take $M = N$, any integer, and infinitely many a_n are greater than N, no matter how large N is. Conversely, if we can choose an M such that *all* members of the sequence S satisfy $|a_n| < M$, then the sequence is bounded. Sequence S_2 is bounded—we could take $M = 2$.

A sequence S is *convergent* if there is a number s such that, given any positive number ε, we can choose an m so that for every $n > m$

$$|a_n - s| < \varepsilon$$

in which case the limit of the sequence is s. We write

$$\lim_{n \to \infty} a_n = s$$

or, more loosely,

$$a_n \to s \quad \text{as } n \to \infty$$

Neither of the sequences S_1 or S_2 is convergent, but the sequence

$$S_3 = \left\{ 1, \frac{1}{2}, \frac{1}{3}, \frac{1}{4}, \ldots, \frac{1}{n}, \ldots \right\}$$

converges to zero. (Take m to be any integer greater than $1/\varepsilon$.)

A sequence that is bounded but not convergent—for example, sequence S_2—is said to oscillate finitely. Another example is

$$S_4 = \left\{ \frac{1}{n} + (-1)^n \right\}$$

Other sequences may be unbounded and not convergent—for example, sequence S_1 above, but also series such as

$$S_5 = \left\{ \sqrt{n} \cos(\pi n) \right\}$$

which is oscillating but unbounded (infinitely oscillating).

Further properties of sequences can be found in texts on calculus (for example, see Stewart, Chapter 11) or on real analysis.

Series

A series is formed by summing the terms of a sequence. Using sequence S_1, we form the series

$$\sum_{n=1}^{\infty} a_n = \sum_{n=1}^{\infty} n \tag{2.40}$$

A second sequence is formed from the partial sums of the series:

$$A_m = \sum_{n=1}^{m} a_n$$

If this *sequence of partial sums* is convergent, then we say that the *series is convergent*. One requirement for convergence is for the successive terms to approach zero; that is, $a_n \to 0$. Conversely, the series diverges if

$$\lim_{n \to \infty} a_n \neq 0$$

This includes the case in which the limit does not exist at all.

Clearly, series (2.40) is not convergent, since $\lim_{n \to \infty} a_n$ does not exist. But what about the following series?

$$\sum_{n=1}^{\infty} \frac{1}{n} \tag{2.41}$$

Here the answer is not so clear. The terms $a_n = 1/n$ do approach zero, but not very fast. It is possible for a series to diverge even though $a_n \to 0$. We'll need to *test* this series for convergence (see below).

If the series in which each a_n is replaced by its absolute value,

$$\sum_{n=1}^{\infty} |a_n|$$

also converges, then the series is said to be *absolutely convergent*.

Series that converge, but not absolutely, are said to be conditionally convergent. These series do not have a well-defined sum, since by rearranging the terms we can achieve any sum that we want (for example, see Stewart, p. 736; Arfken and Weber, Section 5.4). Thus, they must be used with caution. In contrast, an absolutely convergent series has a well-defined sum. In addition, *absolutely* convergent series may be multiplied together to form a double sum that converges to the product of the original two sums.

There are numerous tests for convergence of a series. A number are listed in Gradshteyn and Ryzhik, Section 0.22. (See also Stewart, Section 11.2.) They include the following.

The root test. If

$$\lim_{k \to \infty} |a_k|^{1/k} = q$$

and $q < 1$, the series converges absolutely, but if $q > 1$, the series diverges. If $q = 1$, the test fails.

The ratio test. If

$$\lim_{k \to \infty} \left| \frac{a_{k+1}}{a_k} \right| = q$$

and $q < 1$, the series converges absolutely, but if $q > 1$, the series diverges. If $q = 1$, the test fails, unless the ratio remains greater than 1 as the limit is approached, in which case the series also diverges.

The integral test. If $a_k = f(k)$, where $f(x)$ is defined for $x \geq q \geq 1$, then the series converges or diverges according to whether the integral

$$\int_q^\infty f(x)\,dx$$

converges or diverges.

The comparison test. If $|a_n| < b_n$ and $\sum b_n$ converges, then $\sum a_n$ converges absolutely.

The alternating series test. If the series is *alternating* (that is, successive terms in the series alternate in sign), then the series converges provided that $|a_{k+1}| < |a_k|$ and $|a_k| \to 0$.

Example 2.7.   Show that the series (2.41), $\sum_{n=1}^\infty \frac{1}{n}$, diverges.

Let's apply the integral test. Since $a_k = 1/k$ is defined for $k \geq 1$, we look at

$$\int_1^\infty \frac{1}{x}\,dx = \ln x |_1^\infty$$

which diverges. Thus, the series also diverges.

If the series elements a_n are not just numbers but functions of some variable, say x, then convergence of the series may depend on the value of x chosen. For any particular value of x, we can apply any of the tests listed above. The result determines the *pointwise convergence* of the series at that value of x. For example, the series

$$\sum_{n=1}^\infty x^n$$

converges if $|x| < 1$ but diverges otherwise. A particularly valuable series is one that converges independent of the value of x and converges equally well for all x. Such series are said to be *uniformly convergent*.

A sequence $f_n(x)$ is uniformly convergent to $f(x)$ in an interval I if, for any arbitrarily small positive number ε, we can find a number N such that

$$\left| f_p(x) - f(x) \right| < \varepsilon$$

for all $p > N$, where N is independent of x in I.

If the uniformly convergent sequence $f_n(x)$ is the sequence of partial sums of a series $f_n(x) = \sum_{m=0}^n u_m(x)$, then that series is uniformly convergent in I.

One of the major advantages of uniformly convergent series is that they may be integrated and differentiated term by term, provided that the individual terms are continuous (for integration) or differentiable (for differentiation).

Tests for uniform convergence include the *Weierstrass M test*:

If there is a sequence of numbers M_n such that $|u_n(x)| \le M_n$ for all x in the interval $[a, b]$ and the series $\sum_{n=1}^{\infty} M_n$ converges, then the series $\sum_{n=1}^{\infty} u_n(x)$ converges uniformly and absolutely in the interval $[a, b]$.

Uniform convergence and absolute convergence are independent properties of a series. Thus, the most useful series are those that converge both uniformly *and* absolutely.

2.3.2. Complex Series

A complex sequence is a sequence of complex numbers and thus is actually two sequences of real numbers:

$$z_n = x_n + i y_n$$

The sequence z_n converges if *and only if* **both** the sequences x_n and y_n converge. Similarly, the complex series

$$\sum z_n = \sum x_n + i \sum y_n$$

converges if and only if **both** real series $\sum x_n$ and $\sum y_n$ converge.

We can define absolute convergence for a complex series as we did for a real series: A series converges absolutely if the series of absolute values converges. But for complex series, we are now talking about convergence of a *third* series,

$$\sum |z_n| = \sum \sqrt{x_n^2 + y_n^2}$$

which is different from the previous two. Thus, absolute convergence is a more powerful constraint for complex series than for real series. If a series converges absolutely for z in a range $|z - z_0| < R$, then R is called the radius of convergence for the series.

Let's look at an important example of a complex series: the *geometric series*

$$\sum_{n=0}^{\infty} z^n$$

The partial sums are

$$S_N(z) = 1 + z + z^2 + z^3 + \cdots + z^N$$

$$= \frac{1 - z^{N+1}}{1 - z} = \frac{1}{1 - z} - \frac{z^{N+1}}{1 - z} \tag{2.42}$$

Now if $|z| < 1$, then

$$\left| \frac{z^{N+1}}{1 - z} \right| = \frac{|z|^{N+1}}{|1 - z|} \to 0 \quad \text{as } N \to \infty$$

and so

$$\sum_{n=0}^{\infty} z^n = \frac{1}{1-z} \quad \text{if } |z| < 1 \tag{2.43}$$

The geometric series is uniformly convergent for $|z| \leq r$ and $r < 1$. The limiting magnitude of r (here 1) is the radius of convergence for the series.

2.3.3. The Taylor Series

Suppose that a function $f(z)$ is analytic in a region R: $|z - a| \leq \rho$, centered on the point $z = a$. Then we may express $f(z)$ as a series in powers of $(z - a)$:

$$f(z) = f(a) + (z-a)f'(a) + \frac{(z-a)^2}{2}f''(a) + \cdots + \frac{(z-a)^n}{n!}\frac{d^n f}{dz^n}\bigg|_{z=a} + \cdots \tag{2.44}$$

This series is uniformly convergent within the circle $|z - a| \leq \rho$, where ρ is the radius of convergence.

To prove this result, we'll use the Cauchy formula (equation 2.38) and the geometric series (2.43). First we construct a closed curve Γ that surrounds both the points z and a (Figure 2.15). Then we may write $f(z)$ as an integral around the curve Γ:

$$f(z) = \frac{1}{2\pi i}\oint_{\Gamma}\frac{f(\xi)}{\xi - z}d\xi$$

Next we write

$$\frac{1}{\xi - z} = \left(\frac{1}{\xi - a}\right)\frac{1}{\left(1 - \dfrac{z - a}{\xi - a}\right)}$$

where

$$\left|\frac{z - a}{\xi - a}\right| \leq \eta < 1$$

since z is inside Γ, and thus closer to a than ξ, for any ξ on Γ. (We are free to construct Γ to make sure this is true.) Thus, we can use the geometric series (2.43) to expand the second fraction:

$$\frac{1}{\xi - z} = \left(\frac{1}{\xi - a}\right)\left(1 + \frac{z-a}{\xi - a} + \left(\frac{z-a}{\xi - a}\right)^2 + \cdots\right)$$

$$= \left(\frac{1}{\xi - a}\right)\sum_{n=0}^{\infty}\left(\frac{z-a}{\xi - a}\right)^n$$

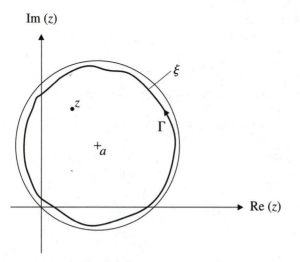

FIGURE 2.15. The curve Γ for evaluating the coefficients in the Taylor series. Γ is entirely enclosed within the region R and encloses the point $z = a$.

The geometric series is uniformly convergent on Γ, so we may integrate term by term with respect to ξ:

$$f(z) = \frac{1}{2\pi i} \sum_{n=0}^{\infty} (z-a)^n \oint_\Gamma \frac{f(\xi)}{(\xi - a)^{n+1}} d\xi \tag{2.45}$$

Then we may use the Cauchy formula (equation 2.39) to express the integral in terms of derivatives of f:

$$f(z) = \sum_{n=0}^{\infty} \frac{(z-a)^n}{n!} \left. \frac{d^n f}{dz^n} \right|_{z=a} \tag{2.46}$$

as we set out to prove.

Now suppose there exists some other series for f about the same point a:

$$f(z) = \sum_{n=0}^{\infty} c_n (z-a)^n$$

What can we say about the coefficients c_n in this series? Let's differentiate the series m times. Then

$$\frac{df}{dz} = \sum_{n=1}^{\infty} n c_n (z-a)^{n-1}$$

$$\frac{d^2 f}{dz^2} = \sum_{n=2}^{\infty} n(n-1) c_n (z-a)^{n-2}$$

$$\frac{d^m f}{dz^m} = m! c_m + \sum_{n=m+1}^{\infty} n(n-1) \cdots (n-m+1) c_n (z-a)^{n-m}$$

Evaluating at $z = a$, we have

$$\left. \frac{d^m f}{dz^m} \right|_{z=a} = m!c_m$$

So if any such series exists, its coefficients are given by equation (2.44).

Example 2.8. Find the Taylor series about the origin for the exponential function

$$f(z) = e^z = e^{x+iy} = e^x e^{iy} = e^x \cos y + i e^x \sin y$$

This function is analytic everywhere:

$$\frac{\partial u}{\partial x} = e^x \cos y = \frac{\partial v}{\partial y}$$

and

$$\frac{\partial v}{\partial x} = e^x \sin y = -\frac{\partial u}{\partial y}$$

The derivatives are

$$\frac{d^n f}{dz^n} = e^z$$

for every n, and so the Taylor series about the origin ($a = 0$) is

$$e^z = 1 + z + \frac{z^2}{2} + \cdots + \frac{z^n}{n!} + \cdots$$

This series is exactly what we might expect, since e^x has the same form when x is real.

2.3.4. The Laurent Series

Suppose that the function f is analytic in an annular region R centered on $z = a$: $\rho_1 < |z - a| < \rho_2$. Then we may still expand f in a series of powers of $(z - a)$, but the series will include negative as well as positive powers. This series is called a *Laurent series*. To find the coefficients formally, we apply the same techniques that we used for the Taylor series.

As before, we construct a curve Γ that surrounds the point z and lies entirely within the region R. This time the curve cannot also surround a, but we can distort it to form a composite curve having four parts (Figure 2.16):

1. The original curve Γ, with two snips in it
2. A circular curve C_1 just inside the inner border of the annulus at $|z - a| = \rho_1 + \varepsilon$
3. A circular curve C_2 just inside the outer border of the annulus at $|z - a| = \rho_2 - \delta$
4. A set of cross cuts linking Γ to C_1 and Γ to C_2

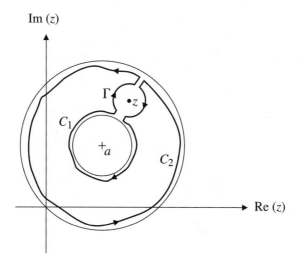

FIGURE 2.16. A curve for finding the Laurent series lies within the annulus but excludes the point z: $C = C_1 + C_2 + \Gamma +$ cross cuts.

Now we proceed as before, writing f as an integral around the curve Γ:

$$f(z) = \frac{1}{2\pi i} \oint_\Gamma \frac{f(\xi)}{\xi - z} d\xi$$

But, when we traverse our composite curve as shown in the diagram, the integrand is analytic everywhere inside and on the constructed composite curve, and thus the integral is zero by the Cauchy theorem.

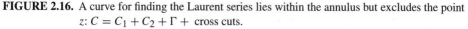

$$\oint_C \frac{f(\xi)}{\xi - z} d\xi = \oint_{C_2} \frac{f(\xi)}{\xi - z} d\xi + \oint_{\Gamma,\text{clockwise}} \frac{f(\xi)}{\xi - z} d\xi + \oint_{C_1,\text{ clockwise}} \frac{f(\xi)}{\xi - z} d\xi$$
$$+ \oint_{\text{cross cuts}} \frac{f(\xi)}{\xi - z} d\xi = 0$$

The sum of the integrals along the cross cuts is zero, since the function is analytic within R, and hence continuous at the cross cuts. We can use a minus sign to change the direction of the clockwise integrals to the usual counterclockwise sense. Thus,

$$\oint_\Gamma \frac{f(\xi)}{\xi - z} d\xi = \oint_{C_2} \frac{f(\xi)}{\xi - z} d\xi - \oint_{C_1} \frac{f(\xi)}{\xi - z} d\xi$$

Now we proceed as we did with the Taylor series.

For ξ on C_2, $|z - a| < |\xi - a|$, and so we expand the denominator as we did before, obtaining the same result (equation 2.45 but not equation 2.46). However, for ξ on C_1, $|z - a| > |\xi - a|$, so we have to expand as follows:

$$\frac{1}{\xi - z} = \frac{-1}{(z - a)} \frac{1}{\left(1 - \dfrac{\xi - a}{z - a}\right)} = \frac{-1}{z - a} \sum_{n=0}^{\infty} \left(\frac{\xi - a}{z - a}\right)^n$$

and thus

$$\oint_{C_1} \frac{f(\xi)}{\xi - z} d\xi = \sum_{n=0}^{\infty} \frac{-1}{(z-a)^{n+1}} \oint_{C_1} f(\xi)(\xi - a)^n \, d\xi$$

which gives a series in negative powers of $z - a$. Writing $p = -(n+1)$, we get

$$\oint_{C_1} \frac{f(\xi)}{\xi - z} d\xi = -\sum_{p=-1}^{-\infty} (z-a)^p \oint_{C_1} f(\xi)(\xi - a)^{-p-1} \, d\xi$$

Here we cannot easily express the coefficents of $(z - a)^p$ in terms of derivatives, as we could for the Taylor series.

Putting together the two parts, we get

$$f(z) = \frac{1}{2\pi i} \oint_{\Gamma} \frac{f(\xi)}{\xi - z} d\xi = \sum_{n=-\infty}^{+\infty} c_n (z - a)^n$$

where

$$c_n = \frac{1}{2\pi i} \oint_C \frac{f(\xi)}{(\xi - a)^{n+1}} d\xi \quad \text{for } -\infty < n < \infty \qquad (2.47)$$

Once we have the result, we can evaluate the integrals over a common circle C centered on a and lying within the annulus, since the integrands in equation (2.47) are analytic everywhere in the annulus.

This expression (2.47) for c_n amply demonstrates the existence of the Laurent series, but it is not generally very useful for *finding* a particular Laurent series. We can resort to a set of "tricks" that work better. The next example illustrates the geometric series method.

Example 2.9.   Find a series about the point $z = 1$ for the function $\dfrac{1}{z^2 - 1}$.

   The function is not analytic at the two points $z = 1$ and $z = -1$. Thus, we cannot draw a circular region centered at $z = 1$ within which the function is analytic. So, there is no Taylor series for this function about the point $z = 1$. However, there are two annuli centered at $z = 1$: $0 < |z - 1| < 2$, with the point $z = -1$ outside the annulus and $z = 1$ inside an infinitesimal hole in the center, and $2 < |z - 1| < \infty$, with both points in the inner hole. Thus, we can find two Laurent series centered at $z = 1$: one in the region $0 < |z - 1| < 2$ and a second for $|z - 1| > 2$ (Figure 2.17).

   In the inner region, we have

$$\frac{1}{z^2 - 1} = \frac{1}{(z-1)(z+1)} = \frac{1}{2}\left(\frac{1}{z-1} - \frac{1}{z+1}\right)$$

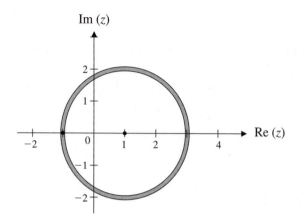

FIGURE 2.17. The function in Example 2.9 is not analytic at $z = \pm 1$. There is a Laurent series valid in each of the annuli $0 < |z - 1| < 2$ and $|z - 1| > 2$.

Now we can expand the second fraction using the geometric series. In the inner annulus, $|z - 1| < 2$, so

$$\frac{1}{z + 1} = \frac{1}{z - 1 + 2} = \frac{1}{2\left(1 + \dfrac{z - 1}{2}\right)} = \frac{1}{2} \sum_{n=0}^{\infty} (-1)^n \left(\frac{z - 1}{2}\right)^n$$

and thus we have

$$\frac{1}{z^2 - 1} = \frac{1}{2}\left[\frac{1}{z - 1} - \frac{1}{2} \sum_{n=0}^{\infty} (-1)^n \left(\frac{z - 1}{2}\right)^n\right]$$

$$= \frac{1}{4} \sum_{n=-1}^{\infty} (-1)^{n+1} \left(\frac{z - 1}{2}\right)^n$$

which is a Laurent series with only one negative power: $(z - 1)^{-1}$.

However, in the outer annulus, $|z - 1| > 2$, so we expand the fraction the other way:

$$\frac{1}{z - 1 + 2} = \frac{1}{(z - 1)\left(1 + \dfrac{2}{z - 1}\right)} = \frac{1}{z - 1} \sum_{n=0}^{\infty} (-1)^n \left(\frac{2}{z - 1}\right)^n$$

and

$$\frac{1}{z^2 - 1} = \frac{1}{2}\left[\frac{1}{z - 1} - \sum_{n=0}^{\infty} (-1)^n \frac{2^n}{(z - 1)^{n+1}}\right]$$

$$= \frac{1}{2} \sum_{n=1}^{\infty} \frac{(-2)^n}{(z - 1)^{n+1}} \qquad (2.48)$$

a series with infinitely many negative powers of $z - 1$ and no positive powers.

2.4. COMPLEX NUMBERS AND LAPLACE'S EQUATION

2.4.1. The Equations of Incompressible Fluid Flow

A fluid may be described by a vector field $\vec{v}(\vec{r})$, giving the velocity of each fluid element, and a scalar field $\rho(\vec{r})$, giving the density at each point. Since the mass within a volume can change only by flow of fluid into or out of the element,[11] we have

$$\frac{d}{dt} \int_V \rho \, dV = - \int_S \rho \vec{v} \cdot \hat{n} \, dA$$

where the minus sign arises because the unit vector $\hat{n}$ normal to S points out of the volume V. Now if we apply this relation to a fixed volume V, we can then apply the divergence theorem[12] and rearrange to get

$$\int_V \left(\frac{\partial \rho}{\partial t} + \vec{\nabla} \cdot (\rho \vec{v}) \right) dV = 0$$

Then since this relation must hold for *any* fixed volume V, we may conclude that

$$\frac{\partial \rho}{\partial t} + \vec{\nabla} \cdot (\rho \vec{v}) = 0 \tag{2.49}$$

which is known as the continuity equation.

Liquids have the property that their density remains almost constant under a very wide range of applied pressure; they are incompressible. (Gases can also behave incompressibly under some circumstances.) For incompressible fluids, equation (2.49) simplifies considerably, since the time and space derivatives of the density are identically zero:

$$\vec{\nabla} \cdot \vec{v} = 0 \tag{2.50}$$

Certain classes of fluid flow are also irrotational,

$$\vec{\nabla} \times \vec{v} = 0 \tag{2.51}$$

which means that there are no swirling motions in the flow (no turbulence).

2.4.2. The Velocity Potential

A vector field that satisfies equation (2.51) may be described as the gradient of a scalar function ϕ, since for any such function

$$\vec{\nabla} \times (\vec{\nabla}\phi) = 0$$

In our fluid flow problem, the function ϕ is called the velocity potential. It is usual to introduce a minus sign, so that

$$\vec{v} = v_x \hat{x} + v_y \hat{y} = -\vec{\nabla}\phi \tag{2.52}$$

---

[11] See Chapter 1, Problem 6.
[12] See Chapter 1, Section 1.2.3.

Then we can put this result into equation (2.50) to get

$$\nabla^2\phi = 0$$

which is Laplace's equation. In this respect, the solutions to irrotational, incompressible fluid flow problems will resemble the solutions to problems in electrostatics, since the governing differential equation is the same.

2.4.3. Analytic Functions as Solutions of Laplace's Equation

An analytic function $w(z) = u + iv$ satisfies the Cauchy-Riemann relations (equations 2.32 and 2.33):

$$\frac{\partial u}{\partial x} = \frac{\partial v}{\partial y} \quad \text{and} \quad \frac{\partial u}{\partial y} = -\frac{\partial v}{\partial x}$$

Differentiating again, we have

$$\frac{\partial^2 u}{\partial x^2} = \frac{\partial^2 v}{\partial x \partial y} = \frac{\partial}{\partial y}\left(-\frac{\partial u}{\partial y}\right)$$

and thus

$$\frac{\partial^2 u}{\partial x^2} + \frac{\partial^2 u}{\partial y^2} = \nabla^2 u = 0 \tag{2.53}$$

That is, u satisfies Laplace's equation in two dimensions. Such functions are said to be *harmonic*. Similarly, we can show that v is also harmonic. Thus, the real part or the imaginary part of any analytic function will be the solution of an incompressible, irrotational fluid flow problem in two dimensions.

Let's define an analytic, complex velocity potential function

$$\Phi = \phi + i\psi$$

The fluid velocity is given by equation (2.52). In particular, if we use the real part ϕ of Φ as our solution for the real velocity potential, then the velocity components are

$$v_x = -\frac{\partial\phi}{\partial x} \quad \text{and} \quad v_y = -\frac{\partial\phi}{\partial y} = \frac{\partial\psi}{\partial x}$$

and so we can form the complex number

$$w = v_x + iv_y = -\left(\frac{\partial\phi}{\partial x} - i\frac{\partial\psi}{\partial x}\right)$$

Then

$$w^* = -\left(\frac{\partial\phi}{\partial x} + i\frac{\partial\psi}{\partial x}\right) = -\frac{d\Phi}{dz}$$

Thus, the velocity components are found from the real and imaginary parts of the derivative of Φ; that is,

$$v_x = -\text{Re}\left(\frac{d\Phi}{dz}\right) \quad \text{and} \quad v_y = \text{Im}\left(\frac{d\Phi}{dz}\right) \tag{2.54}$$

Now any physics problem is described mathematically by one or more differential equations *plus* a set of boundary conditions. For example, fluid cannot flow across a nonporous boundary, such as a solid wall. On such a boundary we have the boundary condition

$$\vec{v} \cdot \hat{n} = -\hat{n} \cdot \vec{\nabla}\phi = 0 \tag{2.55}$$

Thus, to solve such a problem we need only find an analytic function $\Phi = \phi + i\psi$ whose real part satisfies the boundary condition (2.55). The real part ϕ then satisfies both Laplace's equation and the boundary conditions, and the problem is solved.

If the real part of Φ is the velocity potential, what is the imaginary part? Note that

$$\vec{\nabla}\phi \cdot \vec{\nabla}\psi = \frac{\partial\phi}{\partial x}\frac{\partial\psi}{\partial x} + \frac{\partial\phi}{\partial y}\frac{\partial\psi}{\partial y} = \frac{\partial\phi}{\partial x}\left(-\frac{\partial\phi}{\partial y}\right) + \frac{\partial\phi}{\partial y}\frac{\partial\phi}{\partial x} = 0 \tag{2.56}$$

Thus, surfaces of constant ψ are perpendicular to surfaces of constant ϕ. (This is the same relation as that between equipotential surfaces and electric field lines in electrostatics.) The constant ψ surfaces are thus the streamlines of the flow, and from equation (2.55), a solid boundary surface must be a surface of constant ψ.

2.4.4. Steady Irrotational Flow Around an Infinitely Long Cylinder

Suppose that fluid flows from a great distance with velocity $\vec{v} = V_0\hat{x}$ toward an infinitely long solid cylinder of radius a (Figure 2.18). We want to find the flow velocity around the cylinder. To solve this problem, we will use cylindrical coordinates with origin on the cylinder axis. It is a two-dimensional problem with the boundary conditions

$$\vec{v} \to V_0\hat{x} \quad \text{as } r \to \infty \tag{2.57}$$

and since the flow has to be along the surface of the cylinder, streamlines follow the surface at $r = a$ and $\psi = $ constant on the circle $r = a$.

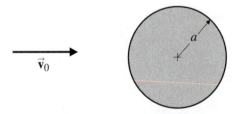

FIGURE 2.18. Fluid flows toward the cylinder (radius a) from infinity with velocity $\vec{v} = V_0\hat{x}$.

First we reformulate the problem in terms of complex functions. The solution for ϕ is the real part of a complex function $\Phi = \phi + i\psi$ that is analytic everywhere outside the cylinder and that satisfies the two boundary conditions. Since Φ is analytic everywhere outside the cylinder (that is, in the region $r > a$), it can be written as a Laurent series[13] centered at the origin:

$$\Phi = \sum_{n=-\infty}^{+\infty} c_n z^n \qquad (2.58)$$

The real part of this function satisfies the differential equation for our problem. Now we also need to satisfy the boundary conditions. At infinity, we have

$$v_x = -\frac{\partial \phi}{\partial x} = V_0 = -\mathrm{Re}\left(\frac{d\Phi}{dz}\right) \quad \text{and} \quad v_y = 0$$

Thus, as $|z| \to \infty$, $\Phi = -V_0 z$. The coefficients c_n, $n \geq 2$, of the positive powers in equation (2.58) are all zero, and the coefficient of z must be $c_1 = -V_0$. The other coefficients c_n, $n < 0$, are not yet determined, since all the negative powers of z go to zero at infinity.

The second boundary condition is $\psi = $ constant on $r = a$, and we may choose that constant to be zero. (As with electrostatic problems, we may add an arbitrary constant to the potential without changing the values of its gradient, here the velocity field.) Now the imaginary part of a complex function may be written as

$$\psi = \frac{1}{2i}(\Phi - \Phi^*)$$

and we want this expression to be identically zero for $|z| = a$. Inserting the series (2.58) for Φ, with both z and c_n in polar form ($z = ae^{i\theta}$ and $c_n = |c_n|e^{i\delta_n}$), we find

$$0 = \frac{1}{2i}\left(-V_0 a(e^{i\theta} - e^{-i\theta}) + \sum_{n=1}^{\infty} c_{-n}a^{-n}e^{-in\theta} - c_{-n}^* a^{-n}e^{in\theta}\right)$$

$$0 = -V_0 a\,\sin\theta - \sum_{n=1}^{\infty} a^{-n}|c_{-n}|\sin(n\theta - \delta_{-n})$$

$$= -V_0 a\,\sin\theta - \frac{|c_{-1}|}{a}\sin(\theta - \delta_{-1}) - \sum_{n=2}^{\infty} a^{-n}|c_{-n}|\sin(n\theta - \delta_{-n})$$

We can[14] satisfy this equation by taking $|c_{-n}| = 0$ for $n > 1$ and $|c_{-1}| = a^2 V_0$, $\delta_{-1} = \pi$. Then

$$\Phi = -V_0\left(z + \frac{a^2}{z}\right) = -V_0\left(z + \frac{a^2 z^*}{zz^*}\right) = -V_0\left(z + \frac{a^2 z^*}{|z|^2}\right)$$

---

[13] See Section 2.3.4.
[14] In Chapter 4, we will show that this is the *only* possible solution.

and on the curve $|z| = a$,

$$\Phi = -V_0(z + z^*) = \Phi^*$$

as we require. Thus, our Laurent series has only two terms, with powers z^1 and z^{-1}.

The velocity components are given by

$$w^* = -\frac{d\Phi}{dz} = V_0\left(1 - \frac{a^2}{z^2}\right)$$

$$= V_0\left(1 - \frac{a^2}{r^2}e^{-2i\theta}\right)$$

$$= v_x - iv_y$$

So

$$v_x = V_0\left(1 - \frac{a^2}{r^2}\cos 2\theta\right); \quad v_y = -V_0\frac{a^2}{r^2}\sin 2\theta$$

The equipotential surfaces are

$$\phi = -V_0\left(r + \frac{a^2}{r}\right)\cos\theta = \text{constant}$$

and the streamlines are given by

$$\psi = -V_0\left(r - \frac{a^2}{r}\right)\sin\theta = \text{constant}$$

The streamlines are shown in Figure 2.19.

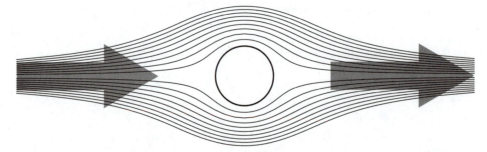

FIGURE 2.19. Streamlines of flow around the cylinder. The lines are given by $\psi = \text{constant}$.

2.5. POLES AND ZEROS

2.5.1. Analytic Continuation

If a function is analytic in a region R, $\rho_1 < |z - a| < \rho_2$, we may describe it using a Laurent series valid in R. We may also be able to find a second Laurent series in a different, but overlapping, region. (See, for example, the discussion of series for the tangent function in Appendix XI.) The second series is an *analytic continuation* of f into the second region.

For example, consider the function

$$f_1(z) = \frac{\sin z}{z - 1}$$

defined in the region $|z| < 1$ (Figure 2.20). We may evaluate the function with a Taylor series:

$$f_2(z) = -\left(z - \frac{z^3}{3!} + \frac{z^5}{5!} + \cdots\right)\left(1 + z + z^2 + z^3 + z^4 + z^5 + \cdots\right)$$

$$= -z - z^2 - \frac{5}{6}z^3 - \frac{5}{6}z^4 - \frac{101}{120}z^5 + \cdots$$

where $f_2(z)$ is valid only in the region $|z| < 1$. To evaluate the function[15] outside this region, we can expand about another point in the original region—for example, $z_1 = -\pi/4$.

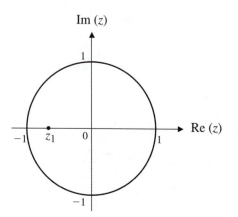

FIGURE 2.20. The function $f_2(z)$ may be analytically continued by expanding in a series about $z_1 = -\pi/4$.

---

[15]You might think that this would be unnecessary, since we already have the closed-form expression $f_1(z)$. But we cannot actually compute $\sin z$ except through its series expansion. See Morse and Feshbach, p. 377, for a good discussion of this point.

For convenience, we define a new variable $w = z - z_1 = z + \pi/4$. Then we obtain the series

$$f_3(z) = \frac{\sin(z + \pi/4 - \pi/4)}{(z + \pi/4 - \pi/4 - 1)} = \frac{\sin(w - \pi/4)}{w - (1 + \pi/4)} = \frac{1}{\sqrt{2}} \left(\frac{\sin w - \cos w}{w - 1.7854} \right)$$

$$= -\frac{1}{\sqrt{2}} \frac{w - w^3/3! + \cdots - (1 - w^2/2! + \cdots)}{1.7854(1 - w/1.7854)}$$

$$= \frac{1}{1.7854\sqrt{2}} \left(1 - w + \frac{w^2}{2!} + \frac{w^3}{3!} + \cdots \right)$$

$$\times \left(1 + \frac{w}{1.7854} + \frac{w^2}{1.7854^2} + \frac{w^3}{1.7854^3} + \cdots \right)$$

$$= 0.39605(1 - 0.4399w + 0.25361w^2 + 0.30871w^3 + \cdots)$$

and this series $f_3(z)$ is valid in the region $|w| < 1 + \pi/4$, or $|z + \pi/4| < 1 + \pi/4 = 1.7854$.

Thus, we have extended the function $f_2(z)$ into the shaded region shown in Figure 2.21. We can now pick a point in the new region and continue to extend the function indefinitely. The expressions $f_1(z)$, $f_2(z)$, and $f_3(z)$ are three different expressions for the same function.

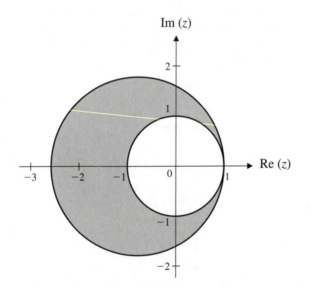

FIGURE 2.21. The function $f_2(z)$ has been continued into the shaded region as $f_3(z)$.

In some circumstances, we can simply use the original series outside of R as well, but it does not necessarily describe the original function in the new region. The new function is also an analytic continuation of f. In this case, we are redefining the function to *be* the series. This result is called *permanence of algebraic form*. For example, suppose we look at the function

$$f(x) = \frac{1}{x} + \frac{1}{2x^2} + \frac{1}{3x^3} + \cdots$$

where x is on the real axis with $|x| > 1$. (This series represents the real function $\ln [x/(x-1)]$.) We can use this series everywhere in the annulus $|z| > 1$ in the complex plane, where it represents a complex function (which we might call the principal branch of $\ln [z/(z-1)]$) that is identical with the original function on the real axis. It is an analytic continuation of the original function.

Notice that we talk about "an" analytic continuation, not "the" analytic continuation. There is more than one way to continue a given function. In any specific application, we must choose the continuation that is best for that particular purpose.

2.5.2. Zeros

The point $z = a$ is a *zero* of a function f if $f(a) = 0$. If f is analytic at a, then we can expand f in a Taylor series in some region $|z - a| < \rho$:

$$f(z) = \sum_{n=0}^{\infty} c_n (z - a)^n$$

If a is a zero of f, then c_0 must be zero. If $c_1 \neq 0$, then a is a *simple zero* of f. If both c_0 and c_1 are zero and $c_2 \neq 0$, then a is a zero of order two. In general, if $c_0, c_1, \ldots, c_{n-1}$ are all zero but c_n is not, then a is a zero of f of order n. Essentially, the order of the zero of a function at $z = a$ is a statement about how fast the function goes to zero as z approaches a.

Example 2.10.   The function $f(z) = \sin (z^2)$ has a zero at $z = 0$. What is its order?
    If we expand in a Taylor series, we get

$$\sin (z^2) = z^2 - \frac{z^6}{3!} + \cdots$$

In this series, both c_0 and c_1 are zero but c_2 is not, so $z = 0$ is a second-order zero of f.

What is the order of the zero $z = 0$ for the function $g = \sin z^2 - z^2$?

2.5.3. Singularities

Isolated Singularities

If $f(z)$ is analytic in the neighborhood of a point $z = a$ but not actually *at $z = a$*, then a is an *isolated singularity* of f.

There are three possible cases.

1. $|f(z)| \to \infty$ as $z \to a$. An example of this case is the function $f(z) = \dfrac{1}{z - a}$. These singularities are called *poles*.

2. $f(z)$ is bounded as $z \to a$. An example of this case is the function $f = \dfrac{\cos(z)}{z - \pi/2}$. Then, using l'Hospital's rule,[16] we have

$$\lim_{z \to \pi/2} \frac{\cos(z)}{z - \pi/2} = \lim_{z \to \pi/2} \frac{-\sin z}{1} = -1$$

These functions present no problem, since we can redefine the function as follows:

$$f(z) = \begin{cases} \dfrac{\cos(z)}{z - \pi/2} & \text{if } z \neq \pi/2 \\ -1 & \text{if } z = \pi/2 \end{cases}$$

The new function is analytic everywhere. This kind of singularity is called *removable*.

3. $f(z)$ oscillates. For example, consider the function

$$f(z) = \exp\left(\frac{1}{z}\right)$$

Along the real axis, $z = x$, we have

$$f(z) \to \infty \quad \text{as } x \to 0 \text{ through positive values}$$
$$f(z) \to 0 \quad \text{as } x \to 0 \text{ through negative values}$$

But along the imaginary axis, $z = iy$,

$$|f(z)| = \left|\exp\left(\frac{1}{iy}\right)\right| = \left|\exp\left(\frac{-i}{y}\right)\right| = 1 \quad \text{for all values of } y$$

Note that the Laurent series

$$\exp\left(\frac{1}{z}\right) = 1 + \frac{1}{z} + \frac{1}{2!z^2} + \cdots$$

which is valid *up to* the singularity at $z = 0$, has infinitely many negative powers. This is a characteristic of this type of singularity, which is called an *essential singularity*. (A Laurent series with infinitely many negative powers that is valid in an annulus with a *finite* inner radius does not necessarily indicate an essential singularity. An example of such a series is in Section 2.3.4, equation 2.48: the series for $\dfrac{1}{z^2 - 1}$ valid for $|z - 1| > 2$. This function has a pole at $z = 1$.)

---

[16]We could get the same result by expanding $\cos z$ in a Taylor series about $z = \pi/2$.

Next we shall investigate singularities of the first type, the *poles*. The function f is analytic in an annular region $0 < |z - a| < \rho$ and may be expanded in a Laurent series centered at $z = a$:

$$f(z) = \sum_{n=-\infty}^{\infty} c_n(z - a)^n$$

If the coefficient $c_{-1} \neq 0$ but all $c_{-m} = 0$ for $m > 1$, then the pole is of order 1 (the pole is *simple*). If c_{-2} is not zero but $c_{-m} = 0$ for $m > 2$, the pole is of order 2 (whether or not c_{-1} is zero). In general, if the series may be written as

$$f(z) = \sum_{n=-m}^{\infty} c_n(z - a)^n$$

with $c_{-m} \neq 0$, the pole is of order m. [An essential singularity (case 3 above) is a pole of infinite order.]

We can test for the order of a pole without finding the Laurent series. The limit

$$\lim_{z \to a} (z - a)^p f(z) = \lim_{z \to a} (z - a)^p \sum_{n=-m}^{\infty} c_n(z - a)^n$$

$$= \lim_{z \to a} \sum_{n=-m}^{\infty} c_n(z - a)^{n+p}$$

will be zero for $p > m$, will be a constant (c_{-m}) for $p = m$, and will not exist for $p < m$. Thus,

---

The order of a pole at $z = a$ is the lowest integer p for which the limit $\lim_{z \to a}(z - a)^p f(z)$ exists.

A function that has well-separated poles as its only singularities is described as *meromorphic*.

Other Kinds of Singularities

Functions may not be analytic for worse reasons than those cited above. For example:

- The function may have a *branch point*. A branch point is a point from which a branch cut emerges (the head of the tadpole, see Sections 2.1.3 and 2.2.1). To determine where the branch points z_p lie, note that we obtain successive branches of the function by increasing the argument of $z - z_p$ by 2π. The function $f = \sqrt{z}$ has a branch point at the origin. The principal branch of the function is not continuous across the branch cut, and therefore the function is not analytic anywhere along the branch cut. Thus, the singularity at $z = 0$ is *not isolated*.

- We may have an infinite set of singularities that converge to a limit point, so that any neighborhood of that point contains infinitely many singularities. The function $f = \tan(1/z)$ has this property. The singularities are at $1/z = n\pi/2$, where n is an odd integer, or

$$z_n = \frac{2}{n\pi}$$

This sequence converges to $z = 0$. No matter how small a neighborhood of the origin we pick, say $|z| < \varepsilon$, there are infinitely many z_n in this neighborhood. To see this, let N be an integer such that $N > 1/\varepsilon$. Then

$$z_N = \frac{2}{N\pi} < \frac{2}{\pi}\varepsilon < \varepsilon$$

is inside the neighborhood, and so are all the z_m, $m > N$. Thus, $z = 0$ is *not an isolated singularity* of f.

2.6. THE RESIDUE THEOREM

2.6.1. Definition of the Residue

If $f(z)$ is analytic in a neighborhood of $z = a$ except perhaps at a, then

the residue of f at a is $\dfrac{1}{2\pi i} \displaystyle\oint_C f(z)\,dz,$ where the closed curve C encloses a.

$$(2.59)$$

By the Cauchy theorem, the residue of f at a is zero if the function f is analytic at a. But the reverse is not necessarily true: A zero residue at a does *not* imply that the function is analytic at a.

Comparison with the expression for the coefficients in the Laurent series (equation 2.47) shows that

the residue of a function f at a is the c_{-1} coefficient
of the Laurent series centered at a.

$$(2.60)$$

Example 2.11.   Find the residue of the function $f = \dfrac{\cos z}{z^2}$ at $z = 0$.

The function may be written as a series:

$$f(z) = \frac{1}{z^2}\left(1 - \frac{z^2}{2!} + \frac{z^4}{4!} + \cdots\right)$$

$$= \frac{1}{z^2} - \frac{1}{2!} + \frac{z^2}{4!} + \cdots$$

The function has a second-order pole at $z = 0$, and since the coefficient $c_{-1} = 0$, the residue there is zero.

Example 2.12. Show that the function $f(z) = \dfrac{\sin z}{z^2} - 2\dfrac{e^z}{z}$ has a pole at the origin, find its order, and find the residue there.

We expand the sine and the exponential in Taylor series:

$$f = \frac{\sin z}{z^2} - 2\frac{e^z}{z}$$

$$= \frac{1}{z^2}\left(z - \frac{z^3}{3!} + \frac{z^5}{5!} - \cdots\right) - \frac{2}{z}\left(1 + z + \frac{z^2}{2!} + \frac{z^3}{3!} + \frac{z^4}{4!} + \cdots\right)$$

$$= -\frac{1}{z} - 2 - \frac{7}{6}z - \frac{1}{3}z^2 - \frac{3}{40}z^3 - \cdots$$

Thus, $f(z)$ has a simple pole at the origin, and the residue there is -1.

2.6.2. The Residue Theorem

If a function f is analytic in a simply connected domain D except for a *finite* number of *isolated* singularities and if curve C is within D, then

$$\oint_C f \, dz = 2\pi i \sum_{n=1}^{N} \operatorname{Res} f(z_n) \tag{2.61}$$

where z_n are the singularities of f contained within C.

To prove this theorem, we deform the curve C so as to exclude each of the singularities z_n, as shown in Figure 2.22. The new curve C' equals the original curve C, plus a set of cross cuts leading to, and a small circle Γ_n around, each singularity. We can do this only if the singularities are isolated, since the function has to be analytic on the cross cuts and the circles.

Then by the Cauchy theorem,

$$\oint_{C'} f \, dz = 0$$

since we have constructed C' so that f is analytic everywhere in and on C'. Since the function is analytic and thus continuous along the cross cuts, they do not contribute to the integral.[17] Thus,

$$\oint_C f \, dz + \sum_{n=1}^{N} \oint_{\Gamma_{n,\text{clockwise}}} f \, dz = 0$$

----

[17]We used this argument previously in Section 2.2.5.

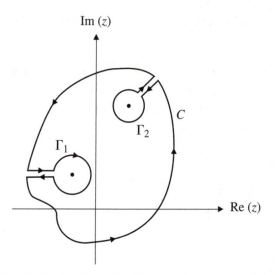

FIGURE 2.22. Contour for proving the residue theorem. The contour excludes the singularities at z_n. Two such points are shown in this diagram.

If the integral around C is in the standard counterclockwise direction, then each of the integrals in the sum is taken clockwise around its circle Γ_n. Now we move the sum to the other side of the equation and use the resulting minus sign to change the direction of integration to the counterclockwise direction. Using the definition of the residue (equation 2.59), we have

$$\oint_C f\,dz = \sum_{n=1}^{N} \oint_{\Gamma_n} f\,dz$$

$$= 2\pi i \sum_{n=1}^{N} \operatorname{Res} f(z_n)$$

and the theorem is proved.

2.6.3. Finding Residues

There are several different methods for finding residues.

1. Find the Laurent series and pick out the coefficient c_{-1}. This is what we did in Section 2.6.1.

2. For a simple pole, the residue at a is

$$\operatorname{Res} f(a) = \lim_{z \to a} (z - a) f(z) \qquad (2.62)$$

To see why, let's insert the Laurent series for f:

$$\lim_{z \to a} (z - a) f(z) = \lim_{z \to a} (z - a) \sum_{n=-1}^{\infty} c_n (z - a)^n$$

$$= \lim_{z \to a} \sum_{n=-1}^{\infty} c_n (z - a)^{n+1}$$

All the terms with positive powers of $(z - a)$ go to zero, leaving c_{-1}, which is the residue.

Example 2.13. Find the residue of the function $f(z) = \dfrac{\tan z}{z^2}$ at the origin.

First we find the order of the pole. Note that the limit

$$\lim_{z \to 0} z \frac{\tan z}{z^2} = \lim_{z \to 0} \frac{\tan z}{z} = \lim_{z \to 0} \frac{\sec^2 z}{1} = 1$$

exists, and thus the pole is simple. The above limit also gives the residue as 1.

3. For a pole of order m, the residue is

$$\operatorname{Res} f(a) = \lim_{z \to a} \frac{1}{(m-1)!} \frac{d^{m-1}}{dz^{m-1}} [(z - a)^m f(z)] \tag{2.63}$$

Again we use the Laurent series to demonstrate the result:

$$\lim_{z \to a} \frac{1}{(m-1)!} \frac{d^{m-1}}{dz^{m-1}} (z - a)^m f(z) = \lim_{z \to a} \frac{1}{(m-1)!} \frac{d^{m-1}}{dz^{m-1}} (z - a)^m \sum_{n=-m}^{\infty} c_n (z - a)^n$$

Since the Laurent series is uniformly convergent, we may differentiate it term by term. All terms with $m + n < m - 1$ (that is, $n < -1$) differentiate to zero.

$$\lim_{z \to a} \frac{d^{m-1}}{dz^{m-1}} \sum_{n=-m}^{\infty} c_n (z - a)^{n+m} = \lim_{z \to a} \sum_{n=-1}^{\infty} c_n (n+m)(n+m-1) \cdots (n+2)(z - a)^{n+1}$$

Again all the positive powers of $(z - a)$ go to zero in the limit, leaving only the $n = -1$ term:

$$\lim_{z \to a} \frac{1}{(m-1)!} \frac{d^{m-1}}{dz^{m-1}} (z - a)^m f(z) = \left(\frac{1}{(m-1)!} (m-1)(m-2) \cdots 1 \right) c_{-1}$$

$$= c_{-1}$$

as required.

Example 2.14. The function $f = \dfrac{\cosh z}{z^2}$ has a pole at the origin. Find the residue there.

First we find the order of the pole. We find that

$$\lim_{z \to 0} z \frac{\cosh z}{z^2} = \lim_{z \to 0} \frac{\cosh z}{z}$$

does not exist. But

$$\lim_{z \to 0} z^2 \frac{\cosh z}{z^2} = \lim_{z \to 0} \cosh z = 1$$

exists, and so the pole is of second order. The residue is

$$\operatorname{Res} f(0) = \lim_{z \to 0} \frac{d}{dz} \cosh z = \lim_{z \to 0} \sinh z = 0$$

We can check this result by finding the Laurent series:

$$f(z) = \frac{1 + z^2/2 + z^4/4! + \cdots}{z^2} = \frac{1}{z^2} + \frac{1}{2} + \frac{z^2}{4!} + \cdots$$

There is no c_{-1} term, and thus the residue is zero, as we found using method 3.

4. For a function of the form

$$f(z) = \frac{g(z)}{h(z)}$$

where $h(z)$ has a simple zero at $z = a$ and $g(z)$ is analytic at a, $f(z)$ has a simple pole at $z = a$, and the residue is given by

$$\operatorname{Res} f(a) = \lim_{z \to a} \frac{g(z)}{h'(z)} \tag{2.64}$$

To understand this result, first write $h(z)$ in a Taylor series centered at $z = a$:

$$h(z) = \sum_{n=1}^{\infty} c_n (z - a)^n$$

There is no c_0 term because $h(z)$ has a simple zero at $z = a$, and the coefficient c_n is given by (equation 2.44)

$$c_n = \frac{1}{n!} \frac{d^n h}{dz^n} \bigg|_{z=a}$$

Now apply method (2):

$$\lim_{z \to a} (z - a) f(z) = \lim_{z \to a} (z - a) \frac{g(z)}{h(z)}$$

$$= \lim_{z \to a} (z - a) \frac{g(z)}{\sum_{n=1}^{\infty} c_n (z - a)^n}$$

$$= \lim_{z \to a} \frac{g(z)}{\sum_{n=1}^{\infty} c_n (z - a)^{n-1}}$$

$$= \frac{g(a)}{c_1} = \frac{g(a)}{h'(a)}$$

as required.

Example 2.15. The function $f(z) = \tan z = \sin z / \cos z$ has a simple pole at $z = \pi/2$. Find the residue there.

By method 4, the residue is

$$\operatorname{Res} f\left(\frac{\pi}{2}\right) = \lim_{z \to \pi/2} \frac{\sin z}{-\sin z} = -1$$

5. Finally, as a last resort, we can actually evaluate the integral in (2.59).

Example 2.16. Evaluate the residue of the function $f(z) = 1/z$ at $z = 0$.

We choose C to be a circle of radius ρ centered at the origin. Then

$$\frac{1}{2\pi i} \oint_C f(z)\, dz = \frac{1}{2\pi i} \int_0^{2\pi} \frac{1}{\rho e^{i\theta}} \rho i e^{i\theta}\, d\theta = \frac{1}{2\pi i} i \int_0^{2\pi} d\theta = \frac{1}{2\pi i} 2\pi i = 1$$

Of course, we could also have obtained this result trivially using method 1.

2.7. USING THE RESIDUE THEOREM

2.7.1. Evaluating the Integral of a Complex Function

The residue theorem is an extremely powerful tool for evaluating integrals. When using the residue theorem to evaluate an integral of the form

$$\oint_C f(z)\, dz$$

I suggest that you always use the steps below.

1. Draw a diagram showing the contour C in the complex plane. Also mark on your diagram any poles or other singularities of the integrand, $f(z)$. If the function has a branch cut, be sure to show it on your diagram too.

2. If there is a branch cut, you will have to deform the contour C so that the entire branch cut is excluded from the contour. The contour and the branch cut may not intersect at any point!

3. Note which poles are *inside* the contour C.

4. Evaluate the residue of f at each of the poles that are inside the contour.

5. Apply the residue theorem.

6. If the contour runs around a branch cut, you will have to evaluate the integral explicitly along both sides of the cut.

Example 2.17.   Evaluate

$$\oint_C \frac{\sin z}{z-1} dz$$

where C is a square with corners at the points $(-i)$, $(2-i)$, $(2+i)$, and $(+i)$.
Step 1: See Figure 2.23.

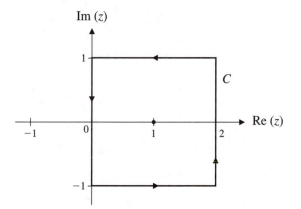

FIGURE 2.23. The contour for Example 2.17. The integrand has a pole at $z = 1$. This pole is inside the contour.

Steps 2 and 3: The integrand has a simple pole at $z = 1$. It is inside the contour. There are no other poles or branch cuts.
Step 4: Using method 1, we have

$$\operatorname{Res} f(1) = \lim_{z \to 1} (z-1) \frac{\sin z}{z-1} = \sin(1)$$

Step 5:

$$\oint_C \frac{\sin z}{z-1} dz = 2\pi i \,(\sin 1) = 5.2871i$$

2.7.2. Using the Residue Theorem to Calculate Real Integrals

Integrals of Trigonometric Functions

Integrals of the form

$$\int_0^{2\pi} f(\sin\theta, \cos\theta)d\theta \tag{2.65}$$

frequently arise in the solution of physical problems. For example, with $\theta = \omega t$, the integral gives 2π times the average value of the function over one period. We may evaluate such integrals around the unit circle in the complex plane. On the unit circle,

$$z = re^{i\theta} = e^{i\theta}, \quad dz = ie^{i\theta}d\theta \Rightarrow d\theta = \frac{dz}{iz} \tag{2.66}$$

and

$$\cos\theta = \frac{e^{i\theta} + e^{-i\theta}}{2} = \frac{1}{2}\left(z + \frac{1}{z}\right) \tag{2.67}$$

Similarly,

$$\sin\theta = \frac{e^{i\theta} - e^{-i\theta}}{2i} = \frac{1}{2i}\left(z - \frac{1}{z}\right) \tag{2.68}$$

These relations allow us to convert the integral over θ to an integral over z.

Example 2.18.   Evaluate

$$\int_0^\pi \frac{1}{3 + \sin^2\theta}d\theta$$

This integral is not in the form (2.65) because the limits are 0 to π rather than 0 to 2π. But notice that the integrand is even [$\sin^2(-\theta) = (-\sin\theta)^2 = \sin^2\theta$] and so

$$\int_{-\pi}^\pi \frac{1}{3 + \sin^2\theta}d\theta = \int_{-\pi}^0 \frac{1}{3 + \sin^2\theta}d\theta + \int_0^\pi \frac{1}{3 + \sin^2\theta}d\theta$$

$$= 2\int_0^\pi \frac{1}{3 + \sin^2\theta}d\theta$$

Then we convert to an integral over the unit circle, using relations (2.66) and (2.68):

$$\int_0^\pi \frac{1}{3 + \sin^2 \theta}\, d\theta = \frac{1}{2}\int_{-\pi}^\pi \frac{1}{3 + \sin^2 \theta}\, d\theta = \frac{1}{2}\oint_{\text{unit circle}} \frac{1}{3 + \left[\dfrac{1}{2i}\left(z - \dfrac{1}{z}\right)\right]^2} \frac{dz}{iz}$$

$$= \frac{1}{2i}\oint_{\text{unit circle}} \frac{1}{\dfrac{7}{2} - \dfrac{1}{4}z^2 - \dfrac{1}{4z^2}} \frac{dz}{z}$$

$$= 2i \oint_{\text{unit circle}} \frac{z}{z^4 - 14z^2 + 1}\, dz$$

Now we can apply the general method for evaluating contour integrals.
Step 1: See Figure 2.24.

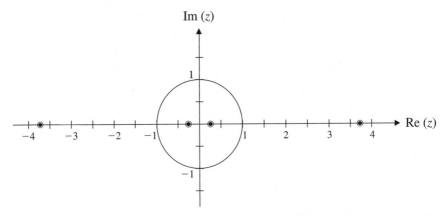

FIGURE 2.24. The contour for Example 2.18 is the unit circle. There are four poles, but only two are inside the contour.

Step 2 is not needed here.
Step 3: The integrand has four poles, given by

$$z_p^4 - 14z_p^2 + 1 = 0$$

or

$$z_p^2 = \frac{14 \pm \sqrt{14^2 - 4}}{2} = 7 \pm 4\sqrt{3} = 7 \pm 6.9282$$

All four roots are real. The values are

$$z_{1,2} = \pm\sqrt{7 + 6.9282} = \pm 3.7321$$

and

$$z_{3,4} = \pm\sqrt{7 - 6.9282} = \pm 0.26796$$

Of these four poles, only the last two are inside the unit circle (Figure 2.24).
  Step 4: The relevant residues are

$$\operatorname{Res} f(z_3) = \lim_{z \to z_3} (z - z_3) \frac{z}{(z - z_1)(z - z_2)(z - z_3)(z - z_4)} = \frac{z_3}{(z_3^2 - z_1^2)(z_3 - z_4)}$$

where we used the fact that $z_2 = -z_1$, and

$$\operatorname{Res} f(z_4) = \lim_{z \to z_4} (z - z_4) \frac{z}{(z - z_1)(z - z_2)(z - z_3)(z - z_4)} = \frac{z_4}{(z_4^2 - z_1^2)(z_4 - z_3)}$$

Thus, since $z_4 = -z_3$, we have

$$\sum_{\text{residues inside}} = \frac{z_3}{(z_3^2 - z_1^2)(z_3 - z_4)} + \frac{z_4}{(z_4^2 - z_1^2)(z_4 - z_3)} = \frac{1}{(z_4^2 - z_1^2)}$$

$$= \frac{1}{(7 - 4\sqrt{3}) - (7 + 4\sqrt{3})} = -\frac{\sqrt{3}}{24}$$

Step 5: Applying the residue theorem, we find

$$\int_0^\pi \frac{1}{3 + \sin^2 \theta} d\theta = 2\pi i (2i) \left(-\frac{\sqrt{3}}{24} \right) = \frac{\sqrt{3}}{6} \pi$$

Let's see how an integral of this type arises from a physics problem.

Example 2.19.   A circular wire loop of resistance R and radius a has its center at a distance $d_0 > a$ from a long straight wire. The loop is oriented so that a diameter of the loop points directly at the long wire, as shown in Figure 2.25. The current in the long wire is increasing at a rate $\alpha = dI/dt$. What is the current in the loop?

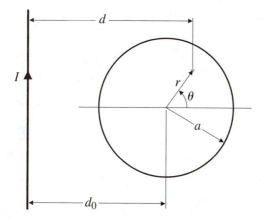

FIGURE 2.25. The physical system in Example 2.19 has a long straight wire carrying current I and
        a circular wire loop whose center is distance d_0 from the wire.

According to Faraday's law, the emf induced in the loop has magnitude

$$\varepsilon = \left| \frac{d\Phi}{dt} \right| = \left| \frac{d}{dt} \int \vec{B} \cdot \hat{n} \, dA \right|$$

where Φ is the magnetic flux through the loop. The magnetic field produced by the wire forms circles centered on the wire, and at a radial distance d from the wire,

$$B_\phi = \frac{\mu_0 I}{2\pi d}$$

Now at a point inside the loop,

$$d = d_0 + r \cos \theta$$

and

$$dA = r \, dr \, d\theta$$

so the flux is

$$\Phi = \int_0^{2\pi} \int_0^a \frac{\mu_0 I}{2\pi \, (d_0 + r \cos \theta)} r \, dr \, d\theta$$

The integral over θ is of the form (2.65), so we may convert it to an integral over the unit circle:

$$\int_0^{2\pi} \frac{1}{(d_0 + r \cos \theta)} d\theta = \oint_{\text{unit circle}} \frac{1}{d_0 + \frac{r}{2} \left(z + \frac{1}{z} \right)} \frac{dz}{iz}$$

$$= -2i \oint_{\text{unit circle}} \frac{1}{2d_0 z + r(z^2 + 1)} dz$$

The poles of the integrand are given by the roots of the quadratic in the denominator:

$$z = \frac{-2d_0 \pm \sqrt{4d_0^2 - 4r^2}}{2r} = -\frac{d_0}{r} \pm \sqrt{\left(\frac{d_0}{r} \right)^2 - 1}$$

Now since $d_0 > a \geq r$, $d_0/r > 1$. So only one of the two poles,

$$z_+ = -\frac{d_0}{r} + \sqrt{\left(\frac{d_0}{r} \right)^2 - 1}$$

is inside the unit circle (Figure 2.26). The relevant residue is

$$\lim_{z \to z_+} -\frac{1}{r} \frac{z - z_+}{(z - z_+)(z - z_-)} = -\frac{1}{r} \frac{1}{(z_+ - z_-)} = -\frac{1}{r} \frac{1}{2\sqrt{\left(\frac{d_0}{r} \right)^2 - 1}} = \frac{1}{2\sqrt{d_0^2 - r^2}}$$

and thus

$$\int_0^{2\pi} \frac{1}{(d_0 + r\cos\theta)}\,d\theta = -2i\,(2\pi i)\frac{1}{2\sqrt{d_0^2 - r^2}} = \frac{2\pi}{\sqrt{d_0^2 - r^2}}$$

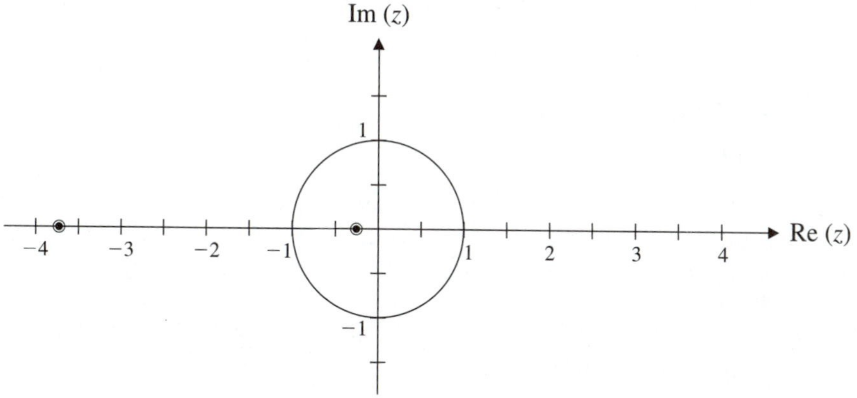

FIGURE 2.26. The integration contour for Example 2.19. Only one of the two poles is inside the contour.

Then the flux is

$$\Phi = \frac{\mu_0 I}{2\pi}\int_0^a \frac{2\pi}{\sqrt{d_0^2 - r^2}}\,r\,dr = \frac{\mu_0 I}{2}\int_{d_0^2}^{d_0^2 - a^2}\frac{-du}{\sqrt{u}}$$

where $u = d_0^2 - r^2$. So

$$\Phi = -\frac{\mu_0 I}{2}\frac{u^{1/2}}{1/2}\Big|_{d_0^2}^{d_0^2 - a^2} = -\mu_0 I\left(\sqrt{d_0^2 - a^2} - d_0\right) \qquad (2.69)$$

and since I is the only time-dependent quantity in this expression, the current in the loop is

$$I = \frac{\varepsilon}{R} = \frac{1}{R}\left|\frac{d\Phi}{dt}\right| = \frac{\mu_0\alpha d_0}{R}\left[1 - \sqrt{1 - \left(\frac{a}{d_0}\right)^2}\right]$$

We can check the result by looking at the limit $a \ll d_0$. We expect the flux to be approximately

$$\Phi \simeq \frac{\mu_0 I}{2\pi d_0}\pi a^2 = \frac{\mu_0 I}{2d_0}a^2$$

Now if we expand the square root in equation (2.69), we get

$$\Phi = \mu_0 I d_0 \left(1 - \sqrt{1 - \frac{a^2}{d_0^2}} \right) \simeq \mu_0 I d_0 \left[1 - \left(1 - \frac{1}{2} \frac{a^2}{d_0^2} \right) \right]$$

$$= \mu_0 I d_0 \left(\frac{1}{2} \frac{a^2}{d_0^2} \right) = \frac{\mu_0 I}{2 d_0} a^2$$

as expected.

2.7.3. Integrals Along the Entire Real Axis

Closing the Contour

An integral of the form $\int_{-\infty}^{+\infty} f(x)\, dx$ may be converted to an integral in the complex plane under some circumstances. The idea is to "close the contour" by adding additional pieces along which the integral is either zero or some multiple of the original integral along the real axis.

The most common way to close the contour is to add a large semicircle at *infinity* (Figure 2.27). For example, the integral

$$I = \int_{-\infty}^{+\infty} \frac{1}{\left(x^2 + a^2 \right)^2}\, dx$$

may be intepreted as an integral along the real axis in the complex plane:

$$I = \int_{\text{real axis}} \frac{1}{\left(z^2 + a^2 \right)^2}\, dz$$

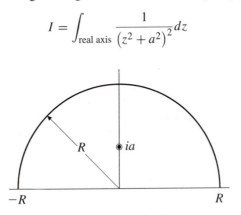

FIGURE 2.27. Closing the contour with a big semicircle of radius $R \to \infty$ in the upper half-plane.

Next we note that the integral along a large semicircle at infinity is zero. First look at the absolute value of the integrand:

$$\left| \frac{1}{\left(z^2 + a^2 \right)^2} \right| = \frac{1}{|z|^4} \frac{1}{\left| 1 + \frac{a^2}{z^2} \right|^2} \leq \frac{1}{|z|^4} \frac{1}{\left(1 - \frac{|a|^2}{|z|^2} \right)^2}$$

where we used the result (2.13) $|z_1 + z_2| \geq ||z_1| - |z_2||$. We can make the large semicircle big enough that $|a| \, / \, |z| < 1/\sqrt{2}$, and then

$$\left| \frac{1}{\left(z^2 + a^2\right)^2} \right| \leq \frac{4}{|z|^4} \quad \text{on the semicircle}$$

Then

$$\left| \int_{\text{semicircle}} \frac{1}{\left(z^2 + a^2\right)^2} dz \right| \leq (\text{length of curve})(\text{max value of } |\text{integrand}| \text{ on curve})$$

$$= (\pi R) \frac{4}{R^4} = \frac{4\pi}{R^3} \to 0 \quad \text{as } R \to \infty$$

Thus,

$$\oint \frac{1}{\left(z^2 + a^2\right)^2} dz = \int_{\text{real axis}} \frac{1}{\left(z^2 + a^2\right)^2} dz + \int_{\text{semicircle}} \frac{1}{\left(z^2 + a^2\right)^2} dz$$

$$= \int_{\text{real axis}} \frac{1}{\left(z^2 + a^2\right)^2} dz + 0$$

Thus, the integral around the closed contour composed of (a) a straight line along the real axis and (b) a large semicircle at infinity equals the integral along the real axis plus zero. We may evaluate the integral around the closed contour using the residue theorem, and the result equals the integral along the real axis.

The integrand has poles at $z = \pm ia$. Each pole is of order 2. Only the pole at $z = +ia$ is inside the contour (Figure 2.27). Using method 3, we find that the residue there is

$$\text{Res} f(ia) = \lim_{z \to ia} \frac{d}{dz} (z - ia)^2 \frac{1}{\left(z^2 + a^2\right)^2}$$

$$= \lim_{z \to ia} \frac{d}{dz} \frac{1}{(z + ia)^2}$$

$$= \lim_{z \to ia} \frac{-2}{(z + ia)^3} = \frac{-2}{(2ia)^3} = \frac{1}{4ia^3}$$

Thus, the value of the integral is

$$\oint \frac{1}{\left(z^2 + a^2\right)^2} dz = 2\pi i \left(\frac{1}{4ia^3} \right) = \frac{\pi}{2a^3}$$

and hence

$$\int_{-\infty}^{+\infty} \frac{1}{\left(x^2 + a^2\right)^2} dx = \frac{\pi}{2a^3}$$

Integrals of the Form $\int_{-\infty}^{+\infty} e^{ikx} f(x)\, dx$

Integrals of the form $\int_{-\infty}^{+\infty} e^{ikx} f(x)\, dx$, where k is a real number, may be evaluated by closing the contour with a semicircle at infinity. Fourier transforms (Chapter 7) are important examples of this class of integrals.

Example 2.20. Evaluate $\int_{-\infty}^{+\infty} \dfrac{e^{ikx}}{x^4+1}\,dx$, where k is real and $k > 0$.

The first step is to close the contour using a large semicircle, as discussed above. On the large semicircle at infinity, $z = R(\cos\theta + i\sin\theta)$, with $0 \le \theta \le \pi$. Thus,

$$\exp(ikz) = \exp(ikR\cos\theta)\exp(-kR\sin\theta) \tag{2.70}$$

and so

$$|\exp(ikz)| = |\exp(ikR\cos\theta)|\exp(-kR\sin\theta) = \exp(-kR\sin\theta) \le 1$$

since $\sin\theta$ is positive for $0 \le \theta \le \pi$.

Thus, we have

$$\left| \int_{\text{semicircle}} \frac{\exp(ikz)}{z^4+1}\,dz \right| \le (\text{length of path}) \max |\text{integrand}|$$

$$= \pi R \max |\exp(ikz)| \left| \frac{1}{z^4+1} \right|$$

$$\le \pi R \frac{2}{R^4} \quad \text{for } R > 2^{1/4}$$

$$= \frac{2\pi}{R^3} \to 0 \quad \text{as } R \to \infty$$

The poles of the integrand are the fourth roots[18] of -1, $\exp(i\pi/4 + 2n\pi i/4) = \exp(i\pi/4 + in\pi/2)$, $n = 0, 1, 2, 3$. Of these poles, only two, $n = 0$ and $n = 1$, are inside our contour (Figure 2.28). The residues are (by method 4)

$$\operatorname{Res} f\left(e^{i\pi/4}\right) = \left.\frac{e^{ikz}}{4z^3}\right|_{z=e^{i\pi/4}} = \frac{\exp\left(\dfrac{ik}{\sqrt{2}}(1+i)\right)}{4e^{3i\pi/4}} = \frac{\sqrt{2}\exp\left(\dfrac{k}{\sqrt{2}}(i-1)\right)}{4(i-1)}$$

and

$$\operatorname{Res} f\left(e^{i3\pi/4}\right) = \frac{\exp\left(\dfrac{ik}{\sqrt{2}}(i-1)\right)}{4e^{9i\pi/4}} = \frac{\sqrt{2}\exp\left(\dfrac{-k}{\sqrt{2}}(1+i)\right)}{4(1+i)}$$

---

[18] See Section 2.1.2.

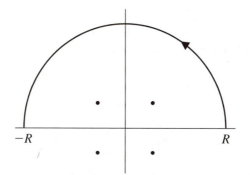

FIGURE 2.28. Contour for the integral in Example 2.20. We close with a contour in the upper half-plane, enclosing two of the four poles.

Thus,

$$\int_{-\infty}^{+\infty} \frac{e^{ikx}}{x^4+1}\,dx = \oint \frac{e^{ikz}}{z^4+1}\,dz = \frac{2\pi i \sqrt{2}}{4}\exp\left(-\frac{\sqrt{2}k}{2}\right)\left(\frac{e^{ki/\sqrt{2}}}{i-1}+\frac{e^{-ki/\sqrt{2}}}{i+1}\right)$$

$$= i\pi \frac{\sqrt{2}}{2}\exp\left(-\frac{\sqrt{2}k}{2}\right)\left(\frac{(i+1)e^{ki/\sqrt{2}}}{(i-1)(i+1)}+\frac{(i-1)e^{-ki/\sqrt{2}}}{(i+1)(i-1)}\right)$$

$$= i\pi \frac{\sqrt{2}}{2}\exp\left(-\frac{\sqrt{2}k}{2}\right)$$

$$\times\left(\frac{i\left(e^{ki/\sqrt{2}}+e^{-ki/\sqrt{2}}\right)+1\left(e^{ki/\sqrt{2}}-e^{-ki/\sqrt{2}}\right)}{-2}\right)$$

$$= \pi \frac{\sqrt{2}}{2}\exp\left(-\frac{\sqrt{2}k}{2}\right)\left(\cos\frac{k}{\sqrt{2}}+\sin\frac{k}{\sqrt{2}}\right)$$

We include real integrals of the form $\int_{-\infty}^{+\infty}\sin kx\, f(x)\,dx$ and $\int_{-\infty}^{+\infty}\cos kx\, f(x)\,dx$ in this class of integrals, since each trigonometric function is a linear combination of e^{ikx} and e^{-ikx}. The sine and cosine *must* be treated as combinations of exponentials because $\int_{\text{semicircle}} f(x)\cos(kx)\,dx$ does not vanish. If you are *sure* that the result of the integration is a finite real value, then you may write

$$\int_{-\infty}^{+\infty}\cos kx\, f(x)\,dx = \text{Re}\int_{-\infty}^{+\infty}e^{ikx}\, f(x)\,dx$$

and

$$\int_{-\infty}^{+\infty}\sin kx\, f(x)\,dx = \text{Im}\int_{-\infty}^{+\infty}e^{ikx}\, f(x)\,dx$$

and evaluate the integral as in Example 2.20. However, if there is any chance that the integral may have an imaginary part (and this does happen in physics problems; see the section below on integrals with poles on the real axis), then you must be more careful and evaluate

$$\int_{-\infty}^{+\infty} \cos kx \; f(x) \, dx = \frac{1}{2} \int_{-\infty}^{+\infty} e^{ikx} f(x) \, dx + \frac{1}{2} \int_{-\infty}^{+\infty} e^{-ikx} f(x) \, dx$$

Example 2.21. Evaluate $\displaystyle\int_{-\infty}^{+\infty} \frac{\cos kx}{x^2 + a^2} dx$, where k is real.

Since the cosine is an even function, we may assume that the real number k is positive. We expect this integral to have a real value, but let's check this assumption by evaluating

$$\int_{-\infty}^{+\infty} \frac{\cos kx}{x^2 + a^2} dx = \frac{1}{2} \int_{-\infty}^{+\infty} \frac{\exp(ikx)}{x^2 + a^2} dx + \frac{1}{2} \int_{-\infty}^{+\infty} \frac{\exp(-ikx)}{x^2 + a^2} dx$$

The first step is to close the contour. Look at the first of the two integrals. On the large semicircle at infinity, $z = R(\cos \theta + i \sin \theta)$, with $0 \le \theta \le \pi$. Using equation (2.70), we have

$$|\exp(ikz)| = |\exp(ikR \cos \theta)| \exp(-kR \sin \theta) = \exp(-kR \sin \theta) \le 1$$

Thus, for $R > a/\sqrt{2}$, we have

$$\left| \int_{\text{semicircle}} \frac{\exp(ikz)}{z^2 + a^2} dz \right| \le (\text{length of path}) \max |\text{integrand}|$$

$$= \pi R \max |\exp(ikz)| \left| \frac{1}{z^2 + a^2} \right|$$

$$\le \pi R \frac{2}{R^2} = \frac{2\pi}{R} \to 0 \quad \text{as } R \to \infty$$

Thus,

$$\int_{-\infty}^{+\infty} \frac{\exp(ikx)}{x^2 + a^2} dx = \oint_{C+} \frac{\exp(ikz)}{z^2 + a^2} dz$$

The integrand has two poles, at $z = \pm ia$, but only the one at $z = +ia$ is inside the contour (Figure 2.29). The pole is simple, and the residue is

$$\text{Res} f(ia) = \lim_{z \to ia} (z - ia) \frac{e^{ikz}}{z^2 + a^2} = \lim_{z \to ia} \frac{e^{ikz}}{z + ia} = \frac{e^{-ka}}{2ia}$$

Using the residue theorem, we have

$$\oint_{C+} \frac{\exp(ikz)}{z^2 + a^2} dz = 2\pi i \left(\frac{e^{-ka}}{2ia} \right) = \frac{\pi}{a} e^{-ka} \tag{2.71}$$

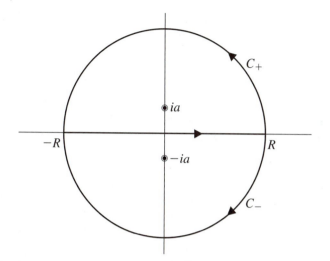

FIGURE 2.29. If the integrand is of the form $f(x)e^{ikx}$, with $k > 0$, we close the contour in the upper half-plane (C_+). If $k < 0$, we close in the lower half-plane (C_-).

Now we turn to the second integral. Evaluating $\exp(-ikz)$ over the big semicircle, we find

$$\exp(-ikz) = \exp(-ikR\cos\theta)\exp(kR\sin\theta)$$

and the real exponential becomes arbitrarily large[19] as $R \to \infty$. (This is why we cannot evaluate the integral of the cosine directly, but instead have to write the cosine as a combination of exponentials.) However, if instead we close the contour *downward* (Figure 2.29), where $\pi \le \theta \le 2\pi$, then $\sin\theta < 0$ and the real exponential is bounded by unity everywhere on this semicircle. Then we can show that

$$\left| \int_{\text{semicircle downward}} \frac{\exp(-ikz)}{z^2 + a^2} dz \right| \to 0 \quad \text{as } R \to \infty$$

Thus,

$$\int_{-\infty}^{+\infty} \frac{\exp(-ikx)}{x^2 + a^2} dx = \oint_{C_-} \frac{\exp(-ikz)}{z^2 + a^2} dz$$

Now the pole at $-ia$ is inside the contour C_-. The residue is

$$\operatorname{Res} f(-ia) = \lim_{z \to -ia} (z + ia) \frac{e^{-ikz}}{z^2 + a^2} = \lim_{z \to -ia} \frac{e^{-ikz}}{z - ia} = \frac{e^{-ka}}{-2ia}$$

---

[19]Although we cannot show that the integral on the upper semicircle is zero, it is not necessarily infinite. Once we have obtained the result for the integral along the real axis using the lower contour, we can use the upper contour to calculate the (finite) integral along the upper semicircle. While possible, this is rarely useful.

As we apply the residue theorem, we have to remember that it applies to contours traversed counterclockwise. This time we are going clockwise, so we have to add a minus sign:

$$\oint_{C_-} \frac{\exp(-ikz)}{z^2 + a^2} dz = -2\pi i \ \text{Res} \ f(-ia) = -2\pi i \left(\frac{e^{-ka}}{-2ia}\right) = \frac{\pi}{a} e^{-ka}$$

which is the same result that we got for the first integral.

Finally, we add the two results to obtain

$$\int_{-\infty}^{+\infty} \frac{\cos kx}{x^2 + a^2} dx = \frac{1}{2}\left(\frac{\pi}{a}e^{-ka} + \frac{\pi}{a}e^{-ka}\right) = \frac{\pi}{a}e^{-ka}$$

The result is a real number. Notice that we could also have obtained this result by taking the real part of equation (2.71).

Jordan's Lemma

Integrals of the type considered above may be converted to complex contour integrals provided that the function $f(x) \to 0$ sufficiently fast as $x \to \infty$. The proof that the integral along the semicircle is zero is facilitated by the use of Jordan's lemma:

If $f(z)$ converges uniformly to zero whenever $z \to \infty$, then

$$\lim_{R \to \infty} \int_{C_R} f(z)e^{ikz} \, dz = 0$$

where k is any positive real number and C_R is the upper half of the circle $|z| = R$.

Note. $f(z)$ converges uniformly to zero if, given any ε, there exists an M such that $|f(z)| < \varepsilon$ whenever $|z| > M$, no matter what the argument of z.

To prove the lemma, choose $R > M$. Then on C_R, $|f(z)| < \varepsilon$ and $z = Re^{i\theta} = R(\cos\theta + i\sin\theta)$.

$$\left|\int_{C_R} f(z)e^{ikz} \, dz\right| \le \varepsilon \left|\int_0^\pi \exp(ikR\cos\theta)\exp(-kR\sin\theta) \ Rie^{i\theta} d\theta\right|$$

$$\le 2\varepsilon R \left|\int_0^{\pi/2} \exp(-kR\sin\theta) \, d\theta\right|$$

Since we need only an upper bound to the integral on the right, we note that $\sin\theta \ge 2\theta/\pi$ throughout the range of integration (Figure 2.30) and thus $e^{-kR\sin\theta} \le e^{-2kR\theta/\pi}$. This latter function may be integrated easily to give

$$\left|\int_{C_R} f(z)e^{ikz} \, dz\right| \le \frac{\varepsilon R\pi}{kR}(1 - e^{-kR}) \le \frac{\varepsilon\pi}{k}$$

Since ε may be chosen as small as we like, the integral goes to zero and the lemma is proved.

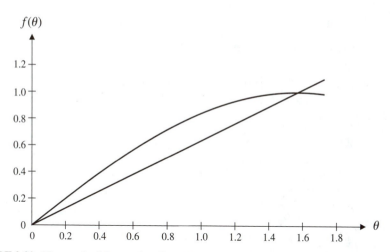

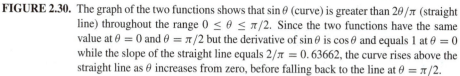

FIGURE 2.30. The graph of the two functions shows that $\sin\theta$ (curve) is greater than $2\theta/\pi$ (straight line) throughout the range $0 \le \theta \le \pi/2$. Since the two functions have the same value at $\theta = 0$ and $\theta = \pi/2$ but the derivative of $\sin\theta$ is $\cos\theta$ and equals 1 at $\theta = 0$ while the slope of the straight line equals $2/\pi = 0.63662$, the curve rises above the straight line as θ increases from zero, before falling back to the line at $\theta = \pi/2$.

Closing the Contour with a Rectangle

Integrands that contain hyperbolic functions do not lend themselves to the methods used above, since the integrand does not go to zero on the contour C_R, whether closed upward or downward. However, closing the contour with a rectangle may work. We choose the height of the rectangle to make the integral along the top side of the rectangle equal to a constant multiple[20] of the integral along the real axis.

Example 2.22.   Evaluate the integral

$$\int_{-\infty}^{+\infty} \frac{1}{\cosh kx}\,dx$$

If we try to close the contour with a semicircle, we find that on C_R,

$$\left|\frac{1}{\cosh kz}\right| = \left|\frac{2}{e^{kz} + e^{-kz}}\right| = \left|\frac{2}{e^{kR\cos\theta}e^{ikR\sin\theta} + e^{-kR\cos\theta}e^{-ikR\sin\theta}}\right|$$

$$\ge \frac{2}{e^{kR\cos\theta} + e^{-kR\cos\theta}}$$

and the lower bound equals 1 on the imaginary axis ($\theta = \pi/2$). So we cannot show that the integral along the semicircle is zero. However, notice that on the line

---

[20] See Problem 31(c) for a slight variant on this theme.

$z = x + i\pi/k,$

$$\cosh kz = \cosh k\left(x + i\frac{\pi}{k}\right) = \frac{1}{2}\left(e^{kx}e^{i\pi} + e^{-kx}e^{-i\pi}\right) = -\cosh kx$$

and so the integral along this line from $+\infty$ to $-\infty$, parallel to the real axis, equals the original integral. Thus, we are led to consider the rectangular contour shown in Figure 2.31.

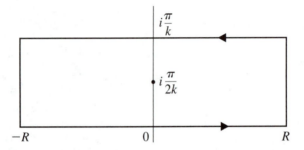

FIGURE 2.31. Closing the contour with a rectangle. In Example 2.22, the upper side is at $\operatorname{Im}(z) = \pi/k$.

The integrand has poles at $z = i(2n+1)\pi/2k$. Only one pole, the one at $z = \pi i/2k$, is inside the contour (Figure 2.31). On the vertical side $x = R$, we have

$$\left|\int_{\text{side}} \frac{1}{\cosh kz}dz\right| = \left|\int_0^{\pi/k} \frac{1}{e^{kR}e^{iky} + e^{-kR}e^{-iky}}i\,dy\right|$$

$$= \frac{1}{e^{kR}}\left|\int_0^{\pi/k} \frac{1}{e^{iky} + e^{-2kR}e^{-iky}}dy\right|$$

$$\leq \frac{1}{e^{kR}}\frac{\pi}{k} \to 0 \quad \text{as } R \to \infty$$

A similar argument holds for the side at $x = -R$. Thus,

$$\oint_{\text{rectangle}} \frac{1}{\cosh kz}dz = \int_{-\infty}^{+\infty} \frac{1}{\cosh kx}dx + \int_{+\infty+i\pi/k}^{-\infty+i\pi/k} \frac{1}{\cosh kz}dz$$

$$= 2\int_{-\infty}^{+\infty} \frac{1}{\cosh kx}dx$$

Using method 4 (Section 2.6.3), we find that the residue at the pole is

$$\operatorname{Res} f\left(\frac{i\pi}{2k}\right) = \lim_{z \to i\pi/2k} \frac{1}{k\sinh(kz)} = \frac{1}{k\sinh(i\pi/2)} = \frac{1}{k}\left(\frac{1}{i\sin\pi/2}\right) = \frac{1}{ik}$$

and thus the integral is

$$\int_{-\infty}^{+\infty} \frac{1}{\cosh kx} dx = \frac{1}{2} \oint_{\text{rectangle}} \frac{1}{\cosh kz} dz = \frac{1}{2} (2\pi i) \left(\frac{1}{ik} \right) = \frac{\pi}{k}$$

Integrals with Poles on the Real Axis

Integrals such as

$$\int_{-\infty}^{+\infty} \frac{\sin kx}{x-2} dx$$

occur in physics applications. The integrand is unbounded at $x = 2$, within the range of integration, so we must carefully state exactly what we mean by this mathematical expression.

In our first definition, we simply remove the offending point from the range of integration. The *principal value* of the integral is defined to be

$$P \int_{-\infty}^{+\infty} \frac{\sin kx}{x-2} dx \equiv \lim_{\varepsilon \to 0} \left(\int_{-\infty}^{2-\varepsilon} \frac{\sin kx}{x-2} dx + \int_{2+\varepsilon}^{+\infty} \frac{\sin kx}{x-2} dx \right) \qquad (2.72)$$

provided that the limit exists.

When evaluating the integral using the residue theorem, we are not allowed to have any singularities *on* the contour, so we must deform the contour so as to avoid the pole. For example, we could put a small semicircle of radius ε over the pole (Figure 2.32). Next we split the integral up into two exponential integrals:

$$\int_{-\infty}^{+\infty} \frac{\sin kx}{x-2} dx = \frac{1}{2i} \left(\int_{-\infty}^{+\infty} \frac{\exp(ikx)}{x-2} dx - \int_{-\infty}^{+\infty} \frac{\exp(-ikx)}{x-2} dx \right)$$

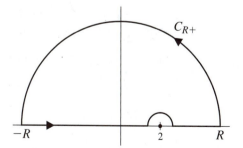

FIGURE 2.32. When there is a pole on the real axis, the contour must be deformed to go around the pole.

(In this discussion, we shall take k to be a positive real number.) To do the first integral, we close the contour upward, as shown in Figure 2.32. (Compare with Example 2.20.) The integral along the big semicircle is zero, by Jordan's lemma, since the function $1/(z-2)$ goes to zero uniformly on the semicircle as $R \to \infty$. There are no poles within the contour, and so

$$\oint_{C_{R+}} \frac{\exp ikz}{z-2} dz = 0$$

For the second term, we have to close downward (Figure 2.33) so that the integral along the big semicircle contributes zero.[21]

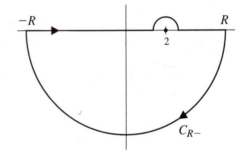

FIGURE 2.33. The contour for the integral of $e^{-ikx}/(x-2)$ is formed by closing with a semicircle in the lower half-plane. The path along the real axis remains unchanged.

Now the pole at $z = 2$ is inside the contour. The residue is

$$\operatorname{Res} f(2) = \lim_{z \to 2} (z-2) \frac{\exp(-ikz)}{z-2} = \exp(-2ki)$$

Thus,

$$\oint_{C_{R-}} \frac{\exp(-ikz)}{z-2} dz = -2\pi i e^{-2ki}$$

Finally, we have to evaluate the integral around the little semicircle at the pole. On this curve, $z = 2 + \varepsilon e^{i\theta}$, where θ varies from π to 0, so

$$\lim_{\varepsilon \to 0} \int_{C_\varepsilon} \frac{\sin kz}{z-2} dz = \lim_{\varepsilon \to 0} \frac{1}{2i} \int_\pi^0 \frac{\exp[ik(2+\varepsilon e^{i\theta})] - \exp[-ik(2+\varepsilon e^{i\theta})]}{\varepsilon e^{i\theta}} \varepsilon i e^{i\theta} d\theta$$

$$= \frac{1}{2}(e^{i2k} - e^{-i2k}) \int_\pi^0 d\theta$$

$$= i \sin(2k)(-\pi)$$

[21] Refer to the discussion in Example 2.21.

The principal value is the integral along the real axis, up to the beginning of the semicircle and continuing from the end of the semicircle, so finally

$$P \int_{-\infty}^{+\infty} \frac{\sin kx}{x-2} dx - \pi i \sin 2k = \int_{-\infty}^{+\infty} \frac{\sin kx}{x-2} dx$$

$$= \frac{1}{2i}[0 - (-2\pi i e^{-2ki})]$$

Thus,

$$P \int_{-\infty}^{+\infty} \frac{\sin kx}{x-2} dx = \pi e^{-2ki} + \frac{\pi i}{2i}(e^{2ki} - e^{-2ki})$$

$$= \frac{\pi}{2}(e^{2ki} + e^{-2ki})$$

$$= \pi \cos 2k$$

In physics problems, however, the physics usually determines the path of integration around the pole. We do not always want the principal value. More often we need the integral along a continuous path from $-\infty$ to $+\infty$. The pole settles onto the real axis as a result of some approximation in the modeling process (zero friction,[22] for example). In such cases, a better interpretation of the integral may be

$$\lim_{\varepsilon \to 0} \int_{-\infty + i\varepsilon}^{+\infty + i\varepsilon} \frac{\sin kx}{x-2} dx$$

where the path of integration passes over the pole,[23] as shown in Figure 2.34.

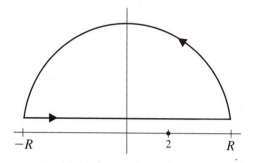

FIGURE 2.34. Integration path that passes above the pole.

[22]We shall have to wait until future chapters for specific examples of this phenomenon. In Chapter 7, Example 7.4, one pole moves onto the real axis as the damping parameter γ approaches zero. The simpler equation may be integrated directly to show that the correct answer is obtained if the path of integration is a straight line that passes over the pole at the origin.

[23]This path is equivalent to a path along the real axis with the pole pushed slightly downward off the real axis.

We may evaluate this integral as we did above, but there is no integral around the little semicircle. Thus,

$$\lim_{\varepsilon \to 0} \int_{-\infty + i\varepsilon}^{+\infty + i\varepsilon} \frac{\sin kx}{x-2} dx = \frac{1}{2i} \left(\lim_{\varepsilon \to 0} \int_{-\infty + i\varepsilon}^{+\infty + i\varepsilon} \frac{\exp(ikx)}{x-2} dx \right.$$

$$\left. - \lim_{\varepsilon \to 0} \int_{-\infty + i\varepsilon}^{+\infty + i\varepsilon} \frac{\exp(-ikx)}{x-2} dx \right)$$

$$= \frac{1}{2i} [0 - (-2\pi i e^{-2ki})] = \pi e^{-2ki} = \pi \cos 2k - \pi i \sin 2k$$

which differs from the previous result by having an additional imaginary part $-\pi i \sin 2k$.

On the other hand, the path may pass below the pole (Figure 2.35). In this case, the integral around the lower curve is zero, but around the upper curve we have

$$\oint_{C_{R+}} \frac{\exp ikz}{z-2} dz = 2\pi i e^{2ki}$$

and so

$$\lim_{\varepsilon \to 0} \int_{-\infty - i\varepsilon}^{+\infty - i\varepsilon} \frac{\sin kx}{x-2} dx = \frac{1}{2i} \left(\lim_{\varepsilon \to 0} \int_{-\infty - i\varepsilon}^{+\infty - i\varepsilon} \frac{\exp(ikx)}{x-2} dx \right.$$

$$\left. - \lim_{\varepsilon \to 0} \int_{-\infty - i\varepsilon}^{+\infty - i\varepsilon} \frac{\exp(-ikx)}{x-2} dx \right)$$

$$= \frac{1}{2i} (2\pi i e^{2ki} - 0) = \pi e^{2ki} = \pi \cos 2k + i\pi \sin 2k$$

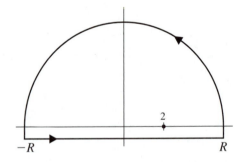

FIGURE 2.35. Integration path that passes below the pole.

Again the imaginary part of this result differs from both of the previous integrals. Thus, the correct choice of path is crucial if we are to obtain physically meaningful results. We'll see examples of these different choices in future chapters.[24]

---

[24]See also Problem 36.

Integrals of Multivalued Functions

We can also evaluate integrals of the form

$$\int_0^\infty x^\alpha f(x)\, dx$$

where α is not an integer. The function z^α has a branch point at the origin. If we choose the branch cut along the positive real axis (that is, $0 \le \theta < 2\pi$), then

$$z^\alpha = (re^{i\theta})^\alpha = r^\alpha e^{i\alpha\theta}$$

and so our real integral is an integral along the top of the branch cut where $\theta = 0$ and $e^{i\alpha\theta} \equiv 1$. We construct a closed contour, as shown in Figure 2.36, so that the branch point at the origin and the entire branch cut are excluded from the interior of the contour.

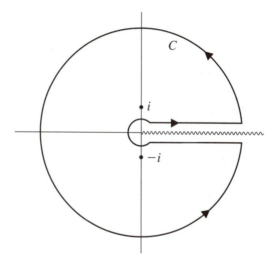

FIGURE 2.36. A contour that excludes the branch cut along the positive real axis.

Example 2.23. Evaluate the integral $I = \displaystyle\int_0^\infty \frac{\sqrt{x}}{x^2+1}\, dx$.

We choose the principal branch of the square root, as above, so I is the integral along the top of the branch cut. Note that the integrand is analytic everywhere inside the keyhole-shaped contour shown in Figure 2.36, except for poles at the two points $z = \pm i$. The residues at the poles are

$$\text{Res } f(+i) = \frac{\sqrt{i}}{2i} = \frac{e^{i\pi/4}}{2i} = \frac{\sqrt{2}}{4i}(1+i)$$

and

$$\text{Res } f(-i) = \frac{\sqrt{-i}}{-2i} = \frac{e^{i3\pi/4}}{-2i} = \frac{\sqrt{2}}{4i}(1-i)$$

(Remember: $0 \leq \theta < 2\pi$ for our chosen branch of the square root function, so $-i = e^{3\pi i/2}$.) Then, using the residue theorem, we find

$$\oint_C \frac{\sqrt{z}}{z^2 + 1} dz = 2\pi i \frac{\sqrt{2}}{4i} 2 = \pi\sqrt{2}$$

Along the big circle of radius R, we have

$$\left| \frac{\sqrt{z}}{z^2 + 1} \right| = \frac{\sqrt{R}}{R^2 \left| 1 + e^{-2i\theta}/R^2 \right|} \leq \frac{1}{R^{3/2} \left(1 - 1/R^2 \right)} \leq \frac{4}{3R^{3/2}} \quad \text{for } R \geq 2$$

Thus,

$$\lim_{R \to \infty} \left| \oint_{C_R} \frac{\sqrt{z}}{z^2 + 1} dz \right| \leq \lim_{R \to \infty} 2\pi R \frac{4}{3R^{3/2}} = \lim_{R \to \infty} \frac{8}{3} \frac{\pi}{\sqrt{R}} = 0$$

Next we investigate the integral around the small circle at the origin. On this circle of radius ε, we set $z = \varepsilon e^{i\theta}$:

$$\oint_{C_\varepsilon} \frac{\sqrt{z}}{z^2 + 1} dz = \int_{2\pi}^{0} \frac{\sqrt{\varepsilon e^{i\theta/2}}}{\varepsilon^2 e^{2i\theta} + 1} \varepsilon i e^{i\theta} \, d\theta$$

$$= i\varepsilon^{3/2} \int_{2\pi}^{0} \frac{e^{i3\theta/2}}{\varepsilon^2 e^{2i\theta} + 1} d\theta \to 0 \quad \text{as } \varepsilon \to 0$$

Finally, look at the integral along the bottom of the branch cut, where $\theta = 2\pi$:

$$\int_{\infty}^{0} \frac{\sqrt{r e^{2\pi i}}}{r^2 e^{4\pi i} + 1} dr = -\int_{0}^{\infty} \frac{\sqrt{r} e^{i\pi}}{r^2 + 1} dr = \int_{0}^{\infty} \frac{\sqrt{r}}{r^2 + 1} dr = I$$

Thus, we have

$$\oint_C \frac{\sqrt{z}}{z^2 + 1} dz = \int_{\text{top of cut}} + \int_{C_R} + \int_{\text{bottom of cut}} + \int_{C_\varepsilon} = 2I = \pi\sqrt{2}$$

and so

$$I = \int_{0}^{\infty} \frac{\sqrt{x}}{x^2 + 1} dx = \frac{\pi\sqrt{2}}{2}$$

2.7.4. Dispersion Relations

We have already noted how powerful the property of analyticity can be. In the 1920s, H. A. Kramers and R. de L. Kronig discovered how to use this property to relate the real and imaginary parts of the dielectric constant of a material, thus deriving relations that relate the dispersive and absorptive properties of a material in its interaction with electromagnetic waves. (See, for example, Jackson, Section 7.10.) The *Kramers-Kronig relations* are a specific example of a more general class of relations called *dispersion relations*. In recent years, these relations have proved important in other branches of physics as well— for example, in particle physics.

Suppose a function $f(z)$ is analytic everywhere in the upper half-plane. Then the first Cauchy formula (2.38) allows us to express the value of the function at a point z_0 in terms of an integral around a curve C that surrounds z_0:

$$f(z_0) = \frac{1}{2\pi i} \oint_C \frac{f(z)}{z - z_0} dz$$

where both z_0 and C are in the upper half-plane. Now if $|f(z)| \to 0$ as $z \to \infty$, we may choose C to be a large semicircle with its flat side along the real axis. The integral along the curved part is zero, and so we find

$$f(z_0) = \frac{1}{2\pi i} \int_{-\infty}^{+\infty} \frac{f(x)}{x - z_0} dx$$

Now we let the point z_0 approach the x-axis from above. The path of integration must remain below the pole, so we put a small semicircle under the pole and obtain

$$f(x_0) = \frac{1}{2\pi i} \left[P \int_{-\infty}^{+\infty} \frac{f(x)}{x - x_0} dx + \lim_{\varepsilon \to 0} \int_{-\pi}^{0} \frac{f\left(x_0 + \varepsilon e^{i\theta}\right)}{\varepsilon e^{i\theta}} i\varepsilon e^{i\theta} d\theta \right]$$

$$= \frac{1}{2\pi i} \left[P \int_{-\infty}^{+\infty} \frac{f(x)}{x - x_0} dx + i\pi f(x_0) \right]$$

Thus,

$$f(x_0) = -\frac{i}{\pi} P \int_{-\infty}^{+\infty} \frac{f(x)}{x - x_0} dx$$

From this it is clear that $f(x)$ has both real and imaginary parts, and they are related by

$$\mathrm{Re}\,[f(x_0)] = \frac{1}{\pi} P \int_{-\infty}^{+\infty} \frac{\mathrm{Im}\,[f(x)]}{x - x_0} dx \tag{2.73}$$

and

$$\mathrm{Im}\,[f(x_0)] = -\frac{1}{\pi} P \int_{-\infty}^{+\infty} \frac{\mathrm{Re}\,[f(x)]}{x - x_0} dx \tag{2.74}$$

Example 2.24. The permittivity of any material approaches ε_0 at very high frequencies,

$$\frac{\varepsilon(\omega)}{\varepsilon_0} = 1 + f(\omega)$$

and $\varepsilon(\omega)$ is analytic in the upper half-plane. If $f = u + iv$, where[25] $f^*(\omega) = f(-\omega)$ and $f(\omega) \to 0$ as $\omega \to \infty$, express the dispersion relations over the (physically meaningful) positive half of the ω-axis.

Since $\varepsilon(\omega)$ does not approach zero at high frequencies, we work instead with the function $f(\omega) = \varepsilon(\omega)/\varepsilon_0 - 1$. The given condition on f shows that

$$u(\omega) - iv(\omega) = u(-\omega) + iv(-\omega)$$

That is, u is an even function while v is odd. Then, from equation (2.73),

$$
\begin{aligned}
u(\omega_0) &= \frac{1}{\pi} P \int_{-\infty}^{+\infty} \frac{v(\omega)}{\omega - \omega_0} d\omega = \frac{1}{\pi} P \left[\int_{-\infty}^{0} \frac{v(\omega)}{\omega - \omega_0} d\omega + \int_{0}^{+\infty} \frac{v(\omega)}{\omega - \omega_0} d\omega \right] \\
&= \frac{1}{\pi} P \left[\int_{0}^{\infty} \frac{v(-\omega)}{-\omega - \omega_0} d\omega + \int_{0}^{+\infty} \frac{v(\omega)}{\omega - \omega_0} d\omega \right] \\
&= \frac{1}{\pi} P \int_{0}^{\infty} v(\omega) \left(\frac{1}{\omega + \omega_0} + \frac{1}{\omega - \omega_0} \right) d\omega = \frac{2}{\pi} P \int_{0}^{\infty} \frac{\omega v(\omega)}{\omega^2 - \omega_0^2} d\omega
\end{aligned}
$$

and, conversely, from equation (2.74),

$$
\begin{aligned}
v(\omega_0) &= -\frac{1}{\pi} P \int_{-\infty}^{+\infty} \frac{u(\omega)}{\omega - \omega_0} d\omega = -\frac{1}{\pi} P \left[\int_{0}^{\infty} \frac{u(-\omega)}{-\omega - \omega_0} d\omega + \int_{0}^{+\infty} \frac{u(\omega)}{\omega - \omega_0} d\omega \right] \\
&= -\frac{2}{\pi} P \int_{0}^{\infty} \frac{\omega_0 u(\omega)}{\omega^2 - \omega_0^2} d\omega
\end{aligned}
$$

These are the Kramers-Kronig relations. Measurements of the absorption properties of a material, $v(\omega)$, determine the dispersion, and vice versa.

2.8. CONFORMAL MAPPING

2.8.1. Definition of a Conformal Transformation

In Section 2.1.3, we noted that complex functions are mappings of the complex plane onto itself. If the function is analytic, then the mapping has some particularly nice properties.

If $f(z)$ is analytic at z_0 and $f'(z_0)$ does not vanish, then the mapping $z \to f(z)$ is *conformal* at z_0.

Conformal mappings have the following properties:

1. Conformal mappings preserve angles.

---

[25]We shall see in Chapter 7 that this condition arises from the fact that $\varepsilon(\omega)$ is the Fourier transform of a real function of time.

2. The magnification is the same for all curves passing through z_0.
3. Infinitesimal circles map to infinitesimal circles.

Let's see how these properties follow from the definition. Consider two sets of curves: $g(z) = $ constant and $h(z) = $ constant (Figure 2.37a). Under the mapping $w = f(z)$, we get two new sets of curves in the w-plane (Figure 2.37b). If we move along the original curves from z_0 by infinitesimal amounts dz_1 and dz_2 in the original plane, this movement corresponds to the displacements dw_1 and dw_2 along the mapped curves. (Remember that addition of complex numbers corresponds to vector addition in the complex plane.)

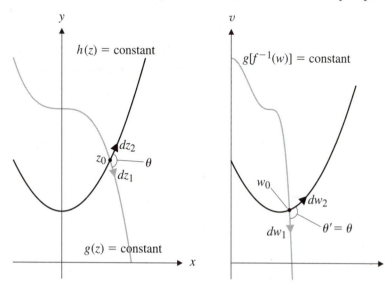

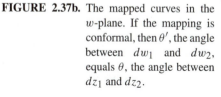

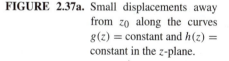

FIGURE 2.37a. Small displacements away from z_0 along the curves $g(z) = $ constant and $h(z) = $ constant in the z-plane.

FIGURE 2.37b. The mapped curves in the w-plane. If the mapping is conformal, then θ', the angle between dw_1 and dw_2, equals θ, the angle between dz_1 and dz_2.

Then

$$dw_1 = \left.\frac{df}{dz}\right|_{z_0} dz_1 \quad \text{and} \quad dw_2 = \left.\frac{df}{dz}\right|_{z_0} dz_2$$

So for $j = 1, 2$, the absolute value of dw_j is

$$|dw_j| = \left|\left.\frac{df}{dz}\right|_{z_0}\right| |dz_j|$$

and the argument is

$$\arg(dw_j) = \arg\left(\left.\frac{df}{dz}\right|_{z_0}\right) + \arg(dz_j)$$

Arg (dw_j) differs from arg (dz_j) by the constant angle arg $\left(\left. \dfrac{df}{dz} \right|_{z_0} \right)$. Thus, the angle between the two curves in the w-plane is

$$\theta' = \arg (dw_1) - \arg (dw_2) = \arg (dz_1) - \arg (dz_2) = \theta$$

and so the mapping preserves angles (property 1).

If a small circle about z_0 has radius $r = |dz|$, then each dw on the mapped curve is a distance

$$r' = |dw| = r \left| \frac{df}{dz} \right|_{z_0}$$

from $w_0 = f(z_0)$, and so the mapped curve is also a circle (property 3), with magnification $\left| df/dz|_{z_0} \right|$ (property 2). Note that this works only if $df/dz|_{z_0}$ is not zero; otherwise, the mapped "circle" reduces to a point.

2.8.2. Some Examples of Mappings

Two trivial cases

$\underline{f(z) = z + a}$

This is a translation of the whole plane by an amount a (Figure 2.38).

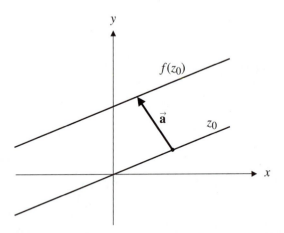

FIGURE 2.38. The mapping $w = z + a$ is a translation of the plane through the vector $\vec{a}$, whose components are the real and imaginary parts of the complex number a.

$\underline{f(z) = az = \rho e^{i\alpha} z}$

This is a rotation by an angle α plus a magnification by ρ (Figure 2.39).

These two examples are not very useful because they do not change the shape of figures in the plane.

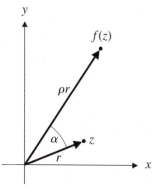

FIGURE 2.39. The mapping $w = az$ corresponds to rotation plus magnification.

More interesting and useful maps

$w = \ln z$

Writing z in polar form, $z = re^{i\theta}$, we find

$$w = u + iv = \ln r + i\theta$$

This function maps circles in the z-plane (constant r) to segments of lines in the w-plane (constant u, $0 \leq v < 2\pi$) and maps radial lines from the origin (θ = constant) to lines parallel to the u-axis (v = constant) (Figure 2.40).

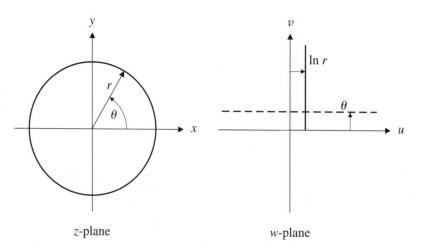

z-plane w-plane

FIGURE 2.40. The function $w = \ln z$ maps circles centered at the origin to straight lines.

$w = \dfrac{2a}{z}$

$$w = \frac{2a}{r} e^{-i\theta} = \frac{2a}{r}(\cos\theta - i\sin\theta) = u + iv$$

What kind of curves in the z-plane map to straight lines in the w-plane? The points that map to the line $v = v_0$ satisfy

$$-\frac{2a}{r}\sin\theta = v_0 \Rightarrow r = -\frac{2a}{v_0}\sin\theta$$

Thus, we have

$$x = r\cos\theta = -\frac{2a}{v_0}\sin\theta\cos\theta = -\frac{a}{v_0}\sin 2\theta$$

and

$$y = r\sin\theta = \frac{-2a}{v_0}\sin^2\theta = -\frac{a}{v_0}(1 - \cos 2\theta)$$

Then

$$\sin^2 2\theta + \cos^2 2\theta = 1$$

$$\left(\frac{xv_0}{a}\right)^2 + \left(\frac{yv_0}{a} + 1\right)^2 = 1$$

$$x^2 + \left(y + \frac{a}{v_0}\right)^2 = \left(\frac{a}{v_0}\right)^2$$

These curves are circles centered at $y = -a/v_0$ with radius $R = |a/v_0|$ (Figure 2.41).

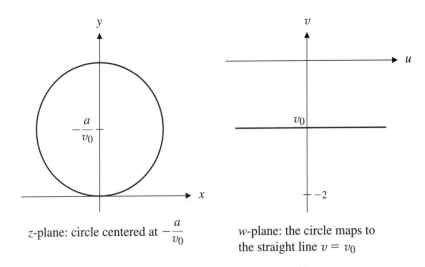

z-plane: circle centered at $-\dfrac{a}{v_0}$

w-plane: the circle maps to the straight line $v = v_0$

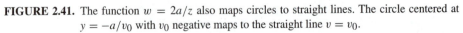

FIGURE 2.41. The function $w = 2a/z$ also maps circles to straight lines. The circle centered at $y = -a/v_0$ with v_0 negative maps to the straight line $v = v_0$.

Example 2.25.   To see how to use this transformation in a physics problem, let the circle represent a metal cylinder of radius R at electric potential V, separated by an insulating strip from a metal plane on the x-axis at zero potential. Find the potential everywhere outside the cylinder.

We use the transformation $w = 2R^2/z$. Then the cylinder maps to the line $v = -R$. The plane at $y = 0$ ($\theta = 0$ and $\theta = \pi$) maps to

$$v = \frac{2R^2}{r} \sin(0) = 0 \quad \text{and} \quad v = \frac{2R^2}{r} \sin(\pi) = 0$$

which is the real axis in the w-plane. Notice, though, that the origin maps to infinity and the points at $y = \pm\infty$ map to the origin. The region outside the cylinder is described by circles of radius $|R^2/v_0|$ greater than R. Each such circle maps to a line $v = -v_0$, with $v_0 < R$. Similarly, the region inside the cylinder maps to points with $v_0 > R$. The region below the plane ($y < 0$) maps to $v > 0$. The physical region of interest (outside the cylinder and above the plane) maps to the region of the w-plane with $0 > v > -R$. Thus, in the w-plane, the potential ϕ equals 0 at $v = 0$ and ϕ equals V at $v = -R$—a parallel plate capacitor! The potential between the plates in the w-plane is given by the function $\phi = -V(v/R)$. We want this to be the real part of an analytic function (see Section 2.4.3), so the complex function we need is

$$\Phi(w) = -\frac{V}{R}\frac{w}{i} = i\frac{V}{R}w$$

Mapping back to the z-plane, we find that the potential is

$$\Phi(z) = i\frac{V}{R}\frac{2R^2}{z} = i\frac{2RV}{r}e^{-i\theta} = \frac{2RV}{r}(i\cos\theta + \sin\theta)$$

The physical potential is the real part of this expression, or

$$\phi = \frac{2RV}{r}\sin\theta$$

And the equipotential surfaces (Figure 2.42) are described by

$$r = \frac{2RV}{\phi}\sin\theta$$

The field lines are described by the function

$$\frac{2RV}{r}\cos\theta = \text{constant}$$

Clearly, the first step in using a conformal transformation in a physics problem is to find the right transformation. This is a nontrivial task! The Schwarz-Christoffel transformation maps the interior of a polygon with N sides to the upper half-plane (Morse and Feshbach, Section 4.7). Circular arcs may be transformed to straight lines using the transformation $w = a^2/(z - z_p)$, where z_p is a point on the circular arc. A few additional cases are explored in the problem set.

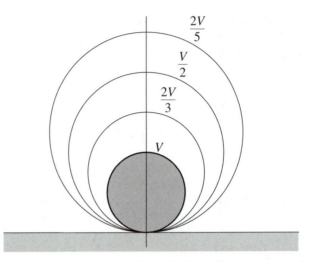

FIGURE 2.42. Equipotential surfaces around a conducting cylinder on a grounded plane (Example 2.25).

2.9. THE GAMMA FUNCTION

The gamma function provides an example of a function that is defined in terms of a complex integral. We will have an opportunity to see how to extend the definition of the function from one with a purely real argument to one with a complex argument—another case of analytic continuation.

The original definition of the gamma function (due to Euler) for x real and $x > 0$ is

$$\Gamma(x) = \int_0^\infty e^{-t} t^{x-1} \, dt \tag{2.75}$$

From this definition we can determine the value of the function for integer arguments. First,

$$\Gamma(1) = \int_0^\infty e^{-t} \, dt = -e^{-t} \Big|_0^\infty = 1$$

Integrating the definition (2.75) by parts yields a recursion relation between $\Gamma(x)$ and $\Gamma(x-1)$:

$$\Gamma(x) = \int_0^\infty e^{-t} t^{x-1} \, dt$$

$$= t^{x-1}(-e^{-t}) \Big|_0^\infty - \int_0^\infty (x-1)(-e^{-t}) t^{x-2} \, dt$$

$$\Gamma(x) = (x - 1)\Gamma(x - 1) \quad (x > 1)$$

(2.76)

Then, when x is an integer,

$$\Gamma(n + 1) = n\Gamma(n) = n(n - 1)\Gamma(n - 1) = \cdots = n(n - 1)(n - 2) \cdots 1\Gamma(1)$$

$$\Gamma(n + 1) = n!$$

(2.77)

For integer arguments, the gamma function is just a factorial.

An important special case of a noninteger argument is

$$\Gamma\left(\frac{1}{2}\right) = \int_0^\infty e^{-t} t^{-1/2} \, dt$$

To evaluate the integral, we change variables. Let $t = u^2$. Then $dt = 2u \, du = 2t^{1/2} \, du$. So

$$\Gamma\left(\frac{1}{2}\right) = \int_0^\infty e^{-u^2} 2 \, du = \int_{-\infty}^\infty e^{-u^2} \, du = \sqrt{\pi}$$

(2.78)

(See Appendix IX for the evaluation of the integral.)

We'd like to extend the definition of the gamma function to all real and complex arguments. The definition (2.75) remains valid for complex arguments z provided that $\mathrm{Re}\,(z) > 0$ to ensure convergence of the integral, but it fails for $x \leq 0$ because the integral diverges at the lower limit. There is another difficulty in that the complex integrand $t^{z-1} e^{-t}$ has a branch point at the origin if z is not an integer. We can deal with both problems if we redefine the function as a contour integral with a well-chosen contour. The new definition must coincide with the original definition when z is a real number greater than zero. To achieve this, we put the branch cut along the positive real axis and create a contour C in the complex t-plane that runs along the top of the cut, around the branch point at the origin, and back to infinity along the bottom of the branch cut (Figure 2.43).

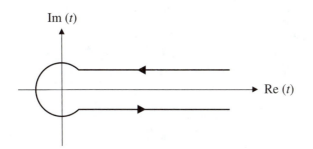

FIGURE 2.43. Contour for evaluating the gamma-function integral.

Let's evaluate the integral along this contour. The integrand is

$$f(z, t) = t^{z-1} e^{-t} = \exp\left(\ln t^{z-1}\right) e^{-t} = \exp\left[(z-1)\ln t - t\right]$$

Recall that $\ln t = \ln r + i\theta$, where $t = re^{i\theta}$. On the top half of contour C, the argument θ equals 0, while on the bottom half, $\theta = 2\pi$. The integrand differs by $\exp\left[2\pi i(z-1)\right] = \exp\left(2\pi i z\right)$ on the two sides. Around the little circle $|t| = \varepsilon$,

$$\int_{C_\varepsilon} t^{z-1} e^{-t}\, dt = \int_0^{2\pi} (\varepsilon e^{i\theta})^{z-1} \exp\left(-\varepsilon e^{i\theta}\right) i\varepsilon e^{i\theta}\, d\theta$$

$$\to \varepsilon^z \int_0^{2\pi} e^{iz\theta} i\, d\theta \quad \text{as } \varepsilon \to 0$$

$$= \frac{\varepsilon^z}{z}(e^{2\pi i z} - 1) \to 0 \quad \text{as } \varepsilon \to 0 \text{ for Re }(z) > 0$$

Thus,

$$\int_C t^{x-1} e^{-t}\, dt = (e^{2\pi i x} - 1)\int_0^\infty t^{x-1} e^{-t}\, dt = (e^{2\pi i x} - 1)\Gamma(x)$$

for $x > 0$. Then we define the gamma function through the integral expression:

$$\boxed{\Gamma(z) = \frac{1}{e^{2\pi i z} - 1}\int_C t^{z-1} e^{-t}\, dt} \tag{2.79}$$

or, equivalently,

$$\Gamma(z) = -\frac{1}{2i\sin \pi z}\int_C (-t)^{z-1} e^{-t}\, dt \tag{2.80}$$

so that the old definition and the new definition are consistent where they are both valid. The contour C is the one shown in Figure 2.43.

The complex function $\Gamma(z)$ defined by equation (2.79) has a singularity where $e^{2\pi i z} - 1 = 0$, or $z = n$, where n is any positive or negative integer. For positive n, we must use a limiting process, since the numerator is also zero. Let $z = n + \delta$. Then

$$\frac{1}{e^{2\pi i z} - 1}\int_C t^{n+\delta-1} e^{-t}\, dt = \frac{1}{e^{2\pi i z} - 1}\left(\int_\infty^0 t^{n+\delta-1} e^{-t}\, dt\right.$$

$$\left. + \int_0^\infty e^{2\pi i(n+\delta)} t^{n+\delta-1} e^{-t}\, dt\right)$$

$$= \frac{e^{2\pi i(n+\delta)} - 1}{e^{2\pi i(n+\delta)} - 1}\int_0^\infty t^{n+\delta-1} e^{-t}\, dt = \int_0^\infty t^{n+\delta-1} e^{-t}\, dt$$

$$\to \Gamma(n) \quad \text{as } \delta \to 0$$

as required. So these singularities are removable. The singularities at negative integer values of z are not removable; they are simple poles.

Properties of the gamma function are summarized in Gradshteyn and Ryzhik, Section 8.3. One useful relation satisfied by the gamma function is

$$\Gamma(x)\Gamma(1-x) = \frac{\pi}{\sin \pi x}$$

from which it follows that

$$\Gamma\left(\frac{1}{2}\right)\Gamma\left(1-\frac{1}{2}\right) = \frac{\pi}{\sin(\pi/2)}$$

$$\left[\Gamma\left(\frac{1}{2}\right)\right]^2 = \pi$$

$$\Gamma\left(\frac{1}{2}\right) = \sqrt{\pi}$$

as we found above by doing the integral. Also,

$$\Gamma\left(-\frac{1}{2}\right)\Gamma\left(1+\frac{1}{2}\right) = \frac{\pi}{\sin\left(-\frac{\pi}{2}\right)} = -\pi$$

So

$$\Gamma\left(-\frac{1}{2}\right) = \frac{-\pi}{\Gamma\left(\frac{3}{2}\right)} = \frac{-\pi}{\frac{1}{2}\Gamma\left(\frac{1}{2}\right)} = -2\sqrt{\pi}$$

where we used relation (2.76).

To compute the gamma function for large real arguments, we may use an asymptotic series due to Stirling[26]:

$$\ln\left[\Gamma(x+1)\right] = \ln(x!) = \frac{1}{2}\ln 2\pi + \left(x+\frac{1}{2}\right)\ln x - x + \frac{1}{12x} + \cdots$$

Finally, we note that the *incomplete gamma functions* are defined by the indefinite integrals

$$\gamma(\alpha, x) = \int_0^x e^{-t} t^{\alpha-1}\, dt$$

$$\Gamma(\alpha, x) = \int_x^\infty e^{-t} t^{\alpha-1}\, dt$$

where $\text{Re}(\alpha) > 0$.

The gamma function has numerous applications in physics and in statistics.

---

[26]This expression may be derived using the method of steepest descent (Optional Topic D).

PROBLEMS

1. If $z_1 = 5 + 2i$ and $z_2 = 3 - 4i$, find z_1/z_2 and $z_1 \times z_2$.

2. Use the polar representation of z to write two expressions for z^3 in terms of r and θ. Use your result to express $\cos 3\theta$ and $\sin 3\theta$ in terms of $\cos \theta$ and $\sin \theta$.

3. Prove De Moivre's theorem:

$$(\cos \theta + i \sin \theta)^n = \cos n\theta + i \sin n\theta$$

4. The equation $(y - y_0)^2 = 4a(x - x_0)$ describes a parabola. Write this equation in terms of $z = x + iy$. *Hint:* Use the geometric definition of the parabola.

5. Show that the equation

$$|z - c| + |z - d| = \alpha$$

represents an ellipse in the complex plane, where c and d are complex constants and α is a real constant. Use geometrical arguments to determine the position of the center of the ellipse and its semi-major and semi-minor axes.

6. Show that the equation

$$z = ae^{i\phi} + be^{-i\phi}$$

represents an ellipse in the complex plane, where a and b are complex constants and ϕ is a real variable. Determine the position of the center of the ellipse and its semi-major and semi-minor axes. *Hint:* Use the geometric interpretation of arithmetic operations in Section 2.1.1 to determine the location of the axes.

7. Find *all* solutions of the equations below. Show your solutions in a plot of the complex plane.

   (a) $z^5 = -1$ (b) $z^4 = 16$

8. Find all solutions of the equations below.

   (a) $\cos z = 100$ (b) $\sin z = 6$

9. Find all solutions of the equation $\cosh z = -5$.

10. Find all complex numbers z such that $z = \ln(-5)$.

11. Investigate the function $w = 1/\sqrt{z}$. Find the functions $u(r, \theta)$ and $v(r, \theta)$, where $w = u + iv$. How many branches does this function have? Find the image of the unit circle under this mapping.

12. The function $w(z) = z^{1/4}$. Find the functions $u(r, \theta)$ and $v(r, \theta)$, where $w = u + iv$. How many branches does this function have? Find the image under this mapping of a square of side 1 centered at the origin.

13. Oblate spheroidal coordinates u, v, w are defined in terms of cylindrical coordinates ρ, ϕ, z by the relations

$$\rho + iz = c \cosh(u + iv), \quad w = \phi$$

Show that the surfaces of constant u and constant v are ellipsoids and hyperboloids, respectively. What values of u and v correspond to the z-axis and the $z = 0$ plane?

14. An AC circuit contains a capacitor C in series with a coil with resistance R and inductance L. The circuit is driven by an AC power supply with emf $\varepsilon = \varepsilon_0 \cos \omega t$.

    (a) Use Kirchhoff's rules to write equations for the steady-state current in the circuit.

    (b) Using the fact that $\cos \omega t = \text{Re} (e^{i\omega t})$, find the current through the power supply in the form

    $$I = \text{Re} (I_0 e^{i\omega t})$$

    and show that $I_0 = \varepsilon_0 / Z$, where Z is the *complex impedance* of the circuit. Express Z in terms of L, R, C, and ω.

    (c) Use the result of (b) to find the amplitude and phase shift of the current.

    (d) How much power is provided by the power supply? Show that the time-averaged power is given by

    $$P = \frac{1}{2}\text{Re} (\varepsilon I^*) = \frac{1}{2}\text{Re} (\varepsilon^* I)$$

    with $\varepsilon = \varepsilon_0 e^{i\omega t}$ and $I = (\varepsilon_0 / Z)e^{i\omega t}$.

15. Small amplitude waves in a plasma are described by the relations

    $$\frac{\partial n}{\partial t} + \frac{\partial}{\partial x} (n_0 v) = 0$$

    $$\varepsilon_0 \frac{\partial E}{\partial x} = -en$$

and

    $$m \frac{\partial v}{\partial t} = -eE - m\nu v$$

    where n_0, e, m, ν, and ε_0 are constants. The constant ν is the collision frequency. Assume that n, E, and v are all proportional to $\exp(ikx - i\omega t)$. Solve the equations for nonzero n, E, and v to show that ω satisfies the equation

    $$\omega^2 + i\nu\omega = \frac{n_0 e^2}{m\varepsilon_0} \equiv \omega_p^2$$

    where ω_p is the plasma frequency. Solve this equation to find the frequency ω and hence show that collisions damp the waves.

16. Write the real and imaginary parts u and v of the complex functions.

    (a) $f = z^2 \sin z$ (b) $f = \dfrac{1}{1+z}$

    In each case, show that u and v obey the Cauchy-Riemann relations. First find the derivative df/dz in terms of x and y, and then express the answer in terms of z. Is the result what you expected?

17. The variables x and y in a complex number $z = x + iy$ may be expressed in terms of z and its complex conjugate z^*:

$$x = \frac{1}{2}(z + z^*)$$

$$y = \frac{1}{2i}(z - z^*)$$

Show that the Cauchy-Riemann relations are equivalent to the condition

$$\frac{\partial f}{\partial z^*} \equiv 0$$

18. One of the functions $u_1 = 2(x - y)^2$ and $u_2 = \dfrac{x^3}{3} - xy^2$ is the real part of an analytic function $w(z) = u + iv$. Which is it? Find the function $v(x, y)$ and write w as a function of z.

19. A very long cylinder of radius a has potential V on one half and $-V$ on the other half. The potential inside the cylinder may be written as a series:

$$\Phi(r, \theta) = \frac{4V}{\pi} \sum_{n=0}^{\infty} \left(\frac{r^2}{a^2}\right)^{2n+1} \frac{\sin [2(2n + 1)\theta]}{2n + 1}$$

Express each term in the sum as the imaginary part of a complex number, and hence sum the series. Show that the result may be expressed in terms of an inverse tangent.

20. The function $f = \sin(z^2)$ (see Example 2.10) also has a zero at $z = \sqrt{\pi}$. What is its order?

21. Find the Taylor series for the following functions about the point specified. In each case, determine the radius of convergence of the series.

(a) $z \cos z$ about $z = 0$

(b) $\ln(1 + z)$ about $z = 0$

(c) $\dfrac{\sin z}{z}$ about $z = \pi/2$

(d) $\dfrac{1}{z^2 - 1}$ about $z = 2$

22. Determine the Taylor or Laurent series for each of the following functions in the immediate neighborhood of the point specified. In each case, determine the radius of convergence of the series.

(a) $\dfrac{\cos z}{z - 1}$ about $z = 1$

(b) $\dfrac{\sin z^2}{z}$ about $z = 0$

(c) $\dfrac{e^z}{z - i\pi}$ about $z = i\pi$

(d) $\dfrac{\ln z}{z-1}$ about $z = 1$

(e) $\tan^{-1}(z)$ about $z = 0$

23. Determine *all* Taylor or Laurent series about the specified point for each of the following functions.

(a) $\dfrac{e^z}{z^2 + 1}$ about the origin

(b) $\dfrac{1}{z^2 + 1}$ about $z = i$

(c) $\dfrac{z}{z^2 - 9}$ about $z = 3$

(d) $\dfrac{1}{z^2 + 9}$ about the origin

24. Find all the singularities of each of the following functions and describe each of them completely.

(a) $\dfrac{e^z}{z} - \sin \dfrac{1}{z}$

(b) $\dfrac{\cos z}{z} - \dfrac{\sin z}{z^2}$

(c) $\dfrac{\tanh z}{z}$

(d) $\ln\left(1 + z^2\right)$

25. Incompressible fluid flows over a thin sheet from a distance X_0 into a corner, as shown in the diagram. The angle between the barriers is $\pi/3$, and at $x = X_0$, $\vec{\mathbf{v}} = V_0 \hat{\mathbf{x}}$. Assuming that the flow is as simple as possible, determine the streamlines and plot them. What is the velocity at $r = X_0/3$, $\theta = \pi/6$?

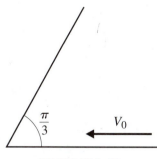

PROBLEM 25

26. Prove the Schwarz reflection principle: If a function $f(z)$ is analytic in a region including the real axis and $f(x)$ is real when x is real, then

$$f^*(z) = f(z^*)$$

Show that the result may be extended to functions that possess a Laurent series about the origin with real coefficients.

(a) Verify the result for the function $f(z) = \cos z$.

(b) Verify the result for the function $f(z) = \tan^{-1}(z)$.

(c) Show that the result does not hold for all z if $f(z) = \ln z$ (the principal branch is assumed).

27. Find the residue of each of the following functions at the point specified.

(a) $\dfrac{z-2}{z^2-1}$ at $z = 1$

(b) $\exp\left(\dfrac{1}{z} - 1\right)$ at $z = 0$

(c) $\dfrac{\sin z}{z^2}$ at the origin

(d) $\dfrac{\cos z}{1/2 - \sin z}$ at $z = \pi/6$

28. Evaluate the following integrals.

(a) $\displaystyle\oint_C \frac{\cos z}{z}\,dz$, where C is a circle of radius 2 centered at the origin

(b) $\displaystyle\oint_C \frac{\sinh z}{z-1}\,dz$, where C is a square of side 4 centered at the origin

(c) $\displaystyle\oint_C \frac{z-1}{z+2}\,dz$, where C is a circle of radius 1 centered at the origin

(d) $\displaystyle\oint_C \frac{z}{4z^2+1}\,dz$, where C is a square of side 1 centered at the point $z = (1+i)/4$

29. Evaluate the following integrals.

(a) $\displaystyle\int_0^{2\pi} \frac{1+\cos\theta}{2-\sin\theta}\,d\theta$

(b) $\displaystyle\int_0^{\pi} \frac{\sin^2\theta}{1+\cos^2\theta}\,d\theta$

(c) $\displaystyle\int_0^{2\pi} \frac{1}{1+\sin^2\theta}\,d\theta$

(d) $\displaystyle\int_0^{\pi} \sin^{2n}\theta\,d\theta$

30. Evaluate each of the following integrals.

(a) $\displaystyle\int_{-\infty}^{+\infty} \frac{1}{x^2+2}\,dx$

(b) $P\displaystyle\int_{-\infty}^{+\infty} \frac{x}{x^3+1}\,dx$

(c) $\displaystyle\int_{-\infty}^{+\infty} \frac{\cos\omega x}{x^2+9}\,dx$

(d) $\displaystyle\int_{-\infty}^{+\infty} \frac{x\sin x}{x^2+2x+2}\,dx$

31. Use a rectangular contour to evaluate the integrals.

(a) $\int_{-\infty}^{+\infty} \dfrac{e^{ax}}{1+e^{bx}}\,dx$, where b is real and $0 < \mathrm{Re}\,(a) < b$

(b) $\int_{-\infty}^{+\infty} \dfrac{\sinh ax}{\sinh 4ax}\,dx$

(c) $\int_{0}^{+\infty} \dfrac{x^2}{\cosh ax}\,dx$

32. Evaluate the integrals

(a) $\int_{0}^{+\infty} \dfrac{x^{3/2}}{1+x^3}\,dx$

(b) $\int_{0}^{\infty} \dfrac{x^{1/3}}{x^2+1}\,dx$

33. Evaluate the integral

$$\int_{0}^{\infty} \dfrac{dx}{1+x^{2N}}$$

by integrating over a pie-slice contour with sides at $\phi = 0$ and at $\phi = \pi/N, 0 \le r < \infty$.

34. Evaluate the integral

$$\int_{0}^{\infty} e^{ix^2}\,dx$$

along the positive real axis by making the change of variable $u^2 = -ix^2$. Take care to discuss the path of integration for the u-integral. Use the Cauchy theorem to show that the resulting u-integral may be reduced to a known integral along the real axis. Hence show that

$$\int_{0}^{\infty} \sin x^2\,dx = \int_{0}^{\infty} \cos x^2\,dx = \frac{1}{2}\sqrt{\frac{\pi}{2}}$$

(The result has numerous applications in physics—for example, in signal propagation.)

35. The power radiated per unit solid angle by a charge undergoing simple harmonic motion is

$$\frac{dP}{d\Omega} = K\sin^2\theta\,\frac{\cos^2\omega t}{(1+\beta\cos\theta\sin\omega t)^5}$$

where the constant $K = e^2 c\beta^4/4\pi a^2$ and $\beta = a\omega/c$ is the speed amplitude/c (see, for example, Jackson, p. 701). Using methods from Section 2.7.2, perform the time average over one period to show that

$$\left\langle \frac{dP}{d\Omega} \right\rangle = \frac{K}{8}\sin^2\theta\,\frac{4+\beta^2\cos^2\theta}{(1-\beta^2\cos^2\theta)^{7/2}}$$

36. **Langmuir waves.** Waves in a plasma may be described by a wave form $n = n_0 \exp(ikx - i\omega t)$ (compare with Section 2.1.4), where the relation between ω and k (the *dispersion relation*) is given by

$$0 = 1 + \frac{\omega_p^2}{k} \int_{-\infty}^{+\infty} \frac{\partial f(v)/\partial v}{\omega - kv} dv$$

where ω_p is the plasma frequency $ne^2/\varepsilon_0 m$ and $f(v)$ is the one-dimensional Maxwellian

$$f(v) = \sqrt{\frac{m}{2\pi k_B T}} \exp\left(-\frac{mv^2}{2k_B T}\right)$$

Notice that the integrand has a singularity at $v = \omega/k$, which is on the real axis, if ω and k are both real. Landau showed that the integral is to be regarded as an integral along the real axis in the complex v-plane, and that the correct integration path passes around and *under* the pole.

(a) Show that the integral may be expressed as

$$\int_{-\infty}^{+\infty} \frac{\partial f(v)/\partial v}{\omega - kv} dv = P \int_{-\infty}^{+\infty} \frac{\partial f(v)/\partial v}{\omega - kv} dv - \frac{i\pi}{k} \left.\frac{\partial f}{\partial v}\right|_{v=\omega/k}$$

(Compare with Section 2.7.3, page 143.)

(b) Evaluate the principal value approximately, assuming $\omega/k \gg v_T = \sqrt{k_B T/m}$. *Hint*: First integrate by parts, and then expand the denominator in a series in powers of kv/ω. Neglect the small effect of the pole, and find the frequency ω as a function of k. Now include the imaginary part due to the pole at ω/k. Show that the wave is damped. The result has been confirmed experimentally.

(c) How would the result change if the path of integration were to pass over, rather than under, the pole?

37. Is the mapping $w = z^2$ conformal? Find the image in the w-plane of the circle $|z - i| = 1$ in the z-plane, and plot it. Comment.

38. Is the mapping $w = z + (1/z)$ conformal? Find the image in the w-plane of

(a) the x-axis   (b) the y-axis   (c) the unit circle in the z-plane

A capacitor plate has a cylindrical bump of radius a on it. The second plate is a distance $d \gg a$ away. One plate is maintained at potential V, and the other is grounded. Find the potential everywhere between the plates.

39. Show that the mapping $z = w + e^w$ is conformal except at a finite set of points in the z-plane. A parallel plate capacitor has plates that extend from $x = -1$ to $x = -\infty$. Find an appropriate scaling that allows you to place the plates at $y = \pm\pi$. Show that the given transformation maps the plates to the lines $v = \pm\pi$. Solve for the potential between the plates in the w-plane, map to the z-plane, and hence find the equipotential surfaces at the ends of the capacitor. Sketch the field lines. This is the so-called fringing field.

40. Two conducting cylinders, each of radius a, are touching. An insulating strip lies along the line at which they touch. One cylinder is grounded, and the other is at potential V. Use one of the mappings from the chapter to solve for the potential outside the cylinders.

41. Show that the mapping $w = 1/(z-2)$ maps to straight line segments the arcs

(a) $|z - 4| = 2$ with endpoints at $z = 3 \pm \sqrt{3}i$

(b) $|z - (2 + i)| = 1$ with endpoints at $z = 3 + i$ and $z = 1 + i$

42. Show that $\Gamma(x) < 0$ for $-1 < x < 0$.

43. Prove *Cauchy's inequality*: If $f(z)$ is analytic and bounded in a region R: $|z - z_0| < R$ and if $|f(z)| < M$ on the circle $|z - z_0| = r$, then the coefficients in the Taylor series expansion of f about z_0 (equation 2.44) satisfy the inequality

$$|c_n| \le \frac{M}{r^n}$$

Hence prove *Liouville's theorem:*

> If $f(z)$ is analytic and bounded in the entire complex plane, then it is a constant.

44. A function $f(z)$ is analytic except for well-separated simple poles at $z = z_n$, $n = 1 - N$, and $z_n \ne 0$. Show that the function may be expanded in a series

$$f(z) = f(0) + \sum_{n=1}^{N} a_n \left(\frac{1}{z_n} + \frac{1}{z - z_n} \right)$$

where a_n is the residue of f at z_n. Is the result valid for $N \to \infty$? Why or why not?
Hint: Evaluate the integral

$$I_N = \frac{1}{2\pi i} \int_{C_N} \frac{f(w)}{w(w - z)} dw$$

where C_N is a circle of radius R_N about the origin that contains the N poles. You may assume that $|f(z)| < \varepsilon R_N$ on C_N for ε a small positive constant.

CHAPTER 3

Differential Equations

Any physical situation can be described by a mathematical problem—usually one involving the solution of a differential equation subject to a set of boundary conditions. Thus, solving a physics problem frequently reduces to solving a differential equation. Because we are concerned with a real physical situation, we expect a solution to exist. In this chapter, we shall review some methods for solving differential equations that occur in physics. Much of the rest of the book is devoted to developing additional techniques. A complete discussion of this topic would require more space than is available here. For important theorems that prove the existence and uniqueness of solutions, see, for example, the texts by Ince or Murray and Miller. Interested students should refer to additional texts listed in the bibliography.

3.1. SOME DEFINITIONS

A differential equation is an equation for a function f of one or more variables that includes derivatives of f with respect to one or more of those variables. The *order* of a differential equation is the order of the highest derivative that appears in the equation. Thus, an equation of the form

$$\frac{d^2y}{dx^2} + 3\frac{dy}{dx} + 2y = 6 \tag{3.1}$$

is a second-order differential equation for the function $y(x)$.

Equation (3.1) is also a *linear* differential equation, because each term in the equation is a linear function of y or its derivatives. On the other hand,

$$y\frac{dy}{dx} = 2 \tag{3.2}$$

is a *nonlinear* differential equation, since it has a term with a product of y and y'.

These two equations are *ordinary* differential equations, which means that the solution y is a function of the one variable x. When we have a function of more than one variable, such as $y(x, t)$, and the equation involves partial derivatives with respect to both x and t,

169

then we have a *partial differential equation*. These equations will be discussed later. (See also Appendix X.)

Equations (3.1) and (3.2) are also *inhomogeneous*, since each equation has a term on the right-hand side that is independent of y and all its derivatives. On the other hand, the equation

$$\frac{d^2y}{dx^2} + 3x^2\frac{dy}{dx} + 2y = 0 \tag{3.3}$$

is a *homogeneous* equation. It is also linear. However, unlike equation (3.1), which has constant coefficients, equation (3.3) has a nonconstant coefficient, with x^2 multiplying the term in dy/dx.

3.2. COMMON DIFFERENTIAL EQUATIONS ARISING IN PHYSICS

Some of the common differential equations that arise in physics are described below.

3.2.1. Newton's Second Law

Newton's second law, $\vec{\mathbf{F}} = m\vec{\mathbf{a}}$, is a second-order differential equation for the position of a particle. If the force is constant, such as the gravitational force on a particle near the Earth's surface, then the equation is very simple. With the y-axis chosen vertically upward, we have

$$m\frac{d^2y}{dt^2} = -mg$$

This equation is quite easy to solve, so we may use it to draw some general conclusions. We may integrate twice to obtain the solution:

$$y = y_0 + \left.\frac{dy}{dt}\right|_0 t - \frac{1}{2}gt^2 \tag{3.4}$$

There are two integration constants, y_0 and $dy/dt|_0 = v_0$, that must be specified to complete the solution.

The functions $y_1(t) = 1$ and $y_2(t) = t$ are solutions of the homogeneous equation $y'' = 0$. Thus, the general solution is a linear combination of these two functions plus a *particular integral*—a function that satisfies the inhomogeneous equation. Here the particular integral is $-\frac{1}{2}gt^2$. Often we obtain the values of y_0 and $dy/dt|_0$ from the specified initial conditions—the values of position and velocity at $t = 0$.

Now let us suppose that there is air resistance that is proportional to the particle's velocity: $\vec{\mathbf{F}}_{ar} = -\alpha\vec{\mathbf{v}}$. Then we get

$$m\frac{d^2y}{dt^2} = -mg - \alpha\frac{dy}{dt} \tag{3.5}$$

This is a second-order, linear, inhomogeneous differential equation with constant coefficients. We shall solve this equation in Example 3.2.

3.2.2. Simple Harmonic Motion

If the force applied to a particle is given by Hooke's law and if y describes the displacement of the particle from its equilibrium position, then we have

$$m\frac{d^2y}{dt^2} = -ky \tag{3.6}$$

which is a second-order, linear, homogeneous differential equation with constant coefficients.

If the system also has a damping force proportional to velocity, then the equation becomes

$$m\frac{d^2y}{dt^2} = -ky - \gamma\frac{dy}{dt}$$

or

$$\frac{d^2y}{dt^2} + 2\alpha\frac{dy}{dt} + \omega_0^2 y = 0 \tag{3.7}$$

where $2\alpha = \gamma/m$ and $\omega_0^2 = k/m$. This is the equation for a *damped harmonic oscillator.*

3.2.3. Bending of a Beam

A beam will bend when a load is placed on it. Let the coordinate x run along the undisplaced beam, and let $y(x)$ be the downward deflection of the beam's center line. The elastic properties of the beam are described by the Young's modulus E of its material. Its response to applied loads is determined by the shape of the cross section, as measured by what engineers call the moment of inertia I of the cross section about the long axis.[1] Then, with origin at the left end of the beam, the deflection $y(x)$ may be computed as follows:

1. The beam bends as a result of the net torque acting on it. In equilibrium, torque due to the internal stresses balances the external torques. Torque balance is expressed by the differential equation[2]

$$\frac{d^2y}{dx^2} = -\frac{1}{EI}m(x) \tag{3.8}$$

   where $m(x)$ is the net counterclockwise torque of all forces acting to the right of point x. The minus sign arises because y increases downward.

2. The torque is due to the shearing force:

$$\frac{dm}{dx} = t(x) \tag{3.9}$$

   where $t(x)$ is the sum of all vertical (downward) components of forces acting to the right of point x.

---

[1] Let x', y', z' be coordinates with origin at the centroid, with y' (vertical) and z' (horizontal) lying in the beam cross section. Then $I = \int (y')^2 dy'\, dz'$.

[2] This is the Bernoulli-Euler law. See, for example, Long, pp. 91–94.

3. The shearing force results from the load per unit length:

$$\frac{dt}{dx} = -q(x) \tag{3.10}$$

Here the minus sign appears because t is $\int_x^L q(x')\,dx'$, where x is the *lower* limit of the integral.

4. Putting these equations together gives us the differential equation satisfied by the beam:

$$\frac{d^4y}{dx^4} = \frac{1}{EI}q(x) \tag{3.11}$$

where $q(x)$ is the load per unit length along the beam. This is one of the few equations in physics that have order greater than 2.

3.2.4. Electric Circuits

Applying Kirchhoff's laws to an electric circuit gives an equation for the charge on a capacitor or the current in the circuit. For a single-loop LRC circuit (Figure 3.1) with appropriate choice of the circuit variables for charge q and current,[3] we find

$$L\frac{d^2q}{dt^2} + R\frac{dq}{dt} + \frac{q}{C} = \mathcal{E} \tag{3.12}$$

where $\mathcal{E}$ is the applied emf. The circuit is a driven, damped harmonic oscillator. Compare equation (3.12) with equation (3.7).

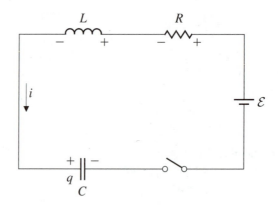

FIGURE 3.1. Circuit diagram that leads to equation (3.12). The algebraic variables i and q are defined here.

Multiloop circuits give rise to a set of coupled linear differential equations. In all such problems, particular care must be taken in the definition of the variables and the consequent relations among them.

---

[3]See, for example, Lea and Burke, p. 990.

3.2.5. Diffusion

Processes involving many random interactions ("collisions") are usually described by a diffusion equation. Heat conduction is an example of this kind of process. Heat transfer along a rod of cross-sectional area A, for example, is described by the heat flux $H = -kA \, \partial T/\partial x$, where k is the thermal conductivity of the rod's material. The minus sign indicates that heat flows toward lower temperature. The rate of change of temperature in a segment of length dx depends on the heat capacity of the segment and the imbalance of heat flow into and out of that segment. Heat flow out of a segment of length dx reduces that segment's temperature:

$$H(x + dx) - H(x) = -mc \, dx \frac{\partial T}{\partial t}$$

where m is the mass per unit length and c is the specific heat. Dividing by dx, we have the differential equation

$$mc \frac{\partial T}{\partial t} = -\frac{\partial H}{\partial x} = kA \frac{\partial^2 T}{\partial x^2} \tag{3.13}$$

$$\frac{\partial T}{\partial t} = \frac{kA}{mc} \frac{\partial^2 T}{\partial x^2} = D \frac{\partial^2 T}{\partial x^2} \tag{3.14}$$

where $D \equiv kA/mc$ is called the diffusion coefficient. This is a partial differential equation, since it contains partial derivatives with respect to x and t. The solution is a function of the two variables x and t.

3.2.6. Waves

The wave equation relates the second time derivative of a function to its space derivative:

$$\frac{\partial^2 y}{\partial t^2} = v^2 \frac{\partial^2 y}{\partial x^2} \tag{3.15}$$

where v is the phase speed of the wave. If we assume a solution of the form $y(x, t) = \mathrm{Re}\, e^{i\omega t} y(x)$ (compare with Chapter 2, Section 2.1.4), then we find

$$-\omega^2 y(x) = v^2 \frac{d^2 y}{dx^2}$$

or

$$\frac{d^2 y}{dx^2} + k^2 y = 0 \tag{3.16}$$

where $k^2 = \omega^2/v^2$. We have reduced our partial differential equation to an ordinary differential equation known as the Helmholtz equation.

3.3. SOLUTION OF LINEAR, ORDINARY DIFFERENTIAL EQUATIONS

A linear, homogeneous, ordinary differential equation of nth order has n linearly independent[4] solutions. The general solution is a linear combination of the n independent solutions. Thus, to completely specify the solution, we require n pieces of information (boundary conditions[5]). For a second-order differential equation, we may know the value of both y and y' at one boundary (say $x = 0$, in which case we have an initial value problem) or perhaps we may know the values of y at two boundaries (say $x = 0$ and $x = L$). In this section, we shall focus on finding the two linearly independent solutions of a second-order equation.

3.3.1. Equations with Constant Coefficients

The easiest class of differential equations to solve is linear, ordinary differential equations with constant coefficients, so we shall start with those.

Homogeneous Equations

A linear, homogeneous differential equation with constant coefficients, no matter what the order, is solved by an exponential function $y = e^{sx}$, where s is a real or complex constant. Simply substitute this solution into the differential equation to obtain an algebraic equation for s. Each distinct solution for s corresponds to a linearly independent solution of the original differential equation. Since a second-order equation gives a quadratic for s, a third-order equation gives a cubic equation, and so on, we can expect to find n solutions to an nth-order equation. (The case of repeated roots is discussed below.)

> **Example 3.1.** Solve equation (3.12) with $\mathcal{E} = 0$ and with the initial conditions $q = Q$ at $t = 0$ and the current $I = dq/dt = 0$ at $t = 0$.
>   Substituting the trial solution e^{st} into the equation, we find
>
> $$Ls^2 e^{st} + Rse^{st} + \frac{e^{st}}{C} = 0$$
>
> and so
>
> $$Ls^2 + Rs + \frac{1}{C} = 0 \qquad (3.17)$$
>
> which has two solutions:
>
> $$s_\pm = -\frac{R}{2L} \pm \sqrt{\left(\frac{R}{2L}\right)^2 - \frac{1}{LC}} = -\alpha \pm i\omega$$

---

[4] A set of N functions $y_n(x)$ is linearly independent if there are no nonzero coefficients a_n that make $\sum_{n=1}^{N} a_n y_n = 0$ for all values of x. See Chapter 1, Section 1.5.2 for a related concept applied to vectors.
[5] In the particular case of the variable t (time) with conditions specified at $t = 0$, the boundary conditions are called *initial conditions*.

where

$$\alpha \equiv \frac{R}{2L}$$

and

$$\omega^2 \equiv \frac{1}{LC} - \left(\frac{R}{2L}\right)^2$$

Both roots are real and negative if $R/2L > 1/\sqrt{LC}$, and both solutions decrease with time. But if $R/2L < 1/\sqrt{LC}$, then the roots form a complex conjugate pair with a negative real part. Since the complex exponential may be expressed in terms of sines and cosines, these solutions are damped oscillations.

The general solution is a linear combination of the solutions e^{s+t} and e^{s-t}:

$$q = \exp(-\alpha t)[A \exp(i\omega t) + B \exp(-i\omega t)]$$

where the values of the constants A and B are given by the initial conditions. At $t = 0$, we have

$$q(0) = Q = A + B$$

and

$$I = \frac{dq}{dt}\bigg|_{t=0} = 0 = (-\alpha + i\omega)A + (-\alpha - i\omega)B$$

So

$$B = A\frac{-\alpha + i\omega}{\alpha + i\omega}$$

and

$$Q = A\left(1 + \frac{-\alpha + i\omega}{\alpha + i\omega}\right) = A\frac{2i\omega}{\alpha + i\omega} \Rightarrow A = \frac{Q}{2}\left(1 - i\frac{\alpha}{\omega}\right)$$

Then

$$B = Q - A = \frac{Q}{2}\left(1 + i\frac{\alpha}{\omega}\right)$$

The solution is then

$$q = \frac{Q}{2}e^{-\alpha t}\left[\left(1 - i\frac{\alpha}{\omega}\right)e^{i\omega t} + \left(1 + i\frac{\alpha}{\omega}\right)e^{-i\omega t}\right]$$

$$= Qe^{-\alpha t}\left(\cos \omega t + \frac{\alpha}{\omega}\sin \omega t\right)$$

A special case arises if $R/2L = 1/\sqrt{LC}$. The square root equals zero, and there is only one solution to the quadratic equation for s: $s = -R/2L$. Thus, we find only one solution for q, $q_1 = e^{st}$. We look for a second solution to the differential equation,[6] of the form $q_2(t) = v(t)e^{st}$, and substitute into the original equation to obtain a new differential equation for the function v:

$$\frac{dq_2}{dt} = v'e^{st} + sve^{st}$$

and

$$\frac{d^2q_2}{dt^2} = v''e^{st} + 2sv'e^{st} + s^2ve^{st}$$

Substituting into the differential equation gives

$$Le^{st}(v'' + 2sv' + s^2v) + e^{st}R(v' + sv) + \frac{ve^{st}}{C} = 0$$

$$Lv'' + v'(2sL + R) + \left(Ls^2 + Rs + \frac{1}{C}\right)v = 0$$

The coefficient of v is zero (equation 3.17). Then, using our solution for s, we have

$$Lv'' + v'\left[2\left(\frac{-R}{2L}\right)L + R\right] + 0 = 0$$

$$v'' = 0$$

Thus, the solution for v is $v = A + Bt$. So the second, independent solution for q is te^{st}. We may readily extend this method to the case where the equation for s has a root repeated n times (see Problem 2).

Inhomogeneous Equations

To find the general solution to a linear, inhomogeneous differential equation, we first find any solution of the inhomogeneous equation [the particular integral $y_p(x)$]. This is not as hard as it might seem, because we can often find the particular integral we need by setting one or more of the derivatives to zero. Next we determine the two solutions $y_1(x)$ and $y_2(x)$ of the homogeneous equation. The general solution is the particular integral plus a linear combination of y_1 and y_2:

$$y(x) = y_p(x) + c_1y_1(x) + c_2y_2(x)$$

We can adjust the constants c_1 and c_2 to satisfy the boundary conditions.

Example 3.2. Find the solution to equation (3.5) (motion under gravity with air resistance) with the initial conditions $y = y_0$ and $dy/dt = v_0$ at $t = 0$.

First we find a solution $y_I(t)$ to the inhomogeneous equation. Set the second derivative to zero. (This means that the particle reaches a terminal velocity at which the

---

[6]This is a general method that we discuss in more detail in Section 3.3.2.

gravitational force is balanced by air resistance and the acceleration is zero.)

$$0 = -mg - \alpha \frac{dy_I}{dt} \Rightarrow y_I(t) = -\frac{mg}{\alpha}t$$

Since the homogeneous equation is linear and has constant coefficients, its solution is of the form e^{st}, where s satisfies the equation

$$ms^2 + \alpha s = 0 \Rightarrow s = 0, \quad -\frac{\alpha}{m}$$

Thus, the general solution $y_H(t)$ to the homogeneous equation is

$$y_H(t) = A + B \exp\left(-\frac{\alpha}{m}t\right)$$

Next we add y_I to y_H to obtain the solution to the complete equation:

$$y(t) = y_H + y_I = A + B \exp\left(-\frac{\alpha}{m}t\right) - \frac{mg}{\alpha}t$$

Finally, we make use of the initial conditions $y = y_0$ and $dy/dt = v_0$ to determine the constants A and B:

$$y_0 = y(0) = A + B$$

and

$$\frac{dy}{dt}\bigg|_{t=0} = -\frac{\alpha}{m}B - \frac{mg}{\alpha} = v_0 \Rightarrow B = -\left(\frac{m}{\alpha}\right)^2 g - \frac{m}{\alpha}v_0$$

Then

$$A = y_0 + \left(\frac{m}{\alpha}\right)^2 g + \frac{m}{\alpha}v_0$$

Thus, the solution is

$$y(t) = y_0 + \left(\frac{m}{\alpha}\right)^2 g\left[1 - \exp\left(-\frac{\alpha}{m}t\right)\right] + \frac{m}{\alpha}\left(v_0\left[1 - \exp\left(-\frac{\alpha}{m}t\right)\right] - gt\right)$$

Check the limit as $\alpha \to 0$ and convince yourself that we get back the solution (3.4). What happens as α becomes very large?

3.3.2. Linear Equations with Nonconstant Coefficients

When the coefficients in a linear equation are not constant, the solution is no longer a simple exponential. Before proceeding to a general method, we need to look at some properties of these equations.

Singular Points

The behavior of the solutions to a homogeneous differential equation is easiest to analyze when the equation is written in *standard form*, where the coefficient of the highest derivative is unity. For example, the standard form of a second-order equation is

$$\frac{d^2y}{dx^2} + f(x)\frac{dy}{dx} + g(x)y = 0 \qquad (3.18)$$

Then the values of y and dy/dx at any point determine the value of the second derivative

$$\frac{d^2y}{dx^2} = -f(x)\frac{dy}{dx} - g(x)y$$

By differentiating this expression, we may obtain the value of all higher derivatives that exist.

> The *singular points* of the differential equation are those points that are singularities[7] of the function $f(x)$ or the function $g(x)$.

All other values of x are *regular points* of the differential equation. At a regular point, y and dy/dx may take on any values, and the resulting value of the second derivative will be finite. We may then use the differential equation to compute the higher-order derivatives, and hence find a Taylor series for y about the regular point. Providing the series has a nonzero radius of convergence, the solution exists.

In contrast, at a singular point, y and dy/dx may take on only a set of special values. The set may be empty (that is, no values exist). Thus, if we are given values of y and dy/dx at the singular point and they are not in the set of special values, the Taylor series will not exist.

The equation

$$x\frac{d^2y}{dx^2} + 2y = 0$$

has a singular point at $x = 0$. [In standard form, $f(x) = 0$ and $g(x) = 2/x$, which has a pole at $x = 0$.] In order for the second derivative to exist at $x = 0$, $y(0)$ must be zero. In fact, y must go to zero as $x \to 0$ at least as fast as x. The value $y(0) = 1$ (or any other finite nonzero constant) is not in the set of special values for this equation, and so there is no analytic solution[8] with $y(0) = 1$.

If the equation is linear, the singular points are fixed. But the singular points of nonlinear equations also depend on the value of y (and perhaps its derivatives). For example, the

---

[7]See Chapter 2, Section 2.5.3.
[8]See Problem 5.

equation

$$y\frac{d^2y}{dx^2} + \left(\frac{dy}{dx}\right)^2 = 0 \tag{3.19}$$

has a singular point where $y = 0$. The solution[9] to this equation, with initial conditions y_0 and y_0', is $y = \sqrt{y_0(2y_0'x + y_0)}$, and thus the singular point occurs at $x = -y_0/2y_0'$, a value that depends on the values of y and y' at $x = 0$. This variability of the singular points accounts for much of the challenge in solving nonlinear equations.

The Wronskian

If y_1 and y_2 are both solutions of a linear, homogeneous differential equation, then any linear combination $Ay_1 + By_2$ is also a solution.

The *Wronskian* of the two solutions is

$$\boxed{W(y_1, y_2) = y_1y_2' - y_2y_1'} \tag{3.20}$$

The two solutions y_1 and y_2 are linearly dependent if there are nonzero coefficients A and B such that $Ay_1 + By_2 = 0$ for all values of x. In this case, the Wronskian is identically zero.

Now observe that

$$\frac{dW}{dx} = y_1y_2'' - y_2y_1''$$

We may evaluate this derivative using the differential equation (3.18). Then

$$\frac{dW}{dx} = y_1(-fy_2' - gy_2) - y_2(-fy_1' - gy_1)$$
$$= -fW \tag{3.21}$$

This is a linear, first-order differential equation for W. To find an expression for W, first rewrite equation (3.21) as

$$\frac{1}{W}\frac{dW}{dx} = \frac{d}{dx}(\ln W) = -f$$

and then integrate to obtain the solution:

$$\boxed{W = W(x_0)\exp\left(-\int_{x_0}^{x} f(\xi)\,d\xi\right)} \tag{3.22}$$

---

[9]You can verify this solution by direct substitution into the equation. See also Problem 12.

Since the exponential function is never zero for real arguments, W is zero for all x if it is zero at any $x = x_0$; otherwise, it is never zero. When the Wronskian is not zero, the two functions y_1 and y_2 are linearly independent.

Equation (3.22) shows that W is a specific function of x. We can evaluate it by using any form of the solutions y_1 and y_2 or by performing the integration in equation (3.22).

Once one solution $y_1(x)$ of the original differential equation is known, we can use the Wronskian to find a second, linearly independent solution. Note that

$$\frac{d}{dx}\left(\frac{y_2}{y_1}\right) = \frac{y_1 y_2' - y_1' y_2}{y_1^2} = \frac{W}{y_1^2}$$

Thus,

$$y_2 = y_1 \int \frac{W}{y_1^2} dx \qquad (3.23)$$

While equation (3.23) provides a formal solution, it is not always a useful one, because the integration may be difficult. It does prove, however, that a second-order differential equation posseses two solutions.

Example 3.3. One solution to the equation $y'' - 4y' + 4y = 0$ is e^{2x} (compare with Section 3.3.1). Use equation (3.23) to show that the second solution is xe^{2x}.

For this equation, the function $f(x) \equiv -4$, and hence, from equation (3.22), the Wronskian is

$$W = W(0)\exp\left(\int_0^x 4\, d\xi\right) = W_0 e^{4x}$$

Then the second solution is

$$y_2(x) = e^{2x}\int \frac{W_0 e^{4x}}{\left(e^{2x}\right)^2} dx = e^{2x} W_0 \int dx = W_0 x e^{2x}$$

and thus the second solution is xe^{2x}. The constant W_0 is arbitrary.

Method of Variation of Parameters

If one solution $y_1(x)$ of the differential equation has been found, we may search for a second solution of the form[10] $y_2(x) = u(x)y_1(x)$. This yields a second, and we hope simpler, differential equation for the function $u(x)$.

---

[10]We did this in Example 3.1.

Example 3.4.   One solution to the Legendre equation

$$(1-x^2)\frac{d^2y}{dx^2} - 2x\frac{dy}{dx} + 2y = 0$$

is $y(x) \equiv P_1(x) = x$. (Verify this by substituting into the differential equation.) Search for a second solution $y_2(x) = xu(x)$.

Differentiating, we find

$$\frac{dy_2}{dx} = u + xu'$$

and

$$\frac{d^2y_2}{dx^2} = 2u' + xu''$$

Now substitute into the differential equation:

$$(1-x^2)(2u' + xu'') - 2x(u + xu') + 2xu = 0$$
$$x(1-x^2)u'' + 2(1-2x^2)u' = 0$$

which is a *first*-order equation for u':

$$\frac{u''}{u'} = -2\frac{1-2x^2}{x(1-x^2)}$$

This equation may be integrated as follows. Let $x^2 = w$. Then $2x\,dx = dw$ and

$$\ln u' = -2\int \frac{1-2x^2}{x(1-x^2)}dx = -\int \frac{1-2w}{(1-w)w}dw$$

$$= -\int\left(\frac{1}{w} - \frac{1}{1-w}\right)dw = -\ln w - \ln(1-w)$$

$$= \ln\left(\frac{1}{x^2(1-x^2)}\right)$$

Taking the exponential of both sides[11] gives

$$u' = \frac{1}{x^2(1-x^2)} = \frac{1}{2}\left(\frac{1}{x^2} + \frac{1}{1-x^2}\right) = \frac{1}{2x^2} + \frac{1}{4}\left(\frac{1}{1-x} + \frac{1}{1+x}\right)$$

Integrating again, we get

$$u = -\frac{1}{2x} + \frac{1}{4}\ln\left(\frac{1+x}{1-x}\right)$$

---

[11]We have omitted the integration constant. You should verify that including it adds a multiple of our first solution $y(x)$ to $y_2(x)$.

Finally,

$$y_2 = xu = -\frac{1}{2} + \frac{x}{4} \ln\left(\frac{1+x}{1-x}\right) = \frac{1}{2} Q_1(x)$$

where in the last step we inserted the usual definition of the Legendre[12] function Q_1.

With minor modifications, this method works for inhomogeneous equations as well. If y_1 and y_2 are solutions of the corresponding homogeneous equation [$h(x) \equiv 0$], the solution to the inhomogeneous equation

$$y'' + f(x)y' + g(x)y = h(x)$$

may be written as

$$y(x) = u_1(x)y_1(x) + u_2(x)y_2(x)$$

The differential equation leads to one equation for the two functions u_1 and u_2, so we are free to impose an additional constraint. The first derivative is

$$y'(x) = u_1'(x)y_1(x) + u_1(x)y_1'(x) + u_2'(x)y_2(x) + u_2(x)y_2'(x)$$

One convenient constraint is

$$u_1'(x)y_1(x) + u_2'(x)y_2(x) = 0 \tag{3.24}$$

in which case the derivative simplifies:

$$y'(x) = u_1(x)y_1'(x) + u_2(x)y_2'(x)$$

The second derivative is then

$$y''(x) = u_1'(x)y_1'(x) + u_1(x)y_1''(x) + u_2'(x)y_2'(x) + u_2(x)y_2''(x)$$

and the differential equation becomes

$$\begin{aligned}
h(x) &= u_1'(x)y_1'(x) + u_1(x)y_1''(x) + u_2'(x)y_2'(x) + u_2(x)y_2''(x) \\
&\quad + f[u_1(x)y_1'(x) + u_2(x)y_2'(x)] + g[u_1(x)y_1(x) + u_2(x)y_2(x)] \\
&= [u_1(x) + u_2(x)][y_1'' + fy_1' + gy_1] + u_1'(x)y_1'(x) + u_2'(x)y_2'(x) \\
&= u_1'(x)y_1'(x) + u_2'(x)y_2'(x) \tag{3.25}
\end{aligned}$$

Using equation (3.24) to eliminate u_2', we have

$$h(x) = u_1'(x)\left[y_1'(x) - \frac{y_1}{y_2}y_2'(x)\right] = -u_1'(x)\frac{W(x)}{y_2}$$

---

[12]The Legendre functions are discussed in some detail in Chapter 8.

where $W(x)$ is the Wronskian (3.20) of the two solutions y_1 and y_2. Thus,

$$u_1(x) = -\int \frac{h(x)y_2(x)}{W(x)} dx \tag{3.26}$$

and, similarly,

$$u_2(x) = \int \frac{h(x)y_1(x)}{W(x)} dx \tag{3.27}$$

Note the similarity between equations (3.23) and (3.27).

Methods of Solution

We have discussed several methods for finding a second solution to a differential equation, but how do we find the first one? Methods for solving an ordinary differential equation include the following:

1. Guess the form of the solution.
2. Form a power series–type solution.
3. Find an asymptotic solution.
4. Recognize the equation, or use a change of variable to transform the equation to a recognizable form.
5. Integrate numerically.

Guessing is not a bad option. A given differential equation with a specified set of boundary conditions has a unique solution (we will not prove this theorem here), so if we can guess the solution, we have the unique solution. Physical intuition and experience are useful guides to the guess.

Later in the text we will identify some equations whose solutions are "named" functions with properties that are well known. Legendre's equation (Example 3.7; Chapter 8, Section 8.3.1) and Bessel's equation (Example 3.9; Chapter 8, Section 8.4.1) are examples. Once we recognize the equation, we may simply write down the solution. Some references (for example, Polyanin and Zaitsev, Murphy) list differential equations and their solutions. These references allow us to look up the solution in the same way that integral tables allow us to find the value of an integral.

Numerical integration gives a numerical solution for a specified set of boundary conditions, but the solution must be repeated for each new set of conditions. Thus, an analytic solution is preferable if one can be found. Numerical methods can be very useful for non-linear equations, however.

Additional methods for solving linear equations, such as expansion in eigenfunctions and transform methods, will be developed in the chapters that follow. Here we shall begin by studying power series solutions.

3.3.3. Power Series Solutions

Solution About a Regular Point

The solution to a linear, homogeneous differential equation can be expressed as a Taylor series about a regular point. Frequently (but not always) we want a solution about $x = 0$, so that the solution is a power series in x. The method is to assume a solution of the form

$$y = \sum_{n=0}^{\infty} a_n x^n \tag{3.28}$$

and substitute into the differential equation. Since a Taylor series is uniformly convergent within its radius of convergence,[13] we may differentiate term by term. Thus,

$$\frac{dy}{dx} = \sum_{n=0}^{\infty} n a_n x^{n-1} \tag{3.29}$$

Notice that the $n = 0$ term disappears because of the factor n multiplying a_n. Then the second derivative is

$$\frac{d^2 y}{dx^2} = \sum_{n=0}^{\infty} n(n-1) a_n x^{n-2} \tag{3.30}$$

where now both the $n = 0$ and $n = 1$ terms are zero.

For example, consider the Helmholtz equation (3.16)

$$\frac{d^2 y}{dx^2} + k^2 y = 0$$

(We can *guess* the solution to this equation: $y = \sin kx$ or $y = \cos kx$. Thus, we can check our result.) This equation has no singular points, so $x = 0$ is a regular point. Substitute in the series (3.28) and (3.30):

$$\sum_{n=0}^{\infty} n(n-1) a_n x^{n-2} + k^2 \sum_{n=0}^{\infty} a_n x^n = 0$$

The algebraic equation that results can be satisfied only if the coefficient of each power of x is separately equal to zero. We always start with the lowest power that appears in the equation.

$\underline{x^0}$:

We obtain this power by taking $n = 2$ in the first term and $n = 0$ in the second term. Thus, we get

$$2 \cdot 1 \cdot a_2 + k^2 a_0 = 0 \Rightarrow a_2 = -k^2 \frac{a_0}{2}$$

---

[13] See Chapter 2, Section 2.3.3.

$\underline{x^1:}$

We take $n = 3$ in the first term and $n = 1$ in the second. Then

$$3 \cdot 2 \cdot a_3 + k^2 a_1 = 0 \Rightarrow a_3 = -k^2 \frac{a_1}{3 \cdot 2}$$

We have to consider special cases until we come to the lowest power of x for which every sum in the algebraic equation contributes to that power. Then we can draw a general conclusion valid for all higher powers of x. Notice that both sums here contribute to the coefficient of x^1, so we are ready for a general statement. Let's look at the coefficient of x^{m-2}.

$\underline{x^{m-2}:}$

We need $n = m$ in the first term and $n = m - 2$ in the second.

$$m(m-1)a_m + k^2 a_{m-2} = 0 \Rightarrow a_m = -k^2 \frac{a_{m-2}}{m(m-1)} \tag{3.31}$$

We can now use the same relation again to express a_{m-2} in terms of a_{m-4}, etc.

$$a_m = \frac{-k^2}{m(m-1)} \frac{-k^2 a_{m-4}}{(m-2)(m-3)} = \frac{(-k^2)^2 a_{m-4}}{m(m-1)(m-2)(m-3)}$$

Continuing in this way, if m is even, we get

$$a_m = (-1)^{m/2} \frac{k^m}{m!} a_0 \tag{3.32}$$

while if m is odd, we get

$$a_m = (-1)^{(m-1)/2} \frac{k^{m-1}}{m!} a_1 \tag{3.33}$$

Equation (3.31) is the *recursion relation* for the series. It relates each coefficient in the series to the preceding one. In this case, the coefficents skip one: a_m is related to a_{m-2} rather than a_{m-1}. Eventually we can relate each coefficient to a_0 or a_1. But we have no relations that determine the first two coefficients a_0 and a_1. These are the two arbitrary constants in the general solution. The two solutions are

$$y_1 = a_0 \left(1 - \frac{(kx)^2}{2} + \frac{(kx)^4}{4!} + \cdots \right) = a_0 \cos kx$$

and

$$y_2 = a_1 \left(x - \frac{k^2 x^3}{3!} + \frac{k^4 x^5}{5!} + \cdots \right) = \frac{a_1}{k} \sin kx$$

in agreement with our guess. Since $(kx)^2$ is positive, the series is alternating, and each term decreases toward zero for $m > |kx| + 1$. Thus, these series converge everywhere.

Example 3.5. Use a power series method to solve Hermite's equation:

$$\frac{d^2y}{dx^2} - 2x\frac{dy}{dx} + 2\alpha y = 0$$

This equation arises in the quantum mechanical treatment of a harmonic oscillator.

The point $x = 0$ is a regular point of this equation, so we expect to be able to find a Taylor series solution about $x = 0$. Substitute in series (3.28) and its derivatives (3.29) and (3.30):

$$\sum_{n=0}^{\infty} n(n-1)a_n x^{n-2} - 2x \sum_{n=0}^{\infty} na_n x^{n-1} + 2\alpha \sum_{n=0}^{\infty} a_n x^n = 0$$

$$\sum_{n=0}^{\infty} n(n-1)a_n x^{n-2} - 2 \sum_{n=0}^{\infty} na_n x^n + 2\alpha \sum_{n=0}^{\infty} a_n x^n = 0$$

Now consider the coefficient of each power of x, starting with the lowest power.
$\underline{x^0}$:

The second term does not contribute. Take $n = 2$ in the first term and $n = 0$ in the third:

$$2 \cdot 1 \cdot a_2 + 2\alpha a_0 = 0 \Rightarrow a_2 = -\alpha a_0$$

$\underline{x^1}$:

All terms contribute:

$$3 \cdot 2 \cdot a_3 - 2a_1 + 2\alpha a_1 = 0 \Rightarrow a_3 = -\frac{\alpha - 1}{3} a_1$$

For all larger powers, all the terms in the equation contribute. So let's look at x^{m-2}:

$$m(m-1)a_m - 2(m-2)a_{m-2} + 2\alpha a_{m-2} = 0$$

and thus

$$a_m = 2a_{m-2} \frac{(m-2) - \alpha}{m(m-1)}$$

Since the recursion relation relates a_m to a_{m-2}, one solution contains only even powers of x while the other contains odd powers. The two series give two independent solutions of Hermite's equation. As in our previous example, the differential operator in Hermite's equation, $\mathcal{D}(x) = d^2/dx^2 - 2xd/dx + 2\alpha = \mathcal{D}(-x)$, is even, and thus the solution must be purely even or purely odd. The general solution is found by choosing values of the two constants a_0 and a_1; that is, it is a linear combination of the two series.

For the even solution, we find

$$a_m = 2\frac{(m-2)-\alpha}{m(m-1)} \times 2\frac{(m-4)-\alpha}{(m-2)(m-3)}a_{m-4}$$

$$= (-2)^{m/2}\frac{\alpha(\alpha-2)(\alpha-4)\cdots(\alpha-m+2)}{m!}a_0$$

Notice that if $\alpha = 2p$ for some integer p, then we have

$$a_{2p+2} = 2a_{2p}\frac{2p-2p}{(2p+2)(2p+1)} = 0$$

Then a_{2p+4}, a_{2p+6}, and so on are all zero also. Thus, the series terminates with the coefficient a_{2p}; it is a polynomial of order $2p$. These solutions are called Hermite polynomials, $H_{2p}(x)$. Since the solution is a polynomial and not an infinite series, the solution exists everywhere.

In this case ($\alpha = 2p$), the second solution to the differential equation does not terminate. With m odd,

$$a_{2n+1} = 2a_{2n-1}\frac{(2n-1)-2p}{(2n+1)(2n)}$$

and the numerator never vanishes. Indeed,

$$a_{2n+1} = (-2)^n\frac{[2(p-n)-1][2(p-n)-3]\cdots(2p-1)}{(2n+1)!}a_1$$

Solution About a Singular Point

In the region surrounding a singular point, a Taylor series may not exist, so we must modify our approach. If the singularity of the *solution* $y(x)$ is isolated, we may express it in terms of a Laurent series

$$\sum_{n=-m}^{\infty} a_n(x-x_0)^n$$

valid for $0 < |x - x_0| < \rho$. However, the singularity may not be isolated; for example, it may be a branch point. Thus, we may have to allow for noninteger powers of x. [It is also possible that, even though x_0 is a singular point of the differential equation, one of the solutions $y(x)$ may be analytic at x_0.]

We can allow for all these possibilities by choosing a series of the form

$$y(x) = (x - x_0)^p \sum_{n=0}^{\infty} a_n(x - x_0)^n \qquad (3.34)$$

where p may be any number, positive or negative, integer or noninteger, real or complex. This is called the Frobenius method.

If the solution is a Taylor or Laurent series, it is uniformly convergent within a region[14] $\rho_1 < |x - x_0| < \rho_2$, and we may differentiate term by term. If p is not an integer, then

$$y' = \left[\frac{d}{dx}(x - x_0)^p\right]\left(\sum_{n=0}^{\infty} a_n(x - x_0)^n\right) + (x - x_0)^p \left(\sum_{n=0}^{\infty} n a_n(x - x_0)^{n-1}\right)$$

The derivative of $(x - x_0)^p$ exists everywhere except on the branch cut. We may differentiate everywhere in a region that is keyhole-shaped—that is, an annulus with a channel cut out at the branch cut. The derivatives are

$$\frac{dy}{dx} = \sum_{n=0}^{\infty}(n + p)a_n(x - x_0)^{n+p-1} \tag{3.35}$$

and

$$\frac{d^2y}{dx^2} = \sum_{n=0}^{\infty}(n + p)(n + p - 1)a_n(x - x_0)^{n+p-2} \tag{3.36}$$

Example 3.6.   Solve the hypergeometric equation

$$(x^2 - x)\frac{d^2y}{dx^2} + \left(2x - \frac{1}{2}\right)\frac{dy}{dx} + \frac{1}{4}y = 0$$

Obtain a solution as a series in powers of x.

The equation has singular points where $x^2 - x = 0$—that is, at $x = 0$ and at $x = 1$. We are looking for a solution about $x = 0$, so we choose a series of the form (3.34). Substituting into the equation, we find

$$0 = \sum_{n=0}^{\infty}(n + p)(n + p - 1)a_nx^{n+p} - \sum_{n=0}^{\infty}(n + p)(n + p - 1)a_nx^{n+p-1}$$

$$+ 2\sum_{n=0}^{\infty}(n + p)a_nx^{n+p} - \frac{1}{2}\sum_{n=0}^{\infty}(n + p)a_nx^{n+p-1} + \frac{1}{4}\sum_{n=0}^{\infty}a_nx^{n+p}$$

The lowest power that appears in this equation is x^{p-1} (with $n = 0$ in the second and fourth terms), and its coefficient is

$$-\left[p(p - 1) + \frac{1}{2}p\right]a_0 = 0$$

Since we want $a_0 \neq 0$ for a nontrivial solution, we must have

$$p\left(p - \frac{1}{2}\right) = 0$$

---

[14]For a Taylor series, $\rho_1 = 0$ and the condition becomes $|x - x_0| < \rho_2$.

This equation is called the *indicial equation*. It has the solutions $p = 0$ and $p = 1/2$. Looking at the coefficient of x^{m+p} gives us the recursion relation for the coefficients:

$$(m + p)(m + p - 1)a_m - (m + p + 1)(m + p)a_{m+1} + 2(m + p)a_m$$

$$-\frac{1}{2}(m + p + 1)a_{m+1} + \frac{1}{4}a_m = 0$$

$$a_{m+1} = \frac{(m + p)(m + p + 1) + 1/4}{(m + p + 1)(m + p + 1/2)}a_m$$

In this case, the two independent solutions arise from the two different possible values of p. The recursion relation relates each a_{m+1} to the preceding a_m.
For $p = 0$,

$$a_{m+1} = \frac{m(m + 1) + 1/4}{(m + 1)(m + 1/2)}a_m = \frac{(m + 1/2)^2}{(m + 1)(m + 1/2)}a_m = \frac{2m + 1}{2(m + 1)}a_m$$

$$= \frac{1}{2^2}\left(\frac{2m + 1}{m + 1}\right)\left(\frac{2m - 1}{m}\right)a_{m-1}$$

$$= \frac{(2m + 1)!!}{2^{m+1}(m + 1)!}a_0$$

and the first solution is

$$y_1 = a_0 \sum_{n=0}^{\infty} \frac{(2n - 1)!!}{2^n n!}x^n$$

This series is well behaved at the origin and converges for $|x| < 1$.
The second solution has $p = 1/2$. Then

$$a_{m+1} = \frac{(m + 1/2)(m + 3/2) + 1/4}{(m + 3/2)(m + 1)}a_m = \frac{m^2 + 2m + 1}{(m + 3/2)(m + 1)}a_m = \frac{m + 1}{m + 3/2}a_m$$

$$= \frac{2(m + 1)}{2m + 3}a_m = \frac{2^{m+1}(m + 1)!}{(2m + 3)!!}a_0$$

and

$$y_2 = \sqrt{x}a_0 \sum_{n=0}^{\infty} \frac{2^n n!}{(2n + 1)!!}x^n$$

Once again the series converges for $|x| < 1$, but this function has a branch point at the origin. The constant a_0 in each solution may be chosen freely; it need not have the same value in the two solutions.

The Frobenius method may fail to provide two independent solutions if the functions $f(x)$ and/or $g(x)$ in the standard form (3.18) are too badly behaved. In particular, the method may fail to provide two solutions about $x = x_0$ unless $(x - x_0) f(x)$ and $(x - x_0)^2 g(x)$ are analytic[15] at $x = x_0$, in which case the singular points are called *regular singular points*. This result is known as *Fuch's theorem*. However, one solution may sometimes be obtained about an irregular singular point, or two solutions may be obtained by choosing a different central point that is a regular singular point.

The Frobenius method gives two independent solutions of equation (3.18) about a regular singular point provided that the indicial equation has two roots that do not differ by an integer. If the equation has a repeated root or two roots that differ by an integer,[16] the method may provide only one solution to the differential equation. This problem arises because the Frobenius method does not provide functions with logarithmic-type singularities. In such cases, the second solution is of the form[17]

$$y_2 = y_1 \ln x + \sum_{n=0}^{\infty} a_n x^{n+p} \tag{3.37}$$

where $y_1(x)$ is the first solution and the coefficients a_n are to be determined.

Example 3.7. Legendre's equation arises in the solution of Laplace's equation in spherical coordinates (Chapter 8). It has the form

$$(1 - x^2)\frac{d^2 y}{dx^2} - 2x\frac{dy}{dx} + l(l + 1)y = 0$$

This equation has singular points at $x = \pm 1$. Let's look at the special case $l = 0$. Then one solution is a constant and is valid everywhere. The second solution may be found as a series in powers of x, but the series converges only for $|x| < 1$. Find a solution valid for $x > 1$.

We can look for a solution about the singular point—that is, a solution in powers of $(x - 1)$. First we change variables to $w = x - 1$. Then we rewrite the equation in terms of w:

$$(1 - x^2)\frac{d^2 y}{dx^2} - 2x\frac{dy}{dx} = 0$$

$$(w + 2)w\frac{d^2 y}{dw^2} + 2(w + 1)\frac{dy}{dw} = 0$$

The equation has singularities at $w = -2$ and at $w = 0$, as expected. We assume a solution of the Frobenius type in powers of w, and substituting into the differential

---

[15] See, for example, Ince, Section 15.3, where this result is also extended to equations of higher order.
[16] See Problem 11. Also see Problem 28(a) for a case in which two independent solutions are obtained even though the roots of the indicial equation differ by an integer.
[17] See Problem 10.

equation, we get

$$0 = \sum_{n=0}^{\infty}(n+p)(n+p-1)a_n w^{n+p} + 2\sum_{n=0}^{\infty}(n+p)(n+p-1)a_n w^{n+p-1}$$

$$+ 2\sum_{n=0}^{\infty}(n+p)a_n w^{n+p} + 2\sum_{n=0}^{\infty}(n+p)a_n w^{n+p-1}$$

We can simplify by adding like powers:

$$0 = \sum_{n=0}^{\infty}(n+p)(n+p+1)a_n w^{n+p} + 2\sum_{n=0}^{\infty}(n+p)^2 a_n w^{n+p-1}$$

The lowest power that appears is w^{p-1}. Its coefficient is

$$2p^2 a_0 = 0$$

and so the only possible nontrivial solution is $p = 0$, a repeated root.
   The coefficient of $w^{m+p} = w^m$ is

$$a_m m(m+1) + 2a_{m+1}(m+1)^2 = 0$$

or

$$a_{m+1} = a_m \frac{-m}{2(m+1)}$$

However, notice that for $m = 0$, we get $a_1 = 0$. There is no series! This is just the constant solution mentioned above.
   The second solution is found by introducing the logarithm:

$$y_2 = y_1 \ln w + w^p \sum_{n=0}^{\infty} a_n w^n$$

where y_1 is the first solution, $y_1 = 1$ in this case. The derivatives are

$$\frac{dy_2}{dw} = \frac{1}{w} + \sum_{n=0}^{\infty}(n+p)a_n w^{n+p-1}$$

and

$$\frac{d^2 y_2}{dw^2} = -\frac{1}{w^2} + \sum_{n=0}^{\infty}(n+p)(n+p-1)a_n w^{n+p-2}$$

The differential equation becomes

$$0 = (w + 2)w \left(-\frac{1}{w^2} + \sum_{n=0}^{\infty} (n + p)(n + p - 1)a_n w^{n+p-2} \right)$$

$$+ 2(w + 1) \left(\frac{1}{w} + \sum_{n=0}^{\infty} (n + p)a_n w^{n+p-1} \right)$$

Rearranging, this becomes

$$\left(-\frac{w + 2}{w} + 2\frac{w + 1}{w} \right) + (w + 2) \sum_{n=0}^{\infty} (n + p)(n + p - 1)a_n w^{n+p-1}$$

$$+ 2(w + 1) \sum_{n=0}^{\infty} (n + p)a_n w^{n+p-1} = 0$$

$$0 = 1 + \sum_{n=0}^{\infty} (n + p)(n + p + 1)a_n w^{n+p} + 2 \sum_{n=0}^{\infty} (n + p)^2 a_n w^{n+p-1}$$

The first term is a constant (w^0 term). Thus, at least one term in the two series must also contribute an equal and opposite constant. Therefore, p cannot be a fraction, since in that case $n + p \neq 0$ for any n. If p were a negative number, $p = -s$, then for $n = 0$ in the last term we would have

$$0 = 2s^2 a_0 w^{-s-1}$$

requiring $a_0 \equiv 0$. Similarly, we would find $a_n \equiv 0$ for each $n < s$. The first nonzero term in the series would be $a_s w^{s-s} = a_s w^0$. Thus, we may as well take $p = 0$ and start the series with a_0. Then the x^0 term is

$$1 + 2a_1 = 0 \Rightarrow a_1 = -\frac{1}{2}$$

The x^m term has coefficient

$$m(m + 1)a_m + 2(m + 1)^2 a_{m+1} = 0, \quad m \geq 1$$

or

$$a_{m+1} = -a_m \frac{m}{2(m + 1)}, \quad m \geq 1$$

$$= (-1)^2 a_{m-1} \frac{m - 1}{2m} \frac{m}{2(m + 1)} = (-1)^2 \frac{a_{m-1}}{2^2} \frac{m - 1}{m + 1}$$

$$= (-1)^m \frac{a_1}{2^m} \frac{1}{(m + 1)} = \frac{(-1)^{m+1}}{2^{m+1}(m + 1)}$$

The constant term a_0 is not determined, since it does not appear in the differential equation. It may be combined with the first solution, $y_1 = $ constant. Thus, the second solution is

$$\ln w + \sum_{m=1}^{\infty} (-1)^{m+1} \left(\frac{w}{2}\right)^{m+1} \frac{1}{m+1} = \ln w - \sum_{m=0}^{\infty} (-1)^m \left(\frac{w}{2}\right)^{m+1} \frac{1}{m+1}$$

The series may be identified as the series for

$$\ln\left(1 + \frac{w}{2}\right) = \ln\left(1 + \frac{x-1}{2}\right) = \ln\left(\frac{x+1}{2}\right)$$

Thus, the second solution is

$$y_2(x) = \ln(x-1) - \ln(x+1) + \ln 2$$

$$= \ln\left(\frac{x-1}{x+1}\right) + \ln 2$$

We may combine the constant $\ln 2$ with the first solution if we wish, as we already argued for the constant a_0. The usual definition of the Legendre function $Q_0(x)$ for $x > 1$ is

$$Q_0(x) = \frac{1}{2} \ln\left(\frac{x+1}{x-1}\right)$$

which differs from our y_2 only by an inconsequential overall factor of $-1/2$.

Indicial Equation with Complex Roots

The indicial equation may have complex roots.

Example 3.8.   Apply the Frobenius method to the Euler differential equation:

$$x^2 y'' + x y' + y = 0$$

Inserting the series (3.34) with $x_0 = 0$, we find

$$\sum_{n=0}^{\infty} (n+p)(n+p-1)a_n x^{n+p} + \sum_{n=0}^{\infty} (n+p)a_n x^{n+p} + \sum_{n=0}^{\infty} a_n x^{n+p} = 0$$

The indicial equation is

$$p(p-1) + p + 1 = 0$$

which has the solutions $p = \pm i$. This problem has a second curiosity as well; the general recursion relation is

$$[(m+p)(m+p-1) + (m+p) + 1]a_m = 0$$

$$[(m \pm i)^2 + 1]a_m = 0$$

$$(m^2 \pm 2im)a_m = 0$$

and the only solution for $m > 0$ is $a_m = 0$. There is no series. Thus, the two solutions of the Euler equation are

$$y_1 = x^i$$

and

$$y_2 = x^{-i}$$

We may rewrite the solutions as follows:

$$y_1 = \exp(\ln x^i) = \exp(i \ln x); \quad y_2 = \exp(-i \ln x)$$

Thus, by taking appropriate linear combinations, we find that for real positive x we have the real solutions $\cos(\ln x)$ and $\sin(\ln x)$.

Asymptotic Methods

Sometimes the equation simplifies for large values of x, allowing us to find the limiting form of the solution.

Example 3.9.   The modified Bessel equation has the form

$$\frac{d^2y}{dx^2} + \frac{1}{x}\frac{dy}{dx} - \left(1 + \frac{m^2}{x^2}\right)y = 0 \tag{3.38}$$

Find the form of the solutions (the modified Bessel functions) for large argument.
   For large x, the terms with powers of $1/x$ become very small, and the equation simplifies to

$$\frac{d^2y_\infty}{dx^2} - y_\infty = 0$$

with the solutions

$$y_\infty = e^{\pm x}$$

The two solutions have exponential behavior at infinity.
   We can then find the complete solution by finding the function $v(x)$, where

$$y(x) = v(x)y_\infty(x) = v(x)e^{\pm x}$$

Then

$$\frac{dy}{dx} = v'e^{\pm x} \pm ve^{\pm x}$$

and

$$\frac{d^2y}{dx^2} = v''e^{\pm x} \pm 2v'e^{\pm x} + ve^{\pm x}$$

Substituting this into the full differential equation (3.38), we have

$$v''e^{\pm x} \pm 2v'e^{\pm x} + ve^{\pm x} + \frac{1}{x}\left(v'e^{\pm x} \pm ve^{\pm x}\right) - \left(1 + \frac{m^2}{x^2}\right)ve^{\pm x} = 0$$

$$v'' + \left(\frac{1}{x} \pm 2\right)v' + \frac{1}{x}\left(\pm 1 - \frac{m^2}{x}\right)v = 0 \quad (3.39)$$

Once again we can simplify. Look at the order of the terms in the equation when x is large:

Term	v''	v'/x	v'	v/x	v/x^2
Order of term[18]	v/x^2	v/x^2	v/x	v/x	v/x^2

Thus, we need to keep only the two terms of order v/x. Then, dividing by $2v$, we have

$$\frac{v'}{v} = -\frac{1}{2x}$$

$$\ln v = -\frac{1}{2}\ln x$$

$$v = \frac{1}{\sqrt{x}}$$

Thus, the general solution for large x is of the form

$$y = \frac{e^{-x}}{\sqrt{x}} \quad \text{or} \quad y = \frac{e^x}{\sqrt{x}} \quad (3.40)$$

These are the asymptotic forms of the modified Bessel functions.

Solution About the "Point at Infinity"

But what if we want more terms in our solution? Since we want a solution valid for large values of x, we change variables to $u = 1/x$, so as to obtain a series in powers of $1/x$. A solution valid in a neighborhood of $u = 0$ is a solution valid about the "point at infinity" in x.

Example 3.10.   Obtain a series expansion for the modified Bessel functions in powers of $1/x$.

We start with equation (3.39) in Example 3.9 and change variables to $u = 1/x$. The derivatives are

$$\frac{dv}{dx} = \frac{dv}{du}\frac{du}{dx} = \frac{dv}{du}\left(-\frac{1}{x^2}\right) = -u^2\frac{dv}{du} \quad (3.41)$$

---

[18]We approximate here by noting that differentiating reduces each power in a series expansion by 1—the same result we obtain by dividing by x. We can (and should) check the approximation once we obtain the solution for v.

and

$$\frac{d^2v}{dx^2} = \frac{d}{du}\left(\frac{dv}{dx}\right)\frac{du}{dx} = -u^2\left(-2u\frac{dv}{du} - u^2\frac{d^2v}{du^2}\right) = u^4\frac{d^2v}{du^2} + 2u^3\frac{dv}{du} \quad (3.42)$$

Substituting in, we get

$$u^4\frac{d^2v}{du^2} + 2u^3\frac{dv}{du} - (u \pm 2)u^2\frac{dv}{du} + u(\pm 1 - m^2u)v = 0$$

$$u^3\frac{d^2v}{du^2} + (u^2 \mp 2u)\frac{dv}{du} + (\pm 1 - m^2u)v = 0$$

This equation has a singular point at $u = 0$, so we look for a series solution of the form (3.34) and substitute in:

$$0 = \sum_{n=0}^{\infty} a_n(n+p)(n+p-1)u^{n+p+1} + \sum_{n=0}^{\infty} a_n(n+p)u^{n+p+1}$$

$$\mp 2\sum_{n=0}^{\infty} a_n(n+p)u^{n+p} \pm \sum_{n=0}^{\infty} a_nu^{n+p} - m^2\sum_{n=0}^{\infty} a_nu^{n+p+1}$$

The lowest power that appears is u^p, and its coefficient is

$$a_0(\mp 2p \pm 1) = 0$$

The indicial equation gives $p = 1/2$ for both the positive and the negative exponential terms. This is the leading term (3.40) that we found in Example 3.9.

Now look at the power u^{k+p+1}. Its coefficient is

$$a_k[(k+p)(k+p-1) + (k+p) - m^2] \pm a_{k+1}[1 - 2(k+1+p)] = 0$$

which gives the recursion relation:

$$a_{k+1} = \pm a_k\frac{(k+1/2)^2 - m^2}{2(k+1)} = \mp a_k\frac{4m^2 - (2k+1)^2}{8(k+1)}$$

Thus, the two solutions are

$$y_1(x) = a_0\frac{e^{-x}}{\sqrt{x}}\left(1 + \frac{(4m^2-1)}{8x} + \frac{(4m^2-9)(4m^2-1)}{2(8x)^2} + \cdots\right)$$

and

$$y_2(x) = a_0\frac{e^{+x}}{\sqrt{x}}\left(1 - \frac{(4m^2-1)}{8x} + \frac{(4m^2-9)(4m^2-1)}{2(8x)^2} + \cdots\right)$$

The leading term is independent of the value of m, as we found above. One solution is exponentially decaying, and one is exponentially growing. These are the functions $K_m(x)$ and $I_m(x)$ (see also Chapter 8, Section 8.4.7).

3.4. NUMERICAL METHODS

The computer is an increasingly useful tool for the solution of physical problems. Many useful pieces of software come with the computer's operating system or may be purchased at a reasonable cost. Spreadsheet programs may be used to solve differential equations numerically. Mathematical packages such as *Mathematica* and *Maple* have more sophisticated routines for solving differential equations. For more complex or unique problems, you may find it necessary to write your own program.

3.4.1. Dimensionless Variables

To prepare a problem for numerical solution or even to plot up the results of an analytic solution, the first step is to write the equation in dimensionless variables. For example, when solving equation (3.11), we would use the variable $u = x/L$. The solution obtained can then be applied to a beam of any length. Then

$$\frac{d}{dx} = \frac{du}{dx}\frac{d}{du} = \frac{1}{L}\frac{d}{du}$$

Rewriting the equation, we have

$$\frac{1}{L^4}\frac{d^4 y}{du^4} = \frac{q(u)}{EI}$$

Scaling the deflection y similarly, we define $v = y/L$. Then

$$\frac{1}{L^3}\frac{d^4 v}{du^4} = \frac{q(u)}{EI}$$

$$\frac{d^4 v}{du^4} = \frac{L^3 q(u)}{EI}$$

The source term q is load, or force per unit length, and I has dimensions of (length)^4 (see footnote 1), so the right-hand side has dimensions

$$L^3 \frac{M}{T^2}\frac{1}{L^4}\frac{1}{[E]}$$

and since $[E] = $ force/area $= M/LT^2$, the right-hand side is also dimensionless.

Example 3.11.   Convert equation (3.5) (motion with air resistance) into dimensionless form.

We would like to divide the space variable y by a characteristic length and the time variable t by a characteristic time. Notice that each term in the equation is a force, so the dimensions of the constant α may be found from

$$\left[\alpha \frac{dy}{dt}\right] = \frac{ML}{T^2} \Rightarrow [\alpha] = \frac{ML}{T^2}\frac{T}{L} = \frac{M}{T}$$

This suggests that we choose our characteristic time to be m/α and the dimensionless time variable to be $\tau = t\alpha/m$. Then the equation becomes

$$m\frac{\alpha^2}{m^2}\frac{d^2y}{d\tau^2} = -mg - \alpha\frac{\alpha}{m}\frac{dy}{d\tau}$$

$$\frac{\alpha^2}{m^2g}\frac{d^2y}{d\tau^2} = -1 - \frac{\alpha^2}{m^2g}\frac{dy}{d\tau}$$

Thus, we choose our characteristic length to be m^2g/α^2 and take the dimensionless space variable to be $u = y\alpha^2/m^2g$. The equation is then

$$\frac{d^2u}{d\tau^2} = -1 - \frac{du}{d\tau}$$

Let's check the dimensions of the characteristic length:

$$\left[\frac{m^2g}{\alpha^2}\right] = \frac{M^2L}{T^2}\frac{T^2}{M^2} = L$$

as required.

In what follows, we shall assume that the equations have been put into dimensionless form.

3.4.2. Difference Equations

The basic idea behind all numerical methods of solving differential equations is to replace the differentials with finite differences. Recall the definition of derivative:

$$\frac{df}{dx} = \lim_{h\to0}\frac{f(x+h) - f(x)}{h}$$

With *finite* differences, we do not take the limit but allow h to be small but finite. Thus, a differential equation is replaced by a difference equation.

If a function is differentiable at the point $x = x_0$, then the derivative may be calculated equally well by taking the limits:

$$\frac{df}{dx}\bigg|_{x_0} = \lim_{h \to 0} \frac{f(x_0) - f(x_0 - h)}{h} = \lim_{h \to 0} \frac{f(x_0 + h/2) - f(x_0 - h/2)}{h}$$

Similarly, we may form finite differences in three different ways:

- *Forward difference:*

$$\Delta f(x) = f(x + h) - f(x)$$

- *Central difference:*

$$\delta f(x) = f\left(x + \frac{h}{2}\right) - f\left(x - \frac{h}{2}\right)$$

- *Backward difference:*

$$\nabla f(x) = f(x) - f(x - h)$$

Higher-order differences may be formed similarly. Thus, the second forward difference is

$$\begin{aligned} \Delta^2 f &= \Delta f(x + h) - \Delta f(x) \\ &= f(x + 2h) - f(x + h) - [f(x + h) - f(x)] \\ &= f(x + 2h) - 2f(x + h) + f(x) \end{aligned}$$

The difference ratio $\Delta f / h$ is an approximation to the derivative. Inaccuracies occur because h is not zero, because of rounding error in the numerical calculation, and because of mistakes in one or more of the values. These inaccuracies increase in opposite directions; making h smaller improves the approximation to the derivative but increases the rounding error. Thus, obtaining an accurate numerical approximation to a derivative is very difficult, and we should attempt it only if it is absolutely unavoidable.

3.4.3. Numerical Solution of a First-Order Differential Equation

Suppose we want to find a solution to a first-order equation

$$y' = f(x, y)$$

that we have been unable to solve analytically. The equation need not be (and usually is not) linear. The output of any numerical method is a set of values for the function y at a set of values x_i for x. We need at least one given value $y_0 = y(x_0)$ (a boundary condition) from which to start.

If the function $f(x)$ can be differentiated analytically, we can use the Taylor series to step away from our given value:

$$y_1 = y(x_0 + h) = y_0 + hy'(x_0) + \frac{h^2}{2} y''(x_0) + \cdots$$

$$= y_0 + hf(x_0, y_0) + \frac{h^2}{2} \frac{df}{dx}\bigg|_{x_0} + \cdots$$

Then

$$y_2 = y_1 + hf(x_1, y_1) + \frac{h^2}{2} \frac{df}{dx}\bigg|_{x_1} + \cdots \tag{3.43}$$

where $x_1 = x_0 + h$, and so on.

As a first approximation, we drop the terms in powers of h greater than 1, effectively expressing the derivative dy/dx using a forward difference. This is called the Euler method. It is said to be first-order accurate, because we kept only the first power in h.

A better approximation is obtained by taking more terms in the Taylor series expansion (3.43). The number of terms needed at each step is determined by the rate at which the series converges, which usually depends on the value of h as well as the particular form of the solution y. If a large number of terms is needed, this method can be inconvenient because of the amount of analytic work that is required before beginning to compute.

Example 3.12.   Solve the differential equation

$$y' = 3xy - 1/3$$

on the range $0 \le x \le 1$, subject to the boundary condition $y(0) = 1$.

The function $f = 3xy - 1/3$ has derivatives

$$\frac{df}{dx} = 3y + 3x\frac{dy}{dx} = 3[y + x(3xy - 1/3)] = 3y(1 + 3x^2) - x$$

$$\frac{d^2 f}{dx^2} = -1 + 18xy + 3(1 + 3x^2)\frac{dy}{dx} = -1 + 18xy + 3(1 + 3x^2)(3xy - 1/3)$$

$$= -2 + 3x(9y + 9x^2 y - x)$$

At this point the labor is getting noticeable, so let's see if we can get away with only these two derivatives. Let's take $h = 0.1$, one-tenth of our interval. We can use a spreadsheet to calculate the various terms and add them together. The spreadsheet is shown in Table 3.1; the numerical values appear in Table 3.2. We can check the result by choosing a value of h equal to one-half of the previous value and repeating the calculation. The maximum difference between the two results is 0.3% (Figure 3.2).

TABLE 3.1. Spreadsheet for Example 3.12

	A	B	C	D
1	**Example 3.12**			
2	**h**	0.1		
3	**x**	**y**	**f = 3xy − 1/3**	**f′ = 3y(1 + 3x^2) − x**
4	0	1	= 3*A4*B4 − 1/3	= 3*B4*(1 + 3*A4^2) − A4
5	0.1	= F4	= 3*A5*B5 − 1/3	= 3*B5*(1 + 3*A5^2) − A5
6	0.2	= F5	= 3*A6*B6 − 1/3	= 3*B6*(1 + 3*A6^2) − A6
7	0.3	= F6	= 3*A7*B7 − 1/3	= 3*B7*(1 + 3*A7^2) − A7
8	0.4	= F7	= 3*A8*B8 − 1/3	= 3*B8*(1 + 3*A8^2) − A8
9	0.5	= F8	= 3*A9*B9 − 1/3	= 3*B9*(1 + 3*A9^2) − A9
10	0.6	= F9	= 3*A10*B10 − 1/3	= 3*B10*(1 + 3*A10^2) − A10
11	0.7	= F10	= 3*A11*B11 − 1/3	= 3*B11*(1 + 3*A11^2) − A11
12	0.8	= F11	= 3*A12*B12 − 1/3	= 3*B12*(1 + 3*A12^2) − A12
13	0.9	= F12	= 3*A13*B13 − 1/3	= 3*B13*(1 + 3*A13^2) − A13
14	1	= F13	= 3*A14*B14 − 1/3	= 3*B14*(1 + 3*A14^2) − A14

	E	F
1		
2		
3	**f″**	**y(n + 1)**
4	= −2 + 3*A4*(9*B4 + 9*A4^2*B4 − A4)	= B4 + $B$2*C4 + $B$2^2/2*D4 + $B$2^3/6*E4
5	= −2 + 3*A5*(9*B5 + 9*A5^2*B5 − A5)	= B5 + $B$2*C5 + $B$2^2/2*D5 + $B$2^3/6*E5
6	= −2 + 3*A6*(9*B6 + 9*A6^2*B6 − A6)	= B6 + $B$2*C6 + $B$2^2/2*D6 + $B$2^3/6*E6
7	= −2 + 3*A7*(9*B7 + 9*A7^2*B7 − A7)	= B7 + $B$2*C7 + $B$2^2/2*D7 + $B$2^3/6*E7
8	= −2 + 3*A8*(9*B8 + 9*A8^2*B8 − A8)	= B8 + $B$2*C8 + $B$2^2/2*D8 + $B$2^3/6*E8
9	= −2 + 3*A9*(9*B9 + 9*A9^2*B9 − A9)	= B9 + $B$2*C9 + $B$2^2/2*D9 + $B$2^3/6*E9
10	= −2 + 3*A10*(9*B10 + 9*A10^2*B10 − A10)	= B10 + $B$2*C10 + $B$2^2/2*D10 + $B$2^3/6*E10
11	= −2 + 3*A11*(9*B11 + 9*A11^2*B11 − A11)	= B11 + $B$2*C11 + $B$2^2/2*D11 + $B$2^3/6*E11
12	= −2 + 3*A12*(9*B12 + 9*A12^2*B12 − A12)	= B12 + $B$2*C12 + $B$2^2/2*D12 + $B$2^3/6*E12
13	= −2 + 3*A13*(9*B13 + 9*A13^2*B13 − A13)	= B13 + $B$2*C13 + $B$2^2/2*D13 + $B$2^3/6*E13
14	= −2 + 3*A14*(9*B14 + 9*A14^2*B14 − A14)	= B14 + $B$2*C14 + $B$2^2/2*D14 + $B$2^3/6*E14

TABLE 3.2. Numerical Values for Example 3.12

	A	B	C	D	E	F
1	**Example 3.12**					
2	**h**	0.1				
3	**x**	**y**	**f = 3xy − 1/3**	**f′ = 3y(1 + 3x^2) − x**	**f″**	**y(n + 1)**
4	0	1	−0.3333	3	−2	0.981333333
5	0.1	0.98	−0.0389	2.93232	0.646096	0.992209283
6	0.2	0.99	0.26199	3.13382319	3.452247331	1.034652997
7	0.3	1.03	0.59785	3.64202792	6.864951309	1.113792731
8	0.4	1.11	1.00322	4.54523973	11.47359534	1.23875299
9	0.5	1.24	1.5248	6.0034532	18.15395671	1.424275531
10	0.6	1.42	2.23036	8.28747931	28.2996385	1.693465796
11	0.7	1.69	3.22294	11.8485815	44.21969028	2.082373136
12	0.8	2.08	4.66436	17.4415887	69.84598597	2.647658296
13	0.9	2.65	6.81534	26.3444039	112.0219549	3.479585048
14	1	3.48	10.1054	40.7550206	182.8975926	4.724385265

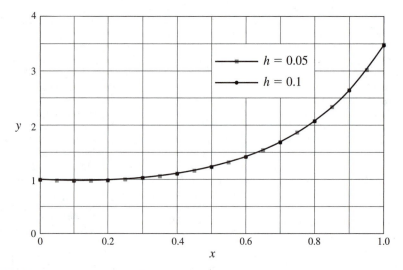

FIGURE 3.2. Numerical solution of the equation $y' = 3xy - 1/3$ with boundary condition $y(0) = 1$ (Example 3.12). The values computed with $h = 0.05$ and $h = 0.1$ do not differ noticeably in this plot.

3.4.4. The Runge-Kutta Method

From Example 3.12, we observe that the derivative of f is

$$\frac{df}{dx} = \frac{\partial f}{\partial x} + \frac{dy}{dx}\frac{\partial f}{\partial y} = \frac{\partial f}{\partial x} + f\frac{\partial f}{\partial y}$$

Thus, the Taylor series is

$$y(x_0 + h) = y(x_0) + hf(x_0, y_0) + \frac{h^2}{2}\left(\frac{\partial f}{\partial x} + f\frac{\partial f}{\partial y}\right) + \cdots = y_0 + \eta$$

where all the terms on the right are evaluated at x_0, y_0. Thus, the increment η in the solution y depends on values of f and all its derivatives at the starting point. We can improve the accuracy of the truncated series if we can use values of f and its derivatives throughout the range x_0 to $x_0 + h$. We can proceed by a process of successive approximations.

The first guess is to use

$$\eta_1 = hf(x_0, y_0)$$

as the increment. This takes us to the point P_1 with coordinates $(x_0 + h, y_0 + \eta_1)$. Now we step back to the midpoint at $(x_0 + h/2, y_0 + \eta_1/2)$. We evaluate the slope $y' = f$ at this point and use this new value to compute an increment η_2, where

$$\eta_2 = hf\left(x_0 + \frac{h}{2}, y_0 + \frac{\eta_1}{2}\right) = hf(x_0, y_0) + h\left(\frac{h}{2}\frac{\partial f}{\partial x} + \frac{\eta_1}{2}\frac{\partial f}{\partial y}\right) + \cdots$$

This takes us to the point P_2 with coordinates $(x_0 + h, y_0 + \eta_2)$. We can now repeat the process to find a third increment:

$$\eta_3 = hf\left(x_0 + \frac{h}{2}, y_0 + \frac{\eta_2}{2}\right) = hf(x_0, y_0) + h\left(\frac{h}{2}\frac{\partial f}{\partial x} + \frac{\eta_2}{2}\frac{\partial f}{\partial y}\right) + \cdots$$

Now we can evaluate the fourth increment by using the values at our last endpoint P_3 with coordinates $(x_0 + h, y_0 + \eta_3)$:

$$\eta_4 = hf(x_0 + h, y_0 + \eta_3) = hf(x_0, y_0) + h\left(h\frac{\partial f}{\partial x} + \eta_3 \frac{\partial f}{\partial y}\right) + \cdots$$

The best result is obtained by taking an appropriate weighted average of all these estimates of the true increment η. That average is

$$\eta = \frac{1}{6}(\eta_1 + 2\eta_2 + 2\eta_3 + \eta_4) \tag{3.44}$$

This increment makes the expression for $y(x + h)$ correct to fourth order in h, and thus the method is called the fourth-order Runge-Kutta method. In practical use, we can calculate the numerical values we need from the function f without evaluating any derivatives analytically.

Example 3.13.   Use the Runge-Kutta method to solve the equation $y' = 3xy - 1/3$ with boundary condition $y(0) = 1$.

Again we can use a spreadsheet to calculate each of the ηs and sum them to compute the increment in y. Our spreadsheet has columns containing the values $x_0, y_0, \eta_1, x_{1/2} = x_0 + h/2, y_1 = y_0 + \eta_1/2, \eta_2, y_2 = y_0 + \eta_2/2, x_1 = x_0 + h, y_3 = y_0 + \eta_3, \eta$, and finally $y_4 = y_0 + \eta$, which becomes the y_0 for the next value x_1 of x. The beginning of the spreadsheet is shown in Table 3.3; the numerical solution appears in Table 3.4. This time we find that the values computed with $h = 0.05$ differ from those with $h = 0.1$ by at most 0.003% (Figure 3.3).

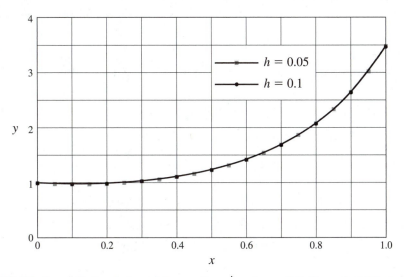

FIGURE 3.3. Runge-Kutta solution of the equation $y' = 3xy - 1/3$ (Example 3.13). Again the solutions for $h = 0.05$ and $h = 0.1$ are not distinguishable.

TABLE 3.3. Spreadsheet for Example 3.13

	A	B	C	D	E	F	G	H
1	Example 3.13							
2	**h**	**n**	**x**	**y**	**x1/2**	**eta1**	**y1**	**eta2**
3	0.1	0	0	1	= C3 + A$3/2	= A$3*(3*C3*D3 − 1/3)	= D3 + F3/2	= A$3*(3*E3*G3 − 1/3)
4		1	= C3 + A$3	= N3	= C4 + A$3/2	= A$3*(3*C4*D4 − 1/3)	= D4 + F4/2	= A$3*(3*E4*G4 − 1/3)
5		2	= C4 + A$3	= N4	= C5 + A$3/2	= A$3*(3*C5*D5 − 1/3)	= D5 + F5/2	= A$3*(3*E5*G5 − 1/3)

TABLE 3.4. Numerical Solution for Example 3.13

	A	B	C	D	E	F	G	H
1	Example 3.13							
2	**h**	**n**	**x**	**y**	**x1/2**	**eta1**	**y1**	**eta2**
3	0.1	0	0	1	0.05	−0.033333	0.983333	−0.018583
4		1	0.1	0.981445	0.15	−0.00389	0.9795	0.010744
5		2	0.2	0.992438	0.25	0.026213	1.005545	0.042083
6		3	0.3	1.035032	0.35	0.05982	1.064941	0.078486
7		4	0.4	1.114386	0.45	0.100393	1.164583	0.123885
8		5	0.5	1.239678	0.55	0.152618	1.315987	0.183805
9		6	0.6	1.425735	0.65	0.223299	1.537384	0.266457
10		7	0.7	1.69581	0.75	0.322787	1.857203	0.384537
11		8	0.8	2.086218	0.85	0.467359	2.319897	0.558241
12		9	0.9	2.6541	0.95	0.683274	2.995737	0.820452
13		10	1	3.490613	1.05	1.01385	3.997538	1.225891

The Runge-Kutta method, or indeed any numerical algorithm, works best if the step size chosen is the largest possible one that will still achieve the desired level of accuracy. Press et al. show how to write routines that adjust the step size automatically to obtain maximum efficiency while retaining accuracy.

3.4.5. Higher-Order Equations

Higher-order differential equations may be reduced to systems of first-order equations through the introduction of additional variables. For example, the second-order equation

$$y'' + 2y'^2 - 6 = 0 \tag{3.45}$$

may be reduced to two first-order equations. First introduce the new variable

$$v = y' \tag{3.46}$$

Then the original differential equation becomes

$$v' = 6 - 2v^2 \tag{3.47}$$

Each of these equations may now be tackled using one of the standard methods.

Example 3.14. Solve equation (3.45) subject to the initial conditions $y(0) = 0.5$ and $y'(0) = 0$. Use the Runge-Kutta method, and compute values for $0 \le x \le 1$.
The differential equations are

$$y' = f(x, y, v) = v$$

TABLE 3.3. *(Continued)*

I	J	K	L	M	N
y2	eta3	y3	eta4	eta	y4
= D3 + H3/2	= A$3*(3*(E3*I3 − 1/3)	= D3 + J3	= A$3*(3*C4*K3 − 1/3)	= (F3 + L3 + 2*(H3 + J3))/6	= D3 + M3
= D4 + H4/2	= A$3*(3*(E4*I4 − 1/3)	= D4 + J4	= A$3*(3*C5*K4 − 1/3)	= (F4 + L4 + 2*(H4 + J4))/6	= D4 + M4
= D5 + H5/2	= A$3*(3*(E5*I5 − 1/3)	= D5 + J5	= A$3*(3*C6*K5 − 1/3)	= (F5 + L5 + 2*(H5 + J5))/6	= D5 + M5

TABLE 3.4. *(Continued)*

I	J	K	L	M	N
y2	eta3	y3	eta4	eta	y4
0.990708	−0.018473	0.981527	−0.003888	−0.018555	0.981445
0.986817	0.011073	0.992518	0.026218	0.010994	0.992438
1.01348	0.042678	1.035116	0.059827	0.042593	1.035032
1.074274	0.079465	1.114497	0.100406	0.079355	1.114386
1.176329	0.125471	1.239857	0.152645	0.125292	1.239678
1.331581	0.186377	1.426056	0.223357	0.186057	1.425735
1.558963	0.270664	1.696399	0.322911	0.270075	1.69581
1.888079	0.391484	2.087294	0.467617	0.390408	2.086218
2.365338	0.569828	2.656046	0.683799	0.567882	2.6541
3.064326	0.84	3.4941	1.014897	0.836512	3.490613
4.103558	1.259288	4.7499	−0.033333	0.991812	4.482425

and

$$v' = g(x, y, v) = 6 - 2v^2$$

We step forward from the boundary at $x = 0$, using an increment h in x. Then we compute the increments in y and v using

$$\eta_1 = hf(x_0, y_0, v_0) = hv_0$$

$$m_1 = hg(x_0, y_0, v_0) = h(6 - 2v_0^2)$$

$$\eta_2 = hf(x_0 + h/2, y_0 + \eta_1/2, v_0 + m_1/2)$$

$$m_2 = hg(x_0 + h/2, y_0 + \eta_1/2, v_0 + m_1/2)$$

and so on. Then we use the increment η given by equation (3.44) with a similar expression for m.

We set up a spreadsheet with columns for x, $x + h/2$, and $x + h$; η_i and m_i, $i = 1, 2, 3, 4$; η and m; and the new values $y(x_0 + h) = y(x_0) + \eta$ and $v(x_0 + h) = v(x_0) + m$. The beginning of the spreadsheet is shown in Table 3.5; the numerical solution appears in Table 3.6. The results for $h = 0.1$ and $h = 0.05$ are shown in Figure 3.4. Splitting h in half gives results that differ by at most 0.004%.

The examples above illustrate the basic ideas involved in finding numerical solutions of differential equations with given initial conditions. The same ideas are used in problems with specified boundary values at two points, but the solution is more difficult and usually involves repeated iteration to converge on a solution. Any one method may prove unsuitable in a given case, and another method will have to be tried. Important issues of stability and convergence have not been discussed here. Details can be found in the references in the bibliography.

TABLE 3.5. Spreadsheet for Example 3.14

	A	B	C	D	E	F	G
1	Example 3.14						
2	h	0.1					
3	x	y	v	eta1	m1	x1	y1
4	0	0.5	0	= $B$3*C4	= $B$3*(6 − 2*C4^2)	= A4 + $B$3/2	= B4 + D4/2
5	0.1	= T4	= V4	= $B$3*C5	= $B$3*(6 − 2*C5^2)	= A5 + $B$3/2	= B5 + D5/2
6	0.2	= T5	= V5	= $B$3*C6	= $B$3*(6 − 2*C6^2)	= A6 + $B$3/2	= B6 + D6/2

	M	N	O	P	Q	R
1						
2						
3	eta3	y3	m3	v3	eta4	m4
4	= $B$3*L4	= B4 + M4	= $B$3*(6 − 2*L4^2)	= C4 + O4	= $B$3*P4	= $B$3*(6 − 2*P4^2)
5	= $B$3*L5	= B5 + M5	= $B$3*(6 − 2*L5^2)	= C5 + O5	= $B$3*P5	= $B$3*(6 − 2*P5^2)
6	= $B$3*L6	= B6 + M6	= $B$3*(6 − 2*L6^2)	= C6 + O6	= $B$3*P6	= $B$3*(6 − 2*P6^2)

TABLE 3.6. Numerical Solution to Example 3.14

	A	B	C	D	E	F	G
1	Example 3.14						
2	h	0.1					
3	x	y	v	eta1	m1	x1	y1
4	0	0.5	0	0.0000	0.6	0.05	0.5000
5	0.1	0.5294	0.5770	0.0577	0.5334	0.15	0.5583
6	0.2	0.6115	1.0386	0.1039	0.3843	0.25	0.6634
7	0.3	0.732	1.3464	0.1346	0.2375	0.35	0.7993
8	0.4	0.8766	1.5275	0.1528	0.1333	0.45	0.9530
9	0.5	1.0348	1.6264	0.1626	0.0709	0.55	1.1162
10	0.6	1.2004	1.6783	0.1678	0.0367	0.65	1.2843
11	0.7	1.3697	1.7049	0.1705	0.0187	0.75	1.4550
12	0.8	1.541	1.7184	0.1718	0.0094	0.85	1.6269
13	0.9	1.7132	1.7252	0.1725	0.0047	0.95	1.7994
14	1	1.8859	1.7286	0.1729	0.0024	1.05	1.9723

	M	N	O	P	Q	R
1						
2						
3	eta3	y3	m3	v3	eta4	m4
4	0.0291	0.5291	0.5831	0.5831	0.0583	0.5320
5	0.0806	0.6100	0.4701	1.0471	0.1047	0.3807
6	0.1187	0.7302	0.3181	1.3568	0.1357	0.2318
7	0.1432	0.8752	0.1900	1.5364	0.1536	0.1279
8	0.1573	1.0339	0.1049	1.6324	0.1632	0.0670
9	0.1650	1.1999	0.0553	1.6818	0.1682	0.0343
10	0.1690	1.3694	0.0285	1.7068	0.1707	0.0174
11	0.1711	1.5408	0.0145	1.7194	0.1719	0.0088
12	0.1721	1.7131	0.0073	1.7257	0.1726	0.0044
13	0.1727	1.8859	0.0037	1.7289	0.1729	0.0022
14	0.1729	2.0588	0.0018	1.7305	0.1730	0.0011

TABLE 3.5. *(Continued)*

H	I	J	K	L
v1	**eta2**	**y2**	**m2**	**v2**
= C4 + E4/2	= $B$3*H4	= B4 + I4/2	= $B$3*(6 − 2*H4^2)	= C4 + K4/2
= C5 + E5/2	= $B$3*H5	= B5 + I5/2	= $B$3*(6 − 2*H5^2)	= C5 + K5/2
= C6 + E6/2	= $B$3*H6	= B6 + I6/2	= $B$3*(6 − 2*H6^2)	= C6 + K6/2

S		T	U		V
eta		**y(n + 1)**	**m**		**v(n + 1)**
= (D4 + 2*I4 + 2*M4 + Q4)/6		= B4 + S4	= (E4 + 2*K4 + 2*O4 + R4)/6		= C4 + U4
= (D5 + 2*I5 + 2*M5 + Q5)/6		= B5 + S5	= (E5 + 2*K5 + 2*O5 + R5)/6		= C5 + U5
= (D6 + 2*I6 + 2*M6 + Q6)/6		= B6 + S6	= (E6 + 2*K6 + 2*O6 + R6)/6		= C6 + U6

TABLE 3.6. *(Continued)*

H	I	J	K	L
v1	**eta2**	**y2**	**m2**	**v2**
0.3	0.03	0.5150	0.582	0.2910
0.84373	0.0843727	0.5716	0.45762	0.8058
1.23075	0.12307497	0.6730	0.29705	1.1871
1.4651	0.14650973	0.8052	0.1707	1.4317
1.59418	0.15941777	0.9563	0.09172	1.5734
1.66191	0.16619128	1.1179	0.04761	1.6502
1.69663	0.16966329	1.2852	0.02429	1.6904
1.71423	0.17142303	1.4554	0.01228	1.7110
1.7231	0.17231005	1.6271	0.00618	1.7215
1.72756	0.17275595	1.7996	0.00311	1.7267
1.7298	0.17297981	1.9724	0.00156	1.7294

S	T	U	V
eta	**y(n + 1)**	**m**	**v(n + 1)**
0.0294	0.5294	0.5770	0.5770
0.0821	0.6115	0.4616	1.0386
0.1205	0.7320	0.3077	1.3464
0.1446	0.8766	0.1811	1.5275
0.1582	1.0348	0.0989	1.6264
0.1655	1.2004	0.0519	1.6783
0.1693	1.3697	0.0266	1.7049
0.1712	1.5410	0.0135	1.7184
0.1722	1.7132	0.0068	1.7252
0.1727	1.8859	0.0034	1.7286
0.1730	2.0588	0.0017	1.7303

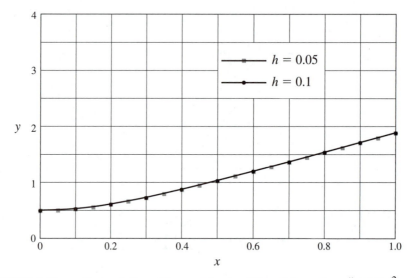

FIGURE 3.4. Runge-Kutta solution of the second-order differential equation $y'' + 0.2y'^2 - 6 = 0$ with $y(0) = 0.5$ and $y'(0) = 0$ (Example 3.14).

3.5. PARTIAL DIFFERENTIAL EQUATIONS: SEPARATION OF VARIABLES

Many of the differential equations that describe physical processes are *partial differential equations*. The solution is a function of three space variables and one time variable, and the equation relates the partial derivatives with respect to the four variables. The diffusion equation (3.13) is an example. Maxwell's equations are also of this form. If we introduce the potentials $\vec{\mathbf{A}}$ and Φ, Maxwell's equations reduce to wave equations for these two potentials. In SI units,

$$\nabla^2 \vec{\mathbf{A}} - \frac{1}{c^2} \frac{\partial^2 \vec{\mathbf{A}}}{\partial t^2} = -\mu_0 \vec{\mathbf{j}} \tag{3.48}$$

In the special case where $\vec{\mathbf{j}}$ is independent of time, the equation simplifies to

$$\nabla^2 \vec{\mathbf{A}} = -\mu_0 \vec{\mathbf{j}}$$

A similar equation holds for Φ:

$$\nabla^2 \Phi = -\frac{\rho}{\varepsilon_0} \tag{3.49}$$

This equation is called Poisson's equation.

The mathematics of partial differential equations is extensive and complicated, and we shall not delve into a general treatment here. Many of the partial differential equations in physics may be solved by the method of separation of variables.[19] In this method, we aim to

---

[19]Additional methods will be discussed in later chapters. See, for example, Chapter 7, Section 7.5.

replace the partial differential equation with a set of coupled ordinary differential equations, one in each of the variables. Then we can make use of any of the techniques we have already discussed for the solution of ordinary differential equations. In particular, series solutions are often used.

For example, let's look at equation (3.49) in the special case $\rho = 0$. This simpler homogeneous equation is called Laplace's equation. In Cartesian coordinates, the Laplacian operator[20] is

$$\nabla^2 \Phi = \frac{\partial^2 \Phi}{\partial x^2} + \frac{\partial^2 \Phi}{\partial y^2} + \frac{\partial^2 \Phi}{\partial z^2}$$

We look for a solution in which Φ is the product of three functions, each a function of one (and only one) of the three variables:

$$\Phi = X(x)Y(y)Z(z)$$

First we substitute into the differential equation:

$$\frac{\partial^2 \Phi}{\partial x^2} + \frac{\partial^2 \Phi}{\partial y^2} + \frac{\partial^2 \Phi}{\partial z^2} = X''YZ + XY''Z + XYZ'' = 0$$

where the prime means differentiation with respect to the argument of the function. Now for a nontrivial solution (Φ not identically zero),[21] we may divide through by Φ to get

$$\frac{X''}{X} + \frac{Y''}{Y} + \frac{Z''}{Z} = 0$$

At this point our equation has separated; each term is a function of *only one* of the three variables x, y, and z. If we now varied x, leaving y and z constant, we could change the value of the first term without changing the second two. The equation would no longer be satisfied. Thus, the only way we can satisfy this equation for *all* values of x, y, and z is if each term separately equals a constant:

$$\frac{X''}{X} = A; \quad \frac{Y''}{Y} = B; \quad \frac{Z''}{Z} = C$$

and

$$A + B + C = 0$$

Thus, the method of separation of variables leads to a set of coupled ordinary differential equations. The constants A, B, and C are called *separation constants*.

---

[20]See Chapter 1, Section 1.2.4, equation (1.50). Solutions in curvilinear coordinate systems will be explored in Chapter 8.

[21]Technically, we need the function $\Phi(\vec{x})$ to be nonzero except on a set of measure zero or, roughly speaking, within a region of zero volume.

In this example, each function is an exponential. Let $A = \alpha^2$ and $B = \beta^2$. Then $C = -\alpha^2 - \beta^2$. The solutions are of the form $X = e^{\pm \alpha x}$, $Y = e^{\pm \beta y}$, and $Z = \exp\left(\pm i\sqrt{\alpha^2 + \beta^2}\right)$. The boundary conditions must be invoked to complete the solution.

Example 3.15. A rectangular copper box of dimensions $l \times w \times h$ has all its walls grounded except for the top, which is at a nonzero potential. (Thin rubber strips separate the top from the other walls.) Find the general solution for the potential everywhere inside the box.

The solution follows a set of steps that we'll want to use for all problems of this type.[22]

1. *Choose an appropriate coordinate system.* Here we choose rectangular coordinates with origin at one corner of the box and $0 \leq x \leq l$, $0 \leq y \leq w$, $0 \leq z \leq h$.

2. *Separate the equation and determine the sets of solutions that exist.* Here the solutions are real and complex exponentials in all three coordinates, as we determined above.

3. *Choose one set of coordinates that has null boundary conditions. Choose the function that satisfies the zero boundary condition at both boundaries.* Here we need $X(0) = 0$ and $X(l) = 0$. We can combine the complex exponentials to give a sine function, which is the solution we want. We reject the hyperbolic sine because it has only *one* zero, at $x = 0$. We need a second zero at $x = l$. Thus, we must choose A to be negative: $A = -k^2$ and $X(x) = \sin kx$.

4. *Choose the value of the separation constant to make the function zero at the second boundary.* Here we need

$$\sin kl = 0$$

which will be satisfied if $kl = n\pi$, or

$$k = \frac{n\pi}{l}$$

This set of constants is the set of *eigenvalues* for the problem. The functions

$$X_n = \sin \frac{n\pi x}{l}$$

are called *eigenfunctions*.

5. *Repeat the procedure for the second set of coordinates:*

$$Y_m = \sin \frac{m\pi y}{w}$$

[22] See also Chapter 8, Section 8.2.

For the last set of coordinates, the separation constant is already *determined* by the two values of A and B that we have chosen in steps 4 and 5. Thus,

$$C = -A - B = \left(\frac{n\pi}{l}\right)^2 + \left(\frac{m\pi}{w}\right)^2$$

and is positive.

6. *Choose the final function to satisfy the one remaining zero boundary condition, given the value of C. In this case, with real exponentials as the possible solutions, we must choose the hyperbolic sine as our solution:*

$$Z = \sinh\left(\sqrt{\left(\frac{n\pi}{l}\right)^2 + \left(\frac{m\pi}{w}\right)^2}\, z\right)$$

Thus, the solution contains terms of the form

$$\sin\frac{n\pi x}{l}\sin\frac{m\pi y}{w}\sinh\left(\sqrt{\left(\frac{n\pi}{l}\right)^2 + \left(\frac{m\pi}{w}\right)^2}\, z\right)$$

7. *Write the general solution as a linear combination of such terms:*

$$\Phi = \sum_{n=1}^{\infty}\sum_{m=1}^{\infty} a_{nm}\sin\frac{n\pi x}{l}\sin\frac{m\pi y}{w}\sinh\left(\sqrt{\left(\frac{n\pi}{l}\right)^2 + \left(\frac{m\pi}{w}\right)^2}\, z\right)$$

This answers the question posed, but there is one final step:

8. *Use the final nonzero boundary condition to determine the coefficients a_{nm}. This is the subject of the next chapter.*

PROBLEMS

1. A vehicle moves under the influence of a constant force $\vec{F}$ and air resistance proportional to velocity (compare with equation 3.5, with $\vec{F}$ replacing the gravitational force). Find the speed of the vehicle as a function of time if it starts from rest at $t = 0$.

2. Find the general solution to the differential equation

$$y''' - 3y'' + 3y' - y = 0$$

Hint: Extend the result for a double root from Section 3.1.1.

3. A capacitor C, inductor L, and resistor R are connected in series with a switch (see the figure). The capacitor is charged by connecting it across a battery with emf $\mathcal{E}$. The battery is disconnected, and then the switch is closed. Find the current in the circuit as a function of time after the switch is closed.

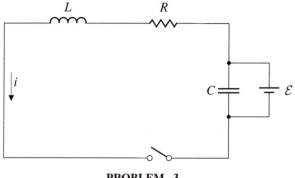

PROBLEM 3

4. The Airy differential equation is

$$y'' - xy = 0$$

Find the two solutions of this equation as power series in x.

5. Solve the equation $xy'' + 2y = 0$ (Section 3.3.2) using the Frobenius method. Show that $y(0)$ cannot equal any nonzero constant, as discussed in Section 3.3.2.

6. Find a solution of Laguerre's differential equation

$$xy'' + (1 - x)y' + \alpha y = 0$$

that is regular at the origin. Show that if α is an integer k, then this solution is a polynomial of degree k.

7. Solve the Bessel equation

$$4x^2 y'' + 4xy' + (4x^2 - 1)y = 0$$

as a Frobenius series in powers of x. Sum the series to obtain closed-form expressions for the two solutions.

8. Solve the hypergeometric equation

$$(x^2 - x)y'' + (3x - 1/2)y' + y = 0$$

as a series

(a) in powers of x (b) in powers of $x - 1$

9. Find two solutions of the Bessel equation

$$x^2 y'' + xy' + \left(x^2 - \frac{9}{4}\right) y = 0$$

as series in x. Verify that your solutions agree with the standard forms

$$\sqrt{\frac{2}{\pi x}} \left(\frac{\sin x}{x} - \cos x\right) \quad \text{and} \quad -\sqrt{\frac{2}{\pi x}} \left(\frac{\cos x}{x} + \sin x\right)$$

10. Consider a linear differential equation of the form

$$x^2 y'' + xf y' + gy = 0$$

Expand the function $f(x)$ in a power series of the form

$$f(x) = f_0 + f_1 x + f_2 x^2 + \cdots$$

and expand the function $g(x)$ similarly. Find the indicial equation. What is the condition on f_0 and g_0 if there is only one root? What is the value of the root in that case? Use the method of variation of parameters to show that the second solution of the differential equation is given by equation (3.37). *Hint:* Show that the equation for v may be reduced to the form

$$\frac{d}{dx}(\ln v') = -\frac{1}{x} + h(x)$$

where $h(x)$ is a series of positive powers of x. Integrate this equation twice to obtain equation (3.37).

11. For a linear differential equation of the form $x^2 y'' + xf y' + gy = 0$, the indicial equation may be written as $h(p) = 0$, where $h(p)$ is a quadratic function. Show that in determining the recursion relation, the coefficient of the c_n term is $h(p + n)$. Hence argue that the method fails to provide two solutions if the solutions of the equation $h(p) = 0$ differ by an integer. Are there any exceptions?

12. Solve the equation $y \dfrac{d^2 y}{dx^2} + \left(\dfrac{dy}{dx} \right)^2 = 0$ (equation 3.19) by writing it in the form

$$\frac{y''}{y'} = -\frac{y'}{y}$$

and integrating twice.

13. Find the two solutions of the equation

$$y'' - y' + \frac{y}{x} = 0$$

as series in powers of x.

14. Determine a solution of the equation

$$(1 + x)y'' + (3 + 2x)y' + (2 + x)y = 0$$

at large x. Hence determine the solutions for all x.

15. Determine the large argument expansion of the Legendre function Q_1 by finding a solution of the equation

$$(1 - x^2)y'' - 2xy' + 2y = 0$$

as a series in powers of $1/x$.

16. Solve the equation

$$x^2 y'' - 4xy' + (6 + x^2)y = 0$$

17. Solve the equation

$$xy'' - y' + 4x^3 y = 0$$

18. The conical functions are Legendre functions with $l = -\frac{1}{2} + i\lambda$.

   (a) Starting from the Legendre equation (Example 3.7), find the differential equation satisfied by the conical functions $P_{-\frac{1}{2}+i\lambda}(x)$ and $Q_{-\frac{1}{2}+i\lambda}(x)$.

   (b) Show that one solution is analytic at the point $x = 1$, and determine a series expansion for the conical function $P_{-\frac{1}{2}+i\lambda}(\cos\theta)$ in powers of $\sin(\theta/2)$. Hence show that this conical function is real.

19. Write the equation $x^4 y'' + y = 0$ in standard form, and use Fuch's theorem to show that the Frobenius method may not give two series-type solutions about $x = 0$. Change to the new variable $u = 1/x$ (as in Example 3.10) and show that the new equation can be solved by the Frobenius method. Obtain the two solutions.

20. Solve the equation

$$y'' + y \cosh x = 0$$

Hint: First expand the hyperbolic cosine in a series, and then use a power series method.

21. The Stark effect describes the energy shift of atomic energy levels due to applied electric fields. The differential equation describing this effect may be written

$$xy'' + y' + \left(k - \frac{1}{4}Ex^2 + \frac{Ux}{2} - \frac{m^2}{4x}\right)y = 0$$

where the term $Ex^2/4$ is the perturbation due to the electric field. Obtain a power series solution for y and obtain explicit expressions for the first four nonzero terms. How many terms are needed before any effect of the electric field is included?

22. Show that the indicial equation for the Bessel equation

$$\frac{d}{dx}\left(x\frac{dy}{dx}\right) + xy = 0$$

has a repeated root. Show that this root leads to only one solution. Find the second solution using equation (3.37). Try to get at least the first three terms in the series.

23. Attempt to solve the equation

$$x^2 y'' + y' = 0$$

using the Frobenius method. Show that the resulting series does not converge for any value of x.

24. Weber's equation is

$$y'' + \left(m + \frac{1}{2} - \frac{x^2}{4} \right) y = 0$$

Show that the substitution $y = \exp(-x^2/4)v(x)$ simplifies this equation. Find two solutions for $v(x)$ as power series in x.

25. The Schrödinger equation in one dimension has the form

$$\frac{\hbar^2}{2m} \frac{d^2\psi}{dx^2} + (E - V)\psi = 0$$

Develop a series solution for ψ in the case where V is the potential due to the interaction of two nucleons:

$$V = C\frac{e^{-\alpha x}}{x}$$

Obtain at least the first three nonzero terms.

26. The Kompaneets equation describes the evolution of the photon spectrum in a scattering atmosphere:

$$\frac{\partial n}{\partial t} = n_e \sigma_T c \frac{kT}{mc^2} \frac{1}{x^2} \frac{\partial}{\partial x} [x^4(n' + n + n^2)]$$

Here n is the photon number density, x is the dimensionless frequency, and σ_T is the Thomson scattering cross section. We may find a steady-state solution ($\partial/\partial t \equiv 0$) when photons are produced by a source $q(x)$ and subsequently escape from the cloud. When n remains $\ll 1$, the Kompaneets equation becomes a linear equation:

$$\frac{1}{x^2} \frac{\partial}{\partial x} [x^4(n' + n)] + q(x) - \frac{4n}{y} = 0$$

where y is the Compton "y" parameter, equal to (fractional energy change per scattering) $\times$ (mean number of scatterings). Assume that $q(x) \simeq 0$ except for $x \ll 1$.

(a) Show that for $x \gg 1$ the solution is an exponential. This is the Wien law.

(b) Show that in the special case $y = 1$, with $q(x) \equiv 0$, the solution is a power law in x.

(c) Verify your answers to (a) and (b) by letting $n = e^{-x}v$ and finding a power series solution for v. Obtain a solution for general y, and then let $y \to 1$ to verify your answer to (b).

27. A particle falls a distance d under gravity. Air resistance is proportional to the square of the particle's speed: $F = kv^2$. Write the differential equation that describes the particle's position as a function of time. Choose dimensionless variables, as in Section 3.4.1, and show that the equation may be put into the form

$$y'' + \alpha y'^2 - \beta = 0$$

Use the Runge-Kutta method to solve this equation with $\alpha = 0.1$. Assume the particle starts from rest. How long does it take for the particle to fall 10 m?

28. In astrophysics, the Lane-Emden equation describes the structure of a star with equation of state $P = K\rho^{(n+1)/n}$. If we define $\rho = \lambda\phi^n$, the equation of hydrostatic equilibrium becomes

$$\frac{1}{x^2}\frac{d}{dx}\left(x^2\frac{d\phi}{dx}\right) + \phi^n = 0$$

where x is a dimensionless distance variable with $x = 0$ at the center of the star. This is the Lane-Emden equation.

(a) Find a series solution for ϕ in the case $n = 1$.

(b) Find the first three nonzero terms in a series solution for ϕ for arbitrary n. Verify that your result agrees with the result of part (a) when $n = 1$. *Hint:* Begin by arguing that the solution contains only even powers of x.

(c) Solve the equation numerically for $n = 2$, $\phi(0) = 1$, and $\phi'(0) = 0$. At what value of x does $\phi(x)$ first equal zero? (This corresponds to the surface of the star.)

29. Investigate the effect of air resistance on the range of a projectile launched with speed v_0. Assume that air resistance is proportional to velocity: $\vec{F}_{res} = -\alpha\vec{v}$. Write the equations for the x- and y-coordinates in dimensionless form. Scale the coordinates with the maximum range $R = v_0^2/g$. What is the dimensionless air resistance parameter? Determine the dimensionless range for values of the air resistance parameter equal to 0, 0.1, 0.2, 0.4, and 0.5. Determine how the maximum range changes, and also determine how the launch angle for maximum range changes as air resistance increases. *Hint:* If there is no air resistance, you can obtain exact expressions for the increments in position and velocity in a time interval Δt. Use the same expressions when $\alpha \neq 0$, but with acceleration computed from the value of $\vec{v}$ at the beginning of your time interval.

30. The equation that describes the motion of a pendulum is

$$y'' = -\frac{g}{\ell}\sin y$$

(a) When y remains small, the equation may be reduced to the harmonic oscillator equation. Solve this equation to obtain the solution $y(t)$.

(b) With the initial conditions $y(0) = \pi/3$, $y'(0) = 0$, solve the nonlinear equation numerically to obtain the period. By how much does the period differ from your result in (a)?

31. Bessel's equation of order ν has the form

$$\frac{d^2y}{dx^2} + \frac{1}{x}\frac{dy}{dx} + \left(1 - \frac{\nu^2}{x^2}\right)y = 0$$

Show that the differential equation

$$\frac{d^2f}{dz^2} + z^r f = 0$$

may be converted to Bessel's equation through the relations

$$f = \sqrt{z}\,y$$

and

$$x = \frac{2}{r+2} z^{1+r/2}$$

What is the order of the resulting Bessel's equation? (The solutions are given in Chapter 8.)

What happens if the plus sign is changed to a minus sign, so that the equation is $d^2 f/dz^2 - z^r f = 0$?

32. Show that the equation

$$u'' + u' + \frac{k}{x^2} u = 0$$

has a solution of the form

$$u = \sqrt{x} e^{-x/2} K_\nu \left(\frac{x}{2} \right)$$

and find the order ν of the modified Bessel function.

CHAPTER 4

Fourier Series

4.1. FOURIER'S THEOREM

The principle of superposition for waves states that when two or more wave disturbances are present at the same time in a system, the total disturbance is the sum of the individual disturbances. This physical principle remains valid so long as the linearity of the system is preserved; that means that neither any of the individual disturbances nor their sum is too large.[1] Mathematically, this principle is due to the linearity of the governing differential equation (3.15). The functional form of the sum of the disturbances need not be a simple sine function—it can be quite complicated and irregular.

This principle is also the idea behind Fourier's theorem, which says that any moderately well-behaved[2] function $f(x)$ defined in a domain $0 \leq x \leq 2\pi$ may be expressed as a sum of sines and cosines:

$$f(x) = \sum_{n=0}^{\infty} (a_n \sin nx + b_n \cos nx) \tag{4.1}$$

Equivalently, we may write the series as a sum of complex exponentials:

$$f(x) = \sum_{n=-\infty}^{+\infty} c_n e^{inx} \tag{4.2}$$

---

[1] If the equation is written in terms of suitable dimensionless variables (see Chapter 3, Section 3.4.1), the necessary condition is $y \ll 1$.

[2] We'll give specifics of how well-behaved the function must be in Section 4.5. For now, you may assume that any function that is the solution to a physics problem is well enough behaved.

219

Here we are using a physical principle (the principle of superposition for waves) to motivate the mathematical result that a function $f(x)$ may be expressed as a superposition of wavelike functions (the sines and cosines) no matter what use we may make of the function f. Result (4.1) or (4.2) can be used in a variety of different physical systems, as we shall see.

Since the sines and cosines are periodic functions, the Fourier series is also periodic. Thus, $f(x \pm 2m\pi) = f(x)$, where m is an integer. We can find a Fourier series for a function that is not periodic, but it will represent the original function only in a finite range. Outside of that range, the original function and its Fourier series representation will differ; the series gives a periodic extension of the function in the original range. In many physical situations, we want a solution valid in a bounded region of space (for example, inside a box of side L), and a Fourier series with period L works well, since the solution does not even exist outside of the range 0 to L.

4.2. FINDING THE COEFFICIENTS

4.2.1. The Real Series

Let's assume that a series of the form (4.1) exists. Then how do we find the coefficients a_n and b_n? First we note that the sine and cosine functions form a set of *orthogonal functions* on the interval $0 \le x \le 2\pi$ (or, in fact, on any interval of length 2π). The functions are orthogonal[3] in the sense that

$$\int_0^{2\pi} \sin nx \sin mx \, dx = 0 \qquad (4.3)$$

and

$$\int_0^{2\pi} \cos nx \cos mx \, dx = 0 \qquad (4.4)$$

unless $m = n$ in both cases, and also

$$\int_0^{2\pi} \sin nx \cos mx \, dx = 0 \qquad (4.5)$$

for all integers m and n. Multiplying two trigonometric functions and integrating over one period is analogous to taking the dot product of two basis vectors.

---

[3]We shall give a more precise definition of the term *orthogonal* in Chapter 8.

To prove relations (4.3)–(4.5), we'll need the following relations:

$$\sin \alpha \cos \beta = \frac{1}{2}[\sin (\alpha + \beta) + \sin (\alpha - \beta)]$$

$$\cos \alpha \cos \beta = \frac{1}{2}[\cos (\alpha + \beta) + \cos (\alpha - \beta)]$$

$$\sin \alpha \sin \beta = \frac{1}{2}[\cos (\alpha - \beta) - \cos (\alpha + \beta)]$$

Thus, we may write the integral in (4.5) as

$$\int_0^{2\pi} \sin nx \cos mx \, dx = \frac{1}{2} \int_0^{2\pi} [\sin (n + m)x + \sin (n - m)x] \, dx$$

$$= \frac{1}{2} \left(-\frac{1}{n+m} \cos (n + m)x - \frac{1}{n-m} \cos (n - m)x \right) \Big|_0^{2\pi}$$

$$= 0 \quad \text{for } m \neq n$$

If $m = n$, the integrand is $\frac{1}{2} \sin 2mx$ and the integral is still zero.

Next we shall prove result (4.3):

$$\int_0^{2\pi} \sin nx \sin mx \, dx = \frac{1}{2} \int_0^{2\pi} [\cos (n - m)x - \cos (n + m)x] \, dx$$

$$= \frac{1}{2} \left[\frac{1}{n-m} \sin (n - m)x - \frac{1}{n+m} \sin (n + m)x \right] \Big|_0^{2\pi}$$

$$= 0 \quad \text{for } m \neq n$$

If $n = m$, the integral is

$$\int_0^{2\pi} \sin^2 nx \, dx = \frac{1}{2} \int_0^{2\pi} (1 - \cos 2nx) \, dx = \frac{1}{2} \left(x - \frac{1}{2n} \sin 2nx \right) \Big|_0^{2\pi} = \pi \quad (4.6)$$

except if $n = m = 0$, in which case the result is trivially zero.

Finally, the integral of the cosines is

$$\int_0^{2\pi} \cos nx \cos mx \, dx = \frac{1}{2} \int_0^{2\pi} [\cos (n - m)x + \cos (n + m)x] \, dx$$

$$= \frac{1}{2} \left[\frac{1}{n-m} \sin (n - m)x + \frac{1}{n+m} \sin (n + m)x \right] \Big|_0^{2\pi}$$

$$= 0 \quad \text{for } m \neq n$$

and again the result is π if $m = n \neq 0$. However, for $m = n = 0$, we now get

$$\int_0^{2\pi} dx = 2\pi$$

Expressing a function $f(x)$ as a Fourier series is akin to expressing a vector in terms of its components. To find a component of a vector, we take the dot product of the vector with the appropriate basis vector. Similarly, here we expect to find the coefficients a_n in the Fourier series (4.1) by taking the "dot product" with the basis function $\sin kx$. That is, we multiply the function $f(x)$ and its series (4.1) by $\sin kx$ for some integer k and integrate from 0 to 2π:

$$\int_0^{2\pi} f(x) \sin kx \, dx = \int_0^{2\pi} \left(\sum_{n=0}^{\infty} a_n \sin nx + b_n \cos nx \right) \sin kx \, dx$$

If the series is uniformly convergent,[4] we can interchange the sum and the integral. (We'll test this hypothesis later.) Then

$$\int_0^{2\pi} f(x) \sin kx \, dx = \sum_{n=0}^{\infty} \left(a_n \int_0^{2\pi} \sin nx \sin kx \, dx + b_n \int_0^{2\pi} \cos nx \sin kx \, dx \right)$$
$$= a_k \pi$$

since all the other terms in the series are zero by relations (4.3) and (4.5). Thus,

$$a_k = \frac{1}{\pi} \int_0^{2\pi} f(x) \sin kx \, dx, \quad k \geq 1 \tag{4.7}$$

Similarly, multiplying by $\cos kx$ gives the result

$$b_k = \frac{1}{\pi} \int_0^{2\pi} f(x) \cos kx \, dx, \quad k \geq 1 \tag{4.8}$$

While for $k = 0$,

$$b_0 = \frac{1}{2\pi} \int_0^{2\pi} f(x) \, dx \tag{4.9}$$

which is the average value of the function over the interval.

Example 4.1.   Find the Fourier series for the step function $f(x) = 0$ for $0 \leq x < \pi$ and $f(x) = 1$ for $\pi \leq x < 2\pi$.
  First we define the series to be

$$f(x) = \begin{cases} 0 & \text{if } 0 \leq x < \pi \\ 1 & \text{if } \pi \leq x < 2\pi \end{cases} = \sum_{n=0}^{\infty} (a_n \sin nx + b_n \cos nx)$$

---

[4]In Chapter 6, we show how to relax this restriction.

Then, using results (4.7) and (4.8), we obtain the coefficients:

$$a_n = \frac{1}{\pi} \int_0^{2\pi} f(x) \sin nx \, dx = \frac{1}{\pi} \int_\pi^{2\pi} \sin nx \, dx$$

$$= \frac{1}{\pi} \frac{1}{n} (-\cos nx)|_\pi^{2\pi} = \frac{1}{n\pi}[-1 + (-1)^n]$$

$$= \begin{cases} 0 & \text{if } n \text{ is even} \\ -2/n\pi & \text{if } n \text{ is odd} \end{cases}$$

and

$$b_n = \frac{1}{\pi} \int_0^{2\pi} f(x) \cos nx \, dx = \frac{1}{\pi} \int_\pi^{2\pi} \cos nx \, dx$$

$$= \frac{1}{\pi} \frac{1}{n} (\sin nx)|_\pi^{2\pi} = 0$$

We have to evaluate b_0 separately:

$$b_0 = \frac{1}{2\pi} \int_0^{2\pi} f(x) \, dx = \frac{1}{2\pi} \int_\pi^{2\pi} dx = \frac{1}{2}$$

As expected, b_0 is the average value of the function over the interval 0 to 2π. Thus, the series is

$$f(x) = \begin{cases} 0 & \text{if } 0 \le x < \pi \\ 1 & \text{if } \pi \le x < 2\pi \end{cases} = \frac{1}{2} - \frac{2}{\pi} \sum_{p=0}^{\infty} \frac{\sin(2p+1)x}{2p+1}$$

Figure 4.1 shows the first four partial sums[5] of this series. Notice that the value of the series at $x = 0$ and at $x = \pi$ is 1/2, not 0 or 1. When a function has a discontinuity, the Fourier series always gives a value equal to the midpoint of the jump. The series is a good representation of the function, but it is not identically equal to the function at every point; there can be significant differences at a finite number of points.

As we include more terms, the sum more closely represents the step function. Just beyond the discontinuity at $x = \pi$, the series overshoots the value $y = 1$. Increasing the number of terms moves the peak of the overshoot closer to $x = \pi$ but does not reduce its magnitude— an effect called the Gibbs phenomenon. The value of the series at the peak of the overshoot is 1.179. (See Appendix V.) The Gibbs phenomenon occurs at every jump of a discontinuous function.

The series we have found actually represents a *square wave*. Figure 4.1 could be repeated indefinitely to both the left and the right. Thus, the jump from 0 to 1 visible at $x = \pi$ in Figure 4.1 also occurs at every odd multiple of π, and, in reverse, the jump from 1 to 0 visible at $x = 0$ occurs at $x = 2m\pi$. Therefore, the series takes on the value $\frac{1}{2} = \frac{1}{2}(0 + 1)$ when x equals any integer multiple of π.

---

[5] See Chapter 2, Section 2.3.1 for the definition of partial sum.

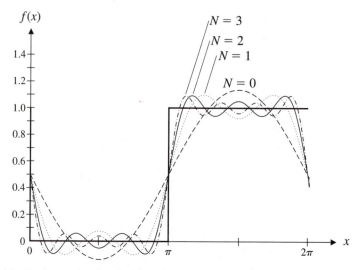

FIGURE 4.1. Fourier series for the step function (heavy line). Shown are the partial sums $\frac{1}{2} - \sum_{n=0}^{N} \frac{\sin(2n+1)x}{2n+1}$ for $N = 0$ (dashed line), $N = 1$ (dotted line), $N = 2$ (solid line), and $N = 3$ (dot-dashed line) terms. The series overshoots the step by 18%. As the number of terms is increased, the overshoot does not decrease, but moves closer to the step. This is the Gibbs phenomenon (see Appendix V).

If the function is defined over a range of values that does not span a range of 2π, then we can proceed by changing to a new variable. For example, if the step function is

$$f(x) = \begin{cases} 0 & \text{if } 0 \le x < 0.5 \\ 1 & \text{if } 0.5 \le x < 1 \end{cases} \tag{4.10}$$

then we define the variable $u = 2\pi x$ so that $0 < u < 2\pi$ corresponds to $0 < x < 1$. Then the series is

$$f(x) = \frac{1}{2} - \frac{2}{\pi} \sum_{n=0}^{\infty} \frac{\sin[(2n+1)2\pi x]}{2n+1} \tag{4.11}$$

In general, if the function is defined for $0 \le x < L$, then we choose the variable $u = 2\pi x/L$ so that u ranges from 0 to 2π and define the series as

$$f(x) = \sum_{n=0}^{\infty} \left[a_n \sin\left(\frac{2n\pi x}{L}\right) + b_n \cos\left(\frac{2n\pi x}{L}\right) \right]$$

With this argument, the orthogonality relation for these sine functions is

$$\int_0^L \sin\left(\frac{2n\pi x}{L}\right) \sin\left(\frac{2m\pi x}{L}\right) dx = \int_0^{2\pi} \sin(nu) \sin(mu) \frac{L}{2\pi} du = \frac{L}{2} \delta_{mn} \tag{4.12}$$

and thus the expression for the coefficient a_n is

$$a_n = \frac{2}{L} \int_0^L f(x) \sin\left(\frac{2n\pi x}{L}\right) dx \qquad (4.13)$$

The factor in front of the integral is one divided by (half of the range of integration). Similarly,

$$b_n = \frac{2}{L} \int_0^L f(x) \cos\left(\frac{2n\pi x}{L}\right) dx, \quad n > 0 \qquad (4.14)$$

For b_0, we get

$$b_0 = \frac{1}{L} \int_0^L f(x) dx \qquad (4.15)$$

Again we find that b_0 is the average value of the function over the interval.

For a function defined on the interval $-L < x < L$, we choose the variable $u = \pi x/L$, which ranges in value from $-\pi$ to $+\pi$, again a range of 2π. The coefficients of the sines in the Fourier series are given by

$$a_n = \frac{1}{L} \int_{-L}^{+L} f(x) \sin\left(\frac{n\pi x}{L}\right) dx \qquad (4.16)$$

The coefficients b_n of the cosines are found similarly.

4.2.2. The Complex Series

Complex exponentials also form a complete set of orthogonal functions. In the orthogonality relation for complex functions, we must multiply one function by the complex conjugate of another:

$$\int_0^{2\pi} y_n y_m^* \, dx = \int_0^{2\pi} e^{inx} e^{-imx} dx = \frac{1}{i(n-m)} e^{i(n-m)x} \Big|_0^{2\pi} = 0, \quad m \neq n$$

while if $n = m$,

$$\int_0^{2\pi} e^{inx} e^{-inx} dx = \int_0^{2\pi} 1 \, dx = 2\pi$$

This expression is valid for $n = m = 0$ also. Thus, the coefficients in the series

$$f(x) = \sum_{n=-\infty}^{+\infty} c_n e^{inx} \tag{4.17}$$

are given by

$$c_n = \frac{1}{2\pi} \int_0^{2\pi} f(x) e^{-inx} dx \tag{4.18}$$

In this case, the number in front of the integral is one divided by (the range of integration). If the function $f(x)$ is real, then

$$c_n^* = \left(\frac{1}{2\pi} \int_0^{2\pi} f(x) e^{-inx} dx \right)^* = \frac{1}{2\pi} \int_0^{2\pi} f(x) e^{inx} dx = c_{-n} \tag{4.19}$$

Let's evaluate the complex series for the step function (4.10):

$$f(x) = \begin{cases} 0 & \text{if } 0 \le x < 0.5 \\ 1 & \text{if } 0.5 \le x < 1 \end{cases} = \sum_{n=-\infty}^{+\infty} c_n e^{in2\pi x}$$

Notice that we have chosen the variable $2\pi x$ as the argument of the exponential, since this variable's value goes from 0 to 2π as x ranges from 0 to 1. Then

$$c_n = \int_0^1 f(x) e^{-i2n\pi x} dx = \int_{1/2}^1 e^{-i2n\pi x} dx = \left(\frac{1}{-2n\pi i} \right) e^{-i2n\pi x} \Big|_{1/2}^1$$

$$= \frac{i}{2n\pi} (e^{-2n\pi i} - e^{-n\pi i}) = \frac{i}{2n\pi}[1 - (-1)^n]$$

Like the a_n and b_n we obtained previously, this coefficient is also zero unless n is odd, in which case the term in parentheses is 2. We must evaluate the integral differently if $n = 0$, since our expression has a zero in the denominator in that case:

$$c_0 = \int_{1/2}^1 dx = \frac{1}{2}$$

Thus, the series is

$$f(x) = \begin{cases} 0 & \text{if } 0 \le x < 0.5 \\ 1 & \text{if } 0.5 \le x < 1 \end{cases} = \frac{1}{2} + \frac{1}{\pi} \sum_{n=-\infty,\, n \text{ odd}}^{+\infty} \frac{i}{n} \exp(i2n\pi x) \tag{4.20}$$

Now notice that the coefficient for $n = -N$, c_{-N}, equals the negative of the coefficient for $n = N$:

$$c_{-N} = -c_N$$

Thus, we may rewrite the series as

$$f(x) = \frac{1}{2} + \frac{1}{\pi} \sum_{n=1,\, n \text{ odd}}^{\infty} \frac{i}{n}(e^{i2n\pi x} - e^{-i2n\pi x})$$

$$= \frac{1}{2} - \frac{2}{\pi} \sum_{n=1,\, n \text{ odd}}^{\infty} \frac{\sin(2n\pi x)}{n} \tag{4.21}$$

where the sum is over odd values of n. The result is the same series (4.11) that we obtained using sines and cosines from the start. One advantage of using the exponential series is that the integrals are often easier to do.

A function $f(x)$ may be represented by a Fourier series of sines and cosines (equation 4.1 and Section 4.2.1) or of complex exponentials (equation 4.2). Since the two series represent the same function,

$$\sum_{n=0}^{\infty} \{a_n \sin nx + b_n \cos nx\} = \sum_{n=-\infty}^{+\infty} c_n e^{inx}$$

Expanding the sines and cosines (Chapter 2, equations 2.6 and 2.7), we get

$$\sum_{n=0}^{\infty} \left\{ a_n \frac{e^{inx} - e^{-inx}}{2i} + b_n \frac{e^{inx} + e^{-inx}}{2} \right\} = \sum_{n=-\infty}^{+\infty} c_n e^{inx}$$

$$\sum_{n=0}^{\infty} \left\{ \frac{-ia_n + b_n}{2} e^{inx} + \frac{ia_n + b_n}{2} e^{-inx} \right\} = \sum_{n=-\infty}^{+\infty} c_n e^{inx}$$

Thus, the relations between the coefficients are

$$c_n = \frac{b_n - ia_n}{2}, \quad c_{-n} = \frac{b_n + ia_n}{2}, \quad n \geq 1 \tag{4.22}$$

and

$$c_0 = b_0$$

4.3. FOURIER SINE AND COSINE SERIES

The function shown in Figure 4.2 is obtained from equation (4.10) by shifting the origin vertically and multiplying by -1. The resulting step function,

$$S_o(x) = \begin{cases} -1/2 & \text{if } -L \leq x < 0 \\ +1/2 & \text{if } 0 \leq x < +L \end{cases} \tag{4.23}$$

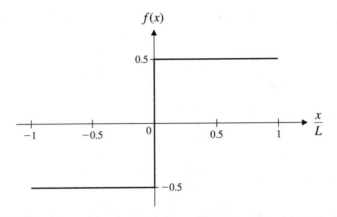

FIGURE 4.2. The step function $S_0(x)$ is an odd function on the interval $(-L, L)$.

is clearly odd in x over the range $-L \leq x \leq L$ and so should be represented by a series of sines, which are also odd in x. We did find that series (4.21) for our previous step function/square wave contained only sines. The origin shift in y is represented by the constant $1/2$ in equation (4.21). Thus,

$$S_0(x) = \frac{2}{\pi} \sum_{n=1,\, n \text{ odd}}^{\infty} \frac{\sin(n\pi x/L)}{n} \tag{4.24}$$

When the function is odd on the range $-L < x < L$, the argument of the sines is $\pi x/L$ because the period is $2L$. [In the previous section, we used the step function (equations 4.20 and 4.21) with $L = 1/2$.] However, we can express the coefficients as an integral over half the period, $0 \leq x \leq L$. The coefficients in a sine series of the form

$$f(x) = \sum_{n=1}^{\infty} a_n \sin \frac{n\pi x}{L}$$

are given by

$$a_n = \frac{1}{L} \int_{-L}^{L} f(x) \sin\left(\frac{n\pi x}{L}\right) dx$$

$$= \frac{1}{L} \left(\int_{-L}^{0} f(x) \sin\left(\frac{n\pi x}{L}\right) dx + \int_{0}^{L} f(x) \sin\left(\frac{n\pi x}{L}\right) dx \right)$$

$$= \frac{1}{L} \left(\int_{L}^{0} f(-x) \sin\left(-\frac{n\pi x}{L}\right) d(-x) + \int_{0}^{L} f(x) \sin\left(\frac{n\pi x}{L}\right) dx \right)$$

Then, since $f(x)$ is odd,

$$a_n = \frac{1}{L} \left(-\int_{L}^{0} -f(x) \left[-\sin\left(\frac{n\pi x}{L}\right) \right] dx + \int_{0}^{L} f(x) \sin\left(\frac{n\pi x}{L}\right) dx \right)$$

$$a_n = \frac{2}{L} \int_0^L f(x) \sin\left(\frac{n\pi x}{L}\right) dx \qquad (4.25)$$

Conversely, if a function is even on the range $-L < x < L$, the function will be represented by a series of cosines with argument $\pi x/L$:

$$f(x) = b_0 + \sum_{n=1}^{\infty} b_n \cos\frac{n\pi x}{L}$$

The coefficients are given by

$$b_n = \frac{2}{L} \int_0^L f(x) \cos\left(\frac{n\pi x}{L}\right) dx \qquad (4.26)$$

and

$$b_0 = \frac{1}{L} \int_0^L f(x)\, dx \qquad (4.27)$$

We can make the odd square wave (4.23) into an even function by shifting it to the left along the x-axis by $L/2$ (Figure 4.3).

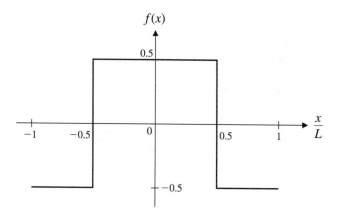

FIGURE 4.3. By shifting the origin to the right, we can create a step function $S_e(x)$ that is an even function. This function is represented by a cosine series.

Then series (4.24) becomes

$$S_e(x) = \frac{2}{\pi} \sum_{n=1,n\ \mathrm{odd}}^{\infty} \frac{1}{n} \sin n\pi \left(\frac{x}{L} + \frac{1}{2} \right)$$

$$= \frac{2}{\pi} \sum_{n=1,n\ \mathrm{odd}}^{\infty} \frac{\sin n\pi \dfrac{x}{L} \cos \dfrac{n\pi}{2} + \cos n\pi \dfrac{x}{L} \sin \dfrac{n\pi}{2}}{n}$$

$$= \frac{2}{\pi} \sum_{n=1,n\ \mathrm{odd}}^{\infty} \frac{(-1)^{(n-1)/2}}{n} \cos \frac{n\pi x}{L} \qquad (4.28)$$

which is a cosine series, as expected.

As with the complete series, if a function is defined *only* in the range $0 \le x \le L$ (Figure 4.4a), then the Fourier sine (or cosine) series will represent a periodic extension of the function outside of the original range (Figure 4.4b). The sine series gives an odd periodic extension, while the cosine series gives an even periodic extension (Figure 4.4, c, d). Each of these series has period $2L$.

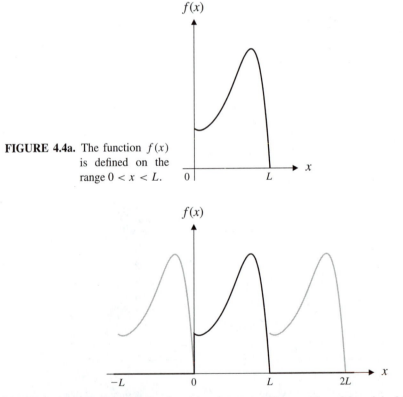

FIGURE 4.4a. The function $f(x)$ is defined on the range $0 < x < L$.

FIGURE 4.4b. The complete Fourier series gives a periodic repetition of the original function. The period is L.

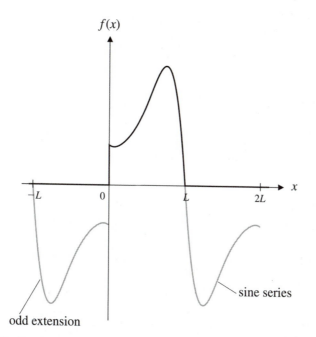

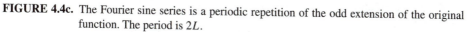

FIGURE 4.4c. The Fourier sine series is a periodic repetition of the odd extension of the original function. The period is $2L$.

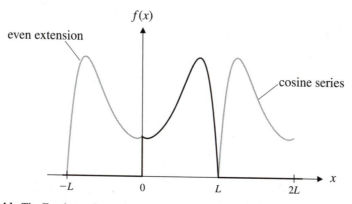

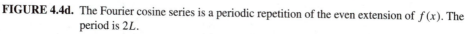

FIGURE 4.4d. The Fourier cosine series is a periodic repetition of the even extension of $f(x)$. The period is $2L$.

Thus, in principle, we can find three different series for a function $f(x)$ defined on the range $0 \le x \le L$:

1. The full series

$$f(x) = \sum_{n=-\infty}^{\infty} c_n e^{i2n\pi x/L}$$

with coefficients $c_n = \dfrac{1}{L} \int_0^L f(x)\, e^{-2in\pi x/L}\, dx$ and period L.

2. The sine series

$$f(x) = \sum_{n=1}^{\infty} a_n \sin \frac{n\pi x}{L}$$

with coefficients $a_n = \dfrac{2}{L} \displaystyle\int_0^L f(x) \sin \dfrac{n\pi x}{L}\, dx$ and period $2L$.

3. The cosine series

$$f(x) = b_0 + \sum_{n=1}^{\infty} b_n \cos \frac{n\pi x}{L}$$

with coefficients $b_n = \dfrac{2}{L} \displaystyle\int_0^L f(x) \cos \dfrac{n\pi x}{L}\, dx$ and b_0 equal to the average value of the function over the range 0 to L. This series also has period $2L$.

Different applications require the use of different series, as we shall see in the next section.

Example 4.2. Find the Fourier sine series and the Fourier cosine series for the function $f(x) = x$, $0 < x < L$.

The sine series $f(x) = \sum a_n \sin n\pi x/L$ represents an odd function on the range $-L < x < L$ (Figure 4.5). The coefficients are (equation 4.25)

$$a_n = \frac{2}{L} \int_0^L x \sin\left(\frac{n\pi x}{L}\right) dx = \frac{2}{L}\left(-x\frac{L}{n\pi}\cos\frac{n\pi x}{L}\Big|_0^L + \int_0^L \frac{L}{n\pi}\cos\frac{n\pi x}{L}dx\right)$$

$$= \frac{2}{L}\left(-\frac{L^2}{n\pi}\cos n\pi + \left(\frac{L}{n\pi}\right)^2 \sin\frac{n\pi x}{L}\Big|_0^L\right) = -2\frac{L}{n\pi}(-1)^n$$

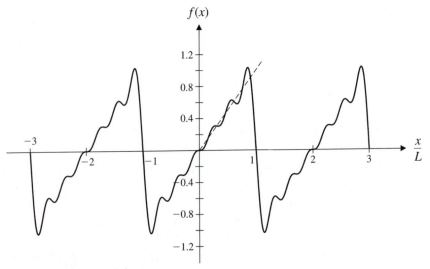

$f(x)$

FIGURE 4.5. The first six terms in the sine series of the function $f(x) = x$ defined for $0 < x < L$.

Thus,

$$x = -\frac{2L}{\pi} \sum_{n=1}^{\infty} \frac{(-1)^n}{n} \sin \frac{n\pi x}{L}$$

The cosine series $f(x) = \sum b_n \cos n\pi x/L$ represents an even function on the range $-L < x < L$ (Figure 4.6). The coefficients are (equations 4.26 and 4.27)

$$b_0 = \frac{1}{L} \int_0^L x \, dx = \frac{L}{2}$$

and

$$b_n = \frac{2}{L} \int_0^L x \cos \frac{n\pi x}{L} dx = \frac{2}{L} \left(x \frac{L}{n\pi} \sin \frac{n\pi x}{L} \Big|_0^L - \int_0^L \frac{L}{n\pi} \sin \frac{n\pi x}{L} dx \right)$$

$$= \frac{2}{n\pi} \frac{L}{n\pi} \cos \frac{n\pi x}{L} \Big|_0^L = \frac{2L}{(n\pi)^2} [(-1)^n - 1]$$

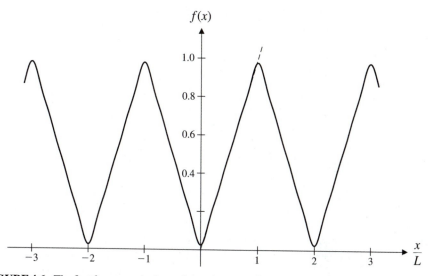

$f(x)$

FIGURE 4.6. The first four terms in the cosine series of the function $f(x) = x$ defined for $0 < x < L$. Compare this graph with Figure 4.5. A few terms of the cosine series represent the function more closely than does the sine series.

Thus, $b_n = 0$ for even n, and $b_n = -4L/(n\pi)^2$ for odd n. Therefore, the cosine series is

$$x = \frac{L}{2} - \frac{4L}{\pi^2} \sum_{p=0}^{\infty} \frac{1}{(2p+1)^2} \cos \frac{(2p+1)\pi x}{L}$$

Compare Figures 4.5 and 4.6; the cosine series converges much faster than the sine series (the terms decrease as $1/n^2$ rather than $1/n$). This happens because the odd extension is a discontinuous function with a jump at $x = \pm L$, whereas the even extension is a continuous function. In Problem 2, you will be asked to look at the full series for this function.

4.4. USE OF FOURIER SERIES TO SOLVE DIFFERENTIAL EQUATIONS

Fourier series may be used to solve differential equations under some circumstances. In this section, we shall look at two of the most common applications.

4.4.1. An Inhomogeneous Linear Equation Has a Periodic Driving Term

When the source term in an inhomogeneous linear equation is periodic, we expect that the solution will also be periodic. One example is an electric circuit connected to a power supply that supplies a periodic emf.

> **Example 4.3.**   A circuit has a 10 Ω resistor, a 2 mH inductor, and a 0.3 μF capacitor connected in series with a power supply that generates a square wave emf $\mathcal{E}(t)$ with a period of $\tau = 0.3$ ms (Figure 4.7). The applied voltage varies between zero and $+1$ V. What is the current in the circuit?

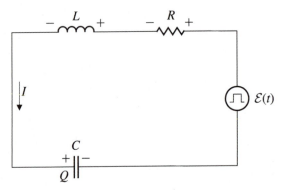

FIGURE 4.7. The series LRC circuit in Example 4.3 with a periodic (square wave) emf. With the current and charge variables (I and Q) as defined in this diagram, $I = dQ/dt$.

The charge and current variables I and Q are defined in Figure 4.7. Kirchhoff's rules applied to the circuit give the differential equation

$$L\frac{dI}{dt} + RI + \frac{Q}{C} = \mathcal{E}(t)$$

where $I = dQ/dt$. If we write the equation in standard form (equation 3.18),

$$\frac{d^2Q}{dt^2} + \frac{R}{L}\frac{dQ}{dt} + \frac{Q}{LC} = \frac{\mathcal{E}}{L}$$

the coefficients of dQ/dt and Q are

$$2\alpha \equiv \frac{R}{L} = \frac{10\ \Omega}{2\ \text{mH}} = 5.0 \times 10^3\ \text{s}^{-1}$$

and

$$\omega_0^2 \equiv \frac{1}{LC} = \frac{1}{(2\ \text{mH})(0.3\ \mu\text{F})} = \frac{1}{6} \times 10^{10}\ \text{s}^{-2}$$

Then

$$\omega_0 = \frac{10^5}{\sqrt{6}}\ \text{rad/s} = 4 \times 10^4\ \text{rad/s}$$

We have already found the series for the square wave emf $\mathcal{E}$ (equation 4.20):

$$\mathcal{E}(t) = \mathcal{E}_0\left(\frac{1}{2} + \frac{1}{\pi} \sum_{n=-\infty,\ n\ \text{odd}}^{+\infty} \frac{i}{n} \exp\left(i\frac{2n\pi t}{\tau}\right)\right)$$

where τ is the period of 0.3 ms and $\mathcal{E}_0 = 1$ V. Next, we write the solution as a Fourier series with period τ:

$$Q(t) = \sum_{n=-\infty}^{\infty} q_n \exp\left(i\frac{2n\pi t}{\tau}\right)$$

(Note: It is much easier to use the series in exponential form here, because the equation has both first and second derivatives. When we differentiate the series, each term in the equation contains simple multiples of the original exponential terms. In contrast, the odd-order derivatives mix sines and cosines: $d \sin x/dx = \cos x$. Instead of one equation for each coefficient c_n, we would have two equations to solve simultaneously for the coefficients of the sines and cosines.)

Now we substitute the series for $\mathcal{E}(t)$ and $Q(t)$ into the differential equation:

$$\sum_{n=-\infty}^{\infty} -\left(\frac{2n\pi}{\tau}\right)^2 q_n e^{i2n\pi t/\tau} + 2\alpha \sum_{n=-\infty}^{\infty} i\frac{2n\pi}{\tau} q_n e^{i2n\pi t/\tau} + \omega_0^2 \sum_{n=-\infty}^{\infty} q_n e^{i2n\pi t/\tau}$$

$$= \frac{\mathcal{E}_0}{L}\left(\frac{1}{2} + \frac{1}{\pi} \sum_{n=-\infty,\ n\ \text{odd}}^{+\infty} \frac{i}{n} e^{i2n\pi t/\tau}\right)$$

Next we make use of the orthogonality of the exponentials by multiplying the whole equation by $\exp[-i(2m\pi t/\tau)]$ and integrating over one period. Only the terms with $n = m$ survive. Since this is true for any integer m, we can equate the coefficients of each exponential separately.

The constant ($n = 0$) term gives

$$\omega_0^2 q_0 = \frac{\mathcal{E}_0}{2L} \Rightarrow q_0 = \frac{\mathcal{E}_0 C}{2}$$

The other terms are given by

$$q_n\left[-\left(\frac{2n\pi}{\tau}\right)^2 + \frac{4\alpha i n\pi}{\tau} + \omega_0^2\right] = \frac{\mathcal{E}_0}{\pi L}\frac{i}{n}$$

$$q_n = \frac{\mathcal{E}_0\tau^2}{\pi L}\frac{i}{n}\frac{1}{[-(2n\pi)^2 + 4\alpha\tau i n\pi + \omega_0^2\tau^2]}$$

$$= \frac{\mathcal{E}_0\tau^2}{\pi L}\frac{4\alpha\tau\pi + [\omega_0^2\tau^2 - (2n\pi)^2]i/n}{[\omega_0^2\tau^2 - (2n\pi)^2]^2 + (4\alpha\tau n\pi)^2}$$

Notice that the real part of q_n is even in n, while the imaginary part is odd. This is exactly what we expect if the resulting series is to be real. The real parts combine to give cosines, while the imaginary parts combine to give sines:

$$Q(t) = \frac{\mathcal{E}_0 C}{2} + \frac{\mathcal{E}_0\tau^2}{\pi L}\sum_{n=-\infty,\, n\neq 0}^{\infty}\frac{4\alpha\tau\pi + [\omega_0^2\tau^2 - (2n\pi)^2]i/n}{[\omega_0^2\tau^2 - (2n\pi)^2]^2 + (4\alpha\tau n\pi)^2}\exp\left(i\frac{2n\pi t}{\tau}\right)$$

$$= \frac{\mathcal{E}_0 C}{2} + \frac{2\mathcal{E}_0\tau^2}{\pi L}\sum_{n=1}^{\infty}\frac{4\alpha\tau\pi\cos\dfrac{2n\pi t}{\tau} - \dfrac{1}{n}[\omega_0^2\tau^2 - (2n\pi)^2]\sin\dfrac{2n\pi t}{\tau}}{[\omega_0^2\tau^2 - (2n\pi)^2]^2 + (4\alpha\tau n\pi)^2}$$

Finally, we can differentiate to get I. Then

$$I(t) = -\frac{4\mathcal{E}_0\tau}{L}\sum_{n=1}^{\infty}\frac{4n\alpha\tau\pi\sin\dfrac{2n\pi t}{\tau} + [\omega_0^2\tau^2 - (2n\pi)^2]\cos\dfrac{2n\pi t}{\tau}}{[\omega_0^2\tau^2 - (2n\pi)^2]^2 + (4\alpha\tau n\pi)^2}$$

The constant in front of the sum is

$$\frac{4\mathcal{E}_0\tau}{L} = \frac{4(1\text{ V})(0.3\text{ ms})}{(2\text{ mH})} = 0.6\,\frac{\text{V}\cdot\text{s}}{\text{H}} = 0.6\,\text{A} = I_0$$

Thus,

$$I(t) = -(0.6\text{ A})\sum_{n=1}^{\infty}\frac{4n\alpha\tau\pi\sin\dfrac{2n\pi t}{\tau} + [\omega_0^2\tau^2 - (2n\pi)^2]\cos\dfrac{2n\pi t}{\tau}}{[\omega_0^2\tau^2 - (2n\pi)^2]^2 + (4\alpha\tau n\pi)^2}$$

Notice that if the natural frequency ω_0 of the circuit times the period τ of the emf is very close to $2n\pi$ for some integer n, then the current will be very large if α is small; there is a resonance at frequency $\omega_n = 2n\pi/\tau$. In our example,

$$\omega_0\tau/\pi = (4.1\times 10^4\text{ s}^{-1})(0.3\times 10^{-3}\text{ s})/\pi = 3.9$$

which is very close to 4. The solution is shown in Figure 4.8. The current is dominated by the $n = 2$ term, which has twice the frequency of the square wave emf.

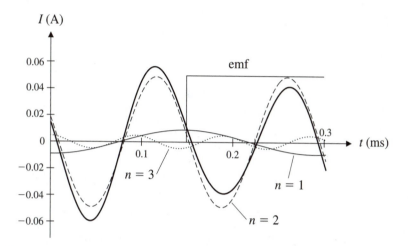

FIGURE 4.8. The first three terms of the solution for the current in Example 4.3. The thick solid line is the sum. The thin solid line is the $n = 1$ term, the dashed line is the $n = 2$ term, and the dotted line is the $n = 3$ term. The solution is dominated by the $n = 2$ term. The square wave emf is shown schematically by a thin solid line. The square wave indicates the phase (but not the amplitude) of the applied emf. The vertical axis does not apply to this curve.

4.4.2. The Solution We Want Equals Zero at Two Boundaries, $x = 0$ and $x = L$

If a physical quantity is described by a function $f(x)$ with $f(0) = f(L) = 0$ and exists only in the interval $0 < x < L$, we can make an odd extension of the function $f(x)$ to the range $-L < x < L$ and represent it as a Fourier sine series. The sine series is *always* zero at the endpoints $x = 0$ and $x = L$ and thus automatically satisfies the boundary conditions.

For example, consider the problem of waves on a string of length L. The differential equation for the string deflection $y(x, t)$ is (equation 3.15)

$$v^2 \frac{\partial^2 y}{\partial x^2} = \frac{\partial^2 y}{\partial t^2}$$

We can solve the equation using separation of variables.[6] Let $y(x, t) = X(x)T(t)$. Then

$$v^2 X''T = XT''$$

Dividing through by $y = XT$, we get

$$v^2 \frac{X''}{X} = \frac{T''}{T}$$

[6]See Chapter 3, Section 3.5.

Since each side of this equation is a function of only one variable, each side must equal a constant if the equation is to remain true for all values of both x and t. Since we want the string displacement y to equal zero at $x = 0$ and $x = L$ for all times t, we want the function X to be a sine, so we choose the separation constant to be negative:

$$\frac{X''}{X} = -k^2 \Rightarrow X = \sin(kx)$$

Next, we choose the constant k to make $\sin(kL) = 0$; that is, $k = n\pi/L$, where n is a positive integer. Then the equation for T is

$$\frac{T''}{T} = -k^2 v^2 = -\left(\frac{n\pi}{L}\right)^2 v^2$$

and the solutions are $T = \sin(n\pi vt/L)$ and $\cos(n\pi vt/L)$. Thus, the general solution may be written

$$y(x, t) = \sum_{n=1}^{\infty} \sin\left(\frac{n\pi x}{L}\right) \left[a_n \sin\left(\frac{n\pi vt}{L}\right) + b_n \cos\left(\frac{n\pi vt}{L}\right)\right] \qquad (4.29)$$

which, for fixed t, is a Fourier sine series in x, as expected. The amplitude of each Fourier component is oscillatory with angular frequency $\omega_n = n\pi v/L$. The initial conditions determine the constants a_n and b_n.

Example 4.4.   Suppose we pull the string up at the middle so that it forms a triangle of height h (Figure 4.9) and then let go. Find the subsequent motion of the string.

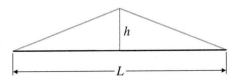

h

L

FIGURE 4.9.   The initial shape of the string ($t = 0$) in Example 4.4.

The solution has the form (4.29). At $t = 0$, we have

$$y(x, 0) = \begin{cases} x(2h/L) & \text{if } 0 \le x < L/2 \\ (L - x)(2h/L) & \text{if } L/2 \le x < L \end{cases} = \sum_{n=1}^{\infty} b_n \sin\left(\frac{n\pi x}{L}\right)$$

and the string velocity at $t = 0$ is zero:

$$\left.\frac{\partial y}{\partial t}\right|_{t=0} = 0 = \sum_{n=1}^{\infty} \frac{n\pi v}{L} \sin\left(\frac{n\pi x}{L}\right) a_n$$

Thus, all of the coefficients a_n are zero, and, from equation (4.25), the coefficients b_n are given by

$$b_n = \frac{2}{L} \int_0^L y(x, 0) \sin \frac{n\pi x}{L} dx$$

$$= \frac{4h}{L^2} \left(\int_0^{L/2} x \sin \frac{n\pi x}{L} dx + \int_{L/2}^L (L - x) \sin \frac{n\pi x}{L} dx \right)$$

To do the first integral, we integrate by parts:

$$\int_0^{L/2} x \sin \frac{n\pi x}{L} dx = x \frac{L}{n\pi} \left(-\cos \frac{n\pi x}{L} \right) \Big|_0^{L/2} + \int_0^{L/2} \frac{L}{n\pi} \cos \frac{n\pi x}{L} dx$$

$$= \frac{L^2}{2n\pi} \left(-\cos \frac{n\pi}{2} \right) + \left(\frac{L}{n\pi} \right)^2 \sin \frac{n\pi}{2}$$

The cosine term is zero unless n is even, and the sine term is zero unless n is odd. The second integral is

$$L \int_{L/2}^L \sin \frac{n\pi x}{L} dx = \frac{L^2}{n\pi} \left(-\cos \frac{n\pi x}{L} \right) \Big|_{L/2}^L = \frac{L^2}{n\pi} \left(\cos \frac{n\pi}{2} - (-1)^n \right)$$

and the last integral is

$$- \int_{L/2}^L x \sin \frac{n\pi x}{L} dx = x \frac{L}{n\pi} \left(\cos \frac{n\pi x}{L} \right) \Big|_{L/2}^L - \int_{L/2}^L \frac{L}{n\pi} \cos \frac{n\pi x}{L} dx$$

$$= \frac{L^2}{2n\pi} \left(2(-1)^n - \cos \frac{n\pi}{2} \right) + \left(\frac{L}{n\pi} \right)^2 \sin \frac{n\pi}{2}$$

Adding the three terms, we get the result

$$b_n = \frac{4h}{n\pi} \left[-\frac{1}{2} \cos \frac{n\pi}{2} + \frac{1}{n\pi} \sin \frac{n\pi}{2} + \cos \frac{n\pi}{2} - (-1)^n \right.$$

$$\left. + \frac{1}{2} \left(2(-1)^n - \cos \frac{n\pi}{2} \right) + \frac{1}{n\pi} \sin \frac{n\pi}{2} \right]$$

$$= \frac{8h}{(n\pi)^2} \sin \frac{n\pi}{2}$$

which is zero unless n is odd, in which case the result is

$$b_n = \frac{8h}{(n\pi)^2} (-1)^{(n-1)/2} = \frac{8h}{(2p + 1)^2 \pi^2} (-1)^p$$

where we wrote $n = 2p + 1$. Then the solution for y is

$$y(x, t) = \frac{8h}{\pi^2} \sum_{p=0}^{\infty} \frac{(-1)^p}{(2p + 1)^2} \sin \left((2p + 1) \frac{\pi x}{L} \right) \cos \left((2p + 1) \frac{\pi vt}{L} \right)$$

Figure 4.10 shows the first four terms of this series. The plot shows the dimensionless variables[7] y/h versus x/L at times $vt/L = 0, 1/4, 1/2$, and 1.

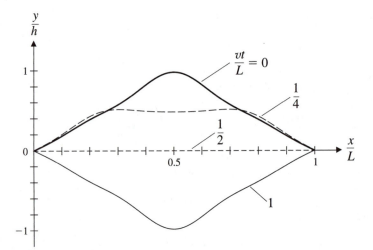

FIGURE 4.10. The string displacement at times $t = L/v$ times 0 (thick solid line), 1/4 (dashed line), 1/2 (short dashes), and 1 (thin solid line). The first four terms of the series were used to make this plot.

4.5. CONVERGENCE OF FOURIER SERIES

Now that we understand why Fourier series are useful, we should worry a little bit about how well they converge.

4.5.1. Pointwise Convergence

A function is very smooth if at least the first two derivatives of the function exist and the second derivative is continuous. It is piecewise very smooth if we can divide the interval $(-L, L)$ into a finite number of subintervals and the derivatives exist within each of the pieces. The step function (4.23) satisfies this requirement, since the derivatives exist within each of the two subintervals $(-L, 0)$ and $(0, L)$, even though the derivatives do not exist at $x = 0$.

The Fourier series of a function $f(x)$ that is piecewise very smooth on an interval $-L < x < L$ converges pointwise to $\frac{1}{2}[f(x + 0) + f(x - 0)]$ for each x in the interval.

---

[7]See Chapter 3, Section 3.4.1.

At the endpoints, we identify $x = -L$ and $x = L$ to obtain the limit $\frac{1}{2}[f(-L+0) + f(L-0)]$.

Fourier series are not, in general, uniformly convergent. They are uniformly convergent only in closed subintervals where $f(x)$ is continuous. But Fourier series *can* be integrated and differentiated term by term. We'll see why in Chapter 6.

4.5.2. Convergence in the Mean

Fourier series converge to the corresponding function $f(x)$ *in the mean*. That is, the square deviation

$$R_N = \int_{-L}^{L} \left| f(x) - \sum_{n=-N}^{N} c_n e^{in\pi x/L} \right|^2 dx \to 0 \quad \text{as } N \to \infty$$

(This idea should remind you of a least-squares fit of a model to a set of data points.) The function and the series may differ substantially at a finite number of isolated points and still converge in the mean. We have already seen this happen with the step function series (Figure 4.1), where the series always takes on the value at the middle of each step, independent of the value of the function at that point.

The Fourier series of any piecewise continuous function converges in the mean. This condition is sufficient but not necessary; that is, there are functions that behave more badly yet whose series also converge in the mean. As physicists, we do not need to concern ourselves with such functions because they do not arise in the solution of physics problems.

Now let's ask if we can do better than the series we have. Is there any set of coefficients c_n for which the remainder R_N is smaller than that given by the Fourier coefficients?

Let

$$R_N = \int_{-L}^{L} \left| f(x) - \sum_{n=-N}^{N} c_n e^{in\pi x/L} \right|^2 dx \tag{4.30}$$

where the c_n are unknown. Then, to minimize R_N, we choose the coefficients c_k so that

$$\frac{\partial R_N}{\partial c_k} = 0$$

$$\frac{\partial}{\partial c_k} \int_{-L}^{L} \left(f(x) - \sum_{n=-N}^{N} c_n e^{in\pi x/L} \right) \left(f^*(x) - \sum_{m=-N}^{N} c_m^* e^{-im\pi x/L} \right) dx = 0$$

If the sum $\sum_{n=-N}^{N} c_n e^{in\pi x/L}$ is real, then $c_m^* = c_{-m}$, and if $f(x)$ is also real, the derivative is

$$\int_{-L}^{L} \left(-2f e^{ik\pi x/L} + 2e^{ik\pi x/L} \sum_{m=-N}^{N} c_{-m} e^{-im\pi x/L} \right) dx = 0$$

$$\int_{-L}^{L} -f e^{ik\pi x/L} dx + \sum_{m=-N}^{N} \int_{-L}^{L} e^{ik\pi x/L} c_{-m} e^{-im\pi x/L} dx = 0$$

In the second term, every term in the sum integrates to zero except the one with $m = k$. Thus, we have

$$\int_{-L}^{L} f e^{ik\pi x/L} dx = 2L c_{-k}$$

and so the value of c_k that minimizes R_N is

$$c_k = \frac{1}{2L} \int_{-L}^{L} f e^{-ik\pi x/L} dx \tag{4.31}$$

which is the usual Fourier coefficient (equation 4.18). Thus, the Fourier coefficients are optimum in the sense that they make the square deviation R_N a minimum.

We can use the result that $R_N \to 0$ to prove another important result about the Fourier coefficients. We begin by expanding expression (4.30):

$$R_N = \int_{-L}^{L} \left(f(x) - \sum_{n=-N}^{N} c_n e^{in\pi x/L} \right) \left(f(x)^* - \sum_{m=-N}^{N} c_m^* e^{-im\pi x/L} \right) dx$$

$$= \int_{-L}^{L} |f(x)|^2 \, dx - \int_{-L}^{L} \sum_{n=-N}^{N} [f(x)^* c_n + f(x) c_{-n}^*] e^{in\pi x/L} dx$$

$$+ \int_{-L}^{L} \sum_{n=-N}^{N} c_n e^{in\pi x/L} \sum_{m=-N}^{N} c_m^* e^{-im\pi x/L} dx$$

The last term may be rearranged to give

$$\sum_{n=-N}^{N} \sum_{m=-N}^{N} c_n c_m^* \int_{-L}^{L} e^{in\pi x/L} e^{-im\pi x/L} dx$$

and the integral is zero unless $m = n$, when it equals $2L$. Thus, the last term is

$$2L \sum_{n=-N}^{N} c_n c_n^* = 2L \sum_{n=-N}^{N} |c_n|^2$$

We can also simplify the middle term, since

$$\int_{-L}^{L} f(x) \sum_{n=-N}^{N} c_{-n}^* e^{in\pi x/L} dx = \sum_{n=-N}^{N} c_{-n}^* \int_{-L}^{L} f(x) e^{in\pi x/L} dx$$

$$= \sum_{n=-N}^{N} c_{-n}^* 2L c_{-n} = 2L \sum_{n=-N}^{N} |c_n|^2$$

and, similarly,

$$\int_{-L}^{L} \sum_{n=-N}^{N} f(x)^* c_n e^{in\pi x/L} dx = 2L \sum_{n=-N}^{N} c_n c_n^* = 2L \sum_{n=-N}^{N} |c_n|^2$$

Combining the last two terms, we have

$$R_N = \int_{-L}^{L} |f(x)|^2 \, dx - 2L \sum_{n=-N}^{N} |c_n|^2$$

Since $R_N \geq 0$ by its definition,

$$\frac{1}{2L} \int_{-L}^{L} |f(x)|^2 \, dx \geq \sum_{n=-N}^{N} |c_n|^2$$

which implies that the series on the right converges. Thus, for a series that converges in the mean, $R_N \to 0$ as $N \to \infty$, and so

$$\frac{1}{2L} \int_{-L}^{L} |f(x)|^2 \, dx = \sum_{n=-\infty}^{\infty} |c_n|^2 \qquad (4.32)$$

This is one version of *Parseval's theorem.*[8]

We can understand this result by thinking about the energy stored in the capacitor in the circuit of Example 4.3. Averaged over one period, the energy stored in the capacitor is

$$U = \frac{1}{\tau} \int_{0}^{\tau} \frac{Q(t)^2}{2C} dt = \frac{1}{2C} \sum_{n} |q_n|^2$$

From this we conclude that the time-averaged energy equals the sum of the energies in each of the individual Fourier components.

PROBLEMS

1. Show that the Fourier series (equation 4.1) for a function $f(x)$ may be written

$$f(x) = \sum_{n=0}^{\infty} k_n \cos(nx + \phi_n)$$

and find expressions for k_n and ϕ_n.

---

[8]See Problem 21 for Parseval's theorem for the sine and cosine coefficients, and see Problem 22 for an additional variant.

2. Develop the full Fourier series for the function $f(x) = x$
   (a) over the range $0 \leq x \leq 1$
   (b) over the range $-1 \leq x \leq 1$
   (c) Make a plot showing the original function and the sum of the first three nonzero terms in each series. Comment on the similarities and differences between the two series.

3. Develop the Fourier series for the function $f(x) = x^2$ over the range $0 \leq x \leq 1$.

4. An odd function $f(x)$ on the range $(-L, L)$ has the additional property that $f(x+L) = -f(x)$.
   (a) Make a sketch showing the important features of this function.
   (b) Which kind of Fourier series (sine series, cosine series, or full series) represents this function on the range $-L \leq x \leq L$?
   (c) Show that the series has only terms of odd order $(n = 2m + 1)$, and find a formula for the coefficients as an integral over the range $0 \leq x \leq L/2$.
   (d) How does your answer change if $f(x + L) = +f(x)$?
   (e) How do your answers change if the function is even, but $f(x + L) = -f(x)$?

5. Which series, the sine series or the cosine series, do you expect will converge more rapidly to the function $f(x) = x^3$ on the range $0 < x < 1$? Give reasons for your answer. Evaluate the first four nonzero terms in the optimum series. How large is the fractional deviation $\left| \dfrac{S_4 - f(x)}{f(x)} \right|$ at $x = 0.5$ and $x = 1$? In this expression, S_4 is the sum of the first four terms.

6. Find the Fourier series on the range $0 \leq x \leq 2\pi$ for the function $f(x) = \sin \alpha x$, where α is *not* an integer. Check your result by evaluating the limit $\alpha \to n$. With the value $\alpha = 0.7$, plot the original function and the first three terms of your series on the range $0 \leq x \leq 2\pi$. Comment.

7. Find an exponential Fourier series for the function $\sinh \alpha x$ on the range $0 \leq x \leq 2\pi$. By combining terms, rewrite your answer as a series in sines and cosines.

8. Find the first four nonzero terms in a Fourier series for the function $\tan x$ on the range $-\pi/4 \leq x \leq \pi/4$.

9. Use numerical integration to find the first ten terms in a Fourier series for the function $\sin x^2$ on the range $0 < x < \pi$. Discuss the percent error between your series and the function $\sin x^2$ over the given range.

10. Find a Fourier series for the ramp function

$$ f(x) = \begin{cases} x & \text{if } 0 < x < 1 \\ 1 & \text{if } 1 < x < 2 \end{cases} $$

on the interval $0 < x < 2$.

11. An electric circuit contains a 3 mH inductor, a 50 μF capacitor, and a 200 Ω resistor in series with a power supply that supplies a rectified sine wave voltage (see the figure) with amplitude 110 V and period 2 ms. Determine the capacitor voltage as a Fourier series.

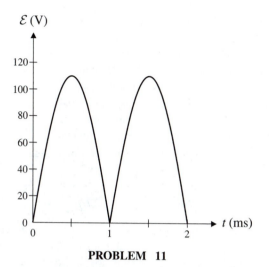

PROBLEM 11

12. A single-loop series *LRC* circuit has resistance $R = 15\ \Omega$, inductance $L = 10$ mH, and capacitance $C = 1.5\ \mu$F. A half-rectified sine wave power supply (see the figure) with period $T = 1.57 \times 10^{-3}$ s is attached to the circuit. Find the voltage across the capacitor as a Fourier series in time once the circuit has reached a steady state.

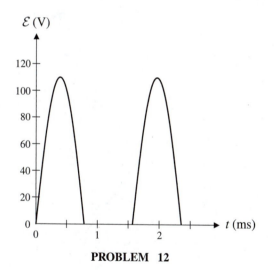

PROBLEM 12

13. A spring-and-dashpot system satisfies the equation $\dfrac{d^2x}{dt^2} + \alpha\dfrac{dx}{dt} + k^2x = f(t)$. The system is driven by a periodic driving force with period T:

$$f(t) = \begin{cases} at & \text{if } 0 < t \le T/2 \\ a(T - t) & \text{if } T/2 \le t < T \end{cases}$$

Find the response of the system $x(t)$ as a Fourier series.

14. A simply supported beam of length L bears a load W that is uniformly distributed over the first $1/4$ of its length. Determine the deflection of the beam as a Fourier series. Make plots showing the first one, two, and three terms of your answer. How many terms are needed to obtain a result accurate to 1%? (The differential equation satisfied by the beam deflection is equation 3.11, and the displacement is zero at the two ends.)

15. A beam rests on supports at its ends, $x = 0$ and $x = L$. The load $q(x)$ varies linearly along the beam: $q = ax$. What are the boundary conditions? Find the displacement of the beam as a Fourier series. Plot your results, and comment.

16. A guitar string of length $L = 65$ cm is plucked by pulling it to the shape

$$y(x, 0) = \begin{cases} ax^2 & \text{if } 0 < x < L/3 \\ (a/4)(L - x)^2 & \text{if } L/3 < x < L \end{cases}$$

and then letting go. Determine the subsequent motion of the string. Which harmonics are excited? Plot the string displacement as a function of x for $t = 0, 0.4$, and 0.8 times L/v. Also plot the original string shape for comparison. Comment.

17. A violin string is plucked to the shape of a triangle with apex one-quarter of the way along the string, as shown in the figure, and then let go. Find the displacement of the string at later times. Plot your result (up to the $n = 10$ term) for $t = L/10v$, $L/5v$, and $L/2v$, where $v = \sqrt{T/\mu}$ is the wave speed. Are all harmonics excited?

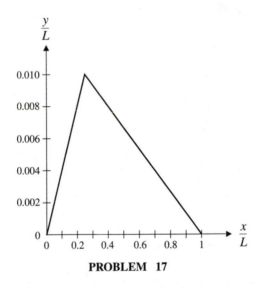

PROBLEM 17

18. A piano string of length L is hit by a hammer of length $\ell = L/10$. The hammer is centered at $x = L/4$, and the impulse it imparts is I. Assume that the impulse is uniformly distributed over the hammer's length. Determine the subsequent displacement of the string as a function of x and t. Which harmonics are excited? Plot the string displacement as a function of x for $t = 0.1, 0.5$, and 0.75 times L/v. Use the first five nonzero terms of the series to make the plot. Comment on your results.

19. Fourier series may be used to evaluate certain series of functions of integers. To illustrate the method, develop the Fourier series for the function x^2 on the range $-\pi$ to π. Set $x = 0$ and hence evaluate

$$\sum_{n=1}^{\infty} \frac{(-1)^{n-1}}{n^2} = 1 - \frac{1}{4} + \frac{1}{9} + \cdots$$

Which sum do you obtain by setting $x = \pi$? Finally, use Parseval's theorem to evaluate

$$\sum_{n=1}^{\infty} \frac{1}{n^4}$$

20. Use the Fourier series for the step function to evaluate the sum

$$\sum_{m=0}^{\infty} \frac{(-1)^m}{2m + 1}$$

(See Problem 19 for the method.) Use Parseval's theorem applied to the same series to obtain the sum

$$\sum_{m=0}^{\infty} \frac{1}{(2m + 1)^2}$$

21. A function $f(x)$ is represented by the Fourier series

$$f(x) = \sum_{n=0}^{\infty} \left(a_n \sin \frac{n\pi x}{L} + b_n \cos \frac{n\pi x}{L} \right)$$

on the range $(-L, L)$. Derive a form of Parseval's theorem (equation 4.32) applicable to this series; that is, express $\int_{-L}^{+L} f(x)^2 dx$ in terms of the coefficients a_n and b_n.

22. If $f(x)$ is represented by the series $\sum f_n e^{inx}$ over the interval $0 < x < 2\pi$, and $g(x) = \sum g_n e^{inx}$ over the same range, prove the generalized Parseval theorem:

$$\frac{1}{2\pi} \int_0^{2\pi} f(x)g(x)\, dx = \sum_{n=-\infty}^{\infty} f_n g_{-n} = \sum_{n=-\infty}^{\infty} f_n g_n^*$$

where the second expression applies when the function $g(x)$ is real.

23. The capacitor shown in the figure on the next page is charged by the battery and discharges through the bulb when the potential across it equals $0.9V$. (See, for example, Lea and Burke, Example 31.3, p. 993.) Assuming that the capacitor discharges very rapidly, show that the potential across the capacitor as a function of time is

$$V_C = V \left(1 - e^{-t/RC} \right), \quad 0 < t < RC \ln 10$$

and repeats periodically with period $T = RC \ln 10$. Find a Fourier series with period T that represents this function.

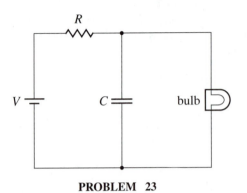

PROBLEM 23

24. A rectangular box of dimensions $a \times b \times a$ has conducting walls. All the walls are grounded except for the one at $y = b$. This wall is separated from the others by a thin insulating strip, and it is at potential V. Using the method illustrated in Chapter 3, Example 3.15, find the potential everywhere inside the box.

25. A rectangular box measuring $a \times b \times c$ has all its walls at temperature T_1 except for the one at $z = c$, which is held at temperature T_2. When the box comes to equilibrium, the temperature function $T(x, y, z)$ satisfies equation 3.14 (extended to three spatial dimensions):

$$\frac{\partial T}{\partial t} = D\nabla^2 T$$

with the time derivative on the left equal to zero. Using the method of Chapter 3, Example 3.15, find the temperature T in the box in the form

$$T(x, y, z) = T_1 + \tau(x, y, z)$$

where τ is expressed in a Fourier series

$$\tau(x, y, z) = \sum_{n,m} a_{nm} \sin \frac{n\pi x}{a} \sin \frac{m\pi y}{b} f(z)$$

Find the function $f(z)$ and the coefficients a_{mn}.

26. An infinitely long conducting tube with circular cross section of radius a is divided into four equal pieces by insulating strips running along its length. One of the four pieces is at potential V, and the other three are grounded. Solve Laplace's equation in two dimensions, using the method of Example 3.15. Evaluate the solution at $\rho = a$ and show that the result is a Fourier series. Determine the coefficients, and hence find the potential everywhere inside the tube.

27. A Fourier series of the form

$$f(x) = \sum c_n e^{inx}$$

may be expressed as a power series

$$f(x) = \sum c_n z^n$$

where $z = \lim_{r \to 1} re^{ix}$ and $r \leq 1$. The function $f(x)$ may be identified by summing the power series. Use this technique to sum the Fourier series

$$\sum_{n=1}^{\infty} \frac{\sin nx}{n}$$

where $0 < x < \pi$. Check your result by evaluating the Fourier sine series of the function you found.

CHAPTER 5

Laplace Transforms

5.1. DEFINITION OF THE LAPLACE TRANSFORM

An integral transform allows us to convert an inhomogeneous linear differential equation to an algebraic equation that is easier to solve. We shall look in detail at two such transforms: the Laplace transform and the Fourier transform (Chapter 7).

The Laplace transform is an integral operator applied to a function $f(t)$ that is defined for $0 < t < \infty$:

$$\mathcal{L}(f) = F(s) = \int_0^\infty f(t)\, e^{-st}\, dt \tag{5.1}$$

The factor e^{-st} in the integrand causes the integral to converge for a very large class of functions f.

A function f is of *exponential order* σ_0 if there exists a real positive constant M such that

$$\left| e^{-\sigma_0 t} f(t) \right| \le M \tag{5.2}$$

for all t, $0 < t < \infty$, and σ_0 is real.

The Laplace transform exists for all piecewise continuous functions of exponential order. The transform is defined for Re $(s) > \sigma_0$.

Let's look at some examples.

1. Exponential functions: $f(t) = e^{\alpha t}$. The function $e^{\alpha t}$ is of exponential order α—we can take $\sigma_0 = \alpha$ and $M = 1$ in equation (5.2). The transform is

$$F(s) = \int_0^\infty e^{\alpha t} e^{-st} dt = \frac{1}{\alpha - s} \left. e^{(\alpha - s)t} \right|_0^\infty$$

The transform exists for Re $(s) > \alpha$, in which case

$$F(s) = \frac{1}{s - \alpha}$$

2. Powers: $f(t) = t^m$. The Taylor series for the exponential function,

$$e^{\varepsilon t} = 1 + \varepsilon t + \frac{\varepsilon^2 t^2}{2} + \cdots + \frac{\varepsilon^m t^m}{m!} + \cdots$$

shows that

$$t^m < \frac{m!}{\varepsilon^m} e^{\varepsilon t}$$

and so

$$\left| t^m e^{-\varepsilon t} \right| < \frac{m!}{\varepsilon^m} e^{\varepsilon t} e^{-\varepsilon t} = \frac{m!}{\varepsilon^m}$$

for any positive real number ε. Thus, these functions are of exponential order ε. With $m = 0$, the transform is

$$\mathcal{L}(1) = \int_0^\infty e^{-st} dt = -\frac{e^{-st}}{s} \Big|_0^\infty = \frac{1}{s} \quad \text{if Re } (s) > 0$$

For $m > 1$, we use integration by parts. For $m = 1$,

$$\mathcal{L}(t) = \int_0^\infty t e^{-st} dt = -\frac{t e^{-st}}{s} \Big|_0^\infty + \frac{1}{s} \int_0^\infty e^{-st} dt$$

$$= -\frac{e^{-st}}{s^2} \Big|_0^\infty = \frac{1}{s^2} \quad \text{if Re } (s) > 0$$

Then, in general,

$$\mathcal{L}(t^m) = \int_0^\infty t^m e^{-st} dt = -\frac{t^m e^{-st}}{s} \Big|_0^\infty + \frac{1}{s} \int_0^\infty m t^{m-1} e^{-st} dt$$

$$= \frac{m}{s} \mathcal{L}(t^{m-1}) = \frac{m(m-1)}{s^2} \mathcal{L}(t^{m-2}) = \cdots = \frac{m!}{s^{m+1}}$$

More powerfully, we can use the definition of the gamma function (equation 2.75). Make a change of variable to $u = st$. Then

$$\mathcal{L}(t^p) = \int_0^\infty t^p e^{-st} dt = \int_0^\infty \left(\frac{u}{s}\right)^p e^{-u} \frac{du}{s}$$

$$= \frac{1}{s^{p+1}} \int_0^\infty u^p e^{-u} du = \frac{\Gamma(p+1)}{s^{p+1}}$$

which is valid for Re $(s) > 0$ and $p > 0$. The power p does not have to be an integer.

3. Sines and cosines. These functions are easily taken care of, since they can be written as linear combinations of exponentials:

$$\mathcal{L}(\cos \omega t) = \int_0^\infty \frac{(e^{i\omega t} + e^{-i\omega t})}{2} e^{-st} dt$$

$$= \frac{1}{2}\left(\frac{1}{s - i\omega} + \frac{1}{s + i\omega}\right) = \frac{s}{s^2 + \omega^2}$$

which is valid for Re $(s) > 0$.

We can now begin to compile a table of transforms (Table 5.1). Each transform exists when the real part of s exceeds some minimum value s_0.

TABLE 5.1. Laplace Transforms

$f(t)$	$F(s)$	s_0	$f(t)$	$F(s)$	s_0
$e^{\alpha t}$	$\dfrac{1}{s - \alpha}$	α	$\cos \omega t$	$\dfrac{s}{s^2 + \omega^2}$	0
1	$\dfrac{1}{s}$	0	$\sin \omega t$	$\dfrac{\omega}{s^2 + \omega^2}$	0
t^p	$\dfrac{\Gamma(p + 1)}{s^{p+1}}$	0			

This table gives transforms of the most common functions, but we will need methods that allow us to find the function $f(t)$ corresponding to any transform $F(s)$. This process is called inverting the transform. Since the Laplace transform is an integral operator, we can change the value of the function $f(t)$ at a finite number of points without changing the value of the integral (5.1). Thus, the inverted transform cannot be unique. This does not usually present any difficulties in the solution of physics problems, as we shall see. Usually we choose a continuous function as the appropriate inverse.

5.2. SOME BASIC PROPERTIES OF THE TRANSFORM

1. The Laplace transform is a *linear operator*. This means that

$$\mathcal{L}(af) = \int_0^\infty af(t) e^{-st} dt = a \int_0^\infty f(t) e^{-st} dt = a\mathcal{L}(f)$$

where a is a constant, and

$$\mathcal{L}(f_1 + f_2) = \int_0^\infty (f_1 + f_2) e^{-st} dt = \int_0^\infty f_1(t) e^{-st} dt + \int_0^\infty f_2(t) e^{-st} dt$$

$$= \mathcal{L}(f_1) + \mathcal{L}(f_2)$$

2. The transform of the derivative df/dt may be found using integration by parts:

$$\mathcal{L}\left(\frac{df}{dt}\right) = \int_0^\infty \frac{df}{dt} e^{-st} dt = f e^{-st}\Big|_0^\infty - \int_0^\infty (-s) f(t) e^{-st} dt$$

If f is of exponential order s_0, the integrated term approaches zero at infinity for $\text{Re}(s) > s_0$. Then

$$\mathcal{L}\left(\frac{df}{dt}\right) = -f(0) + s F(s) \tag{5.3}$$

Then the transform of any higher derivative may be found by iteration:

$$\mathcal{L}\left(\frac{d^2 f}{dt^2}\right) = \mathcal{L}\left[\frac{d}{dt}\left(\frac{df}{dt}\right)\right] = -\frac{df}{dt}\Big|_0 + s\mathcal{L}\left(\frac{df}{dt}\right)$$

$$= -\frac{df}{dt}\Big|_0 - sf(0) + s^2 F(s)$$

and, in general,

$$\mathcal{L}\left(\frac{d^m f}{dt^m}\right) = s^m F(s) - \sum_{n=1}^{m} \frac{d^{m-n} f}{dt^{m-n}}\Big|_0 s^{n-1} \tag{5.4}$$

3. The *attenuation property*: If a function $f(t)$ is attenuated (or reduced) by the factor e^{-at}, then

$$\mathcal{L}(e^{-at} f) = \int_0^\infty e^{-at} f(t) e^{-st} dt = \int_0^\infty f(t) e^{-(s+a)t} dt$$

$$\mathcal{L}(e^{-at} f) = F(s + a) \tag{5.5}$$

An inversion may be simplified if we can recognize the transform as a simpler function of $(s + a)$ for some constant a.

Example 5.1.   Invert the transform $F(s) = 1/(s^2 + 2as + a^2)$.

We recognize the denominator as $(s + a)^2$. The function $F(\sigma) = 1/\sigma^2$ inverts to give $f(t) = t$ (Table 5.1: powers), and so the inverse of $F(s)$ is te^{-at}.

4. The *shifting property* is closely related to the attenuation property. Suppose we shift the function $f(t)$ along the t-axis by an amount t_0. Since the original function is defined only for $t > 0$, we must cut the shifted function off at $t = t_0$ (Figure 5.1) or, equivalently, at $t - t_0 = 0$. (The shifted function equals zero for $t < t_0$.) We can do this by multiplying the shifted function by a step function $S(t - t_0)$:

$$\text{shifted } f = S(t - t_0)f(t - t_0)$$

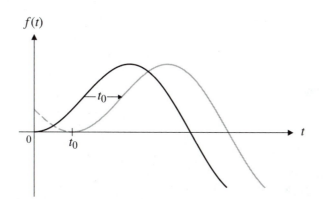

FIGURE 5.1. Shifting a function. The shifted function is zero for $t < t_0$, but $f(t - t_0)$ is not, so we must cut off the dashed portion with the step function $S(t - t_0)$.

Then the transform of the shifted function is given by

$$\mathcal{L}[S(t - t_0)f(t - t_0)] = \int_0^\infty S(t - t_0)f(t - t_0)\, e^{-st}\, dt = \int_{t_0}^\infty f(t - t_0)\, e^{-st}\, dt$$

Now change variables to $u = t - t_0$. The limits become $u = 0$ and ∞.

$$\mathcal{L}[S(t - t_0)f(t - t_0)] = \int_0^\infty f(u)\, e^{-s(u+t_0)}\, du = e^{-st_0}\int_0^\infty f(u)\, e^{-su}\, du$$

$$\mathcal{L}[S(t - t_0)f(t - t_0)] = e^{-st_0} F(s), \quad t_0 > 0 \tag{5.6}$$

Thus, a transform that contains an exponential factor is the transform of a shifted function. In this result, the parameter t_0 must be positive.

5.3. USE OF THE LAPLACE TRANSFORM TO SOLVE A DIFFERENTIAL EQUATION

If we have a linear differential equation of the form[1]

$$A\frac{d^2 y}{dt^2} + B\frac{dy}{dt} + Cy = f(t)$$

---

[1] The method is not limited to second-order equations (see Example 5.3). A second-order equation is used here for illustration.

we may apply the Laplace transform operator to the entire equation. Then we can use the linearity property to write

$$A\mathcal{L}\left(\frac{d^2y}{dt^2}\right) + B\mathcal{L}\left(\frac{dy}{dt}\right) + C\mathcal{L}(y) = F(s)$$

and then use the derivative property:

$$A\left(-\left.\frac{dy}{dt}\right|_0 - sy(0) + s^2Y(s)\right) + B[-y(0) + sY(s)] + CY(s) = F(s)$$

This algebraic equation is easily solved for the transform $Y(s)$:

$$Y(s) = \frac{F(s) + By(0) + A[dy/dt|_0 + sy(0)]}{As^2 + Bs + C} \tag{5.7}$$

The problem of solving the original differential equation is thus reduced to the problem of inverting the transform $Y(s)$. In many cases, it is not too difficult to invert the transform, and this is the advantage of the transform method.

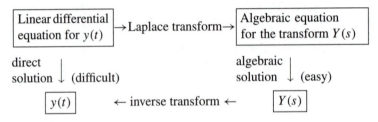

Because of the initial values of y and its derivatives that appear in expression (5.7) for the transform Y, the Laplace transform is well suited to the solution of initial value problems.

Example 5.2. A series LRC circuit has a constant emf $\mathcal{E}$ and a switch. The switch has been open for a long time and is then closed at $t = 0$. What is the current in the circuit for $t > 0$?

If the switch has been open for a long time, we are to assume that the current has dropped to zero and the capacitor is uncharged with $q(0) = 0$. Then we can define the current variable $i(t)$ so that

$$i = \frac{dq}{dt} \tag{5.8}$$

as shown in Figure 5.2. (It is very important that you take care in defining the algebraic variables corresponding to the physical quantities of charge and current. It is easy to make sign errors that will cause your solution to be incorrect. See, for example, Lea and Burke, p. 990.) Then Kirchhoff's loop rule applied to the circuit after the switch is closed gives the differential equation

$$L\frac{di}{dt} + Ri + \frac{q}{C} = \mathcal{E} \tag{5.9}$$

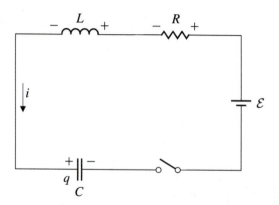

FIGURE 5.2. Definition of charge (i) and current (q) variables in Example 5.1. The switch is closed at $t = 0$.

Now we transform both equations (5.8) and (5.9):

$$I = L\left(\frac{dq}{dt}\right) = sQ - q(0)$$

and

$$L[sI - i(0)] + RI + \frac{Q}{C} = \frac{\mathcal{E}}{s}$$

Use the initial conditions $q(0) = 0$ and $i(0) = 0$, and substitute for $Q(s)$:

$$LsI + RI + \frac{1}{C}\left(\frac{I}{s}\right) = \frac{\mathcal{E}}{s}$$

Solve for I:

$$I = \frac{\mathcal{E}/s}{Ls + R + (1/Cs)}$$

$$= \frac{\mathcal{E}}{L}\left(\frac{1}{\left(s^2 + \frac{R}{L}s + \frac{1}{LC}\right)}\right) \tag{5.10}$$

With the usual definitions $\alpha \equiv R/2L$ and $\omega_0^2 \equiv 1/LC$, we have

$$I(s) = \frac{\mathcal{E}}{L}\left(\frac{1}{s^2 + 2\alpha s + \omega_0^2}\right)$$

There are several possible ways to invert the transform. We could factor the denominator and then use partial fractions. The roots of the denominator are

$$s_\pm = -\alpha \pm \sqrt{\alpha^2 - \omega_0^2} = -\alpha \pm i\sqrt{\omega_0^2 - \alpha^2} = -\alpha \pm i\omega$$

where we have defined $\omega \equiv \sqrt{\omega_0^2 - \alpha^2}$. Then

$$I(s) = \frac{\mathcal{E}}{L}\left(\frac{1}{(s - s_+)(s - s_-)}\right) = -\frac{\mathcal{E}}{L(s_+ - s_-)}\left(\frac{1}{s - s_+} - \frac{1}{s - s_-}\right)$$

From Table 5.1, we can invert each fraction to get an exponential:

$$i(t) = \frac{\mathcal{E}}{2i\omega L}(e^{s_+ t} - e^{s_- t}) = \frac{\mathcal{E}}{2i\omega L}e^{-\alpha t}(e^{+i\omega t} - e^{-i\omega t})$$

$$= \frac{\mathcal{E}}{\omega L}e^{-\alpha t}\sin(\omega t)$$

As t increases from 0, immediately after the switch is closed, the current is positive; that is, it is charging the capacitor.

An alternative method uses the attenuation property (equation 5.5). First we complete the square in the denominator, which may be expressed in terms of $s + \alpha$:

$$I = \frac{\mathcal{E}}{L}\left(\frac{1}{(s + \alpha)^2 + \omega_0^2 - \alpha^2}\right) = \frac{\mathcal{E}}{\omega L}\left(\frac{\omega}{(s + \alpha)^2 + \omega^2}\right)$$

Now, from Table 5.1, we recognize $\omega/(s^2 + \omega^2)$ as the transform of $\sin(\omega t)$. Our transform has $(s + \alpha)^2$ rather than s^2, which indicates that the inverse is the sine function multiplied by $e^{-\alpha t}$. Thus, the solution is

$$i(t) = \frac{\mathcal{E}}{\omega L}e^{-\alpha t}\sin(\omega t)$$

as before.

Even if we do not have values of the function and all its derivatives at $t = 0$, we can sometimes leave the unknown values in the solution and use the remaining boundary conditions to eliminate the unknowns at the end.

Example 5.3.   A simply supported beam of length L is one that rests on a support at each end (Figure 5.3). If the beam supports a load Mg that is uniformly distributed over a distance ℓ at the center of the beam, find the displacement of the beam.

We set up a coordinate system with x-axis along the beam and origin at one end. Then the given boundary conditions are

$$y(0) = 0; \quad y(L) = 0$$

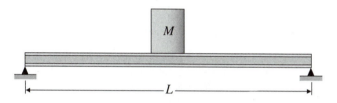

FIGURE 5.3. A simply supported beam of length L carrying a load Mg.

The differential equation satisfied by the beam is equation (3.11):

$$\frac{d^4 y}{dx^4} = \frac{1}{EI} q(x) \tag{5.11}$$

With a fourth-order equation, we need four boundary conditions. We can find the remaining two boundary conditions that we need from equations (3.8)–(3.10). In equilibrium, the torque[2] about any origin must be zero, and thus $m(0) = 0$. Also, $m(L) = 0$ since there are no forces to the right of the right end. Thus, we know two quantities at 0, y and y'', and two at $x = L$.

The load function is

$$q(x) = \begin{cases} Mg/\ell & \text{if } (L - \ell)/2 < x < (L + \ell)/2 \\ 0 & \text{otherwise} \end{cases}$$

Now we transform equation (5.11):

$$s^4 Y - s^3 y(0) - s^2 y'(0) - s y''(0) - y'''(0) = \frac{1}{EI} Q(s)$$

where

$$Q(s) = \int_{\frac{L-\ell}{2}}^{\frac{L+\ell}{2}} \frac{Mg}{\ell} e^{-sx}\, dx = \frac{Mg}{-s\ell} \left[\exp\left(-\frac{L+\ell}{2}s\right) - \exp\left(-\frac{L-\ell}{2}s\right) \right]$$

$$= \frac{Mg}{s\ell} e^{-Ls/2} 2 \sinh \frac{\ell s}{2}$$

Put in the known initial conditions:

$$s^4 Y - s^2 y'(0) - y'''(0) = \frac{Mg}{EI s\ell} e^{-Ls/2} 2 \sinh \frac{\ell s}{2}$$

Thus,

$$Y(s) = \frac{Mg}{s^5 EI \ell} e^{-Ls/2} 2 \sinh \frac{\ell s}{2} + \frac{y'(0)}{s^2} + \frac{y'''(0)}{s^4}$$

[2]Recall that $m(x)$ is the counterclockwise torque due to all forces to the right of x.

The exponentials in the first term suggest that we apply the shifting property (equation 5.6) after inverting $1/s^5$. The powers of s invert to powers of x (Table 5.1). Thus,

$y(x) =$

$$\frac{Mg}{EI\ell}\left(\frac{1}{4!}\right)\left[\left(x - \frac{L-\ell}{2}\right)^4 S\left(x - \frac{L-\ell}{2}\right) - \left(x - \frac{L+\ell}{2}\right)^4 S\left(x - \frac{L+\ell}{2}\right)\right]$$

$$+ y'(0)x + y'''(0)\frac{x^3}{3!}$$

Now we need to apply the remaining boundary conditions to find the unknowns $y'(0)$ and $y'''(0)$. Evaluating the solution at $x = L$, we have

$$y(L) = \frac{Mg}{4!EI\ell}\left[\left(\frac{L+\ell}{2}\right)^4 - \left(\frac{L-\ell}{2}\right)^4\right] + y'(0)L + y'''(0)\frac{L^3}{6}$$

$$= \frac{Mg}{4!EI\ell}\left[\frac{1}{2}L\ell(L^2 + \ell^2)\right] + y'(0)L + y'''(0)\frac{L^3}{6} = 0$$

and

$$\left.\frac{d^2y}{dx^2}\right|_{x=L} = \frac{Mg}{2EI\ell}\left[\left(\frac{L+\ell}{2}\right)^2 - \left(\frac{L-\ell}{2}\right)^2\right] + y'''(0)L$$

$$= \frac{Mg}{2EI\ell}L\ell + y'''(0)L = 0$$

From this result we obtain

$$y'''(0) = -\frac{Mg}{2EI}$$

Then from $y(L) = 0$ it follows that

$$y'(0) = -\frac{Mg}{48EI}(L^2 + \ell^2) - y'''(0)\frac{L^2}{6} = \frac{Mg}{EI}\frac{(3L^2 - \ell^2)}{48}$$

So the solution is

$y(x) =$

$$\frac{Mg}{24EI}\left\{\frac{1}{\ell}\left[\left(x - \frac{L-\ell}{2}\right)^4 S\left(x - \frac{L-\ell}{2}\right) - \left(x - \frac{L+\ell}{2}\right)^4 S\left(x - \frac{L+\ell}{2}\right)\right]\right.$$

$$\left. + \left(\frac{3L^2 - \ell^2}{2}\right)x - 2x^3\right\}$$

The displacement of the beam in the case $\ell = L/2$ is plotted in Figure 5.4.

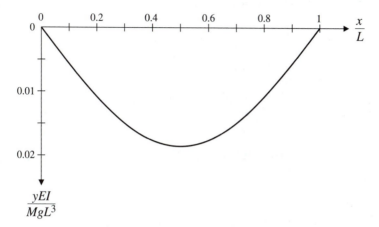

FIGURE 5.4. Deflection of the beam yEI/MgL^3 versus x/L (Example 5.3). Notice the very different scales on the two axes.

5.4. SOME ADDITIONAL USEFUL TRICKS

In this section, we shall compile a list of tricks that may help us to find or to invert a transform.

5.4.1. The Derivative of the Transform

Since the transform $F(s)$ is a smooth function of s, it may be differentiated:

$$\frac{dF}{ds} = \frac{d}{ds} \int_0^\infty f(t)\, e^{-st} dt = \int_0^\infty f(t)\frac{d}{ds} e^{-st} dt$$

$$= -\int_0^\infty tf(t)\, e^{-st} dt$$

Thus,

$$\mathcal{L}(tf) = -\frac{dF}{ds} \tag{5.12}$$

We can repeat this process n times to obtain

$$\mathcal{L}(t^n f) = (-1)^n \frac{d^n F}{ds^n} \tag{5.13}$$

Example 5.4. Find the transform of $t \sin \omega t$.

From Table 5.1, the transform of $\sin \omega t$ is $\omega/(s^2 + \omega^2)$. Thus,

$$\mathcal{L}(t \sin \omega t) = -\frac{d}{ds}\left(\frac{\omega}{s^2 + \omega^2}\right)$$

$$= \frac{2s\omega}{(s^2 + \omega^2)^2}$$

5.4.2. The Integral of the Transform

The transform may also be integrated:

$$\int_s^\infty F(\sigma)\, d\sigma = \int_s^\infty \int_0^\infty f(t)\, e^{-\sigma t}\, dt\, d\sigma$$

Now do the integration over σ first:

$$\int_s^\infty F(\sigma)\, d\sigma = \int_0^\infty f(t)\left.\frac{e^{-\sigma t}}{-t}\right|_s^\infty dt$$

$$= \int_0^\infty \frac{f(t)}{t} e^{-st}\, dt$$

$$\int_s^\infty F(\sigma)\, d\sigma = \mathcal{L}\left(\frac{f}{t}\right) \tag{5.14}$$

It is essential to this derivation that the upper limit of the integral over σ be infinite, so that the exponential goes to zero at the upper limit. It is also necessary that the transform of f/t exist—that is, that the integral converge. For example, we cannot apply this result to the function $f \equiv 1$, since the integral $\int_0^\infty (e^{-st}/t)\, dt$ does not exist.

Example 5.5. Find the transform of the function $f(t) = \dfrac{1 - \cos t}{t}$.

From Table 5.1 and the linearity of the transform, the transform of the function $1 - \cos t$ is

$$\frac{1}{s} - \frac{s}{s^2 + \omega^2}$$

Then

$$\mathcal{L}\left(\frac{1-\cos t}{t}\right) = \int_s^\infty \left(\frac{1}{\sigma} - \frac{\sigma}{\sigma^2+\omega^2}\right) d\sigma$$

$$= \left(\ln\sigma - \frac{1}{2}\ln(\sigma^2+\omega^2)\right)\bigg|_s^\infty$$

$$= \lim_{S\to\infty} \ln\frac{S}{\sqrt{S^2+\omega^2}} - \ln\frac{s}{\sqrt{s^2+\omega^2}}$$

$$= \ln 1 - \ln\frac{s}{\sqrt{s^2+\omega^2}}$$

$$= \ln\frac{\sqrt{s^2+\omega^2}}{s} = \frac{1}{2}\ln\left(1+\frac{\omega^2}{s^2}\right)$$

5.4.3. Periodic Functions

Suppose the function $f(t)$ is periodic with period T (Figure 5.5a). Then we can represent it as

$$f(t) = \begin{cases} g(t) & \text{if } 0 \le t < T \\ g(t-T) & \text{if } T \le t < 2T \\ g(t-2T) & \text{if } 2T \le t < 3T \end{cases}$$

and so on, where $g(t)$ is zero outside of the range $0 \le t < T$ (Figure 5.5b). Equivalently,

$$f(t) = g(t) + S(t-T)g(t-T) + S(t-2T)g(t-2T) + \cdots \qquad (5.15)$$

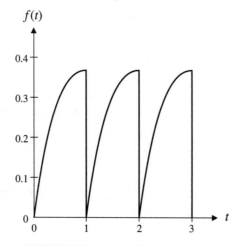

FIGURE 5.5a. A periodic function $f(t)$.

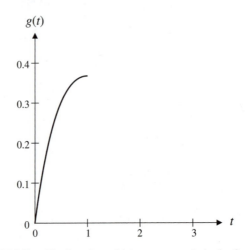

FIGURE 5.5b. The function $g(t)$ is nonzero only in the first period.

Now we can apply the shifting property to find the transform. First we transform g:

$$G(s) = \int_0^T g(t)\, e^{-st}\, dt$$

The upper limit is T, because the function g is zero for $t > T$. Next we use the shifting property (equation 5.6) to evaluate the transform of the second period of the function:

$$\mathcal{L}[S(t-T)g(t-T)] = e^{-sT} G$$

and so on. Thus, from equation (5.15),

$$F(s) = (1 + e^{-sT} + e^{-2sT} + \cdots)G$$

$$\boxed{F(s) = \frac{G}{1 - e^{-sT}}} \qquad (5.16)$$

where we used the known sum of the geometric series (equation 2.43). Thus, a denominator of the form $1 - e^{-sT}$ in the transform $F(s)$ is the signature of a periodic function $f(t)$ with period T.

Example 5.6.   Find the Laplace transform of a square wave of amplitude 1 and period T.

The function $g(t) = 1$ for $0 < t < T/2$ and is zero otherwise. Thus, the transform $G(s)$ is

$$G = \int_0^{T/2} e^{-st}\, dt = \left.\frac{-e^{-st}}{s}\right|_0^{T/2} = \frac{1 - e^{-sT/2}}{s}$$

Note that the upper limit of the integral is $T/2$, since the function $g(t) = 0$ for $T/2 < t < T$. Then, using equation (5.16), we have

$$F(s) = \frac{1}{s}\left(\frac{1 - e^{-sT/2}}{1 - e^{-sT}}\right) = \frac{1}{s(1 + e^{-sT/2})}$$

The denominator $1 - e^{-sT}$ is the signature of a periodic function, but it can be masked by the form of G, as we see here for the square wave. The same thing happens with sines and cosines—the prototypical periodic functions—for which $G \propto 1 - e^{-sT}$, as you should verify.

5.4.4. Discontinuous Functions

Suppose the function f has a discontinuity at $t = t_1$. Then we can find the transform F in the usual way. But we have to be more careful with the transform of the derivative, since the derivative does not exist at t_1. Thus, we divide the range of integration into two pieces and integrate each piece by parts:

$$\mathcal{L}\left(\frac{df}{dt}\right) = \int_0^{t_1} \frac{df}{dt}e^{-st}dt + \int_{t_1}^\infty \frac{df}{dt}e^{-st}dt$$

$$= fe^{-st}\Big|_0^{t_1} + s\int_0^{t_1} fe^{-st}dt + fe^{-st}\Big|_{t_1}^\infty + s\int_{t_1}^\infty fe^{-st}dt$$

$$= e^{-st_1}[f(t_1-) - f(t_1+)] - f(0) + sF(s)$$

$$= sF(s) - f(0) - e^{-st_1}h$$

where $h = f(t_1+) - f(t_1-)$ is the jump in the value of the function at t_1. The transforms of higher derivatives may be evaluated similarly.

5.5. CONVOLUTION

5.5.1. Systems with Memory

Certain physical systems have memory; that is, the behavior of the system at time t depends on what has happened to the system at times less than t. The series LRC circuit is an example of such a system, as we have seen in prior examples. The current in the circuit at time t depends not only on the emf $\mathcal{E}$ applied to the circuit at time t, but also on the past history of the circuit (when the switch was closed, for example). The current in the circuit at any time t is determined both by the structure of the circuit and by the emf $\mathcal{E}(t)$ applied to the circuit over time.

If we choose our current and charge variables as in Example 5.2, so that $i = +dq/dt$, Kirchhoff's loop rule results in the differential equation

$$L\frac{di}{dt} + Ri + \frac{q}{C} = \mathcal{E}(t)$$

Now we transform the equations to get

$$I = sQ - q(0)$$

and

$$L[sI - i(0)] + RI + \frac{Q}{C} = E(s)$$

Let's simplify by choosing $i(0) = q(0) = 0$. Then, combining the equations gives

$$s^2 QL + sRQ + \frac{Q}{C} = E(s)$$

or

$$Q = \frac{E(s)}{s^2 L + sR + 1/C} = E(s)R(s)$$

where

$$R(s) = \frac{1}{L\left(s^2 + \frac{R}{L}s + \frac{1}{LC}\right)} = \frac{1}{L}\left(\frac{1}{(s + \alpha)^2 + \omega^2}\right)$$

The transform of q is the product of the two transforms $E(s)$ and $R(s)$, one being the transform of the input function $\mathcal{E}(t)$ and the other being the transform of the system response function $r(t)$. The inverse of the transform $E(s)R(s)$ is the *convolution* of the functions $\mathcal{E}(t)$ and $r(t)$:

$$\mathcal{L}^{-1}(ER) = \mathcal{E} * r = \int_0^t \mathcal{E}(\tau)r(t - \tau)\, d\tau \qquad (5.17)$$

This result is called the convolution theorem. (See Appendix VI for the proof.)

Physically, this integral represents the input to the system at time τ times the response at time t (that is, a time $t - \tau$ later) to that input, summed over all times τ from 0 to t. The convolution may also be written as

$$\mathcal{E} * r = r * \mathcal{E} = \int_0^t r(\tau)\varepsilon(t - \tau)\, d\tau$$

where τ represents "time ago" — that is, time measured backwards from the present ($\tau = 0$) to the time the clock was started ($\tau = t$).

Thus, the charge on the capacitor in the circuit problem is

$$q(t) = \int_0^t \mathcal{E}(\tau)r(t - \tau)\, d\tau$$

where $r(t)$ is the function whose transform is $R(s)$. We found this function in Example 5.2:

$$r(t) = \frac{1}{L\omega} e^{-\alpha t} \sin \omega t$$

The circuit response is a damped oscillation.

To illustrate the use of the convolution, consider an AC circuit with $\mathcal{E}(\tau) = \mathcal{E}_0 \cos \Omega \tau$. We have

$$q(t) = \int_0^t \frac{\mathcal{E}_0}{L\omega} \cos \Omega \tau \sin \omega (t - \tau) \exp[-\alpha(t - \tau)] d\tau$$

$$= \frac{\mathcal{E}_0}{L\omega} e^{-\alpha t} \int_0^t \cos \Omega \tau \sin \omega (t - \tau) e^{\alpha \tau} d\tau$$

$$= \frac{\mathcal{E}_0}{2L\omega} e^{-\alpha t} \int_0^t \{\sin[\omega t + (\Omega - \omega)\tau] + \sin[\omega t - (\Omega + \omega)\tau]\} e^{\alpha \tau} d\tau$$

The integral is most easily evaluated by writing the sines in terms of exponentials. The result is

$$q(t) = \frac{\mathcal{E}_0}{L\omega} \left(\frac{\alpha[2\Omega\omega \sin \Omega t - (\alpha^2 + \Omega^2 + \omega^2) e^{-\alpha t} \sin \omega t]}{[\alpha^2 + (\Omega - \omega)^2][\alpha^2 + (\Omega + \omega)^2]} \right.$$

$$\left. + \frac{\omega[\alpha^2 - \Omega^2 + \omega^2][\cos \Omega t - e^{-\alpha t} \cos \omega t]}{[\alpha^2 + (\Omega - \omega)^2][\alpha^2 + (\Omega + \omega)^2]} \right)$$

You should check that this solution satisfies the initial conditions $q(0) = 0$ and $i(0) = 0$.

The long-time solution is an AC current. First note that for $t \gg 1/\alpha$,

$$q(t) \to \frac{\mathcal{E}_0}{L\omega} \left(\frac{2\Omega\omega\alpha \sin \Omega t + \omega[\alpha^2 - \Omega^2 + \omega^2] \cos \Omega t}{[\alpha^2 + (\Omega - \omega)^2][\alpha^2 + (\Omega + \omega)^2]} \right)$$

and therefore

$$i(t) = \frac{dq}{dt} \to \frac{\mathcal{E}_0}{L} \Omega^2 \left(\frac{2\alpha \cos \Omega t + \Omega[1 - \omega_0^2/\Omega^2] \sin \Omega t}{[\alpha^2 + (\Omega - \omega)^2][\alpha^2 + (\Omega + \omega)^2]} \right)$$

$$= \frac{\mathcal{E}_0}{Z^2} (R \cos \Omega t + X \sin \Omega t)$$

where Z is the impedance,[3] given by

$$Z^2 \equiv R^2 + X^2 \equiv R^2 + \left(\Omega L - \frac{1}{\Omega C} \right)^2 = R^2 + (\Omega L)^2 (1 - \omega_0^2/\Omega^2)^2$$

$$= L^2 \left[4\alpha^2 + \Omega^2 \left(1 - \frac{\alpha^2 + \omega^2}{\Omega^2} \right)^2 \right]$$

$$= L^2 [\alpha^2 + (\Omega - \omega)^2][\alpha^2 + (\Omega + \omega)^2]/\Omega^2$$

[3] See, for example, Lea and Burke, p. 1022. See also Problem 2.14.

This is the expected long-time behavior. The other term in $q(t)$,

$$-\frac{E_0}{L\omega}\frac{\alpha(\alpha^2+\Omega^2+\omega^2)\sin\omega t+\omega[\alpha^2-\Omega^2+\omega^2]\cos\omega t}{[\alpha^2+(\Omega-\omega)^2][\alpha^2+(\Omega+\omega)^2]}e^{-\alpha t}$$

is an initial *transient* that goes to zero with a time scale $1/\alpha=2L/R$.

5.5.2. The Transform of an Integral

We may apply the convolution theorem in the special case that one of the functions is $g(t)\equiv 1$ with transform $G(s)=1/s$. Then we have

$$\mathcal{L}\left(\int_0^t f(\tau)g(t-\tau)\,d\tau\right)=F(s)G(s)$$

$$\boxed{\mathcal{L}\left(\int_0^t f(\tau)\,d\tau\right)=\frac{F(s)}{s}} \tag{5.18}$$

a result known as the *primitive function theorem*.

Example 5.7.   Find the Laplace transform of the sine integral

$$\text{Si}\,(t)=\int_0^t \frac{\sin u}{u}\,du$$

First we use result (5.14):

$$\mathcal{L}\left(\frac{\sin t}{t}\right)=\int_s^\infty \mathcal{L}(\sin t)\,d\sigma$$

$$=\int_s^\infty \frac{1}{\sigma^2+1}\,d\sigma=\tan^{-1}\sigma\Big|_s^\infty$$

$$=\frac{\pi}{2}-\tan^{-1}s$$

Then, using the primitive function theorem, we have

$$\mathcal{L}[\text{Si}\,(t)]=\frac{\pi}{2s}-\frac{1}{s}\tan^{-1}s$$

or, since $\tan(\pi/2-\theta)=\cot\theta$, we may write the result in the more compact form

$$\mathcal{L}[\text{Si}\,(t)]=\frac{\cot^{-1}s}{s}$$

5.6. THE GENERAL INVERSION PROCEDURE

We have so far amassed a collection of tricks for inverting a given transform $F(s)$ to find the function $f(t)$. We'd like to have a single technique that we can use every time and that does not require recognizing a trick that will work. This procedure is the *Mellin inversion integral:*

$$f(t) = \frac{1}{2\pi i} \int_{\gamma-i\infty}^{\gamma+i\infty} F(s)\, e^{st} ds \qquad (5.19)$$

The integral is along a line parallel to the imaginary axis in the complex s-plane. The line must be positioned so that all of the singularities of the function $F(s)$ are to the left of the line, as shown in Figure 5.6. This line is also called the Bromwich contour.

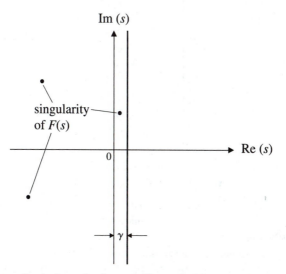

FIGURE 5.6. Contour for the inversion integral. The vertical line must be placed to the right of all the singularities of the transform $F(s)$.

To prove the result (5.19), we work backwards and take the transform of the function $f(t)$ defined by equation (5.19).

$$\mathcal{L}[f(t)] = \int_0^\infty e^{-st} \frac{1}{2\pi i} \int_{\gamma-i\infty}^{\gamma+i\infty} F(\sigma)\, e^{\sigma t} d\sigma\, dt$$

Interchange the order of integration and perform the integration over t:

$$\mathcal{L}[f(t)] = \frac{1}{2\pi i} \int_{\gamma-i\infty}^{\gamma+i\infty} F(\sigma) \int_0^\infty e^{-st+\sigma t} dt\, d\sigma$$

$$= \frac{1}{2\pi i} \int_{\gamma-i\infty}^{\gamma+i\infty} F(\sigma) \left.\frac{e^{-(s-\sigma)t}}{\sigma - s}\right|_0^\infty d\sigma$$

Now, provided that Re $(s) >$ Re $(\sigma) = \gamma$, the exponential approaches zero as $t \to \infty$, and we have

$$\mathcal{L}[f(t)] = \frac{1}{2\pi i} \int_{\gamma-i\infty}^{\gamma+i\infty} \frac{F(\sigma)}{s - \sigma} d\sigma$$

Note that the function $F(\sigma)$ is analytic everywhere to the right of the path of integration, so we may close the contour to the right, as shown in Figure 5.7.

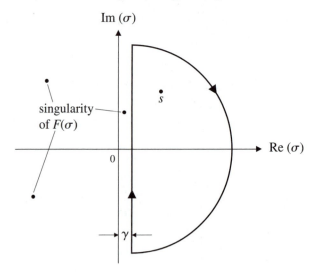

FIGURE 5.7. Contour for proving that equation (5.19) correctly inverts the transform $F(s)$. The contour is closed to the right, enclosing the pole at $\sigma = s$.

Then there is a single pole of the integrand inside the contour,[4] at the point $\sigma = s$. The integrand along the big semicircle is zero, provided that Max $|F(\sigma)|$ approaches zero on the semicircle at least as fast as $R^{-\varepsilon}$, where $\varepsilon > 0$. Notice that we are traversing the contour in the clockwise direction, rather than the standard counterclockwise direction, so we must introduce a minus sign when applying the residue theorem. Then the result is

$$\mathcal{L}[f(t)] = -2\pi i \frac{1}{2\pi i} \lim_{\sigma \to s} (\sigma - s) \frac{F(\sigma)}{s - \sigma} = F(s)$$

If the transform of f is F, then f is the inverse transform of F.

When we *use* the Mellin inversion theorem to evaluate f, we cannot close the contour to the right because the exponential e^{st} causes the integrand to blow up on the large semicircle at infinity. Thus, we extend the function F into the region to the left of the contour, using analytic continuation, and close the contour to the left, thereby enclosing all the poles of F.

Example 5.8.   Invert the transform $F(s) = 3s/(s^2 - 2s - 3)$.
  The denominator factors:

$$s^2 - 2s - 3 = (s + 1)(s - 3)$$

---

[4]Recall that we have already assumed Re $(s) > \gamma$.

Thus, the transform has simple poles at $s = 3$ and at $s = -1$. Using the Mellin inversion integral, we get

$$f(t) = \frac{1}{2\pi i} \int_{\gamma - i\infty}^{\gamma + i\infty} \frac{3se^{st}}{(s+1)(s-3)} ds$$

where $\gamma > 3$. We close the contour with a big semicircle to the left (Figure 5.8). Since $F(s)$ goes uniformly to zero as $R \to \infty$, we can invoke Jordan's lemma[5] to show that the integral along the semicircle goes to zero. The integral along each of the small arcs to the right of the imaginary axis also goes to zero. Let the real part of s on this piece of the contour be x, where $0 \le x \le \gamma$. Then

$$\left| \int_{\text{arc}} \right| \le \text{(length of path)(Max of |integrand| on path)}$$

$$= \gamma 3 e^{\gamma t} \max \left| \frac{x + iR}{(x + 1 + iR)(x - 3 + iR)} \right|$$

$$= \gamma 3 e^{\gamma t} \max \frac{R\sqrt{1 + x^2/R^2}}{R^2 \sqrt{\left(1 - \frac{x^2 - 2x - 3}{R^2}\right)^2 + \frac{4(x-1)^2}{R^2}}} \to 0 \quad \text{as } R \to \infty$$

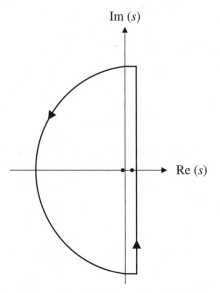

$$\text{Im } (s)$$

$$\text{Re } (s)$$

FIGURE 5.8. Contour for actually performing the inversion (5.19). The contour is closed to the left, thus enclosing all the singularities of $F(s)$. Here the singularities are at $s = -1$ and $s = 3$, so we must choose $\gamma > 3$.

[5]To use the lemma, first make the change of variable $s = i\omega$.

Then applying the residue theorem gives

$$f(t) = 3\left(\frac{-e^{-t}}{-4} + \frac{3e^{3t}}{4}\right) = \frac{3}{4}(3e^{3t} + e^{-t})$$

Example 5.9. Invert the transform $F(s) = 1/\sqrt{s}$.

This function has a branch point at the origin. We must choose the branch cut along the negative real axis so that it does not cross the path of integration, and we must deform our closed contour to avoid the branch cut, as shown in Figure 5.9. Applying the Mellin inversion integral, we have

$$f(t) = \frac{1}{2\pi i} \int_{\gamma-i\infty}^{\gamma+i\infty} \frac{e^{st}}{\sqrt{s}} ds$$

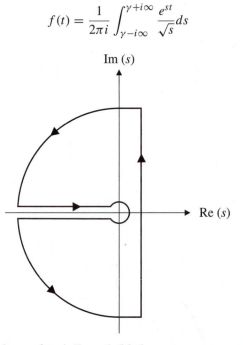

FIGURE 5.9. To invert the transform in Example 5.9, the contour must pass to the right of the branch point at the origin and must exclude the entire branch cut. Here $\gamma > 0$.

The integral around the closed contour is zero, since the integrand is analytic everywhere inside. Thus,

$$\left(\int_{\gamma-i\infty}^{\gamma+i\infty} + \int_{C_R} + \int_{\text{top of branch cut}} + \int_{C_\varepsilon} + \int_{\text{bottom of branch cut}}\right) \frac{e^{st}}{\sqrt{s}} ds = 0$$

and so

$$\int_{\gamma-i\infty}^{\gamma+i\infty} \frac{e^{st}}{\sqrt{s}} ds = -\left(\int_{C_R} + \int_{\text{top of branch cut}} + \int_{C_\varepsilon} + \int_{\text{bottom of branch cut}}\right) \frac{e^{st}}{\sqrt{s}} ds$$

The integral along C_R goes to zero as $R \to \infty$, by Jordan's lemma, since $1/\sqrt{s}$ goes to zero uniformly as $s \to \infty$ on C_R. The integral along the small arcs goes to zero as $R \to \infty$ (and we can also let $\gamma \to 0$ here). The integral around the small circle at the origin is

$$\int_{C_\varepsilon} \frac{e^{st}}{\sqrt{s}} ds = \int_\pi^{-\pi} \frac{\exp{(\varepsilon e^{i\theta} t)}}{\sqrt{\varepsilon e^{i\theta/2}}} i\varepsilon e^{i\theta} d\theta$$

$$= i\sqrt{\varepsilon} \int_\pi^{-\pi} \exp{(\varepsilon e^{i\theta} t)} e^{i\theta/2} d\theta \to 0 \quad \text{as } \varepsilon \to 0$$

We are left with the two integrals along the two sides of the branch cut. On the top side, $s = re^{i\pi}$, and so

$$\int_{\text{top of branch cut}} \frac{e^{st}}{\sqrt{s}} ds = \int_\infty^\varepsilon \frac{\exp{(re^{i\pi} t)}}{\sqrt{re^{i\pi/2}}} e^{i\pi} dr = -\int_\infty^\varepsilon \frac{\exp{(-rt)}}{i\sqrt{r}} dr$$

$$= \int_\varepsilon^\infty \frac{\exp{(-rt)}}{i\sqrt{r}} dr$$

and on the bottom side, $s = re^{-i\pi}$, so

$$\int_{\text{bottom of branch cut}} \frac{e^{st}}{\sqrt{s}} ds = \int_\varepsilon^\infty \frac{\exp{(re^{-i\pi} t)}}{\sqrt{re^{-i\pi/2}}} e^{-i\pi} dr = -\int_\varepsilon^\infty \frac{\exp{(-rt)}}{-i\sqrt{r}} dr$$

$$= \int_\varepsilon^\infty \frac{\exp{(-rt)}}{i\sqrt{r}} dr$$

Thus,

$$\frac{1}{2\pi i} \int_{\gamma-i\infty}^{\gamma+i\infty} \frac{e^{st}}{\sqrt{s}} ds = \frac{-1}{2\pi i} 2 \int_0^\infty \frac{\exp{(-rt)}}{i\sqrt{r}} dr$$

$$= \frac{1}{\pi} \int_0^\infty \frac{\exp{(-rt)}}{\sqrt{r}} dr$$

Now change variables to $u = rt$. Then

$$f(t) = \frac{1}{\pi\sqrt{t}} \int_0^\infty u^{-1/2} e^{-u} du = \frac{1}{\pi\sqrt{t}} \Gamma\left(\frac{1}{2}\right) = \sqrt{\frac{1}{\pi t}}$$

Curiously, the function $f(t) = t^{-1/2}$ is its own transform (except for a multiplicative constant).

5.7. SOME MORE PHYSICS

5.7.1. Modeling Problems

A difficulty that shows up in the mathematical solution of a problem often indicates that we have chosen an inadequate model for the physical system. When the Laplace transform method is applied to the solution of a differential equation, the initial conditions must be included in the solution. In some circumstances, the solution has a different value at $t = 0$, indicating a physical inconsistency. This often happens in circuits containing inductance.

Example 5.10.   The two circuits in Figure 5.10 are coupled by mutual inductance. The switch in the circuit on the right is closed at $t = 0$. Find the current in each circuit for $t > 0$.

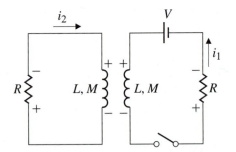

FIGURE 5.10.   These circuits, coupled by mutual induction, exhibit some of the difficulties involved with using an inadequate model for a physical system (Example 5.10). The current variables i_1 and i_2 are defined in this diagram.

As usual, we begin by applying Kirchhoff's loop rule to each circuit. The current variables i_1 and i_2 are defined as shown in the figure. For the circuit on the right, we obtain

$$V = i_1 R + L\frac{di_1}{dt} + M\frac{di_2}{dt}$$

and for the circuit on the left, we obtain

$$0 = i_2 R + L\frac{di_2}{dt} + M\frac{di_1}{dt}$$

with initial conditions that both currents are zero at $t = 0$.
Now we Laplace transform both equations:

$$\frac{V}{s} = I_1 R + sLI_1 + sMI_2$$

and

$$0 = I_2 R + sLI_2 + sMI_1$$

To solve the equations we use a trick. Adding the two equations, we get

$$\frac{V}{s} = XR + sLX + sMX$$

where $X = I_1 + I_2$. Similarly, by subtracting and defining $Y = I_1 - I_2$, we get

$$\frac{V}{s} = YR + sLY - sMY \tag{5.20}$$

Now we can solve:

$$X = \frac{V}{s}\left(\frac{1}{s(L + M) + R}\right)$$

From Table 5.1, the inverse of $\dfrac{1}{(L + M)s + R}$ is $\dfrac{1}{L + M}e^{-\alpha t}$, where $\alpha = R/(L+M)$.
Then, from the primitive function theorem (equation 5.18),

$$x(t) = \frac{V}{L + M}\int_0^t e^{-\alpha t}\,dt = \frac{V}{R}(1 - e^{-\alpha t})$$

Similarly,

$$Y = \frac{V}{s}\left(\frac{1}{s(L - M) + R}\right)$$

$$y(t) = \frac{V}{R}(1 - e^{-\beta t})$$

where $\beta = R/(L - M)$. (β is positive since $L^2 \geq M^2$ is required by fundamental principles of electromagnetic theory.)

Now we can solve for the two currents:

$$i_1 = \frac{1}{2}(x + y) = \frac{V}{2R}(1 - e^{-\alpha t} + 1 - e^{-\beta t})$$

$$= \frac{V}{2R}(2 - e^{-\alpha t} - e^{-\beta t})$$

while

$$i_2 = \frac{1}{2}(x - y) = \frac{V}{2R}[1 - e^{-\alpha t} - (1 - e^{-\beta t})]$$

$$= \frac{V}{2R}(e^{-\beta t} - e^{-\alpha t})$$

Both currents satisfy the initial conditions, and at long times we have

$$i_1 \to \frac{V}{R} \quad \text{and} \quad i_2 \to 0$$

as expected.

Now suppose we make this system by winding the wires for both circuits together into a coil. Then we have $L = M$, and $\beta \to \infty$. We go back to equation (5.20) and notice that it has simplified:

$$\frac{V}{s} = YR + sLY - sMY = YR$$

Thus,

$$Y = \frac{V}{Rs}$$

and, consequently,

$$y(t) = \frac{V}{R}$$

which is a constant. Then the equation for X becomes

$$\frac{V}{s} = XR + s2LX$$

$$X = \frac{V}{s}\left(\frac{1}{2Ls + R}\right)$$

From Table 5.1 and the primitive function theorem (equation 5.18), we obtain

$$x = \frac{V}{2L}\int_0^t e^{-\gamma t}dt$$

$$= \frac{V}{2L\gamma}(1 - e^{-\gamma t})$$

$$= \frac{V}{R}(1 - e^{-\gamma t})$$

where $\gamma = R/2L$. Thus,

$$i_1 = \frac{1}{2}(x + y) = \frac{V}{2R}(2 - e^{-\gamma t})$$

and

$$i_2 = \frac{1}{2}(x - y) = -\frac{V}{2R}e^{-\gamma t}$$

Both solutions still have the correct behavior for long times, but now neither satisfies the correct initial conditions! At the closing of the switch, the currents have instantaneously jumped from zero to $i_1 = V/2R$ and $i_2 = -V/2R$. Of course, this cannot happen. The model fails[6] because it is impossible for L to be *exactly* equal to M.

---

[6]This is, in fact, the exact solution to a somewhat different problem. Imagine combining the circuits by replacing all the inductances with a single inductor L. The second circuit then forms a parallel combination. The solution we have obtained would have zero current through the inductor at $t = 0$, and this is physically possible.

Figure 5.11 shows how i_2 behaves as M approaches L. The absolute value of the current increases more rapidly from zero as $L - M$ approaches zero.

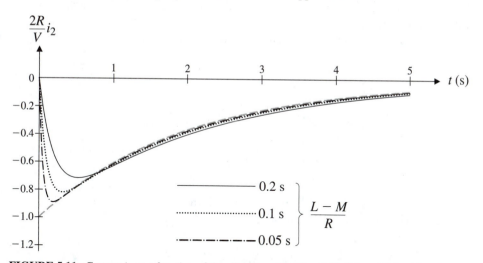

FIGURE 5.11. Current i_2 as a function of time for $(L - M)/R = 0.05, 0.1,$ and 0.2 s. The diagram also shows the limit $L = M$ (dashed line). As $L \to M$, the current increases more and more rapidly from zero. In the limit, the current makes an instantaneous jump to $V/2R$ at the moment the switch is closed. This is the correct mathematical limit, but this solution is physically prohibited.

5.7.2. Nuclear Reaction Networks

A common application of the Laplace transform is to the solution of nuclear reaction problems. For example, in nuclear decay,[7] the number of nuclei of one species decreases, at the same time increasing the number of nuclei of one or more additional daughter species. The daughter nuclei may then in turn decay. For example, in a three-species chain, the decay may be represented as

$$N_1 \to N_2 \to N_3$$

The result is a set of coupled differential equations of the form

$$\frac{dN_1}{dt} = -\lambda_1 N_1, \quad \frac{dN_2}{dt} = -\lambda_2 N_2 + \lambda_1 N_1, \quad \frac{dN_3}{dt} = \lambda_2 N_2$$

which we can more easily solve by converting them to a set of coupled algebraic equations through the use of the Laplace transform:

$$sN_1(s) - N_1(0) = -\lambda_1 N_1(s)$$

$$sN_2(s) - N_2(0) = -\lambda_2 N_2(s) + \lambda_1 N_1(s)$$

$$sN_3(s) - N_3(0) = \lambda_2 N_2(s)$$

[7] See, for example, Lea and Burke, pp. 1161–1165.

Example 5.11. Part of the natural radioactive sequence for the decay of ^{237}Np involves the α decay of ^{217}At into ^{213}Bi, which then β decays into ^{213}Po 98% of the time and α decays into ^{209}Tl 2% of the time. Both species then decay to ^{209}Pb by either an α decay or a β decay. The lead then β decays to ^{209}Bi, which is the stable endpoint of the chain. We may write a set of equations for this reaction chain as follows.

At decays at rate λ_1:

$$\frac{dN_{At}}{dt} = -\lambda_1 N_{At}$$

^{213}Bi, produced by the decay of At, decays in two different ways:

$$\frac{dN_{Bi}}{dt} = \lambda_1 N_{At} - 0.02\lambda_2 N_{Bi} - 0.98\lambda_3 N_{Bi}$$

where λ_2 is the decay rate of ^{213}Bi into Tl and λ_3 is the decay rate of ^{213}Bi into Po. The other relations are obtained similarly:

$$\frac{dN_{Po}}{dt} = 0.98\lambda_3 N_{Bi} - \lambda_4 N_{Po}$$

$$\frac{dN_{Tl}}{dt} = 0.02\lambda_2 N_{Bi} - \lambda_5 N_{Tl}$$

$$\frac{dN_{Pb}}{dt} = \lambda_5 N_{Tl} + \lambda_4 N_{Po} - \lambda_6 N_{Pb}$$

$$\frac{dN_{B2}}{dt} = \lambda_6 N_{Pb}$$

(where B2 refers to ^{209}Bi). These relations transform to the much nicer looking set of equations for the transforms:

$$sN_{At} - N_{At}(0) = -\lambda_1 N_{At} \quad \Rightarrow \quad N_{At} = \frac{N_{At}(0)}{s + \lambda_1}$$

$$sN_{Bi} = -0.02\lambda_2 N_{Bi} - 0.98\lambda_3 N_{Bi} + \lambda_1 N_{At} \quad \Rightarrow \quad N_{Bi} = \frac{\lambda_1 N_{At}}{s + 0.02\lambda_2 + 0.98\lambda_3}$$

$$sN_{Po} = 0.98\lambda_3 N_{Bi} - \lambda_4 N_{Po} \quad \Rightarrow \quad N_{Po} = \frac{0.98\lambda_3 N_{Bi}}{s + \lambda_4}$$

$$sN_{Tl} = 0.02\lambda_2 N_{Bi} - \lambda_5 N_{Tl} \quad \Rightarrow \quad N_{Tl} = \frac{0.02\lambda_2 N_{Bi}}{s + \lambda_5}$$

$$sN_{Pb} = \lambda_5 N_{Tl} + \lambda_4 N_{Po} - \lambda_6 N_{Pb} \quad \Rightarrow \quad N_{Pb} = \frac{\lambda_5 N_{Tl} + \lambda_4 N_{Po}}{s + \lambda_6}$$

and, finally,

$$sN_{B2} = \lambda_6 N_{Pb} \quad \Rightarrow \quad N_{B2} = \frac{\lambda_6 N_{Pb}}{s}$$

The transforms may be inverted by several of the methods we have discussed. For example, if we want to find the abundance of polonium, we first solve for its transform:

$$N_{Po} = \frac{0.98\lambda_3 N_{Bi}}{s + \lambda_4} = \frac{0.98\lambda_3}{(s + \lambda_4)}\frac{\lambda_1 N_{At}}{(s + 0.02\lambda_2 + 0.98\lambda_3)}$$

$$= \frac{0.98\lambda_3}{(s + \lambda_4)}\frac{\lambda_1}{(s + 0.02\lambda_2 + 0.98\lambda_3)}\frac{N_{At}(0)}{(s + \lambda_1)}$$

We may use the Mellin inversion integral to invert the transform. The integrand has three simple poles, each giving an exponential term. The result is

$$N_{Po}(t) = 0.98\lambda_3\lambda_1 N_{At}(0) \left[\begin{array}{c} \dfrac{e^{-\lambda_4 t}}{(0.02\lambda_2 + 0.98\lambda_3 - \lambda_4)(\lambda_1 - \lambda_4)} \\[2mm] + \dfrac{e^{-(0.02\lambda_2 + 0.98\lambda_3)t}}{(\lambda_4 - 0.02\lambda_2 - 0.98\lambda_3)(\lambda_1 - 0.02\lambda_2 - 0.98\lambda_3)} \\[2mm] + \dfrac{e^{-\lambda_1 t}}{(\lambda_4 - \lambda_1)(0.02\lambda_2 + 0.98\lambda_3 - \lambda_1)} \end{array} \right]$$

PROBLEMS

1. Show that the following functions are of exponential order, and find their Laplace transforms.

   (a) $f(t) = \sinh \alpha t$

   (b) $f(t) = \tanh \alpha t$ (*Hint:* Change variables to $e^{-2\alpha t} = u$ and expand the integrand in a series.)

   (c) $f(t) = \sin \sqrt{at}$

   (d) $f(t) = e^{-\alpha t^2}$

   (e) $f(t) = te^{\sqrt{t}}$

   (f) $f(t) = \sin(\omega t + \phi_0)$

   (g) The ramp function $f(t) = \begin{cases} t & \text{if } 0 < t < t_0 \\ t_0 & \text{if } t > t_0 \end{cases}$

2. Using the shifting property or otherwise, find the Laplace transform of the function

$$f(t) = \begin{cases} 0 & \text{if } t < 2 \\ (t - 2)^3 e^{-at} & \text{if } t \geq 2 \end{cases}$$

3. Find the Laplace transform of the function $t \cosh \alpha t$.

4. Find the Laplace transform of the function $f(t) = \dfrac{\sinh \alpha t}{t}$.

5. Find the Laplace transform of

 (a) the triangle wave function with period T:

$$f(t) = \begin{cases} a(t - nT) & \text{if } nT < t < nT + T/2 \\ a[(n+1)T - t] & \text{if } nT + T/2 < t < (n+1)T \end{cases}$$

 (b) the sawtooth function:

$$f(t) = a(t - nT) \quad \text{if } nT < t < (n+1)T$$

6. Use the Mellin inversion integral to invert the following transforms:

 (a) $F(s) = \dfrac{s}{s^2 + 2s + 3}$

 (b) $F(s) = \dfrac{1}{(s^2 + a^2)^2}$

 (c) $F(s) = \dfrac{e^{-\sqrt{s}}}{s}$,

 (d) $F(s) = -\dfrac{\ln(1+s)}{s}$ (Express the answer in terms of the exponential integral

  $\mathrm{Ei}\,(x) = \displaystyle\int_{-\infty}^{x} \dfrac{e^w\,dw}{w}$, where $x < 0$.)

7. Invert the transform $F(s) = \dfrac{1}{\sqrt{s^2 - a^2}}$

 (a) by expanding in a series

 (b) by using the Mellin inversion integral with a branch cut running from $-a$ to a along the real axis (A change of variable to z, where

$$s = \frac{a}{2}\left(z + \frac{1}{z}\right)$$

  may prove useful.)

8. Use the convolution theorem (equation 5.17) to evaluate the inverse transform of

$$F(s) = \frac{\omega s}{s^4 - \omega^4}$$

9. Use the integration rule (equation 5.14) and Table 5.1 to derive the result of Example 5.9 for the transform of $1/\sqrt{t}$.

10. The diagram shows a simplified version of an automobile spark coil circuit. The spark plug itself acts like an open circuit until the potential across it reaches the breakdown voltage for air. Thus, you may ignore that branch of the circuit until the end of the problem (part e). The battery voltage $V = 12$ V, $C = 0.1\,\mu\mathrm{F}$, $R = 10\,\Omega$, and $L = 10$ mH. Assume that the switch (points) have been closed for a long time prior to $t = 0$.

 (a) How long a "long time" is necessary? Write down expressions for the charge on the capacitor and the current through the coil at $t = 0$.

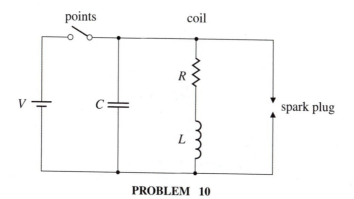

PROBLEM 10

(b) At $t = 0$, the points open. Qualitatively discuss the circuit behavior. What is the expected long-time solution for the charge and current?

(c) Use a Laplace transform method to solve for the potential difference across the spark plug as a function of time.

(d) Plot your solution. What is the maximum potential difference achieved?

(e) If the breakdown voltage of air is 3 MV/m, what spark plug gap would be required with this circuit? Remember that you would like the engine to start even if the battery is a bit low!

11. A beam is supported at one end, as shown in the diagram. A block of mass M and length l is placed on the beam, as shown. Write down the known conditions at $x = 0$. Use the Laplace transform to solve for the beam displacement. Plot your results for $x_0 = 0.6L$ and $l = 0.2L$.

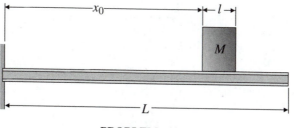

PROBLEM 11

12. Technetium is used in medical procedures as a diagnostic tool. The technetium is obtained as the decay product of ^{99}Mo, which decays to ^{99m}Tc with a half-life of 66.02 h. The technetium in turn decays with a half-life of 6.02 h. A medical radiology department receives a source containing 100 mCi of ^{99}Mo at 9:00 a.m. on Monday morning. Find the amount of technetium present in the sample as a function of time after 9:00 a.m. When is the amount of ^{99m}Tc a maximum?

13. An overdamped harmonic oscillator satisfies the equation

$$\frac{d^2x}{dt^2} + 2\alpha\frac{dx}{dt} + \omega_0^2 x = f(t)$$

where $\alpha^2 > \omega_0^2$ and the driving force is a square wave of period T. Find the displacement $x(t)$ if the initial conditions are $x(0) = dx/dt|_{t=0} = 0$. Plot the result for $\alpha = 2\omega_0$, $\alpha T = 1$, and $0 \le t \le 3T/2$. (*Hint*: Use the result of Example 5.6 and expand the transform in a series.)

14. (a) A harmonic oscillator with resonant frequency ω_0 and no damping is driven by a sinusoidal force $F(t) = F_0 \sin \omega t$. If the initial conditions are $x(0) = 0, dx/dt = 0$ at $t = 0$, use the Laplace transform to find $x(t)$. What happens if $\omega = \omega_0$?

(b) Solve the same problem with the initial conditions $x(0) = 0, dx/dt = a$ at $t = 0$.

15. The two circuits in diagrams (a) and (b) show how we might use a capacitor to prevent sparks across a switch when the switch is opened. Assume that the switch has been closed for a long time and is opened at $t = 0$. For each circuit, use Kirchhoff's rules to solve for the current through the inductor and the charge on the capacitor as a function of time after the switch is opened. Discuss the merits of each of the circuit designs.

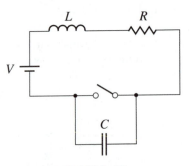

PROBLEM 15a

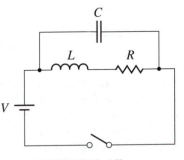

PROBLEM 15b

16. The switch has been in position A in the circuit shown in the diagram for a long time. What are the charge on the capacitor and the current i through the inductor? At time $t = 0$, the switch is moved to position B. What are the charge on the capacitor and the current a long time later? Find the charge on the capacitor as a function of time for $t > 0$. Give your answer in terms of ω_0, where $\omega_0^2 = 1/LC$, $\alpha = R/L$, and $\beta = 1/RC$. You may also find it useful to define $\gamma = (\alpha + \beta)/2$ and $\Omega = \sqrt{2\alpha\beta - \gamma^2}$. You may assume that Ω is real.

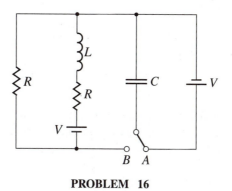

PROBLEM 16

17. The switch in the circuit shown in the diagram has been closed for a long time, and a constant current flows. What is the charge on the capacitor? At time $t = 0$, the switch is opened. What are the charge on the capacitor and the current through the inductor a long time later? Find the current through the inductor as a function of time for $t > 0$. Give your answer in terms of ω_0 and α, where $\omega_0^2 = 1/LC$ and $\alpha = R/2L$.

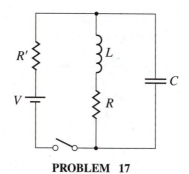

PROBLEM 17

18. The switch in the circuit shown in the diagram has been open for a long time. At $t = 0$, the switch is closed. Find the current through the inductor and the charge on the capacitor

as functions of time for $t > 0$. Give your answer in terms of ω_0, α, and β, where $\omega_0^2 = 1/LC$, $\alpha = R/2L$, and $\beta = \omega_0^2/4\alpha = 1/2RC$. You may also find it useful to define $\Omega = \sqrt{\frac{3}{2}\omega_0^2 - \alpha^2 - \beta^2}$.

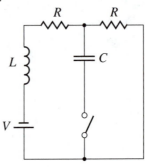

PROBLEM 18

19. In the figure shown, capacitor C_1 has charge Q and capacitor C_2 is uncharged. At $t = 0$, the switch is closed. The two capacitances are equal. Find the voltage across each capacitor as a function of time for $t > 0$.

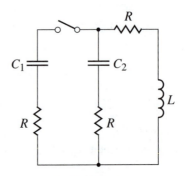

PROBLEM 19

20. The switch in the circuit shown in the diagram has been closed for a long time and is opened at $t = 0$. What are the currents in the circuit for $t < 0$? Use the Laplace transform

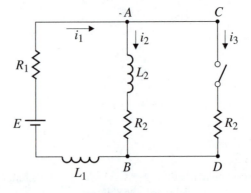

PROBLEM 20

to find the currents in the circuit as a function of time for $t > 0$. Does your answer satisfy the initial conditions?

Rework the problem, leaving the initial value of the current i_2 in arm AB as an unknown to be found. Find the solution for the current i_1 through R_1 and require that it satisfy $i_1(0) = E/(R_1 + R_2/2)$. What value of $i_2(0)$ is required? Give a physical explanation of this result. If $R_1 = R_2 = R$ and $L_1 = L_2 = L$, plot both solutions. Plot current in units of E/R versus time in units of L/R. How long is it before both solutions give the same result to within 1%?

21. The radioactive series that begins with neptunium 93 contains the following decays:

Decay	Type	Half-life
$^{237}\text{Np} \rightarrow {}^{233}\text{Pa}$	α	2.14×10^6 y
$^{233}\text{Pa} \rightarrow {}^{233}\text{U}$	β	27.0 d
$^{233}\text{U} \rightarrow {}^{229}\text{Th}$	α	1.6×10^5 y
$^{229}\text{Th} \rightarrow {}^{225}\text{Ra}$	α	7340 y
$^{225}\text{Np} \rightarrow {}^{225}\text{Ac}$	β	14.8 d
$^{225}\text{Ac} \rightarrow {}^{221}\text{Fr}$	α	10.0 d
$^{221}\text{Fr} \rightarrow {}^{217}\text{At}$	α	4.8 min
$^{217}\text{At} \rightarrow {}^{213}\text{Bi}$	α	0.032 s
$^{213}\text{Bi} \rightarrow {}^{213}\text{Po}$ (98%)	β	47 min
$^{213}\text{Bi} \rightarrow {}^{209}\text{Tl}$ (2%)	α	
$^{213}\text{Po} \rightarrow {}^{209}\text{Pb}$	α	4.2 μs
$^{209}\text{Tl} \rightarrow {}^{209}\text{Pb}$	β	2.2 min
$^{209}\text{Pb} \rightarrow {}^{209}\text{Bi}$	β	3.3 h

If we regard any decay that takes less than one year to be essentially instantaneous, then the chain simplifies to

$$^{237}\text{Np} \rightarrow {}^{233}\text{U} \rightarrow {}^{229}\text{Th} \rightarrow {}^{209}\text{Bi}$$

Write a series of differential equations that describes this decay chain. Apply the Laplace transform to find the fraction of the original ^{237}Np that is in the form of uranium, thorium, and bismuth after 10^5 and 10^6 years. Make a plot showing the amounts of each element as a function of time.

22. Find the Laplace transform of the function $g(t) = t \, df/dt$. Express the result in terms of the transform $F(s)$ of the function $f(t)$. Use the result to solve the differential equation

$$ty' + y = e^{-t}$$

23. Apply the Laplace transform to the differential equation

$$y'' - t^2 y = t^2$$

Does the Laplace transform offer any advantages in solving this equation? Using any method of your choice, solve the original equation or the transformed equation subject to the initial conditions $y(0) = 1$ and $y'(0) = 0$, and comment. (*Hint:* The transform may be expressed as a power series in $1/s$.)

24. Take the Laplace transform of the Bessel equation of order zero,

$$y'' + \frac{1}{x}y' + y = 0$$

and show that

$$(s^2 + 1)Y'(s) + sY(s) = 0$$

Solve for $Y(s)$ and hence find an integral expression for $y(x) = J_0(x)$. [See Chapter 8 for more information on $J_0(x)$.]

CHAPTER 6

Generalized Functions in Physics

6.1. THE DELTA FUNCTION

The particle is an often used and very valuable physical model. We can describe an electron as a particle because its size, as far as we know, is smaller than anything else we might care about. Sometimes we describe something as large as an automobile as a particle because its size and shape are not important to the question we are interested in. When something is modeled as a particle, all of its physical properties, such as mass and charge, are concentrated at a single point. In a similar fashion, an impulsive force is modeled as a force that is concentrated at a point in time. To describe the density of the particle or the force as a function of time, we need a function that has the following properties:

- Its value is zero everywhere except where its argument is zero.
- Its value is infinite where the argument is zero.
- The integral of the function is 1.

This mathematical object was first used in physics by P. A. M. Dirac and is called the delta function $\delta(x)$. Using this function, we write an impulsive force as

$$\vec{F}(t) = \vec{I}\delta(t)$$

The impulse delivered is

$$\vec{I} = \int_{-\infty}^{+\infty} \vec{F}(t)\, dt = \vec{I} \int_{-\infty}^{+\infty} \delta(t)\, dt = \vec{I}$$

To write the density of a particle at the origin, we need a three-dimensional delta function:

$$\rho(\vec{r}) = M\delta(\vec{r}) = M\delta(x)\delta(y)\delta(z) \tag{6.1}$$

The particle's mass is

$$M = \int_{\text{all space}} \rho(\vec{r})\, dV = M \int_{-\infty}^{+\infty} \int_{-\infty}^{+\infty} \int_{-\infty}^{+\infty} \delta(x)\delta(y)\delta(z)\, dx\, dy\, dz = M$$

In both these examples, we can see that the physical dimensionality of the delta function is one over the dimension of its argument:

$$\int_{-\infty}^{+\infty} \delta(t)\, dt = 1 \Rightarrow [\delta(t)] = \frac{1}{[t]} = \frac{1}{T}$$

and, similarly,

$$[\delta(\vec{\mathbf{r}})] \equiv [\delta(x)\delta(y)\delta(z)] = \frac{1}{[x]}\frac{1}{[y]}\frac{1}{[z]} = \frac{1}{L^3}$$

The properties we have listed for the delta function do not describe any proper mathematical function. Thus, we must extend our ideas about functions to include these objects, which are known as *distributions* or *generalized functions*. To begin our study of distributions, we shall study the delta function in one dimension in more detail.

6.1.1. Delta Sequences

The delta function is defined as a mathematical object that possesses a property known as the *sifting property*:

$$\boxed{\int_{-\infty}^{\infty} f(x)\delta(x)\, dx = f(0)} \tag{6.2}$$

That is, if we multiply a function $f(x)$ by the delta function and integrate over the whole real line, we sift out the value of the function at the origin. Let's see how the sifting property follows from the properties listed above:

$$\int_{-\infty}^{\infty} f(x)\delta(x)\, dx = \int_{-\varepsilon}^{\varepsilon} f(x)\delta(x)\, dx \quad \text{because } \delta(x) \text{ is zero except at } x = 0$$

$$= f(x_0) \int_{-\varepsilon}^{\varepsilon} \delta(x)\, dx, \quad \text{where} -\varepsilon < x_0 < \varepsilon$$

$$= f(x_0) \int_{-\infty}^{\infty} \delta(x)\, dx \quad \text{because } \delta(x) \text{ is zero except at } x = 0$$

$$= f(x_0) \quad \text{because the integral of the delta function is 1}$$

$$\rightarrow f(0) \quad \text{as } \varepsilon \rightarrow 0$$

In the second step, we used the mean value theorem[1] for integrals.

Delta functions arise from modeling a physical quantity that is distributed over a small but finite region as being concentrated at a point. Thus, we should expect that the mathematical

---

[1] See Appendix IV for a proof of this result. The application here can be made rigorous by proper use of limiting procedures, as we show below.

quantity $\delta(x)$ would arise as the limit of a set of functions that become increasingly concentrated at a single point. We can construct a sequence of proper functions $\phi_n(x)$ labeled by a parameter n such that as $n \to \infty$ the sequence $\phi_n(x)$ approaches the delta function. What we mean by "ϕ_n approaches the delta function" is that as $n \to \infty$ we obtain the sifting property; that is,

$$\lim_{n \to \infty} \int_{-\infty}^{+\infty} \phi_n(x) f(x)\, dx = f(0)$$

for any continuous function $f(x)$. Such a sequence of functions is called a *delta sequence*.

The simplest delta sequence is a sequence of rectangular blocks with unit area:

$$\phi_n(x) = \begin{cases} n/2 & \text{if } -1/n < x < 1/n \\ 0 & \text{otherwise} \end{cases} \tag{6.3}$$

As n gets larger, the blocks get taller and narrower (Figure 6.1), and so ϕ_n "looks like" the delta function as $n \to \infty$. Next we must test for the sifting property. We multiply $\phi_n(x)$ by a function $f(x)$ that is continuous at the origin and integrate:

$$\int_{-\infty}^{+\infty} \phi_n(x) f(x)\, dx = \frac{n}{2} \int_{-1/n}^{+1/n} f(x)\, dx$$

$$= \frac{n}{2} \frac{2}{n} f(x_0), \quad \text{where } -\frac{1}{n} < x_0 < +\frac{1}{n}$$

$$= f(x_0) \to f(0) \quad \text{as } n \to \infty$$

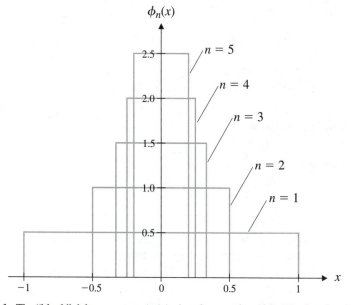

FIGURE 6.1. The "block" delta sequence $\phi_n(x)$ given by equation (6.3). The first five functions are shown here.

(We used the mean value theorem to evaluate the integral.) Since we obtain the sifting property, $\phi_n(x)$ is a delta sequence.

We can use this sequence to determine the sifting property of $\delta(x - a)$. We look at the integral:

$$\int_{-\infty}^{+\infty} \phi_n(x - a) f(x)\, dx = \int_{-\infty}^{+\infty} \phi_n(u) f(u + a)\, du$$

$$= \frac{n}{2} \frac{2}{n} f(u_0 + a), \quad \text{where } -\frac{1}{n} \le u_0 \le +\frac{1}{n}$$

$$\to f(a) \quad \text{as } n \to \infty$$

So we conclude that $\delta(x - a)$ sifts out the value of the function at $x = a$.

Similarly, we can determine how to express $\delta(ax)$. First let $a > 0$. Then

$$\int_{-\infty}^{+\infty} \phi_n(ax) f(x)\, dx = \int_{-\infty}^{+\infty} \phi_n(u) f\left(\frac{u}{a}\right) \frac{du}{a}$$

$$\to \frac{1}{a} f(0) \quad \text{as } n \to \infty$$

where we used the change of variables $u = ax$. But if $a < 0$, the same change of variables gives

$$\int_{-\infty}^{+\infty} \phi_n(ax) f(x)\, dx = \int_{+\infty}^{-\infty} \phi_n(u) f\left(\frac{u}{a}\right) \frac{du}{a}$$

$$= -\int_{-\infty}^{+\infty} \phi_n(u) f\left(\frac{u}{a}\right) \frac{du}{a}$$

$$\to -\frac{1}{a} f(0) \quad \text{as } n \to \infty$$

We can express both results as

$$\boxed{\delta(ax) = \frac{1}{|a|} \delta(x)} \tag{6.4}$$

Sometimes we need a sequence of functions that are better behaved than the block functions $\phi_n(x)$. For example, to investigate the derivative of the delta function, we need a set of functions that are each differentiable. The functions

$$\phi_n(x) = \frac{1}{n\pi} \frac{\sin^2 nx}{x^2} \tag{6.5}$$

are infinitely differentiable and also get peakier around the origin as n increases (Figure 6.2). (The first zero of this function occurs at $x_0 = \pm\pi/n$, and x_0 approaches zero as $n \to \infty$. The value of the function at the origin is n/π and increases linearly with n.) Again these functions "look like" the delta function as $n \to \infty$, but we'll need to test for the sifting property.

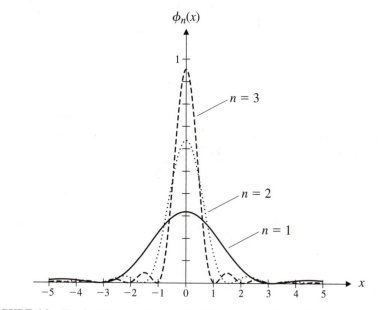

FIGURE 6.2. The first three functions in the delta sequence $\phi_n(x)$ given by equation (6.5).

The derivative is

$$\frac{d\phi_n}{dx} = \frac{1}{n\pi}\left(\frac{2n\sin nx \cos nx}{x^2} - 2\frac{\sin^2 nx}{x^3}\right)$$

$$= \frac{2\sin nx}{n\pi x^3}(nx\cos nx - \sin nx)$$

This clearly exists for all $x \neq 0$. It also exists for $x = 0$, as we can see by expanding the sine and cosine for small x:

$$\frac{d\phi_n}{dx}\bigg|_{x=0} = \lim_{x\to 0}\frac{2nx}{n\pi x^3}\left[nx\left(1 - \frac{n^2x^2}{2}\right) - \left(nx - \frac{n^3x^3}{6}\right)\right]$$

$$= \lim_{x\to 0}\left(-\frac{2}{3}n^3\frac{x}{\pi}\right) = 0$$

This result ($d\phi_n/dx = 0$ at $x = 0$) agrees with the graphs shown in Figure 6.2, which have slope zero at $x = 0$.

Now let's check for the sifting property. We write the sine in terms of complex exponentials:

$$\int_{-\infty}^{+\infty} \frac{1}{n\pi} \frac{\sin^2 nx}{x^2} f(x)\,dx = \int_{-\infty}^{+\infty} \frac{1}{n\pi x^2} \left(\frac{e^{inx} - e^{-inx}}{2i} \right)^2 f(x)\,dx$$

$$= \int_{-\infty}^{+\infty} \frac{1}{n\pi x^2} \left(\frac{e^{2inx} - 2 + e^{-2inx}}{-4} \right) f(x)\,dx$$

$$= \frac{1}{4n\pi} \int_{-\infty}^{+\infty} \left(\frac{1 - e^{2inx}}{x^2} \right) f(x)\,dx$$

$$+ \frac{1}{4n\pi} \int_{-\infty}^{+\infty} \left(\frac{1 - e^{-2inx}}{x^2} \right) f(x)\,dx$$

Now we evaluate each integral along the real axis in the complex plane and use the residue theorem. Suppose the function $f(z)$ is analytic except for a set of isolated simple poles and approaches zero uniformly at infinity. Then the poles of the integrands are the poles of $f(z)$ and of

$$g_{n,\pm}(z) = \frac{1 - e^{\pm 2inz}}{z^2}$$

There is a pole of $g_{n,\pm}(z)$ at the origin. Let's look at the Laurent series:

$$\frac{1 - e^{\pm 2inz}}{z^2} = \frac{1 - \left(1 \pm 2inz - \dfrac{4n^2 z^2}{2} + \cdots \right)}{z^2}$$

$$g_{n,\pm}(z) = \mp \frac{2in}{z} + 2n^2 + \cdots$$

This pole is simple, and the residue of $g_{n,\pm}(z)$ there is $\mp 2in$ (Chapter 2, Section 2.6.3, method 1). We evaluate the first integral by closing the contour with a big semicircle in the upper half-plane. The pole is on the real axis, so we shall evaluate the integral by moving the contour slightly downward so that the path of integration lies below the real axis[2] (Figure 6.3). Then, if the poles of f are simple, the integral[3] is

$$I_1 = \frac{2\pi i}{4n\pi} \left[(-2in) f(0) + \sum_p \operatorname{Res} f(z_p) g_{n,+}(z_p) \right]$$

$$= f(0) + \frac{i}{2n} \sum_p \operatorname{Res} f(z_p) \frac{1 - e^{2inz_p}}{z_p^2} \rightarrow f(0) \quad \text{as } n \rightarrow \infty$$

[2]The result for the complete integral is independent of the method we choose to avoid the pole, provided that $f(z)$ has no poles on the real axis, because the singularity of $(\sin^2 nx)/x^2$ is removable.
[3]There may not be a pole of the integrand at the origin if $f(z)$ is zero at the origin. But then the integral is zero in the limit and so still equals $f(0)$. If the poles of f are not simple, the expression for the residue of the integrand at z_p is more complicated, but the result $I_1 \rightarrow f(0)$ as $n \rightarrow \infty$ still holds. Note that z_p has a positive imaginary part, since the relevant poles are in the upper half-plane, and so the exponential approaches 0 as $n \rightarrow \infty$ faster than any power of n. See Problem 30.

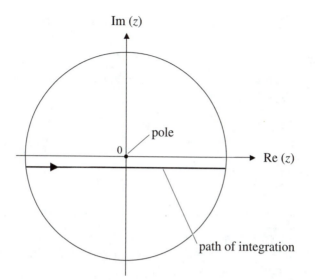

FIGURE 6.3. To demonstrate the sifting property of sequence (6.5), we evaluate the integral using a path that passes below the real axis, under the pole at $z = 0$.

For the second integral, we have to close downward. The pole at $z = 0$ is excluded from this contour, and so the integral is zero in the limit. Thus,

$$\lim_{n \to \infty} \int_{-\infty}^{+\infty} \frac{1}{n\pi} \frac{\sin^2 nx}{x^2} f(x) \, dx = f(0)$$

The sequence ϕ_n possesses the sifting property as $n \to \infty$.

6.1.2. The Derivative of the Delta Function

Now we are ready to determine the sifting property of the derivative of the delta function. Let's use the delta sequence[4] (6.5) to evaluate the integral:

$$\int_{-\infty}^{+\infty} \phi_n'(x) f(x) \, dx$$

We can do the integral by parts:

$$\int_{-\infty}^{+\infty} \phi_n'(x) f(x) \, dx = \phi_n(x) f(x) \Big|_{-\infty}^{+\infty} - \int_{-\infty}^{+\infty} \phi_n(x) f'(x) \, dx$$

---

[4]Since the explicit form of $\phi_n(x)$ is not needed, any differentiable delta sequence will suffice.

The integrated term is zero, provided that $f(x)$ remains bounded as $x \to \pm\infty$. Now if we take the limit as $n \to \infty$, we find

$$\int_{-\infty}^{+\infty} \delta'(x) f(x)\, dx = \lim_{n\to\infty} \int_{-\infty}^{+\infty} \phi'_n(x) f(x)\, dx = -\lim_{n\to\infty} \int_{-\infty}^{+\infty} \phi_n(x) f'(x)\, dx$$

Then, using the sifting property of ϕ_n, we obtain

$$\int_{-\infty}^{+\infty} \delta'(x) f(x)\, dx = -f'(0) \tag{6.6}$$

which is the sifting property for the derivative of the delta function.

We can repeat the steps above to evaluate the sifting property for the nth derivative of the delta function. The result is

$$\int_{-\infty}^{+\infty} \delta^{(n)}(x) f(x)\, dx = (-1)^n f^{(n)}(0) \tag{6.7}$$

Example 6.1. An ideal dipole has dipole moment $\vec{\mathbf{p}} = p\hat{\mathbf{x}}$. Express the charge density in terms of delta functions. Given that the electric potential is

$$\Phi(\vec{\mathbf{x}}) = \frac{1}{4\pi\varepsilon_0} \int \frac{\rho(\vec{\mathbf{x}}')}{|\vec{\mathbf{x}} - \vec{\mathbf{x}}'|} d^3 x'$$

find the electric potential due to a dipole placed at the origin.

We begin by modeling the dipole as two point charges q and $-q$, separated by a distance ℓ along the x-axis. Each point charge has a charge density represented by a delta function (see equation 6.1):

$$\rho(\vec{\mathbf{x}}) = -q\delta(x)\delta(y)\delta(z) + q\delta(x-\ell)\delta(y)\delta(z)$$

$$= -q\ell\delta(y)\delta(z) \left[\frac{\delta(x) - \delta(x-\ell)}{\ell} \right]$$

Now we let $\ell \to 0$ and $q \to \infty$ such that the product $q\ell = p$ remains finite.[5] Then

$$\rho(\vec{\mathbf{x}}) = -p\delta(y)\delta(z) \lim_{\ell\to 0} \left[\frac{\delta(x) - \delta(x-\ell)}{\ell} \right] = -p\delta(y)\delta(z)\delta'(x)$$

---

[5]Since the delta function is not a proper function, we should demonstrate that the limit below behaves as the derivative. The result may be proved using a delta sequence; see Problem 23.

We evaluate the potential using the sifting property of the delta functions:

$$\Phi(\vec{x}) = \frac{1}{4\pi\varepsilon_0} \int \frac{\rho(\vec{x}')}{|\vec{x} - \vec{x}'|} d^3 x'$$

$$= -\frac{p}{4\pi\varepsilon_0} \int_{-\infty}^{+\infty} \int_{-\infty}^{+\infty} \int_{-\infty}^{+\infty} \frac{\delta'(x')\delta(y')\delta(z')}{\sqrt{(x - x')^2 + (y - y')^2 + (z - z')^2}} dx'\, dy'\, dz'$$

$$= -\frac{p}{4\pi\varepsilon_0} \int_{-\infty}^{+\infty} \int_{-\infty}^{+\infty} \frac{\delta'(x')\delta(y')}{\sqrt{(x - x')^2 + (y - y')^2 + z^2}} dx'\, dy'$$

$$= -\frac{p}{4\pi\varepsilon_0} \int_{-\infty}^{+\infty} \frac{\delta'(x')}{\sqrt{(x - x')^2 + y^2 + z^2}} dx'$$

This integral is evaluated using the sifting property of the derivative of the delta function (result 6.6) with $f(x') = 1/\sqrt{(x - x')^2 + y^2 + z^2}$:

$$\Phi(\vec{x}) = -\frac{p}{4\pi\varepsilon_0}[- f'(x')|_{x'=0}] = \frac{p}{4\pi\varepsilon_0} \frac{(x - x')}{[(x - x')^2 + y^2 + z^2]^{3/2}}\Bigg|_{x'=0}$$

$$= \frac{p}{4\pi\varepsilon_0} \frac{x}{r^3}$$

where $r^2 = x^2 + y^2 + z^2$.

6.1.3. The Delta Function of a Function

Sometimes the argument of the delta function is itself a function. We can use any of our delta sequences to determine the sifting property of $\delta[g(x)]$:

$$\int_{-\infty}^{+\infty} \delta[g(x)]f(x)\, dx = \lim_{n\to\infty} \int_{-\infty}^{+\infty} \phi_n[g(x)]f(x)\, dx \qquad (6.8)$$

We make a change of variable to $u = g(x)$. Then $du = g'(x)\, dx$. Next we divide the range of integration into N pieces, where $g'(x) \neq 0$ within each piece (Figure 6.4). Thus, $g'(x)$ is either positive or negative within each piece and is zero only on the boundaries at $x = x_i$. Notice also that there is at most one zero of $g(x)$ in each piece. We label the values of x where $g(x) = 0$ as x_{0i}, $i = 1$ to N. Then the integral on the right-hand side of equation (6.8) is

$$\sum_{i=1}^{N} \int_{x_i}^{x_{i+1}} \phi_n[g(x)]f(x)\, dx = \sum_{i=1}^{N} \int_{g(x_i)}^{g(x_{i+1})} \phi_n(u)f[g^{-1}(u)]\frac{du}{g'(x)}$$

Let's look at one of the integrals in the sum. If $g'(x)$ is positive throughout the range x_i to x_{i+1} then $g(x_i)$ is less than $g(x_{i+1})$. As we take the limit $n \to \infty$, ϕ_n approaches the delta

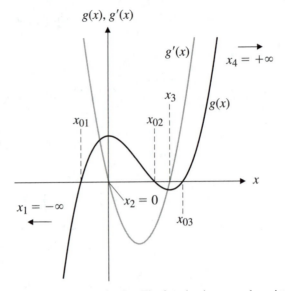

FIGURE 6.4. A function $g(x)$ and its derivative. The function is zero at the points x_{0i}; the derivative is zero at x_2 and x_3. Thus, $g'(x)$ is positive between $-\infty$ and x_2 and between x_3 and ∞, and is negative between x_2 and x_3.

function and thus goes to zero outside any finite range containing $u = 0$. We may expand the range of integration for u without changing the value of the integral[6]:

$$\lim_{n \to \infty} \int_{g(x_i)}^{g(x_{i+1})} \phi_n(u) f[g^{-1}(u)] \frac{du}{g'[g^{-1}(u)]} = \lim_{n \to \infty} \int_{-\infty}^{+\infty} \phi_n(u) f[g^{-1}(u)] \frac{du}{g'[g^{-1}(u)]}$$

$$= \frac{f[g^{-1}(0)]}{g'[g^{-1}(0)]} = \frac{f(x_{0i})}{g'(x_{0i})}$$

where the particular zero x_{0i} of g lies between x_i and x_{i+1}. If there is no zero of g in this range, then the integral is zero.

Now if $g'(x)$ is negative throughout the range x_i to x_{i+1}, then $g(x_i)$ is greater than $g(x_{i+1})$ and we get

$$\lim_{n \to \infty} \int_{g(x_i)}^{g(x_{i+1})} \phi_n(u) f[g^{-1}(u)] \frac{du}{g'[g^{-1}(u)]} = \lim_{n \to \infty} \int_{+\infty}^{-\infty} \phi_n(u) f[g^{-1}(u)] \frac{du}{g'[g^{-1}(u)]}$$

$$= \lim_{n \to \infty} -\int_{-\infty}^{+\infty} \phi_n(u) f[g^{-1}(u)] \frac{du}{g'[g^{-1}(u)]}$$

$$= -\frac{f[g^{-1}(0)]}{g'[g^{-1}(0)]} = \frac{f(x_{0i})}{|g'(x_{0i})|}$$

[6]Strictly, we must also redefine the function $f[g^{-1}(u)]$ to be zero (or any constant) outside the range $u = g(x_i)$ to $u = g(x_{i+1})$.

Adding the results for each of the pieces gives

$$\int_{-\infty}^{+\infty} \delta[g(x)] f(x)\, dx = \sum_{i=1}^{N} \frac{f(x_{0i})}{|g'(x_{0i})|} \tag{6.9}$$

where the sum is over all the zeros of the function $g(x)$: $g(x_{0i}) = 0$, $i = 1$ to N. We may also write the result in the form

$$\delta[g(x)] = \sum_{i=1}^{N} \frac{\delta(x - x_{0i})}{|g'(x_{0i})|} \tag{6.10}$$

Equation (6.9) (or 6.10) has numerous important applications in physics. It is valid only if the sum is over a *finite* number N of zeros and there are no repeated roots. [If there is a repeated root, then $g'(x_{0i}) = 0$ for those roots, and our demonstration fails.]

As a specific example, consider the function $g(x) = x^2 - 4$. There are two roots at $x = \pm 2$, and $g'(x) = 2x$ is zero at $x = 0$. Thus, we divide the range of integration as follows:

$$\int_{-\infty}^{+\infty} \delta(x^2 - 4) f(x)\, dx = \int_{-\infty}^{0} \delta(x^2 - 4) f(x)\, dx + \int_{0}^{+\infty} \delta(x^2 - 4) f(x)\, dx$$

Now we change variables to $u = x^2 - 4$, $du = 2x\, dx$. In the first integral, x is negative, and so $x = -\sqrt{u + 4}$. Thus,

$$\int_{-\infty}^{+\infty} \delta(x^2 - 4) f(x)\, dx$$

$$= \int_{+\infty}^{-4} \delta(u) f(-\sqrt{u + 4}) \frac{du}{-2\sqrt{u + 4}} + \int_{-4}^{+\infty} \delta(u) f(\sqrt{u + 4}) \frac{du}{2\sqrt{u + 4}}$$

Next we extend the range of integration by defining $f(\pm\sqrt{u + 4}) \equiv 0$ for $u < -4$. The precise value chosen is unimportant, since $\delta(u)$ is zero for $u \neq 0$. Thus,

$$\int_{-\infty}^{+\infty} \delta(x^2 - 4) f(x)\, dx$$

$$= \int_{-\infty}^{+\infty} \delta(u) f(-\sqrt{u + 4}) \frac{du}{2\sqrt{u + 4}} + \int_{-\infty}^{+\infty} \delta(u) f(\sqrt{u + 4}) \frac{du}{2\sqrt{u + 4}}$$

Then, using the sifting property, we have

$$\int_{-\infty}^{+\infty} \delta(x^2 - 4) f(x)\, dx = \frac{f(-2)}{4} + \frac{f(2)}{4}$$

which is in agreement with the general result (6.9).

Example 6.2.   A sheet of charge lies in the $z = 0$ plane. The surface charge density is σ_0. Express the volume charge density in spherical coordinates.

The charge is localized to one value of z: $z = 0$. Thus, we may express ρ with a delta function in z:

$$\rho(\vec{\mathbf{x}}) = k\delta(z)$$

To determine k, we integrate over a cylindrical volume of cross-sectional area dA extending from $z = -\infty$ to $z = +\infty$ (Figure 6.5). The cylinder acts like a cookie cutter, cutting out an amount of charge $\sigma_0\, dA$ from the sheet. The amount of charge inside the cylinder may also be computed from the charge density:

$$dq = \int_{\text{cylinder}} \rho(\vec{\mathbf{x}})\, dV = \int_{-\infty}^{+\infty} k\delta(z)\, dz\, dA = k\, dA$$

and thus $k = \sigma_0$:

$$\rho(\vec{\mathbf{x}}) = \sigma_0\delta(z)$$

Now we change to spherical coordinates: $z = r\cos\theta$, and so

$$\rho(\vec{\mathbf{x}}) = \sigma_0\delta(r\cos\theta)$$

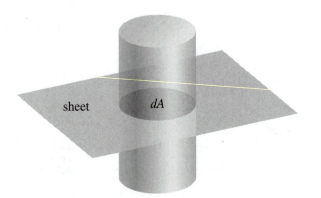

FIGURE 6.5. Cutting out an area dA from the charged sheet (Example 6.2). The charge inside the cylinder is $\sigma_0\, dA$.

To interpret the delta function, we first note that charge exists at all values of r, $0 \le r \le \infty$, so we cannot have a delta function in r. But charge exists at only one value of θ, $\theta = \pi/2$. Thus, we must have a delta function in θ. Then we apply result (6.4) with $a = r$:

$$\rho(\vec{\mathbf{x}}) = \sigma_0\frac{\delta(\cos\theta)}{r}$$

Now we may use result (6.10) to evaluate $\delta(\cos\theta)$. There is only one zero of $\cos\theta$ in the physical range of interest, $0 \le \theta \le \pi$, and that is at $\theta = \pi/2$. Thus,

$$\rho(\vec{\mathbf{x}}) = \frac{\sigma_0}{r} \frac{\delta(\theta - \pi/2)}{|-\sin\theta|_{\theta=\pi/2}} = \frac{\sigma_0}{r}\delta(\theta - \pi/2)$$

We can check the result by calculating the amount of charge inside a spherical shell of thickness dr (Figure 6.6). This shell intersects the sheet in a circle of thickness dr, and so the charge inside is $dq = \sigma_0\, dA = \sigma_0 2\pi r\, dr$. Integrating the volume density and using the sifting property to evaluate the integral over the delta function, we get

$$dq = \int_{\text{shell}} \rho(\vec{\mathbf{r}})\, dV = \int_0^{2\pi} d\phi \int_0^{\pi} \frac{\sigma_0}{r}\delta(\theta - \pi/2)r^2 \sin\theta\, d\theta\, dr$$

$$= 2\pi\sigma_0 r \sin\frac{\pi}{2} dr = \sigma_0 2\pi r\, dr$$

as required.

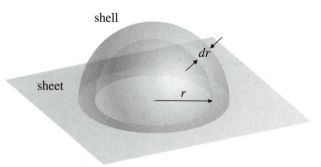

shell

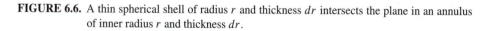

sheet

dr

r

FIGURE 6.6. A thin spherical shell of radius r and thickness dr intersects the plane in an annulus of inner radius r and thickness dr.

6.1.4. The Integral of the Delta Function

To interpret the integral of a delta function, let's integrate our block delta sequence (equation 6.3). First let $x < -1/n$. Then

$$\int_{-\infty}^{x} \phi_n(u)\, du = 0$$

since $\phi_n(u)$ is zero throughout the range of integration. Now let $x > 1/n$. Then

$$\int_{-\infty}^{x} \phi_n(u)\, du = \int_{-1/n}^{1/n} \frac{n}{2}\, du = \frac{n}{2}u\Big|_{-1/n}^{1/n} = 1$$

so

$$\int_{-\infty}^{x} \phi_n(u)\, du = \begin{cases} 0 & \text{if } x < -1/n \\ 1 & \text{if } x > +1/n \end{cases}$$

As we let $n \to \infty$, we have

$$
\int_{-\infty}^{x} \delta(u) \, du = \begin{cases} 0 & \text{if } x < 0 \\ 1 & \text{if } x > 0 \end{cases} = \Theta(x) \tag{6.11}
$$

where $\Theta(x)$ is the *step distribution*.[7] It has the property that

$$
\int_{-\infty}^{+\infty} \Theta(x) f(x) \, dx = \int_{0}^{+\infty} f(x) \, dx
$$

Since $\Theta(x)$ is the integral of $\delta(x)$, we may reasonably conclude that

> The delta function is the derivative of the step distribution:
>
> $$
> \frac{d\Theta}{dx} = \delta(x) \tag{6.12}
> $$

We can check this conclusion by determining that $d\Theta/dx$ possesses the sifting property:

$$
\int_{-\infty}^{+\infty} \frac{d\Theta(x)}{dx} f(x) \, dx = f(x)\Theta(x)\big|_{-\infty}^{+\infty} - \int_{-\infty}^{+\infty} f'(x)\Theta(x) \, dx
$$

$$
= \lim_{x \to \infty} f(x) - 0 - \int_{0}^{+\infty} f'(x) \, dx
$$

$$
= \lim_{x \to \infty} f(x) - f(x)\big|_{0}^{\infty}
$$

$$
= \lim_{x \to \infty} f(x) - \lim_{x \to \infty} f(x) + f(0)
$$

$$
= f(0)
$$

as required.

6.1.5. The Fourier Series of a Delta Function

We may find the Fourier series of a delta function by finding the Fourier series of a delta sequence and taking the limit as $n \to \infty$. Since the Fourier series is periodic, we will actually get a periodic repetition of the delta function. The easiest sequence to handle is the sequence of blocks (6.3). The Fourier series of this delta sequence $\phi_n(x)$ on the range $-L < x < L$ may be written

$$
\phi_n(x) = \sum_{m=-\infty}^{+\infty} c_m e^{im\pi x/L}
$$

---

[7] Previously we used the notation $S(x)$ for the step function. The notation $\Theta(x)$ distinguishes the distribution from the function.

where (equation 4.31)

$$c_m = \frac{1}{2L} \int_{-L}^{L} \phi_n(x) e^{-im\pi x/L} \, dx$$

$$= \frac{1}{2L} \int_{-1/n}^{+1/n} \frac{n}{2} e^{-im\pi x/L} \, dx$$

$$= \frac{n}{4m\pi} \left(\frac{e^{-im\pi/nL} - e^{im\pi/nL}}{-i} \right) = \frac{n}{2m\pi} \sin \frac{m\pi}{nL}$$

Now as $n \to \infty$, $m\pi/nL \to 0$. Expanding the sine in a Taylor series and keeping only the first term, we find

$$c_m \to \frac{n}{2m\pi} \frac{m\pi}{nL} = \frac{1}{2L}$$

Therefore, the Fourier series of the delta function is

$$\delta(x) = \frac{1}{2L} \sum_{m=-\infty}^{+\infty} e^{im\pi x/L} \tag{6.13}$$

which is a very odd-looking series! Since the coefficients do not decrease as m increases, the series does not converge in the usual sense (pointwise or in the mean). That should not surprise us unduly, since $\delta(x)$ is not a proper limit. However, we can check for the sifting property:

$$\int_{-L}^{+L} \frac{1}{2L} \sum_{m=-\infty}^{+\infty} e^{im\pi x/L} f(x) \, dx = \sum_{m=-\infty}^{+\infty} \frac{1}{2L} \int_{-L}^{+L} e^{im\pi x/L} f(x) \, dx$$

The integral on the right gives the coefficient a_{-m} in the Fourier series

$$f(x) = \sum_{p=-\infty}^{+\infty} a_p e^{ip\pi x/L}$$

for the function $f(x)$ on the range $-L$ to $+L$. Then

$$\int_{-L}^{+L} \frac{1}{2L} \sum_{m=-\infty}^{+\infty} e^{im\pi x/L} f(x) \, dx = \sum_{m=-\infty}^{+\infty} a_{-m} = \sum_{m=-\infty}^{+\infty} a_m = f(0)$$

and so the series (6.13) possesses the sifting property. Thus, equation (6.13) represents the delta function on the range $(-L, L)$.

We should perhaps worry about changing the order of the sum and the integral, since the series is not uniformly convergent. In fact, it does not converge at all! The validity of this procedure will be discussed in Section 6.3.

Fourier sine and cosine series may be found similarly.

Example 6.3. An initially stationary string of length L is hit by a hammer blow at $x = L/3$. Determine the subsequent motion of the string. (The speed of waves on the string is $v = \sqrt{T/\mu}$, where T is the tension and μ is the mass per unit length. See also Chapter 4, Section 4.4.2.)

We may model the hammer blow as an impulse I_0 occurring at $t = 0$ and at $x = L/3$. The impulse per unit length is

$$\frac{dI_y(x)}{dx} = I_0 \delta(x - L/3)$$

Using the impulse-momentum theorem,[8] we can determine the initial velocity of the string at $t = 0+$. The change in momentum of a string element of length dx and mass $dm = \mu \, dx$ is

$$\frac{\partial y(x, 0+)}{\partial t} \mu \, dx = \frac{dI_y(x)}{dx} dx$$

$$\frac{\partial y(x, 0+)}{\partial t} = \frac{I_0}{\mu} \delta(x - L/3)$$

which, together with $y(x, 0) = 0$, gives the initial conditions for the string. We expect a solution of the form (equation 4.29)

$$y(x, t) = \sum_{n=1}^{\infty} a_n \sin \frac{n\pi x}{L} \sin \frac{n\pi vt}{L}$$

which satisfies the initial condition of zero string displacement at $t = 0$. Then

$$\frac{\partial y}{\partial t} = \sum_{n=1}^{\infty} a_n \sin \left(\frac{n\pi x}{L} \right) \frac{n\pi v}{L} \cos \frac{n\pi vt}{L}$$

$$\left. \frac{\partial y}{\partial t} \right|_{t=0} = \sum_{n=1}^{\infty} a_n \frac{n\pi v}{L} \sin \frac{n\pi x}{L}$$

Applying the second initial condition, we have

$$\frac{I_0}{\mu} \delta(x - L/3) = \sum_{n=1}^{\infty} a_n \frac{n\pi v}{L} \sin \frac{n\pi x}{L}$$

Now we find the coefficients in the usual way:

$$a_n \frac{n\pi v}{L} = \frac{2}{L} \int_0^L \frac{I_0}{\mu} \delta(x - L/3) \sin \frac{n\pi x}{L} dx = \frac{2I_0}{L\mu} \sin \frac{n\pi}{3} \qquad (6.14)$$

where we used the sifting property. As we expect for a delta function series, the term on the right-hand side does not decrease with increasing n. (Compare equation 6.14

[8] See Lea and Burke, p. 200.

with equation 6.13.) However, the coefficients a_n that describe the string displacement do decrease:

$$a_n = \frac{L}{n\pi v}\frac{2I_0}{L\mu}\sin\frac{n\pi}{3} = \frac{2}{n\pi v}\frac{I_0}{\mu}\sin\frac{n\pi}{3}$$

and the solution for the string displacement is

$$y(x,t) = \sum_{n=1}^{\infty} \frac{2}{n\pi v}\frac{I_0}{\mu}\sin\frac{n\pi}{3}\sin\frac{n\pi x}{L}\sin\frac{n\pi vt}{L}$$

Notice that all the terms with $n = 3m$ and m an integer are zero, since we have $a_{3m} \propto \sin m\pi = 0$ and so every third harmonic is missing. The first few terms are

$$y = \frac{\sqrt{3}I_0}{\pi v\mu}\left(\sin\frac{\pi x}{L}\sin\frac{\pi vt}{L} + \frac{1}{2}\sin\frac{2\pi x}{L}\sin\frac{2\pi vt}{L} - \frac{1}{4}\sin\frac{4\pi x}{L}\sin\frac{4\pi vt}{L} + \cdots\right)$$

Figure 6.7 shows the sum of the series for the string displacement, up to the $n = 10$ term, at times $vt/L = 1/8, 1/4,$ and $1/2$. Notice the peak at $x = L/3$ at the earliest time.

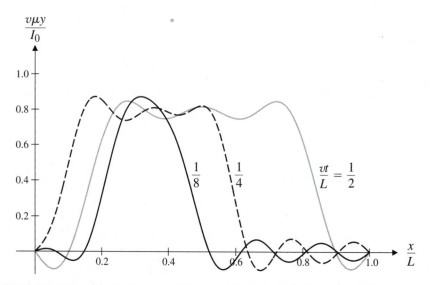

FIGURE 6.7. The string displacement (Example 6.3) at times $vt/L = 1/8, 1/4,$ and $1/2$. The vertical axis shows the dimensionless quantity $v\mu y/I_0$. The dimensionless variable x/L is plotted on the horizontal axis. The hammer blow at $x = L/3$ causes the peak in the displacement at $L/3$ at the earliest time. The displacement spreads along the string.

6.1.6. Integral Representations of the Delta Function

We can find an integral representation of the delta function by finding its Laplace transform. To use this transform, we need a delta sequence defined for $t > 0$. We can use a set of blocks defined as follows:

$$\phi_n(t) = \begin{cases} 0 & \text{if } t \leq 0 \\ n & \text{if } 0 < t < 1/n \\ 0 & \text{if } t > 1/n \end{cases} \tag{6.15}$$

This sequence of functions has essentially the same behavior as our previous block sequence (6.3), and the proof that it is a delta sequence is almost identical. The Laplace transform is

$$\Phi_n(s) = \int_0^\infty \phi_n e^{-st}\,dt = \int_0^{1/n} n e^{-st}\,dt = n \left.\frac{e^{-st}}{-s}\right|_0^{1/n} = \frac{n}{s}(1 - e^{-s/n})$$

Now we take the limit as $n \to \infty$:

$$\mathcal{L}[\delta(t)] = \frac{n}{s}\left\{1 - \left[1 - \frac{s}{n} + \frac{1}{2}\left(\frac{s}{n}\right)^2 + \cdots\right]\right\}$$

$$= 1 - \frac{s}{2n} + \cdots \to 1 \quad \text{as } n \to \infty$$

Thus, the Laplace transform of the delta function is 1.

We could also obtain this result from the relation[9] $\delta(t) = d\Theta/dt$ and the fact that the Laplace transform of the step function is $1/s$. Then

$$\mathcal{L}[\delta(t)] = s\mathcal{L}[\Theta(t)] - \Theta(0) = s/s = 1$$

If we invert the transform using the Mellin inversion integral (equation 5.19), we obtain

$$\delta(t) = \frac{1}{2\pi i}\int_{-i\infty}^{+i\infty} e^{st}\,ds$$

where we may integrate along the imaginary axis since the integrand has no poles.

Now make the change of variable $s = i\omega$. We find

$$\delta(t) = \frac{1}{2\pi i}\int_{-\infty}^{+\infty} e^{i\omega t}\,i\,d\omega = \frac{1}{2\pi}\int_{-\infty}^{+\infty} e^{i\omega t}\,d\omega \tag{6.16}$$

This expression is extremely important and finds many applications in physics. Notice the similarity to equation (6.13). The integral is the extension of the Fourier series representation to the case of a continuous, rather than a discrete, frequency spectrum.[10]

---

[9]This demonstration appears less satisfactory because we have to invoke a value of the step distribution *at* the origin. Using the theory developed in Section 6.2, we can show that the initial condition may be dropped. See Problem 24.

[10]Demonstration of the sifting property may be accomplished in several ways. For example, expression (6.16) may be written as $\lim_{n\to\infty}(1/2\pi)\int_{-\infty}^{+\infty}\exp(-\omega^2/4n^2 + i\omega t)\,d\omega$. The integral is easily evaluated (Chapter 7, Example 7.2), and the result is the delta sequence (6.18).

6.2. DEVELOPING A THEORY OF DISTRIBUTIONS

The delta function (together with its derivatives) is a very useful mathematical object that often simplifies calculations in physical problems. Because it is not a properly defined function, we used sequences of proper functions—the delta sequences—to understand the properties of the delta function. The delta function is mathematically defined by the sifting property (6.2). In this section, we shall begin to formalize and generalize these ideas.

To develop a more rigorous theory of distributions,[11] we start with the definitions of two classes of functions: *core functions* and *test functions*. The properties of these functions determine the particular distribution theory. The example below is illustrative of the techniques but should not be regarded as a description of the only class of distributions that exists.

- **Test functions.** For our illustrative theory, we choose the test functions $f(x)$ to be infinitely differentiable and to go to zero very fast as $x \to \infty$. Taking $f = 0$ outside some finite range will certainly be fast enough. We'll explore this feature further as we develop the theory.

- **Core functions.** For our theory, we choose the core functions $g(x)$ to be infinitely differentiable.

We'll also need to define a new form of convergence, called weak convergence.

- **Weak convergence.** Let $g_n(x)$ be a sequence of core functions. Then the sequence is weakly convergent if

$$\lim_{n \to \infty} \int_{-\infty}^{+\infty} g_n(x) f(x) \, dx$$

exists for any test function $f(x)$.

The definition of a distribution as the weak limit of a sequence of core functions now follows.

- **Definition of a distribution.**

$\phi(x)$ is a distribution if there exists a sequence of core functions $g_n(x)$ that converges weakly to $\phi(x)$, in which case

$$\lim_{n \to \infty} \int_{-\infty}^{+\infty} g_n(x) f(x) \, dx = \int_{-\infty}^{+\infty} \phi(x) f(x) \, dx \qquad (6.17)$$

for any test function $f(x)$.

---

[11] See Lighthill's book for more on this topic. The test functions here are akin to his "good" functions and the core functions to his "fairly good" functions.

Many different sequences of core functions may converge to the same distribution.

The sequence of core functions

$$g_n(x) = \frac{n}{\sqrt{\pi}} e^{-n^2 x^2} \tag{6.18}$$

illustrates some of these concepts. The sequence does not converge pointwise, since $g_n(x) \to 0$ as $n \to \infty$ for $x \neq 0$, but $g_n(x)$ diverges as $n \to \infty$ for $x = 0$. However, the sequence does converge weakly. To see which distribution it converges to, we evaluate the integral (6.17):

$$\lim_{n \to \infty} \int_{-\infty}^{+\infty} \frac{n}{\sqrt{\pi}} e^{-n^2 x^2} f(x)\, dx$$

$$= \lim_{n \to \infty} \left(\int_{-\infty}^{-1/\sqrt{n}} + \int_{-1/\sqrt{n}}^{+1/\sqrt{n}} + \int_{+1/\sqrt{n}}^{+\infty} \right) \frac{n}{\sqrt{\pi}} e^{-n^2 x^2} f(x)\, dx$$

$$= \lim_{n \to \infty} (I_1 + I_2 + I_3)$$

Now the test function $f(x)$ is infinitely differentiable and goes to zero as $x \to \pm\infty$. Thus, $f(x)$ is bounded. Let $|f(x)| \leq M_1$ for $0 \leq x \leq \infty$. Then

$$I_3 \leq \frac{M_1}{2} \int_{+1/\sqrt{n}}^{+\infty} \frac{2n}{\sqrt{\pi}} e^{-n^2 x^2}\, dx = \frac{M_1}{2} \int_{\sqrt{n}}^{+\infty} \frac{2}{\sqrt{\pi}} e^{-u^2}\, du = \frac{M_1}{2} [1 - \Phi(\sqrt{n})]$$

where $\Phi(\sqrt{n})$ is the error function,[12] which approaches 1 as $n \to \infty$.

$$\Phi(x) = \frac{2}{\sqrt{\pi}} \int_0^x e^{-t^2}\, dt = 1 - \frac{1}{\sqrt{\pi}} \frac{e^{-x^2}}{x} + \cdots$$

[The asymptotic expression for $\Phi(x)$ may be found in Appendix IX, equation (29).] Thus, $I_3 \to 0$ as $n \to \infty$. Similarly, we can show that $I_1 \to 0$. However, for I_2, we have

$$I_2 = \int_{-1/\sqrt{n}}^{+1/\sqrt{n}} \frac{n}{\sqrt{\pi}} e^{-n^2 x^2} f(x)\, dx = f(\xi) \int_{-1/\sqrt{n}}^{+1/\sqrt{n}} \frac{n}{\sqrt{\pi}} e^{-n^2 x^2}\, dx$$

$$= f(\xi) \frac{2}{\sqrt{\pi}} \int_0^{\sqrt{n}} e^{-u^2}\, du$$

by the mean value theorem, where $-1/\sqrt{n} \leq \xi \leq +1/\sqrt{n}$ and $u = nx$. Thus,

$$I_2 = f(\xi) \Phi(\sqrt{n})$$

and so

$$\lim_{n \to \infty} \int_{-\infty}^{+\infty} \frac{n}{\sqrt{\pi}} e^{-n^2 x^2} f(x)\, dx = \lim_{n \to \infty} I_2 = f(0) = \int_{-\infty}^{+\infty} \phi(x) f(x)\, dx$$

---

[12]The error function is also called erf (x). See Appendix IX.

The distribution $\phi(x)$, defined as the weak limit of sequence (6.18), exhibits the sifting property, and so it is the delta function. We may write

$$\frac{n}{\sqrt{\pi}} e^{-n^2 x^2} \rightarrow \delta(x) \text{ weakly} \qquad (6.19)$$

Some sequences of core functions may have pointwise limits that are ordinary functions and may also have weak limits that are distributions. It is even possible that $g_n(x) \rightarrow g(x)$ pointwise as $n \rightarrow \infty$, $g_n(x) \rightarrow \phi(x)$ weakly, but

$$\int_{-\infty}^{+\infty} \phi(x) f(x)\, dx \neq \int_{-\infty}^{+\infty} g(x) f(x)\, dx$$

This kind of bizarre sequence is rarely of any importance in physics.

In physics, pointwise convergence is not always important, since a physical quantity cannot be measured *at a point*, but only over some small but finite region. Thus, weak convergence is often all we need.

There is a close relationship between functions and distributions.

The Smudging Theorem. *For every continuous function $h(x)$, we can construct a sequence $g_n(x)$ of core functions such that*

$$\lim_{n \to \infty} |g_n(x) - h(x)| < \varepsilon$$

for any $\varepsilon > 0$ and for all x in any finite interval.[13]

According to this theorem, we can "smooth" the function $h(x)$ over the interval $(x - 1/n, x + 1/n)$ to create an infinitely differentiable $g_n(x)$, and the two functions differ negligibly. Since the sequence of core functions converges to a distribution weakly, this means we can replace the function with a distribution.

> For every continuous function, an equivalent distribution exists.

We cannot make the reverse statement; there is not necessarily a continuous function corresponding to every distribution. The delta function is an example of a distribution for which there is no corresponding continuous function.

The class of distributions is an important addition to our mathematical arsenal. In the next section, we shall learn more about how to use them.

6.3. PROPERTIES OF DISTRIBUTIONS

- Distributions may be added, subtracted, multiplied by constants, or multiplied by infinitely differentiable functions. These assertions may be easily proved using the definition of

---

[13] For the proof of this theorem, see Butkov, p. 243, or Lighthill, p. 21.

a distribution as a weak limit of a sequence of core functions. For example, let $h(x)$ be an infinitely differentiable function, and let $\phi(x)$ be the weak limit of the sequence g_n. Then

$$\int_{-\infty}^{+\infty} [h(x)\phi(x)]f(x)\, dx = \lim_{n\to\infty} \int_{-\infty}^{+\infty} [h(x)g_n(x)]f(x)\, dx$$

$$= \lim_{n\to\infty} \int_{-\infty}^{+\infty} g_n(x)[h(x)f(x)]\, dx$$

$$= \int_{-\infty}^{+\infty} \phi(x)[h(x)f(x)]\, dx$$

since $h(x)f(x)$ is a test function if f is. So we can define a new distribution $\psi(x) = h(x)\phi(x)$. Thus, for example, $x\delta(x)$ is a distribution. Let's see what it does.

$$\int_{-\infty}^{+\infty} [x\delta(x)]f(x)\, dx = \int_{-\infty}^{+\infty} \delta(x)[xf(x)]\, dx = 0 \times f(0) = 0$$

Thus, $x\delta(x) = 0$, the zero distribution.[14]

- Distributions may NOT be multiplied or divided by other distributions depending on the same variable. For example, $[\delta(x)]^2$ is meaningless. [However, $\delta(\mathbf{r}) = \delta(x)\delta(y)\delta(z)$ does make sense because the arguments x, y, and z of the three delta functions are *independent* variables.]
- If $\phi(x)$ is a distribution, so are $\phi(x-a)$ and $\phi(cx)$.
- Distributions are infinitely differentiable, because the core functions are. Moreover, if $g_n(x) \to \phi(x)$ weakly, then $g'_n(x) \to \phi'(x)$:

$$\lim_{n\to\infty} \int_{-\infty}^{+\infty} g'_n(x)f(x)\, dx = \lim_{n\to\infty} \left(f(x)g_n(x)|_{-\infty}^{+\infty} - \int_{-\infty}^{+\infty} g_n(x)f'(x)\, dx \right)$$

The integrated term is zero by the properties of the test functions, and so

$$\boxed{\int_{-\infty}^{+\infty} \phi'(x)f(x)\, dx = -\int_{-\infty}^{+\infty} \phi(x)f'(x)\, dx} \qquad (6.20)$$

This is a property we have already established for the delta function, and it serves to define the derivative $\phi'(x)$. Higher-order derivatives may be computed similarly.

- Since distributions can be differentiated, they can be the solutions of differential equations.

Example 6.4.   Find the solution of the differential equation

$$(x-1)\frac{dy}{dx} + y = 0$$

---

[14]We'll need this result in the next chapter.

We can integrate this equation directly to obtain one solution:

$$\frac{dy}{y} = -\frac{dx}{x-1}$$

$$\ln y = -\ln(x-1) + C$$

$$y = \frac{A}{x-1}$$

This solution is valid for $x > 1$ or for $x < 1$ but not *at* $x = 1$.

Now consider the distribution $y = \delta(x-1)$. If this satisfies the differential equation, then we must have

$$\int_{-\infty}^{+\infty} [(x-1)\delta'(x-1) + \delta(x-1)]f(x)\,dx = 0$$

Let's evaluate the first term using the sifting property of δ' (equation 6.6):

$$\int_{-\infty}^{+\infty} (x-1)\delta'(x-1)f(x)\,dx = -\int_{-\infty}^{+\infty} \delta(x-1)\frac{d}{dx}[(x-1)f(x)]\,dx$$

$$= -\int_{-\infty}^{+\infty} \delta(x-1)[f(x) + (x-1)f'(x)]\,dx$$

$$= -f(1)$$

The second term is

$$\int_{-\infty}^{+\infty} \delta(x-1)f(x)\,dx = f(1)$$

so the differential equation is satisfied. We must choose this delta function solution if the point $x = 1$ is within our domain of interest.

Note that since the differential equation is first order, we expect only one solution. What we learn here is that the character of that solution depends on the domain of validity.

Some differential equations contain a distribution. The solutions of such equations are also distributions.

Example 6.5. Find the solution of Poisson's equation for the electric field due to a sheet of charge with charge density $\rho = \sigma_0\delta(x-a)$:

$$\frac{dE_x}{dx} = \frac{\sigma_0}{\varepsilon_0}\delta(x-a)$$

From equation (6.12), the solution is

$$E_x = A + \frac{\sigma_0}{\varepsilon_0}\Theta(x-a)$$

where Θ is the step distribution (see Section 6.1.4). Since the electric field points directly away from positive charge, we can find the integration constant A by invoking symmetry about the sheet (Figure 6.8):

$$E_x(x < a) = -E_x(x > a)$$

$$A = -\left(A + \frac{\sigma_0}{\varepsilon_0}\right)$$

and so

$$A = -\frac{\sigma_0}{2\varepsilon_0}$$

Thus,

$$E_x = \frac{\sigma_0}{\varepsilon_0}\left(-\frac{1}{2} + \Theta(x - a)\right)$$

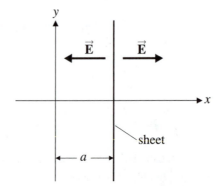

FIGURE 6.8. Electric field due to a sheet of charge (Example 6.5).

6.4. SEQUENCES AND SERIES

We can extend the concept of weak convergence to sequences and series. For example, the function e^{inx} is a core function and therefore may also be regarded as a distribution. This means that *any* Fourier series may be regarded as a series of distributions. The series converges weakly if the sequence of partial sums converges weakly. Once we make this identification, some very useful results follow.

The sequence of functions e^{inx} has no proper limit. But let's see if there's a weak limit:

$$\int_{-\infty}^{+\infty} e^{inx} f(x)\, dx = f(x) \frac{e^{inx}}{in}\Bigg|_{-\infty}^{+\infty} - \int_{-\infty}^{+\infty} \frac{e^{inx}}{in} f'(x)\, dx$$

The integrated term is zero, by the properties of the test functions [$f(x)$ is zero outside some finite range $a \leq x \leq b$]. The absolute value of the remaining term is

$$\left| \int_{-\infty}^{+\infty} \frac{e^{inx}}{in} f'(x)\, dx \right| = \frac{1}{n} \left| \int_{a}^{b} e^{inx} f'(x)\, dx \right|$$

$$\leq \frac{1}{n} \int_{a}^{b} \left| e^{inx} f'(x) \right| dx = \frac{1}{n} \left| e^{in\xi} f'(\xi) \right| (b-a) \to 0 \quad \text{as } n \to \infty$$

by the mean value theorem, where $a \leq \xi \leq b$ and $\left| e^{in\xi} f'(\xi) \right|$ is bounded in the interval $[a, b]$. Thus,

$$e^{inx} \to 0 \text{ weakly}$$

If a sequence of distributions $\phi_n(x)$ converges weakly to $\phi(x)$, $\phi(x) = \lim_{n \to \infty} \phi_n(x)$, then the sequence of derivatives $\phi'_n(x)$ converges to $\phi'(x)$:

$$\phi'(x) = \lim_{n \to \infty} \phi'_n(x) \tag{6.21}$$

To see why, note that

$$\lim_{n \to \infty} \int_{-\infty}^{+\infty} \phi'_n(x) f(x)\, dx = -\lim_{n \to \infty} \int_{-\infty}^{+\infty} \phi_n(x) f'(x)\, dx \quad \text{(result 6.20)}$$

$$= -\int_{-\infty}^{+\infty} \phi(x) f'(x)\, dx$$

$$= \int_{-\infty}^{+\infty} \phi'(x) f(x)\, dx$$

A convergent series of distributions can be differentiated term by term.

To prove this statement, let $S_N(x) = \sum_{n=1}^{N} \phi_n(x)$. Because the sum has a finite number of terms, we may differentiate to obtain $S'_N(x) = \sum_{n=1}^{N} \phi'_n(x)$. Then, taking the limit, we have $S(x) = \lim_{N \to \infty} S_N(x)$, and, by theorem (6.21) above, $S'(x) = \lim_{N \to \infty} S'_N(x)$. This is a powerful result: Series of distributions possess some of the same properties as uniformly convergent series.

If we start with a convergent Fourier series $S(x)$, we can differentiate term by term, thus multiplying each coefficient by in. We can do this m times, so the series with coefficients $(in)^m a_n$ also converges to $d^m S/dx^m$. We have thus shown that *every* Fourier series, considered as a series of distributions, converges (weakly), even when the coefficients do not decrease with increasing n. As a consequence, every Fourier series may be integrated and differentiated term by term.[15]

---

[15] We made use of this result in Chapter 4.

Example 6.6.   Find a Fourier series for a delta function by differentiating the Fourier series for the triangle function

$$f(x) = \begin{cases} x(2h/L) & \text{if } 0 \le x < L/2 \\ (L-x)(2h/L) & \text{if } L/2 \le x < L \end{cases}$$

on the range $(0, L)$ (see Example 4.4):

$$f(x) = \frac{8h}{\pi^2} \sum_{n=0}^{\infty} \frac{(-1)^n}{(2n+1)^2} \sin \frac{(2n+1)\pi x}{L}$$

This series is nicely convergent. If we differentiate once, we get a step distribution (compare with equation 4.28):

$$f'(x) = \begin{cases} 2h/L & \text{if } 0 \le x < L/2 \\ -2h/L & \text{if } L/2 \le x < L \end{cases} = \frac{4h}{L} \left\{ \frac{1}{2} - \Theta(x - L/2) \right\}$$

$$= \frac{8h}{\pi L} \sum_{n=0}^{\infty} \frac{(-1)^n}{(2n+1)} \cos \frac{(2n+1)\pi x}{L}$$

which also converges, but more slowly, and if we differentiate again, we get a delta function:

$$f''(x) = -\frac{4h}{L}\delta(x - L/2) = \frac{4h}{L}\frac{2}{L} \sum_{n=0}^{\infty} (-1)^{n+1} \sin \frac{(2n+1)\pi x}{L} \qquad (6.22)$$

Actually it is a negative delta function, since the step function steps down at $x = L/2$. Compare this result with equation (6.14). Here we have no even terms, because $\sin\left[(n\pi/L)(L/2)\right] = \sin(n\pi/2)$ is zero whenever n is even. The series (6.22) is a weakly convergent series of distributions. The sum of the first twenty terms of the series is plotted in Figure 6.9.

The process of finding the coefficients in a Fourier series of a distribution such as $\delta(x)$ requires a minor modification to our theory. First, the integrals that give the coefficients are not over the correct range $(-\infty, +\infty)$, and second, the functions e^{inx} (or $\sin nx$ or $\cos nx$) are not appropriate test functions as we have defined them. We can solve both problems by multiplying each e^{inx} by a differentiable "fudge function" $u(x)$ such that

$$\frac{1}{2L} \int_{-\infty}^{+\infty} e^{inx} u(x)\phi(x)\, dx = \frac{1}{2L} \int_{-L}^{+L} e^{inx} \phi(x)\, dx \qquad (6.23)$$

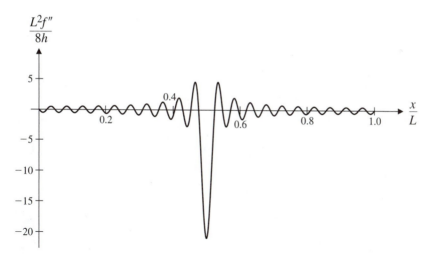

FIGURE 6.9. The first twenty terms in the series (6.22) for the second derivative of the triangle function. The variables are $L^2 f''/8h$ versus x/L. This series represents a negative delta function at $x/L = 0.5$.

Then $e^{inx} u(x)$ is a test function, and the integrals can be evaluated in the usual way.[16] Once we know that such a procedure exists, we can just use the normal expression for the Fourier coefficients—that is, the integral on the right-hand side of equation (6.23). Of course, the resulting series gives a periodic repetition of the original distribution.

6.5. DISTRIBUTIONS IN N DIMENSIONS

The theory of distributions can be extended to N dimensions. In physics, we are frequently concerned with problems in three space dimensions and one time dimension. Sometimes we work in a six-dimensional phase space with three position coordinates and three components of momentum. The modifications are straightforward.

1. The test functions should possess partial derivatives of all orders and go to zero sufficiently fast outside a finite volume of the space.

2. The core functions should possess partial derivatives of all orders.

3. The integrals that define the distributions become volume integrals over the N-dimensional space.

---

[16]Examples of such functions u can be found in Butkov, p. 255, and Lighthill, Section 5.2.

Then all the properties of distributions extend as expected. For example,

$$\int_V \frac{\partial \phi}{\partial x} f(\vec{r}) \, dV = \int \left(\phi f \big|_{-\infty}^{+\infty} - \int_{-\infty}^{+\infty} \phi \frac{\partial f}{\partial x} dx \right) dy \, dz$$

$$= - \int_V \phi \frac{\partial f}{\partial x} dV$$

The delta function in three dimensions is readily expressed in Cartesian coordinates. It is nonzero only at the origin, so

$$\delta(\vec{r}) = \delta(x)\delta(y)\delta(z) \tag{6.24}$$

We may write this as a limit of the core functions (6.18):

$$\delta(\vec{r}) = \lim_{n \to \infty} \frac{n}{\sqrt{\pi}} e^{-n^2 x^2} \frac{n}{\sqrt{\pi}} e^{-n^2 y^2} \frac{n}{\sqrt{\pi}} e^{-n^2 z^2} = \lim_{n \to \infty} \left(\frac{n}{\sqrt{\pi}} \right)^3 e^{-n^2 r^2} \tag{6.25}$$

A particularly useful relation in three dimensions is

$$\nabla^2 \left(\frac{1}{r} \right) = -4\pi \, \delta(\vec{r}) \tag{6.26}$$

First we note that the left-hand side of equation (6.26) has the sort of behavior we expect for a delta function. If $r \neq 0$, then

$$\left(\frac{\partial^2}{\partial x^2} + \frac{\partial^2}{\partial y^2} + \frac{\partial^2}{\partial z^2} \right) \frac{1}{\sqrt{x^2 + y^2 + z^2}} = \frac{\partial}{\partial x} \left(\frac{-x}{r^3} \right) + \frac{\partial}{\partial y} \left(\frac{-y}{r^3} \right) + \frac{\partial}{\partial z} \left(\frac{-z}{r^3} \right)$$

$$= \frac{-3}{r^3} - x \left(-\frac{3x}{r^5} \right) - y \left(-\frac{3y}{r^5} \right) - z \left(-\frac{3z}{r^5} \right)$$

$$= \frac{-3}{r^3} + 3\frac{r^2}{r^5} = 0$$

but as $r \to 0$, $1/r \to \infty$ and its derivatives cannot be computed; the left-hand side of (6.26) is undefined.

Next we must test for the sifting property. Since we have already shown that $\nabla^2 (1/r) = 0$ for $r \neq 0$, the volume integral

$$\int_{\text{all space}} \nabla^2 \left(\frac{1}{r} \right) f(\vec{r}) \, dV = \int_{\text{sphere of radius } \varepsilon} \nabla^2 \left(\frac{1}{r} \right) f(\vec{r}) \, dV$$

reduces to an integral over a small spherical volume surrounding the origin. We write the integrand in terms of a divergence so that we can use the divergence theorem:

$$\int_{\text{sphere of radius } \varepsilon} \nabla^2 \left(\frac{1}{r} \right) f(\vec{r}) \, dV$$

$$= \int_{\text{sphere of radius } \varepsilon} \left[\vec{\nabla} \cdot \left(f(\vec{r}) \vec{\nabla} \frac{1}{r} \right) - \vec{\nabla} \frac{1}{r} \cdot \vec{\nabla} f \right] dV$$

$$= \int_{\text{surface of sphere}} \left(f(\vec{r}) \vec{\nabla} \frac{1}{r} \right) \cdot \hat{n} \, dA - \int_{\text{sphere of radius } \varepsilon} \frac{-\hat{r}}{r^2} \cdot \vec{\nabla} f \, dV$$

$$= \int_{\text{surface of sphere}} \left(f(\vec{r}) \frac{-\hat{r}}{r^2} \right) \cdot \hat{r} r^2 \, d\Omega + \int_{\text{sphere of radius } \varepsilon} \frac{1}{r^2} \frac{\partial f}{\partial r} r^2 \, dr \, d\Omega$$

$$= \int_{\text{sphere of radius } \varepsilon} \frac{\partial f}{\partial r} \, dr \, d\Omega - \int_{\text{surface of sphere}} f(\vec{r}) \, d\Omega$$

$$= \int_{\text{solid angle of sphere}} [f|_{r=0}^{r=\varepsilon} - f(\vec{r})|_{r=\varepsilon}] \, d\Omega$$

$$= -f(0) \int_{\text{sphere}} d\Omega = -4\pi f(0)$$

Thus, since we have obtained the sifting property, relation (6.26) is proved.
  A similar relation holds in two dimensions:

$$\boxed{\nabla_2^2 \left(\ln \frac{\rho}{a} \right) = 2\pi \delta(\vec{\rho})} \qquad (6.27)$$

where ∇_2^2 is the Laplacian operator in two dimensions ($\nabla_2^2 \equiv \partial^2/\partial x^2 + \partial^2/\partial y^2$), a is any constant length, and $\rho = \sqrt{x^2 + y^2}$.

6.6. DESCRIBING PHYSICAL QUANTITIES USING DELTA FUNCTIONS

We often want to use delta functions to describe the mass or charge density of a system that is confined to an infinitesimally thin sheet or line. Let's investigate how to describe such physical quantities using delta functions.

Example 6.7. A ring of charge of radius a lies in the x-y plane. It has a variable line charge density $\lambda_0 \cos \phi$. What is the volume charge density? Use (a) cylindrical coordinates and (b) spherical coordinates.

(a) First we note that the charge exists only at $z = 0$ and at $\rho = a$, so we must have two delta functions:

$$\rho(\vec{\mathbf{x}}) = C\delta(z)\delta(\rho - a)$$

To find C (which may be a function of the coordinates), we integrate over a wedge of space with $-\infty < z < +\infty$, $0 \le \rho < \infty$, and angular extent $d\phi$ (see Figure 6.10a). (This volume is chosen by using the entire range of each coordinate in which we have a delta function, but only a differential range in the third coordinate.) The amount of charge dq in this volume is $\lambda a \, d\phi$. Thus, using the sifting property, we have

$$a\lambda_0 \cos\phi \, d\phi = dq = \int_{\text{wedge}} C\delta(z)\delta(\rho - a)\rho \, d\rho \, dz \, d\phi$$

$$= Ca \, d\phi$$

and so $C = \lambda_0 \cos\phi$:

$$\rho(\vec{\mathbf{x}}) = \lambda_0 \cos\phi \, \delta(z)\delta(\rho - a)$$

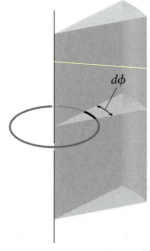

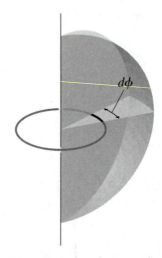

FIGURE 6.10a.
To find an expression for the charge density in Example 6.7, we integrate over a wedge of an infinite cylinder with $-\infty < z < \infty$, $0 \le \rho < \infty$, and angular extent $d\phi$.

FIGURE 6.10b.
To find an expression for the charge density in spherical coordinates, we integrate over the wedge of angular width $d\phi$ that extends throughout the full range of the coordinates θ and r: $0 \le \theta \le \pi$ and $0 \le r < \infty$.

(b) In spherical coordinates, the x-y plane is at $\theta = \pi/2$, so

$$\rho(\vec{x}) = C\delta\left(\theta - \frac{\pi}{2}\right)\delta(r - a)$$

Now we integrate over an "orange-wedge" shape with $0 \le \theta \le \pi$, $0 \le r < \infty$, and extent $d\phi$ in the azimuthal angle (Figure 6.10b):

$$a\lambda_0 \cos\phi\, d\phi = \int_{\text{wedge}} C\delta\left(\theta - \frac{\pi}{2}\right)\delta(r - a)r^2 \sin\theta\, dr\, d\theta\, d\phi$$

$$= Ca^2 \sin\frac{\pi}{2}d\phi = Ca^2 d\phi$$

Thus, $C = \lambda_0 (\cos\phi)/a$, and

$$\rho(\vec{x}) = \lambda_0 \cos\phi\,\delta\left(\theta - \frac{\pi}{2}\right)\frac{\delta(r - a)}{a}$$

Check the dimensions of the result!

This example illustrates a general method for finding densities of linear or planar charge or mass distributions in terms of delta functions in curvilinear coordinates.

First, determine which coordinates take on only a single value, and find the delta functions in these coordinates. (In the example above, charge exists only at $\theta = \pi/2$ and $r = a$.)

Second, determine the function that multiplies the delta functions by integrating over a region of space that extends over the full range of the coordinates in the delta functions, but over only a differential range in the other coordinate or coordinates. The value of your integral should be equal to the value determined by elementary methods. (In our example, we integrated over a wedge that extends from ϕ to $\phi + d\phi$. This wedge slices off a length $a\, d\phi$ of the ring and thus contains an amount $dq = \lambda a\, d\phi$ of charge.) This method guarantees that you will not lose information about the dependence of the coefficient C on the coordinates. Integrating over all space does not work because you will lose the detailed information that you need.

6.7. THE GREEN'S FUNCTION

An important application of distribution theory is the Green's function. The Green's function is actually a distribution that describes the response of a physical system to a unit delta function input (a "point" source). We have already seen an example of this in Example 6.3, where we applied a point impulse to a string. Other examples are the electric field due to a point charge, a point mass on a beam, and a voltage impulse applied to an electric circuit. If the physical system is linear (and thus is described by a linear differential equation), the response of the system to a sum of inputs is just the sum of the responses to the individual inputs. Then if we can model any input as the sum (integral) of a set of point (delta function) inputs, we can use the Green's function to compute the response.

If the impulse per unit length applied to the string is $I(x)$, we can write it as

$$I(x) = \int_0^L I(x')\delta(x-x')\,dx'$$

If the response of the system to the delta function is $G(x, x')$, the response to $I(x)$ is

$$y(x) = \int_0^L I(x')G(x, x')\,dx' \tag{6.28}$$

The Green's function is described in more detail in Optional Topic C, where we show how to find an appropriate Green's function for several different systems and in different geometries.

PROBLEMS

1. Show that the following sequences of functions are delta sequences:

   (a) $\phi_n(x) = \dfrac{n}{2}e^{-n|x|}$

   (b) $\phi_n(x) = \dfrac{n}{\pi}\left(\dfrac{1}{1+n^2x^2}\right)$ (*Hint:* Use contour integration.)

   (c) $\phi_n(x) = \dfrac{1-\cos nx}{n\pi x^2}$

2. Find a Fourier series representation of the delta function $\delta(x)$ in the range $(-L, +L)$ in two ways.

   (a) Start with the Fourier series for a step function (equation 4.24, for example) and differentiate.

   (b) Start with the block functions (equation 6.3) and form the Fourier series. Take the limit as $n \to \infty$.

   Are the results the same? If not, why not? Give a quantitative as well as a qualitative account of any discrepancy.

3. A point load Mg is placed on a beam of length L at a distance $L/3$ from the left-hand end. Find the displacement of the beam

   (a) if the beam is supported at the left end, as in Problem 5.11

   (b) if the beam rests on supports at each end, as in Example 5.3

   Ignore the weight of the beam itself.

4. A damped harmonic oscillator (compare with equation 3.7 and Problem 4.13) has initial conditions $x(0) = x_0$ and $dx/dt|_{t=0} = v_0$. An impulse I is applied at $t = t_0$. Find the motion of the oscillator for $t > 0$.

5. Distributions may be multiplied by infinitely differentiable functions. Do you expect the product

$$\psi(x) = \frac{\delta(x-a)}{x-a}$$

to be a valid distribution? Why or why not? Investigate the properties of this quantity by evaluating the integral

$$\int_{-\infty}^{\infty} \frac{\phi_n(x-a)}{x-a} f(x)\,dx$$

where $\phi_n(x)$ is a delta sequence of your choice and $f(x)$ is a test function. In particular, determine the result for functions that have the property $f(a) = 0$. Is $\psi(x)$ a valid distribution in this case? Can you identify it?

6. Evaluate

(a) $\int_{-\infty}^{\infty} e^{-|x|} \delta(x^2 + 2x - 3)\,dx$

(b) $\int_{-\infty}^{+\infty} e^{-x^2} \delta(x^2 + x - 6)\,dx$

7. A string of length L, with tension T and mass per unit length μ, is hit simultaneously at $t = 0$ at the two points $x = L/3$ and $x = 2L/3$. The impulse delivered at each point is I. Find the subsequent displacement of the string.

8. Using a general curvilinear coordinate system (Chapter 1, Section 1.3) with coordinates u, v, and w, find the charge density due to a point charge q placed at the point $u = u_0$, $v = v_0$, $w = w_0$. *Hint:* Start with the delta sequence (6.25) and note that as $n \to \infty$, only a differential line element ds^2 is needed in the exponent. Then make use of equation (1.61).

9. Show that the sequence of functions

$$f_n(x) = \frac{n}{2a\sqrt{\pi}} \int_{-a}^{+a} \exp[-n^2(x-x')^2]\,dx'$$

converges weakly to the distribution that gives the average value of any test function on the interval $-a$ to a.

10. Show that the sequence of functions

$$f_n(x) = \frac{n}{2\cosh^2 nx}$$

converges weakly to the delta function. *Hint:* Use a method similar to the one used in this chapter for sequence (6.18).

11. According to the properties of distributions in Section 6.3, $e^{-x}\delta'(x)$ is a distribution. Which distribution is it?

12. Starting with the integral (6.16), show that

$$\delta(x) = \frac{1}{\pi} \int_0^{\infty} \cos kx\,dk$$

13. By expanding the cosines in exponentials or otherwise, show that, for x and a both positive real numbers,

$$\delta(x-a) = \frac{2}{\pi} \int_0^{\infty} \cos kx \cos ka\,dk$$

and obtain a similar expression as an integral over sines. Are these results consistent with Problem 12? Discuss.

14. Find the Laplace transform of $\delta(t - a)$. Express the inverse as an integral using equation (5.19) and demonstrate that this integral possesses the sifting property.

15. A uniform disk of radius a and mass M lies in the x-y plane. Express the density in terms of delta functions
   (a) in rectangular Cartesian coordinates
   (b) in cylindrical coordinates
   (c) in spherical coordinates

16. A uniform rod of length ℓ and mass M lies along the x-axis with one end at the origin. Express the density in terms of delta functions
   (a) in rectangular Cartesian coordinates
   (b) in cylindrical coordinates
   (c) in spherical coordinates

17. A line of charge with uniform line charge density λ lies along the z-axis. Find the volume charge density
   (a) in cylindrical coordinates   (b) in spherical coordinates

18. A disk of charge with radius a and surface charge density $\sigma(r) = \sigma_0 r/a$ lies in the x-y plane with center at the origin. Find the volume charge density
   (a) in cylindrical coordinates   (b) in spherical coordinates

19. Current I flows in a circular loop of radius a lying in the x-y plane with its center at the origin. Find an expression for the current density
   (a) in cylindrical coordinates   (b) in spherical coordinates

20. Prove the relation (equation 6.27)

$$\nabla^2 \ln \frac{\rho}{a} = 2\pi \delta(\vec{\rho})$$

where ρ is the radial coordinate in a cylindrical coordinate system, a is a constant length, and $\vec{\rho}$ is the position vector in a plane. Use the result to find the potential due to a line charge λ running parallel to the z-axis at $x = a$, $y = b$.

21. A circuit contains a resistor, a capacitor, and a square wave power supply with period T. Use Kirchhoff's loop rule to write an equation for the current in the circuit in terms of delta functions, and solve it to find the current as a function of time.

22. Starting with the result

$$\Phi = \frac{\vec{p} \cdot \vec{r}}{r^3}$$

for the electric potential due to a dipole placed at the origin (see Example 6.1), calculate the electric field everywhere, including *at* the origin. *Hint*: Show that the electric field contains a delta function term. Use a method similar to that used in Section 6.5 to prove relation (6.26).

23. Using a delta sequence of your choice, show that the limit

$$\lim_{\ell \to 0} \left[\frac{\delta(x) - \delta(x - \ell)}{\ell} \right]$$

exhibits the sifting property of $\delta'(x)$.

24. Use the derivative property (6.20) to show that, for distributions, the Laplace transform of the derivative $\phi'(x)$ equals s times the Laplace transform of ϕ. Show that the Laplace transform of $\ln t$ is $-(\gamma + \ln s)/s$, where γ is Euler's constant, $- \int_0^\infty e^{-x} \ln x \, dx = 0.5772$. Hence show that the Laplace transform of $1/t$ $(t > 0)$, considered as a distribution, is $-\gamma - \ln s$. (See Zemanian, Chapter 8.)

25. Starting from equation (6.16), show that

$$\lim_{R \to \infty} \frac{\sin Rx}{\pi x} = \delta(x)$$

Confirm your result by demonstrating the sifting property. Use contour integration to do the integral.
   Similarly, show that

$$\lim_{R \to \infty} \frac{\cos Rx}{x} = 0$$

if the integral

$$\int_{-\infty}^{+\infty} f(x) \frac{\cos Rx}{x} dx$$

is taken to be the principal value.
   Demonstrate the plausibility of the results by evaluating

$$I_1 = \frac{2}{\pi} \int_\varepsilon^\infty \frac{\sin Nx}{x} dx \quad \text{and} \quad I_2 = \int_\varepsilon^\infty \frac{\cos Nx}{x} dx$$

numerically for a set of values of $\varepsilon \ll 1$ and $N \gg 1$. Show that as N increases, I_1 approaches unity and I_2 decreases toward zero.

26. Show that

$$\frac{d}{dx} \operatorname{sign}(x) = 2\delta(x), \quad \text{where } \operatorname{sign}(x) = \frac{x}{|x|}$$

27. Show that

$$x^m \delta^{(n)}(x) = \begin{cases} 0 & \text{if } n < m \\ (-1)^m [n!/(n - m)!] \delta^{(n-m)}(x) & \text{if } n \geq m \end{cases}$$

Hint: Use proof by induction (Appendix III).

28. The integral $\int_0^\infty x^\alpha f(x)\, dx = \int_{-\infty}^{+\infty} x^\alpha \Theta(x) f(x)\, dx$, where $\alpha < 0$, may be integrated if x^α is interpreted as a distribution. First show that

$$x^\alpha \Theta(x) = \frac{1}{(\alpha+1)(\alpha+2)\cdots(\alpha+n)} \frac{d^n}{dx^n}[x^{\alpha+n}\Theta(x)]$$

where $\alpha + n > 0$ and n is an integer. Use the result to evaluate the integral

$$\int_0^\infty x^{-3/2} e^{-x}\, dx$$

29. A material absorbs light at frequency ν_L due to an atomic transition. The imaginary part of the dielectric constant may be approximated as $\sigma_0 \delta(\nu - \nu_L)$. Use the Kramers-Kronig relations (Chapter 2, Example 2.24) to determine the behavior of the refractive index $n = \sqrt{\mathrm{Re}\,(\varepsilon/\varepsilon_0)}$ as a function of frequency. Comment.

30. Demonstrate the sifting property of the delta sequence (6.5),

$$\phi_n(x) = \frac{1}{n\pi} \frac{\sin^2 nx}{x^2}$$

in the case that $f(x)$ has a second-order pole at $z = z_p$ in the upper half-plane. Can you extend the result to a pole of order m?

CHAPTER 7

Fourier Transforms

7.1. DEFINITION OF THE FOURIER TRANSFORM

The Fourier transform is the second integral transform that we shall study. Like the Laplace transform, it provides a convenient way to solve an inhomogeneous differential equation. But the Fourier transform proves to have many other useful applications in physics, as we shall see.

Recall that a function $f(x)$ may be represented as a Fourier series over any finite range $-L \le x \le +L$:

$$f(x) = \sum_{n=-\infty}^{+\infty} a_n e^{in\pi x/L}$$

Outside of the selected range, we obtain a periodic extension of the original function. The set of coefficients a_n forms a unique representation of the function $f(x)$ over this finite range. Each a_n may be calculated using the integral (equation 4.31)

$$a_n = \frac{1}{2L} \int_{-L}^{+L} f(x) e^{-in\pi x/L} dx$$

Now we'd like to increase the range $(-L, L)$ until f may be represented over the whole real line. But as $L \to \infty$, $n\pi/L \to 0$ for any finite n, so we cannot use equation (4.31) as it stands. We begin by defining

$$k \equiv \frac{n\pi}{L}$$

and $a_n \equiv a(k)$ so that

$$f(x) = \sum_{k=\frac{n\pi}{L}=-\infty}^{+\infty} a(k) e^{ikx}$$

and

$$a(k) = \frac{1}{2L} \int_{-L}^{+L} f(x) e^{-ikx} dx$$

We multiply this expression by $L\sqrt{2/\pi}$ and define

$$F(k) \equiv L\sqrt{\frac{2}{\pi}}a(k) = \frac{1}{\sqrt{2\pi}}\int_{-L}^{+L} f(x)\, e^{-ikx}\, dx$$

Here we may let $L \to \infty$ to obtain

$$F(k) = \frac{1}{\sqrt{2\pi}}\int_{-\infty}^{+\infty} f(x)\, e^{-ikx}\, dx \qquad (7.1)$$

provided that the integral converges.

Now the difference between two neighboring values of k is

$$\Delta k = \frac{(n+1)\pi}{L} - \frac{n\pi}{L} = \frac{\pi}{L}$$

Thus, we may rewrite the original series in terms of $F(k)$ and Δk as

$$f(x) = \sum_k \sqrt{\frac{2}{\pi}}La(k)\, e^{ikx}\sqrt{\frac{\pi}{2}}\frac{1}{L}$$

$$= \frac{1}{\sqrt{2\pi}}\sum_k F(k)\, e^{ikx}\, \Delta k$$

Then, as we let $L \to \infty$ and consequently $\Delta k \to 0$, the sum becomes an integral and we obtain

$$f(x) = \frac{1}{\sqrt{2\pi}}\int_{-\infty}^{+\infty} F(k)\, e^{ikx}\, dk \qquad (7.2)$$

Equations (7.1) and (7.2) define the Fourier transform and its inverse. The factor of $1/\sqrt{2\pi}$ appears in both expressions—this is the *symmetric* Fourier transform. The symmetric transform obeys the following symmetry relation:

> If $F(k)$ is the symmetric Fourier transform of $f(x)$, then $f(-k)$ is the transform of $F(x)$.

An alternative definition of the transform has a factor $1/2\pi$ in front of one integral and unity in front of the other. The product of both factors *must be* $1/2\pi$. It is also possible to define the transform with a plus sign in the exponential in equation (7.1) and a minus sign in the exponential in equation (7.2); in fact, this is usually done when the variable is a time variable.[1]

---

[1] The difference is related to the metric of space-time in special relativity. See Section 7.3.7 for an example.

The Fourier transform and the Laplace transform are closely related. To see how, let $ik = s$ in equation (7.1). Then we have

$$\sqrt{2\pi}\, F(-is) = \int_{-\infty}^{+\infty} f(x)\, e^{-sx}\, dx$$

In the special case that $f(x) \equiv 0$ for $x < 0$, the right-hand side becomes the Laplace transform of f. The inverse (equation 7.2) is

$$f(x) = \frac{1}{\sqrt{2\pi}} \int_{-i\infty}^{+i\infty} F(-is)\, e^{sx}\, d\frac{s}{i}$$

$$= \frac{1}{2\pi i} \int_{-i\infty}^{+i\infty} \sqrt{2\pi}\, F(-is)\, e^{sx}\, ds$$

which is the Mellin inversion integral (equation 5.19) with $\gamma \to 0$.

An important difference between the two transforms is that the Fourier transform may be applied to functions defined over the whole range $-\infty < x < +\infty$. However, the function must approach zero as $x \to \pm\infty$ for the transform to exist.[2] In contrast, the Laplace transform exists for functions that diverge as $x \to \infty$, provided that they do not diverge faster than an exponential (see Chapter 5, Section 5.1). The Laplace transform is restricted to functions defined for positive values of the argument.

7.2. SOME EXAMPLES

Example 7.1. Find the Fourier transform of the function $f(x) = \exp(-\alpha |x|)$, where $\alpha > 0$.

The transform is given by equation (7.1):

$$F(k) = \frac{1}{\sqrt{2\pi}} \int_{-\infty}^{+\infty} e^{-\alpha|x|}\, e^{-ikx}\, dx$$

$$= \frac{1}{\sqrt{2\pi}} \left(\int_{-\infty}^{0} e^{\alpha x} e^{-ikx}\, dx + \int_{0}^{+\infty} e^{-\alpha x} e^{-ikx}\, dx \right)$$

$$= \frac{1}{\sqrt{2\pi}} \left(\frac{e^{(\alpha - ik)x}}{\alpha - ik} \Big|_{-\infty}^{0} - \frac{e^{-(\alpha + ik)x}}{\alpha + ik} \Big|_{0}^{+\infty} \right)$$

$$= \frac{1}{\sqrt{2\pi}} \left(\frac{1}{\alpha - ik} - \frac{-1}{\alpha + ik} \right)$$

$$= \sqrt{\frac{2}{\pi}} \frac{\alpha}{\alpha^2 + k^2} \tag{7.3}$$

---

[2] This restriction can be lifted if we are willing to consider the function $f(x)$ to be a generalized function. See Lighthill's book for an extensive discussion. See also Example 7.3 and Problem 28.

Now let's go backwards and find the function that corresponds to this transform. From equation (7.2),

$$f(x) = \frac{1}{\sqrt{2\pi}} \int_{-\infty}^{+\infty} \sqrt{\frac{2}{\pi}} \frac{\alpha}{\alpha^2 + k^2} e^{ikx} \, dk$$

$$= \frac{1}{\pi} \int_{-\infty}^{+\infty} \frac{\alpha}{\alpha^2 + k^2} e^{ikx} \, dk$$

This integral is of the type that can be evaluated using the residue theorem (Section 2.7.3, Example 2.20). For $x > 0$, we close the contour upward with a big semicircle (Figure 7.1).

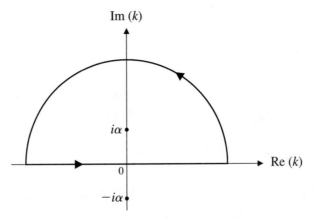

FIGURE 7.1. To invert the transform (7.3), we complete the contour in the upper half-plane when $x > 0$.

The integral along the semicircle is zero, by Jordan's lemma. The integrand has two simple poles at $k = \pm i\alpha$. Only one of the poles is inside the contour. Thus, the integral is

$$\int_{-\infty}^{+\infty} \frac{\alpha}{\alpha^2 + k^2} e^{ikx} \, dk = \oint \frac{\alpha}{\alpha^2 + k^2} e^{ikx} \, dk$$

$$= 2\pi i \left(\frac{\alpha}{2i\alpha}\right) e^{i(i\alpha)x} = \pi e^{-\alpha x}$$

and so

$$f(x) = \frac{1}{\pi} (\pi e^{-\alpha x}) = e^{-\alpha x}$$

For $x < 0$, we must close the contour downward. The pole at $k = -i\alpha$ is inside the contour, and we go around the contour clockwise (Figure 7.2).

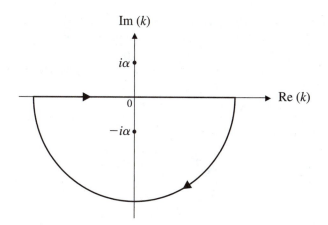

FIGURE 7.2.   For $x < 0$, we close the contour in the lower half-plane.

Thus, the integral is

$$\int_{-\infty}^{+\infty} \frac{\alpha}{\alpha^2 + k^2} e^{ikx}\, dk = \oint \frac{\alpha}{\alpha^2 + k^2} e^{ikx}\, dk$$

$$= -2\pi i \left(\frac{\alpha}{-2i\alpha} \right) e^{i(-i\alpha)x} = \pi e^{\alpha x}$$

and hence $f(x) = e^{\alpha x}$. We may combine our results for $x < 0$ and $x > 0$ to obtain

$$f(x) = \exp\left(-\alpha\, |x|\right)$$

as required.

Example 7.2.   Find the Fourier transform of the Gaussian function $f(x) = Ne^{-\alpha^2 x^2}$.
The transform is

$$F(k) = \frac{1}{\sqrt{2\pi}} \int_{-\infty}^{+\infty} Ne^{-\alpha^2 x^2} e^{-ikx}\, dx$$

We do this integral by completing the square:

$$-\alpha x^2 - ikx = -\alpha^2 \left(x^2 + \frac{ik}{\alpha^2} x - \frac{k^2}{4\alpha^4} \right) - \frac{k^2}{4\alpha^2}$$

$$= -\alpha^2 \left(x + \frac{ik}{2\alpha^2} \right)^2 - \frac{k^2}{4\alpha^2}$$

Now we change variables to $u = \alpha(x + ik/2\alpha^2)$:

$$F(k) = \frac{N}{\sqrt{2\pi}} \exp\left(-\frac{k^2}{4\alpha^2} \right) \int_{-\infty + ik/2\alpha}^{+\infty + ik/2\alpha} e^{-u^2} \frac{du}{\alpha}$$

The path of integration has been moved off the real axis by the amount $k/2\alpha$. However, since the integrand has no poles, the value of the integral is not changed. To see why, construct a rectangular contour from Re $(u) = -R$ to Re $(u) = +R$, with height $k/2\alpha$ (Figure 7.3). The integral along the two vertical pieces at the ends of the rectangle goes to zero as $R \to \infty$:

$$\left| \int_{\text{side}} e^{-u^2}\, du \right| \leq \frac{k}{2\alpha} \max \left| e^{-(\pm R + iy)^2} \right|$$

$$= \frac{k}{2\alpha} \max \left| e^{-R^2 \mp 2i\, Ry + y^2} \right|$$

$$= \frac{k}{2\alpha} \exp \left(\frac{k}{2\alpha} \right)^2 e^{-R^2} \to 0 \quad \text{as } R \to \infty$$

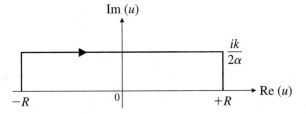

FIGURE 7.3. The transform of the Gaussian is computed using an integral along the upper side of the rectangle. Since there are no poles inside the closed contour and there is no contribution from the short sides, the integral is the same along both long sides.

The integral around the entire rectangle is zero, by the Cauchy theorem, so the integral along the upper path is the same as the integral along the real axis. We know the value of this integral—it is $\sqrt{\pi}$ (Appendix IX). Thus,

$$F(k) = \frac{N}{\alpha\sqrt{2}} \exp \left(-\frac{k^2}{4\alpha^2} \right)$$

Therefore, the transform of a Gaussian is also a Gaussian, but their widths are inversely related. The original function reaches one half its maximum value at $x = 0.83/\alpha$; the transform reaches one half its maximum value at $k = 0.83(2\alpha)$.

This result is closely related to the uncertainty principle in quantum mechanics. The relation between momentum and wave number k is

$$p = \hbar k$$

A particle is described by a wave packet, and the uncertainty in its position is determined by the width of the packet—essentially $1/\alpha$. The width of the corresponding pulse in momentum space is $\hbar\, \Delta k = 2\hbar\alpha$. Thus, the product of the two uncertainties is

$$\Delta x\, \Delta p \simeq \frac{1}{\alpha} 2\hbar\alpha = 2\hbar$$

for a Gaussian wave packet. The quantum mechanical relation is

$$\Delta x \, \Delta p \geq \frac{\hbar}{2}$$

Thus, a particle that is well localized in space is poorly located in momentum, and vice versa.

Example 7.3. Find the Fourier transform of the function $f(x) = 1$.
The transform is

$$F(k) = \frac{1}{\sqrt{2\pi}} \int_{-\infty}^{+\infty} e^{-ikx} \, dx$$

We can recognize this integral as the delta function (equation 6.16):

$$F(k) = \sqrt{2\pi} \, \delta(k)$$

Then, by the sifting property, the inverse is

$$f(x) = \frac{1}{\sqrt{2\pi}} \int_{-\infty}^{+\infty} \sqrt{2\pi} \, \delta(k) e^{ikx} \, dk = 1$$

[Note that e^{ikx} does not approach zero as x approaches infinity, so it is not a proper test function as we defined them in Chapter 6. We must narrow the class of distributions by putting the requirement $g(x) \to 0$ as $x \to \infty$ onto the core functions rather than the test functions. This narrower class still includes the delta function as well as many other useful distributions.]

Examples 7.1–7.3 demonstrate the most commonly used methods for evaluating transforms and their inverses:

1. Integrate an exponential with linear argument.
2. Complete the contour with a large semicircle and use the residue theorem.
3. Complete the square.
4. Identify the delta function or use the sifting property.

7.3. PROPERTIES OF THE FOURIER TRANSFORM

Because the Fourier transform is similar to the Laplace transform, it shares many of the properties that we found in Chapter 5.

7.3.1. Linearity

The Fourier transform operator $\mathcal{F}$ is a linear operator:

$$\mathcal{F}(f + g) = \mathcal{F}(f) + \mathcal{F}(g)$$

$$\mathcal{F}(af) = a\mathcal{F}(f)$$

7.3.2. Complex Conjugate

If the function $f(x)$ is real, then

$$F^*(k) = \left[\frac{1}{\sqrt{2\pi}} \int_{-\infty}^{+\infty} f(x) e^{-ikx} \, dx \right]^* = \frac{1}{\sqrt{2\pi}} \int_{-\infty}^{+\infty} f(x) e^{+ikx} \, dx = F(-k) \quad (7.4)$$

7.3.3. Differentiation

We evaluate the transform of a derivative using integration by parts:

$$\mathcal{F}\left(\frac{df}{dx}\right) = \frac{1}{\sqrt{2\pi}} \int_{-\infty}^{+\infty} \frac{df}{dx} e^{-ikx} \, dx$$

$$= \frac{1}{\sqrt{2\pi}} \left. f e^{-ikx} \right|_{-\infty}^{+\infty} - \frac{1}{\sqrt{2\pi}} \int_{-\infty}^{+\infty} -ikf e^{-ikx} \, dx$$

$$\mathcal{F}\left(\frac{df}{dx}\right) = \frac{ik}{\sqrt{2\pi}} \int_{-\infty}^{+\infty} f e^{-ikx} \, dx = ikF(k) \quad (7.5)$$

The integrated term vanishes, since the transform exists only for functions $f(x)$ that approach zero as $x \to \pm\infty$. Notice that in the case of the Fourier transform, unlike that of the Laplace transform, no initial conditions appear. Extension to higher derivatives is straightforward:

$$\mathcal{F}\left(\frac{d^n f}{dx^n}\right) = (ik)^n F(k) \quad (7.6)$$

The sign that we use in the exponential when forming the transform (equation 7.1) is reflected in the sign that appears in the derivative rule (equations 7.5 and 7.6). If the transform is defined as

$$F(\omega) = \frac{1}{\sqrt{2\pi}} \int_{-\infty}^{+\infty} f(t) e^{+i\omega t} \, dt$$

then the first derivative has transform

$$\mathcal{F}\left(\frac{df}{dt}\right) = -i\omega F(\omega) \quad (7.7)$$

7.3.4. Attenuation and Shifting

Let $g(x) = e^{ax} f(x)$. Then the transform of g is

$$G(k) = \frac{1}{\sqrt{2\pi}} \int_{-\infty}^{+\infty} e^{ax} f(x) e^{-ikx} \, dx$$

$$= \frac{1}{\sqrt{2\pi}} \int_{-\infty}^{+\infty} f(x) e^{-i(ia+k)x} \, dx$$

$$= F(k + ia) \tag{7.8}$$

If $g(x) = f(x - a)$, then

$$G(k) = \frac{1}{\sqrt{2\pi}} \int_{-\infty}^{+\infty} f(x - a) e^{-ikx} \, dx$$

Changing variables to $u = x - a$, we have

$$G(k) = \frac{1}{\sqrt{2\pi}} \int_{-\infty}^{+\infty} f(u) e^{-ik(u+a)} \, du$$

$$= e^{-ika} F(k) \tag{7.9}$$

These relations are analogous to (5.5) and (5.6) for the Laplace transform.

7.3.5. Parseval's Theorem

The integral of the product of two functions may be related to the integral of the product of their transforms.

$$\int_{-\infty}^{+\infty} f(x) g(x) \, dx = \frac{1}{2\pi} \int_{-\infty}^{+\infty} dx \int_{-\infty}^{+\infty} dk \, F(k) e^{ikx} \int_{-\infty}^{+\infty} d\omega \, G(\omega) e^{i\omega x}$$

$$= \int_{-\infty}^{+\infty} F(k) \, dk \int_{-\infty}^{+\infty} G(\omega) \, d\omega \frac{1}{2\pi} \int_{-\infty}^{+\infty} e^{i(k+\omega)x} \, dx$$

$$= \int_{-\infty}^{+\infty} F(k) \, dk \int_{-\infty}^{+\infty} G(\omega) \, \delta(k + \omega) \, d\omega$$

$$= \int_{-\infty}^{+\infty} F(k) \, G(-k) \, dk$$

If the function g is real, then we may use equation (7.4) to write the result as

$$\boxed{\int_{-\infty}^{+\infty} f(x) \, g(x) \, dx = \int_{-\infty}^{+\infty} F(k) \, G^*(k) \, dk} \tag{7.10}$$

This result is called Parseval's theorem.[3] An important special case occurs when $f = g$. Then

$$\int_{-\infty}^{+\infty} [f(x)]^2 \, dx = \int_{-\infty}^{+\infty} |F(k)|^2 \, dk \qquad (7.11)$$

The absolute value signs are necessary since even if $f(x)$ is real, $F(k)$ may not be.[4]

7.3.6. Convolution

Suppose the transform $H(k) = F(k) G(k)$ is the product of the two transforms F and G. Then the inverse is

$$h(x) = \frac{1}{\sqrt{2\pi}} \int_{-\infty}^{+\infty} H(k) \, e^{ikx} \, dk$$

$$= \frac{1}{\sqrt{2\pi}} \int_{-\infty}^{+\infty} F(k) \, G(k) \, e^{ikx} \, dk$$

Now we write the transform $G(k)$ in terms of the function g:

$$h(x) = \frac{1}{2\pi} \int_{-\infty}^{+\infty} F(k) \left[\int_{-\infty}^{+\infty} g(u) \, e^{-iku} \, du \right] e^{ikx} \, dk$$

We then combine the exponentials and interchange the order of integration:

$$h(x) = \frac{1}{\sqrt{2\pi}} \int_{-\infty}^{+\infty} g(u) \, \frac{1}{\sqrt{2\pi}} \int_{-\infty}^{+\infty} F(k) \, e^{ik(x-u)} \, dk \, du$$

$$= \frac{1}{\sqrt{2\pi}} \int_{-\infty}^{+\infty} g(u) \, f(x-u) \, du$$

which is the convolution[5] of the functions g and f. Equivalently, we may write

$$h(x) = \text{the inverse transform of } F(k) \, G(k)$$
$$= \frac{1}{\sqrt{2\pi}} \int_{-\infty}^{+\infty} f(u) \, g(x-u) \, du \qquad (7.12)$$

---

[3] Compare with equation (4.32) for Fourier series.
[4] See Example 7.4.
[5] This convolution differs from the one defined in Chapter 5 only by the numerical factor $1/\sqrt{2\pi}$ and the values of the limits of integration.

7.3.7. Extension to N Dimensions

We may extend the theory of Fourier transforms to as many dimensions as we wish. Frequently, in physics, $N = 4$: three space dimensions and one time dimension. If f is a function of the three Cartesian coordinates x, y, z ($N = 3$), we simply transform with respect to each of the variables, one at a time.

$$F(\vec{k}) = F(k_x, k_y, k_z)$$

$$= \frac{1}{(\sqrt{2\pi})^3} \int_{-\infty}^{+\infty} \int_{-\infty}^{+\infty} \int_{-\infty}^{+\infty} f(x, y, z) \, e^{-ik_x x} \, dx \, e^{-ik_y y} \, dy \, e^{-ik_z z} \, dz$$

$$= \frac{1}{(\sqrt{2\pi})^3} \int_{\text{all space}} f(\vec{x}) \exp(-i\vec{k} \cdot \vec{x}) \, dV$$

It is usual to define the time transform with the opposite signs, so that for $N = 4$,

$$F(\vec{k}, \omega) = \frac{1}{(2\pi)^2} \int_{-\infty}^{+\infty} \int_{\text{all space}} f(\vec{x}, t) \exp[-i(\vec{k} \cdot \vec{x} - \omega t)] \, d^3\vec{x} \, dt$$

Then the function

$$f(\vec{x}, t) = \frac{1}{(2\pi)^2} \int_{-\infty}^{+\infty} \int_{\text{all } k \text{ space}} F(\vec{k}, \omega) \exp[i(\vec{k} \cdot \vec{x} - \omega t)] \, d^3\vec{k} \, d\omega$$

is a sum of plane waves.[6]

The transform of a derivative may be calculated using a method similar to the one we used with the one-dimensional derivatives in Section 7.3.3. For example,

$$\mathcal{F}(\vec{\nabla} f) = \frac{1}{(\sqrt{2\pi})^3} \int_{\text{all space}} [\vec{\nabla} f(\vec{x})] \exp(-i\vec{k} \cdot \vec{x}) \, dV$$

$$= \frac{1}{(\sqrt{2\pi})^3} \int_{\text{all space}} \{\vec{\nabla}[f(\vec{x}) \exp(-i\vec{k} \cdot \vec{x})] - f(\vec{x}) \vec{\nabla} \exp(-i\vec{k} \cdot \vec{x})\} \, dV$$

$$= \frac{1}{(\sqrt{2\pi})^3} \int_{\text{surface at } \infty} f(\vec{x}) \exp(-i\vec{k} \cdot \vec{x}) \hat{n} \, dS$$

$$- \frac{1}{(\sqrt{2\pi})^3} \int_{\text{all space}} -i\vec{k} f(\vec{x}) \exp(-i\vec{k} \cdot \vec{x}) \, dV$$

Now provided that[7] f approaches zero faster than $1/r^2$ as $r \to \infty$, the surface integral is zero, and

$$\mathcal{F}(\vec{\nabla} f) = 0 + i\vec{k} \frac{1}{(\sqrt{2\pi})^3} \int_{\text{all space}} f(\vec{x}) \exp(-i\vec{k} \cdot \vec{x}) \, dV$$

---

[6]Compare with Chapter 2, Section 2.1.4
[7]Again, this condition is also required for the existence of the transform of f.

$$\mathcal{F}(\vec{\nabla} f) = i\vec{k} F(\vec{k}) \tag{7.13}$$

Similarly, we may show that

$$\mathcal{F}(\nabla^2 f) = -k^2 F(\vec{k}) \tag{7.14}$$

$$\mathcal{F}(\vec{\nabla} \times \vec{f}) = i\vec{k} \times \vec{F} \tag{7.15}$$

and so on.

7.4. CAUSALITY

In an initial value problem, the initial conditions appear explicitly in Laplace transform theory. In Fourier transform theory, they appear more subtly. Let's look at an example to see how this occurs.

Example 7.4.   An electron, initially at rest, is acted upon by an electric field $\vec{E}(t) = \vec{E}_0 e^{-\alpha t}$ for $t > 0$, where α is real and positive. The electron is also subject to a damping force $\vec{F}_d = -\gamma \vec{v}$. Find the subsequent motion of the electron.

The equation satisfied by the electron's velocity is

$$m \frac{d}{dt}\vec{v} + \gamma \vec{v} = -e\vec{E}(\vec{x}, t)$$

First notice that the motion will be one dimensional. Then we take the Fourier transform of the entire equation with respect to time. Define the transform by the relations

$$\vec{v}(t) = \frac{1}{\sqrt{2\pi}} \int_{-\infty}^{+\infty} \vec{v}(\omega) e^{-i\omega t} \, d\omega$$

and

$$\vec{v}(\omega) = \frac{1}{\sqrt{2\pi}} \int_{-\infty}^{+\infty} \vec{v}(t) e^{i\omega t} \, dt$$

Then the transform of the electric field is

$$\vec{E}(\omega) = \vec{E}_0 \frac{1}{\sqrt{2\pi}} \int_{0}^{+\infty} e^{-\alpha t} e^{i\omega t} \, dt$$

$$= \vec{E}_0 \frac{1}{\sqrt{2\pi}} \frac{e^{-\alpha t} e^{i\omega t}}{-\alpha + i\omega} \bigg|_0^\infty$$

$$= \vec{E}_0 \frac{1}{\sqrt{2\pi}} \frac{1}{\alpha - i\omega}$$

The range of integration reduces to positive values of t, since $\vec{\mathbf{E}} = 0$ for $t < 0$.

Transforming the differential equation and making use of relation (7.7) for the transform of the derivative, we find

$$-im\omega\vec{\mathbf{v}} + \gamma\vec{\mathbf{v}} = -e\vec{\mathbf{E}}_0 \frac{1}{\sqrt{2\pi}} \frac{1}{(\alpha - i\omega)}$$

and thus

$$\vec{\mathbf{v}}(\omega) = e\vec{\mathbf{E}}_0 \frac{1}{\sqrt{2\pi}} \frac{1}{(\alpha - i\omega)(im\omega - \gamma)}$$

$$= \frac{e}{m}\vec{\mathbf{E}}_0 \frac{1}{\sqrt{2\pi}} \frac{1}{(\omega + i\alpha)(\omega + i\gamma/m)} \tag{7.16}$$

Now we invert the transform:

$$\vec{\mathbf{v}}(t) = \frac{1}{\sqrt{2\pi}} \int_{-\infty}^{+\infty} \frac{e}{m}\vec{\mathbf{E}}_0 \frac{1}{\sqrt{2\pi}} \frac{1}{(\omega + i\alpha)(\omega + i\gamma/m)} e^{-i\omega t} \, d\omega$$

We can use the residue theorem to evaluate the integral if we close the contour with a big semicircle.[8] For $t > 0$ we close the semicircle downward, while for $t < 0$ we must close upward. But for $t < 0$ the electron has felt no force, and so $\vec{\mathbf{v}}$ must remain zero. Thus, causality[9] requires that the transform have *no poles in the upper half-plane*! The poles of our integrand are at $\omega = -i\alpha$ and $\omega = -i\gamma/m$, and both are in the lower half-plane (Figure 7.4). Notice that one of these poles is contributed by the differential equation $(-i\gamma/m)$, and one by the driving force $(-i\alpha)$. It is at this point that the initial conditions enter our solution.

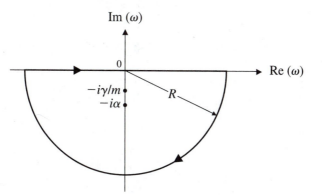

FIGURE 7.4. Contour for inverting the transform (7.16) when $t > 0$. Both poles are in the lower half-plane and thus are inside the contour.

---

[8] See Chapter 2, Section 2.7.3, especially Examples 2.20 and 2.21.

[9] This argument proves important in deciding where to put the integration path when there is a pole *on* the real axis. See Chapter 2, Section 2.7.3.

For $t > 0$ we integrate around the lower contour. The contour is traversed clockwise, so we introduce a minus sign. The solution for $t > 0$ is

$$\vec{v}(t) = \frac{1}{2\pi} \frac{e}{m} \vec{E}_0 (-2\pi i) \left(\frac{e^{-i(-i\alpha)t}}{(-i\alpha + i\gamma/m)} + \frac{e^{-i(-i\gamma/m)t}}{(-i\gamma/m + i\alpha)} \right)$$

$$= \frac{e\vec{E}_0}{\alpha m - \gamma} (e^{-\alpha t} - e^{-\frac{\gamma}{m}t})$$

It is easy to verify that $\vec{v} = 0$ at $t = 0$, and also that $\vec{v} \to 0$ as $t \to \infty$.

7.5. USE OF FOURIER TRANSFORMS IN THE SOLUTION OF PARTIAL DIFFERENTIAL EQUATIONS

The solution of a partial differential equation will be a function of two or more variables. If the equation is linear, we can apply the transform operator in one or more of the variables to obtain either an ordinary differential equation or an algebraic equation. In this section, we show how to apply these techniques to two common partial differential equations.

7.5.1. The Wave Equation

Example 7.5. Suppose an infinitely long cable is pulled up at $x = 0$, so that its shape is described by the function $f(x) = he^{-|x|/a}$, and then let go. What is the subsequent motion of the cable?

The equation of motion for the cable is the wave equation (3.15):

$$\frac{\partial^2 y}{\partial t^2} = v^2 \frac{\partial^2 y}{\partial x^2}$$

Let's transform the equation with respect to space and time. We define the transform as follows:

$$\tilde{y}(k, \omega) = \frac{1}{2\pi} \int_{-\infty}^{+\infty} \int_{-\infty}^{+\infty} y(x, t) e^{-ikx+i\omega t} \, dx \, dt$$

and, conversely,

$$y(x, t) = \frac{1}{2\pi} \int_{-\infty}^{+\infty} \int_{-\infty}^{+\infty} \tilde{y}(k, \omega) e^{ikx-i\omega t} \, dk \, d\omega$$

Then the transformed equation is

$$-\omega^2 \tilde{y} = -v^2 k^2 \tilde{y}$$

or

$$(\omega - vk)(\omega + vk)\tilde{y} = 0$$

One solution to this equation is the trivial solution $\tilde{y} = 0$, but this is not the solution we want. We might be tempted to say that the equation is solved by taking $\omega = \pm vk$, but we want a solution for the transform $\tilde{y}(\omega, k)$. The transform is zero except where $\omega = \pm vk$—a property that is reminiscent of delta functions. The nonzero solution that we need is

$$\tilde{y} = A(k)\delta(\omega - vk) + B(k)\delta(\omega + vk)$$

since we have already established that $x\delta(x)$ is the zero distribution (Chapter 6, Section 6.3). The equation becomes

$$A(k)(\omega + vk)(\omega - vk)\delta(\omega - vk) + B(k)(\omega - vk)(\omega + vk)\delta(\omega + vk)$$
$$= A(k)(\omega + vk) \times 0 + B(k)(\omega - vk) \times 0 = 0$$

as required.

Then the solution for $y(x, t)$ is

$$y(x, t) = \frac{1}{2\pi} \int_{-\infty}^{+\infty} \int_{-\infty}^{+\infty} [A(k)\delta(\omega - vk) + B(k)\delta(\omega + vk)]\, e^{ikx - i\omega t}\, dk\, d\omega$$

$$= \frac{1}{2\pi} \int_{-\infty}^{+\infty} e^{ikx}[A(k)\, e^{-ivkt} + B(k)\, e^{+ivkt}]\, dk$$

This expression shows that the displacement of the cable is a sum of rightward-moving and leftward-moving waves.

To solve for the remaining unknowns, we must use the initial conditions that $\partial y/\partial t = 0$ and $y = he^{-|x|/a}$ at $t = 0$. Thus,

$$y(x, 0) = \frac{1}{2\pi} \int_{-\infty}^{+\infty} e^{ikx}[A(k) + B(k)]\, dk = he^{-|x|/a}$$

and

$$\left.\frac{\partial y}{\partial t}\right|_{t=0} = \frac{1}{2\pi} \int_{-\infty}^{+\infty} ivke^{ikx}[-A(k) + B(k)]\, dk = 0$$

From the second of these relations we conclude that $A = B$, while from the first we may make use of relation (7.3) with $\alpha = 1/a$ to write the transform of the initial condition on y:

$$\sqrt{\frac{2}{\pi}} \frac{ha}{1 + k^2 a^2} = \frac{1}{\sqrt{2\pi}} 2A(k)$$

$$A(k) = \frac{ha}{1 + k^2 a^2}$$

So, finally, our solution is

$$y(x, t) = \frac{ha}{2\pi} \int_{-\infty}^{+\infty} \frac{e^{ikx}}{1 + k^2 a^2} (e^{-ivkt} + e^{+ivkt})\, dk$$

To do this integral, we use the residue theorem:

$$y(x,t) = \frac{h}{2\pi a} \int_{-\infty}^{+\infty} \frac{1}{(k+i/a)(k-i/a)} (e^{ikx-ivkt} + e^{ikx+ivkt})\, dk$$

There are two simple poles on the imaginary axis at $k = \pm i/a$.

1st term: Case I: $x - vt < 0$. We must close the contour downward, thus enclosing the pole at $k = -i/a$. Then the integral along the big semicircle is zero, by Jordan's lemma. We traverse the contour clockwise. Thus,

$$I_1 = -\frac{2\pi i}{2\pi} h \frac{e^{i(-i)(x-vt)/a}}{-2i} = h \frac{e^{(x-vt)/a}}{2}$$

Case II: $x - vt > 0$. We must close the contour upward, enclosing the pole at $k = +i/a$. The result is

$$I_1 = \frac{2\pi i}{2\pi} h \frac{e^{i(i)(x-vt)/a}}{2i} = h \frac{e^{-(x-vt)/a}}{2}$$

2nd term: Case III: $x + vt < 0$. We close downward and get

$$I_2 = -\frac{2\pi i}{2\pi} h \frac{e^{i(-i)(x+vt)/a}}{-2i} = h \frac{e^{x+vt/a}}{2}$$

Case IV: $x + vt > 0$. We close upward. The result is

$$I_2 = \frac{2\pi i}{2\pi} h \frac{e^{i(i)(x+vt)/a}}{2i} = h \frac{e^{-(x+vt)/a}}{2}$$

Finally, we put all this together:

If $x < -vt$ for $t > 0$, then also $x < vt$, so we have Case I and Case III:

$$y(x,t) = h \frac{e^{(x-vt)/a} + e^{(x+vt)/a}}{2} = he^{x/a} \cosh \frac{vt}{a}$$

If $-vt < x < vt$, then we have Case I and Case IV:

$$y(x,t) = h \frac{e^{(x-vt)/a} + e^{-(x+vt)/a}}{2} = he^{-vt/a} \cosh \frac{x}{a}$$

If $x > vt$, then we have Case II and Case IV:

$$y(x,t) = h \frac{e^{-(x-vt)/a} + e^{-(x+vt)/a}}{2} = he^{-x/a} \cosh \frac{vt}{a}$$

This solution is shown in Figure 7.5. The initial peak at $x = 0$ propagates both rightward and leftward along the string and decreases in magnitude with time. The displacement is never negative, so the string does not oscillate as we might have expected. This is a consequence of the infinite length of the string and the fact that the wave is propagating in one dimension (along a line—the string).

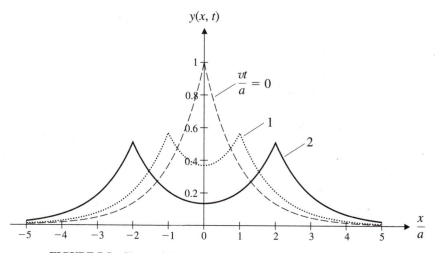

FIGURE 7.5. Shape of the string in Example 7.5 at $vt/a = 0, 1$, and 2.

7.5.2. The Diffusion Equation

Example 7.6. At $t = 0$, a small amount of mud of mass m is introduced at the point $x = \ell$ into an infinitely long pipe of cross-sectional area A, containing fresh water. Determine the distribution of mud in the pipe at times $t > 0$.

The appropriate differential equation is the diffusion equation (equation 3.14):

$$\frac{\partial \rho}{\partial t} = D \frac{\partial^2 \rho}{\partial x^2}$$

where D is the diffusion coefficient for the problem. The boundary conditions are $\rho = 0$ for all x for $t < 0$, and $\rho \to 0$ as $x \to \infty$. At $t = 0$, all the mud is concentrated at one point. We may model the density as a delta function:

$$\rho(x, 0) = \frac{m}{A} \delta(x - \ell)$$

Since the differential equation is first order in time, we may be successful by taking the Fourier transform with respect to x only:

$$\tilde{\rho}(k, t) = \frac{1}{\sqrt{2\pi}} \int_{-\infty}^{+\infty} \rho(x, t) e^{-ikx} \, dx$$

Transforming the whole equation with respect to x, we get the first-order equation

$$\frac{\partial \tilde{\rho}}{\partial t} = -k^2 D \tilde{\rho}$$

which has the solution

$$\tilde{\rho} = \tilde{\rho}_0 \, e^{-k^2 Dt}$$

The initial conditions determine $\tilde{\rho}_0$:

$$\tilde{\rho}_0 = \tilde{\rho}(k, 0) = \frac{1}{\sqrt{2\pi}} \int_{-\infty}^{+\infty} \rho(x, 0) \, e^{-ikx} \, dx$$

$$= \frac{m}{A} \frac{1}{\sqrt{2\pi}} \int_{-\infty}^{+\infty} \delta(x - \ell) \, e^{-ikx} \, dx$$

$$= \frac{m}{A} \frac{e^{-ik\ell}}{\sqrt{2\pi}}$$

Thus, we have

$$\tilde{\rho}(k, t) = \frac{m}{A} \frac{e^{-ik\ell}}{\sqrt{2\pi}} e^{-k^2 Dt}$$

and transforming back gives

$$\rho(x, t) = \frac{m}{A} \frac{1}{2\pi} \int_{-\infty}^{+\infty} \exp\left(ikx - ik\ell - k^2 Dt \right) dk$$

To do the integral, we complete the square:

$$k^2 Dt + ik(\ell - x) = \left(k\sqrt{Dt} + i \frac{(\ell - x)}{2\sqrt{Dt}} \right)^2 + \frac{1}{4} \frac{(x - \ell)^2}{Dt}$$

Thus,

$$\rho(x, t) = \frac{m}{A} \frac{1}{2\pi} \exp\left(-\frac{1}{4} \frac{(x - \ell)^2}{Dt} \right) \int_{-\infty}^{+\infty} \exp\left[-\left(k\sqrt{Dt} + i \frac{(\ell - x)}{2\sqrt{Dt}} \right)^2 \right] dk$$

Now let $u = k\sqrt{Dt} + i(\ell - x)/2\sqrt{Dt}$, and thus $du = \sqrt{Dt} \, dk$. Then we have

$$\rho(x, t) = \frac{m}{A} \frac{1}{2\pi} \exp\left(-\frac{(x - \ell)^2}{4Dt} \right) \int_{-\infty + i\frac{(\ell - x)}{2\sqrt{Dt}}}^{+\infty + i\frac{(\ell - x)}{2\sqrt{Dt}}} e^{-u^2} \frac{du}{\sqrt{Dt}}$$

$$= \frac{m}{A} \frac{1}{2\sqrt{\pi Dt}} \exp\left(-\frac{(x - \ell)^2}{4Dt} \right)$$

Thus, the mud distribution is a Gaussian that spreads with time while the maximum density decreases (Figure 7.6).

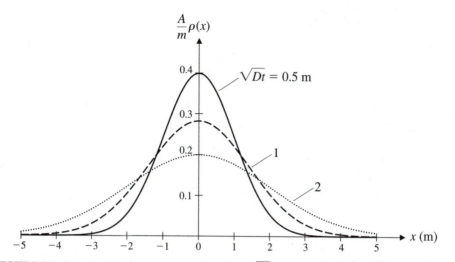

FIGURE 7.6. Distribution of mud in Example 7.6 for $\sqrt{Dt} = 1/2$ m (solid line), 1 m (dashed line), and 2 m (dotted line) and $\ell = 0$.

7.6. FOURIER TRANSFORMS AND POWER SPECTRA

The Fourier transform relates the time dependence of a function $f(t)$ to the frequency dependence of the transform $F(\omega)$. Thus, it is a useful tool for analyzing the frequency components present in a physical system. One of the most common applications is to the electromagnetic power radiated by a system.

An accelerated charge radiates. The power radiated per unit area of wavefront is given by the Poynting flux:

$$\vec{S} = \frac{\vec{E} \times \vec{B}}{\mu_0}$$

Far from the source, $\vec{B}$ is perpendicular to both $\hat{k}$ and $\vec{E}$:

$$\vec{S} = \frac{\vec{E} \times (\hat{k} \times \vec{E})}{\mu_0 c} = \frac{\hat{k}(\vec{E} \cdot \vec{E}) - \vec{E}(\hat{k} \cdot \vec{E})}{\mu_0 c}$$

$$= \frac{E^2}{\mu_0 c} \hat{k}$$

Using $dA = r^2 d\Omega$, we find that the power radiated per unit solid angle is

$$\frac{dP}{d\Omega} = r^2 |\vec{S}| = \frac{r^2 E^2}{\mu_0 c}$$

where E and hence P are functions of time. Then the total energy radiated per unit solid angle is

$$\frac{dW}{d\Omega} = \int_{-\infty}^{+\infty} \frac{r^2}{\mu_0 c} \vec{E}(t) \cdot \vec{E}(t)\, dt$$

Using Parseval's theorem (equation 7.11), we may write this in terms of an integral over the transform of $\vec{E}$:

$$\frac{dW}{d\Omega} = \int_{-\infty}^{+\infty} \frac{r^2}{\mu_0 c} \vec{E}(\omega) \cdot \vec{E}^*(\omega)\, d\omega$$

Thus, the energy radiated per unit solid angle per unit angular frequency is

$$\frac{d^2 W}{d\Omega\, d\omega} = \frac{r^2}{\mu_0 c} \vec{E}(\omega) \cdot \vec{E}^*(\omega) \tag{7.17}$$

which is the *power spectrum*.

The electric field due to an accelerated charge in nonrelativistic motion has two parts: the Coulomb field that decreases as $1/r^2$ and the radiation field that decreases as $1/r$. Thus, at large distances from the charge, the electric field is dominated by the radiation field, which takes the form (Jackson, Chapter 14)

$$\vec{E} = q \frac{\hat{\mathbf{k}} \times (\hat{\mathbf{k}} \times \vec{\mathbf{a}})}{4\pi \varepsilon_0 c^2 r}$$

where $\vec{\mathbf{a}} = d^2\vec{\mathbf{x}}/dt^2$ is the acceleration of the charge and $\hat{\mathbf{k}}$ is a unit vector along the direction of propagation. Therefore, the energy radiated per unit solid angle per unit angular frequency is independent of the distance r to the source, and $d^2 W/d\Omega\, d\omega$ may be expressed in terms of the time transform of the charge's position, since $\vec{\mathbf{a}}(\omega) = -\omega^2 \vec{\mathbf{x}}(\omega)$ (by the derivative property, Section 7.3.3):

$$\frac{d^2 W}{d\Omega\, d\omega} = \frac{q^2}{\mu_0 c (4\pi \varepsilon_0 c^2)^2} \omega^4 |\vec{\mathbf{x}}(\omega)|^2 \sin^2 \theta$$

$$= \frac{q^2}{c^3 (4\pi)^2 \varepsilon_0} \omega^4 |\vec{\mathbf{x}}(\omega)|^2 \sin^2 \theta \tag{7.18}$$

where θ is the angle between $\vec{\mathbf{a}}$ and $\hat{\mathbf{k}}$.

Integrating over the angles, with $\mu \equiv \cos\theta$, we find

$$\frac{dW}{d\omega} = \frac{q^2}{c^3 (4\pi)^2 \varepsilon_0} \omega^4 |\vec{\mathbf{x}}(\omega)|^2 (2\pi) \int_{-1}^{+1} (1 - \mu^2)\, d\mu$$

$$= \frac{8\pi}{3} \frac{q^2}{c^3 (4\pi)^2 \varepsilon_0} \omega^4 |\vec{\mathbf{x}}(\omega)|^2$$

$$= \frac{q^2}{6\pi c^3 \varepsilon_0} \omega^4 |\vec{\mathbf{x}}(\omega)|^2 \tag{7.19}$$

Example 7.7. Find the power spectrum radiated by the electron whose motion we considered in Example 7.4.

Here it is more convenient to write $\vec{\mathbf{a}}(\omega)$ in terms of $\vec{\mathbf{v}}(\omega)$ so that we can use equation (7.16) for the transform of the electron velocity. The electron radiates energy per unit angular frequency:

$$\frac{dW}{d\omega} = \frac{q^2}{6\pi c^3 \varepsilon_0} \omega^2 \, |\vec{\mathbf{v}}(\omega)|^2 = \frac{e^2}{6\pi c^3 \varepsilon_0} \omega^2 \left| \frac{e}{m} \vec{\mathbf{E}}_0 \frac{1}{\sqrt{2\pi}} \frac{1}{(\omega + i\alpha)(\omega + i\gamma/m)} \right|^2$$

$$= \frac{e^4 E_0^2}{12\pi^2 m^2 c^3 \varepsilon_0} \frac{\omega^2}{(\omega^2 + \alpha^2)[\omega^2 + (\gamma/m)^2]} \tag{7.20}$$

The power spectrum (equation 7.20) is shown in Figure 7.7. The spectrum peaks at $\omega = \sqrt{\alpha \gamma / m}$.

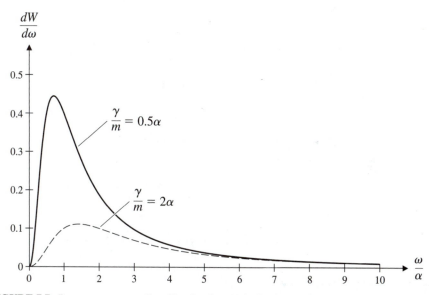

FIGURE 7.7. Power spectrum radiated by the electron in Example 7.7 for $\gamma/m = 0.5\alpha$ (solid line) and $\gamma/m = 2\alpha$ (dashed line). The frequency variable is ω/α.

7.7. SINE AND COSINE TRANSFORMS

In Chapter 4, we saw that the Fourier series of an even function contains only cosines while the series of an odd function contains only sines. Similarly, we can construct sine and cosine transforms for odd and even functions.

Sine and cosine transforms are used when either

1. the function $f(x)$ is known to be even or odd or
2. the function is defined only for $x > 0$.

7.7.1. The Cosine Transform

Let $f(x)$ be defined for $x > 0$, and let us make an even extension of the function to the range $x < 0$. Then the Fourier transform of the extended function is

$$F(k) = \frac{1}{\sqrt{2\pi}} \int_{-\infty}^{+\infty} f(x) e^{-ikx} \, dx$$

$$= \frac{1}{\sqrt{2\pi}} \left(\int_{-\infty}^{0} f(x) e^{-ikx} \, dx + \int_{0}^{+\infty} f(x) e^{-ikx} \, dx \right)$$

$$= \frac{1}{\sqrt{2\pi}} \left(-\int_{+\infty}^{0} f(-u) e^{iku} \, du + \int_{0}^{+\infty} f(x) e^{-ikx} \, dx \right)$$

$$= \frac{1}{\sqrt{2\pi}} \int_{0}^{+\infty} f(x)(e^{ikx} + e^{-ikx}) \, dx$$

$$\boxed{F(k) = \sqrt{\frac{2}{\pi}} \int_{0}^{+\infty} f(x) \cos kx \, dx \equiv F_c(k)} \qquad (7.21)$$

which is the Fourier cosine transform of the original function $f(x)$. Note that the cosine transform is an even function of k:

$$F_c(-k) = \sqrt{\frac{2}{\pi}} \int_{0}^{+\infty} f(x) \cos(-kx) \, dx$$

$$= \sqrt{\frac{2}{\pi}} \int_{0}^{+\infty} f(x) \cos kx \, dx = F_c(k)$$

The inverse transform is given by

$$\boxed{f(x) = \sqrt{\frac{2}{\pi}} \int_{0}^{+\infty} F_c(k) \cos kx \, dk} \qquad (7.22)$$

To verify this expression, we expand out the cosine:

$$f(x) = \sqrt{\frac{2}{\pi}} \int_{0}^{+\infty} F_c(k) \left(\frac{e^{ikx} + e^{-ikx}}{2} \right) dk$$

$$= \frac{1}{\sqrt{2\pi}} \left(\int_{0}^{+\infty} F_c(k) e^{ikx} \, dk + \int_{0}^{+\infty} F_c(k) e^{-ikx} \, dk \right)$$

Now let $\kappa = -k$ in the second term, and use the fact that $F_c(k)$ is an even function of k:

$$f(x) = \frac{1}{\sqrt{2\pi}} \left(\int_0^{+\infty} F_c(k) e^{ikx} \, dk - \int_0^{-\infty} F_c(-\kappa) e^{i\kappa x} \, d\kappa \right)$$

$$= \frac{1}{\sqrt{2\pi}} \int_{-\infty}^{+\infty} F_c(k) e^{ikx} \, dk$$

Thus, equation (7.22) corresponds to the usual inversion relation (7.2).

7.7.2. The Sine Transform

The sine transform is defined similarly by making an odd extension of f. The Fourier transform of the extended function is

$$F(k) = \frac{1}{\sqrt{2\pi}} \int_{-\infty}^{+\infty} f(x) e^{-ikx} \, dx$$

$$= \frac{1}{\sqrt{2\pi}} \left(\int_{-\infty}^{0} f(x) e^{-ikx} \, dx + \int_0^{+\infty} f(x) e^{-ikx} \, dx \right)$$

$$= \frac{1}{\sqrt{2\pi}} \left(-\int_{+\infty}^{0} f(-u) e^{iku} \, du + \int_0^{+\infty} f(x) e^{-ikx} \, dx \right)$$

$$= \frac{1}{\sqrt{2\pi}} \int_0^{+\infty} f(x)(e^{-ikx} - e^{ikx}) \, dx$$

$$= -i \sqrt{\frac{2}{\pi}} \int_0^{+\infty} f(x) \sin kx \, dx = -i F_s(k)$$

where the sine transform is defined by

$$F_s(k) \equiv \sqrt{\frac{2}{\pi}} \int_0^{+\infty} f(x) \sin kx \, dx \qquad (7.23)$$

which is an odd function of k. Again the inverse is

$$f(x) = \sqrt{\frac{2}{\pi}} \int_0^{+\infty} F_s(k) \sin kx \, dk \qquad (7.24)$$

Let's verify equation (7.24):

$$\sqrt{\frac{2}{\pi}} \int_0^{+\infty} F_s(k) \sin kx \, dk = \sqrt{\frac{2}{\pi}} \frac{1}{2i} \int_0^{+\infty} F_s(k)(e^{ikx} - e^{-ikx}) \, dk$$

$$= \sqrt{\frac{2}{\pi}} \frac{1}{2i} \left(\int_0^{+\infty} \frac{F(k)}{-i} e^{ikx} \, dk + \int_0^{-\infty} \frac{F(-\kappa)}{-i} e^{i\kappa x} \, d\kappa \right)$$

$$= \frac{1}{\sqrt{2\pi}} \int_{-\infty}^{+\infty} F(k) \, e^{ikx} \, dk = f(x)$$

as required.

7.7.3. Use of the Sine and Cosine Transforms

Like the Laplace and Fourier transforms, the sine and cosine transforms prove to be useful tools for solving differential equations. Thus, we will need the transforms of derivatives of a function in terms of the transform of the original function. Let's start by finding the cosine transform of df/dx:

$$\mathcal{F}_c\left(\frac{df}{dx}\right) = \sqrt{\frac{2}{\pi}} \int_0^{+\infty} \frac{df}{dx} \cos kx \, dx$$

$$= \sqrt{\frac{2}{\pi}} \left(f \cos kx \big|_0^\infty - (-k) \int_0^{+\infty} f \sin kx \, dx \right)$$

$$= -\sqrt{\frac{2}{\pi}} f(0) + k F_s(k)$$

The transform of the derivative brings in the initial condition $f(0)$, as is the case with the Laplace transform, but it also mixes up the sine and cosine transforms. The Fourier *cosine* transform of df/dx is k times the Fourier *sine* transform of $f(x)$. A similar mixing occurs with the sine transform:

$$\mathcal{F}_s\left(\frac{df}{dx}\right) = \sqrt{\frac{2}{\pi}} \int_0^{+\infty} \frac{df}{dx} \sin kx \, dx$$

$$= \sqrt{\frac{2}{\pi}} \left(f \sin kx \big|_0^\infty - k \int_0^{+\infty} f \cos kx \, dx \right)$$

$$= -k F_c(k)$$

This time no initial conditions appear.

We get back the transform we started with (sine or cosine) when we look at the second derivative:

$$\mathcal{F}_c\left(\frac{d^2 f}{dx^2}\right) = -\sqrt{\frac{2}{\pi}} \left.\frac{df}{dx}\right|_{x=0} + k\mathcal{F}_s\left(\frac{df}{dx}\right)$$

$$= -\sqrt{\frac{2}{\pi}} \left.\frac{df}{dx}\right|_{x=0} + k[-kF_c(k)]$$

$$\mathcal{F}_c\left(\frac{d^2 f}{dx^2}\right) = -\sqrt{\frac{2}{\pi}} \left.\frac{df}{dx}\right|_{x=0} - k^2 F_c(k) \qquad (7.25)$$

and

$$\mathcal{F}_s\left(\frac{d^2 f}{dx^2}\right) = -k\mathcal{F}_c\left(\frac{df}{dx}\right) = -k\left(-\sqrt{\frac{2}{\pi}} f(0) + k F_s(k)\right)$$

$$\mathcal{F}_s\left(\frac{d^2 f}{dx^2}\right) = k\sqrt{\frac{2}{\pi}} f(0) - k^2 F_s(k) \qquad (7.26)$$

These results suggest that the sine and cosine transforms will be most useful in solving differential equations that have only even derivatives or only odd derivatives. The choice of transform will be determined by the initial conditions that are given. For second-order equations:

- If the value of the function is known at $x = 0$, use the sine transform.
- If the value of the first derivative of the function is known at $x = 0$, use the cosine transform.

If we were to use the Laplace transform method, we would need *both* initial conditions. It appears that we can get away with only one condition if we use the sine or cosine transform. But it is implicit in the Fourier transform (sine or cosine) that the function $f(x)$ approach zero as $x \to \infty$, and this provides our second boundary condition. No such constraint is needed with the Laplace transform.

Example 7.8. Use a Fourier transform to solve the equation

$$\frac{d^2 y}{dx^2} - \alpha^2 y = 0$$

with the boundary conditions $y(0) = y_0$ and $y \to 0$ as $x \to \infty$.

Since we are given $y(0)$, we use the sine transform:

$$-k^2 \mathcal{F}_s(y) + k\sqrt{\frac{2}{\pi}} y_0 - \alpha^2 \mathcal{F}_s(y) = 0$$

The solution for the transform is

$$\mathcal{F}_s(y) = \sqrt{\frac{2}{\pi}} \frac{ky_0}{k^2 + \alpha^2}$$

Now the inverse is

$$y(x) = \sqrt{\frac{2}{\pi}} \int_0^\infty \sqrt{\frac{2}{\pi}} \frac{ky_0}{k^2 + \alpha^2} \sin kx \, dk$$

The integrand is even in k, so we can extend the range of integration:

$$y(x) = \frac{1}{\pi} \int_{-\infty}^\infty \frac{ky_0}{k^2 + \alpha^2} \frac{e^{ikx} - e^{-ikx}}{2i} dk$$

We use the residue theorem to evaluate the integral. The integrand has poles at $k = \pm i\alpha$. For the first term, we close the contour upward (remember $x > 0$), and the integral around the big semicircle is zero, by Jordan's lemma. Only the pole at $k = +i\alpha$ is inside the contour (see Figure 7.1), and we get

$$I_1 = \frac{1}{2\pi i} \int_{-\infty}^\infty \frac{ky_0}{(k + i\alpha)(k - i\alpha)} e^{ikx} \, dk = \frac{i\alpha y_0}{2i\alpha} e^{i(i\alpha)x} = \frac{y_0}{2} e^{-\alpha x}$$

Similarly, for the second term we close downward, enclosing the pole at $k = -i\alpha$:

$$I_2 = \frac{1}{2\pi i} \int_{-\infty}^\infty \frac{ky_0}{(k + i\alpha)(k - i\alpha)} e^{-ikx} \, dk = -\frac{-i\alpha y_0}{-2i\alpha} e^{-i(-i\alpha)x} = -\frac{y_0}{2} e^{-\alpha x}$$

where the extra minus sign accounts for the fact that we go around the contour clockwise. Combining the two terms, we find

$$y(x) = I_1 - I_2 = y_0 e^{-\alpha x}$$

which is the expected result.

Let's see how the solution goes using the Laplace transform. We don't know dy/dx at $x = 0$, so let's call it b.

$$s^2 Y(s) - sy_0 - b - \alpha^2 Y = 0$$

$$Y = \frac{sy_0 + b}{s^2 - \alpha^2}$$

Inverting,[10] we get

$$y(x) = y_0 \cosh \alpha x + \frac{b}{\alpha} \sinh \alpha x$$

Now the second boundary condition requires that the positive exponential terms in this result sum to zero, so we need

$$y_0 + \frac{b}{\alpha} = 0 \Rightarrow b = -\alpha y_0$$

Then the solution is

$$y(x) = y_0(\cosh \alpha x - \sinh \alpha x) = y_0\, e^{-\alpha x}$$

as we obtained before.

If we can always use the Laplace transform in this way, why should we bother with sine and cosine transforms? The answer is that we cannot always use the Laplace transform.

Example 7.9. Suppose the system in Example 7.8 is now driven by a function $f(x)$, where the exact functional form is for the moment unknown. Find $y(x)$ in terms of $f(x)$.

In this case, the Laplace transform becomes

$$Y = \frac{s y_0 + b}{s^2 - \alpha^2} + \frac{F}{s^2 - \alpha^2}$$

To invert the second term, we use the convolution theorem (equation 5.17). The inverse is

$$y(x) = y_0 \cosh \alpha x + \frac{b}{\alpha} \sinh \alpha x + \int_0^x f(\xi) \frac{\sinh \alpha(x - \xi)}{\alpha} d\xi$$

Until we evaluate the convolution integral, we cannot make use of the boundary condition at infinity. However, if we use the sine transform, we get

$$\mathcal{F}_s(y) = \sqrt{\frac{2}{\pi}} \frac{k y_0}{k^2 + \alpha^2} - \frac{\mathcal{F}_s(f)}{k^2 + \alpha^2}$$

To invert this expression, we need to work out the convolution theorem for the sine transform:

$$\mathcal{F}_s^{-1}[\mathcal{F}_s(f)\mathcal{F}_s(g)] = \sqrt{\frac{2}{\pi}} \int_0^\infty \mathcal{F}_s(f)\mathcal{F}_s(g) \sin kx\, dk$$

$$= \sqrt{\frac{2}{\pi}} \int_0^\infty \mathcal{F}_s(g) \left(\sqrt{\frac{2}{\pi}} \int_0^\infty f(\xi) \sin k\xi\, d\xi \right) \sin kx\, dk$$

$$= \frac{2}{\pi} \int_0^\infty f(\xi) \int_0^\infty \mathcal{F}_s(g) \frac{1}{2} [\cos k(x - \xi) - \cos k(x + \xi)]\, dk\, d\xi$$

---

[10]Use partial fractions and Table 5.1, or use the Mellin inversion integral.

$$\mathcal{F}_s^{-1}[\mathcal{F}_s(f)\mathcal{F}_s(g)] = \frac{1}{\sqrt{2\pi}} \int_0^\infty f(\xi)[\tilde{g}(x - \xi) - \tilde{g}(x + \xi)] \, d\xi \qquad (7.27)$$

where

$$\tilde{g}(x) = \mathcal{F}_c^{-1}[\mathcal{F}_s(g)]$$

is the inverse *cosine* transform of $\mathcal{F}_s(g)$.

To solve our problem, we need the inverse of $G(k) = 1/(k^2 + \alpha^2)$. This function is even, but fortunately we need its inverse *cosine* transform. The method closely follows what we did in Example 7.8, the only difference being the lack of the factor k in the numerator. (See also Example 7.1.)

$$\begin{aligned}
\tilde{g}(x) = \mathcal{F}_c^{-1}\left(\frac{1}{k^2 + \alpha^2}\right) &= \sqrt{\frac{2}{\pi}} \int_0^\infty \frac{1}{k^2 + \alpha^2} \cos kx \, dk \\
&= \frac{1}{\sqrt{2\pi}} \int_{-\infty}^\infty \frac{e^{ikx} + e^{-ikx}}{2(k^2 + \alpha^2)} \, dk \\
&= \frac{2\pi i}{\sqrt{2\pi}} \left(\frac{e^{i(i\alpha)x}}{2(2i\alpha)} - \frac{e^{-i(-i\alpha)x}}{2(-2i\alpha)}\right) = \sqrt{\frac{\pi}{2}} \frac{e^{-\alpha x}}{\alpha} \qquad (x > 0)
\end{aligned}$$

Although x is always greater than zero, $x - \xi$ is not, because ξ ranges from zero to infinity. However, since we took the inverse cosine transform, the function $\tilde{g}(x)$ must be even. In this example, $\tilde{g}(x) = \sqrt{\pi/2}(e^{-\alpha|x|}/\alpha)$. Then the solution of our differential equation is

$$\begin{aligned}
y(x) &= y_0 e^{-\alpha x} + \frac{1}{2\alpha} \int_0^\infty f(\xi)(e^{-\alpha|x-\xi|} - e^{-\alpha(x+\xi)}) \, d\xi \\
&= y_0 e^{-\alpha x} + \frac{1}{2\alpha} \left(\int_0^x f(\xi) e^{-\alpha(x-\xi)} \, d\xi + \int_x^\infty f(\xi) e^{\alpha(x-\xi)} \, d\xi\right) \\
&\quad - \frac{1}{2\alpha} \int_0^\infty f(\xi) e^{-\alpha(x+\xi)} \, d\xi \\
&= e^{-\alpha x}\left(y_0 + \frac{1}{2\alpha} \int_0^x f(\xi) e^{\alpha\xi} \, d\xi - \frac{1}{2\alpha} \int_0^\infty f(\xi) e^{-\alpha\xi} \, d\xi\right) \\
&\quad + \frac{e^{\alpha x}}{2\alpha} \int_x^\infty f(\xi) e^{-\alpha\xi} \, d\xi
\end{aligned}$$

which clearly obeys the condition $y(x) \to 0$ as $x \to \infty$ no matter what the function f, provided that its Fourier transform exists.

We conclude this discussion by finding the convolution theorem for the cosine transform:

$$\mathcal{F}_c^{-1}[\mathcal{F}_c(f)\mathcal{F}_c(g)] = \sqrt{\frac{2}{\pi}} \int_0^\infty \mathcal{F}_c(f)\mathcal{F}_c(g) \cos kx \, dk$$

$$= \sqrt{\frac{2}{\pi}} \int_0^\infty \mathcal{F}_c(g) \left(\sqrt{\frac{2}{\pi}} \int_0^\infty f(\xi) \cos k\xi \, d\xi \right) \cos kx \, dk$$

$$= \frac{2}{\pi} \int_0^\infty \mathcal{F}_c(g) f(\xi) \frac{1}{2} [\cos k(x - \xi) + \cos k(x + \xi)] \, dk \, d\xi$$

$$\boxed{\mathcal{F}_c^{-1}[\mathcal{F}_c(f)\mathcal{F}_c(g)] = \frac{1}{\sqrt{2\pi}} \int_0^\infty f(\xi)[g(|x - \xi|) + g(x + \xi)] \, d\xi} \qquad (7.28)$$

This time the functions are not mismatched with the transform, but we must take the even extension of g when evaluating $g(x - \xi)$ for $\xi > x$.

PROBLEMS

1. Find the Fourier transform of the following functions, and verify your results by computing the inverse transform.

   (a) $\dfrac{1}{x^2 + 4x + 13}$

   (b) $e^{-\alpha x^2} \cos \beta x$

   (c) $f(x) = \begin{cases} x & \text{if } 0 \le x \le 1 \\ 0 & \text{otherwise} \end{cases}$

   (d) $\dfrac{1}{\cosh ax}$

   (e) $f(t) = \begin{cases} te^{-at} & \text{if } t > 0 \\ 0 & \text{otherwise} \end{cases}$

   (f) $\dfrac{x}{x^2 + a^2}$

2. Invert the following transforms to find the corresponding functions:

   (a) $F(k) = \dfrac{1 - 2ik}{1 + 4k^2}$

   (b) $F(k) = \dfrac{1}{1 + ik^3}$

   (c) $F(k) = \dfrac{1}{i \sinh ak}$

3. If $F(k)$ is the Fourier transform of $f(x)$, show that $id\,F/dk$ is the transform of $xf(x)$. What conditions must $F(k)$ satisfy for this result to hold?

4. Verify Parseval's theorem in the form of equation (7.10) by evaluating the transforms of the functions $f(x) = \cos \beta x$ and $g(x) = e^{-\alpha x^2}$ and evaluating the two integrals in equation (7.10).

5. Verify Parseval's theorem in the form of equation (7.11) by evaluating the transform of

$$f(x) = \begin{cases} 1 & \text{if } -1 \le x \le 1 \\ 0 & \text{otherwise} \end{cases}$$

and evaluating the integrals of $f(x)^2$ and $|F(k)|^2$.

6. Prove that if $F(k)$ is the transform of $f(x)$, then $(1/a)F(k/a)$ is the transform of $f(ax)$. Show that the result is consistent with Parseval's theorem.

7. Find the Fourier transform of the function

$$f(t) = \begin{cases} A \cos \omega_0 t & \text{if } -T < t < T \\ 0 & \text{otherwise} \end{cases}$$

that represents a finite train of data. Plot the Fourier power spectrum $|F(\omega)|^2$ as a function of ωT for the two cases $\omega_0 T = 1$ and $\omega_0 T = 10$. Comment. What happens as T increases toward infinity?

8. Find the Fourier transform of

$$f(t) = \begin{cases} 1 - |t|/T & \text{if } -T < t < T \\ 0 & \text{otherwise} \end{cases}$$

Hence find the transform of the function

$$g(t) = \begin{cases} 1 & \text{if } -T < t < 0 \\ -1 & \text{if } 0 < t < T \\ 0 & \text{otherwise} \end{cases}$$

9. Show that the square deviation between two functions,

$$D = \int_{-\infty}^{+\infty} |f(x) - g(x)|^2 \, dx$$

equals the square deviation between the transforms,

$$D = \int_{-\infty}^{+\infty} |F(k) - G(k)|^2 \, dk$$

10. A spring-and-dashpot system satisfies the equation

$$\frac{d^2x}{dt^2} + 2\alpha \frac{dx}{dt} + \omega_0^2 x = f(t)$$

with $\omega_0 > \alpha$. The driving force per unit mass $f(t)$ is zero for $t < 0$ and

$$f(t) = e^{-\alpha t} \sin \Omega t$$

for $t > 0$. Find $x(t)$ for $t > 0$, and verify that your method gives $x = 0$ for $t < 0$.

11. An electron in an atom may be modeled classically as a damped harmonic oscillator (compare with Problem 10 above). The electron is driven by an incoming EM wave with electric field $E(t) = E_0(\sin \Omega t)/\Omega t$ for $-\infty < t < \infty$. What is the appropriate $f(t)$ for this problem? Solve for the transform $x(\omega)$ of the electron's position.

Use the results of Section 7.6 to determine the power spectrum of the radiated energy. Plot your results in the case $\alpha = \omega_0/10$, $\Omega = 2\omega_0$. Comment.

12. The electric displacement $\vec{D}$ is related to the electric field $\vec{E}$ by the dielectric constant ϵ. In general, ϵ is a function of frequency, so the relationship is one between the Fourier transforms of $\vec{D}$ and $\vec{E}$:

$$\vec{D}(x, \omega) = \epsilon(\omega)\vec{E}(x, \omega)$$

(a) Show that the relationship between $\vec{D}(x, t)$ and $\vec{E}(x, t)$ is

$$\vec{D}(x, t) = \vec{E}(x, t) + \int_{-\infty}^{\infty} G(\tau)\vec{E}(x, t - \tau) \, d\tau$$

and determine an expression for $G(\tau)$ in terms of $\epsilon(\omega)$.

(b) Find $G(t)$ for the one-resonance model

$$\epsilon(\omega) = 1 + \frac{\omega_p^2}{\omega_0^2 - \omega^2 - i\gamma\omega}$$

where ω_p, ω_0, and γ are real positive constants and $\gamma < \omega_0$.

(c) Discuss the physical meaning of your result. Be specific.

13. An electron in an atom may be represented by a damped harmonic oscillator with frequency ω_0 and damping rate Γ. (Compare with Problem 10 with $2\alpha = \Gamma$.) An external electric field $\vec{E}(t)$ acts on the electron. Find the Fourier transform $\vec{x}(\omega)$ of the electron position as a function of time. If the electron loses energy at a rate $P = \vec{j} \cdot \vec{E} = -e\vec{v} \cdot \vec{E}$, use Parseval's theorem to show that the total energy loss is

$$\Delta U = \frac{e^2}{m} \int_{-\infty}^{\infty} \frac{\omega^2 \Gamma |\vec{E}(\omega)|^2}{(\omega_0^2 - \omega^2)^2 + \omega^2 \Gamma^2} \, d\omega$$

Note that the integrand is sharply peaked at $\omega \simeq \omega_0$, while $\vec{E}(\omega)$ is a slowly varying function, and so the integral may be approximated as

$$\Delta U = \frac{e^2}{m} |\vec{E}(\omega_0)|^2 \int_{-\infty}^{\infty} \frac{\omega^2 \Gamma}{(\omega_0^2 - \omega^2)^2 + \omega^2 \Gamma^2} \, d\omega$$

Evaluate the integral by contour integration to show that ΔU is independent of Γ, and hence find ΔU. (In this expression, ω_0 and Γ are real positive constants, and $\omega_0 > \Gamma$.)

14. **The radon problem.** Radon diffuses from the ground into the atmosphere at a rate r (atoms/m^2·s). Model the atmosphere as a semi-infinite medium with boundary (the ground) at $y = 0$. Then the density $\rho(y, t)$ of atmospheric radon is described by the equation

$$\frac{\partial \rho}{\partial t} = D \frac{\partial^2 \rho}{\partial y^2} - \lambda \rho$$

where D is the appropriate diffusion coefficient and λ is the decay rate for radon. The boundary condition at the ground is

$$\frac{\partial \rho}{\partial y}\bigg|_{y=0} = \text{constant} = -\alpha$$

What is the boundary condition at $y \to \infty$? Use the Fourier cosine transform in y to derive an integral expression for $\rho(y, t)$ in the case that $\rho(y, 0) = 0$. Evaluate $\partial \rho / \partial t$ at $t = 0$, and hence determine α in terms of r and D.

Extra credit: Obtain expressions for $\rho(0, t)$ and $\rho(y, \infty)$, and obtain $\rho(y, t)$ as an integral over t.

15. A long copper rod of cross-sectional area $A = 1$ cm^2 is initially at 15°C. At time $t = 0$, one end (at $x = 0$) is placed into a vat of hot oil at 300°C.

   (a) Refer to Chapter 3, Section 3.2.5. Write the equation that describes the change of temperature at position x along the rod at time t.

   (b) Write an expression for the temperature $T(x)$ of the rod immediately after the end is placed in the oil.

   (c) Discuss the use of Fourier and/or Laplace transforms in solving this equation. What determines the best choice of transform for this problem?

   (d) Find the temperature of the rod as a function of position and time for $t > 0$.

   (e) Given the following data for copper, plot the temperature along the first 5 m of the rod at times $t = 0.5, 1.5, 3.0,$ and 6.0 s. Thermal conductivity: 400 W/m·K; specific heat: 385 J/kg·K; density: 8.96 kg/m^3.

16. A long beam is resting on an elastic foundation. The equation satisfied by the beam displacement is

$$EI \frac{d^4 y}{dx^4} = q(x) - \alpha y(x)$$

where $q(x)$ is the load and α is a constant describing the elastic properties of the foundation. If the load is concentrated toward the center of the beam, then we may assume that $y \to 0$ as $x \to \pm\infty$. Transform the equation, and find $Y(k)$ in terms of $Q(k)$. Solve for the beam displacement if

   (a) $q(x) = Mg\delta(x - a)$

   (b) $q(x) = (Mg/L)[S(x - a + L/2) - S(x - a - L/2)]$

17. For the function $e^{-x} \sin x$, find
    (a) the Fourier sine transform
    (b) the Fourier cosine transform
18. For the function $xe^{-\alpha x}$, where α is a real positive constant, find
    (a) the Fourier sine transform
    (b) the Fourier cosine transform
19. Show that the Fourier cosine transform of the function x^{p-1} for $0 < p < 1$ is

$$\sqrt{\frac{2}{\pi}} \frac{1}{k^p} \cos \frac{p\pi}{2} \Gamma(p)$$

Hence show that the function $1/\sqrt{x}$ is its own cosine transform. Obtain similar results for the sine transform. (The results of Chapter 2, Section 2.9 may prove useful.)

20. Determine the form of Parseval's theorem (equation 7.10 and 7.11) that applies to the cosine transform.

21. The magnetic field in a conducting medium diffuses away according to the equation

$$\frac{\partial^2 H(x,t)}{\partial x^2} = \mu \sigma \frac{\partial H(x,t)}{\partial t}$$

Solve this equation by taking the Fourier transform in space. Find $H(x,t)$ if the magnetic field at $t = 0$ is a step function:

$$H(x,0) = \begin{cases} H_0 & \text{if } -d/2 < x < d/2 \\ 0 & \text{otherwise} \end{cases}$$

Express your answer in terms of the error function (Appendix IX).

22. At $t = 0$, the distribution of chlorine in a pipe of water is given by

$$\rho(x,0) = \rho_0 e^{-x^2/a^2}$$

Find the distribution of chlorine for $t > 0$ in terms of the diffusion coefficient D.

23. Develop a three-dimensional version of the convolution theorem. Use the result to obtain the solution of Poisson's equation

$$\nabla^2 \Phi = -\frac{\rho(\vec{r})}{\varepsilon_0}$$

as an integral over space. *Hint:* Use spherical coordinates to do the integration over $\vec{k}$, and use the principal value in the integral over k. Evaluate the resulting integral explicitly if $\rho(\vec{r}) = q\delta(\vec{r})$.

24. Find the three-dimensional Fourier transform of the charge distribution

$$\rho(\vec{r}) = \rho_0 \frac{e^{-r/a}}{4\pi r}$$

This transform is called the form factor of the charged particle.

25. Take the Fourier transform of the three-dimensional wave equation

$$\frac{\partial^2 s}{\partial t^2} - v^2 \nabla^2 s = f(\vec{\mathbf{x}}, t)$$

and solve for the transform $S(\vec{\mathbf{k}}, \omega)$. Show that the introduction of a damping force (through the addition of a term $\alpha\, \partial s/\partial t$ on the left-hand side) moves the poles off the real axis. Invert the transform in the case $\alpha \to 0$ for the cases

(a) $f(\vec{\mathbf{x}}, t) = e^{-r/a}\delta(t)$, where r is distance from the origin

(b) $f(\vec{\mathbf{x}}, t) = \delta(\vec{\mathbf{x}})\delta(t)$

26. At $t = 0$, the distribution of salt in a pipe of fresh water is given by

$$\rho(x, 0) = \rho_0 \left(\frac{\sin \alpha x}{\alpha x} + \frac{1}{4} \right)$$

Solve the diffusion equation (for example, equation 3.14 and Section 7.5.2) to find the salt distribution at $t > 0$ in terms of the diffusion coefficient D. *Hint:* The result of Problem 5 may prove useful.

27. Sum the series

$$\sum_{p=0}^{\infty}(-1)^p \frac{2p + 1}{x^2 + (2p + 1)^2}$$

by taking the Fourier transform of each term, summing the series in the transform space, and then transforming back.

28. Use the derivative rule (7.6) and the symmetry property of Fourier transforms to evaluate the transform of x^n. Check your result by inverting the transform.

29. The differential equation that describes the evolution of neutron density ρ is

$$\frac{\partial \rho}{\partial t} + \frac{\rho}{\tau} - D\nabla^2\rho = S\delta(\vec{\mathbf{r}})\delta(t)$$

where τ is the neutron lifetime and D is the diffusion coefficient. The source is a point source S at $\vec{\mathbf{r}} = 0$, $t = 0$, and the initial condition is $\rho(\vec{\mathbf{r}}, 0) = 0$.

Using any method of your choice, find the density $\rho(\vec{\mathbf{r}}, t)$ of neutrons for $t > 0$.

CHAPTER 8

Sturm-Liouville Theory

8.1. THE STURM-LIOUVILLE PROBLEM

Many differential equations describing physical systems can be reduced to one or more linear ordinary differential equations of the form

$$\frac{d}{dx}\left(f(x)\frac{dy}{dx}\right) - g(x)y + \lambda w(x)y = 0 \tag{8.1}$$

where $w(x) \geq 0$ on the range $a \leq x \leq b$ and the solution $y(x)$ also obeys boundary conditions of the form

and

$$\alpha_1 y + \beta_1 \frac{dy}{dx} = 0 \quad \text{at } x = a$$

$$\alpha_2 y + \beta_2 \frac{dy}{dx} = 0 \quad \text{at } x = b \tag{8.2}$$

We want to determine values of the constant λ for which there are nontrivial solutions $y(x)$. This is called the Sturm-Liouville problem. If $\alpha = 0$ the boundary condition simplifies to $dy/dx = 0$ (Neumann conditions), or if $\beta = 0$ the condition is $y = 0$ (Dirichlet conditions). The constants α and β cannot both be zero.

An example of a Sturm-Liouville problem that we have already seen (Chapter 4, Section 4.4.2) is the problem of waves on a string with fixed ends. Using separation of variables, we reduced the partial differential equation to two ordinary differential equations

$$\frac{d^2 X}{dx^2} + k^2 X = 0 \tag{8.3}$$

357

and

$$\frac{d^2T}{dt^2} + k^2 v^2 T = 0$$

with the boundary conditions in x:

$$X(0) = X(L) = 0$$

The differential equation for X is of the form (8.1) with $f(x) \equiv 1$, $g(x) \equiv 0$, and $w(x) \equiv 1$. Also, the boundary conditions have $\beta_i = 0$, so they are Dirichlet conditions. The solutions to this problem are a set of sine functions $\sin kx$ with a specified set of values for k: $k = n\pi/L$. These are the eigenvalues for the problem, and the resulting functions $X_n(x) \equiv \sin n\pi x/L$ are the eigenfunctions. We also found that these functions have a property called orthogonality (for example, equations 4.3 and 4.12):

$$\int_0^L \sin\frac{n\pi x}{L} \sin\frac{m\pi x}{L} dx = \frac{L}{2}\delta_{mn}$$

We shall see that all problems of Sturm-Liouville type have a similar structure—the general solution is a linear combination of orthogonal eigenfunctions.

8.1.1. Orthogonality of the Eigenfunctions

First, let us show that solutions corresponding to different eigenvalues are orthogonal on the interval (a, b) with respect to the weight function $w(x)$. Suppose that there is a set of solutions $y_m(x)$ to equation (8.1) called eigenfunctions, with corresponding eigenvalues λ_m, and that these solutions also satisfy the boundary conditions (8.2). Then for one solution labeled y_m,

$$\frac{d}{dx}\left(f(x)\frac{dy_m}{dx}\right) - g(x)y_m + \lambda_m w(x)y_m = 0 \tag{8.4}$$

while for a second solution labeled y_n,

$$\frac{d}{dx}\left(f(x)\frac{dy_n}{dx}\right) - g(x)y_n + \lambda_n w(x)y_n = 0 \tag{8.5}$$

Now we multiply equation (8.4) by y_n, multiply equation (8.5) by y_m, and subtract. The term $g(x)y_m y_n$ cancels, and we are left with

$$y_n\frac{d}{dx}\left(f(x)\frac{dy_m}{dx}\right) - y_m\frac{d}{dx}\left(f(x)\frac{dy_n}{dx}\right) + (\lambda_m - \lambda_n)w(x)y_m y_n = 0$$

Next we integrate the whole equation over the range of interest, $x = a$ to $x = b$:

$$\int_a^b \left[y_n\frac{d}{dx}\left(f(x)\frac{dy_m}{dx}\right) - y_m\frac{d}{dx}\left(f(x)\frac{dy_n}{dx}\right)\right] dx = (\lambda_n - \lambda_m)\int_a^b w(x)y_m y_n\, dx$$

$$\tag{8.6}$$

On the left-hand side, we integrate by parts. The first term is

$$\int_a^b y_n \frac{d}{dx}\left(f(x)\frac{dy_m}{dx}\right) dx = y_n f(x)\frac{dy_m}{dx}\bigg|_a^b - \int_a^b \frac{dy_n}{dx} f(x)\frac{dy_m}{dx} dx$$

Clearly the integrated term vanishes if we have Dirichlet ($y_n = 0$) or Neumann ($dy_m/dx = 0$) conditions at each of the two boundaries. In the general case, we may use the boundary conditions (8.2) to eliminate the derivative y' at each boundary, giving

$$f(b)y_n(b)\left(-\frac{\alpha_2}{\beta_2}y_m(b)\right) - f(a)y_n(a)\left(-\frac{\alpha_1}{\beta_1}y_m(a)\right) - \int_a^b f(x)\frac{dy_n}{dx}\frac{dy_m}{dx} dx$$

$$= -\frac{\alpha_2}{\beta_2}f(b)y_n(b)y_m(b) + \frac{\alpha_1}{\beta_1}f(a)y_n(a)y_m(a) - \int_a^b f(x)\frac{dy_n}{dx}\frac{dy_m}{dx} dx$$

All three parts of this expression are symmetric in m and n. We get an identical contribution from the second term on the left-hand side of (8.6), and they subtract to give zero. Thus, the left-hand side reduces to zero, and so the right-hand side must be zero too. There are two possibilities:

- either

$$\lambda_m = \lambda_n \quad \text{and} \quad m = n$$

which[1] means that $y_m \equiv y_n$,

- or

$$\boxed{\int_a^b w(x)y_m y_n \, dx = 0} \tag{8.7}$$

Equation (8.7) is the orthogonality integral that we set out to find. Notice that the eigenfunctions $y_m(x)$ are weighted by the function $w(x)$. [In the case of the string problem (equation 8.3), the weighting function is identically equal to 1.]

There are two other important cases in which the left-hand side of equation (8.6) reduces to zero. If the function $f(x)$ has the value zero at $x = a$ and $x = b$, then the integrated term is zero no matter what the boundary conditions on $y(x)$, provided that $y(x)$ remains finite. We shall see that this is the case for Legendre's equation.[2] Finally, if $yy'f$ has period $(b-a)$, then the integrated term vanishes.

---

[1] See Section 8.1.3 for the degenerate case $\lambda_m = \lambda_n$ with $m \neq n$.
[2] Section 8.3.1.

When equation (8.7) is satisfied, the set of eigenfunctions $y_n(x)$ forms a complete orthogonal set on the interval $[a, b]$. This means that any reasonably well-behaved function $f(x)$ defined for $a \leq x \leq b$ can be expanded in a series of eigenfunctions:

$$f(x) = \sum_{n=0}^{\infty} a_n y_n(x) \tag{8.8}$$

where the coefficients a_n may be found, as we did with the Fourier series (Chapter 4), by multiplying both sides of relation (8.8) by $w(x)y_m(x)$ and integrating over the range a to b. Only the one term in the sum with $m = n$ survives the integration, and

$$a_m = \frac{\int_a^b f(x)y_m(x)w(x)\,dx}{\int_a^b [y_m(x)]^2 w(x)\,dx} \tag{8.9}$$

As we found with the Fourier series in Chapter 4, the sum converges to the function in the mean; that is,

$$\lim_{N \to \infty} \int_a^b \left[f(x) - \sum_{n=0}^{N} a_n y_n(x) \right]^2 w(x)\,dx = 0$$

We can also obtain a useful relation, called the completeness relation, by inserting the expression for a_n (8.9) into equation (8.8):

$$f(x) = \sum_{n=0}^{\infty} \frac{\int_a^b f(x')y_n(x')w(x')\,dx'}{I_n} y_n(x)$$

where

$$I_n = \int_a^b [y_n(x)]^2 w(x)\,dx$$

Interchanging the sum and the integral, we have

$$f(x) = \int_a^b f(x') \sum_{n=0}^{\infty} \frac{y_n(x)y_n(x')w(x')}{I_n}\,dx'$$

Since the quantity multiplying $f(x')$ in the integrand exhibits the sifting property (Chapter 6, equation 6.2), we may conclude that

$$\sum_{n=0}^{\infty} \frac{y_n(x)y_n(x')w(x')}{I_n} = \delta(x - x') \tag{8.10}$$

Equation (8.10) is the completeness relation for the set of eigenfunctions $y_n(x)$.

8.1.2. Reality of the Eigenvalues

Even if the functions $f(x)$, $g(x)$, and $w(x)$ are real, the eigenfunctions may be complex. Again equation (8.3) offers an example, since the complex function e^{ikx} is a solution to this equation. But, provided that the weight function is real and $w(x) \geq 0$ on (a, b), the eigenvalues are real even if the eigenfunctions are not. To prove this result and demonstrate the orthogonality relation, we take the complex conjugate of equation (8.1):

$$\frac{d}{dx}\left(f(x)\frac{dy_n^*}{dx}\right) - g(x)y_n^* + \lambda_n^* w(x)y_n^* = 0 \tag{8.11}$$

Using a method similar to that used in proving orthogonality, we multiply equation (8.1) by y_n^* and equation (8.11) by y_m and subtract. Again the term in $g(x)$ subtracts away, and we have

$$y_m \frac{d}{dx}\left(f(x)\frac{dy_n^*}{dx}\right) - y_n^* \frac{d}{dx}\left(f(x)\frac{dy_m}{dx}\right) + (\lambda_n^* - \lambda_m)w(x)y_n^* y_m = 0$$

Now we integrate over the range $x = a$ to $x = b$, as we did before, integrating the first two terms by parts. We then have

$$\left(y_m f(x)\frac{dy_n^*}{dx} - y_n^* f(x)\frac{dy_m}{dx}\right)\Big|_a^b = (\lambda_m - \lambda_n^*)\int_a^b w(x)y_n^* y_m \, dx$$

We can make the left-hand side zero with appropriate boundary conditions, as we did in the case of real eigenfunctions (Section 8.1.1). Then, if $m = n$,

$$0 = (\lambda_n - \lambda_n^*)\int_a^b w(x)y_n^* y_n \, dx = (\lambda_n - \lambda_n^*)\int_a^b w(x)|y_n|^2 \, dx$$

The integral is > 0 if $w(x) \geq 0$, since $|y_n|^2 \geq 0$ no matter what the form of y_n. Thus,

$$\lambda_n = \lambda_n^*$$

and the eigenvalues are real.

We also obtain the orthogonality integral for complex functions:

$$\int_a^b w(x)y_n^* y_m \, dx = 0, \quad n \neq m \tag{8.12}$$

8.1.3. Degeneracy

The above proofs fail if $\lambda_m = \lambda_n$ for some $m \neq n$. Then we cannot conclude that the corresponding eigenfunctions y_m and y_n are orthogonal (athough in some cases they are). If there are N eigenfunctions that have the same eigenvalue, then we have an N-fold degeneracy. In the case of double degeneracy, we can always construct two linear combinations of the eigenfunctions so that these new functions are orthogonal. Degeneracy reflects a symmetry of the underlying physical system.

Suppose that we are looking for solutions of equation (8.3) with periodic boundary conditions: $y(0) = y(L)$ and $y'(0) = y'(L)$. (Recall that this was the final case we considered in establishing orthogonality in Section 8.1.1.) There are two eigenfunctions, $\sin k_n x$ and $\cos k_n x$, with the same eigenvalue: $k_n = 2n\pi/L$. In this case of double degeneracy, these two eigenfunctions are orthogonal on the range $x = 0$ to $x = L$, as we established in Chapter 4.

Homogeneity of space allows us to place the origin anywhere that we like. If we shift the origin by an amount x_0, then, for example, a new eigenfunction is

$$\sin \frac{2n\pi}{L}(x - x_0) = \sin \frac{2n\pi x}{L} \cos \frac{2n\pi x_0}{L} - \cos \frac{2n\pi x}{L} \sin \frac{2n\pi x_0}{L}$$

which is a linear combination of the previous two functions. The eigenvalues remain the same, as they must, since the physical behavior of the system has not changed.

Example 3.15 (potential in a box) shows what can happen in two dimensions. In that example, the equation for the functions of x and y was found to be

$$\frac{\partial^2 f}{\partial x^2} + \frac{\partial^2 f}{\partial y^2} + k^2 f = 0$$

The eigenfunctions are $f_{nm} = \sin(n\pi x/l) \sin(m\pi y/w)$, with corresponding eigenvalues $k_{nm}^2 = (n^2/l^2 + m^2/w^2)\pi^2$. If the box is square, $l = w$, the eigenfunctions f_{nm} and f_{mn} share the same eigenvalue, reflecting the symmetry of the square.[3]

8.1.4. The Sturm-Liouville Operator as a Self-Adjoint Operator

We may write equation (8.1) in terms of an operator $\mathcal{L}$, where

$$\mathcal{L}y \equiv \frac{d}{dx}\left(f(x)\frac{dy}{dx}\right) - g(x)y$$

---

[3]There are additional degeneracies here, since in some cases we can obtain the same value $m^2 + n^2$ with several values of m and n. See Problem 3 for another example.

so equation (8.1) is

$$\mathcal{L}y + \lambda w(x)y(x) = 0$$

Then the left-hand side of equation (8.6) may be written

$$\int_a^b y_n \mathcal{L} y_m \, dx - \int_a^b y_m \mathcal{L} y_n \, dx$$

We have already shown that if the eigenfunctions y satisfy appropriate boundary conditions, then this difference is zero, or

$$\int_a^b y_n \mathcal{L} y_m \, dx = \int_a^b y_m \mathcal{L} y_n \, dx$$

The operator $\mathcal{L}$ is then said to be *self-adjoint*.

8.2. USE OF STURM-LIOUVILLE THEORY IN PHYSICS

The Sturm-Liouville problem often arises in physics problems as a result of separating variables in a partial differential equation. Thus, there are usually two or more differential equations, each in a different variable, that are linked by a relation between the eigenvalues. The general method[4] of solution is as follows:

1. Determine the set of functions that satisfy each differential equation for arbitrary λ.
2. Determine which coordinates have homogeneous boundary conditions (the function or its derivative is zero) at both ends of the range. These conditions are necessary if we are to have a Sturm-Liouville problem.
3. Choose the eigenfunction that satisfies one of the zero boundary conditions.
4. Choose the eigenvalue to satisfy the second zero boundary condition.
5. Repeat steps 3 and 4 for any other coordinates with two zero boundary conditions.
6. For the remaining coordinate, choose a solution that satisfies a zero boundary condition at one of the boundaries. The remaining condition must be a nonzero condition if the solution is nontrivial.
7. Write the general solution as a linear combination of the eigenfunctions you have found.
8. Use the orthogonality of the eigenfunctions to determine the unknown constants in the linear combination.

Let's see how this plan works in an example.

Example 8.1.   Find the electrostatic potential inside an infinitely long rectangular wave guide with conducting walls. The guide measures $a \times b$. One of the sides of length a is held at potential V; the other sides are grounded.

---

[4] See also Example 3.15.

First we choose a coordinate system that fits the problem. With a rectangular system, we choose Cartesian coordinates and put the origin at one corner. Then the interior of the guide is defined by $0 < x < a$ and $0 < y < b$ (Figure 8.1). The potential is independent of z. The differential equation satisfied by the potential is Laplace's equation:

$$\nabla^2 \Phi = 0$$

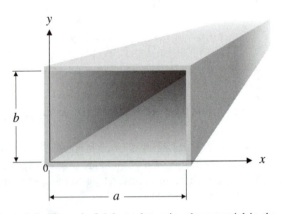

FIGURE 8.1. The task in Example 8.1 is to determine the potential in the region $0 < x < a$, $0 < y < b$. The boundary conditions are that Φ is zero on the sides at $x = 0$, $x = a$, and $y = 0$. The side at $y = b$ has potential V.

Separating variables, we let $\Phi = X(x)Y(y)$, and the differential equation becomes

$$\frac{\partial^2 \Phi}{\partial x^2} + \frac{\partial^2 \Phi}{\partial y^2} = \frac{\partial^2 X}{\partial x^2}Y + X\frac{\partial^2 Y}{\partial y^2} = 0$$

Dividing through by $\Phi = XY$, we have

$$\frac{1}{X}\frac{\partial^2 X}{\partial x^2} + \frac{1}{Y}\frac{\partial^2 Y}{\partial y^2} = 0$$

where the first term is a function of x only and the second is a function of y only. Thus, we can satisfy the equation for *all* values of x and y only if both terms are constants:

$$\frac{1}{X}\frac{\partial^2 X}{\partial x^2} = -\lambda$$

and

$$\frac{1}{Y}\frac{\partial^2 Y}{\partial y^2} = \lambda$$

We have chosen the same constant λ with the opposite sign in the second equation so that the two terms sum to zero, as required.

Each of these equations is of Sturm-Liouville type with $f(x) \equiv 1$, $g(x) \equiv 0$, and $w(x) \equiv 1$. The boundary conditions in x are

$$X(0) = X(a) = 0$$

making the problem in x (differential equation plus boundary conditons) a Sturm-Liouville problem. But the boundary conditions in y are

$$Y(0) = 0; \quad Y(b) = V$$

and thus are not of the correct Sturm-Liouville form.

Now let's follow the steps of the general method.

Step 1. The solutions for X are exponentials of the form $e^{\pm i \sqrt{\lambda} x}$. By choosing the constant λ to be positive, we obtain complex exponentials, or sines and cosines that are linear combinations of the complex exponentials. The y-equation is of the same form and has real exponentials (or $\sinh \sqrt{\lambda} x$, $\cosh \sqrt{\lambda} x$) as its solutions.

Step 2. We have $\Phi = 0$ at two values of x, so we look at the functions of x first.

Step 3. To satisfy the constraint $X(0) = 0$, we need a solution of the form $\sin kx$, where $\lambda = k^2$ has been chosen to be positive.

Step 4. To obtain $X(a) = 0$, we must choose k so that ka is one of the zeros of the sine function—that is, $ka = n\pi$, or $k = n\pi/a$, for some integer n. These are the eigenvalues for the problem. Then the solutions for X are $X_n = \sin n\pi x/a$. These are the eigenfunctions in the variable x.

Step 5 is not needed here.

Step 6. Now that the separation constant λ has been determined—it is $(n\pi/a)^2$—the equation for Y is

$$\frac{\partial^2 Y}{\partial y^2} = \left(\frac{n\pi}{a}\right)^2 Y$$

and the solutions are real exponentials. The function Y must be zero at $y = 0$, so the appropriate linear combination that we need is the hyperbolic sine:

$$Y = \sinh \left(\frac{n\pi}{a} y\right)$$

Notice that the function $Y(y)$ contains the length a of the wave guide in the x-direction. This happens because the equations are coupled through the separation constant that was chosen to fit the boundary conditions in x.

Step 7. The general solution is a linear combination of the eigenfunctions we have found:

$$\Phi(x, y) = \sum_{n=1}^{\infty} c_n \sin \frac{n\pi x}{a} \sinh \frac{n\pi y}{a}$$

Since each eigenfunction $\sin(n\pi x/a)\sinh(n\pi y/a)$ is zero at the three boundaries $x = 0$, $x = a$, and $y = 0$, so is the linear combination.

Step 8. The one remaining boundary condition is the value of Φ at $y = b$:

$$V = \Phi(x, b) = \sum_{n=1}^{\infty} c_n \sin\frac{n\pi x}{a} \sinh\frac{n\pi b}{a}$$

which is a Fourier sine series in x. We find the coefficients c_n by using the orthogonality of the sines. We multiply both sides by $\sin(m\pi x/a)$ and integrate:

$$\int_0^a V \sin\frac{m\pi x}{a}dx = \int_0^a \sum_{n=1}^{\infty} c_n \sin\frac{n\pi x}{a} \sinh\frac{n\pi b}{a} \sin\frac{m\pi x}{a}dx$$

$$V\frac{a}{m\pi}\left(-\cos\frac{m\pi x}{a}\right)\Big|_0^a = \sum_{n=1}^{\infty} c_n \sinh\frac{n\pi b}{a} \int_0^a \sin\frac{n\pi x}{a} \sin\frac{m\pi x}{a}dx$$

The integral on the right-hand side is zero unless $n = m$, in which case it equals $a/2$:

$$V\frac{a}{m\pi}(-\cos m\pi + 1) = \sum_{n=1}^{\infty} c_n \sinh\frac{n\pi b}{a}\left(\frac{a}{2}\delta_{nm}\right)$$

$$V\frac{a}{m\pi}[1 - (-1)^m] = \frac{a}{2}c_m \sinh\frac{m\pi b}{a}$$

Thus,

$$c_m = V\frac{2[1 - (-1)^m]}{m\pi \sinh(m\pi b/a)}$$

and the complete solution is

$$\Phi(x, y) = \frac{2V}{\pi} \sum_{n=1}^{\infty} \frac{[1 - (-1)^n]}{n \sinh(n\pi b/a)} \sin\frac{n\pi x}{a} \sinh\frac{n\pi y}{a}$$

Since only odd values of n contribute $[1 - (-1)^n = 0$ for n even and $1 - (-1)^n = 2$ for n odd], we may write $n = 2m + 1$:

$$\Phi(x, y) = \frac{4V}{\pi} \sum_{m=0}^{\infty} \frac{1}{2m + 1} \sin\left[\frac{(2m + 1)\pi x}{a}\right] \frac{\sinh\dfrac{(2m + 1)\pi y}{a}}{\sinh\dfrac{(2m + 1)\pi b}{a}}$$

Let's see how this solution looks. Figure 8.2 shows the sum of the first ten terms at various values of y, with $b = 2a$. Notice how the solution approaches the value $\Phi/V = 1$ as y approaches b, while maintaining the value zero at the boundaries in x.

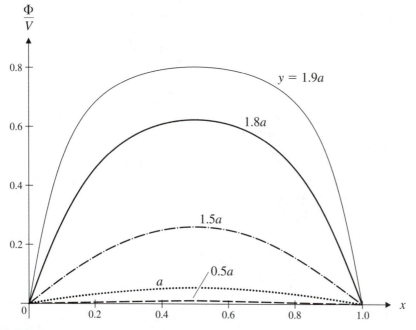

FIGURE 8.2. The solution for the potential with $b = 2a$. The plot shows the first ten terms in the series for Φ versus x for several values of y. The dashed line represents $\Phi(x, y)$ at $y = 0.5a$; the dotted line, $y = a$; the dot-dash line, $y = 1.5a$; the heavy solid line, $y = 1.8a$; and the thin solid line, $y = 1.9a$.

8.3. PROBLEMS WITH SPHERICAL SYMMETRY: SPHERICAL HARMONICS

Suppose our potential problem has spherical boundaries. Then we would like to solve the problem in spherical coordinates. Let's look at Laplace's equation again:

$$\nabla^2\Phi = \frac{1}{r^2}\frac{\partial}{\partial r}\left(r^2\frac{\partial \Phi}{\partial r}\right) + \frac{1}{r^2\sin\theta}\frac{\partial}{\partial \theta}\left(\sin\theta\frac{\partial \Phi}{\partial \theta}\right) + \frac{1}{r^2\sin^2\theta}\frac{\partial^2\Phi}{\partial \phi^2} = 0 \qquad (8.13)$$

We apply the same techniques that we used in the rectangular problem; only the details change. We are looking for a solution of the form

$$\Phi = R(r)P(\theta)W(\phi)$$

Substituting into the differential equation and dividing by Φ, we get

$$\frac{1}{Rr^2}\frac{\partial}{\partial r}\left(r^2\frac{\partial R}{\partial r}\right) + \frac{1}{r^2\sin\theta}\frac{\partial}{\partial\theta}\left(\sin\theta\frac{\partial P}{\partial\theta}\right)\frac{1}{P} + \frac{1}{Wr^2\sin^2\theta}\frac{\partial^2 W}{\partial\phi^2} = 0$$

To separate out an equation for ϕ, we multiply the whole equation by $r^2\sin^2\theta$:

$$\frac{\sin^2\theta}{R}\frac{\partial}{\partial r}\left(r^2\frac{\partial R}{\partial r}\right) + \sin\theta\frac{\partial}{\partial\theta}\left(\sin\theta\frac{\partial P}{\partial\theta}\right)\frac{1}{P} + \frac{1}{W}\frac{\partial^2 W}{\partial\phi^2} = 0$$

Now the last term is a function of ϕ only, while the sum of the first two terms is a function of r and θ only. Thus, if the solution is to satisfy the differential equation for *all* values of r, θ, and ϕ, each of these two pieces must equal a constant.

Often the region of interest is the inside or outside of a complete sphere. In such a region, an increase of ϕ by any integer multiple of 2π corresponds to the same physical point. Thus, the function Φ must have the same value for $\phi = \phi_1$ and $\phi = \phi_1 + 2\pi$; that is, the function W must be periodic with period 2π. We may achieve this behavior if we choose the separation constant so that

$$\frac{1}{W}\frac{\partial^2 W}{\partial\phi^2} = -m^2$$

with m equal to an integer. Then the solutions are the periodic functions

$$W = \begin{cases} \sin m\phi \\ \cos m\phi \end{cases} \quad \text{or } W = e^{\pm im\phi}$$

The equation in r and θ then becomes

$$\frac{\sin^2\theta}{R}\frac{\partial}{\partial r}\left(r^2\frac{\partial R}{\partial r}\right) + \sin\theta\frac{\partial}{\partial\theta}\left(\sin\theta\frac{\partial P}{\partial\theta}\right)\frac{1}{P} - m^2 = 0$$

Next, to separate the r and θ dependences, we divide through by $\sin^2\theta$ to get

$$\frac{1}{R}\frac{\partial}{\partial r}\left(r^2\frac{\partial R}{\partial r}\right) + \frac{1}{\sin\theta}\frac{\partial}{\partial\theta}\left(\sin\theta\frac{\partial P}{\partial\theta}\right)\frac{1}{P} - \frac{m^2}{\sin^2\theta} = 0$$

The first term in this equation is a function of r only, while the sum of the last two terms is a function of θ only. Again both pieces must be constant. The equation has separated:

$$\frac{1}{R}\frac{\partial}{\partial r}\left(r^2\frac{\partial R}{\partial r}\right) = k \tag{8.14}$$

and

$$\frac{1}{\sin\theta}\frac{\partial}{\partial\theta}\left(\sin\theta\frac{\partial P}{\partial\theta}\right)\frac{1}{P} - \frac{m^2}{\sin^2\theta} + k = 0 \tag{8.15}$$

When working in spherical coordinates, changing variables to $\mu = \cos\theta$ is often a useful trick. Then $d\mu = -\sin\theta\,d\theta$, and the theta equation becomes

$$\frac{d}{d\mu}\left((1-\mu^2)\frac{dP}{d\mu}\right) - \frac{m^2}{1-\mu^2}P + kP = 0 \tag{8.16}$$

Equation (8.16) is known as the associated Legendre equation. Let's begin our study of this equation by tackling a special case.

8.3.1. Problems with Axisymmetry: The Legendre Polynomials

If the problem has rotational symmetry about the polar axis, then the function W must be a constant (Φ is independent of ϕ) and so $m = 0$. Then equation (8.16) simplifies:

$$\frac{d}{d\mu}\left((1-\mu^2)\frac{dP}{d\mu}\right) + kP = 0 \tag{8.17}$$

We can solve this equation by looking for a power series solution.[5] The singular points of the equation are at $\mu = \pm 1$, so we should be able to find a solution about $\mu = 0$ of the form

$$y = \sum_{n=0}^{\infty} a_n \mu^n$$

Substituting into the equation, we have

$$\sum_{n=0}^{\infty} n(n-1)a_n\mu^{n-2} - \sum_{n=0}^{\infty} n(n-1)a_n\mu^n - 2\sum_{n=0}^{\infty} na_n\mu^n + k\sum_{n=0}^{\infty} a_n\mu^n = 0$$

where each power of μ must separately equal zero. The constant term in the equation is

$$2a_2 + ka_0 = 0 \Rightarrow a_2 = -\frac{k}{2}a_0$$

and the first power of μ has coefficient

$$3 \times 2a_3 - 2a_1 + ka_1 = 0 \Rightarrow a_3 = a_1\frac{2-k}{3\times 2}$$

For all higher powers, every term in the equation contributes. Looking at μ^p, setting $n = p + 2$ in the first term and $n = p$ in the rest, we find

$$(p+2)(p+1)a_{p+2} - p(p-1)a_p - 2pa_p + ka_p = 0$$

---

[5] See Chapter 3, Section 3.3.3.

and so the recursion relation is

$$a_{p+2} = a_p \frac{p(p-1) + 2p - k}{(p+2)(p+1)} = a_p \frac{p(p+1) - k}{(p+2)(p+1)} \tag{8.18}$$

The first two relations we obtained can also be described by this formula with $p = 0$ and $p = 1$, respectively.

The solution we have obtained is valid for $-1 < \mu < 1$, but the series does not converge for $\mu = \pm 1$. This is a concern, since $\mu = +1$ corresponds to $\theta = 0$ and $\mu = -1$ to $\theta = \pi$. These points are on the z-axis, where usually we do not expect the potential to blow up. Thus, we need a solution that remains valid up to and including these points. We can solve the problem by choosing the separation constant k so that the series terminates after a finite number of terms. In particular, if we choose k to have the value

$$k = l(l+1)$$

for some integer l, then according to the recursion relation (8.18),

$$a_{l+2} = a_l \frac{l(l+1) - l(l+1)}{(l+2)(l+1)} = 0$$

and so every succeeding a_p for $p > l+2$ is also zero. The differential equation (8.17) now takes the form

$$\frac{d}{d\mu}\left((1 - \mu^2) \frac{dP}{d\mu} \right) + l(l+1)P = 0 \tag{8.19}$$

and the corresponding solutions are the Legendre polynomials $P_l(\mu)$. By convention, we choose a_0 (for even l) or a_1 (for odd l) so that

$$P_l(1) \equiv 1 \tag{8.20}$$

The recursion relation becomes

$$a_{p+2} = a_p \frac{p(p+1) - l(l+1)}{(p+2)(p+1)} \tag{8.21}$$

The first few polynomials are found as follows.

$\underline{l = 0}$: The only nonzero coefficient is a_0, which must equal 1 to make $P_0(1) = 1$, so

$$P_0(\mu) = 1 \tag{8.22}$$

$l = 1$: The only nonzero coefficient is a_1, and again we must take $a_1 = 1$ to make $P_1(1) = 1$. Thus,

$$P_1(\mu) = \mu \tag{8.23}$$

$l = 2$: There are two nonzero terms,

$$a_2 = a_0 \left(\frac{-2 \times 3}{2} \right) = -3a_0$$

and the subsequent a_n are all zero. Then

$$P_2(\mu) = a_0(1 - 3\mu^2)$$

Evaluating this at $\mu = 1$, we find

$$P_2(1) = a_0(-2) = 1 \Rightarrow a_0 = -\frac{1}{2}$$

Thus,

$$P_2(\mu) = \frac{1}{2}(3\mu^2 - 1) \tag{8.24}$$

Notice the pattern. We use the recursion relation to determine the nonzero coefficients as multiples of the leading coefficient (a_0 or a_1). Then we evaluate the resulting polynomial at $\mu = 1$ and set the result equal to 1, thus determining the value of the leading coefficient. Let's do one more.

$l = 3$: Applying the recursion relation (8.21) with $l = 3$, we find

$$P_3(\mu) = a_1 \left(\mu + \frac{1 \times 2 - 3 \times 4}{3 \times 2} \mu^3 \right)$$

Evaluating at $\mu = 1$ gives

$$P_3(1) = a_1 \left(1 - \frac{5}{3} \right) = 1 \Rightarrow a_1 = -\frac{3}{2}$$

and so

$$P_3(\mu) = \frac{\mu}{2}(5\mu^2 - 3) \tag{8.25}$$

The first four polynomials are shown in Figure 8.3.

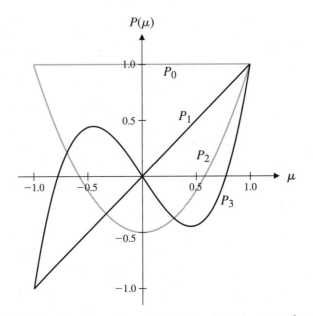

FIGURE 8.3. The first four Legendre polynomials. $P_0(\mu)$ and $P_2(\mu)$ are even functions; $P_1(\mu)$ and $P_3(\mu)$ are odd functions. All the functions are chosen to have the value 1 at $\mu = 1$ (top right).

8.3.2. Solution for the Potential

Now that we have the function of θ, let's return to the potential problem and solve for the function of r. With the separation constant determined, equation (8.14) becomes

$$\frac{\partial}{\partial r}\left(r^2 \frac{\partial R}{\partial r}\right) = l(l+1)R$$

Solutions to this equation are powers of r: r^p, where

$$\frac{\partial}{\partial r}\left(r^2 \frac{\partial r^p}{\partial r}\right) = \frac{\partial}{\partial r}\left(r^2 p r^{p-1}\right) = p(p+1)r^p = l(l+1)r^p$$

One solution has $p = l$; there is a second solution with $p = -(l+1)$. Then $p+1 = -l$, and $p(p+1) = l(l+1)$, as required. We have

$$R = r^l \text{ or } \frac{1}{r^{l+1}} \tag{8.26}$$

Thus, an axisymmetric potential may be expressed as

$$\Phi(r,\theta) = \sum_{l=0}^{\infty} \left(A_l r^l + \frac{B_l}{r^{l+1}} \right) P_l(\mu) \qquad (8.27)$$

where the constants A_l and B_l must be determined by the boundary conditions in r.

8.3.3. Legendre Functions of the Second Kind

Notice that by choosing the separation constant $k = l(l+1)$ we force *one* of the two solutions of equation (8.19) to terminate. The second solution does not terminate. This solution is called the Legendre function of the second kind: $Q_l(\mu)$. For example, with $l = 0$, the second solution begins with a_1 and contains only odd powers of μ. The recursion relation (8.21) is

$$a_{p+2} = a_p \frac{p(p+1) - 0}{(p+2)(p+1)} = a_p \frac{p}{p+2} = a_{p-2} \frac{p-2}{p}\frac{p}{p+2} = a_{p-2}\frac{p-2}{p+2}$$
$$= \frac{a_1}{p+2}$$

and none of the a_n is zero unless a_1 is. Thus, the solution is

$$Q_0(\mu) = a_1 \left(\mu + \frac{\mu^3}{3} + \frac{\mu^5}{5} + \cdots \right)$$
$$= \frac{a_1}{2} \ln \left(\frac{1+\mu}{1-\mu} \right)$$

In this function, a_1 is taken to be 1 so that

$$Q_0(\mu) = \frac{1}{2} \ln \left(\frac{1+\mu}{1-\mu} \right) \qquad (8.28)$$

This expression is valid for $-1 < \mu < 1$.

These functions Q_l are not used in the solution of spherical potential problems, but they do find uses in other situations. For example, they are used when solving potential problems in spheroidal coordinates.[6] Unlike the Legendre polynomials, which diverge at large values

---

[6]See Problem 13 in Chapter 2 and Problem 9 in this chapter.

of the argument, the Legendre functions of the second kind approach zero as the argument tends to infinity. To obtain the limiting form for $Q_l(x)$, we solve the differential equation in inverse powers of x. We find[7]

$$Q_l(x) \rightarrow \sqrt{\pi}\, \frac{\Gamma(l+1)}{\Gamma\left(l+\frac{3}{2}\right)} \frac{1}{(2x)^{l+1}} \quad \text{as } x \rightarrow \infty \tag{8.29}$$

8.3.4. Orthogonality of the Legendre Functions

The Legendre equation (8.19) is of the Sturm-Liouville form (8.1) with

$$f(\mu) \equiv 1 - \mu^2$$

$$g(\mu) \equiv 0$$

and
$$w(\mu) \equiv 1$$

The eigenvalue is $\lambda = l(l+1)$. Even without any boundary conditions specified, except that the function remain finite, the Legendre functions must be orthogonal on the range $[-1, 1]$ because $f(1) = f(-1) = 0$.

$$\int_{-1}^{+1} P_l(\mu) P_{l'}(\mu)\, d\mu = 0 \quad \text{for } l \neq l' \tag{8.30}$$

To make use of this relation in forming series expansions in Legendre polynomials (that is, to use equation 8.9), we will need to find the value of the integral for $l = l'$. In the next few sections, we shall collect some useful tools that will allow us to do that integral.

8.3.5. Properties of Legendre Polynomials

The Generating Function

Suppose we put a point charge q on the polar axis at a distance s from the origin (Figure 8.4). Then the potential[8] at point P is

$$\Phi = \frac{1}{4\pi\varepsilon_0} \frac{q}{D} = \frac{1}{4\pi\varepsilon_0} \frac{q}{\sqrt{s^2 + r^2 - 2rs\cos\theta}}$$

---

[7] See Problem 11.
[8] See, for example, Lea and Burke, Chapter 25, equation 25.9.

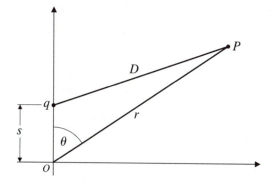

FIGURE 8.4. The potential at P due to a point charge q on the z-axis at $z = s$ may be expressed in terms of Legendre polynomials using equation (8.27). It may also be written as $q/4\pi\varepsilon_0 D$.

which we can also express in the form (8.27). We let $x = r/s$ for convenience, and then for $r < s$ we can expand the denominator to get

$$
\begin{aligned}
\Phi &= \frac{q}{4\pi\varepsilon_0 s} \frac{1}{\sqrt{1 + \dfrac{r^2}{s^2} - 2\dfrac{r}{s}\mu}} = \frac{q}{4\pi\varepsilon_0 s} \frac{1}{\sqrt{1 - 2x\mu + x^2}} \\
&= \frac{q}{4\pi\varepsilon_0 s}\left(1 - \frac{(x^2 - 2x\mu)}{2} + \frac{(-1/2)(-3/2)}{2}(x^2 - 2x\mu)^2 + \cdots\right) \\
&= \frac{q}{4\pi\varepsilon_0 s}\left(1 + x\mu - \frac{x^2}{2}(1 - 3\mu^2) + \cdots\right) \\
&= \frac{q}{4\pi\varepsilon_0 s}\left(1 + x P_1(\mu) + x^2 P_2(\mu) + \cdots\right)
\end{aligned}
$$

which has the form (8.27) with $A_l = q/(4\pi\varepsilon_0 s^{l+1})$ for each l and $B_l \equiv 0$. Thus, we have the identity

$$
\frac{1}{\sqrt{1 - 2x\mu + x^2}} = \sum_{l=0}^{\infty} x^l P_l(\mu) \tag{8.31}
$$

The function

$$
G(x, \mu) \equiv \frac{1}{\sqrt{1 - 2x\mu + x^2}} \tag{8.32}
$$

is called the *generating function* for the Legendre polynomials. We can use it to determine several useful properties of the polynomials.

The Orthogonality Integral

We can obtain the integral we want by integrating the square of the generating function:

$$\int_{-1}^{+1} G^2 \, d\mu = \int_{-1}^{+1} \frac{1}{1 - 2x\mu + x^2} d\mu = \int_{-1}^{+1} \sum_{l=0}^{\infty} x^l P_l(\mu) \sum_{l'=0}^{\infty} x^{l'} P_{l'}(\mu) \, d\mu$$

$$= \sum_{l=0}^{\infty} \sum_{l'=0}^{\infty} x^{l+l'} \int_{-1}^{+1} P_l(\mu) P_{l'}(\mu) \, d\mu$$

The integral of the P_ls is zero unless $l = l'$ (equation 8.30). Thus, evaluating the integral of G^2 by a change of variable to $v = 1 - 2x\mu + x^2$, we have

$$\sum_{l=0}^{\infty} x^{2l} \int_{-1}^{+1} P_l(\mu) P_l(\mu) \, d\mu = \frac{1}{-2x} \int_{(1+x)^2}^{(1-x)^2} \frac{dv}{v}$$

$$= \frac{1}{2x} \ln \frac{(1+x)^2}{(1-x)^2} = \frac{1}{x} \ln \frac{1+x}{1-x}$$

For $x < 1$, we may expand the logarithm:

$$\frac{1}{x} \ln \frac{1+x}{1-x} = \frac{2}{x} \left(x + \frac{x^3}{3} + \frac{x^5}{5} + \cdots + \frac{x^{2l+1}}{2l+1} + \cdots \right)$$

$$= 2 \left(1 + \frac{x^2}{3} + \cdots + \frac{x^{2l}}{2l+1} + \cdots \right) = \sum_{l=0}^{\infty} x^{2l} \int_{-1}^{+1} P_l(\mu) P_l(\mu) \, d\mu$$

Both sides of this equation contain only even powers of x. Equating the coefficients of each power gives

$$\boxed{\int_{-1}^{+1} P_l(\mu) P_l(\mu) \, d\mu = \frac{2}{2l+1}} \tag{8.33}$$

which is the desired result.

Recursion Relations

Our next task is to find a set of relations between different Legendre polynomials. These relations are known as recursion relations. We begin by finding a relation among polynomials with successive values of l: the *pure recursion relation*.[9] First we differentiate the

---

[9] The name "pure" arises from the fact that this relation contains the eigenfunctions only and no derivatives appear.

generating function with respect to x:

$$\frac{\partial G}{\partial x} = \frac{-(x - \mu)}{(1 - 2x\mu + x^2)^{3/2}} = \frac{-(x - \mu)}{(1 - 2x\mu + x^2)} G$$

Rearranging, we obtain

$$(1 - 2x\mu + x^2)\frac{\partial G}{\partial x} = (\mu - x)G$$

We proceed by inserting the expansion of G in Legendre polynomials into this expression,

$$(1 - 2x\mu + x^2)\sum_{l=0}^{\infty} l x^{l-1} P_l(\mu) = (\mu - x)\sum_{l=0}^{\infty} x^l P_l(\mu)$$

and gathering up in powers of x:

$$\sum_{l=0}^{\infty}(l + 1)x^{l+1} P_l(\mu) - \mu\sum_{l=0}^{\infty}(2l + 1)x^l P_l(\mu) + \sum_{l=0}^{\infty} l x^{l-1} P_l(\mu) = 0$$

The coefficient of each power of x must separately equal zero; so, for $l \geq 1$,

$$l P_{l-1}(\mu) - (2l + 1)\mu P_l(\mu) + (l + 1)P_{l+1}(\mu) = 0 \qquad (8.34)$$

Equation (8.34), the pure recursion relation, relates each P_l to the ones above and below it in the sequence. It may be used to determine the P_ls once the first two are known, and it thus provides an alternative to using relation (8.21).

Next we obtain a relation involving the derivatives of P_l. Again we start by differentiating the generating function, this time with respect to μ:

$$\frac{\partial G}{\partial \mu} = \frac{x}{(1 - 2x\mu + x^2)^{3/2}} = \frac{x}{(1 - 2x\mu + x^2)} G$$

Rearranging and inserting the expansion of G gives

$$(1 - 2x\mu + x^2)\sum_{l=0}^{\infty} x^l P_l'(\mu) = x\sum_{l=0}^{\infty} x^l P_l(\mu)$$

$$\sum_{l=0}^{\infty} x^{l+2} P_l'(\mu) - \sum_{l=0}^{\infty} x^{l+1}[P_l(\mu) + 2\mu P_l'(\mu)] + \sum_{l=0}^{\infty} x^l P_l'(\mu) = 0$$

Setting the coefficient of each power of x to zero, we have, for $l \geq 1$,

$$P_l(\mu) = P_{l-1}'(\mu) - 2\mu P_l'(\mu) + P_{l+1}'(\mu) \qquad (8.35)$$

If we differentiate the pure recursion relation (8.34), we obtain

$$l P'_{l-1}(\mu) - (2l + 1) P_l(\mu) - (2l + 1)\mu P'_l(\mu) + (l + 1) P'_{l+1}(\mu) = 0 \tag{8.36}$$

for $l \geq 1$. Using this relation to eliminate P'_{l+1} from equation (8.35), we find

$$P_l(\mu) = P'_{l-1}(\mu) - 2\mu P'_l(\mu) - \frac{l P'_{l-1}(\mu) - (2l + 1) P_l(\mu) - (2l + 1)\mu P'_l(\mu)}{l + 1}$$

$$= \frac{1}{l+1} P'_{l-1}(\mu) - \mu P'_l(\mu)\frac{1}{l+1} + \frac{2l+1}{l+1} P_l(\mu)$$

and so, for $l \geq 1$,

$$\boxed{l P_l(\mu) = \mu P'_l(\mu) - P'_{l-1}(\mu)} \tag{8.37}$$

Similarly, by eliminating P'_{l-1}, we obtain

$$l P_l(\mu) = (2l + 1) P_l(\mu) + (2l + 1)\mu P'_l(\mu) - (l + 1) P'_{l+1}(\mu) - 2\mu l P'_l(\mu) + l P'_{l+1}(\mu)$$

$$(l + 1) P_l(\mu) = P'_{l+1}(\mu) - \mu P'_l(\mu) \tag{8.38}$$

which is valid for $l \geq 0$.

Then, adding equations (8.37) and (8.38), we can express the polynomial P_l entirely in terms of derivatives:

$$\boxed{(2l + 1) P_l(\mu) = P'_{l+1}(\mu) - P'_{l-1}(\mu)} \tag{8.39}$$

which is valid for $l \geq 1$. This expression proves useful in evaluating integrals of the P_ls.

Next we can combine (8.37) with (8.38) to eliminate $P'_{l-1}(\mu)$. Let $l \to l - 1$ in (8.38) to obtain

$$l P_{l-1}(\mu) = P'_l(\mu) - \mu P'_{l-1}(\mu) = P'_l(\mu) - \mu[\mu P'_l(\mu) - l P_l(\mu)]$$

$$\boxed{P_{l-1}(\mu) = \mu P_l(\mu) + \left(\frac{1 - \mu^2}{l}\right) P'_l(\mu)} \tag{8.40}$$

Equation (8.40) is valid for $l \geq 1$. This relation is called a *ladder operator* because it allows us to step down the series in l. If, instead, we eliminate $P'_{l+1}(\mu)$, we obtain the step-up ladder operator:

$$(l + 1) P_{l+1}(\mu) = \mu P'_{l+1}(\mu) - P'_l(\mu) = \mu[(l + 1) P_l(\mu) + \mu P'_l(\mu)] - P'_l(\mu)$$

$$P_{l+1}(\mu) = \mu P_l(\mu) - \left(\frac{1-\mu^2}{l+1}\right) P'_l(\mu) \tag{8.41}$$

These relations prove especially useful in quantum mechanics.

The Rodrigues Formula

The Legendre polynomials may be expressed as

$$P_l(x) = \frac{1}{2^l l!} \frac{d^l}{dx^l} (x^2 - 1)^l \tag{8.42}$$

a relation known as the Rodrigues formula. To demonstrate the validity of equation (8.42), let's evaluate the derivatives. First we expand the quantity $(x^2 - 1)^l$ using the binomial expansion:

$$(x^2 - 1)^l = (-1)^l \left(1 - lx^2 + \frac{l(l-1)}{2} x^4 + \cdots + (-1)^l x^{2l}\right)$$

$$= (-1)^l \sum_{p=0}^{l} (-1)^p \frac{l!}{p!(l-p)!} x^{2p}$$

After differentiating l times, only those terms with $2p \geq l$ remain:

$$\frac{d^l}{dx^l}(x^2 - 1)^l = (-1)^l \sum_{p=[(l+1)/2]}^{l} (-1)^p \frac{l!}{p!(l-p)!} \frac{d^l}{dx^l} x^{2p}$$

$$= (-1)^l \sum_{p=[(l+1)/2]}^{l} (-1)^p \frac{(2p)l!}{p!(l-p)!} \frac{d^{l-1}}{dx^{l-1}} x^{2p-1}$$

$$= (-1)^l \sum_{p=[(l+1)/2]}^{l} (-1)^p \frac{2p(2p-1)l!}{p!(l-p)!} \frac{d^{l-2}}{dx^{l-2}} x^{2p-2}$$

where $[(l+1)/2]$ means the integer part of $(l+1)/2$. Continuing in this way, we find

$$\frac{d^l}{dx^l}(x^2 - 1)^l = (-1)^l l! \sum_{p=[(l+1)/2]}^{l} (-1)^p x^{2p-l} \frac{(2p)!}{(2p-l)! p!(l-p)!} \tag{8.43}$$

Notice that if l is even, this polynomial has $l/2$ even powers of x, while if l is odd, it contains $(l+1)/2$ odd powers of x. The ratio of the coefficient of x^{k+2} to the coefficient of x^k is found by first setting $2p - l = k + 2$ and then setting $2p - l = k$:

$$\frac{a_{k+2}}{a_k} = -\frac{(k+2+l)!}{(k+2)!\left(\dfrac{k+l+2}{2}\right)!\left(\dfrac{l-k}{2}-1\right)!}\frac{k!\left(\dfrac{k+l}{2}\right)!\left(\dfrac{l-k}{2}\right)!}{(l+k)!}$$

$$= -\frac{(l+k+2)(l+k+1)\left(\dfrac{l-k}{2}\right)}{(k+2)(k+1)\left(\dfrac{k+l+2}{2}\right)}$$

$$= -\frac{l^2 - k^2 + l - k}{(k+1)(k+2)} = \frac{k(k+1) - l(l+1)}{(k+1)(k+2)}$$

which is the recursion relation (8.21) for the Legendre polynomials. The numerical factor in front of the derivatives ensures the correct normalization. We can easily check it[10] by looking at the first few polynomials ($l = 0$ is trivial).

$l = 1$:

$$\frac{1}{2^l l!}\frac{d^l}{dx^l}(x^2 - 1)^l = \frac{1}{2}\frac{d}{dx}(x^2 - 1) = x$$

which is correct.

$l = 2$:

$$\frac{1}{2^l l!}\frac{d^l}{dx^l}(x^2 - 1)^l = \frac{1}{2^2 2}\frac{d^2}{dx^2}(x^4 - 2x^2 + 1) = \frac{1}{8}\frac{d}{dx}(4x^3 - 4x)$$

$$= \frac{1}{2}(3x^2 - 1)$$

which is also correct.

As a by-product, we can use equations (8.42) and (8.43) to obtain an explicit formula for P_l:

$$P_l(\mu) = \frac{(-1)^l}{2^l} \sum_{p=[(l+1)/2]}^{l} (-1)^p \mu^{2p-l} \frac{(2p)!}{(2p-l)!p!(l-p)!} \tag{8.44}$$

Or, setting $k = p - l/2$ for l even gives

$$P_l(\mu) = \frac{(-1)^{l/2}}{2^l} \sum_{k=0}^{l/2} (-1)^k \mu^{2k} \frac{(2k+l)!}{2k!\left(k+\dfrac{l}{2}\right)!\left(\dfrac{l}{2}-k\right)!} \tag{8.45}$$

---

[10] See also Problem 7.

and setting $k = p - (l - 1)/2$ for l odd gives

$$P_l(\mu) = \frac{-(-1)^{(l-1)/2}}{2^l} \sum_{k=1}^{(l+1)/2} (-1)^k \mu^{2k-1} \frac{(2k+l-1)!}{(2k-1)! \left(k + \frac{l-1}{2}\right)! \left(\frac{l+1}{2} - k\right)!} \qquad (8.46)$$

We can also use these expressions to evaluate $P_l(0)$. For l odd, the polynomial has no constant term, and so $P_l(0) = 0$. For l even, the constant term $(k = 0)$ in equation (8.45) is

$$P_l(0) = \frac{(-1)^{l/2}}{2^l} \frac{l!}{[(l/2)!]^2} = (-1)^{l/2} \frac{l!}{(l!!)^2} = (-1)^{l/2} \frac{(l-1)!!}{l!!} \qquad (8.47)$$

8.3.6. Solving a Potential Problem: Fourier-Legendre Series

Example 8.2.   A hollow copper sphere of radius a is divided into two halves at the equator by a thin insulating strip (Figure 8.5). The top half of the sphere is held at potential V, and the bottom is grounded. What is the potential everywhere inside?

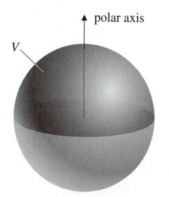

FIGURE 8.5. The potential on the top half of the sphere in Example 8.2 is V. The potential on the bottom half is zero.

Since $\nabla^2 \Phi = 0$ inside the sphere and the system has axisymmetry, the solution must be of the form (8.27):

$$\Phi(r, \theta) = \sum_{l=0}^{\infty} \left(A_l r^l + \frac{B_l}{r^{l+1}} \right) P_l(\mu)$$

We must set all the coefficients B_l to zero, because nothing is at the origin to cause a divergent potential. We find the coefficients A_l by evaluating the potential at the surface $r = a$:

$$\Phi(a, \theta) = \sum_{l=0}^{\infty} A_l a^l P_l(\mu) = \begin{cases} V & \text{if } 1 \geq \mu > 0 \\ 0 & \text{if } 0 > \mu \geq -1 \end{cases}$$

Next we make use of the orthogonality of the P_ls by multiplying both sides of the equation by $P_{l'}(\mu)$ and integrating from -1 to $+1$:

$$\int_{-1}^{+1} \sum_{l=0}^{\infty} A_l a^l P_l(\mu) P_{l'}(\mu) \, d\mu = \int_0^1 V P_{l'}(\mu) \, d\mu$$

On the left-hand side, only the term with $l = l'$ survives the integration. The integral is given by equation (8.33). Then, dropping the primes, we get

$$A_l a^l \frac{2}{2l + 1} = V \int_0^1 P_l(\mu) \, d\mu$$

Now we need to evaluate the integral on the right-hand side. We can make use of equation (8.39), which is valid for $l > 0$:

$$\int_0^1 P_l(\mu) \, d\mu = \frac{1}{2l + 1} \int_0^1 [P'_{l+1}(\mu) - P'_{l-1}(\mu)] \, d\mu$$

Here the right-hand side can be integrated immediately:

$$\int_0^1 P_l(\mu) \, d\mu = \frac{1}{2l + 1} [P_{l+1}(\mu) - P_{l-1}(\mu)]\big|_0^1$$

Since $P_l(1) = 1$ for all values of l,

$$\int_0^1 P_l(\mu) \, d\mu = -\frac{1}{2l + 1} [P_{l+1}(0) - P_{l-1}(0)]$$

$$= -\frac{1}{2l + 1} P_{l+1}(0) \left(1 + \frac{l + 1}{l}\right) = -\frac{P_{l+1}(0)}{l}$$

where we used the pure recursion relation (8.34) to express $P_{l-1}(0)$ in terms of $P_{l+1}(0)$. Then

$$A_l = -\frac{V}{2} \frac{2l + 1}{l a^l} P_{l+1}(0), \quad l \geq 1$$

For the remaining case, $l = 0$, we can easily do the integral, since $P_0(\mu) = 1$ (equation 8.22):

$$\int_0^1 P_0(\mu) \, d\mu = \int_0^1 1 \, d\mu = 1$$

so

$$A_0 = \frac{V}{2}$$

and

$$\Phi(r, \theta) = \frac{V}{2}\left[1 - \sum_{l=1}^{\infty} \frac{2l+1}{l} P_{l+1}(0)\left(\frac{r}{a}\right)^l P_l(\mu)\right]$$

The sum reduces to a sum over odd l, since $P_{l+1}(0) = 0$ if $l+1$ is odd. Then, if we use equation (8.47) for $P_{l+1}(0)$ and set $l = 2n+1$, the potential is

$$\Phi(r, \theta) = \frac{V}{2}\left(1 + \sum_{n=0}^{\infty}(-1)^n \frac{4n+3}{(2n+1)2^{2n+2}} \frac{(2n+2)!}{[(n+1)!]^2}\left(\frac{r}{a}\right)^{2n+1} P_{2n+1}(\mu)\right)$$

The first few terms are

$$\Phi(r, \theta) = \frac{V}{2}\left(1 + \frac{3}{2}\frac{r}{a}\cos\theta - \frac{7}{16}\left(\frac{r}{a}\right)^3 \cos\theta(5\cos^2\theta - 3)\right.$$

$$\left. + \frac{11}{16}\left(\frac{r}{a}\right)^5 \frac{\cos\theta}{8}(63\cos^4\theta - 70\cos^2\theta + 15) + \cdots\right)$$

Figure 8.6 shows the first four terms in the series for the potential versus angle for $r/a = 0.1, 0.2$, and 0.5. Notice that the potential equals $V/2$ at the equator ($\theta = \pi/2$) for all values of r. (This should remind you of a similar property of the Fourier series: The series for a discontinuous function always converges to the midpoint of the jump.)

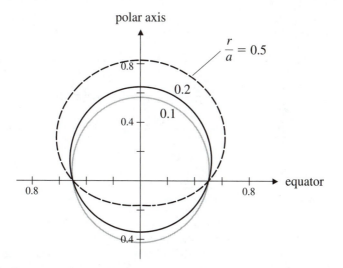

FIGURE 8.6. The potential inside the sphere in Example 8.2 at several values of r: $r/a = 0.1, 0.2$, and 0.5. The value of the potential Φ/V at a specific angle θ is given by the distance of the curve from the origin.

8.3.7. The Associated Legendre Functions and Spherical Harmonics

Orthogonality of Associated Legendre Functions

When a problem does not have symmetry about the polar axis, we need a set of eigen-functions corresponding to nonzero values of the separation constant m. Then the equation for the function of theta is equation (8.16), where we keep the value $k = l(l + 1)$ for that separation constant:

$$\frac{d}{d\mu}\left((1 - \mu^2)\frac{dP}{d\mu} \right) - \frac{m^2}{1 - \mu^2}P + l(l + 1)P = 0 \tag{8.48}$$

Equation (8.48) is of Sturm-Liouville form with $f(\mu) = 1 - \mu^2$, $g(\mu) = m^2/(1 - \mu^2)$, $w(\mu) = 1$, and $\lambda = l(l + 1)$. (Note that m is the eigenvalue for the phi equation.)

The solutions of this equation are the associated Legendre functions $P_l^m(\mu)$. They satisfy the orthogonality relation

$$\int_{-1}^{+1} P_l^m(\mu) P_{l'}^m(\mu)\, d\mu = 0 \quad \text{unless } l = l' \tag{8.49}$$

where the value of m is the same in both functions.

Alternatively, if we disregard the physical origin of this equation, we may identify it as a Sturm-Liouville equation with $g(\mu) = -l(l + 1)$, $w(\mu) = 1/(1 - \mu^2)$, and eigenvalue m^2. In this case, the orthogonality relation is

$$\int_{-1}^{+1} \frac{P_l^m(\mu) P_l^{m'}(\mu)}{1 - \mu^2}\, d\mu = 0 \quad \text{unless } m = m' \tag{8.50}$$

where the value of l is the same in both functions.[11]

Form of the Associated Legendre Functions

To obtain an expression for the function P_l^m, we could solve in a series as we did for the Legendre polynomials. But it is more efficient to relate the solutions P_l^m to the polynomials we have already found. We expand out the differential operator in Legendre's equation (8.19)

---

[11] See Problem 19 for the value of the integral when $m = m'$.

and then differentiate the whole equation m times:

$$(1 - \mu^2)\frac{d^2 P_l}{d\mu^2} - 2\mu\frac{d P_l}{d\mu} + l(l+1)P_l = 0$$

$$(1 - \mu^2)\frac{d^3 P_l}{d\mu^3} - 2 \times 2\mu\frac{d^2 P_l}{d\mu^2} - 2\frac{d P_l}{d\mu} + l(l+1)\frac{d P_l}{d\mu} = 0$$

$$(1 - \mu^2)\frac{d^4 P_l}{d\mu^4} - 2 \times 3\mu\frac{d^3 P_l}{d\mu^3} + [l(l+1) - 2 \times 3]\frac{d^2 P_l}{d\mu^2} = 0$$

Continuing in this way, we get

$$(1 - \mu^2)\frac{d^{m+2} P_l}{d\mu^{m+2}} - 2(m+1)\mu\frac{d^{m+1} P_l}{d\mu^{m+1}} + [l(l+1) - m(m+1)]\frac{d^m P_l}{d\mu^m} = 0 \quad (8.51)$$

so the mth derivative of the Legendre polynomial P_l satisfies the equation

$$(1 - \mu^2)y_m'' - 2(m+1)\mu y_m' + [l(l+1) - m(m+1)]y_m = 0 \quad (8.52)$$

In Appendix VII, we show that the substitution $y_m = (1 - \mu^2)^{-m/2}z(\mu)$ yields the equation

$$(1 - \mu^2)z'' - 2\mu z' + z\left[l(l+1) - \frac{m^2}{1 - \mu^2}\right] = 0$$

and thus $z(\mu)$ satisfies the associated Legendre equation (8.48). Thus, we have shown that $Cz(\mu) = P_l^m(\mu)$ for any constant C. In physics applications, it is common to choose $C = (-1)^m$. Then

$$P_l^m(\mu) = (-1)^m (1 - \mu^2)^{m/2}\frac{d^m}{d\mu^m}P_l(\mu) \quad (8.53)$$

Clearly $P_l^m = 0$ for $m > l$, since the highest power of μ that appears in P_l is μ^l. The function $P_l(\mu)$ is even if l is even and odd if l is odd. The quantity $(1 - \mu^2)^{m/2}$ is an even function of μ, and after m derivatives, the highest power in $(d^m/d\mu^m)P_l(\mu)$ is μ^{l-m}. Thus, $P_l^m(\mu)$ is even if $l + m$ is even and odd if $l + m$ is odd.

Also, since the associated Legendre equation contains m^2, the eigenvalue $-m$ leads to the same differential equation. Thus, P_l^{-m} must be a constant times P_l^m. It is convenient to define

$$P_l^{-m}(\mu) = (-1)^m \frac{(l - m)!}{(l + m)!}P_l^m(\mu) \quad (8.54)$$

as the appropriate solution corresponding to the eigenvalue $-m$. A few of the functions are shown in Figure 8.7 and Table 8.1.

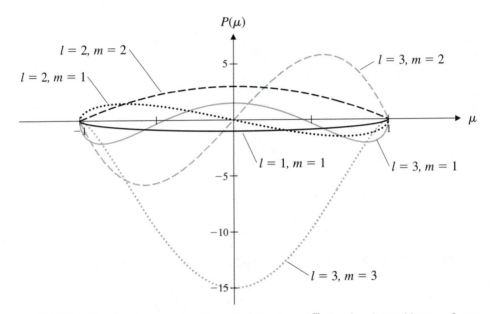

FIGURE 8.7. The first few associated Legendre functions P_l^m. For functions with $m = 0$, see Figure 8.3. Notice that $P_l^m(\pm 1) = 0$ for $m \neq 0$.

TABLE 8.1. Associated Legendre Functions $P_l^m(\mu)$

	$m = 0$	$m = 1$	$m = 2$	$m = 3$
$l = 0$	1	—	—	—
$l = 1$	μ	$-\sqrt{1 - \mu^2}$	—	—
$l = 2$	$\frac{1}{2}(3\mu^2 - 1)$	$-3\mu\sqrt{1 - \mu^2}$	$3(1 - \mu^2)$	—
$l = 3$	$\frac{\mu}{2}(5\mu^2 - 3)$	$-\frac{3}{2}(5\mu^2 - 1)\sqrt{1 - \mu^2}$	$15\mu(1 - \mu^2)$	$-15(1 - \mu^2)^{3/2}$

The Orthogonality Integral

Next we need to find the value of the orthogonality integral when $l = l'$. Using the definition (8.53), we have

$$I_{lm} = \int_{-1}^{+1} P_l^m(\mu) P_l^m \, d\mu = \int_{-1}^{+1} (1 - \mu^2)^m \frac{d^m P_l(\mu)}{d\mu^m} \frac{d^m P_l(\mu)}{d\mu^m} \, d\mu$$

We integrate by parts:

$$
I_{lm} = (1 - \mu^2)^m \frac{d^m P_l(\mu)}{d\mu^m} \frac{d^{m-1} P_l(\mu)}{d\mu^{m-1}} \Big|_{-1}^{+1}
$$
$$
- \int_{-1}^{+1} \frac{d}{d\mu} \left[(1 - \mu^2)^m \frac{d^m P_l(\mu)}{d\mu^m} \right] \frac{d^{m-1} P_l(\mu)}{d\mu^{m-1}} \, d\mu
$$

The integrated term is zero, and the derivative in the integrand is

$$
\frac{d}{d\mu} \left[(1 - \mu^2)^m \frac{d^m P_l(\mu)}{d\mu^m} \right] = -2m\mu (1 - \mu^2)^{m-1} \frac{d^m P_l(\mu)}{d\mu^m} + (1 - \mu^2)^m \frac{d^{m+1} P_l(\mu)}{d\mu^{m+1}}
$$

We can simplify this expression by using equation (8.51) with $m \to m - 1$:

$$
(1 - \mu^2) \frac{d^{m+1} P_l}{d\mu^{m+1}} - 2m\mu \frac{d^m P_l}{d\mu^m} = -[l(l+1) - m(m-1)] \frac{d^{m-1} P_l}{d\mu^{m-1}}
$$

Thus,

$$
I_{lm} = [l(l+1) - m(m-1)] \int_{-1}^{+1} (1 - \mu^2)^{m-1} \left[\frac{d^{m-1} P_l}{d\mu^{m-1}} \right]^2 d\mu
$$
$$
= (l+m)(l-m+1) I_{l,m-1}
$$

Now we may step down to get

$$
I_{lm} = (l+m)(l-m+1)(l+m-1)(l-m+2) I_{l,m-2}
$$
$$
= (l+m)(l+m-1)(l+m-2)(l-m+1)(l-m+2)(l-m+3) I_{l,m-3}
$$
$$
= (l+m)(l+m-1)\cdots(l+1)l\cdots(l-m+2)(l-m+1) I_{l,0}
$$
$$
= \frac{(l+m)!}{(l-m)!} \frac{2}{2l+1}
$$

where we used equation (8.33) for $I_{l,0}$. Thus,

$$
\boxed{\int_{-1}^{+1} P_l^m(\mu) P_{l'}^m \, d\mu = \frac{(l+m)!}{(l-m)!} \frac{2}{2l+1} \delta_{ll'}}
\qquad (8.55)
$$

Spherical Harmonics

The general solution to Laplace's equation in spherical coordinates may be written as

$$\Phi(r,\theta,\phi) = \sum_{l=0}^{\infty} \sum_{m=-l}^{+l} \left(a_{lm} r^l + \frac{b_{lm}}{r^{l+1}} \right) P_l^m(\mu) e^{im\phi} \tag{8.56}$$

[The previous result (equation 8.27) is a special case of equation (8.56) with $a_{lm} = b_{lm} = 0$ unless $m = 0$.] Next we define the combination

$$\sqrt{\frac{2l+1}{4\pi} \frac{(l-m)!}{(l+m)!}} P_l^m(\mu) e^{im\phi} \equiv Y_{lm}(\theta,\phi) \tag{8.57}$$

where the constant has been chosen to make the functions Y_{lm} orthonormal; that is,

$$\int_{-1}^{+1} \int_0^{2\pi} Y_{lm}(\theta,\phi) Y_{l'm'}^*(\theta,\phi) \, d\phi \, d\mu = \delta_{ll'} \delta_{mm'} = \int_{\text{sphere}} Y_{lm}(\theta,\phi) Y_{l'm'}^*(\theta,\phi) \, d\Omega \tag{8.58}$$

The functions $Y_{lm}(\theta,\phi)$ are called *spherical harmonics* (see Figure 8.8 and Table 8.2). They find application not only in potential problems but also in the quantum mechanics of atoms, wave mechanics, oscillations of spheres (for example, the sun), and the structure of the cosmic microwave background.

TABLE 8.2. Spherical Harmonics

	$m=0$	$m=1$	$m=2$	$m=3$
$l=0$	$\dfrac{1}{\sqrt{4\pi}}$	—	—	—
$l=1$	$\sqrt{\dfrac{3}{4\pi}}\cos\theta$	$-\sqrt{\dfrac{3}{8\pi}}\sin\theta e^{i\phi}$	—	—
$l=2$	$\sqrt{\dfrac{5}{4\pi}}\left(\dfrac{3}{2}\cos^2\theta - \dfrac{1}{2}\right)$	$-\dfrac{1}{4}\sqrt{\dfrac{30}{\pi}}\cos\theta\sin\theta e^{i\phi}$	$\dfrac{1}{8}\sqrt{\dfrac{30}{\pi}}\sin^2\theta e^{2i\phi}$	—
$l=3$	$\sqrt{\dfrac{7}{4\pi}}\cos\theta\left(\dfrac{5}{2}\cos^2\theta - \dfrac{3}{2}\right)$	$-\dfrac{1}{8}\sqrt{\dfrac{21}{\pi}}(5\cos^2\theta - 1)\sin\theta e^{i\phi}$	$\dfrac{1}{8}\sqrt{\dfrac{210}{\pi}}\cos\theta\sin^2\theta e^{2i\phi}$	$-\dfrac{1}{8}\sqrt{\dfrac{35}{\pi}}\sin^3\theta e^{3i\phi}$

The defining equation for each figure is

$$\{\mathrm{Re}[Y_l^m(\theta, \phi)]\}^2$$

$m = 0, l = 0$

$m = 0, l = 1$ $m = 1, l = 1$

 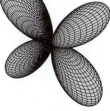

$m = 0, l = 2$ $m = 1, l = 2$ $m = 2, l = 2$

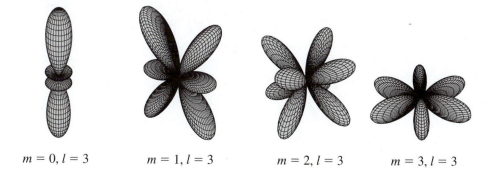

$m = 0, l = 3$ $m = 1, l = 3$ $m = 2, l = 3$ $m = 3, l = 3$

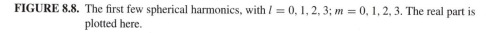

FIGURE 8.8. The first few spherical harmonics, with $l = 0, 1, 2, 3$; $m = 0, 1, 2, 3$. The real part is plotted here.

Recursion Relations for the P_l^m

Here we shall derive a relation between functions with the same l but different m. Let's start with equation (8.51) and multiply by $(-1)^m (1 - \mu^2)^{m/2}$:

$$0 = (-1)^m (1 - \mu^2)^{(m+2)/2} \frac{d^{m+2} P_l}{d\mu^{m+2}} - 2(m+1)\mu(-1)^m (1 - \mu^2)^{m/2} \frac{d^{m+1} P_l}{d\mu^{m+1}}$$

$$+ [l(l+1) - m(m+1)](-1)^m (1 - \mu^2)^{m/2} \frac{d^m P_l}{d\mu^m}$$

Now we use the definition (8.53) to get

$$0 = P_l^{m+2} + 2(m+1)\frac{\mu}{\sqrt{1 - \mu^2}} P_l^{m+1} + [l(l+1) - m(m+1)]P_l^m \qquad (8.59)$$

Numerous similar relations[12] may be obtained by differentiating the relations for the P_l.

The Addition Theorem for Spherical Harmonics

Suppose γ is the angle between two vectors $\vec{x}$ and $\vec{x}'$, as shown in Figure 8.9. The addition theorem allows us to express the Legendre polynomial $P_l(\cos \gamma)$ in terms of spherical harmonics in the angles θ, ϕ, θ', and ϕ' that describe the vectors $\vec{x}$ and $\vec{x}'$.

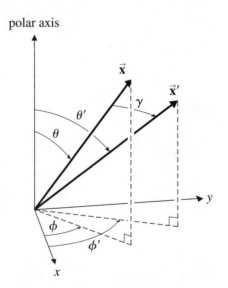

FIGURE 8.9. The angle γ is the angle between the vectors $\vec{x}$ and $\vec{x}'$.

---

[12]Gradshteyn and Ryzhik has an extensive list. See also Problems 15 and 16.

First we note that $r^l P_l(\cos \gamma)$ is a solution of Laplace's equation. Thus, using a "bar" coordinate system with polar axis along the vector $\vec{x}'$, we have

$$\overline{\nabla}^2_{ang} P_l(\cos \gamma) = -\frac{l(l+1)}{r^2} P_l(\cos \gamma)$$

where $\overline{\nabla}^2_{ang}$ is the angular part of the Laplacian operator (the second two terms in equation 8.13). Since $P_l(\cos \gamma)$ is a function of the angles only, the first term in the Laplacian (with derivatives in r) is zero, so we may add it back in to obtain

$$\overline{\nabla}^2 P_l(\cos \gamma) = -\frac{l(l+1)}{r^2} P_l(\cos \gamma)$$

Since ∇^2 is a scalar operator, we may express it in any coordinate system. Returning to our original system, we note that the solutions of

$$\nabla^2 f(\theta, \phi) = -\frac{l(l+1)}{r^2} f(\theta, \phi)$$

are the spherical harmonics Y_{lm} with $-l \le m \le l$. Thus, $P_l(\cos \gamma)$ must be expressible as a sum of these Y_{lm}:

$$P_l(\cos \gamma) = \sum_{m=-l}^{l} A_m(\theta', \phi') Y_{lm}(\theta, \phi) \tag{8.60}$$

In principle, we can find the coefficients A_m in the usual way,

$$A_m = \int P_l(\cos \gamma) Y^*_{lm}(\theta, \phi) \, d\Omega \tag{8.61}$$

but this integral is not easy to do!

Instead, we turn things around and regard $Y^*_{lm}(\theta, \phi)$ as a function of the angles γ and β that describe $\vec{x}$ in a coordinate system with polar axis along $\vec{x}'$. We can expand $Y^*_{lm}(\theta, \phi)$ in spherical harmonics in γ and β, again with a single value of l:

$$Y^*_{lm}(\theta, \phi) = \sum_{m'=-l}^{l} B_{m'} Y_{lm'}(\gamma, \beta) \tag{8.62}$$

If we let $\gamma \to 0$, $\vec{x}$ moves onto the polar axis and we have axisymmetry, so only the $m' = 0$ term remains.[13] Thus, as $\theta \to \theta'$ and $\phi \to \phi'$, we have

$$\lim_{\gamma \to 0} Y^*_{lm}(\theta, \phi) = Y^*_{lm}(\theta', \phi') = B_0 Y_{l0}(0, \beta)$$

$$= B_0 \sqrt{\frac{2l+1}{4\pi}} P_l(1) = B_0 \sqrt{\frac{2l+1}{4\pi}}$$

---

[13] $Y_{lm}(0, \beta) = 0$ unless $m = 0$, because $P_l^m(1) = 0$, as you should verify.

(Of course Y_{l0} is independent of β.) Therefore,

$$B_0 = \sqrt{\frac{4\pi}{2l+1}} Y_{lm}^*(\theta', \phi') \tag{8.63}$$

Alternatively, we can find the coefficients B_m using orthogonality of the Y_{lm} in the (γ, β) coordinate system:

$$B_{m'} = \int Y_{lm}^*(\theta, \phi) Y_{lm'}^*(\gamma, \beta) \, d\Omega_\gamma$$

In particular,

$$
\begin{aligned}
B_0 &= \int Y_{lm}^*(\theta, \phi) Y_{l0}^*(\gamma, \beta) \, d\Omega_\gamma \\
&= \int Y_{lm}^*(\theta, \phi) \sqrt{\frac{2l+1}{4\pi}} P_l(\cos \gamma) \, d\Omega_\gamma \\
&= \sqrt{\frac{2l+1}{4\pi}} A_m(\theta', \phi')
\end{aligned} \tag{8.64}
$$

where we used equation (8.61) in the last step. Since we integrate over the whole sphere, it does not matter whether we use the coordinates θ, ϕ or γ, β as the integration variables.

Finally, we combine equations (8.63), (8.64), and (8.60) to get

$$P_l(\cos \gamma) = \sum_{m=-l}^{l} \frac{4\pi}{2l+1} Y_{lm}(\theta, \phi) Y_{lm}^*(\theta', \phi') \tag{8.65}$$

This is the addition theorem for spherical harmonics.

Note the expected symmetry in the primed and unprimed coordinates. [It does not matter which function has the complex conjugate, since the sum is over positive and negative values of m, and $Y_{l,-m} = (-1)^m Y_{lm}^*$.] We can also check the result by letting θ' approach zero, in which case the left-hand side is just $P_l(\cos \theta)$. Verify that the right-hand side also reduces to $P_l(\cos \theta)$ in this case.

In the special case $\gamma = 0$, we obtain the sum rule for spherical harmonics:

$$\sum_{m=-l}^{l} |Y_{lm}(\theta, \phi)|^2 = \frac{2l+1}{4\pi} \tag{8.66}$$

The addition theorem is useful for performing rotations of the coordinate axes. When combined with the generating function for the Legendre polynomials, it also allows us to express the inverse distance $1/\left|\vec{x} - \vec{x}'\right|$ in terms of spherical harmonics.[14] This is a particularly valuable result, because this inverse distance appears in the integral expressions for electric scalar potential, magnetic vector potential, and Newtonian gravitational potential. The orthogonality of the spherical harmonics often makes the integrals more tractable.

Boundary Value Problems Using Spherical Harmonics

Example 8.3.   Find the potential inside a conducting sphere of radius a with potential $\Phi = V$ on the half $0 \le \phi < \pi$ and zero on the half $\pi \le \phi < 2\pi$.

The potential may be expressed in terms of spherical harmonics, as in equation (8.56). Since there is nothing at the origin to cause a divergent potential, we exclude the negative powers of r. Thus,

$$\Phi(r, \theta, \phi) = \sum_{l=0}^{\infty} \sum_{m=-l}^{l} A_{lm} r^l Y_{lm}(\theta, \phi)$$

To find the constants A_{lm}, we evaluate the potential on the surface and set it equal to the given expression:

$$\Phi(a, \theta, \phi) = \sum_{l=0}^{\infty} \sum_{m=-l}^{l} A_{lm} a^l Y_{lm}(\theta, \phi) = \begin{cases} V & \text{if } 0 \le \phi < \pi \\ 0 & \text{if } \pi \le \phi < 2\pi \end{cases}$$

To evaluate the coefficients, we use the orthogonality of the Y_{lm}. We multiply both sides by $Y^*_{l'm'}(\theta, \phi)$ and integrate over the sphere:

$$\sum_{l=0}^{\infty} \sum_{m=-l}^{l} A_{lm} a^l \int_{\text{sphere}} Y_{lm}(\theta, \phi) Y^*_{l'm'}(\theta, \phi) d\Omega = V \int_{-1}^{+1} \int_{0}^{\pi} Y^*_{l'm'}(\theta, \phi) \, d\mu \, d\phi$$

Only one term on the left-hand side is nonzero. With $l = l'$ and $m = m'$, the integral equals 1. Then

$$A_{l'm'} a^{l'} = V \sqrt{\frac{2l' + 1}{4\pi} \frac{(l' - m')!}{(l' + m')!}} \int_{-1}^{+1} P_{l'}^{m'}(\mu) \, d\mu \int_{0}^{\pi} e^{-im'\phi} \, d\phi$$

Now we may drop the primes:

$$A_{lm} = \frac{V}{a^l} \sqrt{\frac{2l + 1}{4\pi} \frac{(l - m)!}{(l + m)!}} \int_{-1}^{+1} P_l^m(\mu) \, d\mu \left(\frac{1 - (-1)^m}{im} \right)$$

[14] See Problem 20.

The result is zero unless m is odd. Then the integral over μ is zero unless l is also odd, for then $P_l^m(\mu)$ is an even function. Let's call this integral I_{lm}. Then

$$A_{lm} = \frac{V}{a^l} \sqrt{\frac{2l+1}{4\pi} \frac{(l-m)!}{(l+m)!}} I_{lm} \frac{2}{im} \quad \text{for } l \text{ odd and } m \text{ odd}$$

The case $l = m = 0$ is special, for

$$\int_{-1}^{+1} \int_0^\pi Y_{00} = \frac{1}{\sqrt{4\pi}}(2)(\pi) = \sqrt{\pi}$$

giving

$$A_{00} = \frac{V}{2}$$

If $m = 0$ but $l \neq 0$, the result is zero because

$$\int_{-1}^{+1} P_l(\mu)\, d\mu = \int_{-1}^{+1} P_0(\mu)\, P_l(\mu)\, d\mu = 0$$

(equation 8.30). So the potential is

$$\Phi(r, \theta, \phi) = \frac{V}{2} + V \sum_{\substack{l=1, \\ l \text{ odd}}}^{\infty} \sum_{\substack{m=-l, \\ m \text{ odd}}}^{l} \left(\frac{r}{a}\right)^l \sqrt{\frac{2l+1}{4\pi} \frac{(l-m)!}{(l+m)!}} I_{lm} \frac{2}{im} Y_{lm}(\theta, \phi)$$

$$= \frac{V}{2} + V \sum_{\substack{l=1, \\ l \text{ odd}}}^{\infty} \sum_{\substack{m=-l, \\ m \text{ odd}}}^{l} \left(\frac{r}{a}\right)^l \frac{2l+1}{4\pi} \frac{(l-m)!}{(l+m)!} I_{lm} \frac{2}{im} P_l^m(\mu) e^{im\phi}$$

We can demonstrate that the result is real, as expected for a real physical quantity, by combining the terms with $m = M$ and $m = -M$. Using equation (8.54), we have

$$\frac{(l-M)!}{(l+M)!} I_{lM} \frac{2}{iM} P_l^M(\mu) e^{iM\phi} + \frac{(l+M)!}{(l-M)!} I_{l-M} \frac{2}{-iM} P_l^{-M}(\mu) e^{-iM\phi}$$

$$= \frac{(l-M)!}{(l+M)!} I_{lM} \frac{2}{iM} P_l^M(\mu)(e^{iM\phi} - e^{-iM\phi})$$

$$= \frac{(l-M)!}{(l+M)!} I_{lM} \frac{4}{M} P_l^M(\mu) \sin M\phi$$

Thus,

$$\Phi(r,\theta,\phi) = \frac{V}{2} + \frac{V}{\pi} \sum_{\substack{l=1, \\ l \text{ odd}}}^{\infty} \sum_{\substack{m=1, \\ m \text{ odd}}}^{l} \left(\frac{r}{a}\right)^l \frac{2l+1}{m} \frac{(l-m)!}{(l+m)!} I_{lm} P_l^m(\mu) \sin m\phi$$

The result is clearly real and is dimensionally correct. The first few terms are

$$\Phi(r,\theta,\phi) = \frac{V}{2} + \frac{V}{\pi} \left\{ \frac{r}{a} \frac{3}{2} \frac{\pi}{2} \sin\theta \sin\phi + 7 \left(\frac{r}{a}\right)^3 \left[\frac{2}{4!} \frac{3\pi}{16} \frac{3}{2}(5\mu^2 - 1)\sqrt{1-\mu^2} \sin\phi \right.\right.$$

$$\left.\left. + \frac{1}{3} \frac{1}{6!} \frac{45}{8} \pi \, 15 \sin^3\theta \sin 3\phi + \cdots \right] \right\}$$

$$= V \left\{ \frac{1}{2} + \frac{r}{a} \frac{3}{4} \sin\theta \sin\phi + \frac{7}{128} \left(\frac{r}{a}\right)^3 \left[3(5\cos^2\theta - 1)\sin\theta \sin\phi \right.\right.$$

$$\left.\left. + 5 \sin^3\theta \sin 3\phi + \cdots \right] \right\}$$

We can use the addition theorem to show that this result agrees with the result of Example 8.2. Let's compare the two pictures (Figure 8.10).

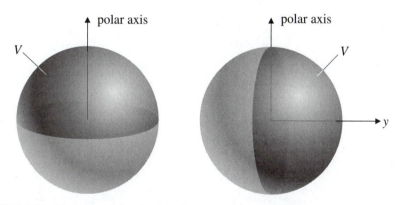

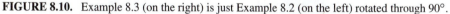

FIGURE 8.10. Example 8.3 (on the right) is just Example 8.2 (on the left) rotated through 90°.

The result of Example 8.2 is

$$\Phi = \frac{V}{2} \left\{ 1 - \sum_{l=1}^{\infty} \frac{2l+1}{l} P_{l+1}(0) \left(\frac{r}{a}\right)^l P_l(\cos\gamma) \right\}$$

where γ is the angle between the position vector $\vec{x}$ and the z-axis in the picture on the left or, equivalently, between the position vector and the y-axis in the picture on the right. To relate the two results, we may use the addition theorem with vector $\vec{x}'$ along the y-axis, so $\theta' = \phi' = \pi/2$.

Let's look at the first few terms:

$$P_1(\cos \gamma) = \cos \theta' = \frac{4\pi}{3}\left[\frac{3}{4\pi}\cos\frac{\pi}{2}\cos\theta + \frac{3}{8\pi}\sin\frac{\pi}{2}\sin\theta(e^{i(\phi-\pi/2)} + e^{-i(\phi-\pi/2)})\right]$$

$$= \frac{4\pi}{3}\left[\frac{3}{8\pi}(\sin\theta)2\cos\left(\phi - \frac{\pi}{2}\right)\right] = \sin\theta\sin\phi$$

and

$$P_3(\cos\gamma) = \frac{4\pi}{7}\left[\frac{7}{4\pi}P_3(\mu)P_3(0) + 2\frac{7}{4\pi}\sum_{m=1}^{3}\frac{(l-m)!}{(l+m)!}P_3^m(\mu)P_3^m(0)\cos m\left(\phi - \frac{\pi}{2}\right)\right]$$

But $P_3^m(0) = 0$ unless $3 + m$ is even, so there is no contribution from $m = 2$. Thus,

$$P_3(\cos\gamma) = 2\left[\frac{2!}{4!}\left(\frac{3}{2}\right)^2(5\mu^2 - 1)\sin\theta(-1)\sin\phi + \frac{1}{6!}(15)^2\sin^3\theta(-\sin 3\phi)\right]$$

$$= -2\left(\frac{3}{16}\left(5\cos^2\theta - 1\right)\sin\theta\sin\phi + \frac{5}{16}\sin^3\theta\sin 3\phi\right)$$

and the potential in Example 8.2, rotated into the new coordinates, is

$$\Phi = \frac{V}{2}\left[1 - \sum_{l=1}^{\infty}\frac{2l+1}{l}P_{l+1}(0)\left(\frac{r}{a}\right)^l P_l(\cos\gamma)\right]$$

$$= \frac{V}{2}\left\{1 - 3\left(-\frac{1}{2}\right)\left(\frac{r}{a}\right)\sin\theta\sin\phi\right.$$

$$\left. - \frac{7}{3}\frac{3}{8}\left(\frac{r}{a}\right)^3\left[-2\left(\frac{3}{16}(5\cos^2\theta - 1)\sin\theta\sin\phi + \frac{5}{16}\sin^3\theta\sin 3\phi\right)\right]\right\}$$

$$= \frac{V}{2}\left\{1 + \frac{3}{2}\left(\frac{r}{a}\right)\sin\theta\sin\phi\right.$$

$$\left. + \frac{7}{64}\left(\frac{r}{a}\right)^3[3(5\cos^2\theta - 1)\sin\theta\sin\phi + 5\sin^3\theta\sin 3\phi] + \cdots\right\}$$

which agrees with Example 8.3.

Now let's try to get a general result. We insert equation (8.65) into the result of Example 8.2 and get

$$\Phi(r, \theta, \phi) = \frac{V}{2} - V\sum_{l=1}^{\infty}\frac{2l+1}{2l}P_{l+1}(0)\left(\frac{r}{a}\right)^l\sum_{m=-l}^{l}\frac{4\pi}{2l+1}Y_{lm}(\theta, \phi)Y_{lm}^*\left(\frac{\pi}{2}, \frac{\pi}{2}\right)$$

$$= \frac{V}{2} - V\sum_{l=1}^{\infty}\frac{2l+1}{2l}P_{l+1}(0)\left(\frac{r}{a}\right)^l\sum_{m=-l}^{l}\frac{(l-m)!}{(l+m)!}P_l^m(\mu)e^{im\phi}P_l^m(0)e^{-im\pi/2}$$

Again we note that $P_l^m(0) = 0$ unless $l + m$ is even. But we also need $l + 1$ to be even so that $P_{l+1}(0) \neq 0$. Thus, l and m are both odd, and $e^{-im\pi/2} = i^{-m} = (-1)^{(m-1)/2}/i$:

$$\Phi(r, \theta, \phi)$$

$$= \frac{V}{2} - V \sum_{\substack{l=1, \\ l \text{ odd}}}^{\infty} \frac{2l+1}{2l} P_{l+1}(0) \left(\frac{r}{a}\right)^l \sum_{\substack{m=-l, \\ m \text{ odd}}}^{l} \frac{(l-m)!}{(l+m)!} P_l^m(\mu) e^{im\phi} P_l^m(0) \frac{(-1)^{(m-1)/2}}{i}$$

$$= \frac{V}{2} \left[1 - 2 \sum_{\substack{l=1, \\ l \text{ odd}}}^{\infty} \sum_{\substack{m=1, \\ m \text{ odd}}}^{l} \frac{2l+1}{l} P_{l+1}(0) \frac{(l-m)!}{(l+m)!} P_l^m(0)(-1)^{(m-1)/2} \left(\frac{r}{a}\right)^l P_l^m(\mu) \sin m\phi \right]$$

Compare this with the result of Example 8.3:

$$\Phi(r, \theta, \phi) = \frac{V}{2} + \frac{V}{\pi} \sum_{\substack{l=1, \\ l \text{ odd}}}^{\infty} \sum_{\substack{m=1, \\ m \text{ odd}}}^{l} \frac{2l+1}{m} \frac{(l-m)!}{(l+m)!} I_{lm} \left(\frac{r}{a}\right)^l P_l^m(\mu) \sin m\phi$$

The results agree if

$$
\boxed{
\begin{aligned}
I_{lm} &= \int_{-1}^{+1} P_l^m(\mu) \, d\mu \\
&= 2 \int_0^1 P_l^m(\mu) \, d\mu \\
&= \pi P_{l+1}(0) P_l^m(0) \frac{m}{l} (-1)^{(m+1)/2}, \quad l, m \text{ odd}
\end{aligned}
}
\tag{8.67}
$$

Thus, we have obtained a useful expression for the integral on the left. In Problem 8.21, you will be asked to verify this result.

8.4. PROBLEMS WITH CYLINDRICAL SYMMETRY: BESSEL FUNCTIONS

8.4.1. Bessel Functions

Bessel functions arise as solutions of potential problems in cylindrical coordinates. Laplace's equation in cylindrical coordinates is (equation 1.69)

$$\frac{1}{\rho} \frac{\partial}{\partial \rho} \left(\rho \frac{\partial \Phi}{\partial \rho} \right) + \frac{1}{\rho^2} \frac{\partial^2 \Phi}{\partial \phi^2} + \frac{\partial^2 \Phi}{\partial z^2} = 0$$

To separate the variables, we let $\Phi = R(\rho)W(\phi)Z(z)$. Then we find

$$\frac{1}{R\rho}\frac{\partial}{\partial\rho}\left(\rho\frac{\partial R}{\partial\rho}\right) + \frac{1}{W\rho^2}\frac{\partial^2 W}{\partial\phi^2} + \frac{1}{Z}\frac{\partial^2 Z}{\partial z^2} = 0$$

The last term is a function of z only, while the sum of the first two terms is a function of ρ and ϕ only. Thus, we take each part to be a constant[15] called k^2. Then

$$\frac{\partial^2 Z}{\partial z^2} = k^2 Z$$

and the solutions are

$$Z = e^{\pm kz}$$

The remaining equation is

$$\frac{1}{R\rho}\frac{\partial}{\partial\rho}\left(\rho\frac{\partial R}{\partial\rho}\right) + \frac{1}{W\rho^2}\frac{\partial^2 W}{\partial\phi^2} + k^2 = 0 \qquad (8.68)$$

Next we multiply through by ρ^2:

$$\frac{\rho}{R}\frac{\partial}{\partial\rho}\left(\rho\frac{\partial R}{\partial\rho}\right) + k^2\rho^2 + \frac{1}{W}\frac{\partial^2 W}{\partial\phi^2} = 0$$

Here the last term is a function of ϕ only, and the first two terms are functions of ρ only. As with spherical coordinates, we often want a solution that is periodic in ϕ with period 2π, so we choose a negative separation constant[16]:

$$\frac{\partial^2 W}{\partial\phi^2} = -m^2 W \Rightarrow W = e^{\pm im\phi}$$

Finally, we have the equation for the function R of ρ:

$$\rho\frac{\partial}{\partial\rho}\left(\rho\frac{\partial R}{\partial\rho}\right) + k^2\rho^2 R - m^2 R = 0$$

To see that this equation is of Sturm-Liouville form, we divide through by ρ:

$$\frac{\partial}{\partial\rho}\left(\rho\frac{\partial R}{\partial\rho}\right) + k^2\rho R - \frac{m^2}{\rho}R = 0 \qquad (8.69)$$

---

[15] Here we choose a positive constant (real k). The solution is rather different if the constant is negative, as we shall see in Section 8.4.7.

[16] In this application, m is usually an integer. Noninteger values are also of interest; see, for example, Section 8.5. In that case, the separation constant is usually written as $-\nu^2$.

Now we have a Sturm-Liouville equation (8.1) with $f(\rho) = \rho$, $g(\rho) = m^2/\rho$, eigenvalue $\lambda = k^2$, and weighting function $w(\rho) = \rho$. Equation (8.69) is Bessel's equation.

It is simpler and more elegant to solve this equation if we change to the dimensionless variable $x = k\rho$. Then

$$k \frac{\partial}{\partial k\rho} \left(k\rho \frac{\partial R}{\partial k\rho} \right) + k^2 \rho R - k \frac{m^2}{k\rho} R = 0$$

$$\frac{d}{dx} \left(x \frac{dR}{dx} \right) + x R - \frac{m^2}{x} R = 0 \qquad (8.70)$$

The equation has a singular point at $x = 0$, so we look for a series solution of the Frobenius type (refer to Chapter 3, Section 3.3.2):

$$R = x^p \sum_{n=0}^{\infty} a_n x^n$$

Then the equation becomes

$$\sum_{n=0}^{\infty} (n+p)(n+p-1) a_n x^{n+p-1} + \sum_{n=0}^{\infty} (n+p) a_n x^{n+p-1}$$

$$+ \sum_{n=0}^{\infty} a_n x^{n+p+1} - m^2 \sum_{n=0}^{\infty} a_n x^{n+p-1} = 0$$

The indicial equation is

$$p(p-1) + p - m^2 = 0 \Rightarrow p = \pm m$$

Thus, one of the solutions (with $p = m$) is analytic at $x = 0$, and one (with $p = -m$) is not. To find the recursion relation, we look at the $k + p - 1$ power of x:

$$(k+p)(k+p-1) a_k + (k+p) a_k + a_{k-2} - m^2 a_k = 0$$

and so

$$a_k = -\frac{a_{k-2}}{(k+p)^2 - m^2} = -\frac{a_{k-2}}{k^2 + 2kp + p^2 - m^2}$$

$$= -\frac{a_{k-2}}{k^2 + 2kp} = -\frac{a_{k-2}}{k(k \pm 2m)}$$

Let's look first at the solution with $p = +m$. We can step down to find each a_k. If we start the series with a_0, then k will always be even, $k = 2n$, and

$$
\begin{aligned}
a_{2n} &= \frac{-1}{2n(2n + 2m)} \frac{-1}{(2n - 2)(2n - 2 + 2m)} a_{2n-4} \\
&= \frac{(-1)^3}{2^3 n(n - 1)(n - 2)2^3(n + m)(n + m - 1)(n + m - 2)} a_{2n-6} \\
&= a_0 \frac{(-1)^n}{2^n n!} \frac{1}{2^n(n + m)(n + m - 1) \cdots (m + 1)}
\end{aligned}
$$

The usual convention is to take

$$
a_0 = \frac{1}{2^m \Gamma(m + 1)} \tag{8.71}
$$

Then

$$
a_{2n} = \frac{(-1)^n}{n! \Gamma(n + m + 1)2^{2n+m}} \tag{8.72}
$$

and the solution is the Bessel function:

$$
\boxed{J_m(x) = \sum_{n=0}^{\infty} \frac{(-1)^n}{n! \Gamma(n + m + 1)} \left(\frac{x}{2}\right)^{m+2n}} \tag{8.73}
$$

The function $J_m(x)$ has only even powers if m is an even integer and only odd powers if m is an odd integer. The series converges for all values of x. Although we have assumed m to be an integer thus far, expression (8.73) is also valid when m is not an integer.

Let's see what the second solution looks like. With $p = -m$, the recursion relation is

$$
a_k = \frac{a_{k-2}}{k(k - 2m)} \tag{8.74}
$$

where again $k = 2n$. Now if m is an integer, we will not be able to determine a_{2m}, because the recursion relation blows up. One solution to this dilemma is to start the series with a_{2m}. Then we can find the succeeding coefficients $a_{2(n+m)}$:

$$
a_{2(n+m)} = \frac{-a_{2(n-1)+2m}}{2^2(n + m)n} = \frac{-a_{2(n-2)+2m}}{2^4(n + m)(n + m - 1)n(n - 1)} = \frac{(-1)^n \Gamma(m + 1)}{n! \Gamma(n + m + 1)2^{2n}} a_{2m}
$$

These coefficients generate the same series that we had before in the case $p = +m$. (Compare the equation above with equation 8.72.) Thus, we do not get a linearly independent solution this way. This dilemma does not arise if the separation constant is taken to be $-\nu^2$

with ν noninteger. In that case, the second recursion relation provides a series $J_{-\nu}(x)$ that is linearly independent of the first.

With $\nu = m$, an integer, we find

$$J_{-m}(x) = \sum_{n=0}^{\infty} \frac{(-1)^n \Gamma(m+1)}{n! \Gamma(n+m+1) 2^{2n}} a_{2m} x^{2(n+m)-m}$$

$$= \sum_{n=0}^{\infty} \frac{(-1)^n \Gamma(m+1) 2^m}{n! \Gamma(n+m+1)} a_{2m} \left(\frac{x}{2}\right)^{m+2n}$$

and if we choose

$$a_{2m} = \frac{(-1)^m}{\Gamma(m+1) 2^m}$$

then

$$\boxed{J_{-m}(x) = (-1)^m J_m(x)} \tag{8.75}$$

With this choice, $J_\nu(x)$ is a continuous function of ν. (Notice that we can also express the series using equation 8.72 for the coefficients, with $m \to -m$, $n \to k+m$, and $a_{2n} \equiv 0$ for $n < m$.)

We still have to determine the second linearly independent solution of the Bessel equation when m is an integer. We can find it by taking the limit as $\nu \to m$ of a linear combination of J_ν and $J_{-\nu}$ known as the Neumann function, $N_\nu(x)$:

$$\boxed{N_\nu(x) \equiv \frac{J_\nu(x) \cos \nu\pi - J_{-\nu}(x)}{\sin \nu\pi}} \tag{8.76}$$

Thus,

$$N_m(x) = \lim_{\nu \to m} N_\nu(x) = \lim_{\nu \to m} \frac{J_\nu(x) \cos \nu\pi - J_{-\nu}(x)}{\sin \nu\pi}$$

$$= \lim_{\varepsilon \to 0} \frac{J_{m+\varepsilon}(x) \cos (m+\varepsilon)\pi - J_{-(m+\varepsilon)}(x)}{\sin (m+\varepsilon)\pi}$$

$$= \lim_{\varepsilon \to 0} \frac{J_{m+\varepsilon}(x) (\cos m\pi \cos \varepsilon\pi - \sin m\pi \sin \varepsilon\pi) - J_{-(m+\varepsilon)}(x)}{\sin m\pi \cos \varepsilon\pi + \cos m\pi \sin \varepsilon\pi}$$

$$= \lim_{\varepsilon \to 0} \frac{J_{m+\varepsilon}(x)(-1)^m \cos \varepsilon\pi - J_{-(m+\varepsilon)}(x)}{(-1)^m \sin \varepsilon\pi}$$

$$= \lim_{\varepsilon \to 0} \frac{J_{m+\varepsilon}(x) - (-1)^m J_{-(m+\varepsilon)}(x)}{\varepsilon\pi}$$

where we expanded the trigonometric functions to first order in ε. Next we expand the Bessel functions in a Taylor series in ν to first order in ε and use relation (8.75):

$$N_m(x) = \lim_{\varepsilon \to 0} \frac{1}{\varepsilon \pi} \left[J_m + \varepsilon \left.\frac{dJ_\nu}{d\nu}\right|_{\nu=m} - (-1)^m \left(J_{-m} + \varepsilon \left.\frac{dJ_{-\nu}}{d\nu}\right|_{\nu=m} \right) \right]$$

$$N_m(x) = \frac{1}{\pi} \left[\left.\frac{dJ_\nu}{d\nu}\right|_{\nu=m} - (-1)^m \left.\frac{dJ_{-\nu}}{d\nu}\right|_{\nu=m} \right] \tag{8.77}$$

The derivative has a logarithmic term[17]:

$$\frac{dJ_\nu}{d\nu} = \frac{d}{d\nu} \left(x^\nu \sum_{n=0}^\infty \frac{(-1)^n}{n!\Gamma(n+\nu+1)} \left(\frac{x}{2}\right)^{2n} \right)$$

$$= \frac{dx^\nu}{d\nu} \sum_{n=0}^\infty \frac{(-1)^n}{n!\Gamma(n+\nu+1)} \left(\frac{x}{2}\right)^{2n} + x^\nu \frac{d}{d\nu} \sum_{n=0}^\infty \frac{(-1)^n}{n!\Gamma(n+\nu+1)} \left(\frac{x}{2}\right)^{2n}$$

and

$$\frac{dx^\nu}{d\nu} = \frac{d}{d\nu} e^{\nu \ln x} = \ln x \, e^{\nu \ln x} = x^\nu \ln x$$

and so $dJ_\nu/d\nu$ has a term containing $J_\nu \ln x$. This term diverges as $x \to 0$, provided that $J_\nu(0)$ is not zero (that is, for $\nu = 0$). The function $N_\nu(x)$ also diverges as $x \to 0$ for $\nu \neq 0$, because it contains negative powers of x. (The series for $J_{-\nu}$ starts with a term $x^{-\nu}$.) But $N_\nu(x)$ does not diverge as $x \to \infty$, because $J_\nu \to 0$ faster than the logarithm approaches infinity.

Two additional functions called Hankel functions are defined as linear combinations of J and N:

$$H_m^{(1)}(x) = J_m(x) + i N_m(x) \tag{8.78}$$

and

$$H_m^{(2)}(x) = J_m(x) - i N_m(x) \tag{8.79}$$

---

[17]This result should not be surprising. We could also find the second solution using the methods of Chapter 3, Section 3.3.3, particularly in the form of equation (3.37). See, for example, Chapter 3, Problem 22.

8.4.2. Properties of the Bessel Functions

The Bessel functions (Js) are well behaved both at the origin and as $x \to \infty$. They have infinitely many zeros. All of the J_m, except J_0, are zero at $x = 0$. The first few functions are shown in Figure 8.11.

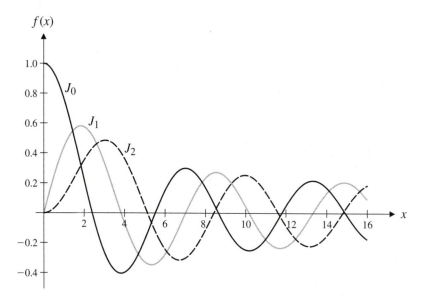

FIGURE 8.11. The first three Bessel functions. All the functions except $J_0(x)$ equal zero at $x = 0$, and all of them approach zero as $x \to \infty$. All of the J_m oscillate with decreasing amplitude.

For small values of the argument, we may approximate the function with the first term in the series:

$$J_m(x) \approx \frac{1}{\Gamma(m+1)} \left(\frac{x}{2}\right)^m \quad \text{for } x \ll 1 \tag{8.80}$$

The Neumann functions are not well behaved at $x = 0$. N_0 has a logarithmic singularity, and for $m > 0$, N_m diverges as an inverse power of x:

$$N_0(x) \approx \frac{2}{\pi} \ln x \quad \text{for } x \ll 1 \tag{8.81}$$

$$N_m(x) \approx -\frac{(m-1)!}{\pi} \left(\frac{2}{x}\right)^m \quad \text{for } x \ll 1 \tag{8.82}$$

For large values of the argument, both J and N oscillate. They are like damped cosine or sine functions:

$$J_m(x) \approx \sqrt{\frac{2}{\pi x}} \cos\left(x - \frac{m\pi}{2} - \frac{\pi}{4}\right) \quad \text{for } x \gg 1, m \qquad (8.83)$$

$$N_m(x) \approx \sqrt{\frac{2}{\pi x}} \sin\left(x - \frac{m\pi}{2} - \frac{\pi}{4}\right) \quad \text{for } x \gg 1, m \qquad (8.84)$$

Thus, the Hankel functions are like complex exponentials[18]:

$$H_m^{(1,2)} \approx \sqrt{\frac{2}{\pi x}} \exp\left[\pm i\left(x - \frac{m\pi}{2} - \frac{\pi}{4}\right)\right] \quad \text{for } x \gg 1, m \qquad (8.85)$$

Notice that if $m > 1$, the large argument expansions apply for $x \gg m$ rather than the usual $x \gg 1$.

8.4.3. Relations Between the Functions

As we found with the Legendre functions, we can determine a set of recursion relations that relate successive $J_m(x)$. Starting with the series representation (8.73), we divide by $(x/2)^m$ and then differentiate:

$$\frac{d}{dx}\left(\frac{2^m J_m(x)}{x^m}\right) = \sum_{n=0}^{\infty} \frac{(-1)^n n}{n!\,\Gamma(n+m+1)}\left(\frac{x}{2}\right)^{2n-1}$$

$$= \sum_{n=1}^{\infty} \frac{(-1)^n}{(n-1)!\,\Gamma(n+m+1)}\left(\frac{x}{2}\right)^{2n-1}$$

Now let $k = n - 1$:

$$\frac{d}{dx}\left(\frac{2^m J_m(x)}{x^m}\right) = -\sum_{k=0}^{\infty} \frac{(-1)^k}{k!\,\Gamma(k+m+1+1)}\left(\frac{x}{2}\right)^{2k+1}$$

$$= -\frac{2^m}{x^m}\sum_{k=0}^{\infty} \frac{(-1)^k}{k!\,\Gamma(k+m+1+1)}\left(\frac{x}{2}\right)^{2k+m+1}$$

$$= -\frac{2^m}{x^m} J_{m+1}(x)$$

[18]Refer to Chapter 2, equation (2.5).

and thus

$$\frac{d}{dx}\left(\frac{J_m(x)}{x^m}\right) = -\frac{J_{m+1}(x)}{x^m} \qquad (8.86)$$

which is valid for $m \geq 0$. In particular, with $m = 0$ we obtain

$$J_1(x) = -J_0'(x) \qquad (8.87)$$

Similarly,

$$\frac{d}{dx}[x^m J_m(x)] = 2^m \frac{d}{dx}\sum_{n=0}^{\infty} \frac{(-1)^n}{n!\Gamma(n+m+1)}\left(\frac{x}{2}\right)^{2m+2n}$$

$$= 2^m \sum_{n=0}^{\infty} \frac{(-1)^n(m+n)}{n!\Gamma(n+m+1)}\left(\frac{x}{2}\right)^{2m+2n-1}$$

$$= x^m \sum_{n=0}^{\infty} \frac{(-1)^n}{n!\Gamma(n+m-1+1)}\left(\frac{x}{2}\right)^{m+2n-1}$$

$$\frac{d}{dx}[x^m J_m(x)] = x^m J_{m-1}(x) \qquad (8.88)$$

which is valid for $m \geq 1$.

Combining relations (8.86) and (8.88), we get

$$J_{m+1} + J_{m-1} = -x^m \frac{d}{dx}\left(\frac{J_m(x)}{x^m}\right) + \frac{1}{x^m}\frac{d}{dx}[x^m J_m(x)]$$

$$= -x^m\left(-m\frac{J_m}{x^{m+1}} + \frac{J_m'}{x^m}\right) + \frac{1}{x^m}\left(mx^{m-1}J_m + x^m J_m'\right)$$

$$= m\frac{J_m}{x} - J_m' + m\frac{J_m}{x} + J_m' = \frac{2m}{x}J_m$$

$$J_{m+1} + J_{m-1} = \frac{2m}{x}J_m \qquad (8.89)$$

which is the pure recursion relation for the Bessel functions. Similarly, by subtracting instead of adding, we find

$$J_{m+1} - J_{m-1} = -2\frac{dJ_m}{dx} \qquad (8.90)$$

The same relations hold for the Ns and the Hs.

8.4.4. Bessel Functions as Solutions of the Helmholtz Equation: The Generating Function

Like Laplace's equation, the Helmholtz equation (equation 3.16 in Chapter 3 extended to two or three dimensions)

$$(\nabla^2 + k^2)\Phi = 0$$

may be written in cylindrical coordinates and separated. Suppose we look for a solution in two dimensions, with Φ independent of z. Then the equation becomes

$$\frac{1}{R\rho}\frac{\partial}{\partial\rho}\left(\rho\frac{\partial R}{\partial\rho}\right) + \frac{1}{W\rho^2}\frac{\partial^2 W}{\partial\phi^2} + k^2 = 0$$

which is the same equation (8.68) that we obtained from Laplace's equation after separating out the z-dependence. Thus, the solutions are of the same form:

$$\Phi = \sum_{m=-\infty}^{+\infty} a_m J_m(k\rho)e^{im\phi}$$

We may exclude the functions $N_m(k\rho)$ if the origin is within our region of interest.

As with the Legendre functions, we may exploit a simple physical situation—here a plane wave—to obtain a generating function for the Bessel functions. A plane wave propagating along the y-direction has the form

$$e^{iky} = e^{ik\rho\sin\phi}$$

which must therefore be expressible in Bessel functions:

$$e^{ik\rho\sin\phi} = \sum_{m'=-\infty}^{+\infty} a_{m'} J_{m'}(k\rho)e^{im'\phi}$$

Now we make use of the orthogonality of the $e^{im\phi}$. We multiply both sides by $e^{-im\phi}$ and integrate over the range 0 to 2π. Only the one term with $m' = m$ survives on the right-hand side:

$$\int_0^{2\pi} e^{ik\rho\sin\phi - im\phi}\,d\phi = 2\pi a_m J_m(k\rho)$$

With $m = 0$, we can evaluate both sides for $\rho = 0$ to obtain $a_0 = 1$. If $m \neq 0$, we expand the function $e^{ik\rho \sin \phi}$ using the exponential series and expand the Bessel function on the right using its series (8.73):

$$\frac{1}{2\pi} \int_0^{2\pi} e^{-im\phi} \left(1 + \sum_{n=1}^{\infty} \frac{(ik\rho \sin \phi)^n}{n!} \right) d\phi = a_m \left(\frac{k\rho}{2} \right)^m \sum_{p=0}^{\infty} \frac{(-1)^p}{p! \Gamma(p + m + 1)} \left(\frac{k\rho}{2} \right)^{2p}$$

With $m \neq 0$, the first term in the integral on the left-hand side integrates to zero. Now for small $k\rho$, we keep only the leading term on the right-hand side. Then we have

$$\frac{1}{2\pi} \sum_{n=1}^{\infty} \left(\frac{ik\rho}{2i} \right)^n \frac{1}{n!} \int_0^{2\pi} e^{-im\phi} (e^{i\phi} - e^{-i\phi})^n \, d\phi = a_m \frac{1}{\Gamma(m + 1)} \left(\frac{k\rho}{2} \right)^m$$

Look first at the case $m = 1$. Keeping only the leading ($n = 1$) term on the left-hand side, we have

$$\frac{k\rho}{4\pi} \int_0^{2\pi} e^{-i\phi} (e^{i\phi} - e^{-i\phi}) \, d\phi = \frac{k\rho}{4\pi} (2\pi - 0) = a_1 \frac{k\rho}{2} \Rightarrow a_1 = 1$$

For $m > 1$, the terms with $n < m$ on the left-hand side are all zero because $(e^{i\phi} - e^{-i\phi})^n = e^{in\phi}(1 - e^{-2i\phi})^n$ has no $e^{im\phi}$ term if $n < m$. (This must be true for consistency, because the lowest power of ρ in the Bessel function series on the right is ρ^m.) Thus, the first nonzero term has $n = m$. In the factor $(e^{i\phi} - e^{-i\phi})^m$, only the term $e^{im\phi}$ survives the integration over ϕ. We have

$$\frac{(ik\rho)^m}{m!(2i)^m} \frac{1}{2\pi} \int_0^{2\pi} e^{-im\phi} e^{im\phi} \, d\phi = a_m \frac{1}{\Gamma(m + 1)} \left(\frac{k\rho}{2} \right)^m \Rightarrow a_m = 1$$

Thus, we have a generating function for the J_m,

$$e^{ik\rho \sin \phi} = \sum_{m=-\infty}^{+\infty} J_m(k\rho) e^{im\phi} \tag{8.91}$$

and the integral representation,

$$J_m(kr) = \frac{1}{2\pi} \int_0^{2\pi} e^{ikr \sin \phi - im\phi} \, d\phi \tag{8.92}$$

Next we write $\sin \phi = (e^{i\phi} - e^{-i\phi})/2i$ and let $e^{i\phi} = t$. Then we have an alternative form for the generating function:

$$\exp\left[\frac{kr}{2}\left(t - \frac{1}{t}\right)\right] = \sum_{m=-\infty}^{+\infty} J_m(kr)t^m \qquad (8.93)$$

8.4.5. Orthogonality of the J_m

Since the Bessel equation is of Sturm-Liouville form, the Bessel functions are orthogonal if we demand that they satisfy boundary conditions of the form (8.2). In particular, suppose the region of interest is $\rho = 0$ to $\rho = a$, and the boundary conditions are that $J_m(ka) = 0$ and $J_m(0)$ is finite. We do not need a more specific boundary condition at $\rho = 0$ because the function $f(\rho) = \rho$ is zero there. Then the eigenvalues are

$$k_{mn} = \frac{\alpha_{mn}}{a}$$

where α_{mn} is the nth zero of J_m. (The zeros are tabulated in standard references such as Abramowitz and Stegun. Programs such as *Mathematica* and *Maple* can also compute them.) Then equation (8.6) becomes

$$J_m(k_{mn'}\rho)\rho\frac{d J_m(k_{mn}\rho)}{d\rho}\bigg|_0^a - J_m(k_{mn}\rho)\rho\frac{d J_m(k_{mn'}\rho)}{d\rho}\bigg|_0^a$$

$$= (k_{mn'}^2 - k_{mn}^2)\int_0^a \rho J_m(k_{mn'}\rho)J_m(k_{mn}\rho)\, d\rho \qquad (8.94)$$

The boundary conditions make the left-hand side zero, and thus, if $n \neq n'$,

$$\int_0^a \rho J_m(k_{mn'}\rho)J_m(k_{mn}\rho)\, d\rho = 0 \qquad (8.95)$$

To determine the value of the integral when $n = n'$, we replace $k_{mn'}$ with an arbitrary value of k, *not* one of the eigenvalues. Then equation (8.94) becomes

$$J_m(ka)a\frac{d J_m}{d\rho}(k_{mn}\rho)\bigg|_{\rho=a} = (k^2 - k_{mn}^2)\int_0^a \rho J_m(k_{mn}\rho)J_m(k\rho)\, d\rho$$

Now we differentiate this expression with respect to k:

$$a\frac{d J_m(ka)}{dka}ak_{mn}\frac{d J_m}{dk_{mn}\rho}(k_{mn}\rho)\bigg|_{\rho=a} = 2k\int_0^a \rho J_m(k_{mn}\rho)J_m(k\rho)\, d\rho$$

$$+ (k^2 - k_{mn}^2)\int_0^a \rho J_m(k_{mn}\rho)\frac{d J_m(k\rho)}{dk}\, d\rho$$

Next let $k \to k_{mn}$. The second term on the right vanishes, and we have

$$k_{mn}[aJ'_m(k_{mn}a)]^2 = 2k_{mn} \int_0^a \rho\, [J_m(k_{mn}\rho)]^2\, d\rho$$

and so

$$\int_0^a \rho\, [J_m(k_{mn}\rho)]^2\, d\rho = \frac{a^2}{2}[J'_m(k_{mn}a)]^2 \qquad (8.96)$$

which is the integral we need.[19]

8.4.6. Solving a Potential Problem

Example 8.4. A cylinder of radius a and height h has its curved surface and its bottom grounded. The top surface has potential V (Figure 8.12). What is the potential inside the cylinder?

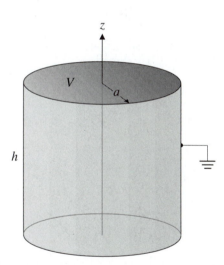

FIGURE 8.12. The top end of the cylinder in Example 8.4 is at potential V. All the other surfaces are grounded.

The potential has no dependence on ϕ, and so only eigenfunctions with $m = 0$ contribute. The potential is zero at $\rho = a$, so the solution we need is $J_0(k\rho)$ with eigenvalues chosen to make $J_0(ka) = 0$. Thus, the eigenvalues are given by

[19] See Problem 29 for the integral with Neumann conditions.

$k_{0n}a = \alpha_{0n}$, where α_{0n} are the zeros of the function J_0. The remaining function of z must be zero at $z = 0$, so we choose the hyperbolic sine. Thus, the potential is

$$\Phi(\rho, z) = \sum_n c_n J_0(k_{0n}\rho) \sinh(k_{0n}z)$$

To find the coefficients c_n, we evaluate this expression at $z = h$:

$$V = \Phi(\rho, h) = \sum_n c_n J_0(k_{0n}\rho) \sinh(k_{0n}h)$$

Next we make use of the orthogonality of the Bessel functions. We multiply both sides by $\rho J_0(k_{0r}\rho)$ and integrate from 0 to a. Only the one term in the sum with $n = r$ survives the integration:

$$V \int_0^a \rho J_0(k_{0r}\rho) \, d\rho = \int_0^a \rho J_0(k_{0r}\rho) \sum_n c_n J_0(k_{0n}\rho) \sinh(k_{0n}h) \, d\rho$$

$$= c_r \int_0^a \rho \, [J_0(k_{0r}\rho)]^2 \, d\rho \sinh(k_{0r}h)$$

$$= c_r \frac{a^2}{2} [J_0'(k_{0r}a)]^2 \sinh(k_{0r}h)$$

To evaluate the left-hand side, we use equation (8.88) with $m = 1$:

$$\int_0^a \rho J_0(k\rho) \, d\rho = \frac{1}{k} \int_0^a \frac{d}{dk\rho} [k\rho J_1(k\rho)] \, d\rho$$

$$= \frac{1}{k} \rho J_1(k\rho)|_0^a = \frac{a}{k} J_1(ka)$$

So

$$c_r = \frac{Va}{k_{0r}} J_1(k_{0r}a) \frac{2}{a^2 [J_0'(k_{0r}a)]^2 \sinh(k_{0r}h)}$$

$$= \frac{V}{k_{0r}a} \frac{2}{J_1(k_{0r}a) \sinh(k_{0r}h)}$$

where we used the result from equation (8.87) that $J_0' = -J_1$. Finally, our solution is

$$\Phi = 2V \sum_{n=1}^{\infty} \frac{J_0(k_{0n}\rho)}{k_{0n}a \, J_1(k_{0n}a)} \frac{\sinh(k_{0n}z)}{\sinh(k_{0n}h)}$$

The first two zeros of J_0 are $\alpha_{01} = 2.4048$ and $\alpha_{02} = 5.5201$, and thus the first two terms in the potential are

$$\Phi = 2V \left(\frac{J_0(2.4048\rho/a)}{2.4048 J_1(2.4048)} \frac{\sinh(2.4048z/a)}{\sinh(2.4048h/a)} + \frac{J_0(5.5201\rho/a)}{5.5201 J_1(5.5201)} \frac{\sinh(5.5201z/a)}{\sinh(5.5201h/a)} \right)$$

These terms are plotted in Figure 8.13 for $h/a = 2$ and $z/a = 0.5$, 1, 1.5, and 1.8. For $z/a = 1.8$, the first two terms do not represent the potential well. The three-term result is also shown for $z/a = 1.5$ and 1.8. While the third term makes little difference at $z/a = 1.5$, it is very significant at $z/a = 1.8$.

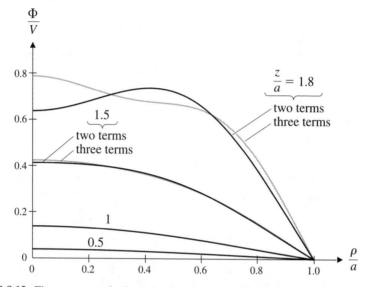

FIGURE 8.13. First two terms in the series for the potential inside the cylinder versus radius for $h/a = 2$ and $z/a = 0.5$, 1, 1.5, and 1.8. As z/a increases, the series converges more slowly and these first two terms do not represent the potential well at $z/a = 1.8$— hence the dip near the axis. The grey lines show the potential when a third term is added to the series. At the smaller values of z, the extra term makes a negligible difference in this diagram.

8.4.7. Modified Bessel Functions

Suppose we change the potential problem in Example 8.4 so that the top and bottom of the cylinder are grounded but the outer wall at $\rho = a$ has a known potential $V(\phi, z)$. Then we need to choose a negative separation constant so that the solutions of the z-equation are trigonometric functions:

$$\frac{\partial^2 Z}{\partial z^2} = -k^2 Z \Rightarrow Z = a \sin kz + b \cos kz$$

At $z = 0$, $Z(z) = 0$, so we need the sine and therefore set $b = 0$. We also need $Z(h) = 0$, so we choose the eigenvalue $k = n\pi/h$.

This change in sign of the separation constant also affects equation (8.69) for the function $R(\rho)$ because the sign of the k^2 term changes here too. The equation becomes

$$\frac{\partial}{\partial \rho}\left(\rho \frac{\partial R}{\partial \rho}\right) - k^2 \rho R - \frac{m^2}{\rho} R = 0$$

or, with variables changed to $x = k\rho$,

$$\frac{\partial}{\partial x}\left(x\frac{\partial R}{\partial x}\right) - xR - \frac{m^2}{x}R = 0 \tag{8.97}$$

which is called the *modified Bessel equation*. The solutions to this equation are $J_m(ik\rho)$. It is usual to define the modified Bessel function $I_m(x)$ by the relation

$$I_m(x) = \frac{1}{i^m}J_m(ix) \tag{8.98}$$

so that the function I_m is always real (whether or not m is an integer). Using equation (8.73), we can write a series expansion for I_m:

$$I_m(x) = \frac{1}{i^m}\sum_{n=0}^{\infty}\frac{(-1)^n}{n!\Gamma(n+m+1)}\left(\frac{ix}{2}\right)^{m+2n}$$

$$I_m(x) = \sum_{n=0}^{\infty}\frac{1}{n!\Gamma(n+m+1)}\left(\frac{x}{2}\right)^{m+2n} \tag{8.99}$$

As with the Js, if m is an integer, I_{-m} is not independent of I_m; in fact,

$$I_{-m}(x) = i^m J_{-m}(ix) = I_m(x)$$

The second independent solution is usually chosen to be

$$K_m(x) = \frac{\pi}{2}i^{m+1}H_m^{(1)}(ix) \tag{8.100}$$

Then these functions have the limiting forms for small argument:

$$I_m(x) \approx \frac{1}{\Gamma(m+1)}\left(\frac{x}{2}\right)^m \quad \text{for } x \ll 1 \tag{8.101}$$

and

$$K_0(x) \approx -0.5772 - \ln \frac{x}{2} \quad \text{for } x \ll 1 \tag{8.102}$$

$$K_m(x) \approx \frac{\Gamma(m)}{2} \left(\frac{2}{x}\right)^m, \quad m > 0, \text{ for } x \ll 1 \tag{8.103}$$

At large x, $x \gg 1, m$, the asymptotic forms are

$$I_m(x) \approx \frac{1}{\sqrt{2\pi x}} e^x \tag{8.104}$$

and

$$K_m(x) \approx \sqrt{\frac{\pi}{2x}} e^{-x} \tag{8.105}$$

(Compare with Chapter 3, Examples 3.9 and 3.10.) These functions, like the real exponentials, do not have multiple zeros and are not orthogonal functions. Note that the Is are well behaved at the origin but diverge at infinity. For the Ks, the reverse is true. They diverge at the origin but are well behaved at infinity (Figure 8.14).

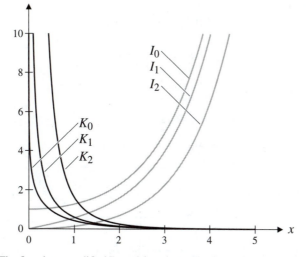

FIGURE 8.14. The first three modified Bessel functions. The functions $K_n(x)$ diverge at the origin, and the functions $I_n(x)$ diverge as $x \to \infty$.

The recursion relations satisfied by the modified Bessel functions are similar to, but not identical to, the relations satisfied by the Js. For the Is, again we can start with the series:

$$\frac{d}{dx}\left(\frac{2^m I_m(x)}{x^m}\right) = \sum_{n=1}^{\infty} \frac{n}{n!\Gamma(n+m+1)}\left(\frac{x}{2}\right)^{2n-1}$$

Now let $k = n - 1$:

$$\frac{d}{dx}\left(\frac{2^m I_m(x)}{x^m}\right) = \sum_{k=0}^{\infty} \frac{1}{k!\Gamma(k+m+1+1)}\left(\frac{x}{2}\right)^{2k+1}$$

$$\frac{d}{dx}\left(\frac{I_m(x)}{x^m}\right) = \frac{I_{m+1}(x)}{x^m} \tag{8.106}$$

Similarly,

$$\frac{d}{dx}(x^m I_m) = x^m I_{m-1} \tag{8.107}$$

Expanding out and combining, we get

$$I'_m(x) - m\frac{I_m}{x} = I_{m+1}$$

$$I'_m + \frac{m I_m}{x} = I_{m-1}$$

Adding, we obtain

$$2I'_m = I_{m+1} + I_{m-1} \tag{8.108}$$

while subtracting, we get

$$\frac{2m}{x}I_m = I_{m-1} - I_{m+1} \tag{8.109}$$

For the Ks, the relations are

$$\frac{d}{dx}(x^m K_m) = -x^m K_{m-1} \quad \text{and} \quad \frac{d}{dx}\left(\frac{K_m(x)}{x^m}\right) = -\frac{K_{m+1}(x)}{x^m}$$

(8.110)

and, consequently,

$$K_{m-1} - K_{m+1} = -\frac{2m}{x} K_m$$

(8.111)

$$K_{m-1} + K_{m+1} = -2K_m'$$

(8.112)

8.4.8. Combining Functions

When solving a physics problem, we start with a partial differential equation and a set of boundary conditions. Separation of variables produces a set of *coupled* ordinary differential equations in the various coordinates. The standard solution method (Section 8.2) requires that we choose the separation constants by fitting the zero boundary conditions first. In a standard three-dimensional problem, once we have chosen the two separation constants we have no more freedom, and the third function is determined.

When we solve Laplace's equation in cylindrical coordinates, the functions couple as follows.

Zero boundary conditions in ρ. The eigenfunctions are of the form

$$J_m\left(\alpha_{mn}\frac{\rho}{a}\right)\left(A_{mn}\sinh\alpha_{mn}\frac{z}{a} + B_{mn}\cosh\alpha_{mn}\frac{z}{a}\right)e^{\pm im\phi}$$

where $J_m(\alpha_{mn}) = 0$. The set of functions $J_m(\alpha_{mn}\rho/a)e^{\pm im\phi}$ forms a complete orthogonal set on the constant z surfaces that bound the region.

Zero boundary conditions in z. The eigenfunctions are of the form

$$\sin\left(\frac{n\pi z}{h}\right)\left[A_{mn}I_m\left(\frac{n\pi\rho}{h}\right) + B_{mn}K_m\left(\frac{n\pi\rho}{h}\right)\right]e^{\pm im\phi}$$

The set of functions $\sin(n\pi z/h)e^{\pm im\phi}$ forms a complete orthogonal set on the boundary surface $\rho = $ constant.

Thus, in solutions of Laplace's equation, the Js in ρ always couple with the hyperbolic sines and cosines (or real exponentials) in z, while the Is and Ks in ρ always couple with the sines and cosines (or complex exponentials) in z.

Example 8.5.   Find the potential inside a cylinder of height h and radius a when the top and bottom are grounded and the curved walls have a potential $V(\phi, z) = V_0$ for $0 < \phi < \pi$ and $-V_0$ for $\pi < \phi < 2\pi$.

As we discussed in Section 8.4.7, the z-functions are of the form $\sin(n\pi z/h)$ so that the potential equals zero at $z = 0$ and at $z = h$. The potential should be finite on the axis at $\rho = 0$, so we exclude the K functions. Thus, the potential is of the form

$$\Phi(\rho, z, \phi) = \sum_{n=1}^{\infty} \sum_{m=-\infty}^{+\infty} A_{nm} \sin\left(\frac{n\pi z}{h}\right) I_m\left(\frac{n\pi}{h}\rho\right) e^{im\phi}$$

Now we evaluate the potential at $\rho = a$:

$$\Phi(a, z, \phi) = \begin{cases} V_0 & \text{if } 0 < \phi < \pi \\ -V_0 & \text{if } \pi < \phi < 2\pi \end{cases} = \sum_{n=1}^{\infty} \sum_{m=-\infty}^{+\infty} A_{nm} \sin\left(\frac{n\pi z}{h}\right) I_m\left(\frac{n\pi}{h}a\right) e^{im\phi}$$

This is a Fourier sine series, and we find the coefficients in the usual way:

$$A_{nm}\frac{h}{2} I_m\left(\frac{n\pi}{h}a\right) 2\pi = V_0 \int_0^h \sin\left(\frac{n\pi z}{h}\right) dz \left(\int_0^\pi - \int_\pi^{2\pi}\right) e^{-im\phi}\, d\phi$$

$$= V_0 \frac{h}{n\pi}(1 - \cos n\pi) \frac{2[1 - (-1)^m]}{im}$$

$$= \begin{cases} V_0 \dfrac{8h}{n\pi m i} & \text{for } n \text{ odd and } m \text{ odd} \\ 0 & \text{otherwise} \end{cases}$$

We must evaluate the terms with $m = 0$ separately. The integral over ϕ is zero, so $A_{n0} = 0$. Thus,

$$\Phi(\rho, z, \phi) = \frac{8V_0}{\pi^2} \sum_{\substack{n=1, \\ n \text{ odd}}}^{\infty} \sum_{\substack{m=-\infty, \\ m \text{ odd}}}^{\infty} \frac{\sin\left(\dfrac{n\pi z}{h}\right)}{n} \frac{I_m\left(\dfrac{n\pi}{h}\rho\right)}{I_m\left(\dfrac{n\pi}{h}a\right)} \frac{e^{im\phi}}{mi}$$

Combining positive and negative m terms, we have

$$\Phi(\rho, z, \phi) = \frac{16V_0}{\pi^2} \sum_{\substack{n=1, \\ n \text{ odd}}}^{\infty} \sum_{\substack{m=1, \\ m \text{ odd}}}^{\infty} \frac{\sin\left(\dfrac{n\pi z}{h}\right)}{n} \frac{\sin m\phi}{m} \frac{I_m\left(\dfrac{n\pi}{h}\rho\right)}{I_m\left(\dfrac{n\pi}{h}a\right)}$$

Figure 8.15 shows the solution for the case $a = h/2$.

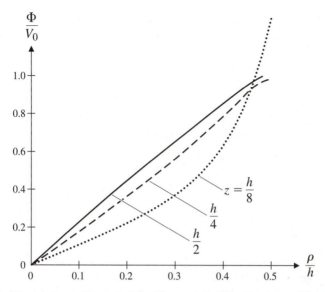

FIGURE 8.15. The solution to Example 8.5 with $a = h/2$. This plot shows $\Phi(\rho)/V_0$ versus ρ/h at $\phi = \pi/2$. The dotted line represents $z = h/8$; the dashed line, $z = h/4$; the solid line, $z = h/2$. The first three values of m and n are included in the sum.

8.4.9. Continuous Set of Eigenvalues: The Fourier-Bessel Transform

In Chapter 7, we approached the Fourier transform by letting the length of the domain in a Fourier series problem become infinite. The orthogonality relation for the exponential functions

$$\frac{1}{2L} \int_{-L}^{L} \exp\left(i\frac{n\pi x}{L}\right) \exp\left(-i\frac{m\pi x}{L}\right) dx = \delta_{mn}$$

becomes

$$\frac{1}{2\pi} \int_{-\infty}^{\infty} e^{ikx} e^{-ik'x} dx = \delta(k - k')$$

That is, the Kronecker delta becomes a delta function, and the countable set of eigenvalues $n\pi/L$ becomes a continuous set of values $-\infty < k < \infty$.

The same thing happens with Bessel functions. With a finite domain in ρ, say $0 \leq \rho \leq a$, we can determine a countable set of eigenvalues from the set of zeros of the Bessel functions J_m, as we did in Example 8.4. If our domain in ρ becomes infinite, then we cannot determine the eigenvalues, and instead we have a continuous set. The orthogonality relation

$$\int_{0}^{a} J_m\left(\alpha_{mn}\frac{\rho}{a}\right) J_m\left(\alpha_{mk}\frac{\rho}{a}\right) \rho \, d\rho = \frac{a^2}{2}[J_m'(\alpha_{mn})]^2 \delta_{nk}$$

becomes

$$\int_0^\infty \rho J_m(k\rho) J_m(k'\rho) \, d\rho = \frac{\delta(k-k')}{k} \tag{8.113}$$

(The proof of this relation is in Appendix VIII.) Then the solution to the physics problem is determined as an integral over k. For example, a solution of Laplace's equation may be written as

$$\Phi(\rho, \phi, z) = \sum_m e^{im\phi} \int_0^\infty A_m(k) f(kz) J_m(k\rho) \, dk$$

where $f(kz)$ depends on the boundary conditions in z. It will be a combination of the exponentials e^{-kz} and e^{+kz}.

Example 8.6.   The potential on a plane at $z = 0$ is given by the function $V(\rho) = V_0(a/\rho) \sin(\rho/a)$. Find the potential above the plane at $z > 0$.

The appropriate function of z is e^{-kz}, chosen so that $\Phi \to 0$ as $z \to \infty$ (a long way from the plane). Then the solution is of the form

$$\Phi(\rho, \phi, z) = \sum_{m=-\infty}^{+\infty} e^{im\phi} \int_0^\infty A_m(k) e^{-kz} J_m(k\rho) \, dk$$

Evaluating Φ on the plane at $z = 0$ gives

$$V(\rho, \phi) = \Phi(\rho, \phi, 0) = \sum_{m=-\infty}^{+\infty} e^{im\phi} \int_0^\infty A_m(k) J_m(k\rho) \, dk$$

Now we can make use of the orthogonality of the $e^{im\phi}$. We multiply both sides by $e^{-im'\phi}$ and integrate over the range 0 to 2π. On the right-hand side, only the term with $m = m'$ survives the integration, and we get

$$\int_0^{2\pi} V(\rho, \phi) e^{-im'\phi} \, d\phi = 2\pi \int_0^\infty A_{m'}(k) J_{m'}(k\rho) \, dk$$

which is a Fourier-Bessel transform. Next[20] we multiply both sides by $\rho J_m(k'\rho)$, integrate from 0 to ∞ in ρ, and use equation (8.113) to get

$$\int_0^\infty \int_0^{2\pi} V(\rho, \phi) e^{-im\phi} \, d\phi \, J_m(k'\rho) \rho \, d\rho$$

$$= 2\pi \int_0^\infty \int_0^\infty A_m(k) J_m(k\rho) J_m(k'\rho) \rho \, d\rho \, dk$$

$$= 2\pi \int_0^\infty A_m(k) \frac{\delta(k-k')}{k} \, dk = 2\pi \frac{A_m(k')}{k'}$$

[20]We also drop the primes on the m' for convenience.

which determines the coefficient $A_m(k')$ in terms of the known function $V(\rho, \phi)$.

In our example, $V(\rho)$ is independent of ϕ, so only $m = 0$ survives the integration over ϕ, leaving

$$A_0(k) = k \int_0^\infty V_0 \frac{a}{\rho} \sin \frac{\rho}{a} J_0(k\rho)\rho \, d\rho = ka V_0 \begin{cases} 0 & \text{if } ka > 1 \\ a/\sqrt{1 - (ka)^2} & \text{if } ka < 1 \end{cases}$$

(The integral is Gradshteyn and Ryzhik formula 6.671#7.) Thus,

$$\Phi(\rho, z) = V_0 a^2 \int_0^{1/a} \frac{k}{\sqrt{1 - (ka)^2}} J_0(k\rho) e^{-kz} \, dk$$

$$= V_0 \int_0^1 \frac{x}{\sqrt{1 - x^2}} J_0\left(x\frac{\rho}{a}\right) e^{-xz/a} \, dx$$

The solution for $\Phi(\rho, z)$ on the z-axis and at $\rho = 2a$ is shown in Figure 8.16.

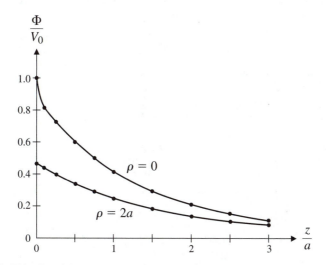

FIGURE 8.16. This plot of the solution to Example 8.6 shows $\Phi(\rho, z)$ on the z-axis ($\rho = 0$) and at $\rho = 2a$. The integration was done numerically.

8.5. SPHERICAL BESSEL FUNCTIONS

8.5.1. The Wave Equation in Spherical Coordinates

In Section 8.4.1, we showed how Bessel functions arise as solutions of Laplace's equation in cylindrical coordinates. Closely related functions—the spherical Bessel functions—arise as solutions of the wave equation in spherical coordinates.

The wave equation (3.15)

$$\nabla^2 F - \frac{1}{c^2} \frac{\partial^2 F}{\partial t^2} = 0$$

may be solved by first Fourier transforming in time. The wave equation is then transformed to the Helmholtz equation (3.16) with $k^2 = \omega^2/c^2$, where ω is the Fourier transform variable (angular frequency of the wave) and c is the wave phase speed. For waves from a point source or in some other spherical geometry, it makes sense to write the ∇^2 operator in spherical coordinates:

$$\frac{1}{r^2}\frac{\partial}{\partial r}\left(r^2\frac{\partial F}{\partial r}\right) + \frac{1}{r^2\sin\theta}\frac{\partial}{\partial \theta}\left(\sin\theta\frac{\partial F}{\partial \theta}\right) + \frac{1}{r^2\sin^2\theta}\frac{\partial^2 F}{\partial \phi^2} + k^2 F = 0 \qquad (8.114)$$

Next we separate variables, as we have done before. First we write $F = R(r)\Theta(\theta)\Phi(\phi)$, multiply the whole equation by $r^2\sin^2\theta$, and divide by F, thus isolating the ϕ dependence:

$$\frac{\sin^2\theta}{R}\frac{\partial}{\partial r}\left(r^2\frac{\partial R}{\partial r}\right) + \frac{\sin\theta}{\Theta}\frac{\partial}{\partial \theta}\left(\sin\theta\frac{\partial\Theta}{\partial \theta}\right) + k^2 r^2 \sin^2\theta + \frac{1}{\Phi}\frac{\partial^2\Phi}{\partial\phi^2} = 0$$

The last term must therefore be a constant, and if our region of interest is $0 \le \phi < 2\pi$, then we must choose the constant to be $-m^2$ so that the solutions are periodic with period 2π. Then

$$\Phi = \sin m\phi, \quad \cos m\phi$$

and, after dividing by $\sin^2\theta$, equation (8.114) becomes

$$\frac{1}{R}\frac{\partial}{\partial r}\left(r^2\frac{\partial R}{\partial r}\right) + k^2 r^2 + \frac{1}{\Theta\sin\theta}\frac{\partial}{\partial \theta}\left(\sin\theta\frac{\partial\Theta}{\partial \theta}\right) - \frac{m^2}{\sin^2\theta} = 0$$

Now we have separated the equation completely, since the first two terms are functions of r only while the last two are functions of θ only. The last two terms are set equal to $-l(l+1)$, giving us the Legendre equation (refer to Section 8.3, equations 8.15 and 8.48) with solutions $P_l^m(\cos\theta)$ and $Q_l^m(\cos\theta)$. The remaining terms give the equation for R:

$$\frac{1}{R}\frac{\partial}{\partial r}\left(r^2\frac{\partial R}{\partial r}\right) + k^2 r^2 - l(l+1) = 0 \qquad (8.115)$$

We proceed by letting $R = Z/\sqrt{r}$. Then

$$\frac{dR}{dr} = -\frac{1}{2}\frac{Z}{r^{3/2}} + \frac{Z'}{r^{1/2}}$$

and

$$\frac{d}{dr}\left(r^2\frac{dR}{dr}\right) = -\frac{1}{4}\frac{Z}{r^{1/2}} + r^{1/2}Z' + r^{3/2}Z''$$

So the equation for Z is

$$-\frac{1}{4}\frac{Z}{r^{1/2}} + r^{1/2}Z' + r^{3/2}Z'' + k^2 r^{3/2}Z - l(l+1)\frac{Z}{r^{1/2}} = 0$$

Dividing by $r^{1/2}$ and gathering terms, we have

$$\frac{d}{dr}(rZ') + k^2 Zr - \left(l + \frac{1}{2}\right)^2 \frac{Z}{r} = 0$$

which is Bessel's equation (8.69) of order $l + 1/2$. Thus, the solution of equation (8.115) that we need is

$$R(r) = \frac{J_{l+1/2}(kr)}{\sqrt{r}}$$

This function is called a *spherical Bessel function*. The usual normalization is

$$j_l(x) = \sqrt{\frac{\pi}{2x}} J_{l+1/2}(x) \qquad (8.116)$$

Thus, the full solution to the wave equation in spherical coordinates is of the form

$$F(r, \theta, \phi, t) = j_l(kr) P_l^m(\cos\theta) e^{im\phi} e^{-ikct} \qquad (8.117)$$

We may define a similar "spherical" analog of each Bessel function. For example, the spherical Neumann function is

$$n_l(x) = \sqrt{\frac{\pi}{2x}} N_{l+1/2}(x) \qquad (8.118)$$

In the large argument limit, the spherical Hankel functions (refer to Section 8.4.1, equations 8.78, 8.79, and 8.85) have the form

$$h_l^{(1)}(kr) \equiv \sqrt{\frac{\pi}{2}} \frac{H_{l+1/2}^{(1)}(kr)}{\sqrt{kr}} \rightarrow \frac{e^{i[kr - \pi(l+1)/2]}}{kr} = (-i)^{l+1} \frac{e^{ikr}}{kr} \quad \text{as } r \to \infty \quad (8.119)$$

Thus, solutions of the form

$$F(r, \theta, \phi, t) = h_l^{(1)}(kr) P_l^m(\cos\theta) e^{im\phi} e^{-ikct} \propto e^{ik(r-ct)}$$

correspond to outgoing waves. Since the wave intensity is proportional to $|F|^2$, and $|F| \sim 1/r$ for large r, this result corresponds to the usual inverse square law. The second Hankel function $h^{(2)}(kr)$ describes incoming waves.

Properties of the spherical Bessel functions follow in a straightforward manner from the properties of the Bessel functions. For example, the recursion relations (8.89) and (8.90) become

$$
j_{l+1}(x) + j_{l-1}(x) = \frac{2(l + 1/2)}{x} j_l(x) = \frac{2l + 1}{x} j_l(x)
\tag{8.120}
$$

and

$$
j_{l+1}(x) - j_{l-1}(x) = -\frac{2}{\sqrt{x}} \frac{d}{dx} J_{l+1/2}(x) = -\frac{2}{\sqrt{x}} \frac{d}{dx} [\sqrt{x}\, j_l(x)]
$$

$$
= -2 \frac{dj_l(x)}{dx} - \frac{j_l(x)}{x}
$$

We can simplify this relation by using equation (8.120):

$$
(2l + 1)[j_{l+1}(x) - j_{l-1}(x)] = -2(2l + 1) \frac{dj_l(x)}{dx} - [j_{l+1}(x) + j_{l-1}(x)]
$$

$$
(l + 1) j_{l+1}(x) - l j_{l-1}(x) = -(2l + 1) \frac{dj_l(x)}{dx}
\tag{8.121}
$$

We may also write the spherical Bessel functions in series expansions. For example, using the series (8.73) for $J_{l+1/2}$, we find

$$
j_l(x) = \sqrt{\frac{\pi}{2x}} \sum_{n=0}^{\infty} \frac{(-1)^n}{n!\,\Gamma\left(n + l + \frac{3}{2}\right)} \left(\frac{x}{2}\right)^{l + \frac{1}{2} + 2n}
$$

$$
j_l(x) = \frac{\sqrt{\pi}}{2} \sum_{n=0}^{\infty} \frac{(-1)^n}{n!\,\Gamma\left(n + l + \frac{3}{2}\right)} \left(\frac{x}{2}\right)^{l + 2n}
\tag{8.122}
$$

In particular, for $l = 0$,

$$
j_0(x) = \frac{\sqrt{\pi}}{2} \sum_{n=0}^{\infty} \frac{(-1)^n}{n!\,\Gamma\left(n + \frac{3}{2}\right)} \left(\frac{x}{2}\right)^{2n}
\tag{8.123}
$$

Let's simplify the Γ function:

$$
\Gamma\left(n + \frac{3}{2}\right) = \left(n + \frac{1}{2}\right) \Gamma\left(n + \frac{1}{2}\right) = \left(n + \frac{1}{2}\right) \left(n - \frac{1}{2}\right) \Gamma\left(n - \frac{1}{2}\right)
$$

$$
= \left(n + \frac{1}{2}\right) \left(n - \frac{1}{2}\right) \left(n - \frac{3}{2}\right) \cdots \frac{1}{2} \Gamma\left(\frac{1}{2}\right)
$$

$$
= \frac{(2n + 1)!!}{2^{n+1}} \sqrt{\pi} = \frac{(2n + 1)!}{(2n)!!} \frac{\sqrt{\pi}}{2^{n+1}} = \frac{(2n + 1)!}{2^n n!} \frac{\sqrt{\pi}}{2^{n+1}}
$$

and so

$$n!\,\Gamma\left(n + \frac{3}{2}\right) = \sqrt{\pi}\,\frac{(2n + 1)!}{2^{2n+1}}$$

Thus,

$$j_0(x) = \sum_{n=0}^{\infty} \frac{(-1)^n}{(2n + 1)!}x^{2n} = \frac{\sin x}{x} \tag{8.124}$$

In fact, each spherical Bessel function may be written in terms of sines and cosines. For example,

$$j_1(x) = \frac{\sin x}{x^2} - \frac{\cos x}{x} \tag{8.125}$$

$$j_2(x) = \left(\frac{3}{x^3} - \frac{1}{x}\right)\sin x - \frac{3}{x^2}\cos x \tag{8.126}$$

(See Figure 8.17.) As l increases, the expressions become more and more complicated.

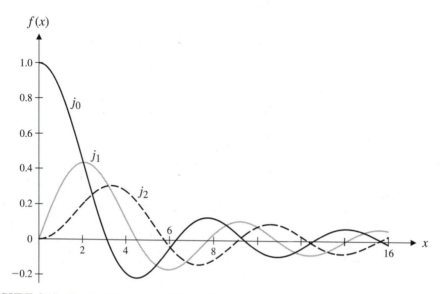

FIGURE 8.17. The first few spherical Bessel functions $j_0(x)$, $j_1(x)$, and $j_2(x)$. Compare with Figure 8.11.

For large arguments, the expressions become simpler[21]:

$$j_n(x) \sim \frac{1}{x}\sin\left(x - n\frac{\pi}{2}\right) \quad \text{for } x \gg 1, n \tag{8.127}$$

8.5.2. Orthogonality

Equation (8.115) is of Sturm-Liouville form with $f(r) \equiv r^2$, $g(r) \equiv l(l+1)$, $\lambda \equiv k^2$, and the weighting function $w(r) \equiv r^2$. Thus, the orthogonality relation is of the form

$$\int_a^b r^2 j_l(kr) j_l(k'r)\,dx = 0 \tag{8.128}$$

unless $k = k'$, provided that boundary conditions of the form (8.2) are satisfied. As with the regular Bessel functions, since $f(0) = 0$, we have

$$\int_0^b r^2 j_l(kr) j_l(k'r)\,dr = 0$$

provided that the boundary conditions (8.2) are satisfied at $r = b$, regardless of the boundary conditions, if any, that apply at $r = 0$. (As always, we do require that the function remain finite when its argument is zero.)

For $k = k' = \alpha_{l+1/2,p}/b$, where $\alpha_{l+1/2,p}$ is the pth zero of $J_{l+1/2}$, and therefore also of j_l, we may use relation (8.96) to obtain

$$\int_0^b r^2 j_l(kr) j_l(kr)\,dr = \frac{\pi}{2k}\int_0^b r J_{l+1/2}(kr) J_{l+1/2}(kr)\,dr = \frac{\pi}{2k}\frac{b^2}{2}[J_{l+1/2}'(\alpha_{l+1/2,p})]^2$$

But

$$j'(x) = \sqrt{\frac{\pi}{2}}\frac{J'}{\sqrt{x}} - \sqrt{\frac{\pi}{2}}\frac{1}{2}\frac{J}{x^{3/2}} = \sqrt{\frac{\pi}{2x}}J' - \frac{1}{2x}j$$

so

$$J' = \sqrt{\frac{2x}{\pi}}\left(j' + \frac{j}{2x}\right)$$

Thus, here we have

$$\int_0^b r^2 j_l(kr) j_l(kr)\,dr = \frac{\pi}{2k}\frac{b^2}{2}\frac{2kb}{\pi}\left[j_l'(\alpha_{l+1/2,p}) + \frac{j_l(\alpha_{l+1/2,p})}{2\alpha_{l+1/2,p}}\right]^2$$

$$= \frac{b^3}{2}[j_l'(\alpha_{l+1/2,p})]^2 \tag{8.129}$$

---

[21] Compare with equation (8.83).

where we used the fact that $\alpha_{l+1/2,p}$ is a zero of j_l. We can write the result in terms of the functions themselves rather than the derivatives[22] by using the recursion relations (8.120) and (8.121):

$$\int_0^b r^2 j_l(kr) j_l(kr)\, dr = \frac{b^3}{2} [j_{l+1}(\alpha_{l+1/2,p})]^2 \qquad (8.130)$$

Spherical Bessel functions find particular application in the study of electromagnetic waves (see Jackson, Chapter 9) and also in the quantum mechanics of a particle in a spherical cavity (Problem 55).

8.6. THE CLASSICAL ORTHOGONAL POLYNOMIALS

The functions we have studied in this chapter have several common features:

- They are orthogonal on an interval $[a, b]$ with respect to a weight function $w(x)$ (for example, relation 8.30).
- They satisfy a set of recursion relations (for example, equations 8.34 through 8.41).
- They may be computed from a generating function (for example, equation 8.32).
- They may be computed from a Rodrigues-type formula (for example, equation 8.42).

These properties are common to a larger class of orthogonal polynomials that arise in physics problems. They are defined by the generalized Rodrigues formula:

$$C_n(x) = \frac{1}{K_n} \frac{1}{w(x)} \frac{d^n}{dx^n} [w(x)s^n] \qquad (8.131)$$

where $w(x)$, the weight function, is real, positive, and integrable on the interval $[a, b]$; $s(x)$ is a polynomial of degree ≤ 2 with real roots; and the product ws satisfies the boundary conditions

$$w(a)s(a) = w(b)s(b) = 0$$

In addition, the first polynomial $C_1(x)$ is a first-degree polynomial in x. The constant K_n serves to normalize the polynomials. [Notice that the Legendre polynomials satisfy these conditions with $w(x) \equiv 1$, $s(x) = x^2 - 1$, $a = -1$, $b = 1$, $K_n = 2^n n!$, and $C_1(x) = x$.]

---

[22]See Problem 48.

We shall begin by showing that the function $C_n(x)$ defined by equation (8.131) is a polynomial of degree n. To construct the proof, we use the symbol $p_k(x)$ to represent any polynomial of degree $\leq k$. Then

$$\frac{d^m}{dx^m}(ws^n p_k) = ws^{n-m} p_{k+m} \tag{8.132}$$

where p_{k+m} is another polynomial of degree $\leq k + m$. To prove this result, we start with the definition of C_1 from equation (8.131) to obtain

$$wK_1C_1 = s\frac{dw}{dx} + w\frac{ds}{dx} \Rightarrow s\frac{dw}{dx} = w\left(K_1C_1 - \frac{ds}{dx}\right)$$

where $K_1C_1 - ds/dx$ is a polynomial of degree ≤ 1. (Remember: We have already specified that C_1 is a first-degree polynomial.) Then

$$\begin{aligned}
\frac{d}{dx}(ws^n p_k) &= \frac{dw}{dx}s^n p_k + nw\frac{ds}{dx}s^{n-1} p_k + ws^n\frac{dp_k}{dx} \\
&= w\left(K_1C_1 - \frac{ds}{dx}\right)s^{n-1} p_k + nw\frac{ds}{dx}s^{n-1} p_k + ws^n\frac{dp_k}{dx} \\
&= ws^{n-1}\left[\left(K_1C_1 + (n-1)\frac{ds}{dx}\right)p_k + s\frac{dp_k}{dx}\right]
\end{aligned}$$

Since $s(x)$ is a polynomial of degree ≤ 2, ds/dx is of degree ≤ 1, and dp_k/dx is of degree $\leq k - 1$, the term in brackets on the right-hand side is a polynomial of degree $\leq k + 1$:

$$\frac{d}{dx}(ws^n p_k) = ws^{n-1} p_{k+1}$$

[Given specific forms for the functions $w(x)$, $s(x)$, and a polynomial p_k, we could use this relation to construct a specific form for the polynomial p_{k+1}, but we don't need to do that.] We can continue to differentiate until we obtain the result (8.132).

Setting $n = m$ and $k = 0$ in equation (8.132) and using the result in equation (8.131), we obtain

$$C_n(x) = p_n$$

that is, C_n is a (yet unspecified) polynomial of degree $\leq n$. This polynomial may be written as a polynomial of degree $\leq n - 1$, with the possible addition of a term in x^n:

$$C_n = p_{n-1} + Ax^n \tag{8.133}$$

Next we shall show that A is not zero, and thus C_n is a polynomial of degree n.

To show that A is not zero, and at the same time establish the orthogonality property of the polynomials C_n, we shall first show that

$$\int_a^b P_m(x)C_n(x)w(x)\,dx = 0 \quad \text{for } m < n \tag{8.134}$$

for any $P_m(x)$. We begin with the defining relation (8.131) and integrate by parts:

$$\int_a^b P_m(x)C_n(x)w(x)\,dx$$

$$= \frac{1}{K_n}\int_a^b P_m(x)\frac{d^n}{dx^n}[w(x)s^n]\,dx$$

$$= \frac{1}{K_n}\left(P_m(x)\frac{d^{n-1}}{dx^{n-1}}[w(x)s^n]\bigg|_a^b - \int_a^b P_{m-1}(x)\frac{d^{n-1}}{dx^{n-1}}[w(x)s^n]\,dx\right)$$

where we used the result that the derivative of a polynomial of degree $\leq m$ is another polynomial of degree $\leq m-1$. Result (8.132) shows that all the derivatives $(d^m/dx^m)(ws^n)$ with $m < n$ vanish at the endpoints $x = a$ and $x = b$ of the interval because ws does, and thus the integrated term vanishes. We may now continue to integrate by parts, reducing the order of the polynomial in the integrand by one each time, until we obtain

$$\int_a^b P_m(x)C_n(x)w(x)\,dx = \frac{1}{K_n}\left((-1)^m\int_a^b P_0\frac{d^{n-m}}{dx^{n-m}}[w(x)s^n]\,dx\right)$$

for $m < n$, and the next integration shows that the integral (8.134) is zero.

Now we multiply equation (8.133) by $C_n w$ and integrate:

$$\int_a^b (C_n)^2\,w(x)\,dx = \int_a^b P_{n-1}C_n w\,dx + A\int_a^b x^n C_n w\,dx$$

$$= 0 + A\int_a^b x^n C_n w\,dx \equiv I_n$$

where we used result (8.134) with $m = n - 1$ to eliminate the first term. The left-hand side is ≥ 0, since the integrand is positive throughout the range of integration, so A must be greater than zero, and C_n is thus a polynomial of order n. Finally, we set $P_m = C_m$ in relation (8.134) to obtain the orthogonality relation we seek:

$$\int_a^b C_m(x)C_n(x)w(x)\,dx = 0 \quad \text{for } m < n \tag{8.135}$$

Properties of some of the polynomials used in physics are listed in Table 8.3.

TABLE 8.3. Classical Orthogonal Polynomials

	$w(x)$	$s(x)$	a	b	K_n	I_n
Legendre	1	$x^2 - 1$	-1	$+1$	$2^n n!$	$\dfrac{2}{2n+1}$
Hermite	e^{-x^2}	1	$-\infty$	$+\infty$	$(-1)^n$	$\sqrt{\pi}\,2^n n!$
Laguerre	$x^\nu e^{-x}$	x	0	$+\infty$	$n!$	$\dfrac{\Gamma(n+\nu+1)}{n!}$
Jacobi	$(1-x)^\nu (1+x)^\mu$	$1-x^2$	-1	$+1$	$(-2)^n n!$	$\dfrac{2^{\nu+\mu+1}\Gamma(n+\nu+1)\Gamma(n+\mu+1)}{(2n+\nu+\mu+1)n!\,\Gamma(n+\nu+\mu+1)}$
Tchebichef	$\dfrac{1}{\sqrt{1-x^2}}$	1	-1	$+1$	$(-1)^n \dfrac{(2n)!}{2^n n!}$	$\dfrac{\pi}{2}$

PROBLEMS

1. Find the eigenfunctions for the one-dimensional Helmholtz equation

$$\frac{d^2 y}{dx^2} + k^2 y = 0$$

subject to the boundary conditions

$$y = 0$$

at $x = 0$ and

$$y' = 0$$

at $x = L$. (This corresponds to a vibrating string with one fixed end and one free end.)

2. Find the eigenfunctions for the Helmholtz equation

$$\frac{d^2 y}{dx^2} + k^2 y = 0$$

subject to the boundary conditions

$$ay + by' = 0$$

at $x = 0$ and

$$\alpha y + \beta y' = 0$$

at $x = L$.

Determine specific forms for the functions, and obtain the eigenvalues for the following cases:

 (i) $a = \alpha$ and $b = \beta$

 (ii) $aL = b$ and $\beta = 2\alpha L$

You should be able to obtain numerical values for the product kL.

3. The displacement of a square vibrating membrane of side L satisfies the two-dimensional Helmholtz equation

$$\frac{\partial^2 s}{\partial x^2} + \frac{\partial^2 s}{\partial y^2} + k^2 s = 0$$

where $k = \omega/v$, ω is the frequency, and v is the speed of waves on the membrane. Suppose the membrane is fixed at its edges at $x = 0, L$ and $y = 0, L$. Separate variables and solve for the eigenfunctions $s(x, y)$. Show that the system exhibits degeneracy— that is, that there is more than one eigenfunction corresponding to a given eigenvalue k^2. In particular, show that there are two eigenfunctions s_1 and s_2 that correspond to the eigenvalue $k^2 = 5\pi^2/L^2$. What symmetry of the physical system causes this degeneracy? (*Hints:* Where are the nodal lines for the two modes? What happens if one side of the membrane is slightly shorter, equal to $L - \varepsilon$?) Any linear combination of the two eigenfunctions is also a solution. Find some of the nodal lines for combinations of the modes—for example, $s_1 + s_2$. How do these modes reflect the symmetry of the system? Can you find an eigenvalue that has threefold degeneracy? If so, what do those modes look like?

4. A set of eigenfunctions $y_n(x)$ satisfies the Sturm-Liouville equation (8.1) with boundary conditions (8.2). The function $g \equiv 0$. Show that the derivatives $u_n(x) = y_n'(x)$ are also orthogonal functions. Determine the weighting function for these functions. What boundary conditions are required for orthogonality? Apply your results to the Legendre equation to determine the orthogonality of the derivatives $P_l'(\mu)$.

5. Use the recursion relations in Section 8.3.5 to show that the derivatives $P_l'(\mu)$ of the Legendre polynomials are orthogonal on the range $(-1, 1)$ with weighting function $(1 - \mu^2)$, in agreement with the results of Problem 4.

6. To obtain Fourier-Legendre series, we often need to evaluate integrals of the form

$$I_l^n = \int_0^1 \mu^n P_l(\mu)\, d\mu$$

 (a) Start by evaluating I_l^0, I_l^1, I_0^n, and I_1^n.

 (b) Next use the recursion relations for the polynomials in Section 8.3.5 to determine recursion relations for the integrals I_l^n. Multiply equation (8.37) by μ^n and integrate by parts to obtain a relation between I_l^n and I_{l-1}^{n-1}.

 (c) Use these results to step down until you can use your results from (a) to obtain an explicit expression for I_l^n. Show that

$$
I_l^n = \begin{cases}
\dfrac{n!}{(n-1)!!(n+l+1)!!} & \text{if } n \geq l \\[3mm]
(-1)^{\frac{l-n-1}{2}} n! \dfrac{(l-n-2)!!}{(n+l+1)!!} & \text{if } n < l \text{ and } l-n \text{ is odd} \\[3mm]
0 & \text{if } n < l \text{ and } l-n \text{ is even}
\end{cases}
$$

7. Verify that the Rodrigues formula gives the correct normalization for every Legendre polynomial. *Hint:* Write $(x^2 - 1)^l = (x-1)^l (x+1)^l$, differentiate, and evaluate the result at $x = 1$.

8. Evaluate the integral

$$
\int_{-1}^{+1} \frac{P_l(x)}{\sqrt{1-x^2}}\, dx
$$

and hence obtain a Fourier-Legendre series for the function $1/\sqrt{1-x^2}$. *Hint:* The recursion relations in Section 8.3.5 may prove useful.

9. Write Laplace's equation in oblate spheroidal coordinates (refer to Chapter 2, Problem 13), separate variables, and hence show that the solution requires Legendre functions in both the coordinates u and v. Argue that the solution exterior to an oblate spheroidal boundary requires the use of the Legendre function of the second kind, Q.

10. Expand the Legendre function $Q_0(x)$ for large values of the argument, and show that your result agrees with the asymptotic form in equation (8.29), modulo a constant.

11. Rewrite the Legendre equation

$$
\frac{d}{dx}\left((1-x^2)\frac{dQ_l}{dx} \right) + l(l+1)Q_l = 0
$$

in terms of the variable $u = 1/x$, and obtain a solution as a series in u. Show that for large x, $Q_l(x)$ goes to zero as $1/x^{l+1}$. Show that for $l = 0$, the solution $Q_0(x)$ may be written as in equation (8.28) but with $x - 1$ in the denominator instead of $1 - x$.

12. Current flow in a conducting sheet is described by the relations $\vec{E} = -\vec{\nabla}\Phi$ and $\vec{j} = \sigma\vec{E}$. Use the charge conservation law ($\partial\rho/\partial t + \vec{\nabla}\cdot\vec{j} = 0$) to show that, in a steady state, Φ satisfies Laplace's equation. Find the eigenfunctions for current flow in a circular copper plate.

Current I flows into and out of a plate of thickness t and conductivity σ through electrodes at $r = a$ extending from $\theta = \pi - \gamma/2$ to $\pi + \gamma/2$ and from $\theta = -\gamma/2$ to $+\gamma/2$. Determine the potential, and plot the current flow lines.

13. A solid sphere of radius a is immersed in a vat of fluid at temperature T_0. Heat is conducted into the sphere according to equation (3.14). If the temperature at the boundary is fixed at T_0 and the initial temperature of the sphere is T_1, find the temperature within the sphere as a function of time.

14. Use the Cauchy formula together with the Rodrigues formula to write $P_l(\mu)$ as a contour integral in the complex plane. Take the contour to be a circle of radius $\sqrt{x^2 - 1}$, and

hence obtain the integral expression

$$P_l(x) = \frac{1}{\pi} \int_0^\pi \left(x + \sqrt{x^2 - 1} \cos \phi\right)^l d\phi$$

15. Starting from the relations in Section 8.3.5, derive the following recursion relations for the associated Legendre functions:

(a) $(l - m + 1)\sqrt{1 - \mu^2} P_l^{m-1} = P_{l-1}^m - \mu P_l^m$

(b) $(2l + 1)\sqrt{1 - \mu^2} P_l^{m-1} = P_{l-1}^m - P_{l+1}^m$

Hint: Start with the pure recursion relation and differentiate.

(c) From relations (a) and (b), derive the following:

$$(2l + 1)\mu P_l^m(\mu) = (l - m + 1) P_{l+1}^m + (l + m) P_{l-1}^m$$

16. Starting from the definition (8.53), obtain the m-raising recursion relation for the associated Legendre functions:

$$P_l^{m+1}(\mu) = -m \frac{\mu}{\sqrt{1 - \mu^2}} P_l^m(\mu) - \sqrt{1 - \mu^2} \frac{d}{d\mu} P_l^m(\mu)$$

Combine this result with equation (8.59) to obtain the m-lowering relation

$$(l + m)(l - m + 1) P_l^{m-1} = \sqrt{1 - \mu^2} \frac{d}{d\mu} P_l^m - m \frac{\mu}{\sqrt{1 - \mu^2}} P_l^m$$

17. Use the results of Problem 15 to show that, for $l + m$ even,

$$P_l^m(0) = (-1)^{(l+m)/2} \frac{(l + m - 1)!!}{(l - m)!!}$$

18. Show by direct substitution into equation (8.15) that $P_m^m(\theta) \propto \sin^m \theta$. Use the value of the orthogonality integral (8.55), together with the result

$$\int_0^{\pi/2} \sin^{2m+1} \theta \, d\theta = \frac{(2m)!!}{(2m + 1)!!}$$

(for example, Gradshteyn and Ryzhik formula 3.621#4), to show that

$$P_m^m(\theta) = (-1)^m \frac{(2m)!}{2^m m!} \sin^m \theta$$

This relation, together with the m-lowering relation (Problem 16), may be used to generate the P_l^m.

19. Verify the result

$$\int_{-1}^{+1} \frac{[P_l^m(\mu)]^2}{1 - \mu^2} d\mu = \frac{1}{m} \frac{(l + m)!}{(l - m)!}$$

for the second orthogonality integral in Section 8.3.7 in the cases

(a) $l = m = 1$ (b) $l = 2, m = 1$ (c) $l = m$

(d) Stepping down in m, use proof by induction (Appendix III) to show that the result is true in general.

20. Using the generating function $G(x, \mu)$ (equation 8.32) and the addition theorem (8.65), derive the expansion

$$\frac{1}{|\mathbf{x} - \mathbf{x}'|} = \sum_{l=0}^{\infty} \sum_{m=-l}^{l} \frac{4\pi}{2l+1} \frac{r_<^l}{r_>^{l+1}} Y_{lm}(\theta, \phi) Y_{lm}^*(\theta', \phi')$$

where $r_<$ and $r_>$ are the lesser and the larger of r and r', respectively.

Hence find the magnetic vector potential due to a circular loop of wire of radius a that is carrying current I. (Refer also to Problem 6.19.)

21. Verify the result (8.67)

$$\int_0^1 P_l^m(\mu) d\mu = (-1)^{(m+1)/2} \frac{\pi}{2} \frac{m}{l} P_{l+1}(0) P_l^m(0)$$

where l and m are both odd.

(a) First evaluate the Legendre functions to show that

$$I_{lm} = \int_{-1}^{+1} P_l^m(\mu) d\mu = -m\pi \frac{(l-2)!! \, (l+m-1)!!}{(l+1)!! \, (l-m)!!}, \quad l, m \text{ odd}$$

(b) Use the expression for $P_m^m(\theta)$ from Problem 18 to show that the result is true for $m = l$. *Hint*: Use contour integration.

(c) Show that the result is true for $m = 1$ and l equal to any odd integer.

(d) Use proof by induction (Appendix III) to show that the result is true for all m, $1 \le m \le l$, with both l and m odd. *Hint*: Use the result from part (b) and step down in m.

22. Find the electrostatic potential inside a hemisphere of radius a with potential $\Phi = 0$ on the flat side and $\Phi = V$ on the curved part.

23. Quantum mechanical treatment of the harmonic oscillator results in the Hermite differential equation

$$y'' - 2xy' + \lambda y = 0$$

Write this equation in standard Sturm-Liouville form. If the boundary conditions are $y(x) \to 0$ as $x \to \pm\infty$, show that the solutions are orthogonal on the range $(-\infty, +\infty)$, and find the weight function $w(x)$. Solve the equation to find a series expansion for the Hermite functions. What value of the eigenvalue λ is required for the functions to remain bounded throughout the interval, including $x \to \pm\infty$? (*Hint*: Experience with Legendre functions should prove useful.) Normalize the solutions by choosing the coefficient of the highest power x^n to be 2^n, and hence determine the first three eigenfunctions.

24. The generating function for Hermite polynomials is

$$G(x, t) = e^{-t^2+2xt} = \sum_{n=0}^{\infty} \frac{t^n}{n!} H_n(x)$$

Use this generating function to establish a pure recursion relation for Hermite polynomials (analogous to equation 8.34 for Legendre polynomials). Also obtain the derivative dH_n/dx in terms of the H_n (analogous to equations 8.40 and 8.41).

Differentiate $G(x, t)$ with respect to t a total of n times to obtain the Rodrigues-type formula

$$H_n(x) = (-1)^n e^{x^2} \frac{d^n}{dx^n} e^{-x^2}$$

25. By using the Rodrigues formula (Problem 24) for the Hermite polynomials or otherwise, obtain the normalization integral:

$$\int_{-\infty}^{+\infty} e^{-x^2} [H_n(x)]^2 \, dx = 2^n n! \sqrt{\pi}$$

26. Starting with relation (8.86), derive the Rodrigues-type formula for Bessel functions:

$$J_n(x) = x^n \left(-\frac{1}{x} \frac{d}{dx} \right)^n J_0(x)$$

27. A drumhead is a circular membrane of radius a. When it is struck, waves propagate across the drumhead. The membrane vibrates with displacement ξ, where $\xi(r, \theta, t) = \eta(r, \theta)e^{-i\omega t}$ and $\eta(r, \theta)$ satisfies the Helmholtz equation

$$\frac{1}{r} \frac{\partial}{\partial r} \left(r \frac{\partial \eta}{\partial r} \right) + \frac{1}{r^2} \frac{\partial^2 \eta}{\partial \theta^2} + k^2 \eta = 0$$

where $k^2 = \omega^2/v^2$ and v is the speed with which waves propagate across the drumhead. (The speed v depends on the tension in the drumhead, among other things.) The boundary condition is that $\eta = 0$ at $r = a$. Separate variables, and find the eigenfunctions. Determine the first three allowable frequencies ω in terms of the drum parameters v and a.

28. Sound waves propagating through a tube may be described by a velocity potential (refer to Chapter 2, Section 2.4) that satisfies the Helmholtz equation

$$\left(\nabla^2 + \frac{\omega^2}{c_s^2} \right) \Phi = 0$$

where c_s is the sound speed in the tube. Assume that for propagation along the length of the tube (in the $+z$-direction), the potential may be written

$$\Phi = \Phi_t e^{ikz}$$

where Φ_t is a function of the transverse coordinates (x and y or r and θ). Because the air cannot move perpendicular to the walls of the tube, the boundary condition is

$$\hat{\mathbf{n}} \cdot \vec{\nabla} \Phi = \frac{\partial \Phi}{\partial n} = 0 \quad \text{on the boundary surface}$$

Write the differential equation and boundary conditions satisfied by Φ_t and hence find the eigenvalues and the set of allowed frequencies ω if

(a) the tube has a rectangular cross section measuring $a \times b$

(b) the tube has a circular cross section of radius a

In each case, show that there is a minimum frequency for waves that propagate along the tube with Φ_t not constant.

29. If γ_{mn} is the nth zero of $J_m'(x)$, show that the Bessel functions satisfy the orthogonality relation:

$$\int_0^a J_m\left(\gamma_{mn}\frac{\rho}{a}\right) J_m\left(\gamma_{mk}\frac{\rho}{a}\right) \rho \, d\rho = \frac{a^2}{2}\left(1 - \frac{m^2}{\gamma_{mn}^2}\right)[J_m(\gamma_{mn})]^2 \delta_{nk}$$

30. Use the generating function (8.93) to show that

(a) $\sin x = 2 \sum_{n=0}^{\infty} J_{2n+1}(x)$

(b) $1 = J_0(x) + 2 \sum_{n=1}^{\infty} J_{2n}(x)$

31. Use the generating function (8.91) to show that

$$J_n(x + y) = \sum_{m=-\infty}^{+\infty} J_m(x) J_{n-m}(y)$$

and hence show that

$$J_0(2x) = J_0^2(x) + 2\sum_{m=1}^{\infty}(-1)^m J_m^2(x)$$

(*Hint:* Use the orthogonality of the complex exponentials.)

32. Starting from the Bessel differential equation, show that

$$\int_0^\infty \frac{J_m(x) J_n(x)}{x} dx = \frac{2}{\pi} \frac{\sin[(m-n)\pi/2]}{m^2 - n^2} \quad \text{for } m + n > 0$$

and that this result equals $1/2n$ when $m = n$. Does this result constitute a second orthogonality relation for the Bessel functions? Why or why not?

33. At time $t = 0$, the surface of the water in a pond has the form $s(\rho, \phi, 0) = A J_0(\alpha\rho)$. By taking the Fourier transform of the wave equation with two spatial dimensions, find the displacement $s(\rho, \phi, t)$ at later times.

34. (a) Show that

$$\int_0^\infty e^{-ax} J_0(x)\, dx = \frac{1}{\sqrt{a^2+1}}$$

Hint: Use the integral expression (8.92) for $J_0(x)$ and perform the integral over x first. Do the integral over the angle using methods from Chapter 2.

(b) Use the same technique to evaluate the integral

$$\int_0^\infty e^{-ax} J_m(x)\, dx$$

35. Show that

$$\int_0^1 x^3 J_0(ax)\, dx = \left(1 - \frac{4}{a}\right) J_1(a)$$

if $J_0(a) = 0$. What is the value of the integral if, instead, $J_1(a) = 0$?

36. Show that the first zero (other than zero) of the Bessel function $J_m(x)$, $x_{m,1}$, is an increasing function of m—that is,

$$x_{0,1} < x_{1,1} < x_{2,1} < x_{3,1}$$

and so on. *Hint:* Use relations (8.86)–(8.90).

37. A pendulum has steadily increasing length $l(t) = l_0 + \alpha t$. Show that the equation that describes small oscillations of this pendulum is

$$l\theta'' + 2\alpha\theta' + g\theta = 0$$

Change variables to $u = (1 + \alpha t/l_0)^{1/2}$ and $y = u\theta$, and hence show that the general solution may be expressed in terms of Bessel functions. Find the solution if the pendulum is released from rest at an angle θ_0 at $t = 0$ (that is, when $l = l_0$).

38. The equation that describes the angular displacement of a vertical pole or column is

$$EI\frac{d^2\theta}{dx^2} + g\lambda x\theta = 0$$

where x increases downward from the top of the pole, E is the Young's modulus, I is the moment of inertia (see also Chapter 3, Section 3.2.3), and λ is the mass per unit length. Make a change of variables to

$$u = \frac{2}{3}\sqrt{\frac{g\lambda}{EI}x^3}, \quad y = \frac{\theta}{\sqrt{x}}$$

and hence show that the solution may be expressed in terms of Bessel functions. Show that there is no solution that fits the boundary conditions $\theta(L) = 0$ and $\theta'(0) = 0$ unless the pole has a minimum length L_{min}. Find an expression for L_{min} in terms of the physical parameters of the pole.

39. Establish the addition theorem for Bessel functions:

$$J_0(kR) = \sum_{m=-\infty}^{\infty} e^{im\phi} J_m(k\rho) J_m(k\rho')$$

where

$$R = \sqrt{\rho^2 + (\rho')^2 - 2\rho\rho' \cos\phi}$$

Hint: Begin by expressing $1/|\vec{x} - \vec{x}'|$ as an expansion in Bessel functions, as in Section 8.4.9, and set $z = z' = 0$ and $\phi' = 0$. Evaluate the constant $A_m(k)$ using the result of Problem 34.

40. Starting from the definitions (8.100), (8.76), and (8.78), show that

$$K_\nu(x) = \frac{\pi}{2} \frac{I_{-\nu}(x) - I_\nu(x)}{\sin \nu\pi}$$

where ν is not an integer. Hence, show that $K_{-\nu}(x) = K_\nu(x)$.

41. The potential on a plane is V_0 for $\rho < a$ and zero for $r > a$. Find an integral expression for the potential everywhere. Evaluate the integral with $\rho = 0$ to find the potential on the z-axis.

42. A cylinder of height h and radius a has the top and bottom grounded. The potential on the wall at $\rho = a$ is V_0. Find the potential inside the cylinder.

43. A cylinder of height h and radius a is grounded except for its base at $z = 0$. On the base the potential is $V_0/\sqrt{1 - \rho^2/a^2}$. Find the potential inside the cylinder.

44. **(a)** Use the series for $J_0(x)$ to show that its Laplace transform is

$$\mathcal{L}[J_0(x)] = \frac{1}{\sqrt{1 + s^2}}$$

(b) Then use the recursion relation (8.87) to find the Laplace transform of $J_1(x)$. Extend the result to show that

$$\mathcal{L}[J_m(x)] = \frac{(\sqrt{s^2 + 1} - s)^m}{\sqrt{s^2 + 1}}$$

Compare with Problem 34.

(c) Use the convolution theorem to establish the relation

$$\int_0^x J_0(x - u) J_0(u)\, du = \sin x$$

45. Obtain expression (8.126) for $j_2(x)$ from the expressions for j_0 and j_1 and the recursion relation (8.120).

46. Starting with the recursion relations (8.86) and (8.88), derive the relations

$$\frac{d}{dx}\left(\frac{j_l(x)}{x^l}\right) = -\frac{j_{l+1}(x)}{x^l}$$

and

$$\frac{d}{dx}[x^{l+1}j_l(x)] = x^{l+1}j_{l-1}(x)$$

47. Use proof by induction (Appendix III) to establish the Rodrigues-type formula

$$j_n(x) = (-1)^n x^n \left(\frac{1}{x}\frac{d}{dx}\right)^n \left(\frac{\sin x}{x}\right)$$

48. Use the recursion relations to show that the orthogonality relation (8.130) is equivalent to (8.129).

49. Starting from the definition (8.76), show that

$$n_l(x) = (-1)^{l+1} j_{-(l+1)}(x)$$

Hence show that

$$n_0(x) = -\frac{\cos x}{x}$$

50. Starting from relations (8.111) and (8.112), establish the recursion relations for the modified spherical Bessel functions $k_l(x) = \sqrt{2/\pi x}\, K_{l+1/2}(x)$:

$$k_{l-1} - k_{l+1} = -\left(\frac{2l+1}{x}\right)k_l$$

and

$$(l+1)k_{l+1} + lk_{l-1} = -(2l+1)\frac{d}{dx}k_l(x)$$

51. The Fresnel integrals

$$S(x) = \sqrt{\frac{2}{\pi}}\int_0^x \sin t^2\, dt$$

and

$$C(x) = \sqrt{\frac{2}{\pi}}\int_0^x \cos t^2\, dt$$

may be expressed as series of spherical Bessel functions. First show that

$$S(x) = \frac{1}{\sqrt{2\pi}}\int_0^{x^2} \sqrt{u}\, j_0(u)\, du$$

and obtain a similar expression for $C(x)$. Use the recursion relation to do the integration and hence establish the result

$$S(x) = \sqrt{\frac{2}{\pi}x}\sum_{n=1}^{\infty} j_{2n-1}(x^2)$$

Determine a similar expression for $C(x)$.

52. Sound waves in a spherical cavity satisfy the differential equation $(\nabla^2 + k^2) F = 0$ for $r < R$ with $\partial F / \partial r = 0$ at $r = R$. Find the eigenvalues k_n for the problem and hence find the allowed frequencies $\omega_n = k_n v$ for sound waves inside the cavity.

53. Electromagnetic waves in a spherical cavity may be described by a mathematical problem similar to that described in Problem 52, with $v = c$, the speed of light. The boundary conditions depend on the polarization. Find the allowed frequencies if the boundary condition is $F(R) = 0$.

54. The modified spherical Bessel functions are defined[23] as

$$ i_l(x) = \sqrt{\frac{\pi}{2x}} I_{l+1/2}(x) \quad \text{and} \quad k_l(x) = \sqrt{\frac{2}{\pi x}} K_{l+1/2}(x) $$

Using expression (8.99) and the result of Problem 8.40, verify the expressions for the modified spherical Bessel functions $i_0(x) = \sinh x / x$ and $k_0(x) = e^{-x}/x$. Use proof by induction to show that

$$ k_l(x) = (-1)^l x^l \left(\frac{1}{x} \frac{d}{dx} \right)^l \frac{e^{-x}}{x} $$

55. We may model the force between particles in an atomic nucleus with a three-dimensional square-well potential

$$ V(r) = \begin{cases} -V_0 & \text{for } r < R \\ 0 & \text{for } r > R \end{cases} $$

Schrodinger's equation for this system takes the form

$$ \left(\nabla^2 - 2\frac{m}{\hbar^2} V(r) \right) \psi = -2\frac{m}{\hbar^2} E \psi $$

Write the differential operator in spherical coordinates and show that the solution may be written in terms of spherical Bessel functions. With $\alpha \equiv 2(m/\hbar^2) V_0 R^2$ and $\varepsilon = -E/V_0$ with $E < 0$, show that the energy levels are determined by the equation

$$ \sqrt{1-\varepsilon} k_l \left(\alpha \sqrt{\varepsilon} \right) j_{l+1} \left(\alpha \sqrt{1-\varepsilon} \right) = \sqrt{\varepsilon} j_l \left(\alpha \sqrt{1-\varepsilon} \right) k_{l+1} \left(\alpha \sqrt{\varepsilon} \right) $$

Find the energy of the lowest energy level for $l = 0$, $\alpha = 10$.

56. The density of neutrons in uranium is described by the equation

$$ \frac{\partial n}{\partial t} = D\nabla^2 n + an $$

where D (the diffusion coefficient) and a (the net production rate) may be taken to be constants in space and time. Solve the equation using separation of variables. Look for a solution with spherical symmetry that satisfies the boundary condition $n = 0$ at $r = R$. Show that the density increases exponentially if R exceeds a critical value R_{crit}, and determine that value in terms of D and a.

---

[23] Some authors choose different normalization.

OPTIONAL TOPIC A

Tensors

A.1. CARTESIAN TENSORS

In Chapter 1, we showed how physical laws, such as Newton's second law (equation 1.8), may be represented as mathematical relations between vectors. In equation (1.8), the force vector $\vec{F}$ and the acceleration vector $\vec{a}$ are parallel. It is sometimes the case that one vector is linearly related to another, as in equation (1.8), but the directions of the vectors are *not* the same. An example from mechanics is the angular momentum $\vec{L}$ of a rigid body, which is determined by the body's angular velocity $\vec{\omega}$ but is not parallel to $\vec{\omega}$ unless the body has sufficient symmetry.[1] In electricity and magnetism, the current density $\vec{j}$ in a magnetized plasma is not necessarily parallel to the electric field $\vec{E}$ that drives it, because the magnetic force causes particles to gyrate. In these cases, the vectors are related by a linear operator that mixes the components. For example, the angular momentum is

$$\vec{L} = \mathbb{I}\vec{\omega}$$

where $\mathbb{I}$ is the inertia tensor.[2] The vector components are related by a matrix. In index notation,[3]

$$L_i = I_{ij}\omega_j \tag{A.1}$$

The matrix $\mathbb{I}$ has $3 \times 3 = 9$ components.[4] It is an example of a rank-two tensor. The rank of a tensor is indicated by the number of indices needed to describe its components.

TABLE A.1. Tensors in Three-Dimensional Space

Object	Notation	Number of Components	Rank of Tensor
scalar	m	$1 = 3^0$	0
vector	v_i	$3 = 3^1$	1
rank-two tensor	I_{ij}	$9 = 3^2$	2

[1] See, for example, Lea and Burke, Chapter 9; Marion and Thornton; Goldstein, Chapter 5.
[2] This is not the identity matrix. Here $\mathbb{I}$ stands for inertia. In Chapter 1, we used the symbol $\mathbb{R}$ for this operator.
[3] Throughout this topic we shall use the summation convention discussed in Chapter 1, Section 1.1.2.
[4] The number 3 here represents the three dimensions of space.

As with Newton's second law, it is essential that a physical relationship such as (A.1) remain true when we change the coordinate system (refer to Section 1.1.1). In the new system,

$$L_i' = I_{ij}' \omega_j'$$ (A.2)

We already know how the vector components transform under coordinate rotations, so let's transform them:

$$L_i' = A_{ik} L_k$$

where A_{ik} is the rotation matrix (equation 1.21). To go in the reverse direction, we multiply on the left by $\mathbb{A}^{-1} = \mathbb{A}^T$:

$$A_{ni}^T L_i' = A_{ni}^T A_{ik} L_k = \delta_{nk} L_k = L_n$$

so (equation 1.25)

$$L_n = A_{in} L_i'$$

Similarly,

$$\omega_m = A_{jm} \omega_j'$$

so equation (A.1) becomes

$$A_{in} L_i' = I_{nm} A_{jm} \omega_j'$$

Now we multiply on the left by $\mathbb{A}$, and we have

$$A_{kn} A_{ni}^T L_i' = A_{kn} I_{nm} A_{jm} \omega_j'$$
$$\delta_{ki} L_i' = L_k' = A_{kn} A_{jm} I_{nm} \omega_j'$$

Comparing with equation (A.2), we find the transformation rule for the tensor I_{ij}:

$$\boxed{I_{kj}' = A_{kn} A_{jm} I_{nm}}$$ (A.3)

Notice how the indices match up in equation (A.3); the repeated indices indicate that we must sum over both n and m.

We can use matrix multiplication to perform the calculation indicated by the indices if we first put the summed indices next to each other.[5] To do this, we have to transpose one of the matrices:

$$I'_{kj} = A_{kn} I_{nm} A^T_{mj}$$

or, in matrix notation,

$$\mathbb{I}' = \mathbb{A} \mathbb{I} \mathbb{A}^T$$

(A.4)

The index notation (A.3) is more general and more powerful. It extends easily to tensors of rank greater than two, whereas the matrix notation does not.

The transformation laws for vectors, scalars, and tensors are compared in Table A.2.

TABLE A.2. Transformation Laws

Object	Transformation Law
scalar	$m' = m$
vector	$v'_i = A_{ij} v_j$
rank-two tensor	$I'_{km} = A_{kn} A_{mj} I_{nj}$

We need one transformation matrix for each index of the tensor. Extending the rule we have found, we get the transformation law for a rank-three tensor:

$$T'_{ijk} = A_{in} A_{jm} A_{kp} T_{nmp}$$

(A.5)

Example A.1. The inertia tensor has components

$$I_{ij} = \int (r^2 \delta_{ij} - x_i x_j)\, dm$$

(A.6)

A uniform square plate of side s is rotating about an axis perpendicular to the plane of the square and through its center. The angular momentum of the square about its center is

$$\vec{L} = \frac{Ms^2}{6} \vec{\omega}$$

Compute the components of the inertia tensor in a coordinate system with origin at the center of the square and x- and y-axes parallel to the sides of the square, and verify the expression for the angular momentum. Next, compute the components of

---

[5] See footnote 8 in Chapter 1.

the tensor in a primed coordinate system with axes rotated by 45° about the original x-axis. Hence find the angular momentum about its center when the square rotates about an axis at 45° to the plane of the square.

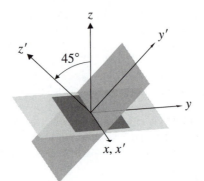

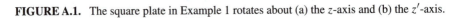

FIGURE A.1.   The square plate in Example 1 rotates about (a) the z-axis and (b) the z'-axis.

Since the square is a planar object and all mass elements are at $z = 0$ (Figure A.1), we may immediately conclude that

$$I_{i3} = I_{3i} = 0$$

unless i also equals 3. Then, with $dm = \sigma \, dA = (M/s^2) \, dx \, dy$ and $r^2 = x^2 + y^2$, we have

$$I_{33} = \frac{M}{s^2} \int_{-s/2}^{s/2} \int_{-s/2}^{s/2} (x^2 + y^2) \, dx \, dy$$

$$= \frac{M}{s^2} \int_{-s/2}^{s/2} \left(\frac{x^3}{3} + xy^2 \right) \Big|_{-s/2}^{s/2} \, dy$$

$$= \frac{M}{s^2} \int_{-s/2}^{s/2} \left(\frac{s^3}{12} + sy^2 \right) \, dy$$

$$= \frac{M}{s^2} \left(\frac{s^3}{12} y + s \frac{y^3}{3} \right) \Big|_{-s/2}^{s/2}$$

$$= \frac{M}{s^2} \frac{s^4}{6} = \frac{Ms^2}{6}$$

The other components are

$$I_{11} = \frac{M}{s^2} \int_{-s/2}^{s/2} \int_{-s/2}^{s/2} (x^2 + y^2 - x^2) \, dx \, dy$$

$$= \frac{M}{s^2} \frac{y^3}{3} \Big|_{-s/2}^{s/2} x \Big|_{-s/2}^{s/2} = \frac{Ms^2}{12}$$

and

$$I_{12} = \frac{M}{s^2} \int_{-s/2}^{s/2} \int_{-s/2}^{s/2} -xy\, dx\, dy = -\frac{M}{s^2} \frac{x^2}{2}\Big|_{-s/2}^{s/2} \frac{y^2}{2}\Big|_{-s/2}^{s/2} = 0$$

Similar calculations give $I_{22} = I_{11}$ and $I_{21} = I_{12}$. Thus, in this system,

$$I_{ij} = \frac{Ms^2}{12} \begin{pmatrix} 1 & 0 & 0 \\ 0 & 1 & 0 \\ 0 & 0 & 2 \end{pmatrix}$$

Then, when the system rotates about the z-axis,

$$L_i = I_{ij}\omega_j = \frac{Ms^2}{12} \begin{pmatrix} 1 & 0 & 0 \\ 0 & 1 & 0 \\ 0 & 0 & 2 \end{pmatrix} \begin{pmatrix} 0 \\ 0 \\ \omega \end{pmatrix} = \frac{Ms^2}{6} \begin{pmatrix} 0 \\ 0 \\ \omega \end{pmatrix}$$

Thus, $\vec{L}$ is parallel to $\vec{\omega}$ in this case.

Now we change to a coordinate system rotated by 45° about the x-axis (Figure A.1). The transformation matrix is

$$\mathbb{A} = \frac{1}{\sqrt{2}} \begin{pmatrix} \sqrt{2} & 0 & 0 \\ 0 & 1 & 1 \\ 0 & -1 & 1 \end{pmatrix}$$

The new components of the inertia tensor are found from the transformation rule (A.3):

$$I'_{ij} = A_{in} I_{nm} A^T_{mj}$$

$$= \frac{1}{2} \begin{pmatrix} \sqrt{2} & 0 & 0 \\ 0 & 1 & 1 \\ 0 & -1 & 1 \end{pmatrix} \frac{Ms^2}{12} \begin{pmatrix} 1 & 0 & 0 \\ 0 & 1 & 0 \\ 0 & 0 & 2 \end{pmatrix} \begin{pmatrix} \sqrt{2} & 0 & 0 \\ 0 & 1 & -1 \\ 0 & 1 & 1 \end{pmatrix}$$

$$= \frac{Ms^2}{24} \begin{pmatrix} \sqrt{2} & 0 & 0 \\ 0 & 1 & 1 \\ 0 & -1 & 1 \end{pmatrix} \begin{pmatrix} \sqrt{2} & 0 & 0 \\ 0 & 1 & -1 \\ 0 & 2 & 2 \end{pmatrix}$$

$$= \frac{Ms^2}{24} \begin{pmatrix} 2 & 0 & 0 \\ 0 & 3 & 1 \\ 0 & 1 & 3 \end{pmatrix}$$

In the new coordinate system, the angular velocity is along the z'-axis and has only one component, so the angular momentum is

$$L'_i = I'_{ij}\omega'_j = \frac{Ms^2}{24} \begin{pmatrix} 2 & 0 & 0 \\ 0 & 3 & 1 \\ 0 & 1 & 3 \end{pmatrix} \begin{pmatrix} 0 \\ 0 \\ \omega \end{pmatrix} = \frac{Ms^2\omega}{24} \begin{pmatrix} 0 \\ 1 \\ 3 \end{pmatrix}$$

and is *not* parallel to $\vec{\omega}$.

The angular velocity has components $\omega(\sqrt{2}/2)(0, -1, 1)$ in the original coordinate system. We may check our result by computing the components of $\vec{L}$ in this frame and transforming the result into the prime frame.

$$L_i = \frac{Ms^2}{12} \begin{pmatrix} 1 & 0 & 0 \\ 0 & 1 & 0 \\ 0 & 0 & 2 \end{pmatrix} \omega \frac{\sqrt{2}}{2} \begin{pmatrix} 0 \\ -1 \\ 1 \end{pmatrix} = Ms^2\omega \frac{\sqrt{2}}{24} \begin{pmatrix} 0 \\ -1 \\ 2 \end{pmatrix}$$

Transforming the resulting vector to the new frame, we have

$$L'_i = A_{ij}L_j = Ms^2\omega \frac{\sqrt{2}}{24} \frac{1}{\sqrt{2}} \begin{pmatrix} \sqrt{2} & 0 & 0 \\ 0 & 1 & 1 \\ 0 & -1 & 1 \end{pmatrix} \begin{pmatrix} 0 \\ -1 \\ 2 \end{pmatrix} = \frac{Ms^2\omega}{24} \begin{pmatrix} 0 \\ 1 \\ 3 \end{pmatrix}$$

which agrees with the result from the transformed inertia tensor.

A.2. INNER AND OUTER PRODUCTS

The components of the inertia tensor (A.6) are made up of components of the position vectors of all the elements of the body. Such definitions are common. In fact, the products $a_i b_j$ of vector components are the components of a tensor called an *outer product* of the two vectors $\vec{a}$ and $\vec{b}$. All such products are tensors. In an outer product, the rank of the resulting tensor is the sum of the ranks of the tensors making up the product.

When we sum over two or more indices, as in the dot product $a_i b_i$ or in the product of a tensor and a vector to give another vector $I_{ij}\omega_j$, the result is an *inner product*. All inner products are also tensors. The process of reducing the rank of a tensor by summing over a pair of indices is called *contraction*. Every contraction reduces the rank of the tensor by two.

It is fairly straightforward to show that an inner or outer product of two tensors obeys the transformation law for a tensor of the appropriate rank.[6] In fact, we may prove a yet more powerful result called the *quotient theorem.*

If a tensor b is the inner or outer product of c and d and if c is a tensor with arbitrary components, then the set of components d is also a tensor.

For example, let

$$b_{ij} = c_{ik}d_{kj}$$

Then if b is a tensor, we can transform the components using equation (A.3):

$$b'_{nm} = A_{ni}A_{mj}b_{ij} = A_{ni}A_{mj}c_{ik}d_{kj}$$

[6]See Problems 4 and 5.

But from the definition of b in the prime frame and the fact that c is also a tensor,

$$b'_{nm} = c'_{np}d'_{pm} = A_{ni}A_{pk}c_{ik}d'_{pm}$$

Setting the two expressions for b'_{nm} equal, we have

$$A_{ni}A_{mj}c_{ik}d_{kj} = A_{ni}A_{pk}c_{ik}d'_{pm}$$

Now we multiply[7] on the left by $A^{-1}_{qn} = A_{nq}$ and use the result $A^{-1}_{qn}A_{ni} = \delta_{qi}$ to obtain

$$A_{mj}c_{qk}d_{kj} = A_{pk}c_{qk}d'_{pm}$$

Next we multiply on the left by $A^{-1}_{rm} = A_{mr}$:

$$(\delta_{rj}d_{kj} - A_{mr}A_{pk}d'_{pm})c_{qk} = 0$$
$$(d_{kr} - A_{mr}A_{pk}d'_{pm})c_{qk} = 0$$

Because the components c_{qk} are arbitrary, the term in parentheses must equal zero, and so we must have

$$d_{kr} = A_{pk}A_{mr}d'_{pm} = A^{-1}_{kp}A^{-1}_{rm}d'_{pm}$$

Thus, the components of d transform as a tensor, as required.

A similar proof works for tensors of any rank.

A.3. PSEUDO-TENSORS AND CROSS PRODUCTS

In Chapter 1, we saw that the vector cross product is not a true vector because it does not possess the proper behavior under reflection of the coordinate axes. Also recall that we may express the cross product using the Levi-Civita symbol (see Chapter 1, equation 1.30):

$$(\vec{\mathbf{u}} \times \vec{\mathbf{v}})_i = \varepsilon_{ijk}u_j v_k$$

Since $\vec{\mathbf{u}}$ and $\vec{\mathbf{v}}$ are both vectors yet $\vec{\mathbf{u}} \times \vec{\mathbf{v}}$ is not, the symbol ε_{ijk} cannot be a tensor. It represents a pseudo-tensor, or a *tensor density*. The transformation law for a tensor density is similar to that for a tensor (Table A.2) except that we must also multiply by the determinant of the transformation matrix.[8] Thus, for the Levi-Civita tensor density,

$$\varepsilon'_{ijk} = \det{(\mathbb{A})}A_{ip}A_{jq}A_{kr}\varepsilon_{pqr}$$

From this result and the fact that $\det{(\mathbb{A})} = \pm 1$, we may easily see that

1. The product of two tensor densities is a tensor.
2. The product of a tensor and a tensor density is another tensor density.

---

[7] We cannot divide out the "factors" A_{ni} and c_{ik} because we are summing over i and k.
[8] See Problem 11 and Appendix I.

We may avoid the difficulties associated with the cross product if instead we represent the components of the cross product as components of an antisymmetric tensor. A rank-two antisymmetric tensor has only three independent components, and these will be the components of the cross product $\vec{u} \times \vec{v}$ if we choose

$$T_{ij} = u_i v_j - v_i u_j$$

In matrix form, the components are

$$\mathbb{T} = \begin{pmatrix} 0 & u_1 v_2 - u_2 v_1 & u_1 v_3 - u_3 v_1 \\ u_2 v_1 - u_1 v_2 & 0 & u_2 v_3 - u_3 v_2 \\ u_3 v_1 - u_1 v_3 & u_3 v_2 - u_2 v_3 & 0 \end{pmatrix}$$

$$= \begin{pmatrix} 0 & (\vec{u} \times \vec{v})_3 & -(\vec{u} \times \vec{v})_2 \\ -(\vec{u} \times \vec{v})_3 & 0 & (\vec{u} \times \vec{v})_1 \\ (\vec{u} \times \vec{v})_2 & -(\vec{u} \times \vec{v})_1 & 0 \end{pmatrix}$$

Then we may also define the vector density:

$$d_i = \frac{1}{2} \varepsilon_{ijk} T_{jk} \tag{A.7}$$

which is known as the *dual* of the tensor T_{ij}. The components of d_i are

$$d_1 = \frac{1}{2}(T_{23} - T_{32}) = T_{23} = u_2 v_3 - u_3 v_2 = (\vec{u} \times \vec{v})_1$$

$$d_2 = \frac{1}{2}(T_{31} - T_{13}) = T_{31} = u_3 v_1 - u_1 v_3 = (\vec{u} \times \vec{v})_2$$

and similarly for d_3. Thus, the vector density $\vec{d}$ is the cross product $\vec{u} \times \vec{v}$.

The inverse dual relationship is

$$T_{ij} = \varepsilon_{ijk} d_k \tag{A.8}$$

We may demonstrate this relationship using equation (1.34) from Chapter 1:

$$T_{ij} = \varepsilon_{ijk} \frac{1}{2} \varepsilon_{klm} T_{lm} = \frac{1}{2}(\delta_{il}\delta_{jm} - \delta_{im}\delta_{jl}) T_{lm}$$

$$= \frac{1}{2}(T_{ij} - T_{ji})$$

Since T_{ij} is antisymmetric, the result follows.

Example A.2. Express the angular velocity and angular momentum of a particle in circular motion as tensors.

We choose to put the origin at the center of the circle. The angular velocity of the particle is given by

$$\vec{\mathbf{v}} = \vec{\boldsymbol{\omega}} \times \vec{\mathbf{r}}$$

Since $\vec{\mathbf{v}}$ and $\vec{\mathbf{r}}$ are vectors, $\vec{\boldsymbol{\omega}}$ is a pseudo-vector. We can find an expression for $\vec{\boldsymbol{\omega}}$ by taking another cross product and using the fact that $\vec{\boldsymbol{\omega}}$ is perpendicular to $\vec{\mathbf{r}}$:

$$\vec{\mathbf{r}} \times \vec{\mathbf{v}} = \vec{\mathbf{r}} \times (\vec{\boldsymbol{\omega}} \times \vec{\mathbf{r}}) = \vec{\boldsymbol{\omega}} r^2$$

Thus,

$$\vec{\boldsymbol{\omega}} = \frac{1}{r^2} \vec{\mathbf{r}} \times \vec{\mathbf{v}}$$

We may express these components using the antisymmetric tensor:

$$\omega_{ij} = \frac{1}{r^2}(r_i v_j - v_i r_j) \equiv \frac{1}{r^2}\Omega_{ij}$$

The angular momentum is

$$\vec{\mathbf{L}} = \vec{\mathbf{r}} \times \vec{\mathbf{p}} = m\vec{\mathbf{r}} \times \vec{\mathbf{v}}$$

We may also define an antisymmetric tensor with components equal to the components of $\vec{\mathbf{L}}$:

$$L_{ij} = m(r_i v_j - v_i r_j) = m\Omega_{ij}$$

Thus, returning to the pseudo-vector representation, we can relate the components of $\vec{\mathbf{L}}$ to the components of $\vec{\boldsymbol{\omega}}$ by

$$L_i = mr^2 \omega_i$$

A.4. GENERAL TENSOR CALCULUS

In a non-Cartesian coordinate system, we need to identify two different classes of vector-like entities that obey different transformation laws. First let's write the transformation matrix for vectors in terms of the old (v^i) and new $(\overline{v}^i)$ components. A differential displacement vector $d\vec{\mathbf{x}}$ has components $d\overline{x}^i$ that are functions of the original coordinates x^i. Thus, we may use the usual rules of partial differentiation to express $d\overline{x}^i$ as

$$d\overline{x}^i = \frac{\partial \overline{x}^i}{\partial x^j} dx^j$$

Since all vectors transform the same way, we obtain the transformation law for any vector $\vec{v}$:

$$\overline{v}^i = \frac{\partial \overline{x}^i}{\partial x^j} v^j \equiv A^i_j v^j$$

where the transformation matrix has components

$$A^i_j = \frac{\partial \overline{x}^i}{\partial x^j} \tag{A.9}$$

You should check that this expression is consistent with our previous result (equations 1.11 and 1.12 in Chapter 1) in terms of rotation angles.

Notice that we are now writing the index on the vector component as an upper, rather than a lower, index. The transformation matrix is written with one upper index and one lower index. The lower index corresponds to the vector in the denominator on the right-hand side of equation (A.9). We'll discuss this convention further below.

Now let's look at the gradient of a scalar function Φ. The gradient has components

$$\vec{\nabla}\Phi = \left(\frac{\partial \Phi}{\partial x}, \frac{\partial \Phi}{\partial y}, \frac{\partial \Phi}{\partial z} \right)$$

or

$$\partial_i \Phi \equiv \frac{\partial \Phi}{\partial x^i} \tag{A.10}$$

The lower index on the left-hand side reminds us that the upper index vector component is in the denominator of this expression. To transform the gradient to a new coordinate system (labeled $\overline{x}^i$), we use the chain rule for derivatives. Since Φ is a scalar, $\overline{\Phi} = \Phi$.

$$\overline{\partial}_i \Phi = \frac{\partial \Phi}{\partial \overline{x}^i} = \frac{\partial x^j}{\partial \overline{x}^i} \frac{\partial \Phi}{\partial x^j}$$

and thus the transformation matrix for the gradient has components

$$B_i^{\ j} = \frac{\partial x^j}{\partial \overline{x}^i} \tag{A.11}$$

The gradient operator is written with a lower index to indicate that it transforms with the matrix $\mathbb{B}$. Upper index and lower index quantities transform differently, in general. The upper index quantities are called *contravariant* tensors, while the lower index quantities are called *covariant* tensors, or *forms*. Thus, the differential displacement vector is a contravariant vector (or just a vector), while the gradient is a covariant vector (or one-form). If the geometrical image for a vector is an arrow with a specified length and direction, then the image for a one-form is a set of level surfaces with a given orientation and separation (Figure A.2).

FIGURE A.2. Level surfaces offer a geometric image for a one-form. The dot product $d\vec{s} \cdot \vec{\nabla}\Phi = d\Phi$. When the vector $d\vec{s}$ is laid across the level surfaces representing $\vec{\nabla}\Phi$, $d\Phi$ is the difference in the value of Φ between the head and the tail of the vector.

In the case of Cartesian coordinates, where the allowed set of transformations is rotations, these distinctions are not necessary, because the two transformation matrices are the same. But let's look at what happens when we transform from Cartesian coordinates to cylindrical coordinates. The relations between the coordinates are

$$x = \rho \cos\phi; \quad y = \rho \sin\phi; \quad z = \bar{z} \tag{A.12}$$

and the inverse relations are

$$\rho = \sqrt{x^2 + y^2}; \quad \phi = \tan^{-1}\frac{y}{x}; \quad \bar{z} = z \tag{A.13}$$

We may differentiate $\tan\phi$ to obtain an expression for $\partial\phi/\partial x$,

$$\tan\phi = \frac{y}{x} \implies \sec^2\phi\frac{\partial\phi}{\partial x} = -\frac{y}{x^2} \implies \frac{\partial\phi}{\partial x} = -\frac{\tan\phi}{\rho\cos\phi}\cos^2\phi = -\frac{\sin\phi}{\rho}$$

and hence find the components of the transformation matrices.

$$A^1_1 = \frac{\partial\rho}{\partial x} = \frac{x}{\rho} = \cos\phi; \quad A^1_2 = \frac{\partial\rho}{\partial y} = \sin\phi; \quad A^1_3 = \frac{\partial\rho}{\partial z} = 0$$

$$A^2_1 = \frac{\partial\phi}{\partial x} = -\frac{\sin\phi}{\rho}; \quad A^2_2 = \frac{\partial\phi}{\partial y} = \frac{\cos\phi}{\rho}; \quad A^2_3 = \frac{\partial\phi}{\partial \bar{z}} = 0$$

and

$$A^3_1 = \frac{\partial\bar{z}}{\partial x} = 0; \quad A^3_2 = \frac{\partial\bar{z}}{\partial y} = 0; \quad A^3_3 = \frac{\partial\bar{z}}{\partial z} = 1$$

while

$$B_1^1 = \frac{\partial x}{\partial\rho} = \cos\phi; \quad B_1^2 = \frac{\partial y}{\partial\rho} = \sin\phi; \quad B_1^3 = \frac{\partial z}{\partial\rho} = 0$$

$$B_2^1 = \frac{\partial x}{\partial\phi} = -\rho\sin\phi; \quad B_2^2 = \frac{\partial y}{\partial\phi} = \rho\cos\phi; \quad B_2^3 = \frac{\partial z}{\partial\phi} = 0$$

and

$$B_3^1 = \frac{\partial x}{\partial\bar{z}} = 0; \quad B_3^2 = \frac{\partial y}{\partial\bar{z}} = 0; \quad B_3^3 = \frac{\partial z}{\partial\bar{z}} = 1$$

Thus, the two matrices[9] are

$$A^i_{\ j} = \begin{pmatrix} \cos\phi & \sin\phi & 0 \\ -\sin\phi/\rho & \cos\phi/\rho & 0 \\ 0 & 0 & 1 \end{pmatrix} \tag{A.14}$$

and

$$B_i^{\ j} = \begin{pmatrix} \cos\phi & \sin\phi & 0 \\ -\rho\sin\phi & \rho\cos\phi & 0 \\ 0 & 0 & 1 \end{pmatrix} \tag{A.15}$$

which are clearly different. In fact,

$$\boxed{\mathbb{B}^{\mathrm{T}} = \mathbb{A}^{-1}} \tag{A.16}$$

Let's verify this result:

$$\mathbb{B}^{\mathrm{T}}\mathbb{A} = \begin{pmatrix} \cos\phi & -\rho\sin\phi & 0 \\ \sin\phi & \rho\cos\phi & 0 \\ 0 & 0 & 1 \end{pmatrix} \begin{pmatrix} \cos\phi & \sin\phi & 0 \\ -\sin\phi/\rho & \cos\phi/\rho & 0 \\ 0 & 0 & 1 \end{pmatrix}$$

$$= \begin{pmatrix} \cos^2\phi + \sin^2\phi & \cos\phi\sin\phi - \sin\phi\cos\phi & 0 \\ \sin\phi\cos\phi - \cos\phi\sin\phi & \cos^2\phi + \sin^2\phi & 0 \\ 0 & 0 & 1 \end{pmatrix} = \begin{pmatrix} 1 & 0 & 0 \\ 0 & 1 & 0 \\ 0 & 0 & 1 \end{pmatrix}$$

Relation (A.16) is true in general. (See Problem 15.)

Since the components of the transformation matrix are not constants, in general, but functions of position, we cannot add vectors defined at different points of space because the sum would not have a well-defined transformation law. For example, we cannot subtract velocity vectors at two different points of space to determine an average acceleration. A finite displacement $\vec{D}_{AB}$ from point A to point B is no longer a true vector. However, operations such as adding the electric field contributions $\vec{E}_1$ and $\vec{E}_2$ at a point P due to two different charge distributions to give the total electric field $\vec{E} = \vec{E}_1 + \vec{E}_2$ remain valid.

Example A.3.   The electric field due to a long straight wire may be written in Cartesian coordinates as

$$\vec{E} = \frac{\lambda}{2\pi\varepsilon_0}\left(\frac{x}{x^2 + y^2}, \frac{y}{x^2 + y^2}, 0\right)$$

Transform the electric field vector into the cylindrical coordinate system.

---

[9]Here, i labels the rows, and j labels the columns. Even though one index is up and one is down, we shall still write one index to the left of the other so that we can maintain the conventions introduced in Chapter 1 for labeling rows and columns.

The electric field is a contravariant vector. Thus, in the new coordinate system, the components are

$$\vec{E}' = \mathbb{A}\vec{E} = \begin{pmatrix} \cos\phi & \sin\phi & 0 \\ -\sin\phi/\rho & \cos\phi/\rho & 0 \\ 0 & 0 & 1 \end{pmatrix} \frac{\lambda}{2\pi\varepsilon_0\rho^2} \begin{pmatrix} x \\ y \\ 0 \end{pmatrix}$$

$$= \frac{\lambda}{2\pi\varepsilon_0\rho^2} \begin{pmatrix} x\cos\phi + y\sin\phi \\ (-x\sin\phi + y\cos\phi)/\rho \\ 0 \end{pmatrix}$$

$$= \frac{\lambda}{2\pi\varepsilon_0\rho^2} \begin{pmatrix} \rho\cos^2\phi + \rho\sin^2\phi \\ (-\rho\cos\phi\sin\phi + \rho\sin\phi\cos\phi)/\rho \\ 0 \end{pmatrix}$$

$$= \frac{\lambda}{2\pi\varepsilon_0\rho^2} \begin{pmatrix} \rho \\ 0 \\ 0 \end{pmatrix} = \frac{\lambda}{2\pi\varepsilon_0\rho} \begin{pmatrix} 1 \\ 0 \\ 0 \end{pmatrix}$$

The electric field has only a radial component. This is the expected result.

A.5. THE METRIC TENSOR

The line element (Chapter 1, Sections 1.1.1 and 1.3.2) is an important example of a scalar that is formed as an inner product of a tensor, called the *metric tensor*, and the differential vector $d\vec{x}$:

$$ds^2 = g_{ij}\,dx^i dx^j \tag{A.17}$$

Since the left-hand side is a scalar and the vector $d\vec{x}$ is a vector whose components may take on any values, we may apply the quotient theorem to conclude that g_{ij} is a tensor. In Cartesian coordinates, $g_{ij} = \delta_{ij}$. In any orthogonal coordinate system, g_{ij} is diagonal. For example, in the cylindrical coordinate system (equation 1.4),

$$ds^2 = d\rho^2 + \rho^2\,d\phi^2 + dz^2$$

and so

$$g_{ij} = \begin{pmatrix} 1 & 0 & 0 \\ 0 & \rho^2 & 0 \\ 0 & 0 & 1 \end{pmatrix} \tag{A.18}$$

We may obtain these components from the Cartesian components by applying the transformation law in the usual way. The lower indices imply that g_{ij} is a covariant tensor, and so we transform with the matrix $\mathbb{B}$:

$$g' = \mathbb{B}g\mathbb{B}^\mathsf{T} = \begin{pmatrix} \cos\phi & \sin\phi & 0 \\ -\rho\sin\phi & \rho\cos\phi & 0 \\ 0 & 0 & 1 \end{pmatrix} \begin{pmatrix} 1 & 0 & 0 \\ 0 & 1 & 0 \\ 0 & 0 & 1 \end{pmatrix} \begin{pmatrix} \cos\phi & -\rho\sin\phi & 0 \\ \sin\phi & \rho\cos\phi & 0 \\ 0 & 0 & 1 \end{pmatrix}$$

$$= \begin{pmatrix} \cos\phi & \sin\phi & 0 \\ -\rho\sin\phi & \rho\cos\phi & 0 \\ 0 & 0 & 1 \end{pmatrix} \begin{pmatrix} \cos\phi & -\rho\sin\phi & 0 \\ \sin\phi & \rho\cos\phi & 0 \\ 0 & 0 & 1 \end{pmatrix}$$

$$= \begin{pmatrix} \cos^2\phi + \sin^2\phi & 0 & 0 \\ 0 & \rho^2\sin^2\phi + \rho^2\cos^2\phi & 0 \\ 0 & 0 & 1 \end{pmatrix} = \begin{pmatrix} 1 & 0 & 0 \\ 0 & \rho^2 & 0 \\ 0 & 0 & 1 \end{pmatrix}$$

in agreement with the result above (equation A.18).

A.6. CONTRACTION

To form an inner product of two tensors of ranks m and n, we must contract the tensors by summing over (at least) two of the indices. The result is a tensor of lower order (by two for each pair of summed indices) than the sum of the ranks of the two tensors in the product, and thus it must transform as a tensor of rank $m + n - 2$. As an example, let's consider the contraction of two vectors to form a dot product—a scalar. If we try to contract by summing over two upper indices, we find that the result is not a scalar.

$$\overline{v}^i \overline{u}^i = A^i{}_j v^j A^i{}_k u^k$$

But the product $A^i{}_j A^i{}_k \neq \delta_{jk}$ except for rotations of a Cartesian coordinate system, and so this product is not a scalar in general. However, if we contract an upper index with a lower index, then

$$\overline{v}^i \overline{u}_i = A^i{}_j v^j B_i{}^k u_k$$

and the product $A^i{}_j B_i{}^k = \delta^k_j$ (equation A.16), so

$$\overline{v}^i \overline{u}_i = v^k u_k$$

and this inner product is a scalar. Thus, we have the following rule:

In general tensor algebra, an upper index may be contracted with a lower index, but not with another upper index.

In order to form an inner product of two vectors, we must first form the lower-index covariant vector that corresponds to the upper-index contravariant vector. The line element (equation A.17) shows us how to do this. Writing the line element as a dot product of $d\vec{\mathbf{x}}$ with itself,

$$dx^i \, dx_i = ds^2 = g_{ij} \, dx^i \, dx^j$$

we observe that

$$dx_i = g_{ij} \, dx^j$$

In this way, the metric tensor maps any contravariant vector to a corresponding covariant vector (or one-form):

$$v_i = g_{ij} v^j \qquad\qquad\text{(A.19)}$$

Similarly, the contravariant metric tensor g^{ij}, whose components form the inverse of the matrix of components g_{ij}, maps a covariant vector to a contravariant vector:

$$v^i = g^{ij} v_j \qquad\qquad\text{(A.20)}$$

Sometimes we say that the metric tensor g^{ij} is used to *raise* an index, while the tensor g_{ij} is used to *lower* an index.

These expressions show that there is a covariant vector corresponding to every contravariant vector, and vice versa. The metric tensor is an isomorphism that maps one space to the other, its "dual" space.

When an inner product is formed, it does not matter which index of a contracted pair is up and which is down. For example,

$$u^i v_i = u^i g_{ij} v^j = u_j v^j$$

A.7. BASIS VECTORS AND BASIS FORMS

The basis vectors in any coordinate system are defined to have components $(1, 0, 0)$, $(0, 1, 0)$, and $(0, 0, 1)$. However, in a non-Cartesian coordinate system, the basis vectors are not necessarily *unit* vectors. We may find the magnitude of a vector from its dot product with itself:

$$|\vec{v}| = \sqrt{v^i v_i}$$

Thus, for the nth basis vector, we have

$$|\vec{e}_{(n)}| = \sqrt{e^i_{(n)} g_{ij} e^j_{(n)}}$$

where (n) labels the basis vector and i and j label the components. But

$$e^i_{(n)} = \delta^i_n$$

and so

$$\boxed{|\vec{e}_{(n)}| = \sqrt{\delta^i_n g_{ij} \delta^j_n} = \sqrt{g_{nn}}} \qquad (A.21)$$

where in these expressions we do *not* sum[10] over n. Thus, in cylindrical coordinates, the second basis vector has magnitude $\sqrt{g_{22}} = \rho$, and the basis vectors in this system are $\hat{\rho}$, $\rho\hat{\phi}$, and $\hat{z}$.

The basis forms are defined similarly. So, for example, the first basis form has components $(1, 0, 0)$, and its magnitude is $\sqrt{g^{11}}$.

Example A.4.   Starting from the components of the velocity vector in Cartesian coordinates, transform to cylindrical coordinates to find the components of $\vec{v}$ in the new system, and hence write the velocity vector in cylindrical coordinates.

First we transform the velocity vector:

$$\bar{v}^i = A^i_j v^j = A^i_j \frac{\partial x^j}{\partial t} = \begin{pmatrix} \cos\phi & \sin\phi & 0 \\ -\dfrac{\sin\phi}{\rho} & \dfrac{\cos\phi}{\rho} & 0 \\ 0 & 0 & 1 \end{pmatrix} \begin{pmatrix} \dfrac{\partial x}{\partial t} \\ \dfrac{\partial y}{\partial t} \\ \dfrac{\partial z}{\partial t} \end{pmatrix}$$

$$= \begin{pmatrix} \dfrac{\partial x}{\partial t}\cos\phi + \dfrac{\partial y}{\partial t}\sin\phi \\ -\dfrac{\partial x}{\partial t}\dfrac{\sin\phi}{\rho} + \dfrac{\partial y}{\partial t}\dfrac{\cos\phi}{\rho} \\ \dfrac{\partial z}{\partial t} \end{pmatrix}$$

---

[10]There is no inconsistency here: n is a label, not an index.

Next we write the derivatives in terms of the new coordinates, using equations (A.12):

$$\bar{v}^i = \begin{pmatrix} \cos\phi\left(\dfrac{\partial\rho}{\partial t}\cos\phi - \rho\sin\phi\dfrac{\partial\phi}{\partial t}\right) + \sin\phi\left(\dfrac{\partial\rho}{\partial t}\sin\phi + \rho\cos\phi\dfrac{\partial\phi}{\partial t}\right) \\[2mm] -\dfrac{\sin\phi}{\rho}\left(\dfrac{\partial\rho}{\partial t}\cos\phi - \rho\sin\phi\dfrac{\partial\phi}{\partial t}\right) + \dfrac{\cos\phi}{\rho}\left(\dfrac{\partial\rho}{\partial t}\sin\phi + \rho\cos\phi\dfrac{\partial\phi}{\partial t}\right) \\[2mm] \dfrac{\partial z}{\partial t} \end{pmatrix}$$

$$= \begin{pmatrix} \dfrac{\partial\rho}{\partial t} \\[2mm] \dfrac{\partial\phi}{\partial t} \\[2mm] \dfrac{\partial z}{\partial t} \end{pmatrix} = \dfrac{\partial \bar{x}^i}{\partial t}$$

as expected from the definition. To write the vector, we must multiply each component by the corresponding basis vector. Then

$$\vec{v} = \frac{\partial\rho}{\partial t}\hat{\mathbf{e}}_{(1)} + \frac{\partial\phi}{\partial t}\hat{\mathbf{e}}_{(2)} + \frac{\partial z}{\partial t}\hat{\mathbf{e}}_{(3)} = \frac{\partial\rho}{\partial t}\hat{\boldsymbol{\rho}} + \frac{\partial\phi}{\partial t}\rho\hat{\boldsymbol{\phi}} + \frac{\partial z}{\partial t}\hat{\mathbf{z}}$$

which is the standard result.

A.8. DERIVATIVES

We have already shown that the gradient of a scalar is a covariant vector. We must be more careful when taking the derivative of tensors of rank one or higher. For the moment, let's discuss vectors. In taking the derivative, we must compare the values of the vector components at neighboring points. But if the coordinates are not Cartesian, then even if the vector is unchanged (its magnitude and direction remain the same), its components at the two neighboring points are not the same (see Figure A.3). Thus, before we can subtract

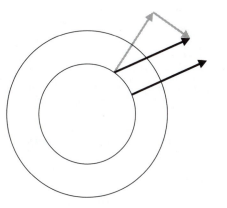

FIGURE A.3. Parallel displacement. When the vector is moved in a non-Cartesian system, its components change even if the magnitude and direction do not. Here a vector that is in the r direction at one point has both r and θ components when moved to a second nearby point.

the two vectors, we must first displace the vector parallel to itself and evaluate its new components. This process is called parallel displacement.

It is customary to start with a covariant vector. Since we are considering differential displacements, the change in the component v_i under parallel displacement is proportional to the displacement. It should also be proportional to the vector components,[11] so we demand that

$$\delta v_i = \Gamma^j_{ik} v_j \, dx^k \tag{A.22}$$

where the components Γ^j_{ik} are called an *affinity* that specifies the *affine connection* of the space. These components are not a tensor. Now, since the dot product $u^i v_i$ is a scalar, we require that it not change under parallel displacement,[12] $\delta(u^i v_i) = 0$. Thus,

$$v_i \, \delta u^i + u^i \, \delta v_i = 0 = v_i \, \delta u^i + u^i \Gamma^j_{ik} v_j \, dx^k$$

Since this result must be true for any covariant vector v_j, we may conclude that

$$\delta u^i = -\Gamma^i_{jk} u^j \, dx^k \tag{A.23}$$

Now if the change du^i in a vector's components is equal to the change δu^i due to parallel displacement, then the vector has not actually changed, and so its derivative is zero. Thus, we define the *covariant derivative* $u^i_{;j}$ in terms of the difference between the actual change in components and the change (A.23) due to parallel transport:

$$du^i - \delta u^i = u^i_{;j} \, dx^j \tag{A.24}$$

And so we obtain an expression for the covariant derivative in terms of the affinity:

$$u^i_{;j} = \frac{\partial u^i}{\partial x^j} + \Gamma^i_{kj} u^k \tag{A.25}$$

Working from equation (A.22), we obtain the result[13]

$$u_{i;j} = \frac{\partial u_i}{\partial x^j} - \Gamma^k_{ij} u_k \tag{A.26}$$

Similarly, by looking at invariant combinations, we may write the covariant derivative of any rank tensor. It turns out that we need one term in the affinity Γ for each index of the

---

[11]This is actually a consequence of requiring that the parallel displacement in general be consistent with the familiar definition in Euclidean space. See Lawden, p. 98.

[12]This is clearly true in Euclidean space. Once again we are requiring that the new ideas be consistent with the old. This is a constraint on the kinds of space we are willing to consider as physically *interesting*.

[13]See also Problem 28.

tensor. It will be added, as in equation (A.25), for an upper index, and subtracted, as in equation (A.26), for a lower index.

The covariant derivative of a vector (equation A.25) is a rank-two tensor. This result follows from the definition (A.24), since the quantity on the left-hand side is the difference between two vectors defined at the same point of space and thus is also a vector. Since the components of dx^j are arbitrary, we may apply the quotient theorem to show that the covariant derivative is a tensor. Similarly, the covariant derivative of any tensor is also a tensor.

It remains to find the components of the affinity.[14] The result we obtain should be valid in any space, including ordinary Euclidean space. But if the space is Euclidean,[15] we may set up a Cartesian coordinate system $\bar{x}^i$ at the original point and compute the Cartesian components $\bar{u}_i$ of the covariant vector u_i. These components do not change under parallel displacement: $\delta \bar{u}_i = 0$. We may obtain the components u_i from $\bar{u}_i$ using the usual transformation matrix:

$$u_i = \frac{\partial \bar{x}^j}{\partial x^i} \bar{u}_j$$

Then

$$\delta u_i = \delta \left(\frac{\partial \bar{x}^j}{\partial x^i} \right) \bar{u}_j + \frac{\partial \bar{x}^j}{\partial x^i} \delta \bar{u}_j$$

$$= \frac{\partial^2 \bar{x}^j}{\partial x^i \partial x^k} dx^k \bar{u}_j + 0$$

$$= \frac{\partial^2 \bar{x}^j}{\partial x^i \partial x^k} dx^k \frac{\partial x^m}{\partial \bar{x}^j} u_m$$

and thus, comparing with equation (A.22), we find

$$\Gamma^m_{ik} = \frac{\partial^2 \bar{x}^j}{\partial x^i \partial x^k} \frac{\partial x^m}{\partial \bar{x}^j} \tag{A.27}$$

This expression shows that the affinity is symmetric in its lower two indices. However, it is not very convenient for calculating Γ.

To obtain a more convenient expression, note that we can obtain the covariant derivative $u_{i;j}$ in two ways:

1. Take the covariant derivative $u^i_{;j}$ and then lower indices.
   or
2. Lower indices and then take the derivative.

---

[14] In their Section 1.5, Morse and Feshbach give a different derivation of this result and provide an expression for the affinity in terms of the metric coefficients h_i, discussed in Chapter 1, Section 1.3.2.

[15] This requirement is more restrictive than necessary, but allows us to present a more transparent argument. In fact, even in non-Euclidean space we can set up a *local* Cartesian system, provided that $\det(g) \neq 0$, and then show that the parallel displacement of the components in this system is zero to first order in the displacement $d\mathbf{x}$.

These two methods must give the same result, and so

$$g_{im}u^m_{;j} = u_{i;j} = (g_{ik}u^k)_{;j} = g_{ik;j}u^k + g_{ik}u^k_{;j}$$

In the second step, we assumed that the product rule applies to the covariant derivative in the same way as for ordinary derivatives.[16] Then it must be true that

$$g_{ik;j} \equiv 0 \tag{A.28}$$

That is, the covariant derivative of the metric tensor is zero, a pleasing and intuitive result. This requirement actually determines the Γ. We write it explicitly as

$$\frac{\partial g_{ik}}{\partial x^j} - \Gamma^m_{kj}g_{im} - \Gamma^m_{ij}g_{mk} = 0$$

We may permute the indices in this expression to obtain two more similar expressions:

$$\frac{\partial g_{ji}}{\partial x^k} - \Gamma^m_{ik}g_{jm} - \Gamma^m_{jk}g_{mi} = 0$$

and

$$\frac{\partial g_{kj}}{\partial x^i} - \Gamma^m_{ji}g_{km} - \Gamma^m_{ki}g_{mj} = 0$$

Now we add the first two, subtract the last, and use the symmetry of the Γs and the g_{ij} to get

$$\frac{\partial g_{ik}}{\partial x^j} - 2\Gamma^m_{kj}g_{im} + \frac{\partial g_{ji}}{\partial x^k} - \frac{\partial g_{kj}}{\partial x^i} = 0$$

Thus,

$$\Gamma^m_{kj}g_{im} = \frac{1}{2}\left(\frac{\partial g_{ik}}{\partial x^j} + \frac{\partial g_{ji}}{\partial x^k} - \frac{\partial g_{kj}}{\partial x^i}\right)$$

Multiplying by g^{in}, we find

$$\boxed{\Gamma^n_{kj} = \frac{g^{in}}{2}\left(\frac{\partial g_{ik}}{\partial x^j} + \frac{\partial g_{ji}}{\partial x^k} - \frac{\partial g_{kj}}{\partial x^i}\right)} \tag{A.29}$$

Expression (A.29) is called the metric affinity. It is also called the Christoffel symbol of the second kind.

We now have the mathematical apparatus that we need to describe physical systems in any coordinate frame.

---

[16]See Problem 29.

Example A.5. The covariant divergence of a vector $u^i_{;i}$ is a scalar. The electric field inside a uniform cylinder of charge is $\vec{E} = \rho_0 \vec{\rho}/2\varepsilon_0$, where ρ_0 is the charge density. Evaluate the divergence of this vector field in Cartesian coordinates ($\vec{\rho} = x\hat{x} + y\hat{y}$). Evaluate the necessary components of the metric affinity in cylindrical coordinates, and hence calculate the divergence of the electric field in cylindrical coordinates. Verify that the results are the same.

In Cartesian coordinates, the Γ are zero, so

$$\vec{\nabla} \cdot \vec{E} = E^i_{;i} = \frac{\rho_0}{2\varepsilon_0} \frac{\partial \overline{E}^i}{\partial \overline{x}^i}$$

$$= \frac{\rho_0}{2\varepsilon_0} \left(\frac{\partial x}{\partial x} + \frac{\partial y}{\partial y} \right) = \frac{\rho_0}{2\varepsilon_0} 2 = \frac{\rho_0}{\varepsilon_0}$$

which is the expected result from Gauss' law.

In cylindrical coordinates, the only derivative of the metric tensor components that is not zero is

$$\frac{\partial g_{22}}{\partial x^1} = 2\rho$$

The divergence is

$$E^i_{;i} = \frac{\partial E^i}{\partial x^i} + \Gamma^i_{ki} E^k$$

$$= \frac{\partial E^i}{\partial x^i} + \frac{g^{in}}{2} \left(\frac{\partial g_{nk}}{\partial x^i} + \frac{\partial g_{in}}{\partial x^k} - \frac{\partial g_{ki}}{\partial x^n} \right) E^k$$

where $E^1 = \rho\rho_0/2\varepsilon_0$, $E^2 = 0$, and $E^3 = 0$. Thus, in this system, $\partial E^i/\partial x^i = \rho_0/2\varepsilon_0$. The metric tensor is diagonal, so in the second term we get contributions from the sum over i only when $i = n$. With the sums put in explicitly, the term becomes

$$\Gamma^i_{ki} E^k = \sum_{n,k} \frac{g^{nn}}{2} \left(\frac{\partial g_{nk}}{\partial x^n} + \frac{\partial g_{nn}}{\partial x^k} - \frac{\partial g_{kn}}{\partial x^n} \right) E^k$$

The terms in g_{nk} are nonzero only if $n = k$, but $\partial g_{nn}/\partial x^n$ (not summed) is zero for each n. Thus, the only nonzero term is

$$\sum_{n,k} \frac{g^{nn}}{2} \frac{\partial g_{nn}}{\partial x^k} E^k = \frac{g^{22}}{2} \frac{\partial g_{22}}{\partial x^1} E^1 = \frac{1}{2\rho^2} (2\rho) \frac{\rho_0}{2\varepsilon_0} \rho = \frac{\rho_0}{2\varepsilon_0}$$

and so

$$\vec{\nabla} \cdot \vec{E} = E^i_{;i} = \frac{\partial E^i}{\partial x^i} + \Gamma^i_{ki} E^k = \frac{\rho_0}{2\varepsilon_0} + \frac{\rho_0}{2\varepsilon_0} = \frac{\rho_0}{\varepsilon_0}$$

as we found in Cartesian coordinates.

Although we have used flat space in our examples here, the maximum benefit of general tensor mathematics accrues in applications such as special and general relativity, where the metric cannot be reduced to the simple form δ_{ij} with $+1$ along the diagonal. Some examples are given in the problem set.

PROBLEMS

1. Determine the velocity of an electron driven by an electric field $\vec{E} = \vec{E}_0 e^{-i\omega t}$ in the presence of a constant uniform magnetic field $\vec{B}_0$. Choose the z-axis along $\vec{B}_0$, but do not make any assumptions about the direction of $\vec{E}_0$. If there are n electrons per unit volume, write the current density in the form $j_i = \sigma_{ij} E_j$ and determine the components of σ_{ij}.

2. Starting with the expression $\vec{v} = \vec{\omega} \times \vec{r}$ for a particle in circular motion, derive the expression (A.6) given in Example A.1 for the inertia tensor.

3. Compute the inertia tensor (equation A.6) for a uniform cylinder of radius R and height h. Hence find its angular momentum when it rotates with angular speed ω about

   (a) an axis through its center and along its length

   (b) an axis through its center and along a diameter

   (c) an axis through its center making a $45°$ angle with each of the axes in (a) and (b)

4. Show that the outer product $a_i b_j$ of two vectors obeys the transformation law for a rank-two tensor.

5. Show that the inner product $a_{ijk} b_k$ of a rank-three tensor and a vector obeys the transformation law for a rank-two tensor.

6. Show that the Kroneker delta tensor δ_{ij} has the same components in every coordinate frame.

7. Show that if a tensor b_{ij} is symmetric in one frame (that is, $b_{ij} = b_{ji}$), then it is symmetric in every frame. Similarly, show that the property of antisymmetry ($b_{ij} = -b_{ji}$) is preserved under coordinate transformations.

8. (a) The following set of components is defined in two-dimensional Cartesian space:

$$A_{ij} = \begin{pmatrix} -y^2 & xy \\ xy & -x^2 \end{pmatrix}$$

Does this set of components transform as a tensor? Why or why not? *Hint:* Try to express the components in terms of inner and outer products of known vectors and tensors. Alternatively, form inner or outer products with vectors and use the quotient theorem.

Repeat the problem for the following sets of components:

(b) $B_{ij} = \begin{pmatrix} -xy & x^2 \\ y^2 & -xy \end{pmatrix}$

(c) $C_{ij} = \begin{pmatrix} x^2 & xy \\ xy & y^2 \end{pmatrix}$

(d) In three-dimensional space,

$$D_{ij} = \begin{pmatrix} 0 & z & -y \\ -z & 0 & x \\ y & -x & 0 \end{pmatrix}$$

9. The magnetic moment tensor has components

$$M_{ik} = \int x_i j_k \, dV$$

where $\vec{j}$ is the current density and the integral is over all space.

(a) Show that M_{ik} is antisymmetric for any steady, finite current distribution.

(b) Show that the corresponding cross product (the dual vector, equation A.7) reduces to the usual magnetic moment vector $\vec{m} = I A \hat{n}$ in the case of a planar current loop.

10. The electric quadrupole tensor is given by

$$Q_{ij} = \int (3x_i x_j - r^2 \delta_{ij}) \rho(\vec{x}) \, dV$$

Calculate the quadrupole tensor for a set of four point charges, two of charge q and two of charge $-q$, at the corners of a square of side a. The charges alternate in sign, so charges of equal sign are at opposite ends of the diagonals of the square. Use a coordinate system with $\bar{x}$- and $\bar{y}$-axes along the diagonals of the square.

The force on a quadrupole charge distribution placed in an external electric field is

$$F_i = \frac{1}{6} Q_{jk} \frac{\partial^2 E_j}{\partial x_i \, \partial x_k}$$

where the derivatives are evaluated at the origin. Find the force on the square when it is placed in an external electric field $\vec{E} = \alpha(2xy, -y^2, x^2 + y^2)$. The square's normal lies in the x-z plane at angle θ to the z-axis, and its center is at the origin.

11. Show that the components of the Levi-Civita symbol ε_{ijk} transform as a tensor density under coordinate rotations and reflections.

12. Starting from the components of the velocity vector in Cartesian coordinates, transform to spherical coordinates to find the components of $\vec{v}$ in the new system, and hence write the velocity vector in spherical coordinates. (Refer to Example A.4.)

13. In a region of space, the electric scalar potential has the form $\Phi = -E_0 z$.

(a) Working in Cartesian coordinates, compute the gradient to obtain the electric field components. Transform to a spherical coordinate system, using the appropriate transformation law from Section A.4.

     (b) Write the potential in spherical coordinates, and compute the gradient using the operator ∂_i (equation A.10).

Confirm that both methods give the same electric field.

14. Write the components of the gradient form in **(a)** cylindrical coordinates and **(b)** spherical coordinates. Use the metric tensor to raise indices, thus mapping to the corresponding vector. Finally, multiply by the basis vectors to obtain the conventional expression for $\vec{\nabla}\Phi$, as derived in Chapter 1 (equation 1.62).

15. Use equations (A.9) and (A.11) in Section A.4 to show that the relation $\mathbb{B}^{\mathrm{T}} = \mathbb{A}^{-1}$ is true in general.

16. If the tensor A^{ij} is symmetric, $A^{ij} = A^{ji}$, show that $A^i{}_j = A_j{}^i$ and that $A_{ij} = A_{ji}$. Can you find a relation between $A^i{}_j$ and $A^j{}_i$? Why or why not?

17. Which of the following relations between tensor components could possibly be true? Say what is wrong with the incorrect expressions.

    (a) $V^i = \varepsilon^{ijk} U_k$

    (b) $T^{ij} = X^{ik} Y_{kj}$

    (c) $V^i = X^{ik} U_k + W^i$

    (d) $V^i = \varepsilon^{ijk} U_i W_j X_k Y^i$

18. In special relativity, space-time is described by four-component vectors v^α. (It is conventional to use Greek letters to label the indices 0, 1, 2, and 3 in four-dimensional space-time.) The coordinates of an event in space-time are $x^\alpha = (ct, x, y, z)$, and the metric[17] is

$$g_{\alpha\beta} = \begin{pmatrix} 1 & 0 & 0 & 0 \\ 0 & -1 & 0 & 0 \\ 0 & 0 & -1 & 0 \\ 0 & 0 & 0 & -1 \end{pmatrix}$$

The Lorentz transformation matrix relating two coordinate systems moving with relative speed v along the x-axis is

$$\Lambda^\mu{}_\nu = \begin{pmatrix} \gamma & -\gamma\beta & 0 & 0 \\ -\gamma\beta & \gamma & 0 & 0 \\ 0 & 0 & 1 & 0 \\ 0 & 0 & 0 & 1 \end{pmatrix}$$

where $\gamma = 1/\sqrt{1 - \beta^2}$ and $\beta = v/c$.

    The electromagnetic potential is described by a four-vector with components $A^\mu = (\Phi, A_x, A_y, A_z)$, where Φ is the electric scalar potential and $\vec{A}$ is the magnetic vector potential. The electromagnetic field tensor has components

$$F^{\mu\nu} = \partial^\mu A^\nu - \partial^\nu A^\mu$$

---

[17] This is the sign convention used by Jackson. There are others.

Show that, in the Gaussian unit system, the components of the field tensor in terms of $\vec{E}$ and $\vec{B}$ are

$$F^{\mu\nu} = \begin{pmatrix} 0 & -E_x & -E_y & -E_z \\ E_x & 0 & -B_z & B_y \\ E_y & B_z & 0 & -B_x \\ E_z & -B_y & B_x & 0 \end{pmatrix}$$

Two particles, each with charge q and mass m, are moving along lines parallel to the x-axis and a distance d apart. Each particle moves with speed $v \ll c$. Start in a reference frame in which the two particles are at rest. Compute the components of the field tensor in this reference frame, and hence find the force acting on each particle. Now transform the field tensor to the lab frame, and again compute the force on each particle. Verify your result to first order in the small quantity $\beta = v/c$ by computing $\vec{E}$ and $\vec{B}$ in the lab frame using Coulomb's law and the Biot-Savart law. (In the Biot-Savart law, the source $I\,d\vec{l}$ should be identified with a point source $q\vec{v}$ at the position of one of the charges.)

19. Using the metric of Lorentz space-time and the electromagnetic field tensor (see Problem 18 above), verify that an electromagnetic wave has the same field structure ($\vec{E} \perp \vec{B}$, $E = B$) (in cgs Gaussian units) in any inertial frame.

20. What invariants can you form from a tensor $T^{\mu\nu}$? Compute these invariants for the electromagnetic field tensor in Lorentz space-time (see Problem 18).

21. In Lorentz space-time, the wave four-vector has components $k^\mu = (\omega/c, k_x, k_y, k_z)$. Use the Lorentz transformation matrix $\Lambda^\mu{}_\nu$ (see Problem 18) to find the components of the wave vector in a second frame moving with velocity $\vec{v} = v\hat{x}$ with respect to the first. What is the result if (a) $\vec{k} = k\hat{x}$ and (b) $\vec{k} = k\hat{y}$? Compare with the nonrelativistic Doppler shift formula, and comment.

22. Use the metric $g_{\mu\nu}$ for Lorentz space-time (see Problem 18) to compute the line element $ds^2 = g_{\mu\nu}\,dx^\mu\,dx^\nu$. The proper time τ is defined by the relation $d\tau = ds/c$. Compute the proper time interval $d\tau$ between two neighboring points on the world line of a particle moving at speed v. (Let the points have coordinates ct, x, y, z and $c(t+dt), x+dx, y+ dy, z+dz$, where $dx = v_x\,dt$ and dy and dz are defined similarly.) Express your result in terms of the time interval dt, $\beta = v/c$, and $\gamma = 1/\sqrt{1-\beta^2}$. Compute the components of the four-velocity $u^\mu = dx^\mu/d\tau$ of a particle and compute the invariant product $u^\mu u_\mu$. Comment.

23. The set of components

$$\varepsilon^{\alpha\beta\gamma\delta} = \begin{cases} +1 & \text{if } \alpha\beta\gamma\delta = 0123 \text{ or an even permutation of this} \\ -1 & \text{if } \alpha\beta\gamma\delta = 1023 \text{ or an even permutation of this} \\ 0 & \text{otherwise} \end{cases}$$

may be used to form the tensor $\mathcal{F}^{\alpha\beta} = \frac{1}{2}\varepsilon^{\alpha\beta\gamma\delta}F_{\gamma\delta}$ dual to F (compare with equation A.7).
(a) Show that the components of $\varepsilon^{\alpha\beta\gamma\delta}$ transform as a tensor in Minkowski space-time.
(b) Find the invariant $\mathcal{F}^{\alpha\beta}F_{\alpha\beta}$ if $F^{\alpha\beta}$ is the electromagnetic field tensor defined in Problem 18. Comment.

24. Use Gauss' law in integral form to find the electric field inside a uniformly charged sphere. Compute the necessary components of Γ^i_{jk}, and hence find the divergence of this electric field in spherical coordinates. Show that the divergence equals the (uniform) charge density divided by ε_0.

25. In two-dimensional flat space described with polar coordinates, a vector $\vec{\mathbf{u}}$ is in the radial direction. Displace the vector to a neighboring point (refer to Figure A.3), and compute the new components in terms of the displacement $(\delta\rho, \delta\theta)$. Compare with relation (A.23) in the text and hence compute the components Γ^1_{ij} of the affinity. Perform the same operations with a vector in the θ direction to find the remaining components of Γ. *Hint:* Remember that the basis vectors are not unit vectors in this system.

26. In three-dimensional space, the affinity Γ^n_{kj} (equation A.29) has $3^3 = 27$ components. Show that the affinity for Euclidean space with cylindrical coordinates has only three nonzero components, and compute them.

    Use equations A.14, A.15, and A.27 to obtain the same result. (Remember that in equation A.27 $\bar{x}$ are the Cartesian coordinates.)

27. The tensor density ε^{ijk} is defined in Cartesian coordinates as in Chapter 1 (Section 1.1.2, equation 1.29). Transform to flat space with cylindrical coordinates, and determine the components of $\bar{\varepsilon}^{ijk}$. Lower indices to find the components of $\bar{\varepsilon}_i{}^{jk}$, $\bar{\varepsilon}_{ij}{}^k$, and $\bar{\varepsilon}_{ijk}$. Compute the cross product $\vec{\boldsymbol{\omega}} \times \vec{\mathbf{v}} = \vec{\mathbf{a}}$ for a particle in uniform circular motion with $\vec{\boldsymbol{\omega}} = \omega\hat{\mathbf{z}}$. *Hint:*

$$(\vec{\boldsymbol{\omega}} \times \vec{\mathbf{v}})^i = \varepsilon^{ijk}\omega_j v_k = \varepsilon^i{}_{jk}\omega^j v^k$$

28. By lowering indices, show that the covariant derivative of a covariant vector may be written

$$u_{i;j} = \frac{\partial u_i}{\partial x_j} - \Gamma^k{}_{ij}u_k$$

as in equation (A.26). Obtain the same result from equation (A.22).

29. Verify the product rule

$$v^i{}_{;j} = (T^{ik}u_k)_{;j} = u_k T^{ik}{}_{;j} + T^{ik}u_{k;j}$$

OPTIONAL TOPIC B

Group Theory

In modern physics, group theory is becoming an increasingly important mathematical tool. As we develop the theory, we shall see that groups form the natural mathematical description of symmetries of physical systems. These symmetries are in turn related to the conservation laws of physics, and thus the mathematics of groups allows us to investigate these laws in a very general way. Space constraints here do not allow us to do more than touch on the basics of group theory. Students who plan to pursue an interest in topics for which group theory is especially important, such as particle physics, will need to consult more advanced texts to supplement the material here.

B.1. DEFINITION OF A GROUP

A *group* G is a set of elements $\{a\}$, together with an operation $*$, that obeys the following set of rules:

1. If a and b are elements of the group, then so is $a * b$.
2. The operation $*$ is associative: $(a * b) * c = a * (b * c)$.
3. There is a unit element 1 in G such that $1 * a = a * 1 = a$ for every element a in G.
4. Every element a in G possesses an inverse element a^{-1}, also in G, which has the property that

$$a * a^{-1} = a^{-1} * a = 1$$

where 1 is the identity element (see rule 3).

In the special case where the operation is commutative ($a * b = b * a$), the group is said to be *abelian*.

The number of elements in the group is called the *order* of the group. The order may be finite or infinite.

465

B.2. EXAMPLES OF GROUPS

The group of order one   The most trivial group is of order one and consists of the single identity element. Convince yourself that this set obeys all of the rules listed above.

The group of order two   There is exactly one group of order two. It consists of the identity and one other element a that is its own inverse, so that $a * a = 1$. This group represents the physical symmetry of reflection in a plane or of rotation through an angle π about a single fixed axis.

The group of order three   There is exactly one group of order three. It contains the identity and two other elements a and b, where b is the inverse of a. We may write a multiplication table for this group, as shown in Table B.1.

Since $b = a^2$, we also have the relation $a * b = a * a^2 = a^3 = 1$. This group and the previous one are examples of *cyclic groups*. A cyclic group is generated by a single element. Every element of the group is obtained as a power of this single element, and in a group of order n, the nth power gives the identity. The cyclic group of order n is called C_n. Every cyclic group is abelian.

**TABLE B.1. Multiplication Table
for the Group of Order Three**

	1	a	b
1	1	a	b
a	a	b	1
b	b	1	a

The cyclic group C_n represents a symmetry of a polygon with n sides. We can think of the element a as a rotation through $2\pi/n$ that rotates the polygon into an identical copy of itself.

The set of integers under the operation of addition   The set of (positive and negative) integers forms a group of infinite order. The operation $*$ is addition:

$$m * n \equiv m + n$$

Then the sum of any two integers is another integer. The operation is associative:

$$n + (m + p) = (n + m) + p$$

The identity element for this group is the integer *zero*:

$$n + 0 = 0 + n = n$$

for any integer n. The inverse of an element n is the negative of n: $-n$. Then

$$n + (-n) = (-n) + n = 0$$

This group is abelian, since the order of the integers in the sum does not matter: $n+m = m+n$ for all integers m and n.

The set of integers does *not* form a group under ordinary multiplication, because the inverse of n, $1/n$, is not in the set of integers.

The set of all rational numbers m/n under the operation of multiplication, where n, $m \neq 0$

$$\frac{m}{n} * \frac{p}{q} \equiv \frac{m}{n} \times \frac{p}{q} = \frac{m \times p}{n \times q}, \quad \text{which is in } G$$

$$\frac{m}{n} \times \left(\frac{p}{q} \times \frac{r}{s} \right) = \left(\frac{m}{n} \times \frac{p}{q} \right) \times \frac{r}{s}$$

The identity element is $1 = 1/1$, and the inverse of m/n is n/m. This group is also abelian and has infinite order.

The rotation group Spatial rotations may be represented by rotation matrices (see, for example, Chapter 1). The identity is a rotation through 0 radians (represented by the identity matrix), and the inverse of a rotation through θ is a rotation through $-\theta$ about the same axis. This group is not abelian. The set of rotations in a plane is given the name SO(2). The rotations are represented by orthogonal 2×2 matrices with determinant $+1$. The O in SO(2) refers to the orthogonality of the matrices, the 2 refers to the fact that they are 2×2 matrices, and the S stands for special and refers to the condition that the determinant is $+1$. This group is abelian.

Notice that in this group the elements themselves *are* "operations" (rotations), represented mathematically by matrices, while the *group operation* is "follow one rotation with another" and is represented mathematically by matrix multiplication.

Permutation groups In these groups, also, the group elements are things we do ("operations"). The *group operation* is "follow one permutation with another." As an example, consider the set of permutations of three objects, S_3. One element of this group is the permutation {interchange first two elements}, which we'll label G_{12}.

$$G_{12}: (abc) \rightarrow (bac)$$

The identity element is {do nothing}.

$$1: (abc) \rightarrow (abc)$$

Consider the two elements $G_{13} = \{$interchange 1st and 3rd$\}$ and $G_{23} = \{$interchange 2nd and 3rd$\}$. The product of these two elements may be written

$$G_{23}G_{13} = \{(abc) \rightarrow (cba)\} \text{ followed by } \{(cba) \rightarrow (cab)\}$$

The result is the group element G_{312}, which means {take the last element and put it first}. (In standard notation, the operations are done in order from right to left. This is consistent with the notation for matrix multiplication.) Now if we multiply by G_{13} again, we find

$$G_{13}G_{23}G_{13} = \{(abc) \to (cba)\}\{(cba) \to (cab)\}\{(cab) \to (bac)\} = G_{12}$$

where G_{12} is the group element {interchange 1st and 2nd}. Each of these three "interchange" elements is its own inverse, since interchanging two elements twice gets us back to where we started. Thus, we can also write this relation as

$$G_{13}G_{23}G_{13}^{-1} = G_{12} \tag{B.1}$$

We say that the element G_{12} is *conjugate to* G_{23}.

The sixth element is G_{132}, which means {take the first element and put it last}, or $1 \to 3 \to 2 \to 1$.

$$G_{132} = \{(abc) \to (bca)\}$$

Using these symbols for the group elements, we can set up a multiplication table[1] listing all of the $3! = 6$ group elements for the group S_3. (See Table B.2.) Note that this group is not abelian, since, for example,[2]

$$G_{132}G_{12} = \{(abc) \to (bac) \to (acb)\} = G_{23}$$

while

$$G_{12}G_{132} = \{(abc) \to (bca) \to (cba)\} = G_{13}$$

TABLE B.2. Multiplication Table for the Group S_3

	1	G_{12}	G_{13}	G_{23}	G_{132}	G_{312}
1	1	G_{12}	G_{13}	G_{23}	G_{132}	G_{312}
G_{12}	G_{12}	1	G_{132}	G_{312}	G_{13}	G_{23}
G_{13}	G_{13}	G_{312}	1	G_{132}	G_{23}	G_{12}
G_{23}	G_{23}	G_{132}	G_{312}	1	G_{12}	G_{13}
G_{132}	G_{132}	G_{23}	G_{12}	G_{13}	G_{312}	1
G_{312}	G_{312}	G_{13}	G_{23}	G_{12}	1	G_{132}

The elements of this group may also be written using the cycle structure, in which G_{13} is written (13)(2): a two-cycle (13) in which the element 1 goes to 3 and 3 goes to 1, and a one-cycle (2), showing that element 2 goes to itself. The one-cycle is sometimes omitted, since it is redundant. The element G_{132} is written (132). The sequence of numbers in the cycle is

---

[1] Since this group is not abelian, the table must be read as follows: Multiply the element in the column on the far left by the element in the top row to obtain each table entry.

[2] *Remember*: We do the operation on the right first.

important, but it does not matter which one we write first, since $2 \to 3 \to 1 \to 2 \to 3$. In this notation, G_{312} is written as the three-cycle $(312) = (123)$.

We can understand how the operations in this permutation group are related to physical symmetries by labeling the vertices of an equilateral triangle $1, 2$, and 3, as shown in Figure B.1. Then each of the two-cycles corresponds to rotation by π about an axis in the plane of the triangle and through one apex. The cycle (12) corrresponds to rotation about axis A in the figure. The three-cycles correspond to rotation by $2\pi/3$ or $4\pi/3$ about an axis through the center of the triangle and perpendicular to the plane of the figure. Each of these operations leaves the triangle unchanged, except for the positions of the labels on the vertices. Once we make this identification, it is easy to see why $(123) * (123) = G_{312}G_{312} = G_{132} = (132)$ and $(123) * (132) = 1$.

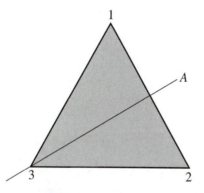

FIGURE B.1. The group S_3 describes the symmetry of this equilateral triangle.

The unitary groups   Another set of groups that is important in physics is the set of unitary groups. The unitary $n \times n$ matrices form the group $U(n)$. A matrix is unitary if its adjoint equals its inverse. The adjoint is formed by taking the complex conjugate and then the transpose:

$$\text{adj}\,(A) = \widetilde{(A^*)}$$

If A is real, then the condition of unitarity becomes $\tilde{A} = A^{-1}$; that is, the matrix is orthogonal. The special group $SU(n)$ is the group of $n \times n$ unitary matrices with determinant $+1$. The general form of a group element of $SU(2)$ is

$$\begin{pmatrix} a & b \\ -b^* & a^* \end{pmatrix} \quad \text{with } aa^* + bb^* = 1 \tag{B.2}$$

For example,

$$\frac{1}{\sqrt{2}} \begin{pmatrix} e^{i\theta} & e^{i\phi} \\ -e^{-i\phi} & e^{-i\theta} \end{pmatrix}$$

would satisfy these constraints.

B.3. CLASSES

If a, b, and c are three elements of a group and

$$b = cac^{-1}$$

(B.3)

then b is said to be conjugate to a (compare with equation B.1).

1. An element is always conjugate to itself, since we can take c to be the identity element. Then

$$cac^{-1} = 1a1 = a$$

2. If a is conjugate to b, then b is conjugate to a:

$$a = cbc^{-1}$$

We multiply on the right by c and on the left by c^{-1}. Then

$$c^{-1}ac = c^{-1}cbc^{-1}c = b = faf^{-1}$$

where $f = c^{-1}$.

3. If a is conjugate to b and c, then b and c are conjugate to each other.

If a is conjugate to both b and c, then there are elements f and g such that

$$a = fbf^{-1} = gcg^{-1}$$

Thus, if $f^{-1}g = h$, then $h^{-1} = (f^{-1}g)^{-1} = g^{-1}f$. Multiplying the above expressions on the left by f^{-1} and on the right by f, we find

$$f^{-1}fbf^{-1}f = b = hch^{-1}$$

and so b is conjugate to c.

The complete set of elements $C = a_1, a_2, \ldots$ that are conjugate to one another is called an equivalence class, or simply a *class*, of the group. The identity element forms a class by itself, since $a1a^{-1} = 1$ for each element a of the group. To find the elements that are conjugate to a, calculate gag^{-1} for each element g of the group. The results may not all be distinct.

For example, in the permutation group S_3, the elements conjugate to G_{12} are computed in Table B.3. The elements in the class are $\{G_{12}, G_{13}, G_{23}\}$. This seems sensible, since these three elements are the "interchange" elements; that is, they do the same kind of thing. Similarly, the two "cycling" elements form a second class for this group, as you should check.

TABLE B.3. Class of Elements Conjugate to G_{12}

$$1G_{12}1^{-1} = G_{12}G_{12}G_{12}^{-1} = G_{12}$$
$$G_{13}G_{12}G_{13}^{-1} = G_{231}G_{13} = G_{23}$$
$$G_{23}G_{12}G_{23}^{-1} = G_{132}G_{23} = G_{13}$$
$$G_{231}G_{12}G_{231}^{-1} = G_{13}G_{132} = G_{23}$$
$$G_{132}G_{12}G_{132}^{-1} = G_{23}G_{231} = G_{13}$$

B.4. SUBGROUPS

Sometimes we can find a subset of the original group that obeys all the rules for a group; this is called a subgroup. For example, the subset $\{1, G_{12}\}$ is a subgroup of S_3 because $(G_{12})^2 = 1$. A second example is the set of all even integers. This set forms a subgroup of the group of integers, since the sum of any two even integers is also even, the subset includes the identity (zero), and each element has its own inverse within the subgroup. This subgroup also has infinitely many elements. The group of all rotations by multiples of $\pi/2$ around the x-axis is a subgroup of the rotation group. This group has only four elements: rotations by $\pi/2 \equiv -3\pi/2$, $\pi \equiv -\pi$, $3\pi/2 \equiv -\pi/2$, and $2\pi \equiv 0$. This last element is the identity element. This subgroup is abelian, while the group of all rotations is not.

If a group G has a subgroup S and element a is *not* in S, then the sets of elements as_i and $s_i a$ are the left and right *cosets* of the subgroup S with respect to a. Note that none of the elements in the cosets are in the subgroup if a is not. We can then write the group elements as

$$G = S + a_1 S + a_2 S + \cdots$$

Thus, the order of the subgroup S is a divisor of the order of the group G.

The set of conjugate elements

$$aSa^{-1}$$

forms a subgroup S'. To see this, note the following:

1. The product of two elements is in the set. If s_1 and s_2 are elements of S, then

$$(as_1a^{-1})(as_2a^{-1}) = as_1(a^{-1}a)s_2a^{-1} = as_1s_2a^{-1} = as_3a^{-1}$$

where $s_3 = s_1s_2$ is also in S, and so as_3a^{-1} is in S'.

2. The associative property is preserved:

$$(as_1a^{-1} * as_2a^{-1}) * as_3a^{-1} = as_1a^{-1} * (as_2a^{-1} * as_3a^{-1})$$

3. The identity element is in S'. If S is a subgroup, it contains the identity. Thus, S' contains the element $a1a^{-1} = aa^{-1} = 1$.

4. The inverse of asa^{-1} is $as^{-1}a^{-1}$:

$$(asa^{-1})(as^{-1}a^{-1}) = a(ss^{-1})a^{-1} = aa^{-1} = 1$$

and

$$(as^{-1}a^{-1})(asa^{-1}) = a(s^{-1}s)a^{-1} = aa^{-1} = 1$$

If the new subgroup S' is the same as the original subgroup S, that subgroup is called *invariant,* normal, or self-conjugate.

Note that the subgroup of even integers is an invariant subgroup, since

$$n + 2m + (-n) = 2m$$

for any integer n. However, the subgroup $\{1, G_{12}\}$ of S_3 is not invariant, because $G_{13}G_{12}G_{13}^{-1} = G_{23}$ (see Table B.3) and G_{23} is not in the subgroup.

B.5. CYCLIC GROUPS

If a is an element of a group G, then a^n is also an element, for any integer n. By the first rule of groups, $a * a = a^2 = b$ must also be in the group. Then $a * b = a^3$ must also be in the group, and so on. If the group has finite order, then $a^n = 1$ for some n.

> If N is the smallest integer for which $a^N = 1$, then N is called the order of the element a.

The set $\{a, a^2, \ldots, a^N = 1\}$ is called the *period* of a. It is a group called a cyclic group C_N. If the order of an element is less than the order of the group, then its period forms a cyclic subgroup of G. In the permutation group of three objects, S_3, each of the interchange elements has order two, while the other two elements (G_{231} and G_{312}) have order three. Then, for example, the period $\{G_{12}, G_{12}^2 = 1\}$ forms a cyclic subgroup of order two. The symmetry of cyclic groups was discussed in Section B.2.

B.6. FACTOR GROUPS AND DIRECT PRODUCT GROUPS

If S is an invariant subgroup of a group G, then we can identify another group, called the factor group G/S, whose elements are the cosets aS. The group operation for the factor group is defined by the rule

$$(aS) * (bS) = (ab)S$$

where the product ab is computed using the group operation of the original group G. The invariant subgroup itself acts as the identity element. The inverse of aS is $a^{-1}S$:

$$(aS) * (a^{-1}S) = (aa^{-1})S = S$$

The group structure of the factor group follows directly from the group structure of G itself. If the subgroup has m elements, then the order of the group is mn for some integer n, and the factor group has order n.

A group may be written as a product of subgroups under some circumstances.

G is a *direct product* of subgroups S_1 and S_2, $G = S_1 \times S_2$, if

(i) all the elements of S_1 commute with all the elements of S_2 and

(ii) every element in G may be written as

$$g = s_1 s_2$$

where s_1 is a member of the subgroup S_1 and s_2 is a member of S_2.

Then the product of two group elements is

$$g_i g_j = s_{1,i} s_{2,i} s_{1,j} s_{2,j}$$
$$= s_{1,i} s_{1,j} s_{2,i} s_{2,j} \quad \text{by property (i)}$$
$$= s_{1,k} s_{2,k} \quad \text{since } S_1 \text{ and } S_2 \text{ are subgroups}$$
$$= g_k$$

and is of the correct form (ii).

B.7. ISOMORPHISM

We can relate the elements of two groups to one another by a mapping. The mapping may be one-to-one, in which one element of G_1 maps to exactly one element of G_2; many-to-one, in which two or more elements of G_1 map to a single element of G_2; or one-to-many, in which a single element of G_1 maps to more than one element of G_2. A mapping is said to be "onto" if the mapping maps all of G_1 onto all of G_2. For example, we can map the elements of the permutation group of three elements, S_3, to the group of integers by assigning each permutation a number, 1 through 6.

$$
\begin{aligned}
1 &\to 1 \\
G_{12} &\to 2 \\
G_{13} &\to 3 \\
G_{23} &\to 4 \\
G_{132} &\to 5 \\
G_{312} &\to 6
\end{aligned}
$$

This mapping is one-to-one but not onto.

We can also construct a mapping from the integers to the permutation group S_3. We divide each integer by 6 and note the remainder. That element maps to the permutation group as follows:

Remainder	maps to	permutation group element
0	$\rightarrow$	G_{312}
1	$\rightarrow$	1
2	$\rightarrow$	G_{12}
3	$\rightarrow$	G_{13}
4	$\rightarrow$	G_{23}
5	$\rightarrow$	G_{132}

This mapping is onto but not one-to-one, since, for example, both 6 and 12 map to G_{312}. It is many-to-one.

In an important class of mappings called *homomorphisms*, the mapping also "preserves the operation." To see what this means, consider a mapping f that maps group G to group H. Then we can write $f(g) = h$. If f is a homomorphism, then

$$f(g_1 * g_2) = f(g_1) * f(g_2) \tag{B.4}$$

For example, the mapping that maps the rational fractions to the integers via

$$\frac{n}{m} \rightarrow nm$$

is a homomorphism, since

$$f\left(\frac{n}{m} * \frac{p}{q}\right) = f\left(\frac{np}{mq}\right) = npmq = (nm)(pq) = f\left(\frac{n}{m}\right) * f\left(\frac{p}{q}\right)$$

This mapping is onto (since every integer is also a rational fraction of the form $n/1$), but not one-to-one ($\frac{3}{2}$ and $\frac{6}{1}$ both map to 6).

A homomorphism that is also one-to-one and onto is called an isomorphism. The group of rotations in a plane is isomorphic to the group of 2×2 orthogonal matrices, since each rotation maps to one matrix, and vice versa.

B.8. REPRESENTATIONS[3]

B.8.1. Reducible and Irreducible Representations

All groups that are isomorphic to one another are the same in a fundamental sense. An isomorphism allows us to represent the elements of an abstract group with the elements of another group. This is a useful idea because the elements of the isomorphic group can be mathematical structures such as the matrices that represent rotations. The group of order two can be represented by the numbers 1 and -1 through the isomorphism $f(1) = 1$ and $f(a) = -1$.

---

[3]In this section, we choose to draw conclusions from examples, rather than by explicit proof. For proofs of the results presented here, the reader should refer to the bibliography.

We can use a homomorphism to construct a representation of an abstract group as follows. We allow the group elements to operate on a vector space of dimension n. For example, to represent the permutation group S_3, we can label the points of a triangle with x and y coordinates, thus allowing the group elements to operate on two-component vectors. Then

A representation of an abstract group G is defined as a homomorphism $f: G \to M$, where M is a group of $n \times n$ matrices. This representation has dimension n. If the mapping is an isomorphism, the representation is said to be *faithful*.

Every group has a trivial one-dimensional representation in which every element of the group is represented by the number 1. This mapping clearly preserves the group operation, but we have lost a good deal of information about the group structure; the representation is not faithful.

Another representation, of dimension equal to the order of the group, may be formed using the group multiplication table. We form n-dimensional vectors whose components are the elements of the group, in some order. We label the group elements $g_1, g_2, \ldots, g_n$. Then we can represent any element g_i with an $n \times n$ matrix M_i that has only one nonzero element in each row and each column. If

$$g_i g_j = g_k$$

then $(M_i)_{jm} = \delta_{mk}$. This matrix represents the multiplication properties of element g_i. For example, the cyclic group C_3 of order three whose multiplication table is given in Table B.1 may be represented by the following matrices:

$$M_1 = \begin{pmatrix} 1 & 0 & 0 \\ 0 & 1 & 0 \\ 0 & 0 & 1 \end{pmatrix}, \quad M_2 = \begin{pmatrix} 0 & 1 & 0 \\ 0 & 0 & 1 \\ 1 & 0 & 0 \end{pmatrix}, \quad M_3 = \begin{pmatrix} 0 & 0 & 1 \\ 1 & 0 & 0 \\ 0 & 1 & 0 \end{pmatrix} \quad \text{(B.5)}$$

where M_2 represents element a and M_3 represents b. Then

$$M_2 \begin{pmatrix} 1 \\ a \\ b \end{pmatrix} = \begin{pmatrix} 0 & 1 & 0 \\ 0 & 0 & 1 \\ 1 & 0 & 0 \end{pmatrix} \begin{pmatrix} 1 \\ a \\ b \end{pmatrix} = \begin{pmatrix} a \\ b \\ 1 \end{pmatrix}$$

which correctly reproduces the second row of the multiplication table. Note also that

$$M_2 M_2 = \begin{pmatrix} 0 & 1 & 0 \\ 0 & 0 & 1 \\ 1 & 0 & 0 \end{pmatrix} \begin{pmatrix} 0 & 1 & 0 \\ 0 & 0 & 1 \\ 1 & 0 & 0 \end{pmatrix} = \begin{pmatrix} 0 & 0 & 1 \\ 1 & 0 & 0 \\ 0 & 1 & 0 \end{pmatrix} = M_3$$

and

$$M_2 M_3 = \begin{pmatrix} 0 & 1 & 0 \\ 0 & 0 & 1 \\ 1 & 0 & 0 \end{pmatrix} \begin{pmatrix} 0 & 0 & 1 \\ 1 & 0 & 0 \\ 0 & 1 & 0 \end{pmatrix} = \begin{pmatrix} 1 & 0 & 0 \\ 0 & 1 & 0 \\ 0 & 0 & 1 \end{pmatrix} = M_1$$

The matrices themselves satisfy the same multiplication table, as required for a representation. This representation is called the *regular representation*.

A one-dimensional representation of the same group may be formed from the complex numbers

$$f_1(1) = 1, \quad f_1(a) = e^{2\pi i/3}, \quad f_1(b) = e^{4\pi i/3} \tag{B.6}$$

Recall from Chapter 2, Section 2.1.1 that multiplication by a complex number represents a rotation. So each of these elements may be visualized as a rotation in the complex plane.

Extending this idea, we may form a two-dimensional representation using the 2×2 rotation matrices:

$$\begin{pmatrix} 1 & 0 \\ 0 & 1 \end{pmatrix}, \quad \frac{1}{2}\begin{pmatrix} -1 & \sqrt{3} \\ -\sqrt{3} & -1 \end{pmatrix}, \quad \frac{1}{2}\begin{pmatrix} -1 & -\sqrt{3} \\ \sqrt{3} & -1 \end{pmatrix} \tag{B.7}$$

We may also use the 3×3 matrices:

$$M_1' = \begin{pmatrix} 1 & 0 & 0 \\ 0 & 1 & 0 \\ 0 & 0 & 1 \end{pmatrix}, \quad M_2' = \begin{pmatrix} -1/2 & \sqrt{3}/2 & 0 \\ -\sqrt{3}/2 & -1/2 & 0 \\ 0 & 0 & 1 \end{pmatrix},$$

$$M_3' = \begin{pmatrix} -1/2 & -\sqrt{3}/2 & 0 \\ \sqrt{3}/2 & -1/2 & 0 \\ 0 & 0 & 1 \end{pmatrix} \tag{B.8}$$

These 3×3 matrices operate on three-component vectors, but they leave the third component unchanged. Labeling the components x, y, and z in the conventional way, we may write any vector in the space as the sum of a vector $\mathbf{u}$ in the x-y plane plus a vector $\mathbf{w}$ parallel to the z-axis: $\mathbf{v} = \mathbf{u} + \mathbf{w}$. The group elements leave $\mathbf{w}$ unchanged.

We may generalize this result. If an n-dimensional space V may be written as the sum of an m-dimensional space and an $(n - m)$-dimensional space, $(V = U + W)$ and if, under the group operations, vectors in W remain in W, then the subspace W is invariant under the group. As we saw in Chapter 1, we can change the basis vectors of the vector space. The entries in the matrix will also change. The matrices with respect to the two bases are related by a similarity transformation (equation 1.78):

$$\boxed{M' = SMS^{-1}} \tag{B.9}$$

The corresponding representations M and M' are said to be *equivalent*. But in the primed basis, the invariance of the subspace W is less obvious.

If we reverse this argument, we can see that invariance of a subspace under a group means that there will be a basis for which the group representation makes the invariance obvious. The representation is said to be *reducible*.

A representation is reducible if g_i is represented with respect to some basis by a matrix M_i that takes the form

$$M_i = \begin{pmatrix} A_i & B_i \\ 0 & D_i \end{pmatrix}$$

where A_i, B_i, and D_i are submatrices of dimensions $m \times m$, $m \times p$, and $p \times p$, respectively, and $n = m + p$ is the order of the representation. Then the group element $g_i g_j$ is represented by

$$\begin{pmatrix} A_i & B_i \\ 0 & D_i \end{pmatrix} \begin{pmatrix} A_j & B_j \\ 0 & D_j \end{pmatrix} = \begin{pmatrix} A_i A_j & A_i B_j + B_i D_j \\ 0 & D_i D_j \end{pmatrix} = \begin{pmatrix} A_k & B_k \\ 0 & D_k \end{pmatrix}$$

where the submatrices A_k, B_k, and D_k are of the specified form and

$$\begin{pmatrix} A_i & B_i \\ 0 & D_i \end{pmatrix} \begin{pmatrix} u \\ w \end{pmatrix} = \begin{pmatrix} A_i u + B_i w \\ D_i w \end{pmatrix}$$

so that the subspace W is invariant under the group. If $B \equiv 0$, then the representation M is completely reducible and is written[4]

$$M = A \oplus D$$

This corresponds to the case in which both subspaces U and W are invariant under the group, as is true for the representation (B.8). For finite groups, every reducible representation is equivalent to a completely reducible representation.

If the submatrices A and/or D are reducible, we may continue this decomposition until M is diagonal or the remaining submatrices are not reducible. The resulting representation is then said to be *irreducible*.

In the representation (B.8), the matrix D is a 1×1 matrix—that is, just a number. The set of matrices D_i is the trivial representation of the group C_3; each $D_i = 1$. We can continue to reduce the representation (B.8) by diagonalizing[5] the matrices A_i. We can do this because A_1 is already diagonal and A_2 and A_3 give rise to the same equation (equation 1.80) for the eigenvalues:

$$\begin{vmatrix} -\frac{1}{2} - \lambda & \pm\sqrt{3}/2 \\ \mp\sqrt{3}/2 & -\frac{1}{2} - \lambda \end{vmatrix} = 0$$

$$\left(-\frac{1}{2} - \lambda\right)^2 + \frac{3}{4} = 0$$

$$-\frac{1}{2} - \lambda = \pm\frac{\sqrt{3}}{2} i$$

---

[4] The symbol $\oplus$ does not imply addition in the usual sense. It cannot, because the matrices A and D may not have the same dimension. They are to be "added" by forming a larger matrix with A and D along the diagonal.
[5] See Chapter 1, Section 1.6.5.

So

$$\lambda = -\frac{1}{2} \pm \frac{\sqrt{3}}{2}i = e^{2\pi i/3} \text{ or } e^{4\pi i/3}$$

This solution is just the representation (B.6) that we found above. There is, in fact, a second representation here:

$$f_2(1) = 1, \quad f_2(a) = e^{4\pi i/3}, \quad f_2(b) = e^{8\pi i/3} = e^{2\pi i/3} \tag{B.10}$$

Thus, the irreducible representations are the trivial representation, (B.6) and (B.10). The regular representation and (B.8) may be reduced to the form

$$M_1'' = \begin{pmatrix} 1 & 0 & 0 \\ 0 & 1 & 0 \\ 0 & 0 & 1 \end{pmatrix}, \quad M_2'' = \begin{pmatrix} 1 & 0 & 0 \\ 0 & e^{2\pi i/3} & 0 \\ 0 & 0 & e^{4\pi i/3} \end{pmatrix},$$

$$M_3'' = \begin{pmatrix} 1 & 0 & 0 \\ 0 & e^{4\pi i/3} & 0 \\ 0 & 0 & e^{2\pi i/3} \end{pmatrix} \tag{B.11}$$

The representation has now been fully reduced and may be written

$$M = T \oplus F_1 \oplus F_2 \tag{B.12}$$

where T is the trivial representation and F_1 and F_2 are as given in (B.6) and (B.10).

In a direct sum of inequivalent, irreducible representations (irreps), as in equation (B.12), a given representation may appear more than once. The representations in the direct sum need not, and usually do not, all have the same dimension.

B.8.2. Orthogonality of the Irreps

Using the representation (B.11), we can form a number of three-component vectors by selecting one component from each of the matrices that represent the group elements. These vectors are orthogonal:

$$\sum_{k=1}^{3} \left(M_k'' \right)_p^* (M_k)_q = K \delta_{pq}$$

where K is a constant. Here the different irreps are labeled by p and q, and k labels the group elements. For example, with $p = 1$ and $q = 2$, we have

$$1 \times 1 + 1 \times e^{2\pi i/3} + 1 \times e^{4\pi i/3} = 1 - \frac{1}{2} + \frac{1}{2}i\sqrt{3} - \frac{1}{2} - \frac{1}{2}i\sqrt{3} = 0$$

while with $p = q = 1$, we find $K = 3$.

Now we want to generalize this result to a group of order m in which one or more of the irreps have dimension greater than one. Let us label the distinct, inequivalent, irreducible representations M_1, M_2, $\ldots$. Within representation i, with dimension n_i, a group element g is represented by the $n_i \times n_i$ matrix $M_i(g)$. The representations satisfy an orthogonality relation that we may write as

$$\sum_g [M_i(g)]_{pq} [M_j(g^{-1})]_{rs} = K \delta_{ij} \delta_{ps}$$

where K depends on q and r. We can determine the value of K by setting $i = j$. Then $\delta_{ij} \to 1$, and

$$\sum_g [M_i(g)]_{pq} [M_i(g^{-1})]_{rs} = K \delta_{ps}$$

Now we take the trace; that is, we set $p = s$ and sum from 1 to n_i. On the left, we have the matrix product:

$$\sum_p [M_i(g^{-1})]_{rp} [M_i(g)]_{pq} = [M_i(g^{-1}) M_i(g)]_{rq} = [M_i(g^{-1}g)]_{rq} = [M_i(1)]_{rq} = \delta_{rq}$$

Since the trace of $\delta_{ps} = n_i$,

$$\sum_g \delta_{rq} = m \delta_{rq} = K n_i$$

and

$$K = \frac{m}{n_i} \delta_{rq}$$

Thus, the orthogonality relation is

$$\boxed{\sum_g [M_i(g)]_{pq} \, [M_j(g^{-1})]_{rs} = \frac{m}{n_i} \delta_{rq} \delta_{ij} \delta_{ps}} \tag{B.13}$$

When the M_i are unitary, we may write this relation as

$$\boxed{\sum_g [M_i(g)]_{pq} [M_j(g)]_{sr}^* = \frac{m}{n_i} \delta_{ij} \delta_{ps} \delta_{qr}} \tag{B.14}$$

We can use this relation to determine a limit on the number and dimensionality of the irreps. As we did with the irreps of C_3, we can think of this relation as the dot product of m-dimensional vectors, where m is the order of the group. Here the vectors are labeled by two indices, p and q, each of which runs over n_i values. There are n_i^2 such vectors, and they are all orthogonal. Similarly, s and r each run over n_j values. Furthermore, these n_j^2

vectors are orthogonal to the first n_i^2 vectors. In total, there are $\sum_i n_i^2$ mutually orthogonal vectors in the m-dimensional vector space. Thus,

$$\sum_i n_i^2 \leq m$$

In fact, it can be shown that the equality holds:

$$\boxed{\sum_i n_i^2 = m} \tag{B.15}$$

We found three irreps for the group C_3, each of dimension one, and so the left-hand side of equation (B.15) equals 3, the order of the group, as required.

B.8.3. Character

Equivalent representations are essentially the same. It would be nice to have a convenient way to characterize representations that labels all equivalent representations the same way. Again we can review the properties of matrices (Chapter 1) to find the one that we need: It is the trace,[6] since

$$\text{Tr}\,(SMS^{-1}) = \text{Tr}\,(M)$$

The *character* of a representation is the set $\{\chi_i\}$ of traces of the matrices that represent the group elements.

The traces are not all distinct. In fact, every element in a class has the same trace. If elements g_i and g_j are represented by M_i and M_j, then

$$\text{Tr}\,[M(g_j g_i g_j^{-1})] = \text{Tr}\,(M_j M_i M_j^{-1}) = \text{Tr}\,(M_i) \tag{B.16}$$

The character of the representation (B.8) of the cyclic group C_3 is the set

$$\{3, 0, 0\}$$

and the character of the regular representation (B.5) is the same. Since the characters are the same, the two representations must be equivalent.

It is most instructive to consider the characters of the irreducible representations. In the case of C_3, or indeed any abelian group, the irreducible representations are all one-dimensional,[7] and the traces are just the numbers themselves. The characters of the three irreducible representations that we have found may be written in the form of m-component

---

[6]See Chapter 1, Problem 40.
[7]See Problem 7.

vectors, where m is the order of the group (equal to 3 for C_3). These vectors are mutually orthogonal in the following sense[8]:

$$\frac{1}{m} \sum_{i=1}^{m} \chi_i^a (\chi_i^b)^* = \delta^{ab}$$

(B.17)

The index i in this expression labels the elements of the group, and a and b label the different characters.

We can obtain a second relation by noting that all elements in the same equivalence class are represented by matrices with the same trace (equation B.16). Thus, equation (B.17) becomes

$$\frac{1}{m} \sum_{j=1}^{N} p_j \chi_j^a (\chi_j^b)^* = \delta^{ab}$$

(B.18)

where N is the number of distinct classes in the group and p_j is the number of elements in the jth class. This constitutes a second orthogonality relation for the N-dimensional vectors $\sqrt{p}\chi$. Since there can be no more than N of these, we find that the number of inequivalent, irreducible representations is less than or equal to the number of classes in the group. Once again it turns out that the equality holds.

Let's check this result for the group C_3. There are three distinct classes, since each element forms its own class. Thus, there are three inequivalent, irreducible representations, as we have found.

We can use the results found so far to determine the character table for any finite group.

Let's find the character table for the permutation group[9] S_3. There are three classes, and so there are three irreps. One is the trivial representation of dimension 1. Then, by equation (B.15), the other two irreps have dimensions that satisfy

$$n_2^2 + n_3^2 = 5$$

Thus, $n_2 = 1$ and $n_3 = 2$. With a one-dimensional representation, the multiplication table for the characters is the same as the group multiplication table. Therefore, for the "interchange" class, we must have $\chi_i^2 = 1$ and thus $\chi_i = \pm 1$. For the "cycling" class, $\chi_c^3 = 1$. We also have $\chi_i \chi_c = \chi_i$, so $\chi_i = -1$ and $\chi_c = 1$. For the last representation, the character of the identity equals the dimension of the representation, here 2. We have now obtained the results shown in Table B.4.

---

[8]For one-dimensional representations, equation (B.14) with $p = q = r = s = 1$ and $n_i = 1$ reduces to equation (B.17). In equation (B.17), the label g has been replaced by i, and i and j have been replaced by a and b.
[9]Recall that this group has order six.

TABLE B.4. Character Table for S_3

Class	Identity	Interchange	Three-cycles
Number of elements in class	1	3	2
Trivial rep T-dimension 1	1	1	1
2nd rep M_2-dimension 1	1	−1	1
3rd rep M_3-dimension 2	2	α	β

Then the orthogonality relation (B.18) gives us two equations for α and β:

$$2 - 3\alpha + 2\beta = 0$$

$$2 + 3\alpha + 2\beta = 0$$

Thus,

$$\beta = -1 \quad \text{and} \quad \alpha = 0$$

We can understand these results by referring back to the symmetries in Figure B.1. Let the triangle be in the x-y plane. Then the second representation has basis $\hat{z}$; the cycling elements correspond to rotations about the z-axis and so leave the z-axis invariant. The interchange elements correspond to rotation about an axis in the plane; these operations invert the z-axis. The third representation must have two basis vectors in the plane of the triangle. With axis A labeled as the x-axis, the interchange elements are represented by the reflection matrices[10]

$$G_{12} = \begin{pmatrix} 1 & 0 \\ 0 & -1 \end{pmatrix}, \quad G_{13} = \begin{pmatrix} -1/2 & -\sqrt{3}/2 \\ \sqrt{3}/2 & 1/2 \end{pmatrix}, \quad G_{23} = \begin{pmatrix} -1/2 & \sqrt{3}/2 \\ -\sqrt{3}/2 & 1/2 \end{pmatrix}$$

while the cycling elements are represented by rotation matrices representing rotations of $2\pi/3$ and $4\pi/3$, respectively:

$$G_{231} = \begin{pmatrix} -1/2 & \sqrt{3}/2 \\ -\sqrt{3}/2 & -1/2 \end{pmatrix} \quad \text{and} \quad G_{132} = \begin{pmatrix} -1/2 & -\sqrt{3}/2 \\ \sqrt{3}/2 & -1/2 \end{pmatrix}$$

Notice that these matrices correctly reproduce the character table. Check that they also satisfy the multiplication table for the group.

The character of a fully reduced representation is the sum of the characters of the irreducible representations that compose it. This result can be helpful in determining the decomposition of any particular representation.

B.8.4. Representations and Physics

Here we give a brief example to illustrate how group representations can simplify our understanding of physical systems. The ammonia molecule is a pyramid with the nitrogen

---

[10]Refer to Chapter 1, Problem 8.

atom at the top and three hydrogen atoms in an equilateral triangle below (see Figure B.2). The group S_3 is the symmetry group for this system, because we can permute the hydrogen atoms but we must leave the nitrogen molecule where it is.

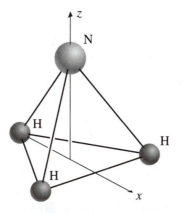

FIGURE B.2. An ammonia molecule is shaped like a tetrahedron with the nitrogen molecule at the top.

We can obtain a physical representation of the symmetry group by using real three-dimensional space as the vector space of the representation and asking how a vector in three-dimensional space transforms under these symmetries. This representation is reducible. To see how it decomposes, we can compute the character. The trace of the representation of the identity is, as always, the dimension of the rep, here 3. We can take one of the interchanges to be reflection in the x-z plane, represented by the matrix

$$M_{\text{int}}^V = \begin{pmatrix} 1 & 0 & 0 \\ 0 & -1 & 0 \\ 0 & 0 & 1 \end{pmatrix} \tag{B.19}$$

with trace $+1$. The cycles are rotations about the z-axis. One of these is represented by

$$M_{\text{cyc}}^V = \begin{pmatrix} -1/2 & -\sqrt{3}/2 & 0 \\ \sqrt{3}/2 & -1/2 & 0 \\ 0 & 0 & 1 \end{pmatrix}$$

with trace zero. Thus, the character is

$$\chi^V = \{3, 1, 0\}$$

Comparing with Table B.4, we see that $\chi^V = \chi_1 + \chi_3$ and thus

$$M^V = T \oplus M_3$$

In particular, the trivial representation is included. This means that we can find a vector that remains invariant under the symmetry group and hence, for example, the molecule can have an electric dipole moment.

A magnetic dipole moment, however, is a pseudo-vector[11] that transforms differently under reflection. The representation (B.19) changes to

$$M_{\text{int}}^{PV} = \begin{pmatrix} -1 & 0 & 0 \\ 0 & +1 & 0 \\ 0 & 0 & -1 \end{pmatrix}$$

with trace -1. The character of the new rep is $\{3, -1, 0\}$, and the decomposition is $M_2 \oplus M_3$. The trivial representation is not included. Thus, the molecule cannot have a permanent magnetic moment. Notice that we did not need to know any of the physical properties of the system—other than its geometrical structure—to draw these conclusions.

B.9. GENERATORS OF GROUPS

The groups we considered in Section B.8 were primarily finite groups. But many of the important groups in physics have infinite order. In groups such as the rotation group SO(3), the group elements depend on a set of continuous parameters—the rotation angles in SO(3)—and we can express the group elements in terms of these parameters. The group elements may be expressed in terms of *generators* that relate each group element to the set of parameters. It is convenient to relate the identity element of the group to the parameter zero.

We can understand the idea behind generators by considering the group of rotations in a plane—that is, SO(2). Each group element may be represented by a matrix of the form

$$R(\theta) = \begin{pmatrix} \cos\theta & \sin\theta \\ -\sin\theta & \cos\theta \end{pmatrix}$$

Here the relevant parameter is the angle θ, and with $\theta = 0$ we obtain the identity element, as required. We can express the cosines and sines in exponential form and then use the Pauli matrices as a basis. These are

$$\sigma_1 = \begin{pmatrix} 0 & 1 \\ 1 & 0 \end{pmatrix}, \quad \sigma_2 = \begin{pmatrix} 0 & -i \\ i & 0 \end{pmatrix}, \quad \text{and} \quad \sigma_3 = \begin{pmatrix} 1 & 0 \\ 0 & -1 \end{pmatrix}$$

Thus,

$$R(\theta) = \cos\theta \begin{pmatrix} 1 & 0 \\ 0 & 1 \end{pmatrix} + \sin\theta \begin{pmatrix} 0 & 1 \\ -1 & 0 \end{pmatrix}$$

$$= \mathbf{1}\cos\theta + i\sigma_2 \sin\theta$$

For very small rotations $\theta \ll 1$ (that is, "near" the identity element), $\cos\theta \approx 1$ and $\sin\theta \approx \theta$, so we may write

$$R(\theta) = \exp(i\sigma_2\theta) \tag{B.20}$$

[11] See Appendix I and Optional Topic A, especially Problem 9.

At this point, you might ask what it means to have a matrix as the argument of an exponential. The meaning is clear if we define the exponential by its series expansion:

$$e^{\mathbb{M}} = 1 + \mathbb{M} + \frac{1}{2}\mathbb{M} * \mathbb{M} + \cdots$$

Note that, in general,

$$e^{\mathbb{A}}e^{\mathbb{B}} = \left(1 + \mathbb{A} + \frac{1}{2}\mathbb{A} * \mathbb{A} + \cdots\right)\left(1 + \mathbb{B} + \frac{1}{2}\mathbb{B} * \mathbb{B} + \cdots\right)$$

$$= 1 + \mathbb{A} + \mathbb{B} + \mathbb{A} * \mathbb{B} + \frac{1}{2}\mathbb{A} * \mathbb{A} * \mathbb{B} + \frac{1}{2}\mathbb{A} * \mathbb{B} * \mathbb{B} + \cdots$$

$$\neq 1 + \mathbb{A} + \mathbb{B} + \mathbb{B} * \mathbb{A} + \frac{1}{2}\mathbb{B} * \mathbb{A} * \mathbb{A} + \frac{1}{2}\mathbb{B} * \mathbb{B} * \mathbb{A} + \cdots$$

$$= e^{\mathbb{B}}e^{\mathbb{A}}$$

Expanding the exponential in the rotation matrix, we have

$$R(\theta) = 1 + i\sigma_2\theta - \frac{1}{2}\sigma_2^2\theta^2 - \frac{1}{6}i\sigma_2^3\theta^3 + \frac{1}{24}\sigma_2^4\theta^4 + \frac{1}{120}i\sigma_2^5\theta^5 - \frac{1}{720}\sigma_2^6\theta^6 + \cdots$$

But

$$\sigma_2^2 = \begin{pmatrix} 0 & -i \\ i & 0 \end{pmatrix}\begin{pmatrix} 0 & -i \\ i & 0 \end{pmatrix} = \begin{pmatrix} 1 & 0 \\ 0 & 1 \end{pmatrix}$$

so we may simplify:

$$R(\theta) = 1 + i\sigma_2\theta - \frac{1}{2}\theta^2 - \frac{1}{6}i\sigma_2\theta^3 + \frac{1}{24}\theta^4 + \frac{1}{120}i\sigma_2\theta^5 - \frac{1}{720}\theta^6 + \cdots$$

$$= 1 - \frac{1}{2}\theta^2 + \frac{1}{24}\theta^4 - \frac{1}{720}\theta^6 + \cdots + i\sigma_2\left(\theta - \frac{1}{6}\theta^3 + \frac{1}{120}\theta^5 + \cdots\right)$$

$$= \cos\theta + i\sigma_2\sin\theta$$

This suggests that the representation

$$R = \exp(i\sigma_2\theta)$$

holds for all values of θ, not just small ones. The matrix

$$\mathbb{S} = i\sigma_2 = \begin{pmatrix} 0 & 1 \\ -1 & 0 \end{pmatrix}$$

is the generator[12] for the rotation matrices.

We can extend this idea to get the generators for the full set of three-dimensional rotations. We need the three matrices

$$\mathbb{S}_1 = \begin{pmatrix} 0 & 0 & 0 \\ 0 & 0 & -1 \\ 0 & 1 & 0 \end{pmatrix}, \quad \mathbb{S}_2 = \begin{pmatrix} 0 & 0 & 1 \\ 0 & 0 & 0 \\ -1 & 0 & 0 \end{pmatrix}, \quad \text{and} \quad \mathbb{S}_3 = \begin{pmatrix} 0 & -1 & 0 \\ 1 & 0 & 0 \\ 0 & 0 & 0 \end{pmatrix}$$

---

[12]Whether $\mathbb{S}$ or σ_2 is taken as the generator is primarily a matter of taste. Jackson uses $\mathbb{S}$.

Then any rotation may be written as

$$R = \exp(-\boldsymbol{\omega} \cdot \mathbf{S}) \tag{B.21}$$

where the dot product is

$$\boldsymbol{\omega} \cdot \mathbf{S} = \omega_1 \mathbb{S}_1 + \omega_2 \mathbb{S}_2 + \omega_3 \mathbb{S}_3 = \begin{pmatrix} 0 & -\omega_3 & \omega_2 \\ \omega_3 & 0 & -\omega_1 \\ -\omega_2 & \omega_1 & 0 \end{pmatrix} \tag{B.22}$$

If the vector $\boldsymbol{\omega}$ has only one component ω_3, we get back our previous result for rotations about the z-axis. The other components generalize the result for a rotation about an arbitrary axis described by the vector $\boldsymbol{\omega}$. Note that since the rotation matrix has determinant $+1$, the matrix $\boldsymbol{\omega} \cdot \mathbf{S}$ is traceless [det $(e^{\mathrm{M}}) = e^{\mathrm{Tr}\,(\mathrm{M})}$].

It is useful to note the relations between the matrices $\mathbb{S}_i$:

$$\mathbb{S}_1 \mathbb{S}_2 = \begin{pmatrix} 0 & 0 & 0 \\ 0 & 0 & -1 \\ 0 & 1 & 0 \end{pmatrix} \begin{pmatrix} 0 & 0 & 1 \\ 0 & 0 & 0 \\ -1 & 0 & 0 \end{pmatrix} = \begin{pmatrix} 0 & 0 & 0 \\ 1 & 0 & 0 \\ 0 & 0 & 0 \end{pmatrix}$$

and

$$\mathbb{S}_2 \mathbb{S}_1 = \begin{pmatrix} 0 & 0 & 1 \\ 0 & 0 & 0 \\ -1 & 0 & 0 \end{pmatrix} \begin{pmatrix} 0 & 0 & 0 \\ 0 & 0 & -1 \\ 0 & 1 & 0 \end{pmatrix} = \begin{pmatrix} 0 & 1 & 0 \\ 0 & 0 & 0 \\ 0 & 0 & 0 \end{pmatrix}$$

So

$$\mathbb{S}_1 \mathbb{S}_2 - \mathbb{S}_2 \mathbb{S}_1 = \begin{pmatrix} 0 & -1 & 0 \\ 1 & 0 & 0 \\ 0 & 0 & 0 \end{pmatrix} = \mathbb{S}_3$$

Similar relations hold for the other products. Since the generators do not commute, neither do the rotations they generate.

The rotation group is a subgroup of the Lorentz group (Jackson, Section 11.7; Arfken and Weber, Section 4.5). The full group has six generators: three for the three components of the relative velocity of the frames and three for the three parameters of the rotation (direction of rotation axis and angle of rotation).

The unitary group SU(2) is generated by the three Pauli matrices and is closely related to the rotation group SO(3). They are homomorphic, not isomorphic. That is, we can find a mapping that maps two elements of SU(2) to one element of SO(3).[13]

Now let's generalize. Suppose that elements near the identity in a group G can be described in terms of N real parameters α_k and that the identity element corresponds to $\alpha_k = 0$ for each k. The identity element is represented by the identity matrix:

$$M(\boldsymbol{\alpha})|_{\alpha=0} = \mathbf{1}$$

---

[13] See Arfken and Weber, Section 4.2, pp. 232–236.

where we have written the N parameters as a vector $\boldsymbol{\alpha}$. Since the parameters describe the matrices continuously, we may use a Taylor series to write elements near the identity as[14]

$$M(d\boldsymbol{\alpha}) = M(0) + \left.\frac{\partial M}{\partial \alpha_k}\right|_{\alpha=0} d\alpha_k + \cdots = M(0) + i\left(\left.-i\frac{\partial M}{\partial \alpha_k}\right|_{\alpha=0}\right) d\alpha_k$$

$$= \mathbf{1} + iX_k \, d\alpha_k \tag{B.23}$$

where the repeated index k implies summation. The quantities

$$X_k = -i\left.\frac{\partial M}{\partial \alpha_k}\right|_{\alpha=0}$$

are the generators of the group. If $\boldsymbol{\alpha}$ has only one component α, we can generate the group element represented by $M(\alpha)$ by repeating the operation represented by $M(\alpha/\beta)$ a total of β times. (You can understand this by thinking about rotations—rotate through an angle θ/β about the same axis β times.)

$$M(\alpha) = \left[M\left(\frac{\alpha}{\beta}\right)\right]^\beta$$

This element must be a member of the group if $M(\alpha/\beta)$ is. Now we let $\beta \to \infty$, α/β becomes differential, we obtain $M(\alpha/\beta)$ from equation (B.23), and we generate $M(\alpha)$ through the limit:

$$M(\alpha) = \lim_{\beta\to\infty}\left[M\left(\frac{\alpha}{\beta}\right)\right]^\beta = \lim_{\beta\to\infty}\left(1 + i\frac{\alpha X_k}{\beta}\right)^\beta$$

Expanding the expression on the right, we have

$$\left(1 + i\frac{\alpha X_k}{\beta}\right)^\beta = 1 + \beta i\frac{\alpha X_k}{\beta} + \frac{\beta(\beta-1)}{2}\left[i\frac{\alpha X_k}{\beta}\right]^2 + \cdots$$

$$\to 1 + i\alpha X_k + \frac{1}{2}(i\alpha X_k)^2 + \cdots \quad \text{as } \beta \to \infty$$

$$\boxed{M(\alpha) = e^{i\alpha X_k}} \tag{B.24}$$

Thus, the structure of the group is defined by the elements near the identity.

B.10. LIE ALGEBRAS

Suppose a group G with elements R_i has generators S_i. Then we can write (compare with equations B.21 and B.24)

$$R = e^{\boldsymbol{\omega}\cdot\mathbf{S}}$$

---

[14]The introduction of i here is to make the resulting generators Hermitian.

In particular, there is a set of elements near the identity for which

$$R_i = e^{\varepsilon_i S_i} = 1 + \varepsilon_i S_i + \frac{1}{2}\varepsilon_i^2 S_i^2 + \cdots$$

with $\varepsilon_i \ll 1$. For such an element,

$$R_i^{-1} = e^{-\varepsilon_i S_i} = 1 - \varepsilon_i S_i + \frac{1}{2}\varepsilon_i^2 S_i^2 + \cdots$$

Now if R is unitary, then

$$\det (R_i) = \exp [\varepsilon_i \operatorname{Tr} (S_i)] = 1$$

and so $\operatorname{Tr} (S_i) = 0$. Let R_j be defined similarly. Then

$$R_j^{-1} R_i R_j = R_j^{-1}\left(1 + \varepsilon_i S_i + \frac{1}{2}\varepsilon_i^2 S_i^2 + \cdots\right) R_j$$

$$= 1 + \varepsilon_i R_j^{-1} S_i R_j + \frac{1}{2}\varepsilon_i^2 R_j^{-1} S_i^2 R_j + \cdots$$

$$= 1 + \varepsilon_i\left(1 - \varepsilon_j S_j + \frac{1}{2}\varepsilon_j^2 S_j^2\right) S_i\left(1 + \varepsilon_j S_j + \frac{1}{2}\varepsilon_j^2 S_j^2\right) + \frac{1}{2}\varepsilon_i^2 S_i^2$$

$$= 1 + \varepsilon_i S_i + \varepsilon_i\varepsilon_j(S_i S_j - S_j S_i) + \frac{1}{2}\varepsilon_i^2 S_i^2$$

where we have dropped terms of third order in the small parameters ε_i and ε_j. Finally, we compute

$$R_i^{-1} R_j^{-1} R_i R_j = \left(1 - \varepsilon_i S_i + \frac{1}{2}\varepsilon_i^2 S_i^2\right)\left(1 + \varepsilon_i S_i + \varepsilon_i\varepsilon_j(S_i S_j - S_j S_i) + \frac{1}{2}\varepsilon_i^2 S_i^2\right)$$

$$= 1 + \varepsilon_i\varepsilon_j(S_i S_j - S_j S_i)$$

to second order in the εs. Now this element of the group can also be expressed in terms of the generators, so we must have

$$1 + \varepsilon_i\varepsilon_j(S_i S_j - S_j S_i) = e^{\boldsymbol{\omega}_{ij}\cdot\mathbf{S}} = 1 + \sum_k \omega_{ij}^k S_k$$

In particular, if we write $\omega_{ij}^k = \varepsilon_i\varepsilon_j\alpha_{ij}^k$, then we must have

$$\boxed{(S_i S_j - S_j S_i) \equiv [S_i, S_j] = \sum_k \alpha_{ij}^k S_k} \tag{B.25}$$

This is a fundamental property of the generators S_i, independent of the particular elements R_i and R_j that we picked. The constants α_{ij}^k are called the structure constants of the group G.

Because of the antisymmetry of the commutator $[S_i, S_j]$ (equation B.25),

$$\alpha_{ij}^k = -\alpha_{ji}^k \tag{B.26}$$

We can now compute the double commutator:

$$
\begin{aligned}
[S_i, [S_j, S_k]] &= S_i(S_j S_k - S_k S_j) - (S_j S_k - S_k S_j)S_i \\
&= S_i S_j S_k - S_j S_k S_i - S_i S_k S_j + S_j S_k S_i \\
&= \left[S_i, \sum_m \alpha_{jk}^m S_m \right] = \sum_m \alpha_{jk}^m [S_i, S_m] = \sum_m \alpha_{jk}^m \sum_n \alpha_{im}^n S_n
\end{aligned}
$$

Then the Jacobi identity

$$[S_i, [S_j, S_k]] + [S_j, [S_k, S_i]] + [S_k, [S_i, S_j]] = 0 \tag{B.27}$$

which follows from the definition of the commutator, gives another constraint on the structure constants:

$$\sum_n \sum_m \alpha_{jk}^m \alpha_{im}^n S_n + \sum_n \sum_m \alpha_{ki}^m \alpha_{jm}^n S_n + \sum_n \sum_m \alpha_{ij}^m \alpha_{km}^n S_n = 0$$

$$\sum_n \sum_m S_n \left(\alpha_{jk}^m \alpha_{im}^n + \alpha_{ki}^m \alpha_{jm}^n + \alpha_{ij}^m \alpha_{km}^n \right) = 0$$

or

$$\sum_m \left(\alpha_{jk}^m \alpha_{im}^n + \alpha_{ki}^m \alpha_{jm}^n + \alpha_{ij}^m \alpha_{km}^n \right) = 0 \tag{B.28}$$

For the rotation group SO(3), we found

$$
S_1 S_2 - S_2 S_1 = \begin{pmatrix} 0 & -1 & 0 \\ 1 & 0 & 0 \\ 0 & 0 & 0 \end{pmatrix} = S_3
$$

so $\alpha_{12}^3 = 1$. Then, from equation (B.26), we have $\alpha_{21}^3 = -1$. In fact, the structure constants for this group are the Levi-Civita symbols

$$\alpha_{ij}^k = \varepsilon_{ijk} \tag{B.29}$$

Although we used a specific representation to calculate the structure constants, they are independent of the particular representation.

We can check that relation (B.28) is also satisfied:

$$\sum_m \left(\alpha_{jk}^m \alpha_{im}^n + \alpha_{ki}^m \alpha_{jm}^n + \alpha_{ij}^m \alpha_{km}^n \right) = \sum_m \left(\varepsilon_{jkm} \varepsilon_{imn} + \varepsilon_{kim} \varepsilon_{jmn} + \varepsilon_{ijm} \varepsilon_{kmn} \right)$$

$$= \sum_m \left(\varepsilon_{jkm} \varepsilon_{nim} + \varepsilon_{kim} \varepsilon_{njm} + \varepsilon_{ijm} \varepsilon_{nkm} \right)$$

$$= \delta_{jn}\delta_{ki} - \delta_{ji}\delta_{kn} + \delta_{kn}\delta_{ij} - \delta_{in}\delta_{jk} + \delta_{in}\delta_{jk} - \delta_{jn}\delta_{ki}$$

$$= 0$$

If we define the commutator as the multiplication operator for the generators, the vector space of generators, with both addition and multiplication defined, becomes a field or *algebra* called the Lie algebra for the group G.

A Lie algebra is a vector space[15] of elements v_i with an operation (the *bracket* or *commutator*) that satisfies the following rules.

1. The bracket is linear:

$$[av_1 + bv_2, v_3] = a[v_1, v_3] + b[v_2, v_3]$$

and

$$[v_1, av_2 + bv_3] = a[v_1, v_2] + b[v_1, v_3]$$

2. The bracket is antisymmetric:

$$[v_1, v_2] = -[v_2, v_1]$$

3. The Jacobi relation is satisfied:

$$[v_1, [v_2, v_3]] + [v_2, [v_3, v_1]] + [v_3, [v_1, v_2]] = 0$$

A group whose generators form a Lie algebra is called a Lie group. Lie groups are tremendously important in physics. The group structure is determined by the Lie algebra of the generators, as expressed by the structure constants. The rotation group, the Lorentz group (which includes the rotation group as a subgroup[16]), and the symmetry groups of particle physics are all Lie groups.

PROBLEMS

1. Show that the set of permutations of two elements is a group. What is its order? Write the multiplication table for this group. How many classes are there? Is the group abelian?

---

[15]See Chapter 1.
[16]See Problem 15.

2. The symmetry group of a square contains those operations that leave the square unchanged. Show that this group has eight elements, and write the multiplication table. What are the classes? Are there any subgroups?

3. Show that the set $\{1, -1, i, -i\}$ forms a group under algebraic multiplication. Write the multiplication table. How many classes are there? Show that this group is isomorphic to a group of *rotations* that preserve the symmetry of a square (compare with Problem 2 and Problem 10).

4. Show that there are two groups of order four, and determine their multiplication tables.

5. Show that any group of order n is isomorphic to a subgroup of S_n.

6. Show that unitary matrices of the form (B.2) form a group under matrix multiplication.

7. Show that in any abelian group, each element forms its own class. Hence show that all of the irreps of an abelian group are one-dimensional.

8. Show that the set of 2×2 matrices of the form

$$\begin{pmatrix} x & y \\ -y & x \end{pmatrix}$$

forms a group under the operations of **(a)** addition and **(b)** matrix multiplication. Show that the group of complex numbers $x + iy$ is isomorphic with this group of matrices in each of the two cases.

9. The *quaternions* are four-dimensional complex numbers of the form $q = a + bi + cj + dk$, where $a, b, c,$ and d are real numbers and the quantities $i, j,$ and k obey the multiplication rules

$$i^2 = j^2 = k^2 = -1$$

and

$$ij = k; \quad ji = -k$$

(a) Show that the set $\{\pm 1, \pm i, \pm j, \pm k\}$ forms a group under this multiplication.

(b) Show that $i, j,$ and k may be represented by the matrices

$$i = \begin{pmatrix} 0 & 1 & 0 & 0 \\ -1 & 0 & 0 & 0 \\ 0 & 0 & 0 & 1 \\ 0 & 0 & -1 & 0 \end{pmatrix}, \quad j = \begin{pmatrix} 0 & 0 & 0 & -1 \\ 0 & 0 & -1 & 0 \\ 0 & 1 & 0 & 0 \\ 1 & 0 & 0 & 0 \end{pmatrix},$$

$$k = \begin{pmatrix} 0 & 0 & -1 & 0 \\ 0 & 0 & 0 & 1 \\ 1 & 0 & 0 & 0 \\ 0 & -1 & 0 & 0 \end{pmatrix}$$

Determine the classes of this group.

(c) Determine the number and dimension of irreps of this group, and find the character table.

(d) Is the representation in (b) reducible? If so, how?

10. Show that the integers 1 through 4 form a group under the operation of multiplication mod 5. Write the multiplication table. What is the identity element? How many classes are there? Is the group abelian? Another group with four elements is a group of rotations that preserve the symmetry of a square—that is, rotations by multiples of $\pi/2$. Are these groups isomorphic? Why or why not?

11. Consider the mapping f that maps the group of rationals to the group of integers by

$$f\left(\frac{m}{n}\right) = m + n$$

Is the mapping a homomorphism? Why or why not?

12. Show that all elements in the same class have the same period.

13. Show that if a set of elements $\{e_i\}$ forms a class of a group G, then the set $\{e_i^{-1}\}$ of the inverses of $\{e_i\}$ is also a class.

14. The *center* Z of a group G is the set of elements that commute with every element in the group. Show that the center is an abelian subgroup of G. Is it possible for a group to have no center?

15. A homomorphism f maps group A to group B. The *kernel* of the homomorphism is the set of all elements of A that map to the identity element of B. Show that the kernel is an invariant subgroup of A.

16. A one-dimensional translation operator T_n translates a function along the x-axis by an amount nd, where d is a fixed step length: $T_n f(x) = f(x + nd)$.

   (a) Show that the set of operators T_n forms a group that may be represented by the complex numbers $T_n \rightleftarrows e^{-iknd}$. What are the corresponding basis functions?

   (b) Work out the orthogonality relation (B.14) for this representation, and comment. You will have to make some changes to account for the infinite order of the group.

   (c) Now let the operator translate by an arbitrary amount x': $T(x')f(x) = f(x + x')$. What are the generators of this group?

17. The water molecule is an isosceles triangle with the oxygen atom at one vertex and the hydrogen atoms symmetrically located on either side. Determine the symmetry group for this molecule. May this molecule possess a permanent electric dipole moment? What about a magnetic moment?

18. The molecule SbS_5 is square-pyramidal. Four S atoms form a square base with the Sb atom at the center. The fifth S atom sits at the top of the pyramid. Determine the symmetry group for this system. What is the order of the group? Work out the multiplication table. How many classes are there? Determine the character table. May this molecule possess a permanent electric dipole moment?

19. The Lorentz group has generators that are 4×4 matrices with mostly zero elements. The matrices K_i are given by

$$K_1 = \begin{pmatrix} 0 & 1 & 0 & 0 \\ 1 & 0 & 0 & 0 \\ 0 & 0 & 0 & 0 \\ 0 & 0 & 0 & 0 \end{pmatrix}$$

and so on. (The nonzero elements of K_i are the ith elements in the top row and the first column, where the first element is labeled with 0, not 1.) Similarly, the generators S_i are given by

$$S_1 = \begin{pmatrix} 0 & 0 & 0 & 0 \\ 0 & 0 & 0 & 0 \\ 0 & 0 & 0 & 1 \\ 0 & 0 & -1 & 0 \end{pmatrix}$$

(The nonzero elements of S_i are given by $a_{jk}(S_i) = \varepsilon_{ijk}$, where a_{jk} is the submatrix formed by removing the top row and left-most column and $j, k = 1, 2, 3$.) Find the group element generated by K_1 and also by S_1.

Compute the product of the two group elements e^{aK_1} and e^{bK_2}. Hence show that the elements generated by K_i do not form a subgroup. Do the elements formed by the S_i form a subgroup? Are there any other subgroups? If so, what are they?

20. Show that the transformations $x' = ax + b$ (where x', x, a, and b are real numbers and $a \neq 0$) form a group. Form a two-dimensional representation of this group that acts on the vectors $(x, 1)$.

21. A homomorphism f maps a group G to a group H. Show that the image $f(G)$ in H is isomorphic to the factor group G/K, where K is the kernel of the homomorphism (Problem 15).

22. A group G has an invariant subgroup S. If element a of group G has period N, where N is prime, and a is not a member of the subgroup S, show that element aS of the factor group G/S also has period N.

OPTIONAL TOPIC C

Green's Functions

Many physical systems are linear and consequently are described by a linear differential equation. An important example is the electromagnetic field, and we shall treat this system in detail in Section C.5. Such systems obey a superposition principle: The response of the system to a sum of inputs is just the sum of the responses to the individual inputs. As a result, these systems can often be analyzed using a Green's function. The Green's function is actually a distribution (Chapter 6) that describes the response of a physical system to a unit delta function input.

In general, any physical system in the region $0 < x < L$ is described by a differential equation for some quantity, say $y(x)$, plus a set of boundary conditions. The boundary conditions provide a sort of shorthand for representing additional sources that exist outside the region of interest. The system is given an input $I(x)$, and we want to find the response. The Green's function is the solution to a similar physical problem, but with the source term replaced by a delta function (a "point source") and with zero (or, at worst, constant) values on the boundary. The Green's function problem is thus easier to solve than the original mathematical problem, which is the advantage of the Green's function method.

We shall begin by discussing one-dimensional problems for which the solution itself is zero at the boundaries. A physical system exists in a region defined by $0 \le x \le L$. If the input to the system is $I(x)$, we can write it as

$$I(x) = \int_0^L I(x')\delta(x - x')\,dx'$$

If the response of the system at x to the delta function at x' is $G(x, x')$, then, by the principle of superposition for linear systems, the response to the input $I(x)$ is

$$y(x) = \int_0^L I(x')G(x, x')\,dx' \qquad \text{(C.1)}$$

The Green's function is symmetric in the two variables:

$$\boxed{G(x, x') = G(x', x)} \qquad \text{(C.2)}$$

If the variable is a time variable, then

$$G(t, t') = G(-t', -t)$$ (C.3)

This sign change is necessary to preserve causality—the response cannot precede the event that caused it. (See Morse and Feshbach, Section 7.3, for a more detailed discussion of this point.)

Because the differential equation satisfied by the Green's function contains a delta function, strictly speaking G is not a function, but a distribution (Chapter 6). Indeed, the physical solution is obtained as an integral over G, as expected if G is a distribution. The method is advantageous provided that the integral (C.1) is relatively easy to do.

The methods used to find the Green's function for a given problem may be classified into three groups: (a) divide the region of interest into two pieces with the point source on the boundary between them, (b) expand the Green's function as a series of eigenfunctions, or (c) use an integral transform.

C.1. DIVISION-OF-REGION METHOD

First we write the differential equation and boundary conditions that describe the system in the absence of sources. (The boundaries may be in space, time, or both.) With no sources, the differential equation will be a homogeneous equation. We determine the solutions of this homogeneous equation.

Next we write the differential equation for the Green's function problem. The source is now a delta function [for example, $\delta(x - x')$] in which the position x' of the source is for the moment considered *fixed*. This source may be imagined as dividing space into two regions, $x < x'$ and $x > x'$. We write down the appropriate solutions of the homogeneous equation that match the boundary condition or conditions in the two regions. There will be one or more unknown constants in these solutions.

We also have to consider the boundary conditions that apply at the intermediate boundary that we have constructed at $x = x'$. We need enough boundary conditions to determine the unknown constants in G.

Once $G(x, x')$ has been determined, we can use it to find the response of the system to any input, using the integral (C.1). In evaluating this integral, we consider x' to be variable and x to be fixed.

To illustrate the method, consider a particle moving under the influence of a driving force $F(t)$ with damping proportional to velocity. The equation[1] describing the system is

$$m\frac{dv}{dt} + \alpha v = F(t)$$

---

[1]Compare with equation (3.5), with mg replaced by $F(t)$.

The boundary conditions in time are $v \rightarrow 0$ as $t \rightarrow \pm\infty$. The equation for the Green's function is

$$m\frac{dG}{dt} + \alpha G = \delta(t - t') \tag{C.4}$$

For $t \neq t'$, the delta function is zero and the equation simplifies to

$$m\frac{dG}{dt} + \alpha G = 0$$

which has the solution (Chapter 3, Section 3.3.1)

$$G = A\exp\left(-\frac{\alpha}{m}t\right)$$

This function approaches zero as $t \rightarrow +\infty$, but blows up as $t \rightarrow -\infty$. Since the appropriate solution for $t < t'$ must approach zero as $t \rightarrow -\infty$, it must be identically zero. The system is set in motion by the impulse applied at $t = t'$. Thus, we can write the solution for $t < t'$ and for $t > t'$:

$$G(t, t') = \begin{cases} 0 & \text{if } t < t' \\ A\exp\left(-\alpha t/m\right) & \text{if } t > t' \end{cases} \tag{C.5}$$

Since G is a function of both t and t', we expect that the "constant" A is actually a function of t'. Our next task is to find this function.

We imagine the delta function input at $t = t'$ dividing the system into two regions, with the point $t = t'$ on the boundary between them. To find an expression for A, we integrate the differential equation (C.4) across the (imaginary) boundary at $t = t'$:

$$\int_{t'-\varepsilon}^{t'+\varepsilon}\left(m\frac{dG}{dt} + \alpha G\right)dt = \int_{t'-\varepsilon}^{t'+\varepsilon}\delta(t - t')\,dt = 1$$

where we used the sifting property to evaluate the right-hand side. On the left-hand side, the first term integrates easily:

$$mG|_{t'-\varepsilon}^{t'+\varepsilon} + \alpha\int_{t'-\varepsilon}^{t'+\varepsilon}G\,dt = 1$$

The remaining integral is bounded:

$$\left|\int_{t'-\varepsilon}^{t'+\varepsilon}G\,dt\right| \leq \max|G|(2\varepsilon) \leq 2\varepsilon|A|$$

Provided G remains finite, this integral approaches zero as $\varepsilon \rightarrow 0$. Now we insert our solution (C.5) for G in the two regions into the resulting equation. At the lower limit, $t < t'$

and $G = 0$; at the upper limit, $t > t'$ and $G = A \exp(-\alpha t/m)$. Thus, as $\varepsilon \to 0$,

$$m\left[A \exp\left(-\frac{\alpha}{m}t'\right) - 0\right] = 1$$

$$A = \frac{1}{m} \exp\left(\frac{\alpha}{m}t'\right)$$

and so

$$G(t, t') = \begin{cases} 0 & \text{if } t < t' \\ \dfrac{1}{m} \exp\left(-\dfrac{\alpha}{m}[t - t']\right) & \text{if } t > t' \end{cases} \tag{C.6}$$

A piecewise-defined solution for G always results from the division-of-region method. Notice that the reciprocity relation (C.3) is satisfied by the function (C.6).

Example C.1.   Find $v(t)$ if the input $F(t)$ is a force that is a constant F_0 between times $t = 0$ and $t = T$ and zero otherwise.

The resulting velocity is (compare with equation C.1)

$$v(t) = \int_{-\infty}^{+\infty} F(t')G(t, t')\, dt' = \int_{-\infty}^{+\infty} \begin{pmatrix} 0 & \text{if } t' < 0 \\ F_0 & \text{if } 0 < t' < T \\ 0 & \text{if } t' > T \end{pmatrix} G(t, t')\, dt'$$

Since the force becomes zero for $t' < 0$ and for $t' > T$, the limits of the integral are 0 to T. Thus,

$$v(t) = 0 \quad \text{if } t < 0$$

and for $t > 0$,

$$v(t) = F_0 \int_0^T G(t, t')\, dt'$$

If $0 < t < T$, since G is zero for $t < t'$ (equation C.6), the upper limit of the integral is t:

$$v(t) = F_0 \int_0^t \frac{1}{m} \exp\left(-\frac{\alpha}{m}[t - t']\right) dt'$$

$$= \frac{F_0}{m} \left. \frac{\exp\left(-\dfrac{\alpha}{m}[t - t']\right)}{\alpha/m} \right|_0^t$$

$$= \frac{F_0}{\alpha} \left[1 - \exp\left(-\frac{\alpha}{m}t\right)\right]$$

On the other hand, if $t > T$,

$$
v(t) = F_0 \int_0^T \frac{1}{m} \exp\left(-\frac{\alpha}{m}[t - t']\right) dt'
$$

$$
= \frac{F_0}{m} \left.\frac{\exp\left(-\frac{\alpha}{m}[t - t']\right)}{\alpha/m}\right|_0^T
$$

$$
= \frac{F_0}{\alpha} \left[\exp\left(-\frac{\alpha}{m}[t - T]\right) - \exp\left(-\frac{\alpha}{m}t\right)\right]
$$

$$
= \frac{F_0}{\alpha} \exp\left(-\frac{\alpha}{m}t\right) \left[\exp\left(\frac{\alpha}{m}T\right) - 1\right]
$$

Once the force ceases, the speed decreases exponentially (see Figure C.1). Notice that both expressions give the same result for $v(t)$ at $t = T$.

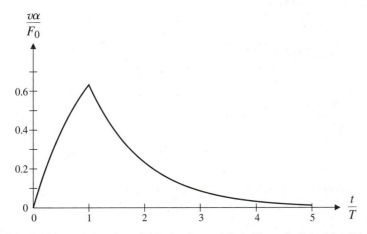

FIGURE C.1. Velocity as a function of time for the particle in Example C.1 with $\alpha T/m = 1$. The velocity decreases exponentially at large times.

C.2. EXPANSION IN EIGENFUNCTIONS

Instead of dividing our region into two pieces and obtaining a piecewise-defined Green's function, we can sometimes expand the Green's function in a series of eigenfunctions.

If the governing differential equation is of the Sturm-Liouville type (equation 8.1) and the boundary conditions are of the form (8.2), then there exists a set of eigenfunctions $y_n(x)$ for the problem, with associated eigenvalues λ_n. We will assume that the eigenfunctions have been normalized so that

$$
\int_a^b w(x) y_m(x) y_n(x) \, dx = \delta_{mn} \tag{C.7}
$$

We require that the differential equation for our Green's function problem be of the same form but with a constant λ that is not equal to *any* of the eigenvalues λ_n. The source is a delta function:

$$\frac{d}{dx}\left(f(x)\frac{dG}{dx}\right) - g(x)G + \lambda w(x)G = -4\pi\delta(x-x') \qquad (C.8)$$

The factor of -4π is traditional—it arises from the use of this method in potential theory.[2]

Then the Green's function may be expanded in eigenfunctions:

$$G(x,x') = \sum_{n=0}^{\infty} \gamma_n\left(x'\right) y_n(x) \qquad (C.9)$$

For the moment we consider the value x' to be *fixed*. We substitute this assumed form (C.9) into equation (C.8):

$$\sum_{n=0}^{\infty} \gamma_n(x')\left[\frac{d}{dx}\left(f(x)\frac{d}{dx}y_n(x)\right) - g(x)y_n(x) + \lambda w(x)y_n(x)\right] = -4\pi\delta(x-x')$$

$$(C.10)$$

Now we know that each eigenfunction satisfies an equation of the form

$$\frac{d}{dx}\left(f(x)\frac{dy_n}{dx}\right) - g(x)y_n + \lambda_n w(x)y_n = 0$$

so we use this to simplify equation (C.10):

$$\sum_{n=0}^{\infty} \gamma_n(x')\left[-\lambda_n w(x)y_n(x) + \lambda w(x)y_n(x)\right] = -4\pi\delta(x-x')$$

Next we make use of the orthogonality of the eigenfunctions by multiplying the whole equation by $y_m(x)$ and integrating over the range $x = a$ to $x = b$:

$$\int_a^b \sum_{n=0}^{\infty}(\lambda-\lambda_n)w(x)\gamma_n(x')y_n(x)y_m(x)\,dx = -4\pi\int_a^b \delta(x-x')y_m(x)\,dx$$

$$\sum_{n=0}^{\infty}(\lambda-\lambda_n)\gamma_n(x')\int_a^b w(x)y_n(x)y_m(x)\,dx = -4\pi y_m(x')$$

$$(\lambda-\lambda_m)\gamma_m(x') = -4\pi y_m(x')$$

where we used the sifting property to evaluate the integral on the right and equation (C.7) to evaluate the integral on the left. Then

$$\gamma_m(x') = 4\pi\frac{y_m(x')}{\lambda_m - \lambda}$$

---

[2] See Section C.6 below.

and the Green's function is

$$G(x, x') = 4\pi \sum_{n=0}^{\infty} \frac{y_n(x')y_n(x)}{\lambda_n - \lambda} \qquad \text{(C.11)}$$

Note the symmetry in x and x', as required by equation (C.2).

Example C.2. Find the Green's function for the one-dimensional Helmholtz equation in the region $0 < x < L$ with $y(x) = 0$ at $x = 0$ and at $x = L$. Hence find $y(x)$ if $f(x)$ equals 1 for $L/4 < x < 3L/4$ and equals 0 outside this range.
The equation for the Green's function is

$$\frac{d^2y}{dx^2} + k^2 y = \delta(x - x') \qquad \text{(C.12)}$$

which is of the form (C.8) with $\lambda \equiv k^2$ and the factor -4π on the right-hand side replaced by 1. Thus, the equation for the eigenfunctions is of the Sturm-Liouville form

$$\frac{d^2 y_n}{dx^2} + \lambda_n y = 0$$

with $\lambda_n \neq k^2$. The solutions to the eigenfunction equation that satisfy the boundary condition at $x = 0$ are

$$y_n = c_n \sin\left(\sqrt{\lambda_n} x\right)$$

To satisfy the second condition at $x = L$, we need

$$\sqrt{\lambda_n} L = n\pi$$

and so the eigenvalues are

$$\lambda_n = \left(\frac{n\pi}{L}\right)^2$$

We still need to normalize the functions. We choose the constant c_n so that

$$\int_0^L y_n(x)y_n(x)\, dx = 1$$

$$c_n^2 \int_0^L \sin^2\left(\frac{n\pi}{L}x\right) dx = c_n^2 \frac{L}{2} = 1$$

$$c_n = \sqrt{\frac{2}{L}}$$

So the normalized eigenfunctions are

$$y_n(x) = \sqrt{\frac{2}{L}} \sin\left(\frac{n\pi}{L}x\right)$$

and the Green's function is (compare with equation C.11 with 4π replaced by -1)

$$G(x, x') = \frac{2}{L} \sum_{n=1}^{\infty} \frac{\sin\left(\frac{n\pi}{L}x'\right)\sin\left(\frac{n\pi}{L}x\right)}{k^2 - (n\pi/L)^2} = \frac{2L}{\pi^2} \sum_{n=1}^{\infty} \frac{\sin\left(\frac{n\pi}{L}x'\right)\sin\left(\frac{n\pi}{L}x\right)}{k^2 L^2/\pi^2 - n^2}$$

Notice that the expression is not valid if $k = n\pi/L$ for any n. Physically, this corresponds to driving the system at a resonant frequency. The displacement becomes arbitrarily large in the absence of damping, and there is no solution.

Then if $f(x) = 1$ for $L/4 < x < 3L/4$,

$$y(x) = \int_0^L G(x, x') f(x') \, dx' = \int_{L/4}^{3L/4} G(x, x') \, dx'$$

$$= \frac{2}{L} \int_{L/4}^{3L/4} \sum_n \frac{\sin\frac{n\pi x}{L} \sin\frac{n\pi x'}{L}}{k^2 - (n\pi/L)^2} \, dx'$$

$$= \frac{2}{L} \sum_n \frac{\sin\frac{n\pi x}{L}}{k^2 - (n\pi/L)^2} \int_{L/4}^{3L/4} \sin\frac{n\pi x'}{L} \, dx'$$

$$= \frac{2}{L} \sum_n \frac{\sin\frac{n\pi x}{L}}{k^2 - (n\pi/L)^2} L \frac{\cos\frac{1}{4}n\pi - \cos\frac{3}{4}n\pi}{n\pi}$$

$$= 2 \sum_{n=1}^{\infty} \frac{\sin\frac{n\pi x}{L}}{(n\pi/L)^2 - k^2} \frac{2\sin\frac{n\pi}{2}\sin\frac{n\pi}{4}}{n\pi}$$

$$= 4 \sum_{n=1,\, n \text{ odd}}^{\infty} (-1)^{(n-1)/2} \left[\frac{\sin\frac{n\pi x}{L}}{(n\pi/L)^2 - k^2} \right] \frac{\sin\frac{n\pi}{4}}{n\pi}$$

The sum is over odd values of n, because $\sin(n\pi/2) = 0$ if n is even. See Figure C.2.

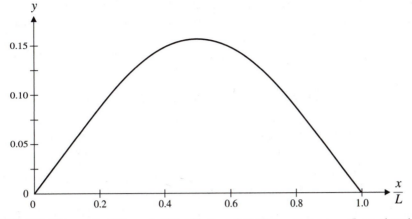

FIGURE C.2. The result of Example C.2 with $k = 2/L$. Terms up to $n = 5$ are plotted here. Including terms up to $n = 9$ does not change the plot noticeably.

C.3. TRANSFORM METHODS

We have already found the Fourier (Chapter 7, Example 7.3) and Laplace (Chapter 6, Section 6.1.6) transforms of the delta function. Thus, we can solve for the Green's function by transforming its defining equation.

The Green's function for the damped harmonic oscillator is given by

$$\frac{d^2 G(t, t')}{dt^2} + 2\alpha \frac{dG(t, t')}{dt} + \omega_0^2 G(t, t') = \delta(t - t') \tag{C.13}$$

We may Fourier transform this equation to get

$$-\omega^2 G(\omega, t') - 2i\omega\alpha G(\omega, t') + \omega_0^2 G(\omega, t') = \frac{1}{\sqrt{2\pi}} e^{i\omega t'}$$

Then the transform of G is

$$G(\omega, t') = \frac{1}{\sqrt{2\pi}} \frac{e^{i\omega t'}}{\omega_0^2 - 2i\alpha\omega - \omega^2}$$

Inverting, we get

$$G(t, t') = \frac{1}{2\pi} \int_{-\infty}^{+\infty} \frac{e^{i\omega t'}}{\omega_0^2 - 2i\alpha\omega - \omega^2} e^{-i\omega t} \, d\omega \tag{C.14}$$

We may evaluate the integral using the residue theorem. The integrand has two simple poles, where the denominator is zero. These are

$$\omega_p = -i\alpha \pm \sqrt{\omega_0^2 - \alpha^2}$$

Both poles are in the lower half-plane, as we expect from causality (Chapter 7, Section 7.4).

For $t < t'$, we must close the contour upward. There are no poles inside the contour, and the result is zero:

$$G(t, t') = 0 \quad \text{for } t < t' \tag{C.15}$$

For $t > t'$, we close downward, enclosing both poles. Then, writing $\sqrt{\omega_0^2 - \alpha^2} \equiv \Omega$, we have

$$G(t, t') = -\frac{1}{2\pi}(-2\pi i)\left(\frac{e^{i(t'-t)(-i\alpha+\Omega)}}{2\Omega} + \frac{e^{i(t'-t)(-i\alpha-\Omega)}}{-2\Omega}\right)$$

$$= \frac{i}{2\Omega}e^{-\alpha(t-t')}\left(e^{-i\Omega(t-t')} - e^{i\Omega(t-t')}\right)$$

$$G(t, t') = e^{-\alpha(t-t')}\frac{\sin\Omega(t - t')}{\Omega} \quad \text{for } t > t' \tag{C.16}$$

The result is of the same form that we would get from the division-of-region method (Section C.1).

C.4. EXTENSION TO N DIMENSIONS

Problems defined in a region with two or more dimensions are handled in a similar manner. The Green's function is defined in the same region of space (and/or time) as the original function f, but both the differential equation and the boundary conditions are simpler.

1. The source term on the right-hand side of the original differential equation is replaced by a delta function representing a point source.
2. The boundary conditions are replaced by
   (a) Dirichlet conditions: $G(\vec{\mathbf{x}}, \vec{\mathbf{x}}') = 0$ on S
   or
   (b) Neumann conditions: $\hat{\mathbf{n}} \cdot \vec{\nabla} G(\vec{\mathbf{x}}, \vec{\mathbf{x}}') = \text{constant}^3$ on S
   where S is the surface that bounds the volume of interest, V.
   In case (a), we have the Dirichlet Green's function; in case (b), we have the Neumann Green's function.

The method used to find the Green's function will be a combination of the methods used in the one-dimensional case.

We'll begin with an example in two dimensions with homogeneous boundary conditions. We'll consider inhomogeneous boundary conditions later.

A potential problem is defined in a two-dimensional rectangular region of dimensions $a \times b$. The boundaries are conducting and are grounded. We'll choose a coordinate system

---

[3]The constant may be zero in some cases. See Section C.6 below.

with x- and y-axes parallel to the sides of the region and the origin in one corner. Then the potential satisfies Poisson's equation:

$$\frac{\partial^2 \Phi}{\partial x^2} + \frac{\partial^2 \Phi}{\partial y^2} = -4\pi\sigma(x, y)$$

where the term on the right-hand side describes the sources for the potential (the charge distribution in the region).

The Green's function is the solution to the similar, but simpler, equation

$$\frac{\partial^2 G(\vec{\mathbf{x}}, \vec{\mathbf{x}}')}{\partial x^2} + \frac{\partial^2 G(\vec{\mathbf{x}}, \vec{\mathbf{x}}')}{\partial y^2} = -4\pi\delta(\vec{\mathbf{x}} - \vec{\mathbf{x}}') \tag{C.17}$$

where for the moment we consider $\vec{\mathbf{x}}'$ to be *fixed*.

Method 1 Our goal is to reduce equation (C.17) to a one-dimensional equation so that we may integrate across the discontinuity, as we did in Section C.1. We do this by expanding in eigenfunctions. This is equivalent to expressing the delta function in one of the coordinates using the completeness relation (8.10).

We imagine the delta function dividing the rectangular region into two pieces. We can choose the interior boundary through the point $\vec{\mathbf{x}}'$ to be parallel to the x-axis or parallel to the y-axis. In this example, we'll choose to divide in y (Figure C.3). Then *within* each region there are no sources, and the differential equation for G is simpler:

$$\frac{\partial^2 G(\vec{\mathbf{x}}, \vec{\mathbf{x}}')}{\partial x^2} + \frac{\partial^2 G(\vec{\mathbf{x}}, \vec{\mathbf{x}}')}{\partial y^2} = 0 \tag{C.18}$$

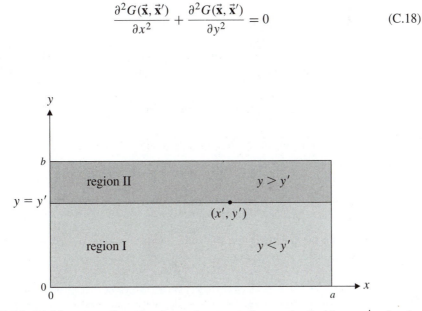

FIGURE C.3. Dividing a two-dimensional space into two regions: region I with $y < y'$ and region II with $y > y'$. The source at (x', y') is on the boundary between the two regions.

We begin by solving this equation by separation of variables, subject to the boundary conditions at $x = 0$ and $x = a$. Let $G = X(x)Y(y)$. Then

$$X''Y + XY'' = 0$$

We divide through by $G = XY$:

$$\frac{X''}{X} + \frac{Y''}{Y} = 0$$

To satisfy the equation at *all* values of x and y, we must have

$$\frac{X''}{X} = -k^2$$

and

$$\frac{Y''}{Y} = k^2$$

where we chose the separation constant to be negative so that the solution for X is a function with more than one zero. (Compare with Chapter 8, Section 8.2.) Here we need $X = \sin kx$. Then to satisfy the boundary condition at $x = a$, we need to choose the eigenvalue $k_n = n\pi/a$ so that $\sin k_n a = 0$. The equation for Y becomes

$$Y'' = \left(\frac{n\pi}{a}\right)^2 Y$$

with solutions[4]

$$Y = \sinh \frac{n\pi y}{a} \quad \text{and} \quad Y = \cosh \frac{n\pi y}{a}$$

In region I, $0 \le y < y'$, we must choose the hyperbolic sine function so that $Y(0) = 0$. Then our solution has the form

$$G_I(\vec{\mathbf{x}}, \vec{\mathbf{x}}') = \sum_{n=1}^{\infty} C_n(x', y') \sin \frac{n\pi x}{a} \sinh \frac{n\pi y}{a} \quad \text{for } y < y'$$

In region II, $b \ge y > y'$, we must choose a solution that is zero at $y = b$. The appropriate choice is $\sinh[n\pi(b - y)/a]$. Then we have

$$G_{II}(\vec{\mathbf{x}}, \vec{\mathbf{x}}') = \sum_{n=1}^{\infty} D_n(x', y') \sin \frac{n\pi x}{a} \sinh \frac{n\pi(b - y)}{a} \quad \text{for } y > y'$$

Now G represents the potential due to a unit "point" charge[5] placed at $\vec{\mathbf{x}}'$, and so the potential it produces must be continuous everywhere in our region. This gives our first

---

[4]Alternatively, we could choose the functions $\exp(\pm n\pi y/a)$, but the hyperbolic functions are more useful here.

[5]The word "point" here refers to a source located at a point in the two-dimensional space. Equivalently, this solution also applies to a region that is infinite in the third (z) dimension, with no z-dependence, in which case our source would be a line charge.

boundary condition at the internal boundary at $y = y'$:

$$G_{\mathrm{I}}(x, y': x', y') = G_{\mathrm{II}}(x, y': x', y') \tag{C.19}$$

$$\sum_{n=1}^{\infty} C_n(x', y') \sin \frac{n\pi x}{a} \sinh \frac{n\pi y'}{a} = \sum_{n=1}^{\infty} D_n(x', y') \sin \frac{n\pi x}{a} \sinh \frac{n\pi (b - y')}{a}$$

Thus, by orthogonality of the eigenfunctions $\sin(n\pi x/a)$,

$$C_n(x', y') \sinh \frac{n\pi y'}{a} = D_n(x', y') \sinh \frac{n\pi (b - y')}{a}$$

Defining the new "constant" $E_n(x', y')$, we have

$$E_n \equiv \frac{C_n(x', y')}{\sinh [n\pi (b - y')/a]} = \frac{D_n(x', y')}{\sinh (n\pi y'/a)}$$

(The "constants" are functions of x' and y', but remember that for the moment we are considering these values fixed.) Then the Green's function becomes

$$G_{\mathrm{I}} = \sum_{n=1}^{\infty} E_n(x', y') \sin \frac{n\pi x}{a} \sinh \frac{n\pi y}{a} \sinh \frac{n\pi (b - y')}{a} \tag{C.20}$$

and

$$G_{\mathrm{II}} = \sum_{n=1}^{\infty} E_n(x', y') \sin \frac{n\pi x}{a} \sinh \frac{n\pi (b - y)}{a} \sinh \frac{n\pi y'}{a} \tag{C.21}$$

We need one more condition to find the unknowns E_n. We obtain it by returning to the differential equation for G (equation C.17). Writing G in the form $\sum g_n(y, x', y') \sin(n\pi x/a)$, we have

$$\sum_{n=1}^{\infty} \left\{ -\left(\frac{n\pi}{a}\right)^2 g_n \sin \frac{n\pi x}{a} + \frac{\partial^2 g_n}{\partial y^2} \sin \frac{n\pi x}{a} \right\} = -4\pi \delta(x - x') \delta(y - y')$$

We begin by reducing this relation to an equation in the one coordinate y that we chose to divide the region. To make use of the orthogonality of the sines, we multiply both sides by $\sin(n'\pi x/a)$ and integrate from 0 to a. Only the one term in the sum with $n = n'$ survives the integration. Then, dropping the prime on n, we have

$$\frac{a}{2}\left[-\left(\frac{n\pi}{a}\right)^2 g_n + \frac{\partial^2 g_n}{\partial y^2}\right] = -4\pi \sin \frac{n\pi x'}{a} \delta(y - y') \tag{C.22}$$

Thus, $E_n \propto \sin(n\pi x'/a)$. We can incorporate this result by writing $E_n = F_n \sin(n\pi x'/a)$ and, equivalently, $g_n = \gamma_n(y, y') \sin(n\pi x'/a)$. Our expression for G now has complete symmetry in the primed and unprimed coordinates if F_n is a constant. Next we integrate equation (C.22) across the boundary at $y = y'$:

$$\int_{y'-\varepsilon}^{y'+\varepsilon} \frac{a}{2} \left[-\left(\frac{n\pi}{a}\right)^2 \gamma_n + \frac{\partial^2 \gamma_n}{\partial y^2} \right] dy = -4\pi \int_{y'-\varepsilon}^{y'+\varepsilon} \delta(y - y') \, dy$$

The first term goes to zero as we let $\varepsilon \to 0$, since $G(\vec{x}, \vec{x}')$, and hence γ_n, is continuous at $y = y'$. We evaluate the right-hand side using the sifting property. Then we have

$$\frac{a}{2} \frac{\partial \gamma_n}{\partial y} \Big|_{y'-\varepsilon}^{y'+\varepsilon} = -4\pi \tag{C.23}$$

At the upper limit, $y = y' + \varepsilon > y'$, so $G = G_{\text{II}}$ and, using result (C.21), we have

$$\frac{a}{2} \frac{\partial \gamma_n}{\partial y} \Big|_{y'+} = -\frac{a}{2} \frac{n\pi}{a} F_n \cosh \frac{n\pi(b - y')}{a} \sinh \frac{n\pi y'}{a}$$

At the lower limit, $y = y' - \varepsilon < y'$, so $G = G_{\text{I}}$ (equation C.20), and so

$$\frac{a}{2} \frac{\partial \gamma_n}{\partial y} \Big|_{y'-} = \frac{a}{2} F_n \sinh \frac{n\pi(b - y')}{a} \left(\frac{n\pi}{a} \cosh \frac{n\pi y'}{a} \right)$$

Thus, equation (C.23) becomes

$$-\frac{n\pi}{2} F_n \left(\cosh \frac{n\pi(b - y')}{a} \sinh \frac{n\pi y'}{a} + \sinh \frac{n\pi(b - y')}{a} \cosh \frac{n\pi y'}{a} \right) = -4\pi$$

The term in parentheses simplifies, leaving

$$F_n = \frac{8}{n \sinh (n\pi b/a)}$$

and so the Green's function (equations C.20 and C.21) is

$$G(\vec{x}, \vec{x}') = \sum_{n=1}^{\infty} \frac{8}{n \sinh (n\pi b/a)} \sin \frac{n\pi x'}{a} \sin \frac{n\pi x}{a} \sinh \frac{n\pi y}{a} \sinh \frac{n\pi(b - y')}{a} \tag{C.24}$$

for $y < y'$ and

$$G(\vec{x}, \vec{x}') = \sum_{n=1}^{\infty} \frac{8}{n \sinh (n\pi b/a)} \sin \frac{n\pi x'}{a} \sin \frac{n\pi x}{a} \sinh \frac{n\pi y'}{a} \sinh \frac{n\pi(b - y)}{a} \tag{C.25}$$

for $y > y'$.

Here again we note the symmetry of the function: $G(\vec{x}, \vec{x}') = G(\vec{x}', \vec{x})$. The symmetry may be made more obvious by writing the function as a single expression. We define

$$y_< = \min (y, y')$$

and

$$y_> = \max(y, y')$$

(compare with Jackson, *Classical Electrodynamics*) so that we may write

$$G(\vec{x}, \vec{x}') = \sum_{n=1}^{\infty} \frac{8}{n \sinh(n\pi b/a)} \sin\frac{n\pi x'}{a} \sin\frac{n\pi x}{a} \sinh\frac{n\pi y_<}{a} \sinh\frac{n\pi(b - y_>)}{a} \quad (C.26)$$

Once again we find that division of the region leads to a piecewise-defined Green's function, but this time each piece is expressed as a series of eigenfunctions in the undivided coordinate.

Method 2 An alternative method allows us to expand the Green's function in a double sum, using eigenfunctions in both x and y. But eigenfunctions of which equation? Observe that the differential equation (C.18) is a special case of the equation

$$\frac{\partial^2 G(\vec{x}, \vec{x}')}{\partial x^2} + \frac{\partial^2 G(\vec{x}, \vec{x}')}{\partial y^2} + \lambda_{mn} G = 0 \quad (C.27)$$

with the constant $\lambda \equiv 0$. Thus, the eigenfunctions we need are solutions of equation (C.27) with λ_{mn} *not* equal to zero. This is the Helmholtz equation (compare with equation 3.16 or Chapter 8, Section 8.4.4). The solutions of this equation are again found by separation of variables:

$$\frac{X''}{X} + \frac{Y''}{Y} + \lambda_{mn} = 0$$

Again we choose the separation constant for the X equation to be negative and then choose the constant to be $-(n\pi/a)^2$ so as to obtain the eigenfunctions $\sin(n\pi x/a)$. Then the equation for Y is

$$\frac{Y''}{Y} + \lambda_{mn} - \left(\frac{n\pi}{a}\right)^2 = 0$$

This time we also want eigenfunctions in y that are zero at both boundaries, so we choose

$$\lambda_{mn} - \left(\frac{n\pi}{a}\right)^2 = \left(\frac{m\pi}{b}\right)^2$$

Then

$$Y'' = -\left(\frac{m\pi}{b}\right)^2 Y$$

and

$$Y = \sin\frac{m\pi y}{b}$$

Thus, the eigenfunctions are

$$XY = C_{mn} \sin \frac{n\pi x}{a} \sin \frac{m\pi y}{b} \tag{C.28}$$

where we must choose the coefficient C_{mn} to normalize the functions—that is,

$$\int_0^a \int_0^b \left(C_{mn} \sin \frac{n\pi x}{a} \sin \frac{m\pi y}{b} \right)^2 dx\, dy = 1$$

Thus,

$$C_{mn} = \frac{2}{\sqrt{ab}}$$

The corresponding eigenvalue is

$$\lambda_{mn} = \left(\frac{n\pi}{a} \right)^2 + \left(\frac{m\pi}{b} \right)^2 \tag{C.29}$$

Now we make use of the general result (C.11) with $\lambda = 0$ to get

$$G(\vec{\mathbf{x}}, \vec{\mathbf{x}}') = 4\pi \sum_{n,m} \frac{4}{ab} \frac{\sin \frac{n\pi x}{a} \sin \frac{m\pi y}{b} \sin \frac{n\pi x'}{a} \sin \frac{m\pi y'}{b}}{(n\pi/a)^2 + (m\pi/b)^2}$$

$$= \frac{16\pi}{ab} \sum_{n,m} \frac{\sin \frac{n\pi x}{a} \sin \frac{m\pi y}{b} \sin \frac{n\pi x'}{a} \sin \frac{m\pi y'}{b}}{(n\pi/a)^2 + (m\pi/b)^2} \tag{C.30}$$

Here again the symmetry in the primed and unprimed variables is obvious.

C.5. INHOMOGENEOUS BOUNDARY CONDITIONS

In the examples we have seen so far, we have imposed the value zero on our solution at the boundaries. This is not always the case. As an example, let's look at a problem involving diffusion in one dimension. Suppose heat is conducted along a rod that extends from $x = 0$ to $x = \infty$. The value of the temperature at $x = 0$ is a specified function of time,

$$T(0, t) = \tau(t)$$

and $T(x, 0) = 0$ for $x > 0$. The differential equation for the problem is equation (3.14):

$$\frac{\partial T}{\partial t} - D \frac{\partial^2 T}{\partial x^2} = 0$$

We can convert this problem into a one-dimensional problem by using a transform. The sine transform applied to the space variable is useful because $T(x, t)$ is defined only for positive values of x, our equation has only second derivatives in space, and we know the value of our function T at $x = 0$. (Refer to Chapter 7, Section 7.7.3.) When we apply the transform,

$$\tilde{T}(k, t) = \sqrt{\frac{2}{\pi}} \int_0^\infty T(x, t) \sin kx \, dx$$

our equation becomes

$$\frac{\partial \tilde{T}}{\partial t} - D\left(k\sqrt{\frac{2}{\pi}}\tau(t) - k^2\tilde{T}\right) = 0$$

or

$$\frac{\partial \tilde{T}}{\partial t} + k^2 D\tilde{T} = Dk\sqrt{\frac{2}{\pi}}\tau(t) \tag{C.31}$$

Equation (C.31) is a one-dimensional inhomogeneous differential equation that we can solve using methods already discussed.

From Section C.1, with the correspondence $v \to \tilde{T}$, $\alpha/m \to k^2 D$, the appropriate Green's function is (equation C.6)

$$\tilde{G}(k, t, t') = \begin{cases} 0 & \text{if } t < t' \\ \exp[-k^2 D(t - t')] & \text{if } t > t' \end{cases} \tag{C.32}$$

Thus, the solution for the transform $\tilde{T}(k, t)$ is

$$\tilde{T}(k, t) = \int_0^\infty \tilde{G}(k, t, t')Dk\sqrt{\frac{2}{\pi}}\tau(t') \, dt'$$

$$= Dk\sqrt{\frac{2}{\pi}} \int_0^t \exp[-k^2 D(t - t')]\tau(t') \, dt'$$

Transforming back, we have

$$T(x, t) = \frac{2}{\pi}D \int_0^\infty k \sin kx \int_0^t \exp[-k^2 D(t - t')]\tau(t') \, dt' \, dk \tag{C.33}$$

The expression (C.33) is most easily evaluated by doing the k-integration first and noting that $k \sin kx = -(d/dx)\cos kx$:

$$T(x, t) = -D\frac{2}{\pi}\frac{d}{dx} \int_0^t \tau(t') \int_0^\infty \cos kx \exp[-k^2 D(t - t')] \, dk \, dt'$$

We write the cosine in terms of exponentials and complete the square to obtain

$$T(x, t) = -D\frac{d}{dx}\frac{2}{\sqrt{\pi}} \int_0^t \frac{\tau(t')}{\sqrt{4D(t - t')}} \exp\left(-\frac{x^2}{4D(t - t')}\right) dt'$$

$$= \int_0^t \frac{x\tau(t')}{\sqrt{4\pi D(t - t')^3}} \exp\left(-\frac{x^2}{4D(t - t')}\right) dt' \tag{C.34}$$

This last expression[6] has the form that we expect for a Green's function–type solution, but now it is the boundary condition rather than the source function that appears in the integral. In the next section, we generalize this result to show how both the boundary conditions and the sources can contribute to a solution.

C.6. GREEN'S THEOREM

Here we shall develop a general theorem applicable to solutions of Poisson's equation[7] in a region where the solution is not necessarily zero on the boundaries.

Let Φ and Ψ be scalar functions of position defined in a volume V. Then

$$\vec{\nabla} \cdot (\Phi \vec{\nabla} \Psi) = \vec{\nabla} \Phi \cdot \vec{\nabla} \Psi + \Phi \nabla^2 \Psi$$

and also

$$\vec{\nabla} \cdot (\Psi \vec{\nabla} \Phi) = \vec{\nabla} \Psi \cdot \vec{\nabla} \Phi + \Psi \nabla^2 \Phi$$

Now we subtract and integrate over the volume V and use the divergence theorem (Chapter 1, equation 1.44):

$$\int_V [\vec{\nabla} \cdot (\Phi \vec{\nabla} \Psi) - \vec{\nabla} \cdot (\Psi \vec{\nabla} \Phi)] \, dV = \int_S (\Phi \vec{\nabla} \Psi - \Psi \vec{\nabla} \Phi) \cdot \hat{\mathbf{n}} \, dS$$

where S is the surface of the volume V and $\hat{\mathbf{n}}$ is the outward normal. Then

$$\int_V \left(\Phi \nabla^2 \Psi - \Psi \nabla^2 \Phi \right) dV = \int_S \left(\Phi \vec{\nabla} \Psi - \Psi \vec{\nabla} \Phi \right) \cdot \hat{\mathbf{n}} \, dS \qquad \text{(C.35)}$$

This relation is known as Green's theorem. It is valid for all scalar functions with derivatives that exist within the volume V and on the surface S.

Now we choose Φ to be the required solution of the Poisson equation

$$\nabla^2 \Phi = -\frac{\rho(\vec{\mathbf{x}})}{\varepsilon_0}$$

in the volume V. We choose Ψ to be the Green's function that satisfies the related equation

$$\nabla^2 G(\vec{\mathbf{x}}, \vec{\mathbf{x}}') = -4\pi \delta(\vec{\mathbf{x}} - \vec{\mathbf{x}}') \qquad \text{(C.36)}$$

where $\vec{\mathbf{x}}$ and $\vec{\mathbf{x}}'$ are within V. Physically, the Green's function $G(\vec{\mathbf{x}}, \vec{\mathbf{x}}')$ is a constant times the potential at $\vec{\mathbf{x}}$ due to a unit point charge at $\vec{\mathbf{x}}'$. In SI units,[8] the constant is $4\pi \varepsilon_0$. Then

---

[6] It appears that $T(0, t) = 0$ in this expression. This is an artifact of our use of the sine transform. The resulting function is necessarily odd and thus equals zero at $x = 0$. To obtain the correct boundary condition, we must take the limit as $x \to 0$ for positive values of x. See Problem 10.

[7] See also Problem 5.

[8] In Gaussian units, the constant is 1, a more pleasing result.

equation (C.35) becomes

$$\int_V (\Phi \nabla^2 G - G \nabla^2 \Phi)\, dV = \int_S (\Phi \vec{\nabla} G - G \vec{\nabla} \Phi) \cdot \hat{\mathbf{n}}\, dS$$

And using the differential equations satisfied by Φ and G, we get

$$\int_V \left[\Phi[-4\pi \delta(\vec{\mathbf{x}} - \vec{\mathbf{x}}')] - G\left(-\frac{\rho(\vec{\mathbf{x}})}{\varepsilon_0}\right) \right] dV = \int_S (\Phi \vec{\nabla} G - G \vec{\nabla} \Phi) \cdot \hat{\mathbf{n}}\, dS$$

The integrals are over the variable $\vec{\mathbf{x}}$, with $\vec{\mathbf{x}}'$ held fixed. Thus, using the sifting property, we have

$$-\Phi(\vec{\mathbf{x}}') + \frac{1}{4\pi \varepsilon_0} \int_V G(\vec{\mathbf{x}}, \vec{\mathbf{x}}') \rho(\vec{\mathbf{x}})\, dV = \frac{1}{4\pi} \int_S (\Phi \vec{\nabla} G - G \vec{\nabla} \Phi) \cdot \hat{\mathbf{n}}\, dS$$

giving the formal solution for Φ:

$$\Phi(\vec{\mathbf{x}}') = \frac{1}{4\pi \varepsilon_0} \int_V G(\vec{\mathbf{x}}, \vec{\mathbf{x}}') \rho(\vec{\mathbf{x}})\, dV - \frac{1}{4\pi} \int_S (\Phi \vec{\nabla} G - G \vec{\nabla} \Phi) \cdot \hat{\mathbf{n}}\, dS$$

The boundary conditions specify either Φ or $\vec{\nabla}\Phi \cdot \hat{\mathbf{n}}$ on the surface, but not both. (See Appendix X.)

(a) If Φ is specified on the surface (Dirichlet conditions), we choose $G_D(\vec{\mathbf{x}}, \vec{\mathbf{x}}') = 0$ on S and obtain

$$\Phi(\vec{\mathbf{x}}') = \frac{1}{4\pi \varepsilon_0} \int_V G_D(\vec{\mathbf{x}}, \vec{\mathbf{x}}') \rho(\vec{\mathbf{x}})\, dV - \frac{1}{4\pi} \int_S \Phi \vec{\nabla} G_D \cdot \hat{\mathbf{n}}\, dS \qquad (C.37)$$

This solution[9] is consistent with our previous expressions in the special case $\Phi = 0$ on S.

(b) If $\vec{\nabla}\Phi \cdot \hat{\mathbf{n}}$ is specified on the surface (Neumann conditions), we choose $\vec{\nabla} G_N(\vec{\mathbf{x}}, \vec{\mathbf{x}}') \cdot \hat{\mathbf{n}}$ = constant = K on S and obtain

$$\Phi(\vec{\mathbf{x}}') = \frac{1}{4\pi \varepsilon_0} \int_V G_N(\vec{\mathbf{x}}, \vec{\mathbf{x}}') \rho(\vec{\mathbf{x}})\, dV + \frac{1}{4\pi} \int_S G_N \vec{\nabla}\Phi \cdot \hat{\mathbf{n}}\, dS - \frac{1}{4\pi} \int_S (\Phi K)\, dS$$

In general, we cannot choose $K = 0$, because

$$\int_V \nabla^2 G\, dV = -4\pi \int_V \delta(\vec{\mathbf{x}} - \vec{\mathbf{x}}')\, dV = -4\pi$$

But according to the divergence theorem,

$$\int_V \nabla^2 G\, dV = \int_S \vec{\nabla} G(\vec{\mathbf{x}}, \vec{\mathbf{x}}') \cdot \hat{\mathbf{n}}\, dS = KA$$

[9]Compare this result with equation (C.34). See Problem 18.

where A is the area[10] of surface S. Thus,

$$K = -4\pi/A$$

and so the solution for Φ becomes

$$\Phi(\vec{\mathbf{x}}') = \frac{1}{4\pi\varepsilon_0} \int_V G_N(\vec{\mathbf{x}}, \vec{\mathbf{x}}')\rho(\vec{\mathbf{x}})\,dV + \frac{1}{4\pi} \int_S G_N\vec{\nabla}\Phi \cdot \hat{\mathbf{n}}\,dS + \frac{1}{A} \int_S \Phi\,dS$$

$$\Phi(\vec{\mathbf{x}}') = \frac{1}{4\pi\varepsilon_0} \int_V G_N(\vec{\mathbf{x}}, \vec{\mathbf{x}}')\rho(\vec{\mathbf{x}})\,dV + \frac{1}{4\pi} \int_S G_N\vec{\nabla}\Phi \cdot \hat{\mathbf{n}}\,dS + \langle\Phi\rangle \qquad \text{(C.38)}$$

where $\langle\Phi\rangle$ is the average value of Φ on the surface S. This additional constant does not affect the physics, which is governed by the gradient of Φ, and for most purposes can be removed by redefining the reference point for Φ.

C.7. THE GREEN'S FUNCTION FOR POISSON'S EQUATION IN A BOUNDED REGION

C.7.1. Symmetry

The Dirichlet Green's function is symmetric in the primed and unprimed coordinates. To see why, we begin with equation (C.35) with $\Phi = G(\vec{\mathbf{u}}, \vec{\mathbf{x}})$ and $\Psi = G(\vec{\mathbf{u}}, \vec{\mathbf{x}}')$. Then

$$\int_V \left[G(\vec{\mathbf{u}}, \vec{\mathbf{x}})\nabla_u^2 G(\vec{\mathbf{u}}, \vec{\mathbf{x}}') - G(\vec{\mathbf{u}}, \vec{\mathbf{x}}')\nabla_u^2 G(\vec{\mathbf{u}}, \vec{\mathbf{x}}) \right] d^3\vec{\mathbf{u}}$$

$$= \int_S \left[G(\vec{\mathbf{u}}, \vec{\mathbf{x}})\vec{\nabla}_u G(\vec{\mathbf{u}}, \vec{\mathbf{x}}') - G(\vec{\mathbf{u}}, \vec{\mathbf{x}}')\vec{\nabla}_u G(\vec{\mathbf{u}}, \vec{\mathbf{x}}) \right] \cdot \hat{\mathbf{n}}\,dS_u$$

The Dirichlet Green's function is zero on S, so the right-hand side is zero. On the left, we use the defining differential equation (C.36):

$$-4\pi \int_V \left[G(\vec{\mathbf{u}}, \vec{\mathbf{x}})\delta(\vec{\mathbf{u}} - \vec{\mathbf{x}}') - G(\vec{\mathbf{u}}, \vec{\mathbf{x}}')\delta(\vec{\mathbf{u}} - \vec{\mathbf{x}}) \right] d^3\vec{\mathbf{u}} = 0$$

Thus, using the sifting property, we have

$$G(\vec{\mathbf{x}}', \vec{\mathbf{x}}) = G(\vec{\mathbf{x}}, \vec{\mathbf{x}}')$$

which is the symmetry relation we need (compare with equation C.2).

For the Neumann case, the right-hand side becomes

$$-\frac{4\pi}{A} \int_S [G(\vec{\mathbf{u}}, \vec{\mathbf{x}}) - G(\vec{\mathbf{u}}, \vec{\mathbf{x}}')]\,dS_u$$

---

[10]Note that A is the area of the *total* bounding surface of the volume V. If the surface comprises several separated pieces, A is the sum of the areas of all the pieces.

Thus, symmetry may be proved only for infinitely large bounding surfaces ($A \rightarrow \infty$). However, symmetry can always be imposed as an additional condition on the Neumann Green's function.

C.7.2. The Division-of-Region Method

Because G satisfies equation (C.36), the solution is of the form

$$G(\vec{\mathbf{x}}, \vec{\mathbf{x}}') = \frac{1}{|\vec{\mathbf{x}} - \vec{\mathbf{x}}'|} + \psi(\vec{\mathbf{x}}, \vec{\mathbf{x}}')$$

where $\nabla^2(1/|\vec{\mathbf{x}} - \vec{\mathbf{x}}'|) = -4\pi\delta(\vec{\mathbf{x}} - \vec{\mathbf{x}}')$ (equation 6.26) and $\nabla^2\psi = 0$. The function ψ is required for G to satisfy the boundary conditions. This form of solution is not always useful, however, because it can be hard to integrate. Instead we can expand the Green's function in a set of eigenfunctions appropriate to the region of interest and make use of the orthogonality of the eigenfunctions when performing the integrations in equations (C.37) and (C.38).

Here we outline the divison-of-region method for calculating the Green's function in a bounded region of space. The method works for either the Dirichlet or the Neumann Green's function. In the Neumann case, it produces the symmetric Green's function. Remember that in *calculating* the Green's function we take the coordinates of $\vec{\mathbf{x}}'$ to be fixed.

1. Draw the volume under consideration.
2. Choose a suitable coordinate system. (We'll call the coordinates u, v, w.) Each boundary of your volume should correspond to a constant value of one of the coordinates.
3. Place a point charge of magnitude 1 at an *arbitrary* point $\vec{\mathbf{x}}'$ within the volume.
4. Determine the solutions of Laplace's equation $\nabla^2\Phi = 0$ in your coordinate system. Note which of the functions are orthogonal functions.
5. Choose one of the coordinates, say u, and use it to divide your volume into two regions, with the point charge on the boundary between those two regions—that is,

$$\text{Region I: } u < u'$$

$$\text{Region II: } u > u'$$

It is important that the functions in the other two coordinates (v, w) be orthogonal functions. This means that if you are using spherical coordinates, you may *only* divide the space in r, since the only functions of r in the solution of $\nabla^2\Phi = 0$ are powers and are not orthogonal functions. With other systems, you have some choice.

6. The equation for the Green's function is equation (C.36). Thus, in either of your two regions (but *not* on the boundary between them), $\nabla^2 G = 0$. Thus, you may expand G in the eigenfunctions identified in step 4. Use the orthogonal functions in v and w. You must choose the eigenvalues and eigenfunctions so as to satisfy the boundary conditions on G at the edges of your volume. Remember: For the Dirichlet Green's function, $G_D = 0$ on the boundary, and for the Neumann problem, $\partial G_N/\partial n = -4\pi/A$ on the boundary.

You will have two different linear combinations of the (nonorthogonal) eigenfunctions in u: one valid in region I ($u < u'$) and one valid in region II ($u > u'$). At this point, your function will have two sets of unknown constants.

7. Use continuity of the potential across the inner boundary at $u = u'$:

$$G(u = u'_-, v, w) = G(u = u'_+, v, w)$$

Equivalently, you may invoke the symmetry of the Green's function: $G(\vec{x}, \vec{x}') = G(\vec{x}', \vec{x})$. This condition will allow you to express one set of unknown constants in terms of the other. Now you should have only one set of unknown constants remaining.

8. Substitute your expression into the differential equation for G and use orthogonality to reduce to an equation in u.

9. Use discontinuity of the field at $\vec{x} = \vec{x}'$ by integrating the differential equation across the boundary at $u = u'$. The last set of constants is thereby found.

10. Check for symmetry, correct dimensions, and so forth.

C.7.3. Dirichlet Green's Function for the Region Outside a Sphere

Let's find the Dirichlet Green's function for the region outside a sphere of radius a.

Step 1. The diagram is shown in Figure C.4.

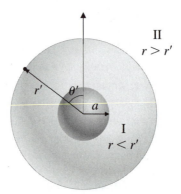

FIGURE C.4. The sphere's boundary is at $r = a$. The region outside the sphere is divided into two regions, $r < r'$ and $r > r'$.

Step 2. In spherical coordinates, the inner boundary of the region is at $r = a$, and the outer boundary is $r \to \infty$.

Step 3. We put a point charge at $\vec{x}'$ with coordinates r', θ', ϕ'. Then we need to solve the equation

$$\nabla^2 G(\vec{x}, \vec{x}') = -4\pi \delta(\vec{x} - \vec{x}') \tag{C.39}$$

with the boundary condition

$$G(\vec{x}, \vec{x}') = 0 \quad \text{for } r = a \text{ and for } r \to \infty \tag{C.40}$$

Note: At this point, $\vec{x}$ is the variable and $\vec{x}'$ is *fixed*.

Step 4. Within each region, we have the simpler equation

$$\nabla^2 G(\vec{x}, \vec{x}') = 0$$

with solutions (equation 8.56)

$$G(\vec{x}, \vec{x}') = \sum_{l,m} \left(A_{lm} r^l + \frac{B_{lm}}{r^{l+1}} \right) Y_{lm}(\theta, \phi)$$

in each region.

Step 5. The functions $Y_{lm}(\theta, \phi)$ are orthogonal functions, but the powers of r are not, so we divide the space in r:

$$\text{Region I: } a < r < r'$$

$$\text{Region II: } r' < r < \infty$$

Step 6. Next we apply the boundary conditions at $r = a$ and $r \to \infty$.
In region II, as $r \to \infty$, G must approach zero, so $A_{lm} = 0$,

$$G_{\mathrm{II}}(\vec{x}, \vec{x}') = \sum_{l,m} \frac{B_{lm}}{r^{l+1}} Y_{lm}(\theta, \phi) \qquad (C.41)$$

while at $r = a$, equation (C.40) requires

$$A_{lm} a^l + \frac{B_{lm}}{a^{l+1}} = 0$$

and so

$$B_{lm} = -a^{2l+1} A_{lm}$$

Thus,

$$G_{\mathrm{I}}(\vec{x}, \vec{x}') = \sum_{l,m} A_{lm} \left(r^l - \frac{a^{2l+1}}{r^{l+1}} \right) Y_{lm}(\theta, \phi) \qquad (C.42)$$

Step 7. Next we look at the boundary at $r = r'$. The potential (G) must be continuous across this boundary, so

$$\sum_{l,m} A_{lm} \left((r')^l - \frac{a^{2l+1}}{(r')^{l+1}} \right) Y_{lm}(\theta, \phi) = \sum_{l,m} \frac{B_{lm}}{(r')^{l+1}} Y_{lm}(\theta, \phi)$$

Now we use the orthogonality of the Y_{lm}. We multiply both sides by $Y^*_{l'm'}(\theta, \phi)$ and integrate over the whole range of θ and ϕ, to show that the coefficients of each Y_{lm} must be separately equal. Then

$$B_{lm} = A_{lm}[(r')^{2l+1} - a^{2l+1}]$$

Inserting this result into equation (C.41), we have

$$G_{\mathrm{II}}(\vec{\mathbf{x}}, \vec{\mathbf{x}}') = \sum_{l,m} \frac{A_{lm}[(r')^{2l+1} - a^{2l+1}]}{r^{l+1}} Y_{lm}(\theta, \phi)$$

while G_{I} is given by (C.42). We can make this look more symmetric by writing $A_{lm} = \alpha_{lm}/(r')^{l+1}$. Then

$$G_{\mathrm{I}}(\vec{\mathbf{x}}, \vec{\mathbf{x}}') = \sum_{l,m} \frac{\alpha_{lm}}{(r')^{l+1} r^{l+1}} (r^{2l+1} - a^{2l+1}) Y_{lm}(\theta, \phi)$$

$$G_{\mathrm{II}}(\vec{\mathbf{x}}, \vec{\mathbf{x}}') = \sum_{l,m} \frac{\alpha_{lm}[(r')^{2l+1} - a^{2l+1}]}{(r')^{l+1} r^{l+1}} Y_{lm}(\theta, \phi)$$

which exhibits the symmetry of G in r and r'.

Step 8. Now we have one set of constants left to find, the α_{lm}, and one remaining boundary condition at $r = r'$. To see what it is, we go back to the differential equation (C.39). Writing $G = \sum g_{lm} Y_{lm}$ and expressing the delta function in spherical coordinates (see Chapter 6, especially Problem 8), we have

$$\sum_{l,m} \left(\frac{1}{r} \frac{\partial^2 r g_{lm}}{\partial r^2} Y_{lm} + \nabla^2_{\mathrm{ang}} Y_{lm} g_{lm} \right) = -4\pi \frac{\delta(r - r')}{r^2} \delta(\mu - \mu') \delta(\phi - \phi')$$

$$\sum_{l,m} \left(\frac{1}{r} \frac{\partial^2 r g_{lm}}{\partial r^2} Y_{lm} - \frac{l(l+1)}{r^2} Y_{lm} g_{lm} \right) = -4\pi \frac{\delta(r - r')}{r^2} \delta(\mu - \mu') \delta(\phi - \phi')$$

Again we multiply both sides by $Y^*_{l'm'}$ and integrate over the full range of θ and ϕ. On the left-hand side, we get zero, except when $l = l'$ and $m = m'$. We use the sifting property to evaluate the integral on the right. Then, dropping the primes on l and m, we get

$$\frac{1}{r} \frac{\partial^2 r g_{lm}}{\partial r^2} - \frac{l(l+1)}{r^2} g_{lm} = -4\pi \frac{\delta(r - r')}{r^2} Y^*_{lm}(\theta', \phi')$$

Now we can relabel again: Let $g_{lm} = f_{lm} Y^*_{lm}(\theta', \phi')$ and, correspondingly, $\alpha_{lm} = \beta_{lm} Y^*_{lm}(\theta', \phi')$. Then the Green's function is symmetric in all three coordinates if β_{lm} is a constant, independent of all the coordinates. With this relabeling,

$$\frac{1}{r} \frac{\partial^2 r f_{lm}}{\partial r^2} - \frac{l(l+1)}{r^2} f_{lm} = -4\pi \frac{\delta(r - r')}{r^2}$$

Step 9. Now we multiply this equation by r and then integrate across the boundary at $r = r'$:

$$\int_{r'-\varepsilon}^{r'+\varepsilon} \left(\frac{\partial^2 r f_{lm}}{\partial r^2} - \frac{l(l+1)}{r} f_{lm} \right) dr = -4\pi \int_{r'-\varepsilon}^{r'+\varepsilon} \frac{\delta(r - r')}{r} dr$$

On the right-hand side, we use the sifting property to evaluate the integral:

$$\frac{\partial r f_{lm}}{\partial r}\bigg|_{r'-\varepsilon}^{r'+\varepsilon} - \int_{r'-\varepsilon}^{r'+\varepsilon} \frac{l(l+1)}{r} f_{lm}\, dr = -\frac{4\pi}{r'}$$

Next we let $\varepsilon \to 0$. We have already ensured that f_{lm} is continuous at $r = r'$, and r is also continuous, so the second term on the left-hand side goes to zero. Then we have

$$\frac{\partial r f_{lm}}{\partial r}\bigg|_{r'-}^{r'+} = -\frac{4\pi}{r'} \tag{C.43}$$

At the upper limit, $r = r'+$, we are in region II, and so

$$\frac{\partial r f_{lm}}{\partial r}\bigg|_{r'+} = \frac{\partial}{\partial r} \frac{\beta_{lm}\left[(r')^{2l+1} - a^{2l+1}\right]}{(r')^{l+1}r^l}\bigg|_{r'+} = -l\frac{\beta_{lm}\left[(r')^{2l+1} - a^{2l+1}\right]}{(r')^{l+1}(r')^{l+1}}$$

$$= -l\frac{\beta_{lm}\left[(r')^{2l+1} - a^{2l+1}\right]}{(r')^{2l+2}} = -l\beta_{lm}\left(\frac{1}{r'} - \frac{a^{2l+1}}{(r')^{2l+2}}\right)$$

At the lower limit, we are in region I, and so

$$\frac{\partial r f_{lm}}{\partial r}\bigg|_{r'-} = \frac{\partial}{\partial r} \frac{\beta_{lm}}{(r')^{l+1}}\left(r^{l+1} - \frac{a^{2l+1}}{r^l}\right)\bigg|_{r'+} = \frac{\beta_{lm}}{(r')^{l+1}}\left((l+1)(r')^l + l\frac{a^{2l+1}}{(r')^{l+1}}\right)$$

$$= \beta_{lm}\left(\frac{l+1}{r'} + l\frac{a^{2l+1}}{(r')^{2l+2}}\right)$$

Then equation (C.43) becomes

$$-l\beta_{lm}\left(\frac{1}{r'} - \frac{a^{2l+1}}{(r')^{2l+2}}\right) - \beta_{lm}\left(\frac{l+1}{r'} + l\frac{a^{2l+1}}{(r')^{2l+2}}\right) = -\frac{4\pi}{r'}$$

$$-\beta_{lm}\frac{2l+1}{r'} = -\frac{4\pi}{r'}$$

The r' cancels, which it must, since β_{lm} should be a constant, not a function of either set of coordinates. Then

$$\beta_{lm} = \frac{4\pi}{2l+1}$$

and so finally we have

$$G_{\mathrm{I}}(\vec{x}, \vec{x}') = \sum_{l,m} \frac{4\pi}{2l+1} \frac{(r^{2l+1} - a^{2l+1})}{(r')^{l+1}r^{l+1}} Y_{lm}(\theta, \phi) Y_{lm}^*(\theta', \phi')$$

$$G_{\mathrm{II}}(\vec{x}, \vec{x}') = \sum_{l,m} \frac{4\pi}{2l+1} \frac{\left[(r')^{2l+1} - a^{2l+1}\right]}{(r')^{l+1}r^{l+1}} Y_{lm}(\theta, \phi) Y_{lm}^*(\theta', \phi')$$

In a more compact form,

$$G(\vec{x}, \vec{x}') = \sum_{l,m} \frac{4\pi}{2l+1} \frac{(r_<^{2l+1} - a^{2l+1})}{r_<^{l+1} r_>^{l+1}} Y_{lm}(\theta, \phi) Y_{lm}^*(\theta', \phi') \tag{C.44}$$

where $r_<$ and $r_>$ are the lesser and the greater of r and r', respectively.

Step 10. Since G is $4\pi\varepsilon_0$ times the potential due to a unit point charge, it should have dimensions of 1/length, as our expression does. This series should converge nicely for $r > a$.

As an example of using this result, let's find the potential due to a ring of charge of radius $b > a$ and uniform linear charge density λ, lying in the equatorial plane of a grounded sphere of radius a. Using equation (C.37), we have

$$\Phi(\vec{x}) = \frac{1}{4\pi\varepsilon_0} \int_V G_D(\vec{x}, \vec{x}')\rho(\vec{x}')\, dV' - \frac{1}{4\pi} \int_S \Phi(\vec{x}')\vec{\nabla}' G_D \cdot \hat{n}\, dS'$$

When we are doing the integration, the coordinates $\vec{x}$ are fixed while $\vec{x}'$ is the integration variable. With $\Phi = 0$ on S, the second integral is zero:

$$\Phi(\vec{x}) = \frac{1}{4\pi\varepsilon_0} \int_V \lambda \frac{\delta(r' - b)\delta(\mu')}{b}$$

$$\times \sum_{l,m} \frac{4\pi}{2l+1} \frac{(r_<^{2l+1} - a^{2l+1})}{r_<^{l+1} r_>^{l+1}} Y_{lm}(\theta, \phi) Y_{lm}^*(\theta', \phi')(r')^2\, dr'\, d\mu'\, d\phi'$$

$$= \frac{\lambda}{\varepsilon_0} \sum_{l,m} \frac{1}{2l+1} Y_{lm}(\theta, \phi) b \frac{(r_<^{2l+1} - a^{2l+1})}{r^{l+1} b^{l+1}} \sqrt{\frac{2l+1}{4\pi} \frac{(l-m)!}{(l+m)!}} P_l^m(0) \int_0^{2\pi} e^{-im\phi'}\, d\phi'$$

where now $r_<$ and $r_>$ are the lesser and the greater of r and b, respectively.

The integral over ϕ' yields zero unless $m = 0$. Then

$$\Phi(\vec{x}) = \frac{\lambda b}{2\varepsilon_0} \sum_l \frac{(r_<^{2l+1} - a^{2l+1})}{r^{l+1} b^{l+1}} P_l(0) P_l(\mu)$$

Now we check dimensions: $2\pi\lambda b$ is the total charge on the ring, and the term in r has dimensions of 1/length, so the answer is of order charge/($\varepsilon_0 \times$ length), which is correct for potential.

Note that $P_l(0) = 0$ for l odd, so only even l appear in the sum. The first few terms for $0 < r < b$ are

$$\Phi(\vec{x}) = \frac{Q}{4\pi\varepsilon_0 b} \left(\frac{r-a}{r} - \frac{1}{4} \frac{r^5 - a^5}{r^3 b^2}(3\cos^2\theta - 1) \right.$$

$$\left. + \frac{3}{64} \frac{r^9 - a^9}{r^5 b^4}(35\cos^4\theta - 30\cos^2\theta + 3) + \cdots \right)$$

while for $r > b$,

$$\Phi(\vec{x}) = \frac{Q}{4\pi \varepsilon_0 b} \sum_l \frac{(b^{2l+1} - a^{2l+1})}{r^{l+1} b^{l+1}} P_l(0) P_l(\mu)$$

$$= \frac{Q}{4\pi \varepsilon_0 r} \left(1 - \frac{a}{b}\right) - \frac{Q}{4\pi \varepsilon_0 b}$$

$$\times \left[\left(b^3 - \frac{a^5}{b^2}\right) \frac{(3\cos^2\theta - 1)}{4r^3} - \frac{3}{64} \frac{b^9 - a^9}{r^5 b^4}(35\cos^4\theta - 30\cos^2\theta + 3) + \cdots \right]$$

The first term is the potential due to a point charge, and this term dominates for large r. The magnitude of the point charge is the sum of the charge on the ring plus the charge induced on the surface of the sphere. Thus, the total induced charge on the sphere is $Q_{\text{ind}} = -Qa/b$. The solution in the equator and on the polar axis is shown in Figure C.5.

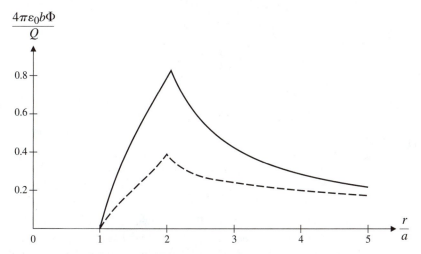

FIGURE C.5. Potential Φ times $4\pi \varepsilon_0 b/Q$ in the equatorial plane ($\theta = \pi/2$, solid line) and on the polar axis ($\theta = 0$, dashed line) versus r/a. The first three terms in the series are shown here, with $b = 2a$.

C.7.4. Dirichlet Green's Function for the Interior of a Cylinder

Again we follow the steps of the standard method. Our region is an infinitely long cylindrical tube of radius a.

Step 1. The volume is shown in Figure C.6.

Step 2. We use cylindrical coordinates. The volume is bounded by the surface $\rho = a$.

Step 3. We place a point charge at $\vec{x}'$ with coordinates ρ', ϕ', z'.

Step 4. The solutions of Laplace's equation in cylindrical coordinates are given in Chapter 8, Section 8.4.8. We may choose to divide the space in any one of the three coordinates.

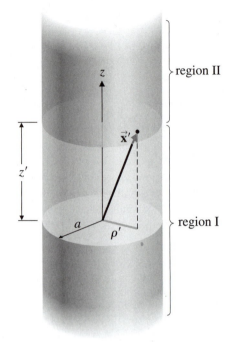

FIGURE C.6. The interior of the tube is divided at $z = z'$.

Step 5. Here we'll divide the volume in z:

$$\text{Region I: } z < z'$$

$$\text{Region II: } z > z'$$

Step 6. The eigenfunctions in ρ that take the value zero at $\rho = a$ are the functions $J_m(x_{mn}\rho/a)$, where x_{mn} is the nth zero of J_m. We exclude the Ns because they are not well behaved on the cylinder axis at $\rho = 0$. The appropriate functions in z are the real exponentials, $e^{-x_{mn}z/a}$ for large positive z and $e^{+x_{mn}z/a}$ for large negative z. Thus, we have

$$G_{\text{I}}(\vec{\mathbf{x}}, \vec{\mathbf{x}}') = \sum_{m,n} A_{mn} J_m\left(x_{mn}\frac{\rho}{a}\right) e^{x_{mn}z/a} e^{im\phi} \quad \text{for } z < z'$$

and

$$G_{\text{II}}(\vec{\mathbf{x}}, \vec{\mathbf{x}}') = \sum_{m,n} B_{mn} J_m\left(x_{mn}\frac{\rho}{a}\right) e^{-x_{mn}z/a} e^{im\phi} \quad \text{for } z > z'$$

Step 7. Setting the two functions equal at the inner boundary, $z = z'$, we have

$$\sum_{m,n} A_{mn} J_m\left(x_{mn}\frac{\rho}{a}\right) e^{x_{mn}z'/a} e^{im\phi} = \sum_{m,n} B_{mn} J_m\left(x_{mn}\frac{\rho}{a}\right) e^{-x_{mn}z'/a} e^{im\phi}$$

Since the functions $J_m(x_{mn}\rho/a)e^{im\phi}$ form an orthogonal set, we may equate the coefficients of each term in the sum separately:

$$A_{mn}e^{x_{mn}z'/a} = B_{mn}e^{-x_{mn}z'/a} \equiv \alpha_{mn}$$

Thus,

$$G_{\mathrm{I}}(\vec{\mathbf{x}}, \vec{\mathbf{x}}') = \sum_{m,n} \alpha_{mn} J_m\left(x_{mn}\frac{\rho}{a}\right) e^{x_{mn}(z-z')/a} e^{im\phi}$$

$$G_{\mathrm{II}}(\vec{\mathbf{x}}, \vec{\mathbf{x}}') = \sum_{m,n} \alpha_{mn} J_m\left(x_{mn}\frac{\rho}{a}\right) e^{x_{mn}(z'-z)/a} e^{im\phi}$$

Step 8. Next we substitute into the differential equation for G, which we now write as

$$G = \sum g_{mn}(z) J_m\left(x_{mn}\frac{\rho}{a}\right) e^{im\phi} \quad \text{with } g_{mn}(z) = \alpha_{mn} \exp\left(\frac{x_{mn}}{a}(z_< - z_>)\right) \quad \text{(C.45)}$$

with $z_<$ and $z_>$ equal to the lesser and greater of z and z', respectively. Then, with the delta function expressed in cylindrical coordinates, we have

$$\sum J_m\left(x_{mn}\frac{\rho}{a}\right) e^{im\phi} \left[-\frac{x_{mn}^2}{a^2} g_{mn}(z) + \frac{d^2 g_{mn}(z)}{dz^2}\right] = -\frac{4\pi}{\rho}\delta(\rho - \rho')\delta(\phi - \phi')\delta(z - z')$$

We multiply both sides by $\rho J_{m'}(x_{m'n'}\rho/a)\,e^{-im'\phi}$ and integrate over ρ and ϕ. On the left, we use the orthogonality of the eigenfunctions and equation (8.96); on the right, we use the sifting property.

$$\frac{a^2}{2}\left[J_m'(x_{mn})\right]^2 (2\pi) \left[-\frac{x_{mn}^2}{a^2} g_{mn}(z) + \frac{d^2 g_{mn}(z)}{dz^2}\right] = -4\pi J_m\left(x_{mn}\frac{\rho'}{a}\right) e^{-im\phi'}\delta(z - z')$$

Now we rewrite:

$$g_{mn}(z) = \gamma_{mn} \exp\left(\frac{x_{mn}}{a}(z_< - z_>)\right) J_m\left(x_{mn}\frac{\rho'}{a}\right) e^{-im\phi'}$$

Step 9. The next step is to integrate across the boundary at $z = z'$ to find the γ_{mn}:

$$a^2\left[J_m'(x_{mn})\right]^2 \pi \int_{z'-\varepsilon}^{z'+\varepsilon} \left[-\frac{x_{mn}^2}{a^2}\gamma_{mn}e^{\frac{x_{mn}}{a}(z_<-z_>)} + \frac{d^2\gamma_{mn}e^{\frac{x_{mn}}{a}(z_<-z_>)}}{dz^2}\right] dz'$$

$$= -4\pi \int_{z'-\varepsilon}^{z'+\varepsilon} \delta(z - z')\,dz'$$

As $\varepsilon \to 0$, the first term on the left vanishes, and the equation simplifies to

$$a^2\left[J_m'(x_{mn})\right]^2 \left.\frac{d\gamma_{mn}e^{\frac{x_{mn}}{a}(z_<-z_>)}}{dz}\right|_{z'-\varepsilon}^{z'+\varepsilon} = -4$$

At the upper limit, $z > z'$, so the derivative is

$$\frac{d}{dz} \exp\left(\frac{x_{mn}}{a}(z'-z)\right)\bigg|_{z=z'} = -\frac{x_{mn}}{a}$$

while at the lower limit, $z < z'$, so

$$\frac{d}{dz} \exp\left(\frac{x_{mn}}{a}(z-z')\right)\bigg|_{z=z'} = \frac{x_{mn}}{a}$$

Thus,

$$a^2[J_m'(x_{mn})]^2 \gamma_{mn} \frac{x_{mn}}{a}(-2) = -4$$

and so

$$\gamma_{mn} = \frac{2}{x_{mn}a[J_m'(x_{mn})]^2}$$

Inserting this value into the expression (C.45) for G gives

$$G(\vec{x}, \vec{x}') = \frac{2}{a} \sum_{m,n} \frac{1}{x_{mn}[J_m'(x_{mn})]^2} J_m\left(x_{mn}\frac{\rho}{a}\right) J_m\left(x_{mn}\frac{\rho'}{a}\right) e^{im(\phi-\phi')} \exp\left(\frac{x_{mn}}{a}(z_< - z_>)\right)$$

Step 10. The result exhibits the necessary symmetry in $\vec{x}$ and $\vec{x}'$ and has dimensions of inverse length, as required.

What is the potential if the cylinder has a band with potential V_0 between $z = -a$ and $z = +a$ and is grounded everywhere else?

This time only the second integral in equation (C.37) is nonzero. Remember that $\vec{n}$ is the outward normal, here $\hat{\rho}$.

$$\Phi(\rho, \phi, z) = -\frac{1}{4\pi} \int_0^{2\pi} \int_{-a}^{+a} V_0 \frac{\partial G}{\partial \rho'}\bigg|_{\rho'=a} a\, d\phi'\, dz'$$

$$= -\frac{V_0}{4\pi} \int_0^{2\pi} \int_{-a}^{+a} \frac{2}{a} \sum_{m,n} \frac{1}{a[J_m'(x_{mn})]^2} J_m\left(x_{mn}\frac{\rho}{a}\right) J_m'(x_{mn}) e^{im(\phi-\phi')}$$

$$\times \exp\left(\frac{x_{mn}}{a}(z_< - z_>)\right) a\, d\phi'\, dz'$$

Only the $m = 0$ terms survive the integration over ϕ', as expected for a system with azimuthal symmetry. Thus,

$$\Phi(\rho, z) = -\frac{V_0}{a} \sum_n \frac{J_0(x_{0n}\rho/a)}{J_0'(x_{0n})} \int_{-a}^{+a} \exp\left(\frac{x_{0n}}{a}(z_< - z_>)\right) dz'$$

For $z > a$, this becomes

$$\Phi(\rho, z) = \frac{V_0}{a} \sum_n \frac{J_0(x_{0n}\rho/a)}{J_1(x_{0n})} \int_{-a}^{+a} \exp\left(\frac{x_{0n}}{a}(z' - z)\right) dz'$$

$$= V_0 \sum_n \frac{J_0(x_{0n}\rho/a)}{J_1(x_{0n})} \frac{2}{x_{0n}} \sinh x_{0n} \exp\left(-x_{0n}\frac{z}{a}\right)$$

On the other hand, for $-a < z < a$,

$$\Phi(\rho, z) = V_0 \sum_n \frac{J_0(x_{0n}\rho/a)}{a J_1(x_{0n})} \left[\int_z^{+a} \exp\left(\frac{x_{0n}}{a}(z - z')\right) dz' + \int_{-a}^z \exp\left(\frac{x_{0n}}{a}(z' - z)\right) dz'\right]$$

$$= V_0 \sum_n \frac{J_0(x_{0n}\rho/a)}{x_{0n} J_1(x_{0n})} \left[e^{x_{0n}\frac{z}{a}}(-e^{-x_{0n}} + e^{-x_{0n}\frac{z}{a}}) + e^{-x_{0n}\frac{z}{a}}(e^{x_{0n}\frac{z}{a}} - e^{-x_{0n}})\right]$$

$$= 2V_0 \sum_n \frac{J_0(x_{0n}\rho/a)}{x_{0n} J_1(x_{0n})} \left[1 - e^{-x_{0n}} \cosh\left(x_{0n}\frac{z}{a}\right)\right]$$

This solution is plotted in Figure C.7, where we show Φ/V_0 versus ρ/a. A few terms in the series represent the solution well for $z > a$, but many terms are required for $|z| < a$, and even then the expression fails[11] at $\rho = a$.

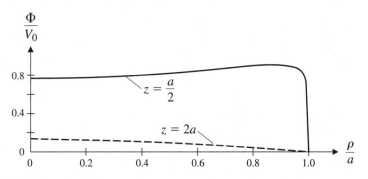

FIGURE C.7. Solution for the potential $\Phi(\rho)$ inside the tube. The solid line shows the potential at $z = a/2$ (42 terms). The dashed line shows the potential at $z = 2a$ (10 terms). For $z > a$, fewer terms are needed in the sum to obtain an accurate result. The additional terms decrease exponentially.

C.7.5. Expansion in Eigenfunctions

An alternative method of solution produces a single expression for $G(\vec{x}, \vec{x}')$ throughout the region. We write G as an expansion in eigenfunctions, as in Sections C.2 and C.4, method 2. The penalty we pay for having a single expression is that we have a triple sum rather than

---

[11]We have seen this phenomenon before in Section C.5. Here our eigenfunctions are zero at $\rho = a$, and so the solution, a superposition of the eigenfunctions, is also zero on the boundary. The solution has the correct limit as $\rho \to a$ from below. In Figure C.7, the solid curve $\Phi(\rho)$ approaches V_0 as $\rho \to a$, but dives toward zero at $\rho = a$.

the double sum we obtained using the division-of-region method. The eigenfunctions we need are solutions of the Helmholtz equation:

$$(\nabla^2 + k^2) f(\vec{\mathbf{x}}) = 0 \tag{C.46}$$

To demonstrate the method, let's look at the Dirichlet problem in the interior of a cylinder of radius a and height h. The solutions of equation (C.46) that satisfy the boundary conditions $G = 0$ at $\rho = a$ and $z = h$ are (Chapter 8, Section 8.4.4)

$$f(\vec{\mathbf{x}}) = J_m\left(x_{mn}\frac{\rho}{a}\right) \sin\frac{p\pi z}{h} e^{im\phi}$$

and the eigenvalues are

$$k_{mnp}^2 = \left(\frac{x_{mn}}{a}\right)^2 + \left(\frac{p\pi}{h}\right)^2$$

where x_{mn} is the nth zero of $J_m(x)$. We must normalize the eigenfunctions, using equations (8.96) and (4.12). Then, from equation (C.11) extended to three dimensions, with $\lambda = k^2 = 0$,

$$G(\vec{\mathbf{x}}, \vec{\mathbf{x}}') = 4\pi \sum_{m,n,p} \frac{J_m\left(x_{mn}\frac{\rho}{a}\right) \sin\frac{p\pi z}{h} e^{im\phi} J_m\left(x_{mn}\frac{\rho'}{a}\right) \sin\frac{p\pi z'}{h} e^{-im\phi'}}{\frac{a^2}{2}\left[J_m'(x_{mn})\right]^2 \frac{h}{2} 2\pi \left[\left(\frac{x_{mn}}{a}\right)^2 + \left(\frac{p\pi}{h}\right)^2\right]}$$

$$= \frac{8}{h} \sum_{m=-\infty}^{+\infty} \sum_{n=1}^{\infty} \sum_{p=1}^{\infty} \frac{J_m\left(x_{mn}\frac{\rho}{a}\right) J_m\left(x_{mn}\frac{\rho'}{a}\right) \sin\frac{p\pi z}{h} \sin\frac{p\pi z'}{h} e^{im(\phi-\phi')}}{[J_{m+1}(x_{mn})]^2 \left[x_{mn}^2 + \left(\frac{p\pi a}{h}\right)^2\right]}$$

The dimensions of this expression are 1/length, as required.

PROBLEMS

1. Use the division-of-region method to find the Green's function for a damped harmonic oscillator. Hence find the response of the oscillator to the input $f(t) = 1 - t/T$ for $0 < t < T$ and zero otherwise.

2. Find the Green's function for a beam supported at one end (refer to Chapter 5, Problem 11) using a Laplace transform method.

3. Find the Green's function for a wave on a string of length L,

$$\frac{\partial^2 G}{\partial t^2} - v^2 \frac{\partial^2 G}{\partial x^2} = \delta(x - x')\delta(t - t')$$

by taking the Fourier transform of the equation in time and using the divison-of-region method for the space dependence. Give your answer as an integral over the Fourier transform variable ω.

4. Find the Green's function for the diffusion equation

$$\frac{\partial G}{\partial t} - D\frac{\partial^2 G}{\partial x^2} = \delta(x - x')\delta(t - t')$$

by taking the Fourier transform in space and using the division-of-region method in time. Take $-\infty < x < \infty$ and $0 \le t < \infty$.

Use the result to compute the function $f(x, t)$ that satisfies the diffusion equation

$$\frac{\partial f}{\partial t} - D\frac{\partial^2 f}{\partial x^2} = S(x, t)$$

with the source function

$$S(x, t) = e^{-x^2/a^2}\delta(t)$$

5. Show that we can use Green's theorem (Section C.6) to obtain a solution for Φ, where Φ satisfies the Helmholtz equation

$$(\nabla^2 + k^2)\Phi(\vec{\mathbf{x}}) = S(\vec{\mathbf{x}})$$

with a source function $S(\vec{\mathbf{x}})$. Determine the solution for Φ in terms of the Green's function when Φ satisfies the Dirichlet boundary conditions $\Phi(\vec{\mathbf{x}}) = F(\vec{\mathbf{x}})$, a known function, on the boundary surface S.

6. Find the Green's function for the one-dimensional Poisson equation

$$\frac{d^2\Phi}{dx^2} = -\rho(x)$$

with boundary conditions $\Phi(x) = 0$ at $x = 0$ and $x = a$. Hence find the solution for Φ when

(a) $\rho(x) = \sin(\pi x/a)$
(b) $\rho(x) = x^2(a^2 - x^2)$

7. Use a division-of-region method to find the Green's function for the one-dimensional Helmholtz equation

$$\frac{d^2 y}{dx^2} + k^2 y = f(x)$$

in the region $0 < x < a$, with $y(x) = 0$ at $x = 0$ and $x = a$. Find the Fourier sine series for $G(x, x')$ and hence show that your result agrees with the result of Example C.2.

8. Sometimes we may expand the Green's function as a series of eigenfunctions even if the differential equation is not of the Sturm-Liouville form. The governing differential equation for the displacement of a beam is equation (3.11):

$$\frac{d^4 y}{dx^4} = \frac{q(x)}{EI}$$

A beam of length L rests on a support at each end so that the boundary conditions are $y(0) = y(L) = 0$. Show that the Green's function may be expanded in a series of eigenfunctions, and determine the form of the Green's function. Use it to find the beam displacement when it is subjected to a load $q(x) = ax/L$. Compare with Chapter 4, Problem 15.

9. Find the Green's function for heat transfer along a rod with insulated ends. The relevant differential equation is

$$\frac{\partial T}{\partial t} - D\frac{\partial^2 T}{\partial x^2} = Q(x, t)$$

and the boundary conditions are $\partial G/\partial x = 0$ at $x = 0$ and at $x = L$; $G(x, 0) = 0$. Treat the problem as a two-dimensional problem, and use method 1 in Section C.4, dividing the region in time.

10. Verify that equation (C.34) gives the correct result $\tau(t)$ in the limit $x \to 0$ from above. *Hint*: Expand $\tau(t')$ in a Taylor series about $t' = t$.

11. Find the Green's function for the wave equation in three space dimensions, using spherical coordinates. The wave equation is

$$v^2 \nabla^2 G - \frac{\partial^2}{\partial t^2}G = \delta(\vec{x} - \vec{x}')\delta(t - t')$$

where $G(\vec{x}, \vec{x}', t, t')$ is the displacement and ∇^2 is the Laplacian operator in three dimensions. Transform the equation in space and time, and solve for G. Hence find the displacement $f(\vec{r}, t)$ if the source is $h\delta(t)e^{-a^2r^2}$. Compare with Example 7.5.

12. Find the Dirichlet Green's function for the two-dimensional Helmholtz equation in a circular region of radius a. Obtain the result as a double sum over appropriate eigenfunctions.

13. Find the Neumann Green's function for the one-dimensional Poisson equation

$$\frac{d^2\Phi}{dx^2} = -\frac{\rho(x)}{\varepsilon_0}$$

with boundary conditions $dG/dx = 0$ at $x = 0$ and $x = L$. Express your answer as a series of eigenfunctions. Hence find the potential Φ if

$$\rho(x) = \begin{cases} \rho_0 x/L & \text{if } 0 < x < L/2 \\ 0 & \text{otherwise} \end{cases}$$

and $d\Phi/dx = 0$ at $x = 0$ and $x = L$.

14. Use a Fourier transform method and cylindrical coordinates to find the Green's function for the wave equation in two dimensions.

15. Using the division-of-space method, find (a) the Dirichlet and (b) the Neumann Green's function for Poisson's equation in the interior of a sphere of radius a.

16. (a) Find the Dirichlet Green's function for Poisson's equation in the half-space $z > 0$ using Cartesian coordinates.

(b) Evaluate the integral to show that G may be interpreted as the potential due to a point charge and its image in the plane $z = 0$. *Hint:* Use polar coordinates to evaluate the integral, and use contour integration to evaluate the integral over the angular variable, as in Chapter 2, Section 2.7.2.

17. Find the Dirichlet Green's function for Poisson's equation in the interior of a sphere of radius a as a triple sum over appropriate eigenfunctions.

18. Obtain a relation analogous to equation (C.37) for the diffusion equation

$$\frac{\partial f}{\partial t} - D\nabla^2 f = S(\vec{x}, t)$$

Define the Green's function through the equation

$$\frac{\partial G}{\partial t} - D\nabla^2 G = D\delta(\vec{x} - \vec{x}')\delta(t - t')$$

Hint: Note that since $G(t, t') = G(t - t')$, $\partial G/\partial t' = -\partial G/\partial t$. Integrate over both space and time, and let $G = 0$ on the bounding surface.

Apply your result to the example in Section C.5. Apply the sine transform in space to the equation

$$\frac{\partial G}{\partial t} - D\frac{\partial^2 G}{\partial x^2} = D\delta(x - x')\delta(t - t')$$

and obtain the Green's function. Show that the solution (C.34) may be expressed in terms of $\partial G/\partial x'$ and that this solution is consistent with the general result you found above.

19. Find the Dirichlet Green's function for Poisson's equation in the interior of a hemisphere of radius a.

(a) Choose $0 \le \phi \le \pi$ and $0 \le \theta \le \pi$.

(b) Choose $0 \le \theta \le \pi/2$ and $0 \le \phi \le 2\pi$.

(c) Using one of the two Green's functions, (a) or (b), evaluate the potential inside the hemisphere if $\Phi = 0$ on the spherical surface and $\Phi(r) = V_0(1 - r/a)$ on the flat face.

20. Obtain the Green's function inside a cylindrical tube (Section C.7.4) by dividing space in ρ.

OPTIONAL TOPIC D

Approximate Evaluation of Integrals

D.1. THE METHOD OF STEEPEST DESCENT

Often functions are defined in terms of integrals in the complex plane. The gamma function (Chapter 2) is one example that we have seen. Another example is the Hankel function[1]

$$H_\nu^{(1)}(\xi) = \frac{1}{\pi i} \int_C \exp\left[\frac{\xi}{2}\left(z - \frac{1}{z}\right)\right] \frac{dz}{z^{\nu+1}} \tag{D.1}$$

where the contour C runs from the origin to $-\infty$, as shown in Figure D.1.

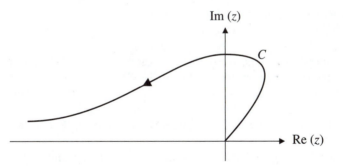

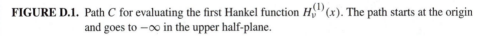

FIGURE D.1. Path C for evaluating the first Hankel function $H_\nu^{(1)}(x)$. The path starts at the origin and goes to $-\infty$ in the upper half-plane.

Such integrals may be written in the form

$$I(\xi) = \int_C g(z) \exp\left[\xi f(z)\right] dz \tag{D.2}$$

---

[1]This function is introduced in Chapter 8 from a different point of view.

where ξ is real, $f(z)$ is an analytic function, and C is a specified contour in the complex plane. We wish to evaluate the function $I(\xi)$ in the limit that ξ becomes large. In this case, the exponent

$$\xi f(z) = \xi[u(x, y) + iv(x, y)]$$

has a large absolute value. The imaginary part causes the integrand to oscillate wildly, contributing very little to the integral, except where v is approximately constant. The integrand is maximized where the real part u is maximum.

We know that we can deform the contour C without changing the value of the integral, so long as we do not cross any singularities of the integrand.[2] We can evaluate the integral most easily if we deform the contour so that it passes through a point z_0 such that

1. v is approximately constant near z_0 and
2. u is a maximum at z_0 and decreases rapidly along the deformed contour C' as we move away from z_0.

Then the integral will be dominated by a small region of the path very near z_0.

To make u a maximum, we need

$$\frac{df}{dz} = 0 \quad \text{at } z = z_0 \tag{D.3}$$

since $\partial u/\partial x$ and $\partial u/\partial y$ are then both zero at z_0. (Refer to Chapter 2, Section 2.2.2.) Now if the function $f(z)$ is analytic, then the function $u(x, y)$ is harmonic; that is,

$$\nabla^2 u = \frac{\partial^2 u}{\partial x^2} + \frac{\partial^2 u}{\partial y^2} = 0$$

Thus, if u is a maximum with respect to x ($\partial^2 u/\partial x^2 < 0$), it must be a minimum with respect to y ($\partial^2 u/\partial y^2 > 0$). Thus, the point z_0 that we seek is a saddle point. In order that the contribution of the integrand to the total integral be concentrated in the vicinity of z_0, we must choose our path C' so that it crosses the saddle in such a way that u decreases from its value at z_0 at the maximum rate. (C' looks like the rider's legs on the saddle.)

We also know that for analytic functions $\vec{\nabla}u$ is perpendicular to $\vec{\nabla}v$ (Chapter 2, Section 2.4.3). Thus, by choosing the path that maximizes the change in u, we are also choosing the path along which the imaginary part v remains almost constant. This is exactly what we want.

Now if the function $g(z)$ varies relatively slowly in the neighborhood of z_0, we can approximate the integral as follows:

$$I(\xi) = \int_C g(z) \exp[\xi f(z)]\, dz$$

$$\simeq g(z_0) \int_{\text{path near } z_0} \exp\left[\xi\left(f(z_0) + \frac{1}{2}(z - z_0)^2 f''(z_0)\right)\right] dz$$

$$= g(z_0) \exp[\xi f(z_0)] \int_{\text{path near } z_0} \exp\left[\xi\frac{1}{2}(z - z_0)^2 f''(z_0)\right] dz \tag{D.4}$$

[2] See Chapter 2, Section 2.2.4. We may not cross a branch cut either.

where we used the Taylor series to express $f(z)$ near z_0 and truncated after the third term. The second term [involving $f'(z_0)$] is zero. To choose the path, we set $z - z_0 = re^{i\phi}$ and $f''(z_0) = ae^{i\alpha}$, where a and α are real constants. Then the Taylor series becomes

$$f(z) = f(z_0) + \frac{1}{2}r^2 e^{2i\phi} a e^{i\alpha}$$

so

$$u = u(x_0, y_0) + \frac{1}{2}ar^2 \cos(2\phi + \alpha)$$

and

$$v = v(x_0, y_0) + \frac{1}{2}ar^2 \sin(2\phi + \alpha)$$

To keep v approximately constant, we choose the path so that $2\phi + \alpha = n\pi$; that is,

$$\phi = \phi_0 = -\frac{\alpha}{2} + \frac{n}{2}\pi \tag{D.5}$$

This path is a straight line in the neighborhood of z_0. Then

$$u = u(x_0, y_0) + \frac{1}{2}ar^2 \cos(n\pi) = u(x_0, y_0) + (-1)^n \frac{1}{2}ar^2$$

Now we want u to decrease as we move away from z_0, so we choose $n = \pm 1$. Then

$$u = u(x_0, y_0) - \frac{1}{2}ar^2$$

The choice of $n = +1$ or -1 is made by requiring that r go from a negative value[3] before passing through z_0 to a positive value after. The integral (D.4) is then reduced to an integral over r, with the value of r increasing throughout the range of integration:

$$I(\xi) \simeq g(z_0) \exp[\xi f(z_0)] \int_{\text{path near } z_0} \exp\left(-\frac{\xi ar^2}{2}\right) dr\, e^{i\phi_0}$$

$$= g(z_0) \exp[\xi f(z_0)] e^{i\phi_0} \int_{r \text{ small and negative}}^{r \text{ small and positive}} \exp\left(-\frac{\xi ar^2}{2}\right) dr$$

---

[3] In the usual polar representation of a complex number, the amplitude r is a positive real number, and negative real parts are described by phase angles between $\pi/2$ and $3\pi/2$. But here we want to describe the straight line path with a constant phase angle ϕ_0, and so we must allow r to take negative values.

The exponent $\xi a r^2/2$ becomes very large for $r \neq 0$ because ξ is very large. Thus, we may extend the limits in r to $\mp\infty$ without appreciably changing the value of the integral[4]:

$$I(\xi) \simeq g(z_0) \exp\left[\xi f(z_0)\right] e^{i\phi_0} \int_{-\infty}^{+\infty} \exp\left(-\frac{\xi a r^2}{2}\right) dr$$

$$I(\xi) = g(z_0) \exp\left[\xi f(z_0)\right] e^{i\phi_0} \sqrt{\frac{2\pi}{\xi a}} \qquad\qquad (D.6)$$

The result (D.6) is the asymptotic form of the integral (D.2).

Let's use this method to evaluate the asymptotic form of the Hankel function (D.1). Comparing equation (D.1) with the standard form (D.2), we find that the function $f(z)$ is

$$f(z) = \frac{1}{2}\left(z - \frac{1}{z}\right)$$

and is analytic except at $z = 0$. Similarly,

$$g(z) = \frac{1}{z^{\nu+1}}$$

We want the path of integration to pass through z_0, where

$$\left.\frac{df}{dz}\right|_{z=z_0} = \frac{1}{2}\left(1 + \frac{1}{z_0^2}\right) = 0 \Rightarrow z_0 = \pm i$$

Of these choices, only $+i$ lies near the original path C. So we deform C to pass through the point $z_0 = i$. Then

$$f''(z_0) = -\left.\frac{1}{z^3}\right|_i = -i = 1e^{-i\pi/2}$$

So $a = 1$ and $\alpha = -\pi/2$. Thus, the new path C' must pass through z_0 at an angle $\phi_0 = \pi/4 + \pi/2 = 3\pi/4$ (equation D.5). Before z_0, the difference $z - z_0$ has a positive real part on this path (see Figure D.2) and $e^{3i\pi/4} = -\frac{1}{2}\sqrt{2} + \frac{1}{2}i\sqrt{2}$, so r is negative, as required.

Then

$$f(z_0) = f(i) = \frac{1}{2}\left(i - \frac{1}{i}\right) = i$$

[4] See Appendix IX for the evaluation of the integral.

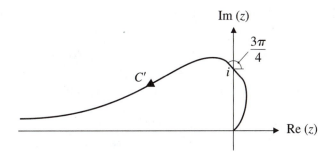

FIGURE D.2. Path C' for evaluating the first Hankel function $H_\nu^{(1)}(x)$. The path passes through the point $z = i$ at $45°$ to the imaginary axis.

Thus, applying the result (D.6), we have

$$H_\nu^{(1)}(\xi) \simeq \frac{1}{\pi i} \frac{1}{i^{\nu+1}} e^{i\xi} e^{i3\pi/4} \sqrt{\frac{2\pi}{\xi}} = \sqrt{\frac{2}{\pi\xi}} \exp i \left(\xi + \frac{3\pi}{4} - (\nu + 2) \frac{\pi}{2} \right)$$

$$= \sqrt{\frac{2}{\pi\xi}} \exp i \left(\xi - \frac{\pi}{2}\nu - \frac{\pi}{4} \right)$$

which is the expression given in Chapter 8, Section 8.4.2 (equation 8.85).

D.2. THE METHOD OF STATIONARY PHASE

A similar method applies to integrals of the form

$$I(x) = \int_{-\infty}^{+\infty} g(x) \exp [i\phi(x)] dx \tag{D.7}$$

when the function $\phi(x)$ (the phase) is real-valued and $g(x)$ is a slowly varying function. Integrals of this type arise through the use of Fourier transforms, for example, and are of importance in determining the signal that arrives after propagation through a dispersive medium (see, for example, Jackson, Chapter 7). The major contribution to the integral comes from the neighborhood of the point (or points) where $\phi(x)$ is stationary—hence the name of the method. Away from the stationary points, the integrand oscillates wildly and there is very little contribution to the integral.[5]

Again we expand the phase in a Taylor series in the vicinity of the stationary point x_s:

$$\phi(x) = \phi(x_s) + \frac{1}{2} (x - x_s)^2 \phi''(x_s) \tag{D.8}$$

---

[5]A more precise statement may be made in the language of generalized functions. See Chapter 6, especially Section 6.4.

Then, since $g(x)$ varies much more slowly than $e^{i\phi}$,

$$I(x) \simeq g(x_s)e^{i\phi(x_s)} \int_{x_s-\varepsilon}^{x_s+\varepsilon} \exp\left(\frac{i}{2}(x-x_s)^2\phi''(x_s)\right) dx$$

Here also we can extend the range of integration to $(-\infty, +\infty)$ with negligible change in the value of the integral.

To evaluate the integral, we make a change of variables to

$$u = (x - x_s)\sqrt{\frac{-i}{2}\phi''(x_s)}$$

The limits of the integral also change correspondingly. Let's assume for the moment that $\phi''(x_s)$ is positive[6] and equals $2A^2$. Then

$$u = (x - x_s)\sqrt{\frac{\phi''(x_s)}{2}}e^{-i\pi/4} = A(x-x_s)e^{-i\pi/4}$$

and the integral becomes

$$I(x) \simeq g(x_s)e^{i\phi(x_s)}\frac{e^{i\pi/4}}{A}\int_C e^{-u^2} du$$

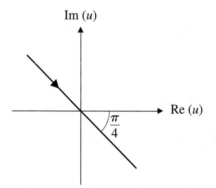

FIGURE D.3. The contour for the u-integral runs from $-\infty$ to $+\infty$ but at a $45°$ angle to the real axis.

The path of integration C for u is at $45°$ to the real axis (see Figure D.3). However, since the integrand has no poles and the contribution to the integral from $u = \pm\infty$ is zero, the value of the Gaussian integral (equal to $\sqrt{\pi}$ when evaluated along the real axis)

---

[6]If $\phi''(x_s)$ is negative, the path is rotated to cross the real axis at an angle of $+\pi/4$ rather than $-\pi/4$. The result is the same.

is not changed by this path shift. Thus,

$$I(x) \simeq g(x_s)e^{i\phi(x_s)}\frac{e^{i\pi/4}}{A}\int_{-\infty}^{+\infty}e^{-u^2}\,du$$

$$I(x) \simeq g(x_s)e^{i\phi(x_s)}e^{i\pi/4}\sqrt{\frac{2\pi}{\phi''(x_s)}} \qquad\qquad\text{(D.9)}$$

If there is more than one stationary point, then the integral is the sum of the contributions from each of the stationary points.

PROBLEMS

1. Use the method of steepest descent to evaluate the asymptotic form of the gamma function

$$\Gamma(\xi) = \int_0^\infty t^{\xi-1}e^{-t}\,dt \simeq \sqrt{2\pi}\xi^{\xi-1/2}e^{-\xi}$$

where the path of integration is along the real axis. This is Stirling's formula.

2. The modified Bessel function $K_\nu(\xi)$ has an integral representation

$$K_\nu(\xi) = \frac{1}{2}\int_0^\infty \exp\left[-\frac{\xi}{2}\left(s+\frac{1}{s}\right)\right]\frac{ds}{s^{1-\nu}}$$

with path of integration along the real axis. Use the method of steepest descent to find the asymptotic form of $K_\nu(\xi)$ as $\xi \to \infty$.

3. The Bessel function may be represented by an integral

$$J_\nu(\xi) = \frac{1}{2\pi}\int_C \exp\left(-i\xi\sin z + i\nu z\right)dz$$

where the contour C is as shown in the figure. Use the method of steepest descent or stationary phase, as appropriate, to derive the asymptotic form (8.83).

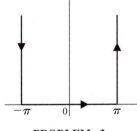

PROBLEM 3

4. Expand the function $g(x)$ in equation (D.7) in a Taylor series about the stationary point. Show that there is no contribution to the integral from the second (linear) term in the series if the expansion of the phase ϕ is truncated at the quadratic term, as in equation (D.8).

5. The function $H_\nu^{(2)}(x)$ has the integral expression

$$H_\nu^{(2)}(\xi) = \frac{1}{\pi i} \int_C \exp\left[\frac{\xi}{2}\left(z - \frac{1}{z}\right)\right] \frac{dz}{z^{\nu+1}}$$

where the path of integration goes from $-\infty$ to zero along a path in the lower half-plane that is the mirror image of the path in Figure D.1. Verify the asymptotic form (8.85) for this function.

6. An alternative integral expression for the Bessel functions is

$$F_\nu(x) = k \int_C \exp\left(i\nu u - ix\sin u\right) du$$

where

(a) for $H^{(1)}$ use contour C_1 and $k = 1/\pi$

(b) for $H^{(2)}$ use contour C_2 and $k = 1/\pi$

(c) for J use contour C_3 and $k = 1/2\pi$

and the contours are as shown in the figure. Evaluate the integrals for large values of x and verify the asymptotic forms in Chapter 8.

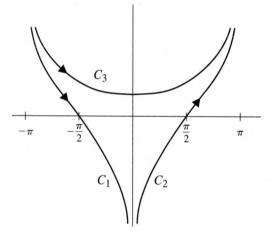

PROBLEM 6

7. The Airy integral

$$Ai(x) = \frac{1}{2\pi} \int_{-\infty}^{\infty} \exp i\left(tx + \frac{t^3}{3}\right) dt$$

arises in the study of diffraction. The path of integration lies slightly *above* the real axis.[7] Use the method of stationary phase to show that

$$Ai(x) \sim \frac{1}{2\sqrt{\pi}x^{1/4}} \exp\left(-\frac{2}{3}x^{3/2}\right) \quad \text{for large, positive } x$$

8. The amplitude of a signal arriving from a distant source after propagation through a dispersive medium may be written as a Fourier integral of the form

$$s(x, t) = \int_{-\infty}^{\infty} A(\omega) \exp\left[ik(\omega)x - i\omega t\right] d\omega$$

where $k(\omega)$ is the dispersion relation for the medium (see, for example, Jackson, Chapter 7). Use the method of stationary phase to show that, at time t, the largest amplitude signal is contributed by frequencies with group speed $d\omega/dk = D/t$, where D is the distance from source to receiver. Find an approximate expression for the amplitude at time t. Obtain an explicit form for the solution if $ck(\omega) = \sqrt{\omega^2 - \omega_p^2}$ and ω_p is a constant.

[7]See Jeffreys and Jeffreys, Section 17.07.

OPTIONAL TOPIC E

Calculus of Variations

E.1. INTEGRAL PRINCIPLES IN PHYSICS

Physical systems are often characterized by being an extremum (maximum, minimum, or point of inflection) of some physical property.[1] For example, a system in equilibrium is at an extremum of the potential energy. For stable equilibrium, the potential energy is a minimum. According to Fermat's principle, light rays follow the path of minimum time. The physical quantity being minimized in these examples (energy or time) is often expressed as an integral over the system. For example, the time for light to travel a distance $d\ell$ is $dt = d\ell/v$, where v is the light speed, and thus the total time for light to travel from point A to point B is

$$t = \int_A^B \frac{d\ell}{v} \tag{E.1}$$

where the integral is taken along the path from A to B. The path from A to B is described by one or more parameters, and the path actually followed by the light between A and B will cause the value of the integral (E.1) to be a minimum.

If we want to find the true path between points A and B, we must adjust the parameters so as to make the integral (E.1) an extremum. To see how to do this, first recall that if a function $f(x)$ has an extremum at the point x_0, then $df/dx|_{x_0} = 0$, and the change in f due to a small change δx in x is

$$\delta f = \frac{df}{dx}\bigg|_{x_0} \delta x + \text{terms of order } (\delta x)^2$$

$$= 0 \quad \text{to first order in } \delta x$$

Similarly, if the integral (E.1) is an extremum, its value will not change when we make small changes in the parameters. That is, the integral along paths that lie very close to the true path will be the same as the integral along the true path, to first order in changes in the parameters that describe the path. The difference δI between the value of the integral I along the true path and that along a neighboring path is called the *variation* in the integral I.

---

[1] Note the language carefully; "a" maximum is not the same as "the" maximum. We are interested in local extrema.

541

Example E.1. As a simple first example, consider the path taken by light moving from point A in a medium with refractive index n_1 to point B in a medium with refractive index n_2 (see Figure E.1). For now, let's take it for granted that the light travels in a straight line within each medium. Then the path meets the boundary surface at point P, and we can parameterize the path by the distance s of the point P along the boundary surface, as shown. Since the speed of light in each medium is $v = c/n$, the total time taken for light to travel from A to B is

$$t = \int_A^P \frac{n_1}{c} d\ell + \int_P^B \frac{n_2}{c} d\ell$$

$$= \frac{n_1}{c} \sqrt{s^2 + h_1^2} + \frac{n_2}{c} \sqrt{(w-s)^2 + h_2^2}$$

where in this case the integrals are trivial. Now we compute the change δt in t due to a small change δs in s:

$$\delta t = \frac{\partial t}{\partial s} \delta s$$

$$= \left(\frac{n_1}{c} \frac{s}{\sqrt{s^2 + h_1^2}} - \frac{n_2}{c} \frac{w-s}{\sqrt{(w-s)^2 + h_2^2}} \right) \delta s$$

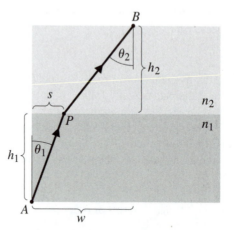

FIGURE E.1. The path of a light ray from point A in medium 1 to point B in medium 2 passes through point P on the boundary between the two materials. We use Fermat's principle to determine the location of point P and also to show that line AP is straight.

This change must be zero for any δs when the value of s corresponds to the true path. Thus, the value of s we need is given by

$$\frac{n_1}{c} \frac{s}{\sqrt{s^2 + h_1^2}} - \frac{n_2}{c} \frac{w-s}{\sqrt{(w-s)^2 + h_2^2}} = 0$$

or, in a more familiar form,

$$n_1 \sin \theta_1 = n_2 \sin \theta_2$$

which is Snells' law.

Now let's ask whether the light does travel in a straight line from A to P. Let the path be described by the function $y(x)$, where $y(0) = 0$ at point A and $y(s) = h_1$ is point P. A differential piece of the path has length

$$d\ell = \sqrt{dx^2 + dy^2} = \sqrt{1 + \left(\frac{dy}{dx}\right)^2}\, dx \qquad \text{(E.2)}$$

and the total time taken for light to travel from A to P is

$$t = \frac{n_1}{c} \int_0^s \sqrt{1 + \left(\frac{dy}{dx}\right)^2}\, dx$$

Here the integrand is a function of the derivative y' of y. Our goal is to find the function $y(x)$ that minimizes the time t, subject to the constraints that the endpoints $y(0)$ and $y(s)$ are fixed. The quantity t is a functional; that is, it is a function of the function $y(x)$.

Now as the function $y(x)$ changes, so does its derivative $y'(x)$. If the derivative changes by a small amount $\delta y'(x)$ (the variation in y'), then the integral changes by an amount δt, where

$$t + \delta t = \frac{n_1}{c} \int_0^s \sqrt{1 + (y' + \delta y')^2}\, dx$$

$$= \frac{n_1}{c} \int_0^s \sqrt{1 + (y')^2 + 2y'\, \delta y'}\, dx \quad \text{to first order in } \delta y'$$

$$= \frac{n_1}{c} \int_0^s \sqrt{1 + (y')^2} \left(1 + \frac{y'\, \delta y'}{1 + (y')^2}\right) dx$$

$$= t + \frac{n_1}{c} \int_0^s \frac{y'\, \delta y'}{\sqrt{1 + (y')^2}}\, dx$$

When $y(x)$ describes the true path, $\delta t = 0$. To obtain δt in terms of δy, we integrate by parts:

$$\delta t = \frac{n_1}{c} \frac{y'\, \delta y}{\sqrt{1 + (y')^2}}\Bigg|_0^s + \frac{n_1}{c} \int_0^s \left[\frac{y''}{\sqrt{1 + (y')^2}} - \frac{1}{2} \frac{2(y')^2 y''}{\left(1 + (y')^2\right)^{3/2}}\right] \delta y\, dx$$

With the endpoints fixed, the variation δy must equal zero at $x = 0$ and $x = s$. Thus, the integrated term is zero. The remaining integral must equal zero *no matter what* the variation $\delta y(x)$ if $y(x)$ describes the true path. Thus, the term in square brackets must be zero. This term simplifies to

$$\frac{y''}{[1+(y')^2]^{3/2}} = 0$$

and thus the second derivative $y'' = 0$. This means that the slope of the line, $y'(x)$, must be a constant, and thus the path is a straight line.

E.2. THE EULER EQUATION

Let's review the general procedure that we used in Example E.1. We have an integral

$$I(y) = \int_a^b f(x, y, y')\, dx \qquad (E.3)$$

evaluated along some path between the points a and b. That path is described by the function $y(x)$. The endpoints a and b are fixed, and thus the function $y(x)$ has fixed values $y(a) = Y_1$ and $y(b) = Y_2$ at the endpoints. Our goal is to find the function $y(x)$ such that the integral $I(y)$ is an extremum.

We begin by expressing the variation δI in the integral I in terms of the variation in the function y and its derivative y':

$$\delta I = \int_a^b \left(\frac{\partial f}{\partial y} \delta y + \frac{\partial f}{\partial y'} \delta y' \right) dx$$

To express the second term in terms of the variation δy, we integrate by parts:

$$\int_a^b \left(\frac{\partial f}{\partial y'} \delta y' \right) dx = \frac{\partial f}{\partial y'} \delta y \Big|_a^b - \int_a^b \frac{d}{dx} \left(\frac{\partial f}{\partial y'} \right) \delta y\, dx$$

The integrated term[2] is zero because $\delta y = 0$ at $x = a$ and $x = b$. Thus,

$$\delta I = - \int_a^b \left[\frac{d}{dx} \left(\frac{\partial f}{\partial y'} \right) - \frac{\partial f}{\partial y} \right] \delta y\, dx$$

We want the variation δI to be zero for *arbitrary*[3] small variations $\delta y(x)$ from the true function $y(x)$, and so the term in square brackets must be zero:

$$\frac{d}{dx} \left(\frac{\partial f}{\partial y'} \right) - \frac{\partial f}{\partial y} = 0 \qquad (E.4)$$

---

[2] In the case that the value of y is *not* fixed at the endpoints, we obtain a second condition: $\partial f / \partial y' = 0$ at $x = a$ and $x = b$.

[3] Some restrictions apply, most importantly that δy be differentiable.

The arbitrariness of δy is essential to this argument. We can perhaps find a specially contrived δy to make the integral zero, but if δy must be arbitrary, then we are forced to make the square bracket zero. Then equation (E.4) is the desired equation for the function $y(x)$. This equation is known as the Euler-Lagrange equation.

An alternative form of equation (E.4) may be derived by noting that the total derivative

$$\frac{df}{dx} = \frac{\partial f}{\partial x} + \frac{\partial f}{\partial y}\frac{dy}{dx} + \frac{\partial f}{\partial y'}\frac{d^2 y}{dx^2}$$

Solving for $\partial f/\partial y$ and substituting into equation (E.4), we obtain

$$\frac{d}{dx}\left(\frac{\partial f}{\partial y'}\right) - \frac{\partial f}{\partial y} = \frac{d}{dx}\left(\frac{\partial f}{\partial y'}\right) - \frac{1}{y'}\left(\frac{df}{dx} - \frac{\partial f}{\partial x} - \frac{\partial f}{\partial y'}\frac{d^2 y}{dx^2}\right) = 0$$

and so

$$\frac{df}{dx} - \frac{\partial f}{\partial x} - \frac{\partial f}{\partial y'}y'' - y'\frac{d}{dx}\left(\frac{\partial f}{\partial y'}\right) = 0$$

We may simplify by noting that

$$\frac{d}{dx}\left(y'\frac{\partial f}{\partial y'}\right) = y''\frac{\partial f}{\partial y'} + y'\frac{d}{dx}\left(\frac{\partial f}{\partial y'}\right)$$

So finally we have

$$\frac{d}{dx}\left(f - y'\frac{\partial f}{\partial y'}\right) - \frac{\partial f}{\partial x} = 0 \qquad (E.5)$$

This version is particularly useful when f does not depend explicity on x ($\partial f/\partial x \equiv 0$), since then we may integrate once to obtain

$$\boxed{f - y'\frac{\partial f}{\partial y'} = \text{constant}} \qquad (E.6)$$

E.2.1. Application to Mechanics

The Lagrangian of a mechanical system is

$$\mathcal{L} = T - V$$

where T is the kinetic and V the potential energy of the system. These quantities are expressed in terms of the generalized coordinates q_i that describe the position and velocity of all of the parts of the system. For a system of particles, the coordinates q_i are just the ordinary coordinates of the particles in the system. Hamilton's principle[4] says that the motion of the system will be such as to minimize the integral

$$\mathcal{I} = \int_{t_1}^{t_2} \mathcal{L}(q_i, \dot{q}_i, t)\, dt$$

Thus, we can determine the motion of the system [that is, determine the $q_i(t)$] using the calculus of variations.

Example E.2.   A particle of mass m slides down a wedge whose sloping side of length L makes an angle θ with the horizontal (see Figure E.2). The wedge of mass M is free to slide horizontally. How long does the particle take to reach the ground?

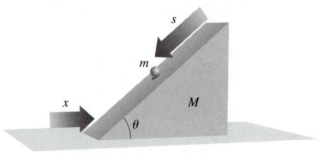

FIGURE E.2. A particle of mass m slides down a wedge of mass M that is itself free to slide horizontally.

It is convenient to choose our generalized coordinates to be the distance s that the particle has slid along the wedge surface and the distance x that the wedge has slid over the ground. With these coordinates, the kinetic energy of the wedge is $\frac{1}{2}M\dot{x}^2$. The velocity of the particle *with respect to the ground* is

$$\vec{v} = \left(\frac{dx}{dt} - \frac{ds}{dt}\cos\theta\right)\hat{x} - \frac{ds}{dt}\sin\theta\,\hat{y}$$

and thus the total kinetic energy is

$$T = \frac{1}{2}M\left(\frac{dx}{dt}\right)^2 + \frac{1}{2}m\left[\left(\frac{dx}{dt} - \frac{ds}{dt}\cos\theta\right)^2 + \left(\frac{ds}{dt}\sin\theta\right)^2\right]$$

$$= \frac{1}{2}(M+m)\left(\frac{dx}{dt}\right)^2 - m\frac{dx}{dt}\frac{ds}{dt}\cos\theta + \frac{1}{2}m\left(\frac{ds}{dt}\right)^2$$

With reference level at the top of the wedge, the potential energy is

$$V = -mgs\sin\theta$$

[4] See, for example, Goldstein, Chapter 7, Section 5.

Thus, the Lagrangian is

$$\mathcal{L} = T - V = \frac{1}{2}(M + m)\left(\frac{dx}{dt}\right)^2 - m\frac{dx}{dt}\frac{ds}{dt}\cos\theta + \frac{1}{2}m\left(\frac{ds}{dt}\right)^2 + mgs\sin\theta$$

The Euler-Lagrange equations for this system are

$$\frac{d}{dt}\left(\frac{\partial\mathcal{L}}{\partial\dot{x}}\right) - \frac{\partial\mathcal{L}}{\partial x} = 0$$

$$\frac{d}{dt}\left[(M + m)\left(\frac{dx}{dt}\right) - m\frac{ds}{dt}\cos\theta\right] - 0 = 0 \tag{E.7}$$

and

$$\frac{d}{dt}\left(\frac{\partial\mathcal{L}}{\partial\dot{s}}\right) - \frac{\partial\mathcal{L}}{\partial s} = 0$$

$$\frac{d}{dt}\left(-m\frac{dx}{dt}\cos\theta + m\frac{ds}{dt}\right) - mg\sin\theta = 0 \tag{E.8}$$

Equation (E.7) may be integrated immediately to get

$$(M + m)\left(\frac{dx}{dt}\right) - m\frac{ds}{dt}\cos\theta = C_1 \tag{E.9}$$

which may be recognized as conservation of momentum in the x-direction. Thus, if the system starts from rest, the integration constant $C_1 = 0$. Integrating again, we find

$$(M + m)x - ms\cos\theta = C_2$$

and again, if $x = s = 0$ at $t = 0$, then $C_2 = 0$. Equation (E.8) may also be integrated to get

$$\left(-\frac{dx}{dt}\cos\theta + \frac{ds}{dt}\right) - gt\sin\theta = 0$$

where again we used the information that the system starts from rest. We may now eliminate dx/dt using equation (E.9) and integrate again with the initial condition $s(0) = 0$:

$$\frac{ds}{dt}\left(1 - \frac{m}{M + m}\cos^2\theta\right) = gt\sin\theta$$

$$s = \left(\frac{M + m}{M + m\sin^2\theta}\right)\frac{gt^2}{2}\sin\theta$$

The particle reaches the bottom of the wedge ($s = L$) in a time

$$t = \sqrt{\frac{2L}{g \sin \theta} \left(\frac{M + m \sin^2 \theta}{M + m} \right)}$$

We may check our result by letting $M \to \infty$ (fixed wedge) to obtain $t = \sqrt{2L/g \sin \theta}$, the correct result for a particle accelerating at $g \sin \theta$.

E.3. VARIATION SUBJECT TO CONSTRAINTS

Sometimes the variations cannot be completely arbitrary because of additional constraints on the physical system. If we want to find an extremum of an integral I (equation E.3) with the constraint that a second integral $J = \int_a^b g(x, y, y') \, dx$ remain constant, we proceed by introducing a constant (but as yet unknown) multiplier λ to form the combination

$$K = I + \lambda J$$

Then if I is stationary and J is constant, K must also be stationary. Proceeding as in Section E.2, we obtain the Euler-Lagrange equation (E.4 or E.6) but with the function f replaced by the function $h = f + \lambda g$.

> **Example E.3.** A uniform cable of length L hangs between two points at the same height and a distance D apart. Find the shape of the curve described by the cable. The curve is called a catenary.
>
> In this problem, the relevant physical quantity is the gravitational potential energy of the cable, which will be a minimum when the cable is in its equilibrium configuration. If we choose the reference level at the height of the two endpoints of the cable and let $y(x)$ describe the distance of the cable below this level at coordinate x, $0 \le x \le D$, the potential energy is
>
> $$U(y) = \int -y(x)g \, dm = -\mu g \int_0^D y \sqrt{1 + (y')^2} \, dx \qquad \text{(E.10)}$$
>
> where μ is the mass per unit length of the cable, $dm = \mu \, d\ell$, and we express the length element $d\ell$ of the cable as in Example E.1 (equation E.2). At the endpoints, $y(0) = y(D) = 0$.
>
> However, the length of the cable is a constant, so we may not vary $y(x)$ in a completely arbitrary manner. We must have
>
> $$L = \int d\ell = \int_0^D G(y') \, dx = \int_0^D \sqrt{1 + (y')^2} \, dx \qquad \text{(E.11)}$$
>
> [Here we choose to call the integrand $G(y')$ to avoid confusion with the acceleration due to gravity, g.] None of the integrands contain x explicitly, so we may use equation (E.6), including the constraint. We may absorb the constants μ and g into the

constant multiplier λ:

$$f + \lambda G - y' \frac{\partial(f + \lambda G)}{\partial y'} = \text{constant}$$

$$(y + \lambda)\sqrt{1 + (y')^2} - y' \frac{\partial}{\partial y'}(y + \lambda)\sqrt{1 + (y')^2} = C$$

$$(y + \lambda)\sqrt{1 + (y')^2} \left(1 - \frac{(y')^2}{1 + (y')^2} \right) = C$$

$$y + \lambda = C\sqrt{1 + (y')^2}$$

Solving for the derivative, we find

$$\frac{dy}{dx} = \pm\sqrt{\frac{1}{C^2}(y + \lambda)^2 - 1}$$

As the cable hangs, the slope dy/dx is positive on the first half of the cable and negative on the second. Now we may integrate to relate the coordinates (X, Y) of a point on the cable. For $0 \le x \le D/2$,

$$\int_0^Y \frac{dy}{\sqrt{\frac{1}{C^2}(y + \lambda)^2 - 1}} = \int_0^X dx$$

Let $y + \lambda = C \cosh\theta$, $dy = C \sinh\theta \, d\theta$. Then

$$\int_{\cosh^{-1}(\lambda/C)}^{\cosh^{-1}(Y+\lambda)/C} \frac{C \sinh\theta \, d\theta}{\sinh\theta} = X$$

$$C \left[\cosh^{-1}\left(\frac{Y + \lambda}{C} \right) - \cosh^{-1}\frac{\lambda}{C} \right] = X \quad \text{for } 0 < X < D/2 \qquad \text{(E.12)}$$

Rearranging, we can solve equation (E.12) for Y on the first half of the cable[5]:

$$y = C \cosh\left(\frac{x}{C} + \cosh^{-1}\frac{\lambda}{C} \right) - \lambda \qquad \text{(E.13)}$$

The slope of the cable is then

$$\frac{dy}{dx} = \sinh\left(\frac{x}{C} + \cosh^{-1}\frac{\lambda}{C} \right)$$

---

[5]Having completed the integration, we no longer need to distinguish between Y and y.

At the midpoint, $x = D/2$, the slope is zero:

$$0 = \sinh\left(\frac{D}{2C} + \cosh^{-1}\frac{\lambda}{C}\right)$$

Thus,

$$\frac{D}{2C} = -\cosh^{-1}\frac{\lambda}{C}$$

It is convenient to choose C to be negative, $C = -\gamma$, and then

$$\lambda = -\gamma\cosh\frac{D}{2\gamma}$$

Our solution (E.13) becomes

$$y = \gamma\left[\cosh\frac{D}{2\gamma} - \cosh\left(\frac{D/2 - x}{\gamma}\right)\right]$$

with slope

$$\frac{dy}{dx} = \sinh\left(\frac{D/2 - x}{\gamma}\right)$$

Next we apply the length constraint (E.11) to find γ. Again it is easiest to work with the first half of the cable.

$$\frac{L}{2} = \int_0^{D/2}\sqrt{1 + (y')^2}\,dx = \int_0^{D/2}\sqrt{1 + \sinh^2\left(\frac{D/2 - x}{\gamma}\right)}\,dx$$

$$= \int_0^{D/2}\cosh\left(\frac{D/2 - x}{\gamma}\right)dx = -\gamma\sinh\left(\frac{D/2 - x}{\gamma}\right)\Big|_0^{D/2}$$

$$= \gamma\sinh\frac{D}{2\gamma}$$

which is a transcendental equation for γ that can be solved numerically. If the cable length L is twice the distance D between the supports, then

$$\frac{D}{\gamma} = \sinh\frac{D}{2\gamma}$$

with solution $D/\gamma = 4.355$, and hence $\gamma = 0.2296D$. The shape of the cable is shown in Figure E.3.

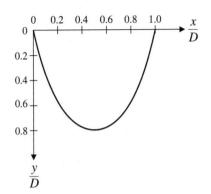

FIGURE E.3. The cable's shape is such as to minimize its potential energy. Here we see a cable with length $L = 2D$. Remember that y increases downward.

E.4. EXTENSION TO FUNCTIONS OF MORE THAN ONE VARIABLE

The theory of calculus of variations may easily be extended to integrals over more than one variable. The integral is of the form

$$I = \int f(x_i; u)d^n\vec{\mathbf{x}}$$

where x_i, $i = 1 - n$, are the independent variables and u is a function of the variables x_i. Upon taking the variation, we have

$$\delta I = \int \left(\frac{\partial f}{\partial u}\delta u + \sum_{i=1}^{n} \frac{\partial f}{\partial(\partial u/\partial x_i)}\delta\left(\frac{\partial u}{\partial x_i}\right) \right) d^n\vec{\mathbf{x}}$$

Once again, we integrate by parts to convert the variation in each derivative to a variation in u. We obtain the Euler equation:

$$\sum_{i=1}^{n} \frac{\partial}{\partial x_i}\left(\frac{\partial f}{\partial(\partial u/\partial x_i)}\right) - \frac{\partial f}{\partial u} = 0 \tag{E.14}$$

In this expression, the derivative $\partial/\partial x_i$ is a partial derivative in the sense that we are holding all the other coordinates fixed, but it is a total derivative in the sense that we must include the implicit as well as the explicit dependence upon x. For example, if $n = 3$,

$$\left.\frac{\partial F}{\partial x}\right|_{\text{const } y,z} = \left.\frac{\partial F}{\partial x}\right|_{\text{const } u,y,z,u_x,u_y,u_z} + \frac{\partial F}{\partial u}\frac{\partial u}{\partial x} + \frac{\partial F}{\partial(\partial u/\partial x)}\frac{\partial(\partial u/\partial x)}{\partial x}$$

$$+ \frac{\partial F}{\partial(\partial u/\partial y)}\frac{\partial(\partial u/\partial y)}{\partial x} + \frac{\partial F}{\partial(\partial u/\partial z)}\frac{\partial(\partial u/\partial z)}{\partial x}$$

If we already knew the function $u(x, y, z)$, we could obtain the same result by explicitly writing out the coordinate-dependence of the function u and all its derivatives to obtain $F(x, y, z)$ and then evaluating the usual partial derivative $\partial F/\partial x$. But in cases where we want to use equation (E.14), the function u is the solution that we seek, and so the explicit dependence on x is not known *a priori*.

Example E.4.   The electrostatic field in a volume V may be described as the gradient of a potential: $\vec{E} = -\vec{\nabla}\phi$, where the potential $\phi(x, y, z)$ takes on specified values on the bounding surface S. Show that the function ϕ that minimizes the electric energy in the volume V satisfies Laplace's equation: $\nabla^2\phi = 0$.
    The electric energy is given by the integral

$$U = \frac{\varepsilon_0}{2} \int_V E^2 \, dV = \frac{\varepsilon_0}{2} \int_V (\vec{\nabla}\phi)^2 dV$$

$$= \frac{\varepsilon_0}{2} \int_V \left[\left(\frac{\partial\phi}{\partial x}\right)^2 + \left(\frac{\partial\phi}{\partial y}\right)^2 + \left(\frac{\partial\phi}{\partial z}\right)^2 \right] dV$$

The integrand $f = (\partial\phi/\partial x)^2 + (\partial\phi/\partial y)^2 + (\partial\phi/\partial z)^2$, and so the Euler equation (E.14) is

$$\frac{\partial}{\partial x}\left(\frac{\partial\phi}{\partial x}\right) + \frac{\partial}{\partial y}\left(\frac{\partial\phi}{\partial y}\right) + \frac{\partial}{\partial z}\left(\frac{\partial\phi}{\partial z}\right) = \nabla^2\phi = 0$$

PROBLEMS

1. The speed of waves in a medium varies with y as $v = v_0(1 + y)$. What is the path of a ray? This model describes the propagation of seismic waves through the Earth's outer layers. If the waves start at $y = 0$, $x = 0$, find the value of x at which the waves return to the surface as a function of the initial slope of the ray $dy/dx = m = \tan\theta$. Also determine the total time of travel for each ray.

2. Find the path of a light ray through a medium whose refractive index increases linearly with depth: $n = 1 + x/a$. Assume that the path starts from the origin, $x = y = 0$, and $dy/dx = 1$ at $x = 0$.

3. Rework Problem 2, reversing the roles of the labels x and y and describing the time as an integral over the new x. Is the result the same? Which method is easier?

4. Repeat Problem 2 with refractive index function $n(x) = n_0 e^x$.

5. **The brachistochrone.** A smooth wire runs between two fixed points $A(0, 0)$ and $B(X, Y)$. Find the shape of the wire such that a particle sliding without friction on the wire reaches point B in minimum time. Assume that the particle starts from point A with speed v_0. *Hint:* Set $y' = \tan\theta$ and obtain x and y as functions of θ. Show that the coordinates of the two fixed points are sufficient to determine the initial and final values of θ and the two integration constants. Determine an explicit solution in the case $v_0 = 0$, $X = Y = -1$. Plot the shape of the wire in this case.

6. Show that the Sturm-Liouville equation (8.1) arises from the problem of finding the extremum of the integral

$$I = \int_a^b [f(y')^2 + gy^2]\,dx = \int_a^b F\,dx$$

subject to the constraint

$$J = \int_a^b wy^2\,dx = \int_a^b G\,dx = \text{constant}$$

7. Show that the catenary (Example E.3) is symmetric about the midpoint—that is, $y(D - x) = y(x)$.

8. Show that the curve that encloses the greatest area with a fixed perimeter is a circle.

9. Investigate the problem of finding an extremum of the integral

$$I = \int \psi H \psi^*\,dx$$

subject to the constraint

$$J = \int \psi \psi^*\,dx = 1$$

where H is the Hamiltonian operator

$$H = -\frac{\hbar^2}{2m}\frac{d^2}{dx^2} + V(x)$$

and $V(x)$ is a known function. Show that the resulting differential equation is the Schrödinger equation.

   Hint: First integrate by parts to eliminate the second derivative.

10. Using polar coordinates, write the Lagrangian for a particle moving in the potential $V = -GMm/r$, and form the Euler-Lagrange equations. Show that the equation in the angular coordinate indicates conservation of angular momentum.

11. A spherical pendulum is a mass free to move on the end of a string of fixed length ℓ. Write the Lagrangian in terms of the spherical angles θ and ϕ, and hence find the equations of motion. Show that one possible motion is the *conical pendulum* with constant θ. What is the value of $d\phi/dt$ in this case?

12. Consider the one-dimensional motion of a particle with potential energy $V(x)$ that is independent of time. Show that the Euler-Lagrange equations may be written in the form of equation (E.6). Give a physical interpretation of this equation.

13. The Lagrangian for a vibrating string may be written

$$\mathcal{L} = \int_0^L \left[\frac{1}{2}\mu \left(\frac{\partial y}{\partial t}\right)^2 - \frac{1}{2}T \left(\frac{\partial y}{\partial x}\right)^2 \right] dx$$

where $y(x, t)$ is the displacement of the string, μ is the mass per unit length, and T is the tension. The first term in the integrand is the kinetic energy density, and the second is the potential energy density. Determine the Euler-Lagrange equations for the system, and comment.

14. As an alternative approach to Problem 13, we may expand the displacement $y(x, t)$ as a Fourier series in x (refer to Chapter 4, Section 4.2):

$$y(x, t) = \sum_{n=0}^{\infty} a_n(t) \sin \frac{n\pi x}{L}$$

Write the Lagrangian as a function of the generalized coordinates $a_n(t)$ and the time t. What are the Euler-Lagrange equations now?

15. A volume V is formed by rotating a curve $y(x)$ defined for $-a < x < a$ around the x-axis. Given that the curve is symmetric about the y-axis, $y(a) = y(-a) = 0$, and $y'(0) = 0$, show that the curve that gives the maximum volume for a given surface area is a circle and the corresponding volume is a sphere.

16. The Lagrangian for a particle moving under the influence of electromagnetic fields is

$$\mathcal{L} = \frac{1}{2}mv^2 - q\phi + q\vec{A} \cdot \vec{v}$$

where ϕ and $\vec{A}$ are given functions of position and time. Find the equations of motion, and hence show that the force acting on the particle is the Lorentz force $\vec{F} = q(\vec{E} + \vec{v} \times \vec{B})$.

17. Show that the shortest distance between two points on the surface of a sphere is a great circle. *Hint:* You may place the polar axis through one of the points.

APPENDICES

I. TRANSFORMATION PROPERTIES OF THE VECTOR CROSS PRODUCT

In Chapter 1, we noted that the cross product of two vectors does not transform as a vector (Section 1.1.2). Here we shall investigate the transformation properties of the cross product.[1] In a second, primed coordinate system, the definition of the cross product in terms of components (Chapter 1, equation 1.30) is unchanged. We can then express the vector components in the primed system in terms of the original (unprimed) components, using the transformation law for vectors:

$$\left(\vec{u}' \times \vec{v}'\right)_i = \varepsilon_{ijk} u'_j v'_k = \varepsilon_{ijk} A_{jl} u_l A_{km} v_m \tag{1}$$

We want to compare this expression with the usual transformation law (1.23) for vectors:

$$A_{ip}(\vec{u} \times \vec{v})_p = A_{ip} \varepsilon_{pjk} u_j v_k$$

These expressions are not the same.

As a first step toward finding the correct transformation law, we write the determinant[2] of the transformation matrix using the Levi-Civita symbol:

$$\det \mathbb{A} = \begin{vmatrix} A_{11} & A_{12} & A_{13} \\ A_{21} & A_{22} & A_{23} \\ A_{31} & A_{32} & A_{33} \end{vmatrix}$$

$$= A_{11}(A_{22}A_{33} - A_{23}A_{32}) - A_{12}(A_{21}A_{33} - A_{23}A_{31}) + A_{13}(A_{21}A_{32} - A_{22}A_{31})$$

$$= A_{1j}\varepsilon_{jkl}A_{2k}A_{3l} = -A_{2i}\varepsilon_{ijk}A_{1j}A_{3k}$$

Similar results are obtained for all permutations of the indices 1, 2, and 3. For $p = 1, q = 2$, and $r = 3$, the first expression for det A gives directly

$$\varepsilon_{pqr} \det \mathbb{A} = \varepsilon_{jkl} A_{pj} A_{qk} A_{rl} \tag{2}$$

[1] See also Optional Topic A, Section A.3.
[2] See also Chapter 1, Section 1.6.2.

This relation is confirmed for $p = 2$, $q = 1$, and $r = 3$ using the second expression for det $\mathbb{A}$ and the index name changes $i \rightarrow j$, $j \rightarrow k$, $k \rightarrow l$. Similarly, we may verify the relation for each set of values of p, q, and r.

Now we can make use of this relation to see how the cross product transforms. We start with the definition (1.30), apply the transformation matrix, and then write the result in terms of the components in the primed system:

$$A_{ip}(\vec{\mathbf{u}} \times \vec{\mathbf{v}})_p = A_{ip}\varepsilon_{pjk}u_j v_k = A_{ip}\varepsilon_{pjk}A_{jl}^{-1}u_l' A_{km}^{-1}v_m'$$

Since the inverse of the transformation matrix $\mathbb{A}$ equals its transpose, we have

$$A_{ip}(\vec{\mathbf{u}} \times \vec{\mathbf{v}})_p = \varepsilon_{pjk}A_{ip}A_{lj}A_{mk}u_l' v_m'$$

Now we use equation (2):

$$A_{ip}(\vec{\mathbf{u}} \times \vec{\mathbf{v}})_p = \varepsilon_{ilm}u_l' v_m' \det \mathbb{A} = (\vec{\mathbf{u}}' \times \vec{\mathbf{v}}')_i \det \mathbb{A} \tag{3}$$

The usual transformation law for vectors does not lead to the correct expression (1) for the cross product, so we multiply both sides of equation (3) by det $\mathbb{A}$ and use the result that det $\mathbb{A} = \pm 1$ to obtain

$$(\vec{\mathbf{u}}' \times \vec{\mathbf{v}}')_i = (\det \mathbb{A}) A_{ip}(\vec{\mathbf{u}} \times \vec{\mathbf{v}})_p \tag{4}$$

which is the transformation law we seek. Because of the determinant in equation (4),

$$(\vec{\mathbf{u}}' \times \vec{\mathbf{v}}') = \mathbb{A}(\vec{\mathbf{u}} \times \vec{\mathbf{v}})$$

for rotations, but

$$(\vec{\mathbf{u}}' \times \vec{\mathbf{v}}') = -\mathbb{A}(\vec{\mathbf{u}} \times \vec{\mathbf{v}})$$

for reflections.[3]

II. PROOF OF THE HELMHOLTZ THEOREM

The Helmholtz theorem states that any vector field $\vec{\mathbf{F}}$ may be decomposed into the sum of two vectors: $\vec{\mathbf{F}} = \vec{\mathbf{u}} + \vec{\mathbf{v}}$, where $\vec{\mathbf{u}} = \vec{\nabla}\Phi$ and $\vec{\mathbf{v}} = \vec{\nabla} \times \vec{\mathbf{A}}$.

The proof takes the form of a demonstration of how to construct the two functions $\Phi(x, y, z)$ and $\vec{\mathbf{A}}(x, y, z)$. First we construct the vector function:

$$\vec{\mathbf{w}}(x, y, z) = \int_{\text{all space}} \frac{\vec{\mathbf{F}}(x', y', z')}{4\pi R} \, dx' \, dy' \, dz' \tag{5}$$

[3]For additional information, see Optional Topic A.

where

$$R = |\vec{\mathbf{x}} - \vec{\mathbf{x}}'|$$

and the integral is over all space. This function satisfies the equation

$$\nabla^2 \vec{\mathbf{w}} = -\vec{\mathbf{F}} \tag{6}$$

Let's demonstrate this. Since the operator $\vec{\nabla}$ differentiates with respect to x, y, z but not x', y', z', we may move it inside the integral, where it operates only on the function $1/R$:

$$\nabla^2 \vec{\mathbf{w}} = \nabla^2 \int \frac{\vec{\mathbf{F}}(x', y', z')}{4\pi R} \, dx' \, dy' \, dz' = \int \vec{\mathbf{F}}(x', y', z') \nabla^2 \frac{1}{4\pi R} \, dx' \, dy' \, dz'$$

$$= \int \vec{\mathbf{F}}(x', y', z')[-\delta(\vec{\mathbf{x}} - \vec{\mathbf{x}}')] \, dV' = -\vec{\mathbf{F}}(x, y, z)$$

where we used equation (6.26) and the sifting property. Thus, we have verified equation (6).

But from equation (1.51),

$$\nabla^2 \vec{\mathbf{w}} = \vec{\nabla}(\vec{\nabla} \cdot \vec{\mathbf{w}}) - \vec{\nabla} \times (\vec{\nabla} \times \vec{\mathbf{w}})$$

and so

$$\vec{\mathbf{F}} = \vec{\nabla} \times (\vec{\nabla} \times \vec{\mathbf{w}}) - \vec{\nabla}(\vec{\nabla} \cdot \vec{\mathbf{w}})$$

Then we define the scalar function Φ by

$$\Phi = -\vec{\nabla} \cdot \vec{\mathbf{w}} \tag{7}$$

and the vector function $\vec{\mathbf{A}}$ by

$$\vec{\mathbf{A}} = \vec{\nabla} \times \vec{\mathbf{w}} \tag{8}$$

so that

$$\vec{\mathbf{F}} = \vec{\nabla} \times \vec{\mathbf{A}} + \vec{\nabla}\Phi \tag{9}$$

as required by the theorem.

Next we can find explicit expressions for Φ and $\vec{\mathbf{A}}$. Let's start with Φ (equations 7 and 5):

$$\Phi(x, y, z) = -\vec{\nabla} \cdot \vec{\mathbf{w}} = -\vec{\nabla} \cdot \int \frac{\vec{\mathbf{F}}(x', y', z')}{4\pi R} \, dx' \, dy' \, dz'$$

$$= -\int \vec{\nabla} \cdot \frac{\vec{\mathbf{F}}(x', y', z')}{4\pi R} \, dx' \, dy' \, dz'$$

$$= -\int \vec{\mathbf{F}}(x', y', z') \cdot \vec{\nabla} \frac{1}{4\pi R} \, dx' \, dy' \, dz'$$

Then we use the result

$$\vec{\nabla}\frac{1}{R} = -\frac{(x-x')}{R^3}\hat{x} - \frac{(y-y')}{R^3}\hat{y} - \frac{(z-z')}{R^3}\hat{z} = -\vec{\nabla}'\frac{1}{R}$$

where $\vec{\nabla}'$ is the grad operator with respect to the primed coordinates. Thus,

$$\Phi = \int \mathbf{F}(x', y', z') \cdot \vec{\nabla}'\frac{1}{4\pi R}\, dx'\, dy'\, dz'$$

Now we move the grad operator back through the vector $\mathbf{F}$ again, using the result $\vec{\nabla} \cdot (\psi\vec{u}) = \psi\vec{\nabla} \cdot \vec{u} + \vec{u} \cdot \vec{\nabla}\psi$ with $\vec{u} = \mathbf{F}$ and $\psi = 1/R$:

$$\Phi = \int \left[\vec{\nabla}' \cdot \left(\frac{\mathbf{F}(x', y', z')}{4\pi R} \right) - \frac{\vec{\nabla}' \cdot \mathbf{F}}{4\pi R} \right] dx'\, dy'\, dz' \qquad (10)$$

We can rewrite the first term using the divergence theorem (equation 1.44):

$$\Phi = \int_{S_\infty} \frac{\mathbf{F}(x', y', z') \cdot \hat{n}}{4\pi R}\, dA' - \int \frac{\vec{\nabla}' \cdot \mathbf{F}}{4\pi R}\, dV'$$

The surface integral is zero provided that $\mathbf{F} \to 0$ on the surface at infinity at least as fast as $R^{-(1+\varepsilon)}$. Thus,

$$\boxed{\Phi = -\int \frac{\vec{\nabla}' \cdot \mathbf{F}(x', y', z')}{4\pi R}\, dV'} \qquad (11)$$

We can obtain an expression for $\vec{A}$ similarly:

$$\vec{A} = \vec{\nabla} \times \vec{w} = \vec{\nabla} \times \int \frac{\mathbf{F}(x', y', z')}{4\pi R}\, dx'\, dy'\, dz'$$

First we move the curl inside the integral, where it operates only on the function $1/R$:

$$\vec{A} = \int \vec{\nabla} \times \frac{\mathbf{F}(x', y', z')}{4\pi R}\, dx'\, dy'\, dz' = \int \left(\vec{\nabla}\frac{1}{4\pi R} \right) \times \mathbf{F}(x', y', z')\, dx'\, dy'\, dz'$$

Next we convert to a derivative with respect to the prime variables and use the result of Problem 1.23:

$$\vec{A} = -\int \left(\vec{\nabla}'\frac{1}{4\pi R} \right) \times \mathbf{F}(x', y', z')\, dx'\, dy'\, dz'$$

$$= -\int \left[\vec{\nabla}' \times \left(\frac{\vec{F}}{4\pi R} \right) - \frac{1}{4\pi R}\vec{\nabla}' \times \vec{F} \right] dx'\, dy'\, dz'$$

Now we use a variant of the divergence theorem (Problem 1.30b) to convert the first integral to a surface integral:

$$\vec{A} = -\int_{S_\infty} \hat{n}' \times \left(\frac{\vec{F}}{4\pi R}\right) dA' + \int \frac{\vec{\nabla}' \times \vec{F}}{4\pi R} dx'\, dy'\, dz'$$

$$\boxed{\vec{A} = \int \frac{\vec{\nabla}' \times \vec{F}}{4\pi R} dx'\, dy'\, dz'} \tag{12}$$

again provided that $\vec{F} \to 0$ on the surface at infinity at least as fast as $R^{-(1+\varepsilon)}$.

Equations (11) and (12) allow us to calculate the vectors $\vec{u} = \vec{\nabla}\Phi$ and $\vec{v} = \vec{\nabla} \times \vec{A}$.

III. PROOF BY INDUCTION: THE CAUCHY FORMULA

To prove the Cauchy formula (Chapter 2, equation 2.39),

$$\oint_C \frac{f(z)}{(z-a)^n} dz = \frac{2\pi i}{(n-1)!} \left. f^{(n-1)}(z)\right|_{z=a} \tag{13}$$

we use the idea of proof by induction. First we assume that the result is true for some value of n, say $n = m$. Then we shall show that the result must also be true for $n = m + 1$. Since we have already shown that the result holds for $n = 1$ (Chapter 2, equation 2.38), it must be true for all values of $n \geq 1$.

We start with $n = m + 1$ and express the derivative on the right-hand side of equation (13) in terms of the $(m - 1)$st derivative:

$$\left. f^{(m)}(z)\right|_{z=a} = \lim_{h \to 0} \frac{1}{h} \left(\left. f^{(m-1)}(z)\right|_{z=a+h} - \left. f^{(m-1)}(z)\right|_{z=a} \right)$$

Now we use the assumed result for $n = m$ to express the $(m - 1)$st derivatives:

$$\left. f^{(m)}(z)\right|_{z=a} = \lim_{h \to 0} \frac{(m-1)!}{2\pi i h} \left(\oint_C \frac{f(z)}{[z-(a+h)]^m} dz - \oint_C \frac{f(z)}{(z-a)^m} dz \right)$$

$$= \lim_{h \to 0} \frac{(m-1)!}{2\pi i h} \oint_C f(z) \left(\frac{1}{(z-a-h)^m} - \frac{1}{(z-a)^m} \right) dz$$

$$= \lim_{h \to 0} \frac{(m-1)!}{2\pi i h} \oint_C f(z) \frac{(z-a)^m - (z-a-h)^m}{(z-a-h)^m (z-a)^m} dz$$

We expand the second term in the numerator of the fraction in the integrand, using the binomial series:

$$\frac{(z-a)^m - (z-a-h)^m}{(z-a-h)^m(z-a)^m}$$

$$= \frac{(z-a)^m - \left[(z-a)^m - mh(z-a)^{m-1} + \frac{m(m-1)}{2}(z-a)^{m-2}h^2 + \cdots\right]}{(z-a-h)^m(z-a)^m}$$

$$= \frac{mh(z-a)^{m-1} - \frac{m(m-1)}{2}(z-a)^{m-2}h^2 + \cdots}{(z-a-h)^m(z-a)^m}$$

Then, since $z - a \neq 0$ anywhere on C, we may divide out a factor of $(z-a)^{m-1}$ to get

$$mh\frac{1 - \frac{(m-1)}{2}(z-a)^{-1}h + \cdots}{(z-a-h)^m(z-a)}$$

Then the mth derivative becomes

$$f^{(m)}(z)\Big|_{z=a} = \lim_{h\to 0}\frac{(m-1)!mh}{2\pi ih}\oint_C f(z)\frac{1 - \frac{(m-1)}{2}(z-a)^{-1}h + \cdots}{(z-a-h)^m(z-a)}dz$$

$$= \frac{m!}{2\pi i}\oint_C \frac{f(z)}{(z-a)^{m+1}}dz$$

which shows that the result (13) is also true for $n = m + 1$. Hence the general result (13) is proven.

IV. THE MEAN VALUE THEOREM FOR INTEGRALS

If the functions $f(x)$ and $g(x)$ are continuous on the closed interval $[a, b]$ and $g(x) > 0$ on the open interval (a, b), then

$$\int_a^b f(x)g(x)\,dx = f(c)\int_a^b g(x)\,dx \tag{14}$$

for some c in the interval (a, b).

The proof uses the mean value theorem for derivatives[4] applied to the function

$$F(x) = \int_a^x f(t)g(t)\,dt - \frac{A}{B}\int_a^x g(t)\,dt \qquad (15)$$

where

$$A = \int_a^b f(x)g(x)\,dx$$

is the left-hand side of equation (14) and

$$B = \int_a^b g(x)\,dx$$

is the integral on the right-hand side of equation (14). Notice that if both f and g are continuous, then F is differentiable. Then there exists some c between a and b such that

$$F'(c) = \frac{F(b) - F(a)}{b - a}$$

But $F(b) = F(a) = 0$, so $F'(c) = 0$. Next we evaluate the derivative directly from equation (15):

$$F'(x) = f(x)g(x) - \frac{A}{B}g(x)$$

Setting $x = c$, we obtain

$$f(c)g(c) - \frac{A}{B}g(c) = 0$$

Then, since $g(x) > 0$ for all x in (a, b), we may divide out the factor $g(c)$ to obtain

$$f(c)B = A$$

or

$$f(c)\int_a^b g(x)\,dx = \int_a^b f(x)g(x)\,dx$$

which is the result we set out to prove.

In the special case $g(x) \equiv 1$, then $B = b - a$, and we obtain the simpler result

$$\int_a^b f(x)\,dx = (b - a)f(c), \quad \text{where } a < c < b \qquad (16)$$

[4] See, for example, Stewart, p. 289.

The area under the curve $f(x)$ between $x = a$ and $x = b$ equals the area of a rectangle of width $b - a$ and height $f(c)$; see Figure 1.

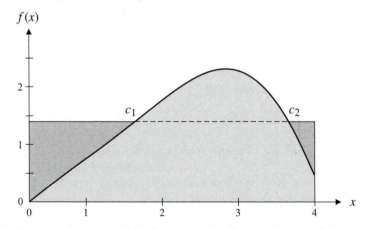

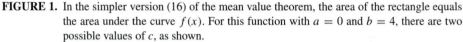

FIGURE 1. In the simpler version (16) of the mean value theorem, the area of the rectangle equals the area under the curve $f(x)$. For this function with $a = 0$ and $b = 4$, there are two possible values of c, as shown.

V. THE GIBBS PHENOMENON

The Fourier series of a function that has a discontinuity always converges to the midpoint of the discontinuity, and it overshoots as it makes the jump. To investigate these phenomena further, we will concentrate our attention on the step function.

We will denote the sum of the first N terms of a Fourier series by S_N:

$$S_N = \sum_{n=0}^{N} (a_n \sin nx + b_n \cos nx)$$

where (equation 4.7)

$$a_n = \frac{1}{\pi} \int_0^{2\pi} f(x) \sin nx \, dx$$

and b_n is defined similarly (equation 4.8). Then we can write

$$a_n \sin nx + b_n \cos nx = \frac{1}{\pi} \int_0^{2\pi} f(u)(\sin nu \sin nx + \cos nu \cos nx) \, du$$

$$= \frac{1}{\pi} \int_0^{2\pi} f(u) \cos n(u - x) \, du$$

for $n \geq 1$, and (equation 4.9)

$$b_0 = \frac{1}{2\pi} \int_0^{2\pi} f(u) \, du$$

We'll also need the combination

$$a_n \cos nx - b_n \sin nx = \frac{1}{\pi} \int_0^{2\pi} f(u) \sin n(u - x) \, du$$

In this case, the term with $n = 0$ is identically zero. The series

$$T(x) = \sum_{n=0}^{\infty} (a_n \cos nx - b_n \sin nx)$$

is called the *allied series*. The sum of its first N terms is denoted by T_N. Then

$$S_N + iT_N = \frac{1}{\pi} \int_0^{2\pi} f(u) \left\{ \frac{1}{2} + \sum_{n=1}^{N} e^{in(u-x)} \right\} du$$

where the first term in the braces (1/2) is the $n = 0$ term. We can evaluate the sum (refer to Chapter 2, Section 2.3.2, equation 2.42 with $z = e^{i(u-x)}$) to get

$$S_N + iT_N = \frac{1}{\pi} \int_0^{2\pi} f(u) \left\{ -\frac{1}{2} + \frac{1 - e^{i(N+1)(u-x)}}{1 - e^{i(u-x)}} \right\} du$$

$$= \frac{1}{\pi} \int_0^{2\pi} f(u) \left\{ \frac{e^{-i(u-x)/2} - e^{i(N+1/2)(u-x)}}{e^{-i(u-x)/2} - e^{i(u-x)/2}} - \frac{1}{2} \right\} du$$

The quantity in braces may be rewritten as follows:

$$\frac{\cos\left(\frac{u-x}{2}\right) - i \sin\left(\frac{u-x}{2}\right) - \cos\left[\left(N + \frac{1}{2}\right)(u - x)\right] - i \sin\left[\left(N + \frac{1}{2}\right)(u - x)\right]}{-2i \sin\left(\frac{u-x}{2}\right)} - \frac{1}{2}$$

$$= i \frac{\cos\left(\frac{u-x}{2}\right) - \cos\left[\left(N + \frac{1}{2}\right)(u - x)\right]}{2 \sin\left(\frac{u-x}{2}\right)} + \frac{\sin\left[\left(N + \frac{1}{2}\right)(u - x)\right]}{2 \sin\left(\frac{u-x}{2}\right)}$$

Thus, the real part of this equation is

$$S_N(x) = \frac{1}{2\pi} \int_0^{2\pi} f(u) \frac{\sin\left[(N + \frac{1}{2})(u - x)\right]}{\sin\left(\dfrac{u - x}{2}\right)} du \qquad (17)$$

Now let's apply this result to the step function $f(u) = 1$ for $0 < u < \pi$ and $f(u) = 0$ for $\pi < u < 2\pi$. For this function, the upper limit of the integral in equation (17) becomes π:

$$S_N(x) = \frac{1}{2\pi} \int_0^{\pi} \frac{\sin\left[\left(N + \frac{1}{2}\right)(u - x)\right]}{\sin\left(\dfrac{u - x}{2}\right)} du$$

Now let $N + \frac{1}{2} = \eta$, and $(u - x)\eta = v$:

$$S_N(x) = \frac{1}{2\pi} \int_{-x\eta}^{(\pi-x)\eta} \frac{\sin v}{\sin(v/2\eta)} \frac{dv}{\eta}$$

$$= \frac{1}{2\pi\eta} \int_{-x\eta}^{x\eta} \frac{\sin v}{\sin(v/2\eta)} dv + \frac{1}{2\pi\eta} \int_{x\eta}^{(\pi-x)\eta} \frac{\sin v}{\sin(v/2\eta)} dv$$

For $0 < x < \pi$, the first term is much larger than the second, since $\sin v/2\eta$ is not close to zero in the second range of integration. In the first integral, the integrand is even, and so

$$S_N(x) \simeq \frac{1}{\pi\eta} \int_0^{x\eta} \frac{\sin v}{\sin v/2\eta} dv \quad \text{for } 0 < x < \pi$$

Thus, as we let N, and hence η, become large, $\sin v/2\eta \simeq v/2\eta$ and we have

$$S_N(x) \simeq \frac{2}{\pi} \int_0^{x\eta} \frac{\sin v}{v} dv \quad \text{for } 0 < x < \pi \tag{18}$$

The integral is zero[5] for $x = 0$ and increases steadily as x increases until $\eta x = \pi$, because the integrand is positive throughout this range. For $v > \pi$, the integrand becomes negative and the integral begins to decrease. Further cycles of the sine cannot increase the integral back to its value at $x\eta = \pi$ because of the larger denominator. As $\eta \to \infty$, we find[6]

$$S(x) = \frac{2}{\pi} \int_0^\infty \frac{\sin v}{v} dv = \frac{2}{\pi} \frac{\pi}{2} = 1 \quad \text{for } 0 < x < \pi$$

as expected. The peak of the overshoot is thus greater than one and is given by[7]

$$\boxed{\frac{2}{\pi} \int_0^\pi \frac{\sin v}{v} dv = \frac{2}{\pi} \text{Si}(\pi) = 1.179} \tag{19}$$

The peak occurs at

$$x = \frac{\pi}{N + 1/2}$$

and gets closer to $x = 0$ as N increases (see Figure 2).

---

[5] When $x = 0$, the second integral becomes $\frac{1}{\pi} \int_0^{\pi\eta} \frac{\sin v}{v} dv \to \frac{1}{2}$ as $\eta \to \infty$, so we retrieve the expected result that the series converges to the midpoint of the jump (Chapter 4, Section 4.2.1).

[6] See Chapter 2, Section 2.7.3, for the evaluation of the integral.

[7] The integral may be evaluated numerically. See also Abramowitz and Stegun, Table 5.1.

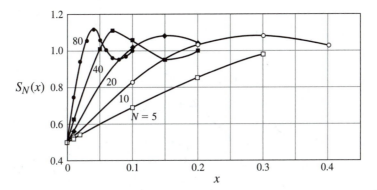

FIGURE 2. The function $S_N(x)$ (equation 18) for $N = 5, 10, 20, 40$, and 80. The peak moves closer to $x = 0$ as N increases, while the height of the peak remains approximately constant for large N.

VI. THE LAPLACE TRANSFORM AND CONVOLUTION

To prove the convolution theorem (5.17) for Laplace transforms, let's calculate the transform of the convolution of the functions f and g:

$$\mathcal{L}(f * g) = \int_0^\infty e^{-st} \int_0^t f(\tau)g(t - \tau)\, d\tau\, dt = \lim_{T \to \infty} \int_0^T \int_0^t e^{-st} f(\tau)g(t - \tau)\, d\tau\, dt$$

The integral is taken over the half-space in the t-τ plane bounded by the t-axis and the line $\tau = t$. The integral is evaluated by summing over vertical strips, as shown in Figure 3a.

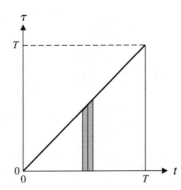

FIGURE 3a. The Laplace transform of the convolution is an integral over the triangular area below the diagonal line $\tau = t$, summed over the vertical strips shown.

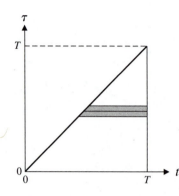

FIGURE 3b. We can equally well compute the area as a sum of the horizontal strips shown here.

We can equally well evaluate the integral by summing over horizontal strips, as shown in Figure 3b. Then the integral becomes

$$\mathcal{L}(f * g) = \lim_{T \to \infty} \int_0^T \int_\tau^T e^{-st} f(\tau) g(t - \tau) \, dt \, d\tau$$

Now we change variables to $u = t - \tau$:

$$\mathcal{L}(f * g) = \lim_{T \to \infty} \int_0^T d\tau \int_0^{T-\tau} du \, e^{-s(u+\tau)} f(\tau) g(u)$$

$$= \lim_{T \to \infty} \int_0^T \left(\int_0^{T-\tau} g(u) e^{-su} \, du \right) f(\tau) e^{-s\tau} \, d\tau$$

The integral is over a triangular region of the τ-u plane, as shown in Figure 4. As $T \to \infty$, the integral over the square gives

$$\lim_{T \to \infty} \int_0^{T/2} g(u) e^{-su} \, du \int_0^{T/2} f(\tau) e^{-s\tau} \, d\tau$$

$$= \int_0^\infty g(u) e^{-su} \, du \int_0^\infty f(\tau) e^{-s\tau} \, d\tau = G(s)F(s)$$

We can show that the integrals over the two smaller triangles go to zero in the limit:

$$\left| I_{\text{top triangle}} \right| = \left| \int_{T/2}^T f(\tau) e^{-s\tau} \, d\tau \int_0^{T-\tau} g(u) e^{-su} \, du \right|$$

Both f and g are of exponential order, so, from equation (5.2) in Chapter 5, max $|g(u)| \le M_1 e^{\sigma_1 u}$ on the range $0 < u < T/2$, and a similar relation holds for f: $|f(\tau)| \le M_2 e^{\sigma_2 \tau}$

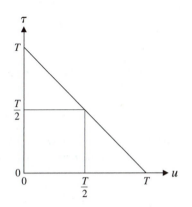

FIGURE 4. In the u-τ plane, we integrate over the area to the left of the diagonal line. We may divide this area into the square and the two smaller triangles shown here.

for $T/2 < \tau < T$. Thus,

$$\left|I_{\text{top triangle}}\right| \leq M_1 M_2 \left|\int_{T/2}^{T} e^{(\sigma_2 - s)\tau}\, d\tau \int_{0}^{T-\tau} e^{(\sigma_1 - s)u}\, du\right|$$

$$= M_1 M_2 \int_{T/2}^{T} e^{(\sigma_2 - s)\tau} \frac{1 - e^{(\sigma_1 - s)(T-\tau)}}{s - \sigma_1}\, d\tau$$

$$= \frac{M_1 M_2}{s - \sigma_1} \left[\int_{T/2}^{T} \left(e^{(\sigma_2 - s)\tau} - e^{(\sigma_1 - s)T} e^{(\sigma_2 - \sigma_1)\tau}\right) d\tau\right]$$

$$= \frac{M_1 M_2}{s - \sigma_1} \left[\frac{e^{(\sigma_2 - s)T} - e^{(\sigma_2 - s)T/2}}{\sigma_2 - s} - e^{(\sigma_1 - s)T} \frac{e^{(\sigma_2 - \sigma_1)T} - e^{(\sigma_2 - \sigma_1)T/2}}{\sigma_2 - \sigma_1}\right]$$

$$= \frac{M_1 M_2}{s - \sigma_1} \left[\frac{e^{(\sigma_2 - s)T/2} - e^{(\sigma_2 - s)T}}{s - \sigma_2} - e^{-sT} \frac{e^{\sigma_2 T} - e^{(\sigma_2 + \sigma_1)T/2}}{\sigma_2 - \sigma_1}\right]$$

which approaches zero as $T \to \infty$ provided that[8] Re $(s) > \max(\sigma_1, \sigma_2)$. We can use a similar argument to show that the integral over the lower triangle also goes to zero. Then we have

$$\mathcal{L}(f * g) = F(s)G(s)$$

and, equivalently,

$$\mathcal{L}^{-1}(FG) = f * g$$

---

[8]The argument must be modified slightly if $\sigma_1 = \sigma_2$. The result is unchanged.

VII. PROOF THAT $P_l^m(\mu) = (-1)^m (1-\mu^2)^{m/2} \dfrac{d^m}{d\mu^m} P_l(\mu)$

In Chapter 8, we showed that the mth derivative of the Legendre polynomial P_l satisfies the equation (8.52):

$$(1-\mu^2)y_m'' - 2(m+1)\mu y_m' + [l(l+1) - m(m+1)]y_m = 0 \tag{20}$$

Now let $y_m(\mu) = (1-\mu^2)^r z(\mu)$. We will be able to show that $z(\mu)$ satisfies the associated Legendre equation with a suitable choice of the power r. We begin by differentiating y_m:

$$y_m' = (1-\mu^2)^r z' + r(-2\mu)(1-\mu^2)^{r-1} z$$

and

$$
\begin{aligned}
y_m'' &= (1-\mu^2)^r z'' - 4r\mu(1-\mu^2)^{r-1} z' - 2r(1-\mu^2)^{r-1} z \\
&\quad - 2r\mu[-2(r-1)\mu](1-\mu^2)^{r-2} z \\
&= (1-\mu^2)^{r-1}\left[(1-\mu^2)z'' - 4r\mu z' - \frac{2rz}{1-\mu^2}(1+\mu^2 - 2\mu^2 r)\right]
\end{aligned}
$$

Substituting this into the differential equation (20), we get

$$
\begin{aligned}
0 &= (1-\mu^2)(1-\mu^2)^{r-1}\left[(1-\mu^2)z'' - 4r\mu z' - \frac{2rz}{1-\mu^2}(1+\mu^2 - 2\mu^2 r)\right] \\
&\quad - 2(m+1)\mu[(1-\mu^2)^r z' - 2r\mu(1-\mu^2)^{r-1} z] \\
&\quad + [l(l+1) - m(m+1)](1-\mu^2)^r z
\end{aligned}
$$

We can divide out a factor of $(1-\mu^2)^r$ to get

$$
\begin{aligned}
0 &= \left[(1-\mu^2)z'' - 4r\mu z' - \frac{2rz}{1-\mu^2}(1+\mu^2 - 2\mu^2 r)\right] - 2(m+1)\mu\left(z' - \frac{2r\mu}{(1-\mu^2)}z\right) \\
&\quad + [l(l+1) - m(m+1)]z
\end{aligned}
$$

Gathering up terms, we have

$$
\begin{aligned}
0 &= (1-\mu^2)z'' - 2\mu(2r + m + 1)z' \\
&\quad + z\left[l(l+1) - m(m+1) - \frac{2r}{1-\mu^2}(1+\mu^2 - 2\mu^2 r - 2(m+1)\mu^2)\right] \tag{21}
\end{aligned}
$$

We want this equation to become equation (8.48). We'll get the correct coefficient of z' if we choose $r = -m/2$. Inserting this value into equation (21), we get

$$
\begin{aligned}
0 &= (1-\mu^2)z'' - 2\mu z' + z\left[l(l+1) - m(m+1) + \frac{m}{1-\mu^2}(1-\mu^2 - \mu^2 m)\right] \\
&= (1-\mu^2)z'' - 2\mu z' + z\left[l(l+1) - \frac{m^2}{1-\mu^2}\right]
\end{aligned}
$$

and we have the equation we seek. Thus, we have shown that $Cz = P_l^m$ for any constant C. In physics applications, it is usual to choose $C = (-1)^m$ and then

$$P_l^m(\mu) = (-1)^m (1 - \mu^2)^{m/2} \frac{d^m}{d\mu^m} P_l(\mu)$$

as in equation (8.53).

VIII. PROOF OF THE RELATION $\int_0^\infty \rho J_m(k\rho) J_m(k'\rho)\, d\rho = \dfrac{\delta(k - k')}{k}$

To prove the delta function relation above (equation 8.113), we start with the Bessel differential equation (8.69):

$$\frac{\partial}{\partial \rho} \left(\rho \frac{\partial J_m(k\rho)}{\partial \rho} \right) + k^2 \rho J_m(k\rho) - \frac{m^2}{\rho} J_m(k\rho) = 0$$

With a different eigenvalue k', the equation is

$$\frac{\partial}{\partial \rho} \left(\rho \frac{\partial J_m(k'\rho)}{\partial \rho} \right) + (k')^2 \rho J_m(k'\rho) - \frac{m^2}{\rho} J_m(k'\rho) = 0$$

Following the usual procedure for proving orthogonality (Section 8.1.1), we multiply the first equation by $J_m(k'\rho)$, multiply the second by $J_m(k\rho)$, and subtract:

$$J_m(k'\rho) \frac{\partial}{\partial \rho} \left(\rho \frac{\partial J_m(k\rho)}{\partial \rho} \right) - J_m(k\rho) \frac{\partial}{\partial \rho} \left(\rho \frac{\partial J_m(k'\rho)}{\partial \rho} \right) + [k^2 - (k')^2]\rho J_m(k\rho) J_m(k'\rho) = 0$$

Now we integrate from 0 to ∞. We integrate the first two terms on the left-hand side by parts, and the integrals cancel, leaving only the integrated terms:

$$\rho \left[J_m(k'\rho) \frac{\partial J_m(k\rho)}{\partial \rho} - J_m(k\rho) \frac{\partial J_m(k'\rho)}{\partial \rho} \right]\Bigg|_0^\infty = [(k')^2 - k^2] \int_0^\infty \rho J_m(k\rho) J_m(k'\rho)\, d\rho \tag{22}$$

The integrated terms vanish at the lower limit $\rho = 0$. To evaluate these terms at the upper limit, we use the asymptotic form (8.83) for $J_m(k\rho)$ and the recursion relation (8.90). The first term is

$$\lim_{\rho \to \infty} \frac{k\rho}{2} J_m(k'\rho) \left[J_{m-1}(k\rho) - J_{m+1}(k\rho) \right]$$

$$= \lim_{\rho \to \infty} \frac{k\rho}{2} \frac{2}{\pi\rho\sqrt{kk'}} \cos\left(k'\rho - \frac{m\pi}{2} - \frac{\pi}{4} \right)$$

$$\times \left[\cos\left(k\rho - \frac{(m-1)\pi}{2} - \frac{\pi}{4} \right) - \cos\left(k\rho - \frac{(m+1)\pi}{2} - \frac{\pi}{4} \right) \right]$$

Using the identity

$$\cos A \cos B = \frac{1}{2}[\cos (A + B) + \cos (A - B)]$$

we find the first term to be

$$\lim_{\rho \to \infty} \frac{1}{2\pi} \sqrt{\frac{k}{k'}} \left\{ \begin{array}{l} \cos [(k + k')\rho - m\pi] + \cos [(k - k')\rho + \pi/2] \\ - \cos [(k + k')\rho - m\pi - \pi] - \cos [(k - k')\rho - \pi/2] \end{array} \right\}$$

$$= \lim_{\rho \to \infty} \frac{1}{\pi} \sqrt{\frac{k}{k'}} \left[(-1)^m \cos (k + k')\rho - \sin (k - k')\rho \right]$$

The second integrated term is found by interchanging k and k' in this expression. Thus, the difference of the two integrated terms is

$$\lim_{\rho \to \infty} \frac{1}{\pi} \sqrt{\frac{k}{k'}} \left[(-1)^m \cos (k + k')\rho - \sin (k - k')\rho \right]$$

$$- \frac{1}{\pi} \sqrt{\frac{k'}{k}} \left[(-1)^m \cos (k + k')\rho - \sin (k' - k)\rho \right]$$

$$= \lim_{\rho \to \infty} \frac{1}{\pi} \left[\frac{k - k'}{\sqrt{kk'}} (-1)^m \cos (k + k')\rho + \frac{k + k'}{\sqrt{kk'}} \sin (k' - k)\rho \right]$$

Inserting this result into equation (22), we have

$$\lim_{\rho \to \infty} \frac{1}{\pi} \left[\frac{(-1)^{m+1}}{\sqrt{kk'}} \frac{\cos (k + k')\rho}{k + k'} + \frac{1}{\sqrt{kk'}} \frac{\sin (k' - k)\rho}{k' - k} \right] = \int_0^\infty \rho J_m(k\rho) J_m(k'\rho) \, d\rho \tag{23}$$

Now we already know[9] from Chapter 6 (equation 6.16) that

$$\lim_{R \to \infty} \frac{1}{2\pi} \int_{-R}^{R} e^{ikr} \, dr = \delta(k)$$

$$\lim_{R \to \infty} \frac{e^{ikR} - e^{-ikR}}{2ik\pi} = \lim_{R \to \infty} \frac{1}{\pi} \frac{\sin kR}{k} = \delta(k)$$

and thus the sine term on the left-hand side of equation (23) is

$$\frac{1}{\sqrt{kk'}} \delta(k' - k) = \frac{\delta(k' - k)}{k}$$

The cosine term is zero in the limit. (See Chapter 6, Problem 25.) We can understand this result by looking at plots of the two functions; see Figure 5.

Thus, we have the desired result:

$$\int_0^\infty \rho J_m(k\rho) J_m(k'\rho) \, d\rho = \frac{1}{k} \delta(k' - k)$$

---

[9] See also Lighthill, p. 29.

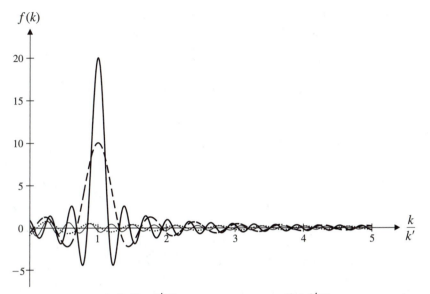

FIGURE 5. The functions $\dfrac{\sin (k - k')R}{k - k'}$ (heavy lines) and $\dfrac{\cos (k + k')R}{k + k'}$ (thin lines) versus k/k' for $Rk' = 20$ (solid line) and 10 (dashed line). While the function with the sine has a prominent peak, the function with the cosine has no peak for k and k' positive.

IX. THE ERROR FUNCTION

The Gaussian function

$$\exp\left(-\frac{(x - x_0)^2}{a^2}\right) \tag{24}$$

occurs frequently in statistics. For example, it describes the distribution of a set of measurements around the mean at $x = x_0$, provided that the number of data points is large and the errors are random. As a consequence, it also appears in physics—for example, in the Maxwellian velocity distribution, where, for each component,

$$f(v_x) = \left(\frac{m}{2\pi kT}\right)^{1/2} \exp\left(-\frac{mv_x^2}{2kT}\right)$$

The function (24) has a maximum value of 1 at $x = x_0$. It reaches one-half its maximum value where

$$\frac{(x - x_0)^2}{a^2} = -\ln\left(\frac{1}{2}\right) = \ln 2 = 0.69315$$

or at

$$x = x_0 \pm a\sqrt{\ln 2} = x_0 \pm 0.83255a$$

The spread $x - x_0 = 0.83255a$ is called the half-width half-maximum, while the spread $2 \times 0.83255a = 1.6651a$ is called the full-width half-maximum (Figure 6).

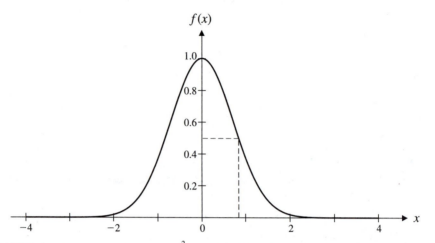

FIGURE 6. The Gaussian function e^{-x^2}. The dashed line shows the half-width half-maximum.

Frequently we need values of the area under the curve of this function. The function

$$\Phi(x) = \frac{2}{\sqrt{\pi}} \int_0^x e^{-u^2}\, du \tag{25}$$

is called the error function or the probability integral. It is also sometimes written as erf (x). The normalization arises from the value of the integral

$$\int_0^\infty e^{-u^2}\, du = \frac{1}{2} \int_{-\infty}^{+\infty} e^{-u^2}\, du = \frac{\sqrt{\pi}}{2} \tag{26}$$

To evaluate this integral (26), we begin with its square:

$$I^2 = \int_{-\infty}^{+\infty} e^{-x^2}\, dx \int_{-\infty}^{+\infty} e^{-y^2}\, dy = \int_{-\infty}^{+\infty} \int_{-\infty}^{+\infty} e^{-(x^2+y^2)}\, dx\, dy$$

This is an integral over the whole x-y plane. We may rewrite this integral in polar coordinates. The area element is

$$dx\, dy = dA = r\, dr\, d\theta$$

and

$$x^2 + y^2 = r^2$$

Thus,

$$I^2 = \int_0^\infty \int_0^{2\pi} e^{-r^2} r \, dr \, d\theta = 2\pi \int_0^\infty e^{-w} \frac{dw}{2}$$

where we have made the change of variable $w = r^2$. Then

$$I^2 = \pi(-e^{-w}\big|_0^\infty) = \pi$$

giving

$$I = \int_{-\infty}^{+\infty} e^{-x^2} \, dx = \sqrt{\pi}$$

The complementary error function

$$\text{erfc}\,(x) = 1 - \Phi(x) = \frac{2}{\sqrt{\pi}} \int_x^\infty e^{-u^2} \, du \qquad (27)$$

represents the area under the Gaussian curve from the point x out to infinity, appropriately normalized.

For small x, we may usefully approximate the error function using the series expansion:

$$\Phi(x) = \frac{2}{\sqrt{\pi}} \sum_{k=0}^\infty (-1)^k \frac{x^{2k+1}}{(2k+1)k!} \qquad (28)$$

which may be easily derived by integrating the series expansion for e^{-x^2}. For large values of x, the function approaches the value 1 exponentially,

$$\Phi(x) \simeq 1 - \frac{1}{\sqrt{\pi}x} e^{-x^2} \left(1 - \frac{1}{2x^2} + \frac{3}{4x^4} - \frac{15}{8x^6} + \cdots \right) \qquad (29)$$

and the complementary function likewise approaches zero:

$$\text{erfc}\,(x) \simeq \frac{1}{\sqrt{\pi}x} e^{-x^2} \left(1 - \frac{1}{2x^2} + \cdots \right) \qquad (30)$$

Values of the function erf (x) are shown in Table 1 and Figure 7.

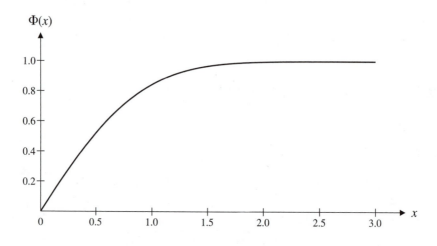

FIGURE 7. The error function erf (x).

TABLE 1. Values of the Error Function

x	erf (x)	x	erf (x)
0	0	0.7	0.6778
0.1	0.1125	0.8	0.7421
0.2	0.2227	0.9	0.7969
0.3	0.3286	1.0	0.8427
0.4	0.4284	1.5	0.9661
0.5	0.5205	2.0	0.9953
0.6	0.6039	2.5	0.9996

X. CLASSIFICATION OF PARTIAL DIFFERENTIAL EQUATIONS

Here we shall give a brief summary of the types of partial differential equations most frequently encountered in physics. For a more extensive discussion, see, for example, Morse and Feshbach, Chapters 2 and 6. The discussion here is presented in two dimensions, but the extension to higher dimensions is relatively straightforward.

Suppose we have a function $f(x, y)$ that satisfies a linear partial differential equation of the form

$$A\frac{\partial^2 f}{\partial x^2} + 2B\frac{\partial^2 f}{\partial x\,\partial y} + C\frac{\partial^2 f}{\partial y^2} = g\left(x, y, f, \frac{\partial f}{\partial x}, \frac{\partial f}{\partial y}\right) \tag{31}$$

where A, B, and C are each functions of the coordinates. The function f exists in a specified region with boundaries defined parametrically by the curves $x = \xi(\tau)$, $y = \theta(\tau)$. The type of boundary conditions that we need in order to find a solution depends on the form of the differential equation, as specified by the relative values of the functions A, B, and C, and also on the form of the boundary. The boundary conditions are classified as follows:

1. **Dirichlet conditions**: The value of the function f is specified on the boundary.
2. **Neumann conditions:** The value of the normal derivative of f is specified on the boundary. For example, if one of the boundaries is at $x = a$, we would specify $\partial f / \partial y$ at $x = a$.
3. **Cauchy conditions:** Both f and its derivative are specified on the boundary.

The boundary is closed if it forms a single closed curve, part of which may be at infinity, and values are specified everywhere on the boundary, including at infinity. It is open if it extends to infinity in some region and no values are specified on the part at infinity.

The differential equation (31) is classified according to the values of the functions A, B, and C, as shown in Table 2. This table also shows examples of each class from the text, along with specific values for A, B, and C for each example.

TABLE 2. Classification of PDEs

	Name		
	Hyperbolic	**Parabolic**	**Elliptic**
Definition	$B^2 > AC$ everywhere	$B^2 = AC$ everywhere	$B^2 < AC$ everywhere
Example	Wave equation (3.15)	Diffusion equation (3.14)	Poisson's equation (3.49)
Variables	x, t	x, t	x, y
	$B = 0,\, AC = -v^2$	$A = D,\, B = C = 0$	$A = C = 1,\, B = 0$

To determine what kind of boundary conditions we need for a solution (Table 3), we step away from the boundary using a Taylor series. We find that this is possible given Cauchy

TABLE 3. Existence of Solutions for Given Boundary Conditions

Boundary Condition	Boundary Type	Hyperbolic Equation	Parabolic Equation	Elliptic Equation
Dirichlet	open	insufficient to determine solution	**unique solution exists** in one direction (Example 7.6)	insufficient to determine solution
Dirichlet	closed	solution not unique	no solution; overdetermined	**unique solution exists** (Examples 8.1, 8.2, 8.3)
Neumann	open	insufficient to determine solution	**unique solution exists** in one direction (Problem 7.14)	insufficient to determine solution
Neumann	closed	solution not unique	no solution; overdetermined	**unique solution exists** (Problem 8.12)
Cauchy	open	**unique solution exists** (Examples 4.4, 7.5)	no solution; overdetermined	unstable, unphysical solution
Cauchy	closed	no solution; overdetermined	no solution; overdetermined	no solution; overdetermined

conditions, provided that the boundary does not coincide with one of the characteristic curves for the equation, which are specified by

$$A \, dy = \left(B \pm \sqrt{B^2 - AC} \right) dx \tag{32}$$

To understand this result, consider the wave equation (3.15). Here we have $A = v^2$, $B = 0$, and $C = -1$, and the characteristics satisfy

$$\frac{dx}{dt} = \pm v$$

with solutions

$$x \pm vt = \text{constant}$$

These are the wave fronts. It is possible to find a solution for a wave on a string, for example, if we know both $f(x)$ and $\partial f / \partial t$ at $t = 0$ (see Example 4.4). But the line $x - vt = u = $ constant is a wave front. Any function of $x - vt$ satisfies the differential equation. Infinitely many functions $f(u)$ have a specified value and a specified derivative at one given point, and so the solution is not determined if the boundary is a line $x - vt = $ constant.

The directionality of the parabolic equations is well illustrated by our example: the diffusion equation. We can integrate forward in time, but not backward. Physically, this is related to the increase in entropy with time. The solutions smooth out as time increases, and it is not possible, in general, to determine the initial state from a given final state.

XI. THE TANGENT FUNCTION: A DETAILED INVESTIGATION OF SERIES EXPANSIONS

The tangent function offers us the opportunity to investigate Taylor and Laurent series in some detail. While the Taylor series for $\tan x$ is familiar and easily found in reference books, the Laurent series for the function $\tan z$ are rarely seen. They do have uses, however. We'll use this function to discuss the relation of these less familiar series to the Taylor series and to discuss some methods for finding them.

First let's see whether the function $\tan z$ is analytic by checking the Cauchy-Riemann relations. We start by finding the functions u and v:

$$\begin{aligned}
\tan z &= \frac{\sin z}{\cos z} = \frac{e^{iz} - e^{-iz}}{2i} \frac{2}{e^{iz} + e^{-iz}} \\
&= -i \frac{e^{i(x+iy)} - e^{-i(x+iy)}}{e^{i(x+iy)} + e^{-i(x+iy)}} = -i \left(\frac{e^{ix} e^{-y} - e^{-ix} e^{y}}{e^{ix} e^{-y} + e^{-ix} e^{y}} \right) \left(\frac{e^{-ix} e^{-y} + e^{ix} e^{y}}{e^{-ix} e^{-y} + e^{ix} e^{y}} \right) \\
&= -i \frac{e^{-2y} + e^{2ix} - e^{-2ix} - e^{2y}}{e^{-2y} + e^{2ix} + e^{-2ix} + e^{2y}} = -i \frac{2i \sin 2x - 2 \sinh 2y}{2 \cos 2x + 2 \cosh 2y} \\
&= \frac{\sin 2x + i \sinh 2y}{\cos 2x + \cosh 2y} = u + iv
\end{aligned}$$

Thus, for this function,

$$u(x, y) = \frac{\sin 2x}{\cos 2x + \cosh 2y}$$

and

$$v(x, y) = \frac{\sinh 2y}{\cos 2x + \cosh 2y}$$

You might want to check that these expressions are correct in the case $y = 0$.

Now let's compute the partial derivatives to check the Cauchy-Riemann conditions:

$$\frac{\partial u}{\partial x} = \frac{2\cos 2x}{\cos 2x + \cosh 2y} - \sin 2x \frac{-2\sin 2x}{(\cos 2x + \cosh 2y)^2}$$

$$= \frac{2\cos 2x(\cos 2x + \cosh 2y) + 2\sin^2 2x}{(\cos 2x + \cosh 2y)^2} = 2\frac{1 + \cos 2x \cosh 2y}{(\cos 2x + \cosh 2y)^2}$$

and

$$\frac{\partial v}{\partial y} = \frac{2\cosh 2y}{\cos 2x + \cosh 2y} - \sinh 2y \frac{2\sinh 2y}{(\cos 2x + \cosh 2y)^2}$$

$$= \frac{2\cosh 2y(\cos 2x + \cosh 2y) - 2\sinh^2 2y}{(\cos 2x + \cosh 2y)^2} = 2\frac{1 + \cos 2x \cosh 2y}{(\cos 2x + \cosh 2y)^2} = \frac{\partial u}{\partial x}$$

Similarly,

$$\frac{\partial u}{\partial y} = \frac{-2\sin 2x \sinh 2y}{(\cos 2x + \cosh 2y)^2}$$

and

$$\frac{\partial v}{\partial x} = \frac{-2(-\sin 2x)\sinh 2y}{(\cos 2x + \cosh 2y)^2} = \frac{2\sin 2x \sinh 2y}{(\cos 2x + \cosh 2y)^2} = -\frac{\partial u}{\partial y}$$

So the Cauchy-Riemann relations are satisfied.

Where, if ever, does this fail? It fails if the numerator blows up ($y \to \infty$) or if the denominator goes to zero. The denominator is zero where

$$\cos 2x + \cosh 2y = 0$$

which can only happen if $y = 0$ and $\cos 2x = -1$, or $2x = (2n + 1)\pi$; that is, $x = \left(n + \frac{1}{2}\right)\pi$. We already know that the tan function approaches infinity (for real arguments) when x equals an odd number times $\pi/2$. So this is consistent.

Now let's find some series representations for tan z and look at the relations between them.

Series 1 We can find a Taylor series about the origin in the region $|z| < \pi/2$ (see Figure 8):

$$\tan z = \sum_{n=0}^{\infty} a_n z^n$$

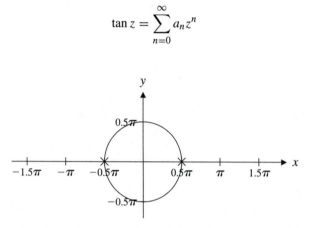

FIGURE 8. The region within which we can find a Taylor series for the tangent function: $|z| < \pi/2$.

The coefficients are given by (equation 2.44)

$$a_n = \frac{1}{n!} \frac{d^n}{dz^n} (\tan z) \Big|_0$$

Thus,

$$a_0 = \tan (0) = 0$$

$$a_1 = \frac{1}{1} \frac{d}{dz} \tan z \Big|_0 = \sec^2 (0) = 1$$

$$a_2 = \frac{1}{2} \frac{d^2}{dz^2} \tan z \Big|_0 = \frac{1}{2} 2 \sec z (\sec z \tan z) |_0 = 0$$

Before proceeding, it helps to simplify this derivative:

$$\frac{d^2}{dz^2} \tan z = 2 \frac{\sin z}{\cos^3 z}$$

Then

$$a_3 = \frac{1}{3!} \frac{d^3}{dz^3} \tan z \Big|_0 = \frac{1}{3} \frac{d}{dz} \frac{\sin z}{\cos^3 z} \Big|_0 = \frac{1}{3} \left(\frac{\cos z}{\cos^3 z} - 3 \frac{\sin z (-\sin z)}{\cos^4 z} \right) \Big|_0$$

$$= \frac{1}{3} \frac{\cos^2 z + 3 \sin^2 z}{\cos^4 z} \Big|_0 = \frac{1}{3} \frac{1 + 2 \sin^2 z}{\cos^4 z} \Big|_0 = \frac{1}{3}$$

$$a_4 = \frac{1}{4!}\frac{d^4}{dz^4}\tan z\Big|_0 = \frac{1}{4\times 3}\frac{d}{dz}\frac{1+2\sin^2 z}{\cos^4 z}\Big|_0$$

$$= \frac{1}{4\times 3}\frac{4\sin z\cos^2 z - 4(1+2\sin^2 z)(-\sin z)}{\cos^5 z}\Big|_0$$

$$= \frac{1}{3}\frac{\sin z(\cos^2 z + 1 + 2\sin^2 z)}{\cos^5 z}\Big|_0 = \frac{1}{3}\frac{\sin z\,(2+\sin^2 z)}{\cos^5 z}\Big|_0 = 0$$

and

$$a_5 = \frac{1}{5!}\frac{d^5}{dz^5}\tan z\Big|_0 = \frac{1}{5\times 3}\frac{d}{dz}\frac{2\sin z + \sin^3 z}{\cos^5 z}\Big|_0$$

$$= \frac{1}{5\times 3}\frac{(2\cos z + 3\cos z\sin^2 z)\cos z - 5(-\sin z)(2\sin z + \sin^3 z)}{\cos^5 z}\Big|_0$$

$$= \frac{2}{15}$$

and so on. Thus, we have

$$\tan z = z + \frac{1}{3}z^3 + \frac{2}{15}z^5 + \cdots$$

Let's have the computer package *Maple* check this. Asked for the series expansion of $\tan z$, it outputs

$$\tan z = z + \frac{1}{3}z^3 + \frac{2}{15}z^5 + \frac{17}{315}z^7 + \frac{62}{2835}z^9 + O(z^{10})$$

So the program agrees with our calculations (and gives us some more terms, as well).

Series 2   Now let's look in the annulus $|z - \pi/2| < \pi$ that surrounds the singularity at $z = \pi/2$. The radius $\rho = \pi$ is determined by the position of the neighboring singularities at $z = -\pi/2$ and at $z = 3\pi/2$ (see Figure 9). To minimize confusion, we change to a new variable $w = z - \pi/2$. Then

$$\tan z = \frac{\sin z}{\cos z} = \frac{\sin\,(w + \pi/2)}{\cos\,(w + \pi/2)} = \frac{\cos w}{-\sin w}$$

Now we use the previous series for $\tan w$:

$$\tan z = -\frac{1}{\tan w} = -\frac{1}{w + \frac{1}{3}w^3 + \frac{2}{15}w^5 + \frac{17}{315}w^7 + \frac{62}{2835}w^9 + O(w^{10})}$$

$$= -\frac{1}{w\left(1 + \frac{1}{3}w^2 + \frac{2}{15}w^4 + \frac{17}{315}w^6 + \frac{62}{2835}w^8 + \cdots\right)}$$

$$= -\frac{1}{w}\left(1 - \frac{1}{3}w^2 - \frac{2}{15}w^4 - \frac{17}{315}w^6 + \cdots + \frac{(-1)(-2)}{2}\left(\frac{1}{3}w^2 + \frac{2}{15}w^4 + \frac{17}{315}w^6 + \frac{62}{2835}w^8 + \cdots\right)^2 \right.$$
$$\left. + \frac{(-1)(-2)(-3)}{3!}\left(\frac{1}{3}w^2 + \frac{2}{15}w^4 + \frac{17}{315}w^6 + \frac{62}{2835}w^8 + \cdots\right)^3 + \cdots\right)$$

$$= -\frac{1}{w}\left[1 - \frac{1}{3}w^2 - \frac{2}{15}w^4 - \frac{17}{315}w^6 + \cdots + \left(\frac{1}{9}w^4 + \frac{4}{45}w^6 + \cdots\right) - \frac{1}{27}w^6\right] + \cdots$$

$$= -\frac{1}{w}\left(1 - \frac{1}{3}w^2 - \frac{1}{45}w^4 - \frac{2}{945}w^6 + \cdots\right)$$

$$= -\frac{1}{w} + \frac{1}{3}w + \frac{1}{45}w^3 + \frac{2}{945}w^5 + \cdots$$

$$= -\frac{1}{(z - \pi/2)} + \frac{1}{3}\left(z - \frac{\pi}{2}\right) + \frac{1}{45}\left(z - \frac{\pi}{2}\right)^3 + \frac{2}{945}\left(z - \frac{\pi}{2}\right)^5 + \cdots$$

which is the series we set out to find. It is a Laurent series with only one negative power and is valid for $0 < |z - \pi/2| < \pi$.

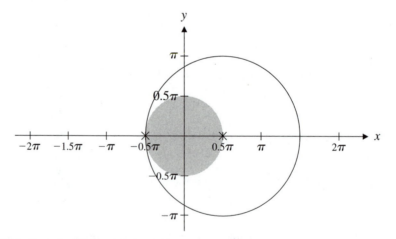

FIGURE 9. Both Series 1 and Series 2 for $\tan z$ are valid in the shaded region, so they must be identical there.

Now in the region of overlap, the two series are supposed to be identical. Let's check. The overlap region is $0 < |z| < \pi/2$ (see Figure 9). In this region, we can expand the second series. We begin by writing the first term as a function of $2z/\pi$, whose absolute value is less than 1 in the overlap region. Then

$$\tan z = -\frac{1}{(z-\pi/2)} + \frac{1}{3}\left(z-\frac{\pi}{2}\right) + \frac{1}{45}\left(z-\frac{\pi}{2}\right)^3 + \frac{2}{945}\left(z-\frac{\pi}{2}\right)^5 + \cdots$$

$$= \frac{2}{\pi}\frac{1}{(1-2z/\pi)} + \frac{1}{3}z - \frac{1}{6}\pi + \frac{1}{45}z^3 - \frac{1}{30}z^2\pi + \frac{1}{60}z\pi^2 - \frac{1}{360}\pi^3$$

$$+ \frac{2}{945}z^5 - \frac{1}{189}z^4\pi + \frac{1}{189}z^3\pi^2 - \frac{1}{378}z^2\pi^3 + \frac{1}{1512}z\pi^4 - \frac{1}{15120}\pi^5 + \cdots$$

$$= \frac{2}{\pi}\left(1 + \frac{2}{\pi}z + \frac{4}{\pi^2}z^2 + \frac{8}{\pi^3}z^3 + \frac{16}{\pi^4}z^4 + \cdots\right) - \frac{1}{6}\pi - \frac{1}{360}\pi^3$$

$$- \frac{1}{15120}\pi^5 + \frac{1}{3}z + \frac{1}{60}z\pi^2 + \frac{1}{1512}z\pi^4 - \frac{1}{30}z^2\pi - \frac{1}{378}z^2\pi^3 + \frac{1}{45}z^3$$

$$+ \frac{1}{189}z^3\pi^2 - \frac{1}{189}z^4\pi + \frac{2}{945}z^5 + \cdots$$

$$= \frac{2}{\pi} - \frac{1}{6}\pi - \frac{1}{360}\pi^3 - \frac{1}{15120}\pi^5 + \frac{4}{\pi^2}z + \frac{1}{3}z + \frac{1}{60}z\pi^2 + \frac{1}{1512}z\pi^4 + \frac{8}{\pi^3}z^2$$

$$- \frac{1}{30}z^2\pi - \frac{1}{378}z^2\pi^3 + \frac{16}{\pi^4}z^3 + \frac{1}{45}z^3 + \frac{1}{189}z^3\pi^2 + \frac{32}{\pi^5}z^4 - \frac{1}{189}z^4\pi + \cdots$$

$$= 6.65\times 10^{-3} + 0.97z + 7.1\times 10^{-2}z^2 + 0.24z^3 + 8.8\times 10^{-2}z^4 + \cdots$$

Each term in the expanded series is itself a series, of which we have evaluated only the first few terms. We can compare with the first series:

$$\tan z = z + \frac{1}{3}z^3 + \frac{2}{15}z^5 + \cdots$$

The a_0 term should be zero; with four terms, we have 0.006 and are decreasing toward zero. The a_1 term should be 1; we have 0.97. If we could add enough terms, we should be able to get back the first series *exactly*.

Series 3 Next we'll look at the series in the annulus $\pi/2 < |z| < 3\pi/2$. This region is centered on the origin, but excludes the singularities at $z = \pm\pi/2$ (see Figure 10). We have a Laurent series with coefficients (equation 2.47)

$$a_n = \frac{1}{2\pi i}\oint_C \frac{\tan z}{z^{n+1}}\,dz$$

The contour C must lie within the annulus. Thus, it contains three poles, at $z = 0, \pm\pi/2$. Using the residue theorem, we have

$$a_n = \sum_{p=1}^{3} \text{Res}\,(z_p)$$

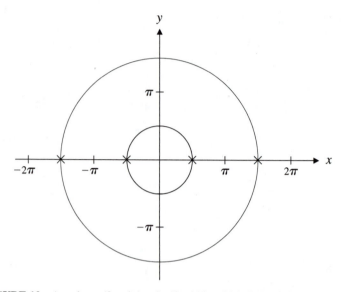

FIGURE 10. Annulus $\pi/2 < |z| < 3\pi/2$ within which Series 3 for $\tan z$ is valid.

The residue at $z = 0$ is a'_n, the nth coefficient of the Taylor series valid in the immediate neighborhood of $z = 0$. To see this, note that near $z = 0$,

$$\frac{\tan z}{z^{n+1}} = \frac{\sum_m a'_m z^m}{z^{n+1}} = \sum_m \frac{a'_m}{z^{n+1-m}}$$

The residue is the coefficient of the z^{-1} term in this series ($m = n$)—that is, the coefficient a'_n.

To evaluate the residues at the other poles, we use the series valid about each of them. For example, near $z = \pi/2$, we found

$$\tan z = -\frac{1}{(z - \pi/2)} + \frac{1}{3}\left(z - \frac{\pi}{2}\right) + \frac{1}{45}\left(z - \frac{\pi}{2}\right)^3 + \cdots$$

The integrand has a simple pole at $z = \pi/2$, so the residue there, by method 1, is

$$\text{Res}\left(\frac{\pi}{2}\right) = \lim_{z \to \pi/2} \frac{(z - \pi/2)}{z^{n+1}}\left(-\frac{1}{(z - \pi/2)} + \frac{1}{3}\left(z - \frac{\pi}{2}\right) + \frac{1}{45}\left(z - \frac{\pi}{2}\right)^3 + \cdots\right)$$

$$= \frac{-1}{(\pi/2)^{n+1}} = -\left(\frac{2}{\pi}\right)^{n+1}$$

In the neighborhood of $-\pi/2$, let $u = z - (-\pi/2) = z + \pi/2$. Then

$$\tan z = \frac{\sin z}{\cos z} = \frac{\sin(u - \pi/2)}{\cos(u - \pi/2)} = \frac{-\cos u}{\sin u}$$

and thus the series for $\tan z$ near $z = -\pi/2$ is Series 2, with w replaced by $u = z + \pi/2$. The residue at $z = -\pi/2$ is then

$$\mathrm{Res}\left(-\frac{\pi}{2}\right) = \frac{-1}{(-\pi/2)^{n+1}} = -\left(-\frac{2}{\pi}\right)^{n+1} = (-1)^n \left(\frac{2}{\pi}\right)^{n+1}$$

These results are valid for both positive and negative values of n. Thus, the integral that gives the Laurent coefficients is

$$a_n = \frac{1}{2\pi i}\oint_C \frac{\tan z}{z^{n+1}}dz = \sum_{i=1}^{3}\mathrm{Res}\,(z_i) = a_n' + \left(\frac{2}{\pi}\right)^{n+1}[-1 + (-1)^n]$$

where $a_n' \equiv 0$ for negative n. The term $-1 + (-1)^n$ is zero for even values of n and equals -2 for odd values of n (positive or negative). Thus, Series 3 is

$$\tan z = \cdots - 2\left(\frac{\pi}{2}\right)^4 \frac{1}{z^5} - 2\left(\frac{\pi}{2}\right)^2 \frac{1}{z^3} - \frac{2}{z} + z\left(1 - \frac{8}{\pi^2}\right)$$

$$+ z^3\left[\frac{1}{3} - \frac{4}{\pi}\left(\frac{2}{\pi}\right)^3\right] + z^5\left[\frac{2}{15} - \frac{4}{\pi}\left(\frac{2}{\pi}\right)^5\right] + \cdots$$

$$= \cdots - 12.176\frac{1}{z^5} - 4.9348\frac{1}{z^3} - \frac{2}{z} + z(0.18943)$$

$$+ z^3(4.8219 \times 10^{-3}) + z^5(1.9266 \times 10^{-4}) + \cdots$$

This is a Laurent series with infinitely many positive and negative powers, valid for $\pi/2 < |z| < 3\pi/2$.

Let's evaluate this series at $z = \pi$ and compare with the known value $\tan \pi = 0$. The sum of the six terms listed above gives

$$\tan \pi = \cdots - \frac{1}{2^3 \pi} - \frac{1}{2\pi} - \frac{2}{\pi} + \pi - \frac{8}{\pi} + \frac{\pi^3}{3} - \frac{2^5}{\pi} + \frac{2}{15}\pi^5 - \frac{2^7}{\pi} + \cdots$$

$$= \cdots - \frac{1365}{8\pi} + \pi + \frac{1}{3}\pi^3 + \frac{2}{15}\pi^5 + \cdots = -3.2 \times 10^{-2}$$

The next two terms are

$$-2\left[\left(\frac{2}{\pi}\right)^6 \frac{1}{z^7} + \left(\frac{2}{\pi}\right)^8 z^7\right] + \frac{17}{315}z^7 = 2.5246 \times 10^{-2} \quad \text{at } z = \pi$$

Thus, the sum of eight terms of the series is $-3.1981 \times 10^{-2} + 2.5246 \times 10^{-2} = -6.735 \times 10^{-3}$. The series is approaching the correct value, but rather slowly.

At $z = 3\pi/4$, the value should be -1. With six terms, we get

$$\tan\frac{3\pi}{4} = \cdots - 12.176\left(\frac{4}{3\pi}\right)^5 - 4.9348\left(\frac{4}{3\pi}\right)^3 - \frac{8}{3\pi} + \frac{3\pi}{4}(0.18943)$$

$$+ \left(\frac{3\pi}{4}\right)^3(4.8219 \times 10^{-3}) + \left(\frac{3\pi}{4}\right)^5(1.9266 \times 10^{-4})$$

$$= \cdots - 0.87 + \cdots$$

These series clearly converge more slowly than the standard ones we are used to using, but they do have their uses. Notice how we used the first two series to obtain the coefficients for the third; this is a good demonstration of the idea of analytic continuation.

The region of validity for the third series overlaps with that for the second. Again, the two series are identical in the region of overlap.

Bibliography

Abramowitz, Milton, and Stegun, Irene A. *Handbook of Mathematical Functions*, 9th printing. National Bureau of Standards, Applied Math Series 55, 1970.

Anton, Howard, and Rorres, Chris. *Elementary Linear Algebra*, 6th edition. Wiley, New York, 1991.

Arfken, George B., and Weber, Hans J. *Mathematical Methods for Physicists*, 4th edition. Academic Press, San Diego, 1995.

Bauerle, G. *Studies in Mathematical Physics, Volume 1: Finite and Infinite Dimensional Lie Algebras and Applications in Physics*. North Holland, Amsterdam, 1990.

Bowman, Frank. *Introduction to Bessel Functions*. Dover Publications, New York, 1958.

Bradbury, Ted Clay. *Mathematical Methods with Applications to Problems in the Physical Sciences*. Wiley, New York, 1984.

Butkov, Eugene. *Mathematical Physics*. Addison-Wesley, Reading, 1968.

Chattopadhyay, P. K. *Mathematical Physics*. Wiley, New York, 1990.

Chow, Tai L. *Mathematical Methods for Physicists: A Concise Introduction*. Cambridge University Press, Cambridge, 2000.

Cohen, Harold. *Mathematics for Scientists and Engineers*. Prentice Hall, Englewood Cliffs, 1992.

Cook, David. *Computation and Problem Solving in Undergraduate Physics*. Brooks/Cole, Pacific Grove, CA, in press.

Courant, R., and Hilbert, D. *Methods of Mathematical Physics*. Wiley, New York, 1953.

Dennery, Philippe, and Krzywicki, André. *Mathematics for Physicists*. Harper and Row, New York, 1967.

Edwards, C. H., Jr., and Penney, David E. *Differential Equations and Boundary Value Problems*. Prentice Hall, Englewood Cliffs, 1996.

Ford, Lester R. *Differential Equations*. McGraw-Hill, New York, 1955.

Garcia, Alejandro L. *Numerical Methods for Physics*. Prentice Hall, Upper Saddle River, 2000.

Georgi, Howard. *Lie Algebras in Particle Physics*. Perseus Books, Reading, 1999.

Geroch, Robert. *Mathematical Physics*. The University of Chicago Press, Chicago, 1985.

Goldstein, Herbert. *Classical Mechanics*. Addison-Wesley, Reading, 1959.

Golub, Gene H., and Van Laon, Charles F. *Matrix Computations*. The Johns Hopkins University Press, Baltimore, 1983.

Gradshteyn, I. S., and Ryzhik, I. M. *Table of Integrals, Series and Products*, corrected and enlarged edition. Academic Press, Orlando, 1980.

Halmos, Paul R. *Finite Dimensional Vector Spaces*. Springer-Verlag, New York, 1974.

Ince, E. L. *Ordinary Differential Equations*. Dover, New York, 1956.

Jackson, J. D. *Classical Electrodynamics*, 3rd edition. Wiley, New York, 1999.

Jeffreys, Sir Harold, and Jeffreys, Bertha Swirles. *Mathematical Physics*. Cambridge University Press, Cambridge, 1962.

Jones, H. F. *Groups, Representations and Physics,* 2nd edition. Institute of Physics Publishing, London, 1998.

Lawden, Derek. *Tensor Calculus and Relativity*. Science Paperbacks, Methuen Publishing Co., 1962.

Lea, S. M., and Burke, J. R. *Physics: The Nature of Things*. Brooks/Cole, Pacific Grove, CA, 1997.

Lighthill, M. J. *Introduction to Fourier Analysis and Generalised Functions*. Cambridge University Press, Cambridge, 1958.

Long, Robert R. *Mechanics of Solids and Fluids*. Prentice Hall, Englewood Cliffs, 1963.

Margenau, H., and Murphy, G. M. *Methods of Mathematical Physics*. Van Nostrand, Princeton, 1956.

Marion, J., and Thornton, S. *Classical Dynamics of Particles and Systems*, 4th edition. Saunders, Fort Worth, 1995.

Mathews, J., and Walker, R. L. *Mathematical Methods of Physics*. W. A. Benjamin, New York, 1964.

Morse, Philip M., and Feshbach, Herman. *Methods of Theoretical Physics*. McGraw-Hill, New York, 1953.

Murphy, G. M. *Ordinary Differential Equations and Their Solutions*. Van Nostrand, Princeton, 1960.

Murray, Francis J., and Miller, Kenneth S. *Existence Theorems for Ordinary Differential Equations*. New York University Press, New York, 1954.

Polyanin, A. D., and Zaitsev, V. F. *Handbook of Exact Solutions for Ordinary Differential Equations*. CRC Press, Boca Raton, 1995.

Press, W. H., Teukolsky, S. A., Vetterling, W. T., and Flannery, B. P. *Numerical Recipes*, 2nd edition. Cambridge University Press, Cambridge, 1992.

Schiff, Leonard I. *Quantum Mechanics*, 2nd edition. McGraw-Hill, New York, 1955.

Stewart, James. *Calculus*, 4th edition. Brooks/Cole, Pacific Grove, CA, 1999.

Temme, Nico M. *Special Functions: An Introduction to the Classical Functions of Mathematical Physics*. Wiley, New York, 1996.

Tinkham, Michael. *Group Theory and Quantum Mechanics*. McGraw-Hill, New York, 1964.

Tucker, Alan. *Linear Algebra*. Macmillan, New York, 1993.

Tung, Wu-Ki. *Group Theory in Physics*. World Scientific, Philadelphia, 1985.

Wallace, Philip. *Mathematical Analysis of Physical Problems*. Dover, New York, 1984.

Wilkes, M. V. *A Short Introduction to Numerical Analysis*. Cambridge University Press, London, 1966.

Zemanian, A. H. *Distribution Theory and Transform Analysis*. Dover, New York, 1965.

Zwillinger, Daniel. *Handbook of Differential Equations*. Academic Press, San Diego, 1998.

Index

The main reference or definition is given in **boldface**.